XXXI-st SIAR International Congress of Automotive and Transport Engineering "Automotive and Integrated Transport Systems" (AITS 2021)

IOP Conference Series: Materials Science and Engineering
Volume 1220

Chisinau, Moldova
28 - 30 October 2021

Editors:

Viorel Bostan
Victor Ceban
Cornel Stan
Anghel Chiru
Adrian Clenci
Stefan Tabacu

ISBN: 978-1-7138-4679-6
ISSN: 1757-8981

Printed from e-media with permission by:

Curran Associates, Inc.
57 Morehouse Lane
Red Hook, NY 12571

proceedings.com

Some format issues inherent in the e-media version may also appear in this print version.

Institute of Physics
Dirac House, Temple Back
Bristol BS1 6BE UK

Phone: 44 1 17 929 7481
Fax: 44 1 17 920 0979

techtracking@iop.org

Additional copies of this publication are available from:

Curran Associates, Inc.
57 Morehouse Lane
Red Hook, NY 12571 USA
Phone: 845-758-0400
Fax: 845-758-2633
Email: curran@proceedings.com
Web: www.proceedings.com

TABLE OF CONTENTS

IOP Conference Series:
Materials Science and Engineering

The XXXI-st SIAR International Congress of Automotive and Transport Engineering "Automotive and Integrated Transport Systems – AITS 2021"

Chisinau, Republic of Moldova
28 – 30 October 2021

Editors: Viorel BOSTAN and Victor CEBAN (Technical University of Moldova),
Cornel STAN (West Saxon University of Zwickau),
Anghel CHIRU (Transilvania University of Brasov),
Adrian CLENCI and Stefan TABACU (University of Pitesti)

IOP Conf. Series: Materials Science and Engineering 1220 (2022) 011001 doi:10.1088/1757-899X/1220/1/011001

Foreword

The Congress of Automotive and Transport Engineering is an annual scientific event initiated in 1990 by the Society of Automotive Engineers of Romania (SIAR) and various Romanian universities with an interest in Automotive and Transport Engineering. This Congress is endorsed by FISITA (Fédération Internationale des Sociétés d'Ingénieurs des Techniques de l'Automobile) and EAEC (European Automotive Engineers Cooperation).

The XXXI-st edition of SIAR's International Congress with the theme "Automotive and Integrated Transport Systems – AITS 2021" was organized for the first time by the Department of Transport of the Faculty of Mechanical Engineering and Transport from the Technical University of Moldova with the support of the State Agrarian University of Moldova and Society of Automotive Engineers (SAE International) as co-host.

The Congress purpose is to gather members from academia, industry and government and present their possibilities for investigations and research, in order to establish new future collaborations in the automotive engineering and transport domain.

Due to the Government restrictions caused by the COVID 19 pandemic, by the General Meeting of the SIAR, it was decided to hold the AITS 2021 Congress on October 28-29th 2021 in mixed mode (online and reduced physical presence) in compliance with all the necessary conditions. Using modern technologies and the Microsoft Teams platform, more than 200 participants from different countries, such as Germany, Austria, USA, Bulgaria, Romania, Moldova attended the Congress and presented their scientific achievements within plenary sessions and three types of panel discussions (Road Vehicles, Automotive Technology and Transportation, Road Safety).

Thus, the Congress participants had the opportunity to:

- gain insight into the cutting edge automotive and transport industry and ideas for future developments;
- update their skills and knowledge by attending focused technical sessions;
- network with potential new partners, clients and suppliers;
- view the latest technology products and services in the technical exhibition.

The papers included in this volume were selected by the Scientific Committee amongst the ones presented within the Congress and are covering the latest issues such as green vehicles, advanced powertrain and propulsion, integrated transport systems, automotive design and testing, advanced engineering methods, road safety, mobility sustainability, materials, automotive technology and maintenance.

We hope that these papers will be of interest to experts from academia and industry around the world and we wish them to create a favourable framework for addressing key issues in the field of Automotive and Road Transport Engineering that concern contemporary society today.

Yours faithfully,
AITS 2021 Proceedings Editors

IOP Conf. Series: Materials Science and Engineering 1220 (2022) 011001 doi:10.1088/1757-899X/1220/1/011001

Acknowledgement

HONORARY COMMITTEE

Cristian-Nicolae ANDREESCU - Professor, Politehnica University of Bucharest, Romania
Ion BOSTAN - Academician of the Academy of Sciences of Moldova
Viorel BOSTAN - Rector of the Technical University of Moldova
Nicolae BURNETE - President of Society of Automotive Engineers of Romania
Vasile CHIRTOACA - President of DAAC „Hermes", Republic of Moldova
Anghel CHIRU - Professor, Transilvania University of Brasov, Romania
Adrian CLENCI - Professor, Former SIAR President, University of Pitesti, Romania
Eugen DATCO - Executive Director of the Union of Carriers and Roads, Republic of Moldova
Radu DINESCU - IRU President, General Secretary of the National Union of Road Hauliers from Romania
Gheorghe FRATILA- Professor, Politehnica University of Bucharest, Romania
Radu GAIGINSCHI - Professor, Technical University „Gheorghe Asachi" of Iasi, Romania
Carles GRASAS - FISITA Vice-President Europe
Günther HOHL - EAEC Pres ident and Former FISITA Vice-President, Austria
Murli IYER - Executive Global Advisor of SAE International, US
Stefan KANYA - General Manager of AVL Romania
Eugen-Mihai NEGRUS - Society of Automotive Engineers of Romania Honorary President
Victor OTAT - Professor, University of Craiova, Romania
Attila PAPP - Executive Director, Magic Engineering Brasov, Romania
Gheorghe-Alexandru RADU - Professor, Transilvania University of Brasov, Romania
Alexander SIMIONESCU - General Manager, Renault Technologie Roumanie
Cornel STAN - Professor, West Saxon University of Zwickau, Germany
Ion TABACU - Professor, University of Pitesti, Romania
Vasile TRONCIU - Pro-Rector for Research and Doctoral Studies, Technical University of Moldova
Radian TUFĂ - Romanian Automotive Register
Liviu VOLCONOVICI - Rector of the State Agrarian University of Moldova

SCIENTIFIC AND TECHNICAL COMMITTEE

Presidents of the Scientific and Technical Committee
Prof. Dr. Eng. Cornel STAN - West Saxon University of Zwickau, Germany
Prof. Dr. Eng. Anghel CHIRU - Transilvania University of Brasov, Romania

Board Members
Prof. Dr.-Eng. Horia ABAITANCEI - Transilvania University of Brasov, Romania
Assoc.Prof. Dr.-Eng. Grigori AMBROSI - Technical University of Moldova
Prof. Dr.-Eng. Gabriel ANGHELACHE - Politehnica University of Bucharest, Romania
Prof. Dr.-Eng. Csaba ANTONYA - Transilvania University of Brasov, Romania
Assoc.Prof. Dr.-Eng. Asen ASENOV - University of Ruse, Bulgaria
Prof. Dr.-Eng. Istvan BARABAS - Technical University of Cluj-Napoca, Romania
Prof. Dr.-Eng. Alexandru BOROIU - University of Pitesti, Romania
Prof. Dr. hab. Larisa BUGAIAN - Technical University of Moldova
Prof. Dr.-Eng. Radu CHIRIAC - Politehnica University of Bucharest, Romania
Prof. Dr.-Eng. Adrian Constantin CLENCI - University of Pitesti, Romania
Prof. Dr.-Eng. Corneliu COFARU - Transilvania University of Brasov, Romania

IOP Conf. Series: Materials Science and Engineering 1220 (2022) 011001 doi:10.1088/1757-899X/1220/1/011001

Assoc.Prof. Dr.-Eng. Adrian SACHELARIE - Technical University „Gheorghe Asachi" of Iasi, Romania
Prof. Dr.-Eng. Corina SANDU - Virginia Tech, U.S.A.
Assoc.Prof. Dr.-Eng. Loreta SIMNICEANU - University of Craiova, Romania
Prof. Dr.-Eng. Adrian ŞOICA - Transilvania University of Brasov, Romania
Prof. Dr.-Eng. Victor ŞONTEA - Technical University of Moldova
Prof. Dr.-Eng. Ştefan TABACU - University of Pitesti, Romania
Prof. Dr.-Eng. Mihail TODOROV - Technical University of Sofia, Bulgaria
Assoc.Prof. Dr.-Eng. Adrian-Ioan TODORUT - Technical University of Cluj-Napoca, Romania
Prof. Dr.-Eng. Bogdan VARGA - Technical University of Cluj-Napoca, Romania
Prof. Dr.-Eng. Teodor VIRGIL - „Dunarea de Jos" University of Galati, Romania
Prof. Dr.-Eng. Mate ZOLDY - Budapest University of Tehnology and Economics, Hungary

ORGANIZING COMMITTEE

Congress Chairmen
Assoc.Prof. Dr.-Eng. Victor CEBAN - Head of the Transport Department, Faculty of Mechanical Engineering and Transport, Technical University of Moldova, SIAR Vice-President

Grigore AMBROSI
Oleg FRUNZE
Olivian PĂDURE
Igor ROTARU
Vasile PLĂMĂDEALĂ
Dumitru CEBAN
Elena BIVOL
Vadim NANTOI
Oxana RADU

Mihail GADIBADI
Dumitru NOVOROJDIN
Igor BEŞLEAGA
Vitalie ARNAUT
Ilie MANOLI
Natalia ŞESTENCO-DIACEK
Nina GARŞTEA
Oleg PETROV

Galina VOLOŞCENCO
Ion COTRUȚA
Margareta VÎRCOLICI
Aliona LÎSÎI
Ilie BEIU
Vladimir GOIAN
Iurie TEZEC
Anatolie CORPOCEAN

SPONSORS
The organizing committee would like to thank the following companies for their support:

Peer review declaration

All papers published in this volume of *IOP Conference Series: Materials Science and Engineering* have been peer reviewed through processes administered by Editors. Reviews were conducted by expert referees to the professional and scientific standards expected of a proceedings journal published by IOP Publishing.

- Type of peer review: *single-blind* (the identity of reviewers is not known by the authors)
- Criteria used by reviewers:
  The papers were reviewed for scientific content, the novelty of information, language correctness and presentation. The reviewers were instructed to evaluate the papers according to a review form with the following sections: originality, relationship to literature, organization of manuscript (including quality of communication, clarity of results), information included in the abstract, English usage, and if the manuscript is free of commercialism. Then the reviewers had to fill in a box with comments and suggestions for authors. In addition, the reviewers had to propose the paper to be published, revised or rejected. The comments were sent to authors (by the track director), and the authors had to update their papers according to the reviewers' comments and upload them again. The conference track director took the final decision of acceptance (or rejection) only after the authors had corrected their papers.
- Conference submission management system: Open Conference System (www.siarcongress.eu).
- Number of submissions received: 199
- Number of submissions sent for review: 98
- Number of submissions accepted: 84
- Acceptance Rate: 42%
- Average number of reviews per paper: 2
- Total number of reviewers: 71
- Plagiarism check system: Sistemantiplagiat.ro (papers were uploaded on the server, then the report was downloaded)
- Contact person for queries: Dinu COVACIU, Transilvania University of Brasov, Romania, email: dinu.covaciu@unitbv.ro

IOP Conf. Series: Materials Science and Engineering 1220 (2022) 012001 doi:10.1088/1757-899X/1220/1/012001

Researches regarding the animal fats use at a truck diesel engine

A Nicolici[1], C Pana[1], N Negurescu[1], A Cernat[1], G Lazaroiu[1], L Mihaescu[1] and C Nutu[2]

[1]Department of Thermotechnics, Engines, Thermal and Frigorific Equipment, University Politehnica of Bucharest, Bucharest, Romania
[2]Department of Automotive Engineering, University Politehnica of Bucharest, Bucharest, Romania

cernatalex@yahoo.com

Abstract. The paper objectives answer to European Commission requires which decided for transport sector a 90% reduction of pollutants emissions till the year 2050. Alternative fuels like animal fats can represent a viable solution for diesel engines due to their good combustion properties, pollution reduction, and the possibility of raw state use in blends with diesel fuel or as biodiesel. A D2156 truck diesel engine is fueled with preheated raw animal fats, blend with diesel fuel in 5% and 10%. At the 55% engine load and 1450 rpm engine speed, for 10% animal fats-diesel fuel blend, the pollutant emissions levels decrease with 21% for NO_x, with 50% for HC, with 66% for CO, with 30% for smoke due to the decrease of carbon content and increase of oxygen content. The CO_2 emission level decrease by 14%. An important novelty aspect is a good correlation between engine running regime, diesel fuel, and animal fats cyclic dose, pollutants emissions level and exhaust gases temperature. The use of animal fats is a good opportunity to improve the environmental protection from pollutants emissions and greenhouse gases for diesel engines. The use of raw animal fats does not require significant changes to the diesel engine design.

1. Introduction

The European Commission decided that the EU will become greenhouse gas neutral by 2050 and the transport sector must reduce its pollutants emissions by 90% by 2050. The present paper objective answers this request because the use of the alternative fuel can be considered an efficient solution for the thermal efficiency increase and for the pollutants emissions decrease at the diesel engine, engines that have an important spread in the field of heavy transport (truck, busses, etc.). Reduction of the greenhouse gases and pollutant emissions from the diesel engine exhaust is an actual concern of researchers in the automotive industry, a preoccupation in concordance with the necessity of a clean environment and decreasing tendency of the classic fuel reserves and implementation of the new energy sources. Animal fats are biodegradable, non-toxic, and free of sulphur and have the advantage offered by large reserves and suitable combustion properties, like the raised energetically value which makes them suitable for biodiesel production. Animal fats represent waste material that remains from the processes developed in food and in tanneries plants. Their use constitutes a significant alternative for fossil fuels replace and for the number of waste materials reduce from the food and tanning industry. Animal fats have a higher cetane number and oxygen content versus diesel fuel, with a calorific power closer to diesel fuel, being suitable to be used as fuel for diesel engines. The value of the caloric power of the animal fats is slightly lower compared to diesel fuel, but the higher oxygen content allows at the same air-fuel ratio to increase the fuel dose per cycle in order to establish the engine power. High oxygen and lower carbon content of animal fats may have a favourable impact on the emission level of some exhaust emissions like unburned hydrocarbons, carbon monoxide, and smoke. The higher cost of transesterification and emulsification with alcohols rises the price of the fuel and may represent a disadvantage. A much cheaper and easier way of using animal fats as fuel is their use in a raw state, but their higher viscosity at normal temperature requires preheating. Thru preheating the animal fats becomes liquid and can be mixed with diesel fuel in a stable and homogeneous blend, a blend that can be used to fuel a diesel engine without major constructive modifications. Researchers use animal fat to produce biodiesel or to mix with classic fuel or biodiesels in order to develop new fuels that can be used for diesel engines fuelling for better performance. Alptekin [1] uses animal fat biodiesels at a direct injection diesel engine for various engine loads at speed of 1400 rev/min. The brake specific fuel consumption (BSFC) was 16% higher compared to classic fuel. The in-cylinder pressure recorded 1.2% higher peak pressures in some individual combustion cycles versus diesel fuel. Also, for biodiesel, the combustion starts earlier per cycle versus diesel fuel. At animal fat use in biodiesel, the CO and HC emissions levels decrease, but the NO_x emissions increase. Behçet [2] use a single cylinder diesel engine of 7,5 kW in an experimental investigation of the waste animal fats (from fish and chicken) use as fuel, after the animal fat was blended with 50% of diesel fuel. Depending on the type of animal fat, the BSFC was increased only at high engine speeds in the range of 1500…2500 rev/min, with 2%...4.5%. At low speeds range of 1000…1300 rev/min the BSFC is similar for animal fat and diesel fuel. The NO_x emission level rises by 3%...23%, but the level for emissions of CO and HC was decreased at animal fat use. The CO emission level drops by ~14% and the HC emission reduces its level by ~10%. In a test, Lapuerta [3] uses two biodiesels, the first one based on 100% animal fat and the second one produced with 50% animal fat and 50% oil from soybean. Also, Lapuerta uses this biodiesel in blends with classic diesel fuel in a percentage of 30% or 70% in volume. Lapuerta shows that for all tested fuels, compared to diesel fuel, the brake thermal efficiency remains unchanged and only the fuel consumption slightly rises. The emission level of HC and smoke was decreased, but the NO_x emission level rises at the use of biodiesel constituted from 50% animal fat and 50% soybean oil. Only at the use of simple animal fat, the NO_x emission level was decreased. Altun [4] obtains Inedible animal tallow Methyl ester by processing Inedible animal tallow and methanol and uses this biodiesel to fuel a Mitsubishi diesel engine. The BSFC increases for biodiesel use versus diesel fuel, but CO and NO_x emissions levels decrease for all tested speeds, from 1000 to 2250 rev/min. At speed of 1500 rev/min the CO emission level decrease by 7% and the NO_x emission level drops by 17% comparative to diesel fuel. Kumar [5] uses animal fat or emulsions of animal fat with different alcohols to fuel a Lister-Petter diesel engine. At 1500 rev/min engine speed, the maximum pressure during combustion decreases by 10%, and the maximum rate of pressure rise decreases by 30% at animal fat use versus diesel fuel. The duration of the combustion increases by 7°CA at full load and by 3 °CA at 60% load at animal fat use. At animal fat use the peak of heat release rate decreases with 38%...46% and appears later on cycle comparative to diesel fuel, the specific energy consumption being increased by ~34%. The smoke emission decreases by 46% at full load but slightly increases at 60% load. The NO_x emission level decreases by 33% at full load and by 45% at 60% load. The CO and HC emissions levels are increased at animal fat use versus diesel fuel. Barrios [6] affirms that biodiesel produced from animal fat can be a good substitution for classic diesel fuel. Barrios uses 6 different animal fats as a base for biodiesel, which are blended with diesel fuel in different proportions, from 10% to 50%. For a Euro 4 diesel engine, versus to classic fuel, at all used blends the NO_x emission level increases, and the particle emission level decreases. Barrios [6] observes a correlation between the animal fat – biodiesel quantity and reduction of particle emission, for biodiesel based on animal fat the smaller quantities of particles are registered comparative to biodiesel based on soybeans. Selvam [7] uses animal fats Methyl ester in blends with diesel fuel to fuel a Kirloskar diesel engine for all loads and 1500 rev/min regime. Comparative to diesel fuel the use of fuel blends leads to a slight increase of BSFC and NOx emission levels, but the emission levels of CO,

HC, and smoke density are decreased. The general objective of the current research is the animal fats use at an automotive truck diesel engine for the decrease of the pollutants emissions level, especially for nitrogen oxides (NO_x) and smoke. By fulfilling these objectives, the paper brings an important contribution to solving pollution problems in large urban areas for diesel engines, the solution of diesel fuel-animal fat blends being easily used at old or new diesel engines with benefits of pollution reduction. The paper presents some experimental aspects of animal fat use, in raw state preheated and blend it with diesel fuel, to fuel a truck diesel engine. The influences of animal fat percentage on the combustion process are presented and analysed.

2. Research methodology

The experimental investigation was carried out on a 6 cylinders truck diesel engine, turbocharged, 10 litre displacement, manufactured by Roman Brasov Plant under the serial D2156 MTN8. A test bed schema is shown in the following figure.

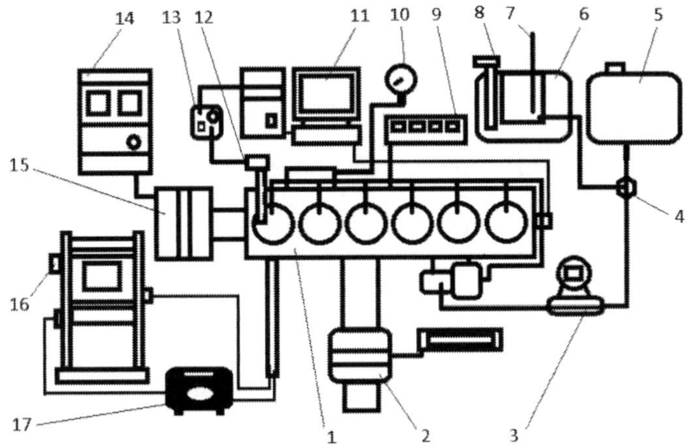

Figure 1. D2156MTN8 truck diesel engine test bed.

The experimental test bed, presented in figure 1, has the following components: 1-D2156 MTN diesel engine, 2-Meriam air flowmeter, 3- Optimass fuel flowmeter, 4- fuelling line tap, 5- diesel fuel tank, 6- animal fat-diesel fuel blend tank, 7- thermometer, 8-electric resistance, 9-Shimaden temperature indicators, 10-supercharging pressure manometer, 11-AVL91 data acquisition computer, 12- Kistler pressure transducer, 13-AVL charge amplifier, 14-dyno control cabinet, 15-Hofmann eddy current dyno, 16-AVL DiCom 4000 gas analyser, 17- AVL opacimeter. Before the start of the experimental investigation procedure, all the equipment was calibrated. The animal fat, in its raw state, was preheated at 40°C and blended with diesel fuel in a special design tank which is equipped with an electric resistance in order to maintain the fuels blend at a high temperature thus the animal fat be perfectly miscible in diesel fuel. The electric resistance is supplied from the laboratory's electrical circuit. The blend temperature is set up by a thermometer. The fuelling system is equipped with a tap used to change the fuelling line from diesel fuel to animal fat-diesel fuel blend. For further investigations, the engine can be equipped with a special design fuelling system that uses the heat from the cooling system to preheat the animal fats at the necessary working temperature. Almost 25% of combustion heat is released to the cooling system and only a small fraction from it is necessary for animal fats preheating, without affecting the engine power. During the experiment, firstly, the engine was fuelled only with diesel fuel and the reference of the experimental investigation was determined. Secondly, the engine was fuelled with different blends of diesel fuel and animal fats, the percentage of animal fat in blends with diesel fuel being 5% and 10%, defined as an energetic substitute ratio xc[%]. At a stable operating regime of 1450 rev/min engine speed and 55% engine load, 250 consecutive cycles were acquisitioned with 1 °CA resolution and get analysed for diesel fuelling and for animal fat-diesel fuel blend use. The engine power and the injection timing were maintained constant. In order to assure the best engine performance at animal fat-diesel fuel blends use, the combustion process is controlled through an optimal tune between engine load, injection timing, exhaust gases temperature, and animal fat content in blend with diesel fuel, a tune procedure that defines the novelty aspect of the experimental research.

3. Results

At animal fat use in blend with diesel fuel, the brake specific fuel consumption (BSEC), figure 2a, decreases by 4.6% for xc=5% and by 1.3% for xc=10%.

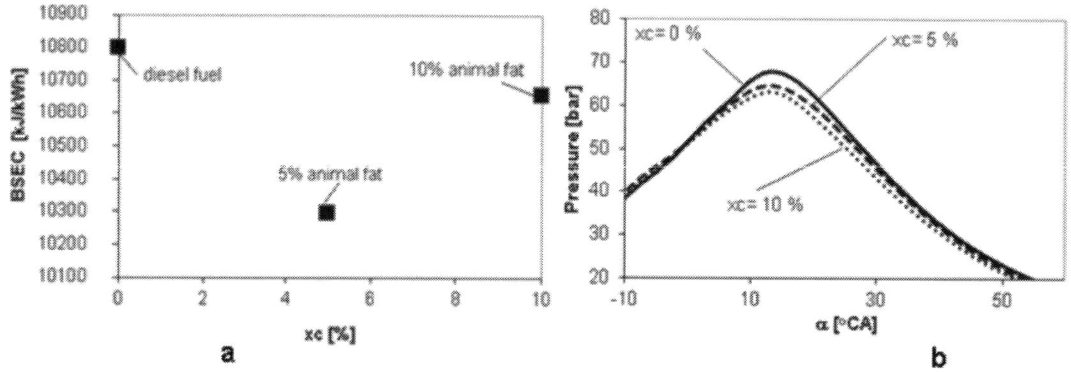

Figure 2. a) BSFC versus xc, b) Pressure diagrams for different xc.

Due to the higher cetane number and higher oxygen content of animal fat, the combustion process is improved for small quantities of animal fat, but the influence is reduced. The increase of animal fat percentage in diesel fuel blends at higher values leads to the increase of specific fuel consumption. For a higher percentage of animal fats biodiesel in blends, similar results were obtained by different researchers [7]. Pressure diagrams presented in figure 2b are obtained by averaging 250 individual pressure diagrams. The animal fats use influences the cycle heat release comparative to diesel fuel, the pressure diagrams being modified once with the rise of animal fats content, figure 3a, the maximum decrease of peak pressure being ~6.9%. The maximum pressure of 68 bar registered for diesel fuelling (xc=0%) decrease to 64.5 bar for xc=5% and to 63.2 bar for xc=10%, figure 3a, in correlation the heat release rate and its maximum values per cycle start to decrease with the rise of animal fats content in blend with diesel fuel because the calorific power of the animal fats is lower versus diesel fuel, figure 4a.

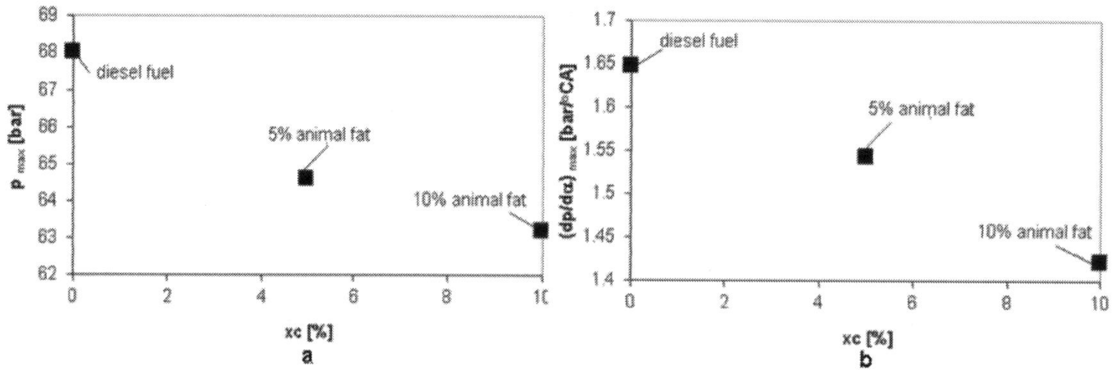

Figure 3. a) Maximum pressure versus xc, b) Maximum rate of pressure rise versus xc.

The maximum rate of pressure rise, figure 3b, decreases with 6.5% for xc=5 and with 13.7% for xc=10% the combustion noise being reduced at animal fats use. The reduction of the maximum rate of pressure rise and of the maximum pressure during combustion is related to the mechanical stress of the engine. The heat release rate diagrams for diesel fuel use and for diesel fuel-animal fats blends use are presented in figure 4a. Because the lower heating value of animal fats is 9.3% smaller versus diesel fuel, the heat release rate also decreases with the rise of animal fats content in the fuel blend. The maximum value of heat release decreases with ~14% for xc=5% and with 16.5% at xc=10% comparative to diesel fuel. Also, the use of animal fats influences the moment when those maximum values are achieved, the angle of maximum heat release rate appears later on cycle with 1 °CA and 2 °CA for xc=5% and xc=10%, respectively, versus classic fuelling. The reduction of the heat release rate and of the heat quantity at animal fats use is in correlation with the reduction of the maximum pressure and maximum rate of pressure rise, figures 3a and 3b.

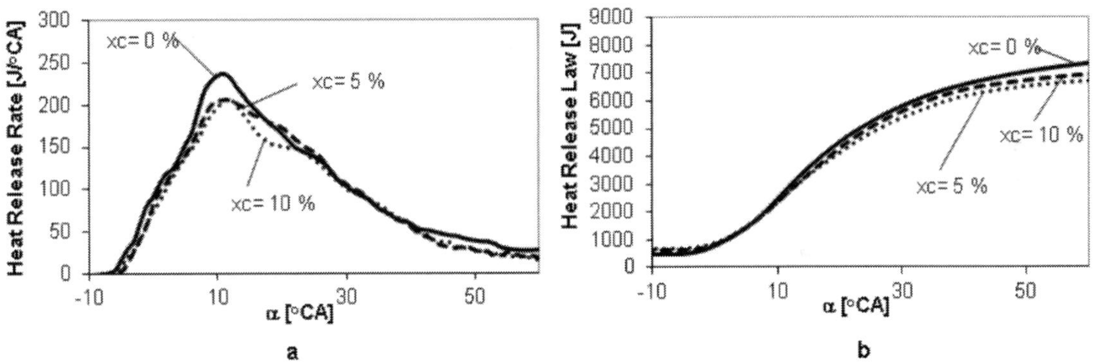

Figure 4. a) Heat release rate diagrams, b) combustion law diagrams for different xc.

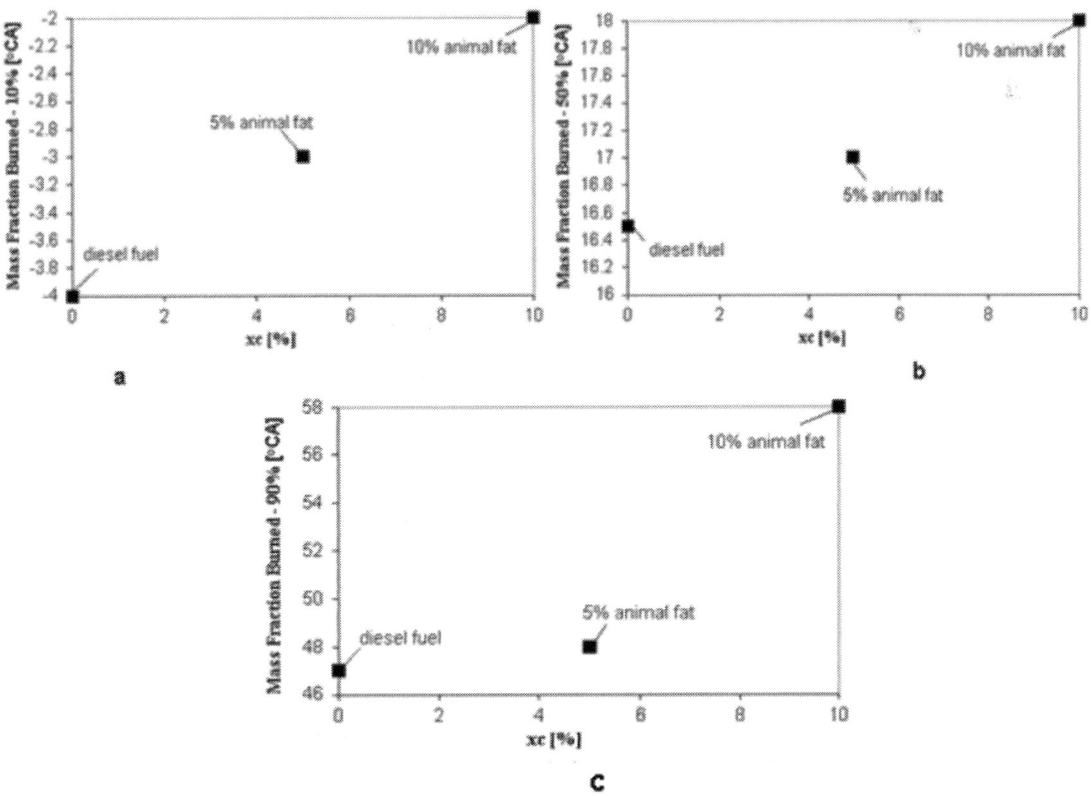

Figure 5. a) MFB-10% versus xc, b) MFB-50% versus xc, c) MFB-90% versus xc.

Comparative to diesel fuelling, the heat release quantity decreases with 5% at xc=5% and with 8.5% at xc=10%, figure 4. Modification of the heat release rate at the use of animal fats influence the moment per cycle when a specific percent from the combustion heat is released, the 10%, 50% and 90% mass fraction burned (MFB), figure 5. The high viscosity of animal fats leads to the deterioration of the animal fats-diesel fuel blend atomization thru injector nozzle once with the increase of the animal fats content and comparative to diesel fuel the 10% from heat release appears later on cycle, with 1 °CA or 2 °CA for MFB-10%, figure 5a. The decrease of heat release during the earlier stage of combustion and of its release rate affects the main stage of heat release during combustion and leads to a later achievement per cycle of the MFB-50%, the angle of MFB-50% being increased with 0.5-1 °CA, figure 5b. A later start of combustion and a later achievement of 50% from heat release, which are in correlation with the decrease of maximum pressure and maximum rate of pressure rise, affects the end of combustion, MFB-90%, figure 5c.

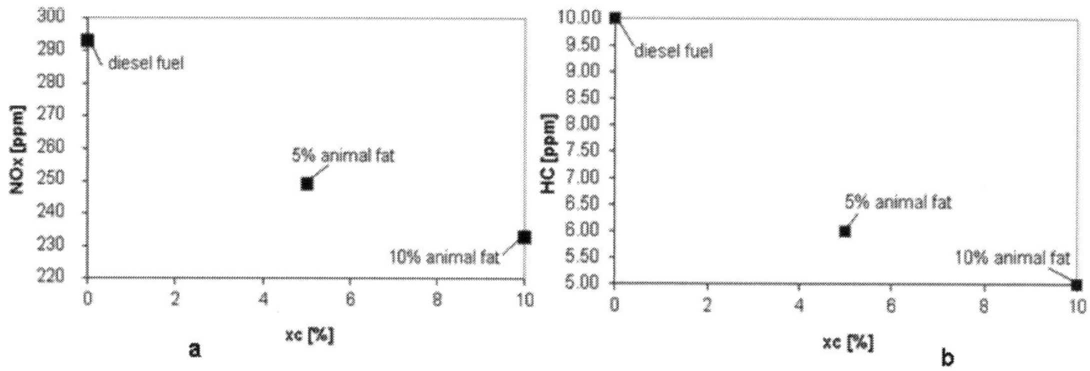

Figure 6. a)NO_x emission level versus xc, b) HC emission level versus xc.

For classic fuelling the conventional end of combustion, MFB-90%, appears at 47 °CA, at 48 °CA for xc=5%, and at 58 °CA for xc=10%, the combustion process being ended with 11 °CA later versus TDC for maximum xc. Thus, the combustion duration $\Delta_{10\% - 90\%}$ = 51 °CA for classic fuelling (xc=0%) is not affected by small percent's of animal fats, xc=5%, the values being similar for diesel fuel and xc=5%. animal fats in blend with diesel fuel. If the percentage of animal fats is doubled then the influence is visible, the $\Delta_{10\% - 90\%}$ = 60 °CA (xc=10%) being increased. The increase of the combustion duration from 51 °CA to 60 °CA is in correlation with the decrease of the maximum rate of pressure rise and of the heat release rate at animal fats use. An injection timing tune may correct this issue. In further research, the injection timing will be optimized according to the substitute ratio of diesel fuel with animal fats to assure the optimum values of specific energetic consumption, NO_x and smoke emissions minimum level. The use of animal fats in mixture with diesel fuel leads to the reduction of pollutant emissions and greenhouse gases emission. At animal fat use the NO_x emissions level decrease with 15% for xc=5% and with ~ 21% for xc=10%, figure 6a. The NO_x emission level may be explained by the reduction of the pressure and temperature during combustion due to the lower value of the caloric power for the animal fats versus diesel fuel. The smoke emission level is evaluated by opacity and number K, figure 7. At animal fats use the smoke opacity decrease with 26% at xc=5% and with 28% at xc=10%, figure 7a. The smoke number decrease with 26% at xc=5% and with 30% at xc=10%, figure 7b. The smoke emission level is reduced due to the decrease of carbon content and by the increase of the oxygen content at the rise of xc. The HC emissions level decrease with 40% at xc=5% and with 50% for xc=10%, as figure 6b shows. The HC reduction is related to a lower carbon content of animal fats and the higher oxygen content in animal fats comparative to diesel fuel improves the HC oxidation. The CO emission level decreases by 66% when the animal fats-diesel fuel blends are used compared to diesel fuel, figure 8a. The CO emission level is decreased due to a much higher oxygen content of animal fats versus diesel fuel (beef fat 12.5 *(% by mass)* versus diesel fuel 0.3 *(% by mass)*, [5]). Also, the carbon content in animal fats is lower compared to diesel fuel and for xc=10% the carbon content in animal fats-diesel fuel blend is 1.2% less compared to classic fuel, xc=0%. In terms of greenhouse gases emission, the CO_2 emission level decrease by 7.14% at xc=5% and by 14% at xc=10%, figure 8b, due to the lower carbon content of animal fats that substitute the diesel fuel.

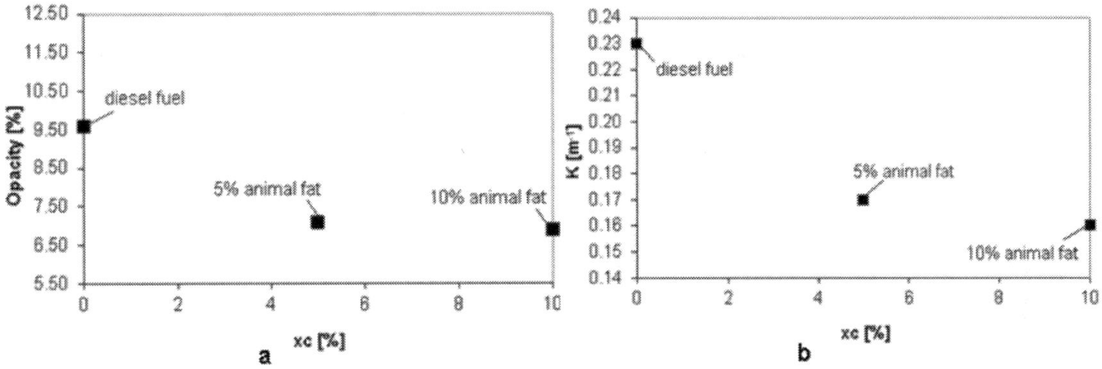

Figure 7. a) Smoke opacity versus xc, b) Smoke number versus xc.

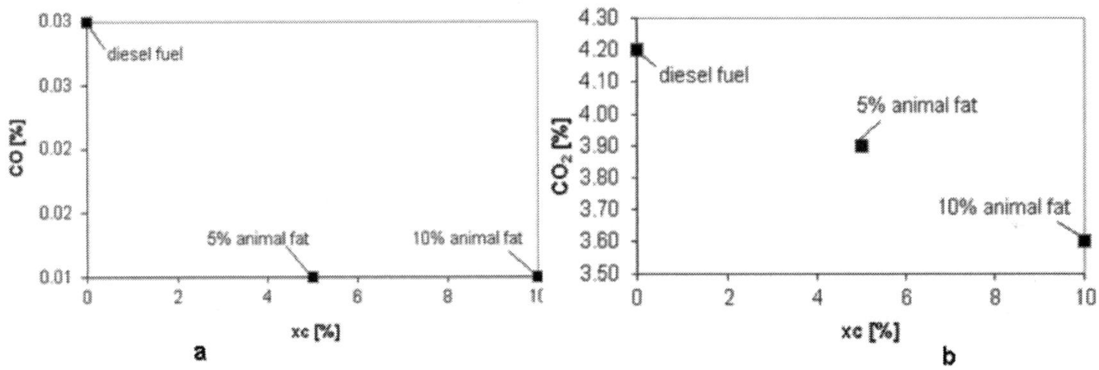

Figure 8. a) CO emission level versus xc, b) CO_2 emission level versus xc.

4. Conclusions

The research regarding the use of animal fats in blends with diesel fuel to fuel a truck diesel engine, at a regime of 1450 rev/min and 55% load, leads to the following main conclusions: at animal fat use in blend with diesel fuel the brake specific fuel consumption decrease with 4.6% for xc=5% and with 1.3% for xc=10%, due to higher oxygen content and cetane number of animal fat which improve the combustion process, for small animal fats quantities. The maximum pressure decreases at animal fat use, from 68 bar (xc=0%) to 64.5 bar (xc=5%) and to 63.2 bar (xc=10%) because of heat release rate decrease with the rise of animal fats content. The maximum rate of pressure rise decreases with 6.5% (xc=5%) and 13.7% (xc=10%) at animal fats use, the combustion noise and the mechanical stress of the engine being reduced. The heat release rate decreases for all xc because the lower heating value of animal fats is 9.3% smaller compared to diesel fuel. The peak values of heat release rate decreases with ~14% (xc=5%) and 16.5% (xc=10%) comparative to diesel fuel, the moment per cycle of peak values occurs with 1… 2 °CA, depending on xc. The reduction of heat release rate and of cyclic heat quantity (up to 8.5%) is in correlation with maximum pressure and maximum rate of pressure rise reductions. The MFB-10% from heat release appears later on cycle, with 1 °CA or 2 °CA, because of the high viscosity of animal fats which deteriorate the fuel blend atomization once with the increase of the animal fats content. The main stage of heat release, defined by the angle of MFB-50%, is increased by 0.5…1 °CA in concordance with the decrease of maximum pressure and maximum rate of pressure rise.

This phenomenon affects the end of combustion, MFB-90%, which appears with 1 or 11 °CA later per cycle comparative to diesel fuel. The combustion duration $\Delta_{10\% - 90\%}$ = 51 °CA for classic fuelling (xc=0%) is not affected by small percent's of animal fats, like xc=5%, the values being similar for diesel fuel, but for xc=10% , the combustion duration is increased at $\Delta_{10\% - 90\%}$ = 60 °CA. This increase in combustion duration is related to the reduction of the maximum rate of pressure rise and of the heat release rate at animal fats use. The NO_x emissions level decrease by ~ 21% at maximum xc, due to the lower value of animal fats caloric power that assures the reduction of pressure and temperature during combustion. The smoke emission level decrease by 28% (for opacity) and 30% (for smoke number) at animal fats use due to the decrease of carbon content and the increase of the oxygen content at the rise of xc. The HC emissions level decrease by 50% for maximum xc due to lower carbon and higher oxygen contents of animal fats comparative to diesel fuel. The CO emission level decrease by 66% due to a much higher oxygen content of animal fats versus diesel fuel. The CO_2 emission level maximum decrease of 14% is assured at xc=10% due to the lower carbon content of animal fats. The use of animal fats in truck diesel engines represents a good opportunity to improve environmental protection. The pollutants emissions level and the greenhouse gases level decreases compared to diesel fuelling. The use of these animal fats does not require significant changes in the diesel engine design.

Acknowledgements
The authors address special thanks to AVL GmbH Graz Austria for providing the necessary equipment. This work has been funded by the University Politehnica of Bucharest, through the "Excellence Research Grants" Program, UPB-GEX 2017. Identifier: UPB- GEX2017, Ctr. No. TR 27-17-06/2017". Grant of the Romanian Ministry of Research and Innovation, CCCDI-UEFISCDI, project number PN-III-P1-1.2-PCCDI-2017-0404, within PNCDI III. Sectoral Operational Programme Human Resources Development 2007–2013 of the Ministry of European Funds through the Financial Agreement POSDRU/159/1.5/S/134398. This work has been funded by the European Social Fund from the Sectoral Operational Programme Human Capital 2014-2020, through the Financial Agreement with the title "Scholarships for entrepreneurial education among doctoral students and postdoctoral researchers (Be Antreprenor!)", Contract no. 51680/09.07.2019-SMIS code: 124539.

References

[1] Alptekin E, Canakci M, Ozsezen A N, Turkcan A, Sanli H 2015 Using waste animal fat based biodiesels–bioethanol–diesel fuel blends in a DI diesel engine *Fuel* **157** pp 245-254 (https://doi.org/10.1016/j.fuel.2015.04.067)

[2] Behçet R, Çakmak A 2017 Experimental Investigation of Using Animal Fat Based Biodiesel-Diesel Blends in a Diesel Engine *8th International Science and Technology Conference* (Berlin: July 17-19 2017, Germany)

[3] Lapuerta M, Rodríguez-Fernández J, Oliva F, Canoira L 2009 Biodiesel from Low-Grade Animal Fats: Diesel Engine Performance and Emissions *Energy Fuels* **23** pp121-129 (https://doi.org/10.1021/ef800481q)

[4] Altun S 2011 Performance and exhaust emissions of a DI diesel engine fueled with waste cooking oil and inedible animal tallow methyl esters *Turkish J. Eng. Env. Sci.* **35** pp107 – 114 (doi:10.3906/muh-1004-108)

[5] Senthil Kumar M, Kerihuel A, Bellettre J, Tazerout M 2006 A Comparative Study of Different Methods of Using Animal Fat as a Fuel in a Compression Ignition Engine *Journal of Engineering for Gas Turbines and Power* **128** pp 907-914

[6] Barrios C C, Domínguez-Sáez A, Martín C, Álvarez P 2014 Effects of animal fat based biodiesel on a TDI diesel engine performance, combustion characteristics and particle number and size distribution emissions *Fuel* **117** pp 618-623 https://doi.org/10.1016/j.fuel.2013.09.037

[7] Selvam D J P, Vadivel K 2012 Performance and emission analysis of DI diesel engine fuelled with methyl esters of beef tallow and diesel blends Procedia Engineering **38** pp 342-358 (doi: 10.1016/j.proeng.2012.06.043)

Effect of injection system characteristics on performance, efficiency and emissions of agricultural tractor Diesel engines

Andrei Laurentiu Niculae[1], Radu Chiriac[1,3], Alexandru Racovitza[1] and Vlad Ungureanu[2]

[1]Faculty of Mechanical Engineering and Mechatronics, University Politehnica of Bucharest, Splaiul Independentei 313, 060042, Romania
[2]Faculty of Engineering in Foreign Languages, University Politehnica of Bucharest, Splaiul Independentei 313, 060042, Romania
[3]EA7341 CMGPCE of Conservatoire National des Arts et Metiers, Paris, France

racovitzaalexandru@gmail.com

Abstract. The development and the control of Diesel engines combustion process are fundamentally influenced by the parameters of the injection system, namely by the fuel injection characteristic and the injection rate shape. This work comparatively evaluates, by numerical simulation the influences of the injection system characteristics on performance, efficiency, and emissions of two tractor engine types, with four and three cylinders, with similar energetic configuration and same piston displacement. Therefore, the simulation applied two AVL-BOOST computation models, each of them using the AVL-MCC combustion equations system, with the calibration of the models being based on experimental data obtained by testbed investigations. The corresponding specification of the injection systems, together with those of the fuel spray characteristics simulated by the AVL Hydsim tool and the appropriate injection parameters have been adapted for both engines. The results highlighted that for the 3-cylinders engine, presenting lesser total displacement compared to the 4-cylinders one, it is possible to achieve in the condition of similar performance per cylinder, higher efficiency and lower Soot emissions.

1. Introduction

A large number of research studies has proved that diesel engine operation could be improved by correlating the use of the last generation biofuels with a new approach of the high-pressure injection strategy. Thus, various fueling biodiesel blends offer a certain opportunity in limiting the specific fuel consumption together with diesel engine emissions following the configuration of the injection characteristic in terms of timing, number of pulses or shape of the rate of injection (ROI). It is thus emphasized that one of the top priorities in the development of Diesel engines is the optimization of injection system parameters [1].

Many studies revealed that by increasing the injection fragmentation number and ensuring a fine-tuning of the start of injection (SOI), the above-presumed aims could be achieved. Therefore, a study involving a medium-duty common-rail 4-cylinder diesel engine using B20 and B50 fuels and tested at 2000 rpm and 60 Nm load, operated under modified injection strategies, with one, two and three injection pulses have been developed [2]. The findings highlighted a significantly lower level of NO_x, under 100 ppm, together with a slight decrease of smoke concentrations, when delaying the start of injection, in case of using B50 fuel, when applying the three-pulse split injection characteristic.

Another parametric study [3] developed on a 1.5 l Renault common rail DI diesel engine, fueled with diesel and B20 fuels proved the same trend of decreasing NO_x and particles emissions when using one pilot injection versus classic configuration, at maximum power and torque operating regimes, as an effect of the main-injection ignition delay and the maximum heat release values reduction.

Furthermore, the use of B40 in the operation of a modern single-cylinder common-rail direct injection engine has been proved to prevail in use with some other oxygenated additives. Testing the engine at its maximum load (3.5 kW) and speed (1500 rpm), using an injection pressure of 600 bar and starting the injection at latest 21 °CA BTDC, the results concerning the main emissions compared to the reference were down to 60% for B40 versus 7% when using n-butanol and DMC, operating under a rate of 15% EGR [4]. The recommendation to use lower EGR fractions, no higher than 10% when fueling the engine with biodiesel to reduce NO_x emissions without influencing other emissions or combustion efficiency is also expressed by other research [5]. Also, the injection pressure has an important influence on engine performance and efficiency. An increase from 200 to 350 bar leads to a decrease with 7% in brake specific fuel consumption at high load, and a decrease in CO, HC and smoke emissions with 7.5%, 21% and 35.5% for pine oil biodiesel blend P30 [6].

Another similar research program centred on a twin-cylinder turbocharged diesel engine, running at 2250 rpm, with correspondent BMEP within 2 and 4.15 bar established that increasing the injection pressure from 700 to 1100 bar is beneficial when fueling with B20 in terms of reducing by 30% Soot and NO_x emissions, especially when using multiple injections as compared to single injection [7]. The benefits of the injection pressure increase when fueling diesel engines with biodiesel blends are highlighted too by other studies [8-10]. As another main conclusion, it was revealed that multiple injections lead to an increase of the engine gross indicated efficiency and further investigations should point out a suitable connection between the level of the generated NO_x emissions and the dwell between the pilot and the main injections. The post injection of a slight amount of fuel in case of a medium dwell between main and post injections, could also play a significant role in balancing the NO_x and Soot emissions, without affecting the engine performance and efficiency [7].

For increasing the thermodynamic efficiency, the engine manufacturers developed different solutions to correlate the compression ratio with the combustion chamber architecture, the air excess and the injection system characteristics, therefore making the compression ignition engines more efficient and reliable. This raised the problem of self-ignition control of the air-fuel blend and the subsequent development of the combustion process. The latter one can be optimized using the link between the combustion process characteristics in conjunction with the organized fuel motions inside the combustion chamber and the particularities of the injection process.

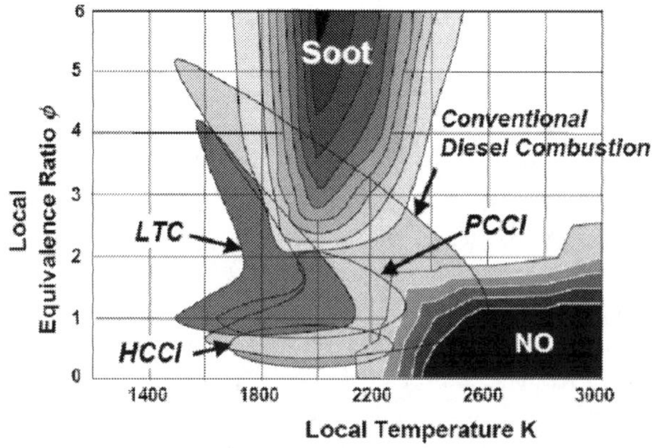

Figure 1. Comparison between LTC, HCCI, PCCI and CDC [11]

Different new technologies (figure 1) were proposed as combustion methods to maximize the engine performance with a reduced level of pollutant emissions and fuel consumption in comparison to conventional Diesel combustion (CDC). All these new technologies are related to the idea that low-temperature combustion (LTC) in Diesel engines offers the best tradeoff between low Soot and NO_x levels but conserving its high efficiency. Other such new technologies as homogeneous charge compression ignition (HCCI) or premixed charge compression ignition (PCCI) are still under research.

According to Shijin Shuai et al. [12], during the last decades, there was a rising interest in low-temperature combustion manifested because of the advantages regarding the low levels of NO_x and particulate matter (PM). Comparing to conventional Diesel combustion, this technique reduces NO_x and PM emissions at the same time by realizing a fuel blend before ignition maintaining the engine performance. For comparison, the homogeneous charge compression ignition technology uses the fuel blended with air and compressed towards the self-ignition point.

Starting from numerical simulation, this work is proposing to evaluate the influences of the injection system characteristics on the performance, efficiency and emissions of two tractor engines, with three and four cylinders, with similar energetic configuration and same piston displacement volume. The simulation is based on the application of two AVL-BOOST computation models, using the AVL-MCC combustion equations. The model calibration was made considering the experimental data. The AVL Hydsim software was used to study the injection system parameters, thus the injection characteristics were successively adapted for both engines.

The injection characteristics of a radial pump have been considered. The start of the injection and fuel delivery were mechanically and hydraulically controlled while a solenoidal valve being provided for fuel cutting [13].

Simulation results highlighted that for the engine with lesser total displacement volume, higher efficiency and lower Soot emissions is possible, with similar performance per cylinder.

2. Materials and Methods

2.1. Simulation of injection system characteristics using AVL Hydsim

For this project, the selected fuel injection pump was Delphi DP200. This unit is designed for direct injection Diesel engines that develop a power of no more than 25 kW per cylinder, with a configuration of two, three, four or six cylinders [14]. From a practical point of view, this pump model is typically used for agricultural traction engines. Figure 2 shows the model which was developed using the producer specifications.

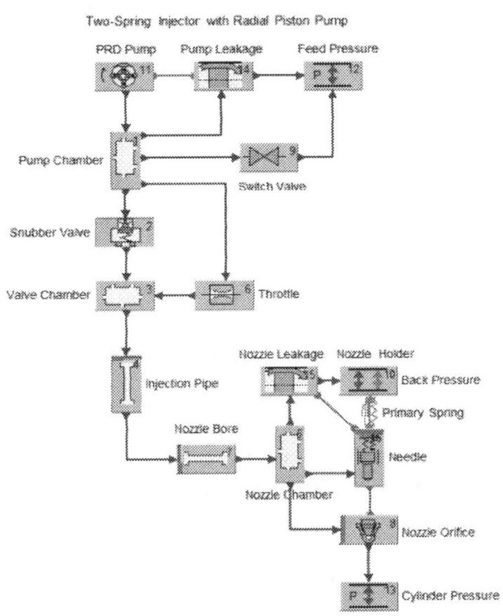

Figure 2. The Delphi DP200 simulation model developed using the AVL Hydsim software.

The engine speed was considered from 1200 to 2400 rpm, thus allowing the rotational pump velocity to vary from 600 to 1200 rpm. For higher engine speeds, the period of needle lift is longer, so a larger fuel quantity is injected. In the upper part of the speed domain, the start of injection is usually slightly advanced to avoid the shifting of the combustion process towards expansion. Figure 3 shows the simulation results as the variation of injection rate and the needle lift by crankshaft position for different engine speeds as they were obtained from the AVL Hydsim model. It can be observed that the injection rate profile and its duration and similarly the needle lift duration are changed in correlation with the engine speed. As the engine speed is increasing the duration of the needle lift, respectively of the injection rate profile are enlarged.

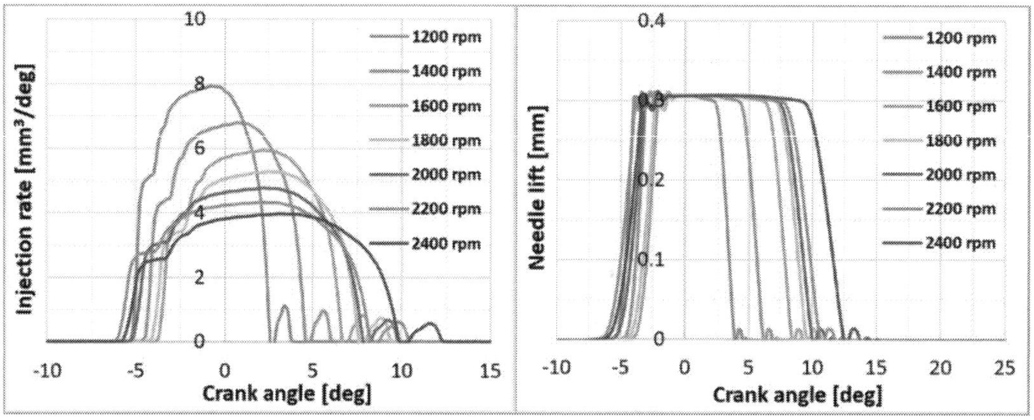

Figure 3. The injection rate and needle lift variation based on the crank angle.

Having these results, the delay corresponding to each operating condition was found and consequently, the injection rate was introduced in the engine model developed under AVL BOOST as the corresponding normalized rate of injection (figure 4).

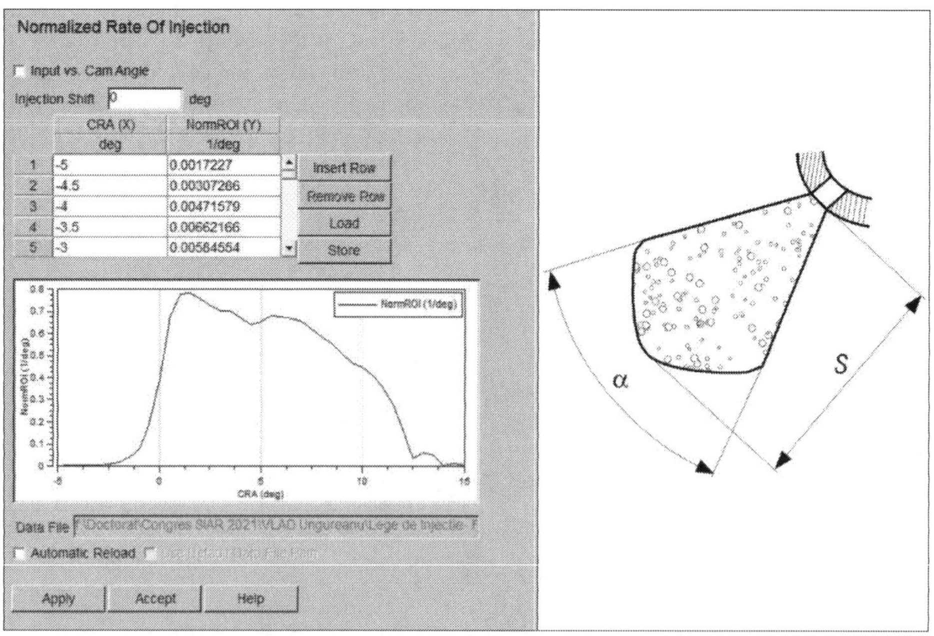

Figure 4. The normalized rate of injection and the representation of the spray cone angle (α) and fuel jet penetration (S).

The computation of the fuel quantity injected per cycle and cylinder was done considering the trapezoidal rule:

$$\int_a^b f(x)dx \approx \sum_{k=1}^{N} \frac{f(x_{k-1}) + f(x_k)}{2} \Delta x_k \tag{1}$$
$$= \frac{\Delta x}{2}(f(x_0) + 2f(x_1) + 2f(x_2) + \cdots + 2f(x_{N-1}) + f(x_N))$$

where: Δx is the difference between two consecutive crank angles and $f(x)$ represents the injected fuel quantity expressed in mm3/deg. The results obtained were thus registered, for instance, considering the engine speed of 2400 rpm this was 54.15 mm^3/ (cycle cylinder).

The fuel jet characteristics, as the spray cone angle α and the fuel jet penetration S (figure 4) resulted from the Hydsim model are slightly influenced by the rotational speed of the injection pump (engine speed).

Droplets of different sizes are contained by the fuel jet. The Sauter mean diameter $\bar{d}_{pS}$ is the average diameter of the individual droplet diameters d_{pi} which is used for the calculation of fuel vaporization, efficiency, and mass transfer. It can be defined as the equivalent diameter $\bar{d}_{32}$ of an average sphere with the same ratio volume by surface $\frac{6V}{S}$ of all droplets in the fuel jet which are split into *N* different classes, each *i* class containing n_i droplets [1, 15].

$$\bar{d}_{32} = \bar{d}_{pS} = \frac{6V}{S} \Rightarrow \bar{d}_{32} = \bar{d}_{pS} = \frac{\left(\sum_{i=1}^{N} n_i \cdot d_{pi}^3\right)}{\left(\sum_{i=1}^{N} n_i \cdot d_{pi}^2\right)} \tag{2}$$

The computation of the spray cone angle (α) and penetration (S) is based on the Arai, Tabata and Hyroyasu [16]:

$$\alpha = 0.05 \cdot \left(\frac{\Delta p \cdot d_{hole}^2}{\rho_G \cdot \upsilon_G^2}\right)^{0.25} \tag{3}$$

$$t_B = 28.65 \cdot \frac{\rho_D \cdot d_{hole}}{(\rho_G \cdot \Delta p)^{0.5}} \tag{4}$$

$$if\ t < t_B \qquad S = 0.39 \cdot t \cdot \sqrt{\frac{2}{\rho_D} \cdot \Delta p} \tag{5}$$

$$if\ t \geq t_B \qquad S = 2.95 \cdot \left(\frac{\Delta p}{\rho_G}\right)^{0.25} \cdot \sqrt{d_{hole} \cdot t} \tag{6}$$

where: Δp – pressure difference; d_{hole} – noozle hole diameter; ρ_G – gas density; υ_G – kinematic viscosity of gas; ρ_D – liquid density; t – time from injection start; t_B – breakup time.

Figure 5 shows the variation of the spray cone angle and Sauter mean diameter during the injection process which appears similar for different engine speeds. However, based on the simulation results several remarks can be undertaken:

- Regardless of the engine speed, the spray cone angle achieves finally approximately the same value of 8 degrees;

- The spray angle is slightly variable during the injection process displaying higher values at the beginning of injection and lower values at its ending;
- The Sauter mean diameter is variable during the injection process, lower values of 31 microns are registered at the beginning of injection and considerably higher values of 108 microns at the end;
- The average value of the Sauter mean diameter is 35 microns. This value could be characteristic for the main part of the injection process.

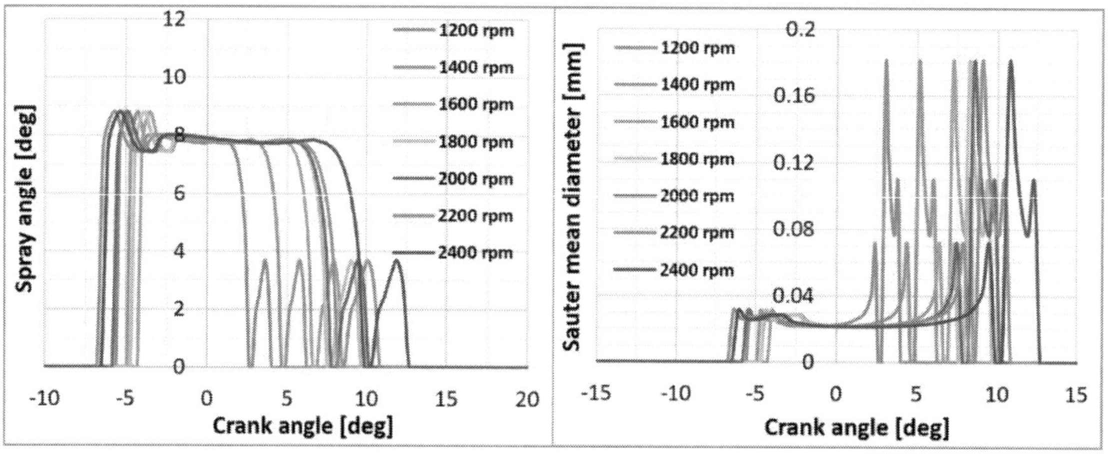

Figure 5. The spray angle and Sauter diameter relative to the crank angle.

One of the key parameters for defining the efficiency of how a direct injection Diesel engine is using the air existing in the combustion chamber is the fuel spray penetration S. This can be a piece of valuable information in the study of engine performance optimization. Figure 6 reveals the variation of fuel jet penetration and the injection pressure by the crank angle at different engine speeds. It can be noticed that regardless of the engine speed the maximum penetration is about 15.5 mm which could represent the piston bowl radius. The maximum injection pressure in the injector nozzle does not overpass 700 bar.

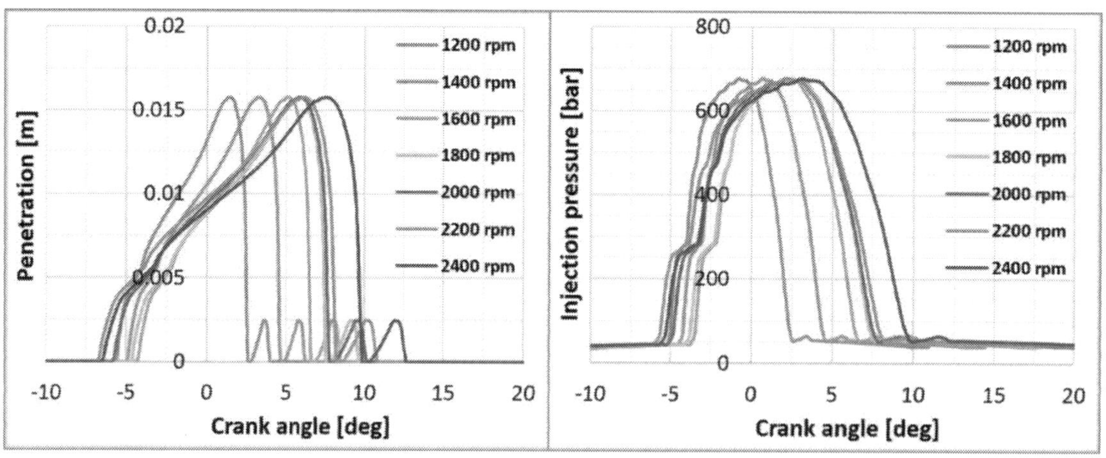

Figure 6. The spray penetration and the injection pressure relative to the crank angle.

2.2. *Engine implementation using AVL BOOST*

A comparative analysis between two agricultural tractor engines has been realized. The main criterion for choice was based on the compatibility with the Delphi DP200 Diesel fuel injection pump. Table 1 presents the technical specifications of these two Diesel engines considered.

Table 1. Engine datasheet

	Engine type	
	2302.053	2404.056
Number of strokes	4	
Bore [mm]	102	
Stroke [mm]	115	
Number of cylinders	3 vertical in-line cylinders	4 vertical in-line cylinders
Engine displacement [cm^3]	2819	3759
Rated power [kW] / Speed [rpm]	38 / 2300	56 / 2400
Maximum torque [Nm] / Speed [rpm]	170 / 1600	278 / 1400
Compression ratio	17.5	
Description of the combustion system	Direct injection, Waste-Gate Turbocharger	

Using the AVL BOOST software, a controlled virtual environment for test-benching the engines can be created. Figure 7 shows the models used for this study.

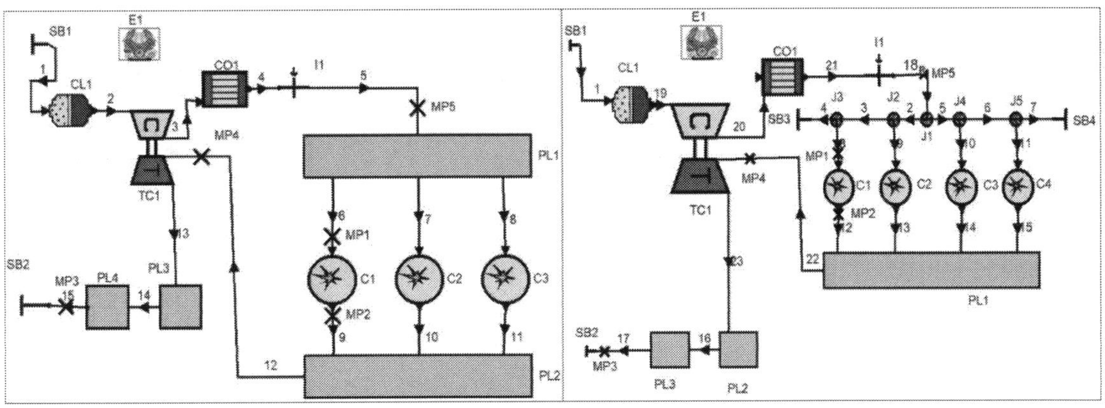

Figure 7. The AVL BOOST models developed for 3 and 4-cylinder engines.

The engine performance parameters have been calculated and they are plotted relative to the engine speed as follows in Figure 8. The variations of the effective power and effective torque for both engine types at full load within the considered range of speeds, obtained by tests and numerical simulation are presented. The simulation results are validated by the experimental data for the entire engine speed characteristic. The maximal rating power for the 4-cylinder engine generally corresponds to 56 kW at 2400 rpm and almost 38 kW was achieved for the maximal rating power in the case of the 3-cylinder engine at 2300 rpm. Maximum torque values were confirmed with the levels of 280 Nm at 1400 rpm for the 4-cylinder engine and 170 Nm at 1600 rpm for the 3-cylinder engine. Confirming the effective power and torque values is highly important in analyzing the further efficiency and emissions behaviour of the engines. As expected, higher values for the 4-cylinder engine were obtained, therefore an engine with a greater number of cylinders is preferable for high torque applications

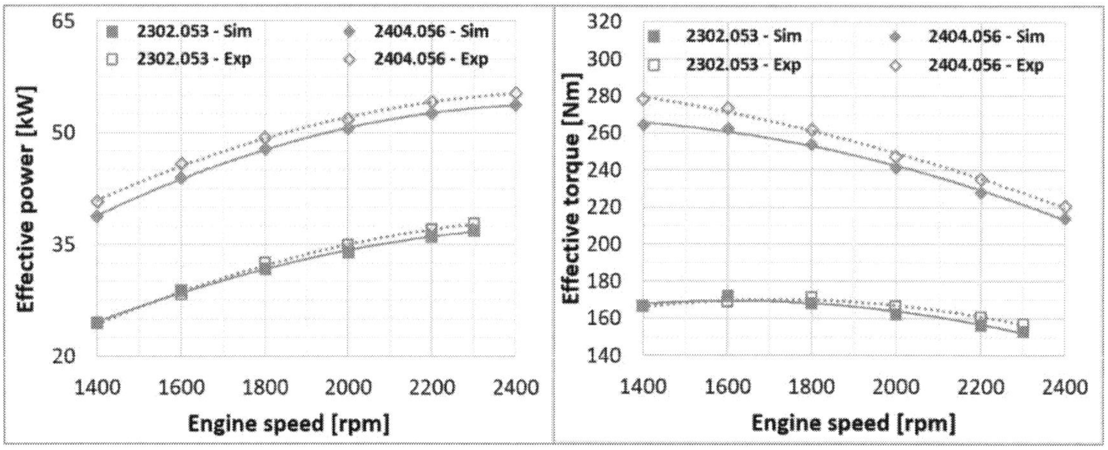

Figure 8. The effective power (P_e) and effective torque (M_e) variations relative to the crank angle.

Figure 9 depicts the variation of brake specific fuel consumption (BSFC) values obtained experimentally and by simulation relative to the engine speed. For both engines, the efficiency proves to be higher at low speed. The lowest specific consumption value is reached at about 220 g/kWh in the case of 2302.053 engine type. Noticing that all the BSFC value points for the 3-cylinder engine are below the correspondent values of the 4-cylinder engine, a first assumed issue of the ongoing research has been attended. The BSFC at a low speed of 1400 rpm has roughly the same value of 221 g/kWh for both engines, while at a high speed of 2300 rpm there is a gain in efficiency of 6.4% for the 3-cylinder engine, with 233 g/kWh in respect to 248 g/kWh for the 4-cylinder engine. The simulation results keep the same trend as experimental data but with the BSFC values shifted upwards by approximately 3.5%.

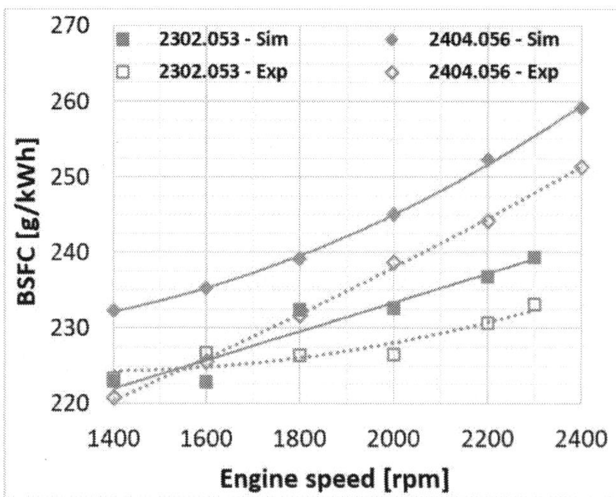

Figure 9. The brake specific fuel consumption (BSFC) variation relative to the crank angle.

The maximum values of in-cylinder temperature and pressure are plotted in Figure 10. Given the reduced number of cylinders, for a 4-stroke engine, the 2302.053 TCI moves its pistons faster to complete a full cycle, thus more heat and higher pressures and temperatures are achieved.

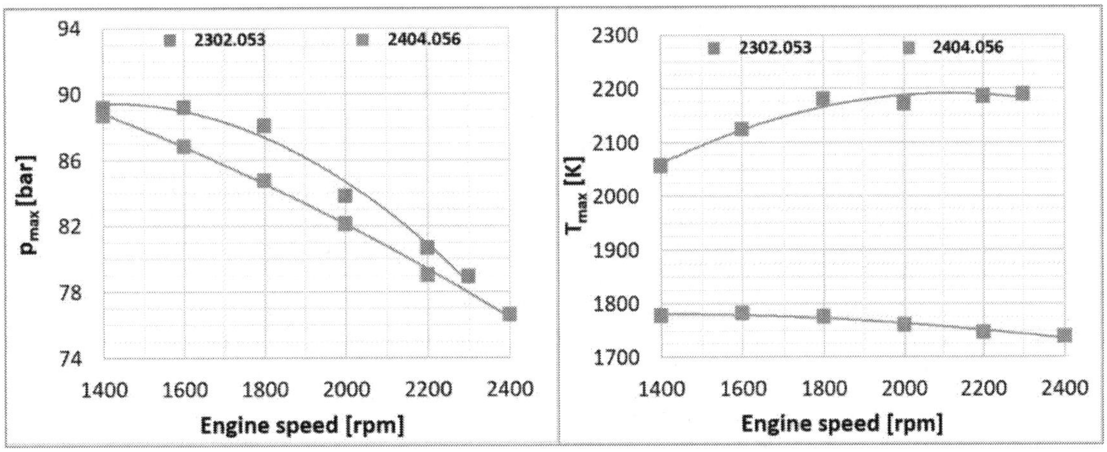

Figure 10. The maximum in-cylinder temperature and pressure values relative to the crank angle.

The pollutant emissions values of nitrogen oxides (NO_x) and Soot obtained by simulation for both engines are shown in Figure 11. Over the whole speed domain, higher NO_x and lower Soot levels are achieved for the 3-cylinder engine-while for the 4-cylinder engine the situation is reversed. Once again, it is emphasized that is not easy to find the balance between NO_x and Soot and if the efficiency is also considered, the 3-cylinder engine seems to be more efficient and less pollutant. Such results justify the actual trend existing for the engine's development of reducing total engine displacement.

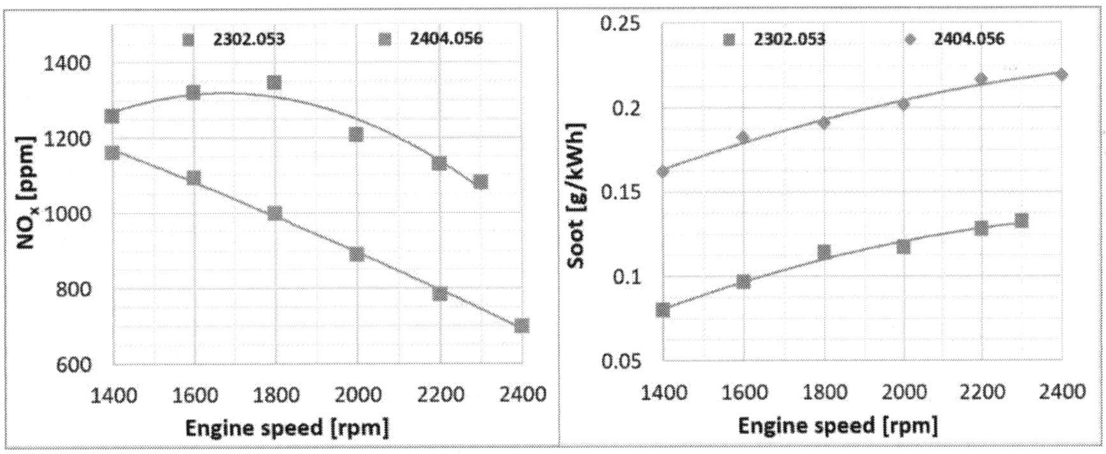

Figure 11. The nitrogen oxides (NO_x) and Soot values relative to the engine speed.

3. Conclusion

The simulation models have been calibrated using the experimental data recorded for the operation of both engines on pure diesel fuel.

The simulation results show that by decreasing the engine total displacement by 25% from 4 to 3 cylinders it is possible to achieve an improvement in the engine efficiency of 5% under the same operating conditions while maintaining the supercharging pressure and injection characteristics.

At the same time for the 3-cylinder engine, an average decrease in Soot emissions values by 44% and an average increase in NO_x by 31% is obtained.

The study will be extended to the same engines fueled with biodiesel blends B20 and B30 figuring out the possibility to reduce NO_x emissions by correlating their energetic formula with the injection characteristics and the fuel properties.

Acknowledgements
The authors of this paper acknowledge the AVL-AST team from AVL List GmbH for the special support offered under the University Partnership Program on performing the simulation activity.

References
[1] Chiriac R 2012 Internal Combustion Engines. Basic Operation Principles ed A.G.I.R. Bucharest
[2] How H G, Masjuki H H, Kalam M A and Teoh Y H 2018 Influence of injection timing and split injection strategies on performance, emissions, and combustion characteristics of diesel engine fueled with biodiesel blended fuels *Fuel* **213** 106–14
[3] Plamondon E and Seers P 2019 Parametric study of pilot–main injection strategies on the performance of a light-duty diesel engine fueled with diesel or a WCO biodiesel–diesel blend *Fuel* **236** 1273–81
[4] Jayabal R, Thangavelu L and Subramani S 2020 Combined effect of oxygenated additives, injection timing and EGR on combustion, performance and emission characteristics of a CRDi diesel engine powered by sapota biodiesel/diesel blends *Fuel* **276** 118020
[5] Jaliliantabar F, Ghobadian B, Carlucci A P, Najafi G, Mamat R, Ficarella A, Strafella L, Santino A and De Domenico S 2020 A comprehensive study on the effect of pilot injection, EGR rate, IMEP and biodiesel characteristics on a CRDI diesel engine *Energy* **194**
[6] Saravanan C G, Raj Kiran K, Vikneswaran M, Rajakrishnamoorthy P and Yadav S P R 2020 Impact of fuel injection pressure on the engine characteristics of CRDI engine powered by pine oil biodiesel blend *Fuel* **264** 116760
[7] Chacko N, Rajkumar S and Thangaraja J 2021 Experimental and modeling analysis of multiple-injection strategies with B20 operation in a CRDI engine *Fuel* **293** 120433
[8] Shameer P M and Ramesh K 2018 Assessment on the consequences of injection timing and injection pressure on combustion characteristics of sustainable biodiesel fuelled engine *Renew. Sustain. Energy Rev.* **81** 45–61
[9] Rajan K, Sujith V, Ganesan M, Peer Haroon M, Mathivanan S D and Elumalai R 2020 Performance and emissions characteristics of DI diesel engine using biodiesel blend with different injection pressures *Mater. Today Proc.* **33** 4699–702
[10] Deokar A J and Harari P A 2021 Effect of injection pressure, injection timing and nozzle geometry on performance and emission characteristics of diesel engine operated with Thevetia peruviana biodiesel *Mater. Today Proc.* Article in press
[11] Neely G D, Sasaki S, Huang Y, Leet J A and Stewart D W 2005 New diesel emission control strategy to meet US tier 2 emissions regulations *SAE Tech. Pap.*
[12] Shuai S, Abani N, Yoshikawa T, Reitz R D and Park S W 2009 Evaluation of the effects of injection timing and rate-shape on diesel low temperature combustion using advanced CFD modeling *Fuel* **88** 1235–44
[13] Diesel-Engine Management: An Overview Bosch 2003
[14] Delphi DP 200 Workshop Manual 2008
[15] Azzopardi, Barry J Sauter Mean Diameter https://dx.doi.org/10.1615/AtoZ.s.sauter_mean_diameter
[16] AVL Hydsim Theory 2019

IOP Conf. Series: Materials Science and Engineering 1220 (2022) 012003 doi:10.1088/1757-899X/1220/1/012003

Experimental studies on the influence of diesel quality and additive for fuel on the smoke emission of diesel cars

S M Burciu

Engineering Faculty, Thermal Systems and Automotives Department, "Dunarea de Jos" University of Galati, Galati, Romania

Mugurel.Burciu@ugal.ro

Abstract. The paper presents experimental determinations that highlight the substantial reduction of smoke in the exhaust gases for a car with a diesel engine with pollution norm Euro 4 in the case of using OMV MaxxMotion diesel and Super Diesel additive from Liqui Moly. For the load determinations when running on a horizontal road (by simulation on the chassis dynamometer V-Tech Dyno VT-2/B1), in case of using diesel with CN55 and Super Diesel additive from Liqui Moly, was found a smoke reduction of up to almost 50%, the larger reductions being at speeds above 110 km/h, which indicates a substantial reduction in smoke when running in extra-urban mode. At the same time, only the superior quality of diesel with cetane number 55 compared to that with cetane number 51, led to better combustion, so that the smoke emission was reduced by up to 33% at high speeds, this reduction being quite important up to 27% for lower speeds below 90 km/h.

1. Introduction

Exhaust gases from internal combustion engines that perform propulsion cars contain significant amounts of carbon monoxide, carbon dioxide, unburned or partially oxidized hydrocarbons, nitrogen oxides and smoke. For the compression ignition engine, according to the current anti-pollution legislation, it is important to measure the degree of smoke, the concentrations of nitrogen oxides, hydrocarbons and carbon dioxide emissions [3, 4, 6, and 7]. The composition of the exhaust gases depends on: engine speed and load, air/fuel ratio, engine wear, fuel system adjustment/ deregulation, turbocharger technical condition and, of course, the anti-pollution norm of the system implemented on the car [6-8]. All vehicles equipped with a diesel engine have been required to significantly reduce their nitrogen oxide emissions with the entry into force of the Euro 6 anti-pollution standard. For example, nitrogen oxide emissions from motor vehicles must be limited to 80 mg/km (meaning an additional reduction of 56% compared to Euro 5, respectively a reduction of 68% compared to Euro 4). Smoke emissions from diesel vehicles must also be reduced to be limited to 5 mg/km (meaning an additional 80% reduction compared to the Euro 4 standard) [3-5].

2. Test equipment used to determine the opacity coefficient

The gas analyzer with opacimeter module type 'Ultima 600', existing within the department of Thermal Systems and Automotive Engineering of the Faculty of Engineering from Galati, allows measurement of the following sizes and pollutant emissions, with the intervals and accuracy indicated in table 1.

IOP Conf. Series: Materials Science and Engineering 1220 (2022) 012003 doi:10.1088/1757-899X/1220/1/012003

Table 1. Opacimeter 'Ultima 600' characteristics.

Emission, size	Measuring range	Precision
Diesel		
Opacity coefficient (smoke index) K	0 – 9,99 m^{-1}	0,01 m^{-1}
Engine rotation	300 –7000 rpm	10 rpm
Lubricating oil temperature	0–150 degrees C	1 degree C
Exhaust gases temperature	0 –150 degrees C	1 degree C

To measure the smoke index, the 'Ultima 600' opacimeter uses the absorption measurement method. The evaluation of the smoke opacity is made by determining the degree of reduction of the light intensity emitted by a source when crossing a pipe of a given length in which the analysed gases circulate. The opacity coefficient (smoke index) K results from the relation that expresses the amount of light that reaches the photoelectric cell:

$$\Phi = \Phi_0 \cdot e^{-K \cdot L} \Rightarrow K\left[m^{-1}\right] = \frac{1}{L} \cdot \ln \frac{\Phi_0}{\Phi} \tag{1}$$

Where: Φ_0 – the luminous flux emitted by the light source, which has a constant value; Φ – the luminous flux that reaches the photoelectric cell; L – the distance between the light source and the photoelectric cell, which has a constant value.

The experimental determinations were performed using the *Renault Laguna III 1.5 dCi* car (110 HP/ 4000 rpm, 2009), Euro 4 pollution standard, having a fuel tank of 66 liters, 6-speed manual transmission for moving forward, a car that is not equipped with a particulate filter (FAP).

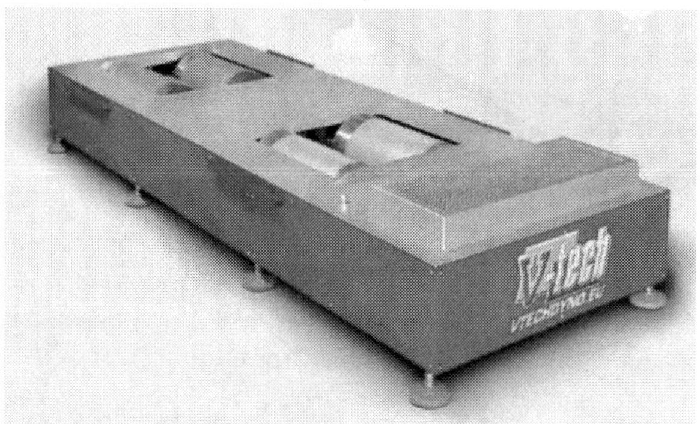

Figure 1. Chassis dynamometer VT-2M/B1- Dyno.

The load determinations were performed using the chassis dynamometer VT-2M/B1- Dyno type (figure 1) using the road method, and for measuring the smoke index was used the Ultima 600 Johnson Controls Automotive Electronic opacimeter indicated above.

The following diesel fuels and additives were used:

OMV diesel fuel (cetane number CN51) having the price per liter 1.122 Euro (at the end of May 2021): With OMV Diesel's unique additive package, the engine benefits from extra protection against corrosion and deposits [1].

OMV MaxxMotion diesel fuel (cetane number CN 55) having the price per liter 1.205 Euro (at the end of May 2021) [1]: Facilitates the optimal operation of the car at very low temperatures up to -38 degrees Celsius.

Unique cleaning molecules help keep the engine clean by removing long-term deposits from the injectors.

Prevents the corrosion phenomenon that occurs over time when using a regular fuel and protects engine subassemblies.

Protection against corrosion of the fuel system and the tank contributes to the optimal operation of the engine for a longer time.

Liqui Moly Super Diesel Additive, identification code (LM 8379), having the price 4.878 Euro for 250 ml dose of additive [2]:

Ensures cleanliness and prevents the formation of deposits in the fuel system and in the combustion chamber. Keeps the injectors clean, ensures better combustion and thus lower specific fuel consumption, increasing engine power. Increases the cetane number of diesel, ensuring an efficient combustion process.

Increases the lubrication effect of low sulfur diesel (according to DIN EN 590) and protects the injection pump against wear. Contains antioxidants and prevents corrosion.

3. Experimental results

The smoke index was determined at load on the chassis dynamometer VT-2M/B1-Dyno type using the road method.

Determinations for the smoke index were performed at various stabilized running speeds, specific to the movement of the car on an urban and extra-urban road, using the indicated equipment, figure 2. The determinations were performed considering the running of the car on a horizontal road.

Figure 2. Determinations on chassis dynamometer.

Based on the values of the measured smoke index, the following graphs were made, figure 3 and figure 4:

- Variation of the smoke index depending on the speed for running in the 5th gear of the manual gearbox, between speeds 70 km/h and 150 km/h, variation presented in figure 3;
- Variation of the smoke index depending on the engine speed for its operation between 1750 rpm and 3700 rpm for running in 5th gear of the manual gearbox, the variation shown in figure 4.

Following the load determinations on the chassis dynamometer VT-2M/B1-Dyno type, using the road method, the following aspects can be indicated:

- In the case of using diesel fuel with CN55 (OMV MaxxMotion) and Super Diesel additive from Liqui Moly identification code (LM 8379), compared to diesel fuel CN51 (OMV

Diesel), there was a reduction of smoke by 35-48%, the biggest reductions being at speeds over 110 km/h, an aspect that indicates a substantial reduction of smoke when driving outside the cities;

- In the case of using only OMV MaxxMotion diesel fuel with CN55, compared to normal OMV Diesel CN51, a reduction of smoke was found by 26 -33%, the larger reductions being at speeds over 110 km/h, an aspect that indicates a substantial reduction of smoke when driving out of town.

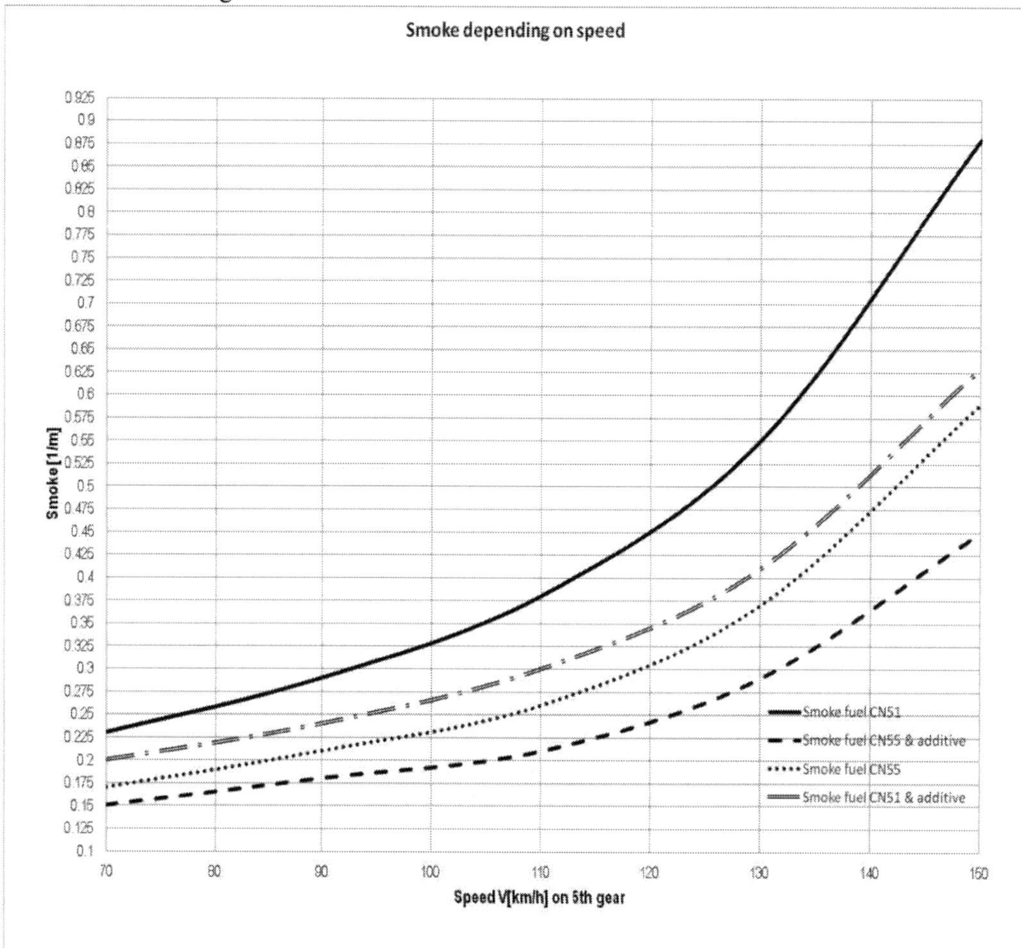

Figure 3. Variation of the smoke index depending on the car speed.

In the case of using diesel fuel with CN51 (OMV Diesel) and Super Diesel additive from Liqui Moly identification code (LM 8379), compared to diesel fuel CN51 (OMV Diesel), there was a reduction of smoke by 13-28%, the biggest reductions being at speeds over 130 [km/h], an aspect that indicates an important reduction of smoke when running outside the cities.

In the case of using diesel fuel with CN55 (OMV MaxxMotion) and Super Diesel additive from Liqui Moly identification code (LM 8379) compared to OMV MaxxMotion diesel fuel with CN55, a reduction of smoke was found by 11-25%, the larger reductions being at speeds over 110 [km/h], an aspect that indicates a substantial reduction of smoke when running in extra-urban mode.

For all four fuel variants used and indicated above, the smoke level is well below the limit of 1.5 [m-1], [3-4], thus indicating a good condition of the engine mechanism and a proper operation of the fuel and supercharging system.

For driving with approximately constant speeds (stationary speeds) up to 100 km/h, that means driving inside of cities, a reduction of smoke emission between 10% and 40% was found, the positive influence only of diesel quality being up to 30% and that of the additive up to 20%.

In the case of using diesel with CN55 and the Super Diesel additive, there was a significant reduction up to almost 50% in the amount of smoke compared to the situation in which the classic diesel with CN51 was used.

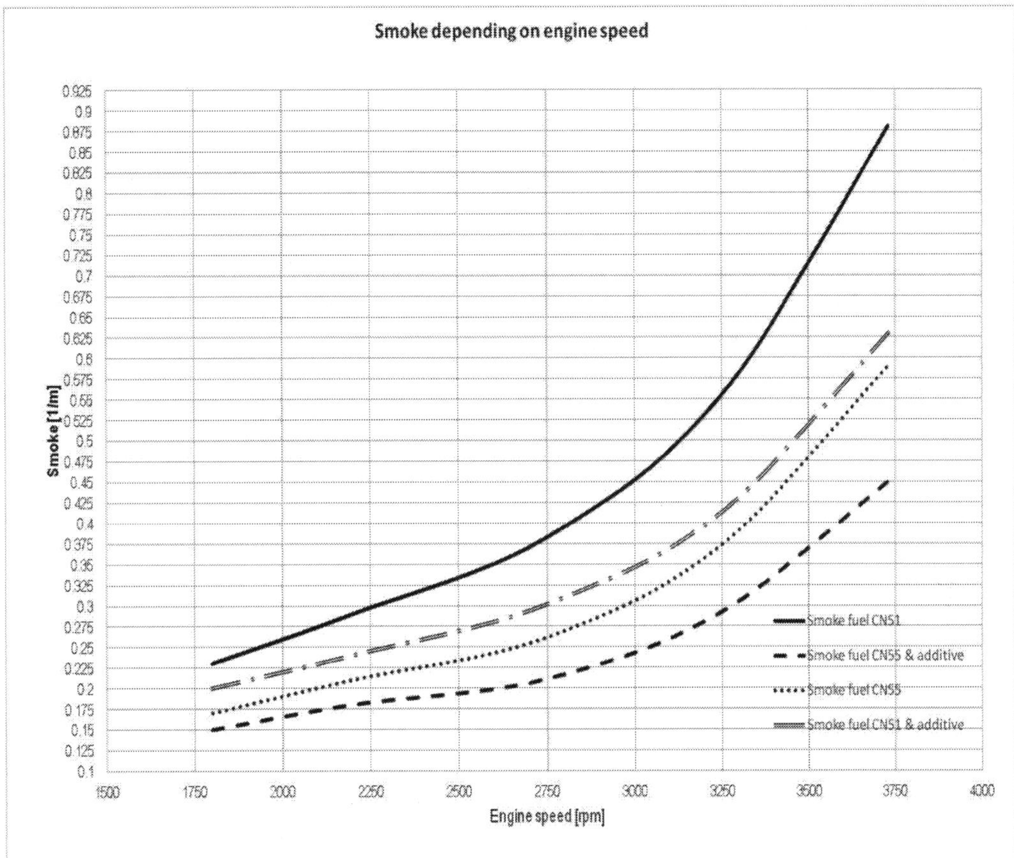

Figure 4. Variation of the smoke index depending on the engine speed.

We can indicate that for this variant which implies the largest price difference between the two types of fuel (OMV MaxxMotion diesel fuel with CN55 completed with the 250 ml dose of Super Diesel additive comparted to OMV Diesel CN51) for only an additional cost of *0.16 Euro / liter of fuel (i.e. a 14% increase in price)*, can be achieved a substantial reduction of pollution produced by the engine when running in urban and extraurban respectively, under appropriate technical conditions of the engine mechanism, of the injection and supercharging system.

4. Conclusion

In the cases presented, the greatest reduction in smoke up to almost 50%, was obtained when used OMV MaxxMotion diesel with CN 55 together with the Liqui Moly Super Diesel Additive, identification code (LM 8379) 250 ml dose. Substantial reduction of the smoke index for the car with the indicated Euro 4 pollution norm (which does not have a particulate filter FAP in the anti-pollution system) in case of using OMV MaxxMotion diesel with CN 55 and Super Diesel additive from Liqui Moly or in case of use only a better quality of diesel fuel, can be explained by improving the combustion process in the sense of increasing the combustion speed and combustion efficiency of the fuel mixture in the combustion chamber, which causes the reduction of the unburned HC chains (which lead to the formation of smoke).

We can indicate that for the variant which implies the largest price difference between the two types of fuel (OMV MaxxMotion diesel fuel with CN55 completed with the 250 ml dose of Super Diesel additive compared to OMV Diesel CN51) for only an additional cost of *0.16 Euro / liter of fuel (i.e. a 14% increase in price)*, can be achieved a substantial reduction of pollution produced by the engine when running in urban and extraurban respectively (smoke emission reduction of up to 48%), under appropriate technical conditions of the engine mechanism, of the injection and supercharging system.

As it is shown in figure 3 and figure 4, even for approximately constant speeds (stationary speeds) of driving up to 100 km/h, a general reduction of smoke emission between 10% and 40% was found, the positive influence only of diesel quality being up to 30% and that of the additive up to 20%.

The better quality of diesel (diesel with cetane number 55 compared to that with cetane number 51) will determine a better burning speed and combustion efficiency of the fuel mixture in the combustion chamber, which will reduce smoke emission by 26% up to 33% which is an important reduction even for lower speeds (up to 90 km/h).

For the load determinations when driving on a horizontal road (by simulation on the chassis dynamometer V-Tech Dyno VT-2/B1), in case of using MaxxMotion diesel fuel with CN55 and the Super Diesel additive from Liqui Moly, was found a smoke reduction by 35-48%, the reductions more being at speeds over 110 km/h, an aspect that indicates a substantial reduction in smoke when driving outside the cities.

References

[1] https://www.omv.ro/ro-ro/statii-de-distributie-carburant/carburanti-statii/omv-maxxmotion-carburanti-de- performanta/omv-maxxmotion-performance-diesel.
[2] https://limorom.ro/produs/aditiv-super-diesel-liqui-moly-1900-1973-8379-250-ml/.
[3] ***, RNTR 7, RAR.
[4] ***, RAR, http://www.rarom.ro.
[5] S.M.Burciu – Laboratory work: Part I - Car dynamics. Part II - Car diagnostics, *GUP Galati* 2017.
[6] S. M. Burciu, G Coman, Experimental study of NO_x and CO_2 emissions at load regimes for a passenger car powered by direct injection diesel engine, Available: https://iopscience.iop.org/article/10.1088/1757-899X/595/1/012035, 2019.
[7] S.M. Burciu, G Coman, A comparative study of smoke emissions between idling regimes and high load regimes for a standard diesel direct injection car, Available: https://iopscience.iop.org/article/10.1088/1757-899X/444/7/072018.
[8] Gelu Coman, Mugurel Salvadore Burciu, Nicuşor Baroiu, Vehicles emissions under different driving conditions in urban areas, DOI 10.37358/rc.19.2.6930, *Revista de Chimie* 2019, ISSN: 0034-7752.

IOP Conf. Series: Materials Science and Engineering 1220 (2022) 012004 doi:10.1088/1757-899X/1220/1/012004

The study on the influence of utilizing n-butanol at fuelling spark ignition engines

Cristian Sandu[1], Constantin Pană[1], Niculae Negurescu[1], Alexandru Cernat[1], Cristian Nuțu[2] and Rareș Georgescu[1]

[1]Thermotechnics, Engines, Thermical and Frigorific Equipment Department, University Politehnica of Bucharest, Bucharest, Romania
[2]Automotive Engineering Department, University Politehnica of Bucharest, Bucharest, Romania

sandu_cristian2007@yahoo.com

Abstract. The ever-increasing emissions restrictions on internal combustion engines have led researchers into the study of alternative fuels solutions. As hybrid and electric vehicles started becoming more and more present in the markets worldwide, the popularity of conventional internal combustion engines started to decrease (we see this especially with diesel fueled engines). For spark ignition engines, alcohols and gasoline-blended mixes have proven to be an attractive solution in the last years. Out of these alcohols, we mention ethanol, methanol and butanol. Butanol is a promising solution because of its high oxygen content with the possibility of improving the combustion process and thus even reducing emissions. Butanol also has a high combustion speed and can potentially reduce the combustion duration while improving the overall thermal efficiency. High miscibility is another important aspect of butanol, allowing a higher percentage volume of butanol to be mixed with gasoline. The additional oxygen content may also improve combustion stability thus reducing cyclic variability. The objective of this study is to determine what is the impact of fueling a spark ignition engine with a blend of 10% vol. n-butanol and 90% vol. gasoline. The study will look at combustion stability, variability, thermal efficiency and emissions. A baseline was established at fueling the engine with pure gasoline at an engine speed of 2500 min^{-1} and an engine load of 55%. After the baseline, the same measurements were done at fueling with a blended mix of n-butanol and gasoline.

1. Introduction

Globally we are faced with a large fossil fuel consumption, which is not sustainable for more than a few more decades. This alarming use of fossil fuel has led researchers into the study of alternative fuel solutions. Large fuel consumptions have led to high emissions of greenhouse gases such as CO_2 but also other pollutants, especially the transportation sector, such as carbon monoxides CO, unburnt hydrocarbons HC and nitrogen oxides (NO_x). Regarding the transportation sector, special attention was given to the study of alcohol blends such as methanol, butanol, ethanol, etc. According to studies done by Gautam M. and Martin II D.W [1, 2] and Yacoub Y. et al. [3], the use of blended conventional fuels such as gasoline or diesel with alcohols may improve the thermal efficiencies of the internal combustion engines and reduce the fuel consumption. A study conducted by Wu. M. et al, [4] concluded that the use of butanol, produced from corn, in the transportation sector may lead to a decrease of gasoline consumption by 39-56% and greenhouse emissions by up to 48% on a life-cycle basis.

The analysis of butanol alcohols indicates certain advantages over short-chain alcohols such as lower viscosity, better blending properties with conventional fuels and higher energy density, [5]. Butanol is also less corrosive than ethanol making it ideal for classic fuel distribution systems, [6]. Two studies conducted by Liu et al. [7, 8] concluded that the use of n-butanol has led to lower soot concentrations when compared to biodiesel. Also, Liu et al. [9] have studied multiple fuels and their effect on pollutant emissions and have concluded that oxygenated blends have little impact on engine emissions. A different study by Yuqiang Li et al. [10] analyzed the effects of fueling a spark ignition engine with isopropanol-n-butanol-ethanol and gasoline. The blending percentage was 30% IBE (isopropanol-n-butanol-ethanol) and 70% gasoline. Results show an improvement of the thermal efficiency by 4,3% and lower pollutant emissions, 4% for carbon monoxide, 15-20% reduction of unburnt hydrocarbons and in some cases a reduction of nitrogen oxides by up to 18,6%. When analyzing the combustion process at fueling with butanol blends, J. Serras-Pereira et al. [11] concluded that butanol is more difficult to combust in cold engine conditions. They also noted the initial combustion phase has a longer duration between 5-7°CA than conventional gasoline. In a study by Alasfour et al. [14] a 7% power loss was observed when fueling a spark ignition engine with a blend of 30% butanol and 70% gasoline. Alasfour et al. [15] also studied the effect of the 30% butanol-gasoline blend on emissions, concluding a decrease in pollutants like CO and HC for λ between 0,9 and 1,05 and also a decrease by 9% for NO_x. Increasing the engine spark timing will lead to higher NO_x emissions was observed by Alasfour in his study *NOx emission from a spark-ignition engine using 30% iso-butanol – gasoline* [16].

The objective of this paper is to determine the effects of fueling a 4-cylinder spark ignition engine with a 10% n-butanol-gasoline blend on energetic performances, emissions and cyclic variability over 250 engine cycles. The engine is turbocharged and was firstly fueled with pure gasoline in order to set up the reference at the engine load χ=55%, engine speed of n=2500 min^{-1} and different air-fuel ratios (λ). After setting the reference the engine was fueled with n-butanol-gasoline (10% vol. butanol - 90% vol. gasoline) with the same engine adjustments.

2. Experimental test bench

The engine used is a 4 cylinder, turbocharged, 1,5l type Cielo Nubira A15MF. The engine is fueled by only gasoline and the engine load of χ=55% and speed of 2500 min^{-1}; different parameters are recorded such as air and water temperature, intake air temperature, exhaust gases temperature, fuel consumption, air consumption, boost pressure and emissions level. In-cylinder pressure is measured across 250 engine cycles and the mean values of these measurements are used to determine the maximum pressure (p_{max}), maximum pressure rise rate $(dp/d\alpha)_{max}$, heat release rate ($dQ/d\alpha$) and total heat released (Q). In order to determine the cycle-to-cycle variation during the combustion process, the coefficients of variability (COV) of maximum pressure, mean effective pressure (IMEP), the angles of 5%, 50% and 90% of mass fraction burnt (MFB) are calculated.

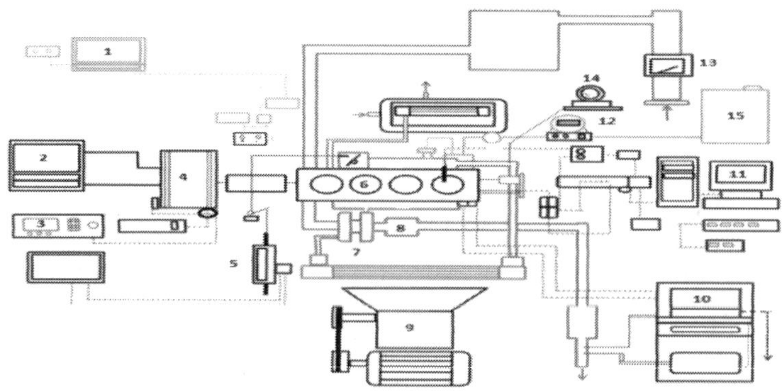

Figure 1. Test bench schema.

The main components of the test bench are: 1 – ECU, 2 – dyno cooling system, 3 – dyno control cabinet, 4 – Eddy current dyno, 5 – throttle body actuator, 6 – A15MF engine, 7 – turbocharger, 8 – catalytic converter, 9 – radiator cooling system, 10 – gas analyzer, 11 – desktop, 12 – fuel mass flow transducer, 13 – air flow transducer, 14 – boost pressure transducer, 15 – fuel reservoir.

3. Results

In the graphs below the energetic performances of the engine are presented (figures 2-11) at fueling with normal gasoline (in blue) and at fueling with 10% n-butanol-gasoline blend (in red).

The pressure diagram is presented in figure 2 for both fuels. The engine adjustments were kept the same at fueling with both fuels so for the blended fuel the initial combustion phase was longer and moved to the expansion (to the right of the TDC) as illustrated below. The maximum pressure is higher for conventional gasoline vs blended fuel. The maximum pressure, for both fuels at different air-fuel ratios, is presented in figure 3.

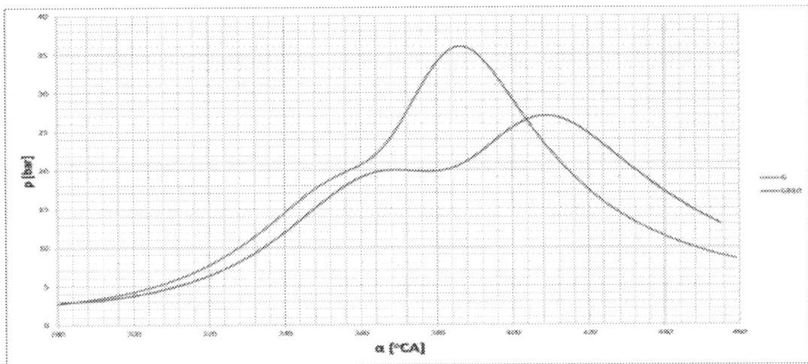

Figure 2. Pressure diagrams.

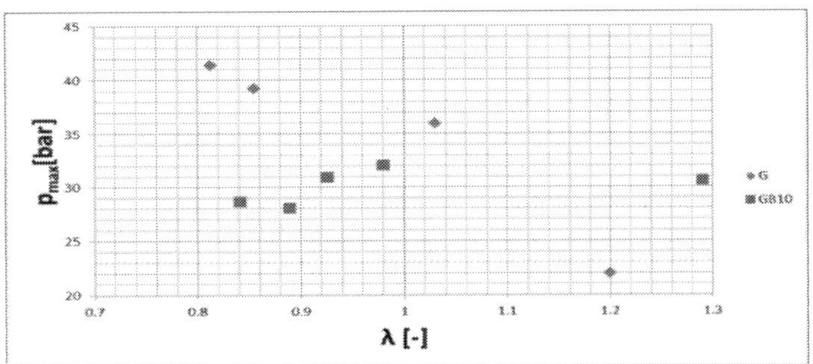

Figure 3. Maximum cylinder pressure.

Higher maximum pressures are observed for pure gasoline (blue) at all air-fuel ratios. At rich air-fuel ratios, gasoline has higher maximum pressures by up to 25% vs blended mix. At leaner mixtures, the gap decreases because of the additional oxygen content of the butanol. For $\lambda \cong 1{,}3$ the n-butanol-gasoline mix has a higher maximum pressure vs pure gasoline. Using the in-cylinder pressure measurements, the maximum pressure rise rate $(dp/d\alpha)_{max}$ was determined and is presented in figure 4. The same observation is available, pure gasoline has higher pressure rise rates at rich λ vs blended mix but lower values at lean values.

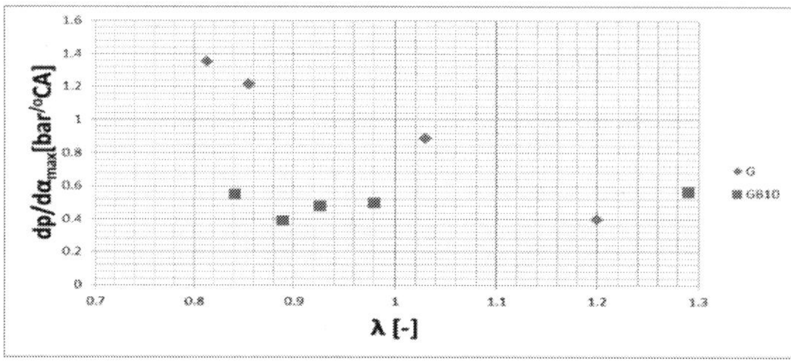

Figure 4. Maximum pressure rise rate.

The heat release rate was calculated with the relation:

$$\frac{dQ}{d\alpha} = \frac{V}{k-1} \cdot \frac{dp}{d\alpha} + \frac{k}{k-1} \cdot p \cdot \frac{dV}{d\alpha} \tag{1}$$

where:

- $\frac{dU}{d\alpha}$ -the internal energy variation;
- $\frac{dL}{d\alpha}$ - the mechanical work variation;
- k - adiabatic exponent;
- V, p – the volume and pressure.

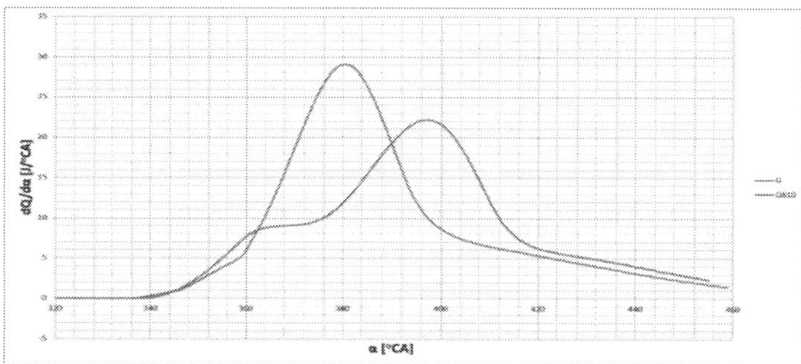

Figure 5. Heat release rate.

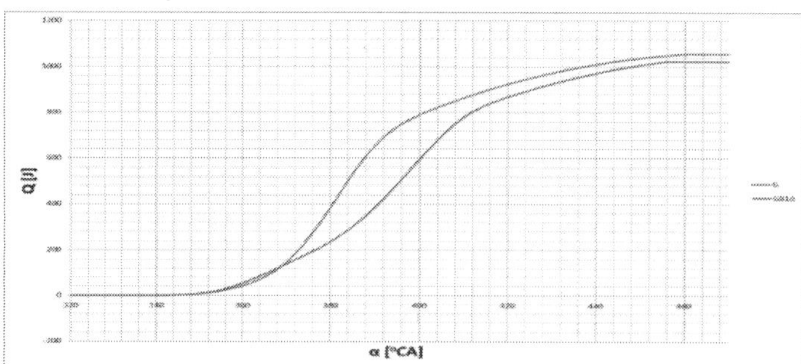

Figure 6. Heat release laws.

The heat release rate and heat release laws are presented in figures 5 and 6. It is observed that gasoline has a faster and higher release rate vs the blended mix. Because the same engine adjustments are used (e.g., same spark timing), the blended mix moves further into the expansion during the combustion process. By integrating the first relation (1) the heat release laws were obtained (figure 6). It is observed that the combustion of pure gasoline occurs faster than the blended mix but also that the total heat released at the end is almost similar in both cases.

It is observed in the figure below that the brake specific fuel consumption for the n-butanol-gasoline blend is higher compared to standard gasoline. There are two reasons for this, 1) the same spark timing was used at fueling with the blended mix, the initial combustion phase increased and the entire process moved into the expansion and the thermal efficiency decreased (figure 8), 2) Butanol has a lower calorific heat compared to gasoline.

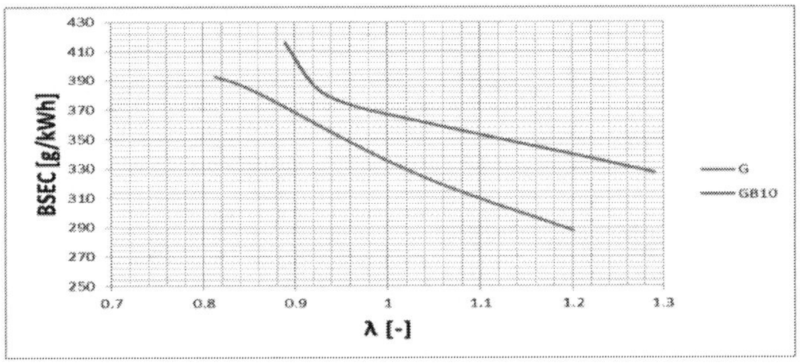

Figure 7. Brake Specific Energy Consumption (BSEC).

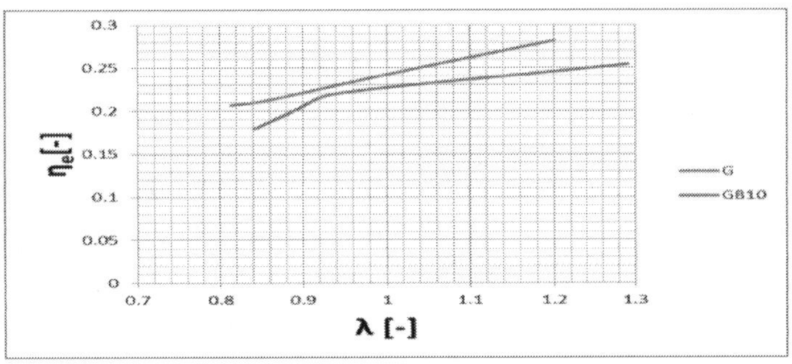

Figure 8. Brake thermal efficiency.

The engine operation becomes more stable in the lean mixtures zone at fueling with the blended mix. In figure 8, it is observed that for λ=1,29 (GB10) the thermal efficiency of the engine is the same as the engine operating with pure gasoline at λ=1,06 (G).

The maximum pressure values over 250 consecutive engine cycles for both fuels are presented in figure 9. The graph indicates a larger dispersion in values at fueling with gasoline (greater variation) vs n-butanol-gasoline blend. The coefficient of variability (COV) for maximum pressure will be determined, according to [17] with the following formula:

$$COV_{p_{Max}} = \frac{\sqrt{\frac{1}{N_c - 1} \sum_{k=1}^{N_c} \left(p_{Max_k} - p_{Max_m} \right)^2}}{p_{Max_m}} \tag{2}$$

$$p_{Max_m} = \frac{\sum_{k=1}^{N} p_{Max_k}}{N_c} \tag{3}$$

where: (2) represents the standard deviation and (3) the average value for maximum pressure.

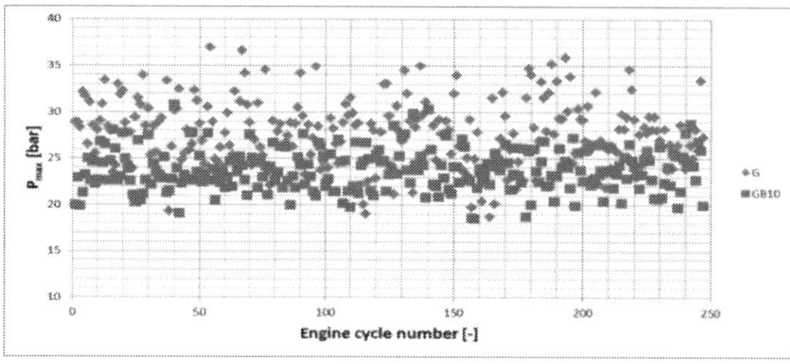

Figure 9. Maximum pressure dispersion over 250 consecutive cycles.

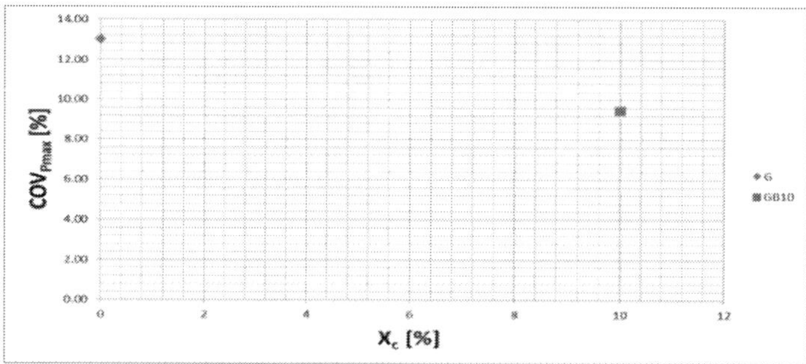

Figure 10. COV for maximum pressure.

The COV for maximum pressure for gasoline and the blended mix is presented in figure 10. After the calculation, it is observed that the variation is higher at fueling with gasoline 13% and it's lower for n-butanol-gasoline blend at 9,47%. The substitute ratio is x_c, in this case, 10% n-butanol. There is a reduction of almost 4% in variation at fueling with the blended mix, which indicates an improvement in the combustion process due to the additional oxygen available for burning. The maximum pressure for the n-butanol-gasoline mix is lower by an average of 17%.

The dispersion for the indicated mean effective pressure is presented below. The dispersion is not as noticeable as for maximum pressure but it is noticed that gasoline has a higher dispersion again. The coefficient of variation for pure gasoline is approximately 4,5% while for n-butanol-gasoline is 2,73%. The addition of butanol seems to improve the combustion slightly. The IMEP for the blended mix is reduced by around 5% vs pure gasoline.

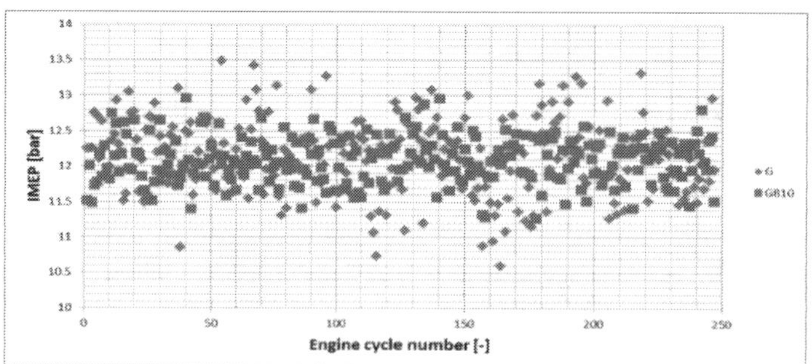

Figure 11. IMEP dispersion over 250 consecutive cycles.

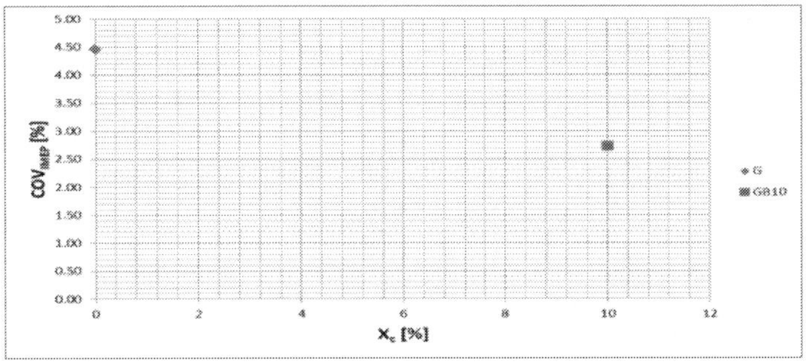

Figure 12. COV for IMEP.

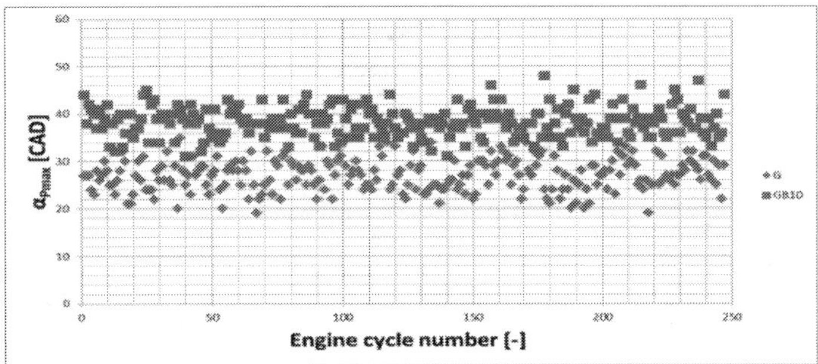

Figure 13. Crank angle at which the maximum pressure is achieved.

Figure 13 is represented by the crank angle at which the maximum pressure in the cylinder is achieved. It is interesting to observe that the n-butanol-gasoline blend achieves maximum pressure later than pure gasoline. The average α_{pmax} for gasoline is 27 CAD vs 38 CAD for the blended mix. The late ignition of the blended fuel is influenced by the air-fuel mixture, especially during the initial and final combustion phases [17]. A slower initial combustion phase will determine maximum pressure to be achieved further from the top dead center (TDC).

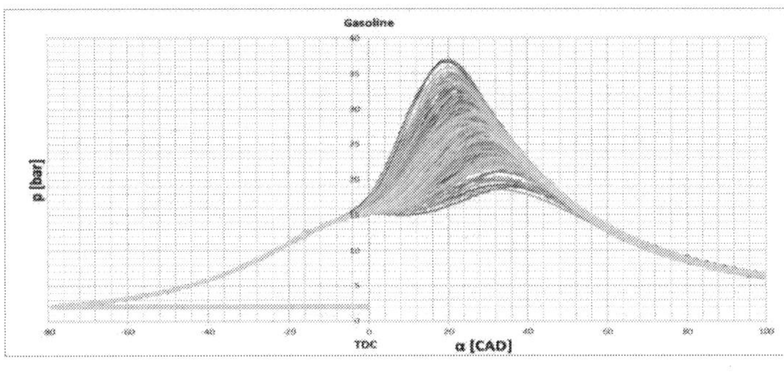

Figure 14. Indicated diagram for gasoline over 250 consecutive cycles.

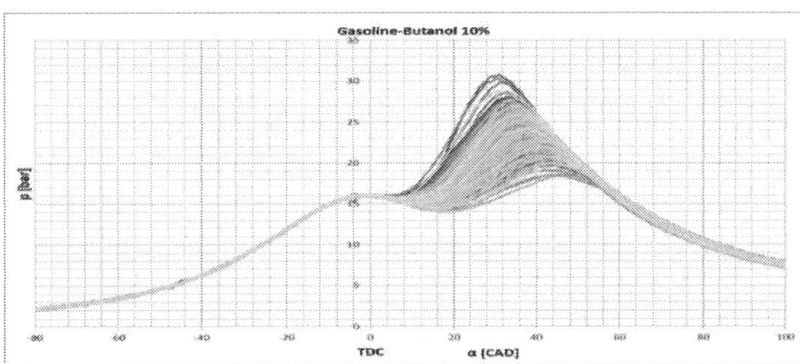

Figure 15. Indicated diagram for n-butanol-gasoline blend over 250 consecutive cycles.

The indicated diagrams for both fuels are presented in figures 14-15. As a comparison between the two diagrams, it is observed that there is lower dispersion in values at fueling with the blended mix indicating a small improvement in combustion but also lower maximum pressure values. The combustion of the blended mix moved further away into the expansion. The same engine tuning/adjustments were used in both cases and should be taken into consideration.

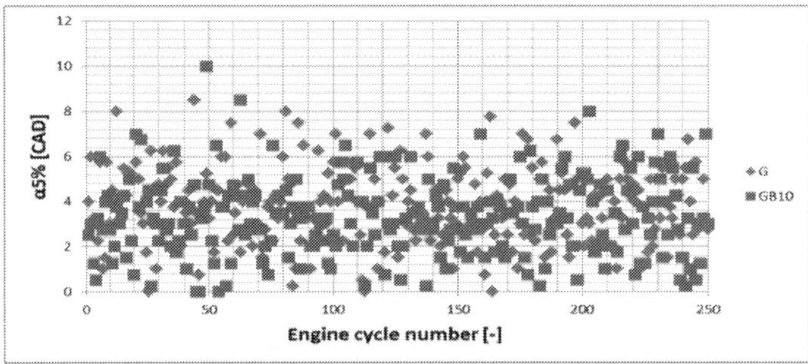

Figure 16. Angle of 5% MFB.

The mass fraction burned (MFB) for 5% heat release is represented in figure 16 where it is observed a high dispersion for both fuels. Calculating the coefficient of variation, COV_G=44,83% and COV_{GB10}=56,54%, where G = gasoline and GB10 = gasoline-butanol 10%. The addition of butanol to the blend doesn't improve the initial burning phase, it only causes a small delay in fuel combustion.

For the mass fraction burned for 50% heat release, the addition of oxygen greatly improves the combustion stability as shown in figure 17. The release rate is also faster for the blended mix, with an average of 16 CAD for GB10 vs 23 CAD for gasoline. Dispersion in values is lower for gasoline-butanol, as seen in the image below but also indicated by the $COV_{GB10}\cong$14% vs $COV_G\cong$17%.

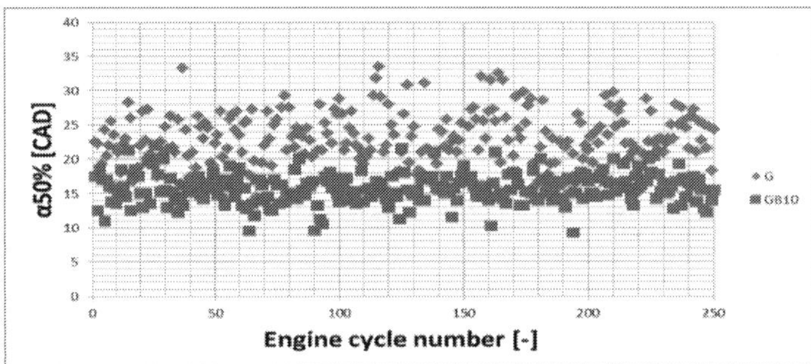

Figure 17. Angle of 50% MFB.

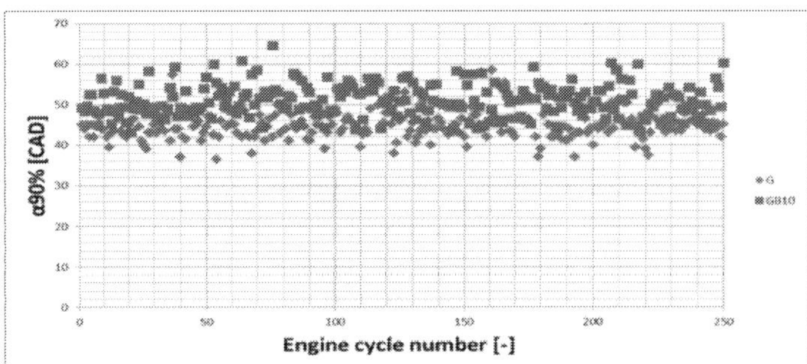

Figure 18. Angle of 90% MFB.

Gasoline releases 90% of the heat slightly faster than the blended mix, with an average angle of 45 CAD for gasoline and 51 CAD for the blended mix. Dispersion values are low for both fuels, the coefficient of variability being around 0,5%.

From the images above and COV calculations, it can be concluded that the addition of butanol doesn't have a good influence over the initial combustion phase (causing a slight delay), improves the combustion during the main phase of burning and has little to no impact during the final phase. The reduction in variability during the main combustion phase is around 3% better for the blended mix. The overall variation coefficients are presented in figure 19, where at 90% MFB, $COV_G = COV_{GB10}$.

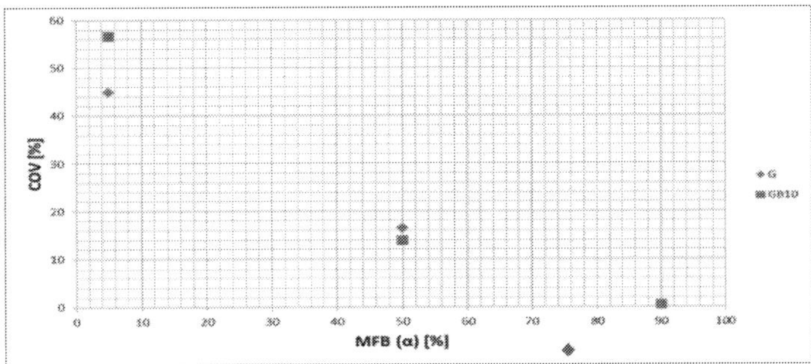

Figure 19. COV variation for 5%, 50% and 90% MFB.

Carbon monoxide is an indicator of incomplete combustion. High levels of carbon monoxide are usually achieved at rich fuel mixtures because there is a lack of oxygen for complete combustion. In figure 20, it is observed that the blended mix produces more carbon monoxide at rich mixtures (λ<0,95) compared to gasoline. The blended mix has higher vaporization heat and the gases temperatures from the cylinder reduce the combustion speed. At stoichiometric mixtures, it is noticed that the blended mix slightly reduces the carbon monoxide emissions. At lean mixtures, there are no noticeable differences between the two fuels.

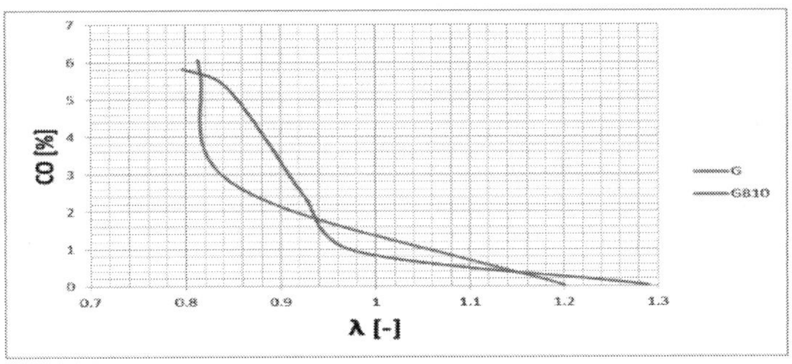

Figure 20. CO emissions.

Unburnt hydrocarbons are also a product of incomplete combustion, high values are achieved at rich mixtures. Similar to carbon monoxide, HC emissions are slightly higher for the blended mix vs gasoline, but there is a clear improvement at stoichiometric and lean mixtures due to combustion improvement.

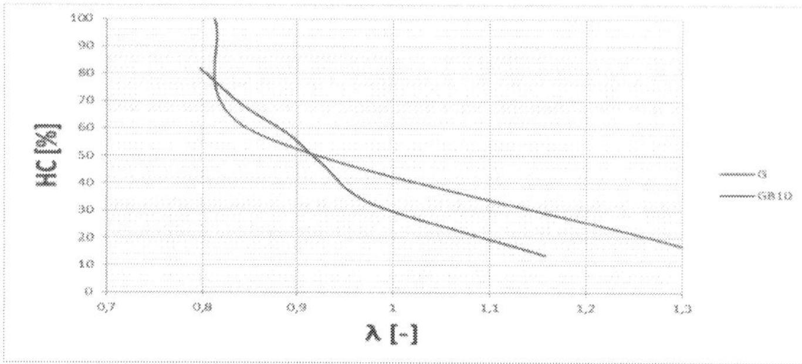

Figure 21. HC emissions.

The formation of nitrogen oxides is an endothermic process and depends on temperatures and oxygen content. The NOx emissions are presented in figure 22. Higher NOx emissions are achieved for all air excess coefficients (λ) because of the additional oxygen content of butanol. Emissions for the blended mix can be higher with around 50% vs pure gasoline. The trend can be reduced by using very lean mixtures. Increasing the spark timing of the engine may lead to an increase in NOx emissions, as stated by Alasfour [16].

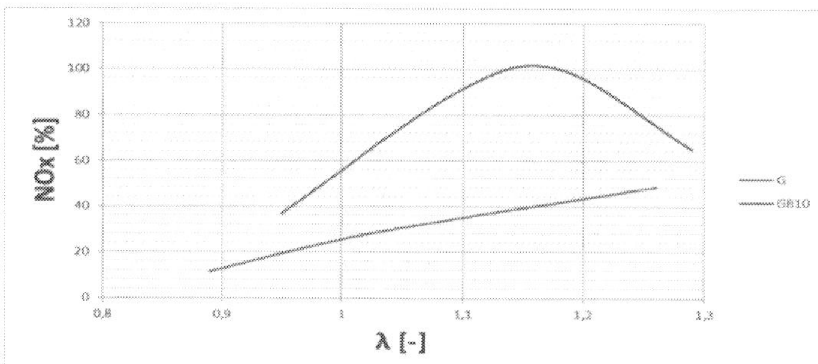

Figure 22. NOx emissions.

Carbon dioxide is a product of complete combustion. The blended mix improves the combustion process and CO_2 emissions are lower due to the lower carbon content. In the range of λ=1,06…1,29, HC, CO and CO_2 emissions level with the blended mix is lower than gasoline.

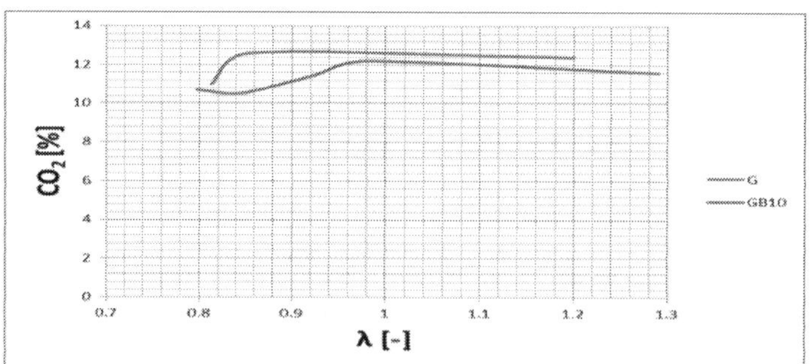

Figure 23. CO_2 emissions.

4. Conclusions

At the investigated engine regime, the addition of 10% -vol butanol to gasoline shows an improvement of the combustion process. The variation in maximum pressure improves from 13% for pure gasoline to 9,47% for the blended mix, however, there is a drop in maximum pressure for the blended mix by around

17%. The variation of the indicated mean effective pressure was also reduced from 4,5% for gasoline to 2,73% for the blended mix. A more stable IMEP can lead to better drivability properties for vehicles. The blended mix achieves maximum pressure at a later angle vs gasoline because of the initial combustion phase delay induced by the addition of butanol. The coefficients of variation for both fuels are high in the first phase of combustion, gasoline being a little more stable. During the main combustion phase, the blended mix improves the process by achieving faster combustion by around 5 CAD vs pure gasoline. The COV for 50% heat release was improved from 17% for gasoline to 14% for the blended mix. The blended mix has no impact on the final combustion phase, the variation being similar for both fuels around 0,5%.

The maximum pressure and maximum pressure rise rates are lower for the blended mix but with beneficial effects for the engine strengths. Maximum pressure is achieved later for the blended mix vs gasoline because of the initial delay and no changes to engine adjustments. The total heat released is similar for both fuels. Carbon monoxide (CO) emissions have decreased by 28% at fueling with the blended mix. HC emissions were also lower by 20% compared to pure gasoline. NO_x however is higher at the use of the blended mix due to the additional oxygen content; this effect can be reduced by using very lean mixtures. The brake specific energetic consumption is higher for the blended mix, but it may be improved through engine optimization.

Overall, butanol can be considered a good alternative fuel for spark ignition engines without requiring any constructive modifications. The researches will continue for the engine adjustments and butanol fraction in mixture with gasoline optimization.

Acknowledgements

The authors would like to thank AVL List GmbH Graz, Austria, for providing the possibility to use the Simulation Software AVL BOOST.

This work has been funded by:

- the European Social Fund from the Sectoral Operational Programme Human Capital 2014-2020, through the Financial Agreement with the title "Scholarships for entrepreneurial education among doctoral students and postdoctoral researchers (Be Entrepreneur!)", Contract no. 51680/09.07.2019 - SMIS code: 124539;
- a grant of the Romanian Ministry of Research and Innovation, CCCDI – UEFISCDI, project number PN-III-P1-1.2-PCCDI-2017-0404 / 31PCCDI/2018, within PNCDI III.

Annexes

Abbreviations:

- G – gasoline
- GB10 – gasoline – butanol mixture (10% vol. butanol)
- TDC – top dead center
- COV – coefficient of variation
- IMEP – indicated mean effective pressure
- BSEC – brake specific energy consumption
- CAD – crank angle degree
- MFB – mass fraction burned

References

[1] Gautam M., Martin II D.W. (2000) Combustion characteristics of higher-alcohol/gasoline blends, Proceedings of the IMechE Part A: J. Power and Energy 214, 5, 497-511.Yoon J, Ryu J and Lim K B 2005 A novel reconfigurable ankle rehabilitation robot for various exercises Proceedings of the 2005 IEEE International Conference on Robotics and Automation ICRA 2005 pp 2290-2295

[2] Gautam M., Martin II D.W., Carder D. (2000) Emission characteristics of higher-alcohol/gasoline blends, Proceedings of the IMechE Part A: J. Power and Energy 214, 2, 165-182.

[3] Yacoub Y., Bata R., Gautam M. (1998) The performance and emission characteristics of C1-C5 alcohol-gasoline blends with matched oxygen content in a single-cylinder spark-ignition engine, Proceedings of the IMechE Part A: J. Power and Energy 212, 5, 363-379.
[4] Wu M, Wang M, Liu J, Huo H. Assessment of potential life-cycle energy and greenhouse gas emission effects 293 from using corn-based butanol as a transportation fuel. Biotechnol Progr 2008;24:1204-1214
[5] Jin C, Yao M, Liu H, Lee CF, Ji J. Progress in the production and application of n-butanol as a biofuel, Renew Sust Energ Rev 2011;15:4080-4106.
[6] Doğan O. The influence of n-butanol/diesel fuel blends utilization on a small diesel engine performance and emissions. Fuel 2011;90:2467–2472.
[7] H, Huo M, Liu Y, Wang X, Wang H, Yao M, Lee CF. Time-resolved spray, flame, soot quantitative measurement fueling n-butanol and soybean biodiesel in a constant volume chamber under various ambient temperatures. Fuel 2014;133:317-325.
[8] Liu H, Bi X, Huo M, Lee CF, Yao M. Soot Emissions of Various Oxygenated Biofuels in Conventional Diesel Combustion and Low-Temperature Combustion Conditions. Energ Fuel 2012;26:1900-1911.
[9] Liu H, Li S, Zheng Z, Xu J, Yao M. Effects of n-butanol, 2-butanol, and methyl octynoate addition to diesel fuel on combustion and emissions over a wide range of exhaust gas recirculation (EGR) rates. Appl Energ 2013; 112:246-256.
[10] Yuqiang L, Lei M, Karthik N, Timoty H L, Yilu L, Shengming L, Chia-Fon L, Combustion, performance and emissions characteristics of a spark-ignition engine fueled with isopropanol-n-butanol-ethanol and gasoline blends. Elsevier 2016; 184: 864-872.
[11] Serras-Pereira J, Aleiferis P.G, Richardson D. An Analysis of the Combustion Behavior of Ethanol, Butanol, Iso-Octane, Gasoline, and Methane in a Direct-Injection Spark-Ignition Research Engine. Combustion Science and Technology (2013), 185:3, 484-513.
[12] He B-Q, Yuan J, Liu M-B, Zhao H. Combustion and emission characteristics of a n-butanol HCCI engine. Fuel 2014;115:758–64.
[13] Zheng M, Han X, Asad U, Wang J. Investigation of butanol-fuelled HCCI combustion on a high efficiency diesel engine. Energy Convers Manage 2015;98:215–24.b
[14] Alasfour F.N. (1997) Butanol -A single-cylinder engine study: Availability analysis, Appl. Therm. Eng. 17, 6, 537-549.
[15] Alasfour F.N. (1998) NOx emission from a spark-ignition engine using 30% iso-butanol – gasoline blend: Part 1: Preheating inlet air, Appl. Therm. Eng. 18, 5, 245-256.d
[16] Alasfour F.N. (1998) NOx emission from a spark-ignition engine using 30% iso-butanol – gasoline
[17] N. Negurescu, C. Pană, M.G. Popa, Spark ignition engines. Processes, Matrix ROM, Bucharest 2013

IOP Conf. Series: Materials Science and Engineering 1220 (2022) 012005 doi:10.1088/1757-899X/1220/1/012005

The effect of added ethanol to diesel fuel on performance and combustion of a Diesel engine

B C Benea

Department of Automotive and Transportation, Mechanical Engineering Faculty, Transilvania University of Brasov, Brasov, Romania

b.benea@unitbv.ro

Abstract. Diesel, as a fossil fuel, is widely used in industrial and agricultural fields. Added biofuels to fossil fuels affect the performance and combustion characteristics in the internal combustion engines. Ethanol is a renewable fuel, which is produced from plant, sugary and starchy biomass. Ethanol as an important additive to gasoline and diesel fuel can improve engine performance and reduce emissions. In this study, ethanol was added to pure diesel fuel with concentrations of 5, 8, 10, and 15%. The produced torque and power, brake thermal efficiency (BTE), and combustion were evaluated. The tests were made on a four cylinders CI engine at the six rotational speeds of 1200, 1700, 2200, 2700, 3200, and 3700 rpm, under full load mode. The results showed that the torque and power increase on average by 4.2% at fuel blend with a concentration of 8% ethanol (D92E8) as compared with those of pure diesel fuel. Although the power increases for the D92E8 fuel mixture, some irregularities are observed in the engine performance, and the engine runs more unevenly. Increasing the ethanol concentration by more than 8% in diesel increases the ignition delay and the engine runs erratically.

1. Introduction

Compression ignition (CI) engines are used in industry and agriculture for mobile and stationary applications due to their high reliability, durability, and efficiency [1]. Due to increasingly stringent pollution rules, manufacturers have adopted various strategies to reduce NOx and soot emissions [2]. The most used fuels were biomass (renewable) fuels. Renewable fuels are usually used in a blend with fossil fuels (diesel) [3].

Bioethanol as an important renewable element fuel is mainly used to power spark-ignition engines. Ethanol contains oxygen in its molecular structure, so it needs less oxygen to burn compared to diesel. Ethanol can be produced by hydrolysis of cellulose and fermentation of sugar and starch [4, 5].

Adding ethanol to diesel reduces the viscosity of the mixture, which leads to a finer spray of fuel in the combustion chamber [6].

2. Method

The tests were made on a Renault K9K engine. A schematic diagram of the engine test bed is presented in figure 1. The engine specifications are presented in table 1.

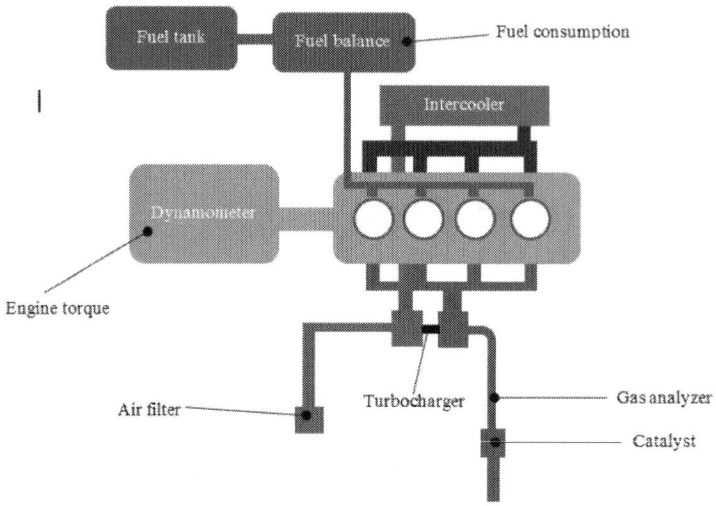

Figure 1. Schematic diagram of the engine test bed.

Table 1. Engine properties.

Engine type	Renault K9K four stroke
Number of cylinders	4
Bore (mm)	76
Stroke (mm)	80.5
Total displacement (cm^3)	1451
Compression ratio	15.3
Maximum power	72 kW at 3700 rpm
Maximum torque	200 Nm at 2700 rpm
Fuelling	Common-rail direct injection

The engine was mounted on a Horiba Titan 250 test bench. The engine test bed is equipped with an electric Dynas3 LI250 dynamometer, which is designed for operation within a range of 0-8000 rotations per minute. It can measure engine power up to 250 kW with an accuracy of ±2%.

The ethanol and pure diesel were blended by volume. Five fuel blends of D100, D95E5, D92E8, D90E10 and D85E15 were used. The characteristics of fuel are presented in table 2.

The experiments were made on a full-load engine condition, for five rotational speeds of the engine: 1200, 1700, 2200, 2700, 3200 and 3700 rpm. The tests were made at an ambient temperature of 20°C and relative humidity of 58%.

Table 2. Fuels properties.

Properties	Diesel	Ethanol	D95E5	D92E8	D90E10	D85E15
Density (at 20°C), kg/m^3	840.2	796.3	837.1	835.7	834.8	832.6
Viscosity (at 20°C), mm^2/s	5.34	3.53	5.21	5.15	5.11	4.96
Cetane number	51.1	8.4	48.7	47.9	46.8	44.9
Oxygen content, %	0	34.78	1.65	2.64	3.3	4.95
Net calorific value, MJ/kg	43.16	24.89	42.05	41.58	41.04	40.12

The in-cylinder pressure was measured with Kistler 6005 pressure sensor installed at the glow plug hole of the cylinder (Figure 2) and the crankshaft angle was measured using AVL 365 sensor that is mounted on the crankshaft pulley (Figure 3). This is an optic sensor that uses the beam of light passing through the slots.

Figure 2. In-cylinder pressure sensor.

Figure 3 AVL 365C position sensor.

3. Results

3.1. Brake power

Figure 2 shows the influence of increasing ethanol concentration on engine power. By increasing the ethanol content, the engine power increases. The power developed by the engine when using the D92E8 fuel blend. Engine power decreases as the ethanol content increases by more than 8% in blends. The decrease in power is due to insufficient mixing of diesel with ethanol at high concentrations of ethanol, leading to an inhomogeneous mixture.

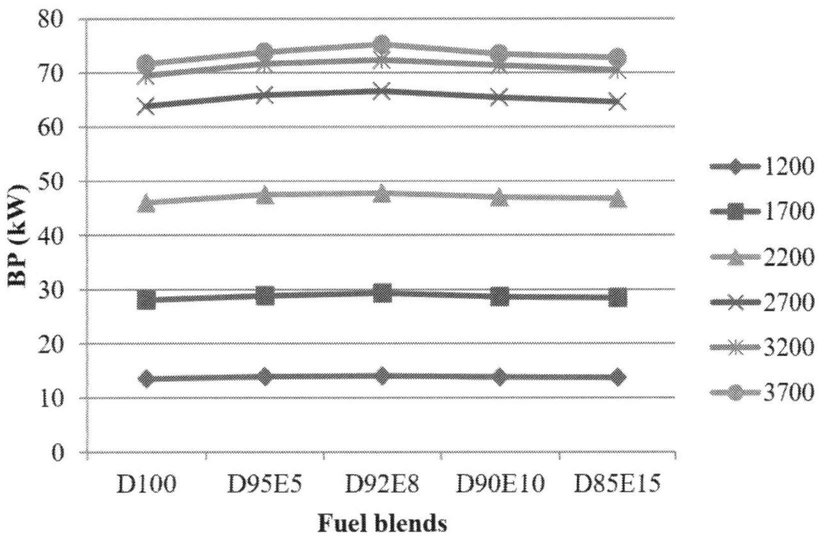

Figure 4. Brake power.

3.2. Engine torque

Figure 3 shows the effect of increasing the concentration of ethanol added to diesel fuel on the engine torque. The engine develops the maximum torque when it was fuelled with the D92E8 blend. The engine torque is dependent on the break mean effective pressure. The torque increases with the increase of the ethanol concentration up to 8%, after which the torque begins to decrease. The increase in torque occurs due to the increase in heat released in the cylinder due to the presence of additional oxygen and better combustion. The maximum torque developed by the engine fuelled by D92E8 is 4.8% higher than the torque developed by the engine fuelled by pure diesel.

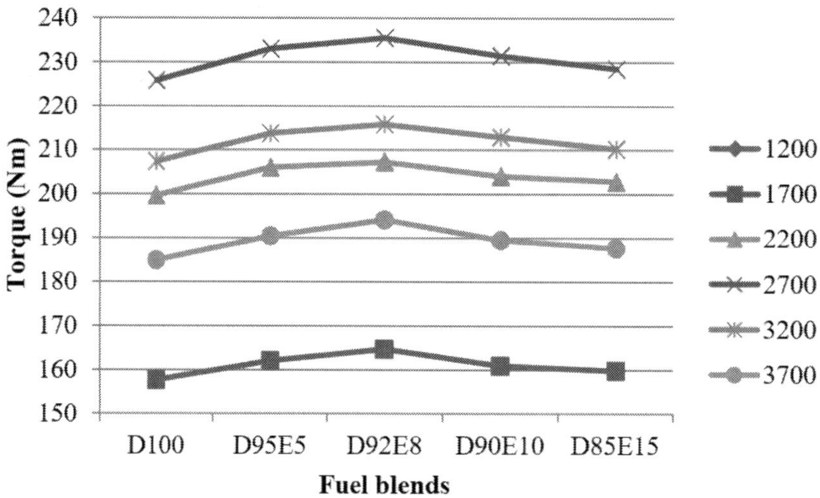

Figure 5. Engine torque.

3.3. Brake thermal efficiency

The increase in engine power is associated with an increase in energy produced in the cylinder as a result of burning fuel. By using an oxygenated fuel, combustion is improved.

Pure ethanol has one mole of oxygen in its molecular structure (34.8% of the total mass). The oxygen content for each type of mixture is shown in Table 2 and varies between 0% for D100 and 4.95% for D85E15.

The power developed by the engine increases with the ethanol content. The reason for the increase is more complete combustion due to the presence of additional oxygen compared to pure diesel.

At the concentration of 15% ethanol, the combustion quality does not improve due to the tendency of the mixture to form two phases, compared to the 8% ethanol mixture.

Ethanol and diesel are not perfectly miscible. In higher concentrations, ethanol and diesel are immiscible, the percentages varying depending on the temperature of the mixture.

Figure 4 shows the BTE according to the fuel blend for the maximum torque.

The higher values of BTE are obtained for D100 and D92E8. For D90E10 and D85E15 the BTE is reduced due to the lower quality of fuel blends.

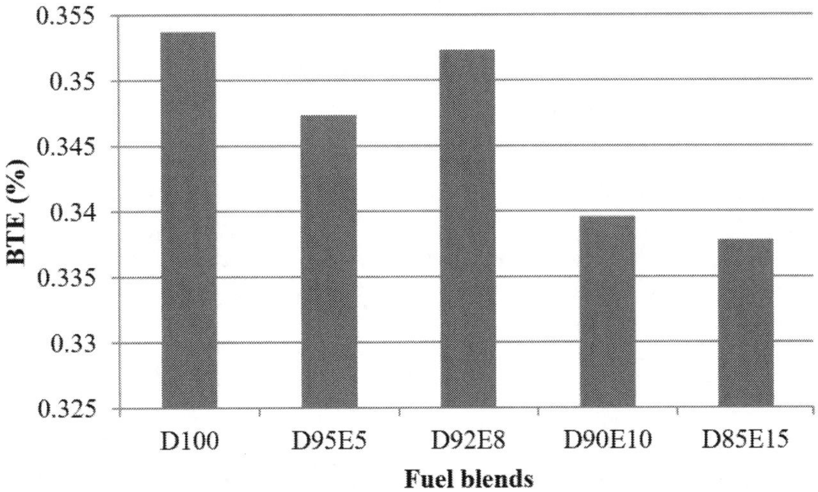

Figure 6. The mean BTE.

3.4. Combustion process
The in-cylinder pressures of tested fuels are shown in Figure 7. The pressure variation in the cylinder indicates the quality of combustion in the combustion chamber. As the fuel oxidizes, it releases heat and then the gas mixture expands inside the combustion chamber and increases the in-cylinder pressure. The maximum pressure was obtained when the engine was running with the D82E8 blend, indicating a better quality of fuel burning.

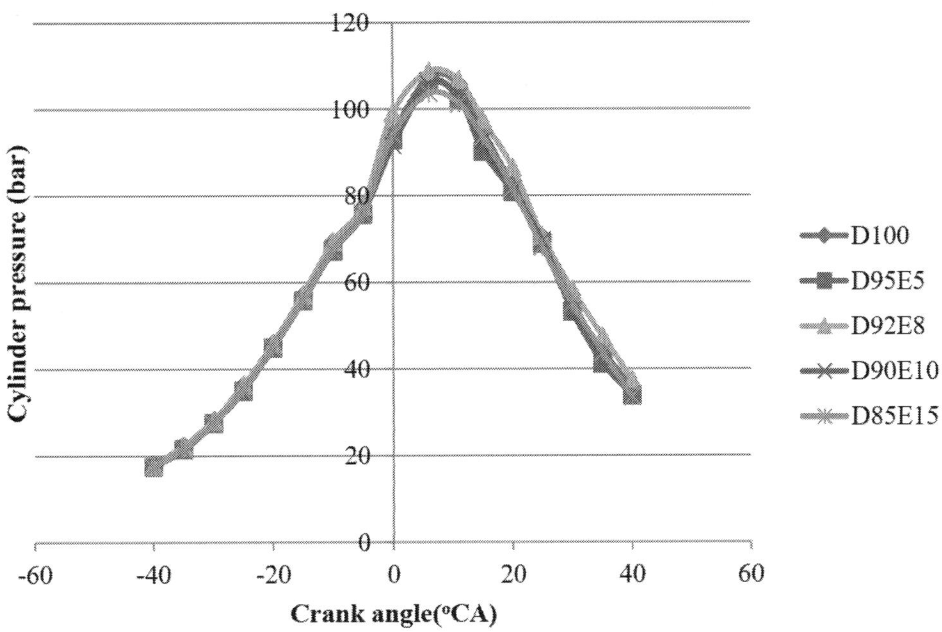

Figure 7. Cylinder pressure.

4. Conclusions

This paper has investigated the variation of power and momentum developed by the engine powered by pure diesel and mixtures of diesel with ethanol in proportions of 5, 8, 10, and 15%. The tests were performed on a 4-cylinder engine under full load at speeds of 1200, 1700, 2200, 2700, 3200, and 3700 rpm.

The results show the maximum engine power increases by 4.2% for an 8% ethanol concentration in diesel. This is the result of the higher pressure developed in the cylinder after the combustion of the D92E8 mixture.

The operation of the engine depends on the homogeneity of the fuel, the amount of oxygen, the quality of the fuel spray in the cylinder, its vaporization and the ignition delay. The addition of ethanol over 10% in the mixture influences the injectors spraying, the vaporization of the mixture and the ignition delay, which causes the worsening of the engine parameters

References

[1] Shahir S A, Masjuki H H, Kalam M A, Imran A and Ashraful A M 2015 Performance and emission assessment of diesel – biodiesel – ethanol/bioethanol blend as a fuel in diesel engines: a review. *Renew. Sustain. Energy Rev.* **48**, 62–78

[2] Zoeldy M 2011 Ethanol-biodiesel-diesel blends as a diesel extender option on compression ignition engines. *Transport* **26**, 303-309

[3] Park S H, Cha J and Lee C S 2012 Impact of biodiesel in bioethanol blended diesel on the engine performance and emissions characteristics in compression ignition engine. *Appl. Energy* **99**, 334-343

[4] Ahmad T-A, Abbas R-E 2016 The effect of added ethanol to diesel fuel on performance, vibration, combustion and knocking of a CI engine. *Fuel* **185**, 718-733

[5] Hansen A C, Zhang Q and Lyne P W 2005 Ethanol-diesel fuels blends – a review. *Bioresource technology* **96**, 277-285

[6] Chauhan B S et al 2011 Experimental studies on fumigation of ethanol in small Diesel engine. *Energy* **36**, 1030-38

Study regarding the influence of geometrical characteristics of the distribution system on truck engines efficiency parameters

L Barothi[1], R M Stoica[1], D Voicu[1], R Vilau[1], L Deleanu[2]

[1]Department of Military Vehicles and Transportation, Military Technical Academy "Ferdinand I ", 39-49 George Coşbuc Blvd., Bucharest, Romania
[2]Department of Mechanical Engineering, "Dunărea de Jos" University of Galati, 47 Domneasca Street, Galaţi, Romania

laszlo.barothi@mta.ro

Abstract. The present paper aims to analyze the influence of certain parameters specific to the distribution systems on a truck engine performance. The research is focused on how the change in the shape of lobes of a camshaft can influence the vehicle functional parameters. It was considered a non-variable fuel distribution system which represents by itself the controlling part, needed to move the poppet valves in order to ensure engine operating cycles. There have been compared results obtained in the case of different profiles of lobes, taking into account the accelerations during movements, in order to obtain minimum inertia of the distribution system components. By increasing the total height of the lobes, it was observed an increase in engine power and torque. Also, the intake air efficiency was improved by approximately 4%. Furthermore, engine fuel consumption reached higher levels in the case of the modified camshaft, in order to increase engine power.

1. Introduction

The present paper presents a theoretical method to increase engine power and torque by modifying a pre-existent engine distribution system, by analyzing the influence of tappet profile and tappet fallower vertical lift.

The engine submitted to analysis is MAN D2066 LF36. This engine equips heavy duty trucks, both of military and civilian use, and has the following characteristics:

- euro emission rating: Euro 4 with common rail/EDC 7;
- six inline cylinders with turbocharger;
- four valves per cylinder;
- wet engine mass: 975 kg;
- engine power: 324 kW at 1500-1900 rev/min;
- engine torque: 2100 Nm at 1000-1400 rev/min;
- cylinder bore:120 mm;
- stroke: 160 mm;
- firing order: 1-5-3-6-2-4.

There are performed modifications at the command of the distribution system in order to increase engine power and torque [1, 2]. Considering that the engine is not equipped with a variable distribution system, the command part, meaning the lift and drop movement of valves, is exclusively realized through the cam lobes. It is considered that the intake air has a laminar flow.

2. Study of camshaft lobes kinematics

The existing distribution system mounted on the engine has the following parameters:

- $\alpha_i = 284°$ RAC – opening position of intake valve;
- $\alpha_e = 280°\ RAC$ – opening position of exhaust valve;
- $h_{cmax} = 12.5\ mm$ – maximum valve lift;
- $r_0 = 21\ mm$ – radius of camshaft lobe base circle;
- $r_2 = 2 \ldots 12.5\ mm$ – radius of camshaft lobe head circle;
- r_1 – circle radius of the continuous profile;
- $r_r = 13\ mm$ – radius of roller tappet;
- $n_{max} = 2400\ rev/min$;
- a– adjacent leg;
- b– hypotenuse.

In figure 1 and figure 2, there are shown kinematic characteristics for a camshaft lobe of different profiles (two of the profiles are detailed in figure 1 and figure 2: the convex and concave profile, respectively).

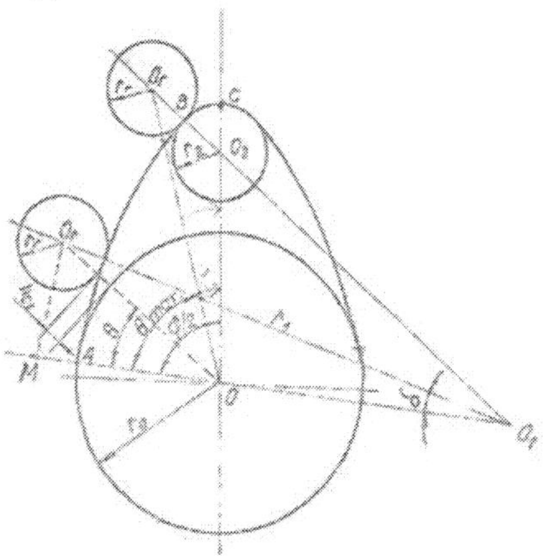

Figure 1. Kinematic characteristics of camshaft lobe, convex profile [1]

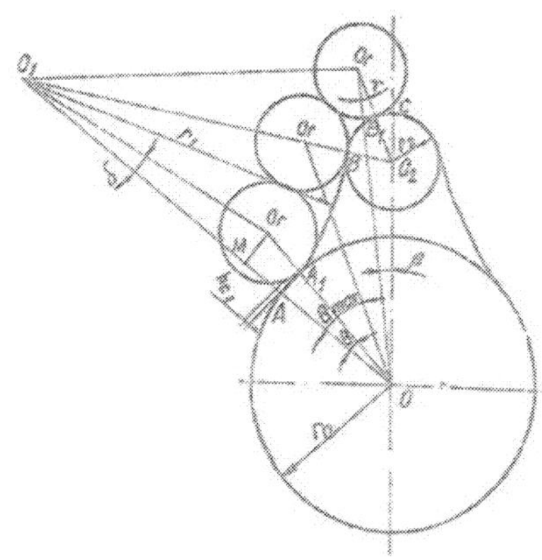

Figure 2. Kinematic characteristics of camshaft lobes, concave profile [1]

First, it was verified whether it is beneficial to modify the camshaft lobe profile. For that purpose, it was performed a kinematic analysis on different solutions of the camshaft lobe, by varying the radius of the base and head circles. Kinematic analysis was performed with the maximum engine speed (2400 rev/min), because the accelerations and stresses are highest. Regarding the head circle radius, it was observed that an increase of r_2 radius determines an increase of acceleration during the intake and exhaust periods. This result hinders engine functioning due to the fact that engine parts are submitted to increased stress. Following a preliminary analysis, it was determined that camshaft lobes with r_2 values higher than 12.5 mm have the disadvantage of generating higher loads so that in subsequent calculations the radius of the camshaft lobe head circle was set to 12.5 mm.

Hence, according to specialized literature [3, 4], the mathematical formulas presented further can be used to plot graphs illustrating the behaviour of some constituents of the distribution system, as can be observed from figure 3 and figure 4.

2.1. Tappet kinematics for convex profile

Maximum lift of roller tappet is:

$$h_{t1} = (r_1 - r_0)\cdot[-\cos(\theta) + \frac{1}{a}\sqrt{1 - a^2\cdot\cos(\theta)} - (r_0 - r_r)][mm] \tag{1}$$

where: θ – represents the rotation degree of the camshaft.

Considering that the angular velocity of the camshaft can be expressed with formula $\omega_C = \frac{\pi \cdot n}{30}$, further there are detailed expressions for different parameters of the distribution system.

Speed of tappet lift is:

$$w_{t1} = \omega_C \cdot (r_1 - r_0) \cdot [\sin(\theta) - \frac{a \cdot \sin(2 \cdot \theta)}{2 \cdot \sqrt{2 \cdot (1 - a^2 \cdot \sin^2(\theta))}}][\frac{m}{s}] \tag{2}$$

Tappet acceleration is defined as:

$$a_{t1} = \omega_C^2 \cdot (r_1 - r_0) \cdot \left[\cos(\theta) - \frac{a \cdot \cos(2 \cdot \theta) + a^3 \cdot \sin^4(\theta)}{(1 - a^2 \cdot \sin^2(\theta))^{\frac{3}{2}}} \right] [\frac{m}{s^2}] \tag{3}$$

2.2. Tappet kinematics for concave profile

Tappet displacement on r_1 is defined as:

$$h_{t2} = (r_1 - r_0) \cdot [-\cos(\theta) + \frac{1}{b}\sqrt{1 - b^2 \cdot \sin^2(\theta)}][mm] \tag{4}$$

Tappet lift speed on r_1 circular curve is:

$$w_{t2} = (r_1 - r_0) \cdot \omega_C \cdot [\frac{a \cdot \sin(2 \cdot \theta)}{2 \cdot \sqrt{2 \cdot (1 - b^2 \cdot \sin^2(\theta))}} - \sin(\theta)][\frac{m}{s}] \tag{5}$$

Tappet acceleration on r_1 circular curve is:

$$a_{t1} = (r_1 - r_0) \cdot \omega_C^2 \cdot \left[\frac{a \cdot \cos(2 \cdot \theta) + a^3 \cdot \sin^4(\theta)}{(1 - a^2 \cdot \sin^2(\theta))^{\frac{3}{2}}} - \cos(\theta) \right] [\frac{m}{s^2}] \tag{6}$$

Geometrical characteristics of classical camshaft lobes profiles can be studied with the aid of mathematical formulae inferable from geometrical conditions, which implies forming a continuous profile with the primitive circles as requisite. Therefore, the difference between all three analyzed profiles (concave, convex and planar) is made by the value of radius r_1. The small circle radius is between 2 and 5.5 mm.

Within the graphs in figures 3 and 4, curves 1 represent the convex profile, 2 is characteristic to the planar profile and 3 represents the concave profile. Also, on the y-axis is plotted the tappet lift height against crankshaft rotation angle (represented on the x-axis).

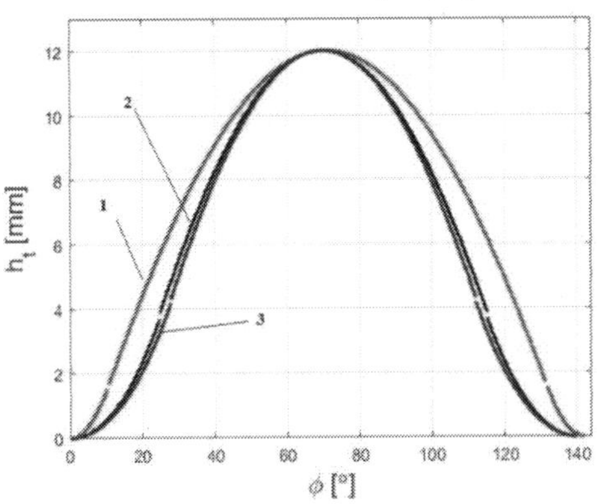

Figure 3. Lift height of roller tappet in case of three profiles, intake valve.

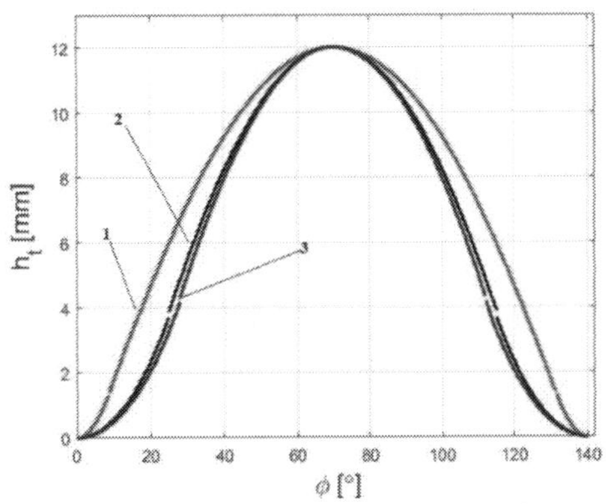

Figure 4. Lift height of roller tappet in case of three profiles, exhaust valves.

As it can be observed from figure 3 and figure 4, the two graphs depicting the lift height of intake and exhaust camshaft lobes are similar. The main difference consists in the angular interval, expressed in crankshaft degrees when the valves are opened. Hence, the exhaust valve opens and closes in a narrower period of time (within 140 crankshafts degrees).

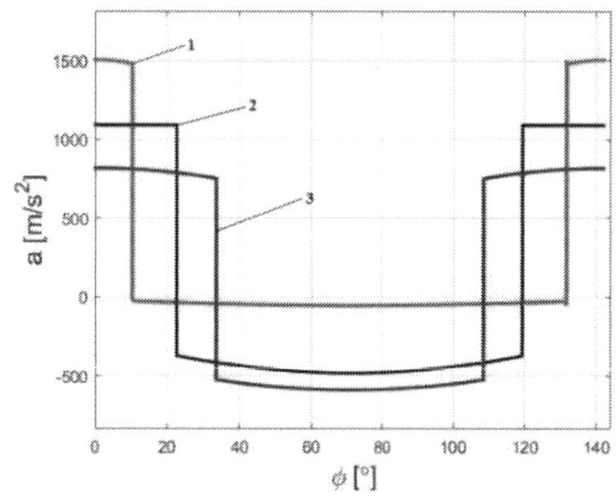

Figure 5. Graphs of tappet accelerations for different cam lobes profiles, during valve displacement intake valve.

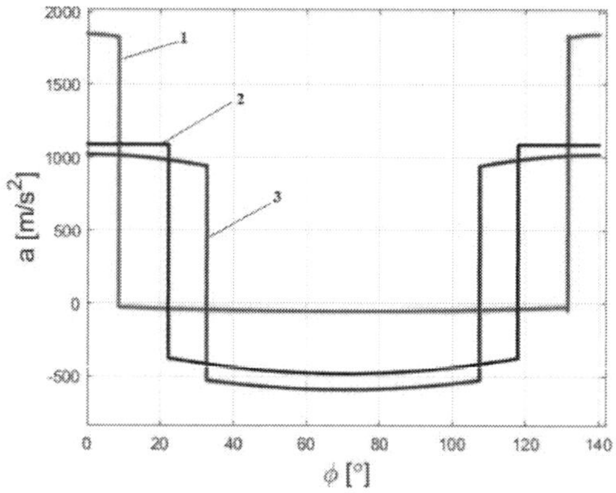

Figure 6. Graphs of tappet accelerations for different cam lobes profiles, during displacement, exhaust valves.

From figures 5 and 6, it can be observed that accelerations of exhaust valves, in the case of all lobe profiles, are higher than the acceleration of intake valves. This result can be explained by the fact that exhaust valves have to reach their maximum height in a shorter period of time.

Following the kinematic study, there have been obtained several results regarding maximum speed, maximum and minimum acceleration of roller tappet (table 1).

Table 1. Kinematic results of roller tappet at 2400 rev/min.

	Convex profile		Planar profile		Concave profile	
	Intake valve	Exhaust valve	Intake valve	Exhaust valve	Intake valve	Exhaust valve
r_1 [mm]	116.29	137.88	Inf.	Inf.	-269.82	-335.71
Maximum speed [m/s]	2.14	2.19	2.27	2.27	2.15	2.18
Maximum acceleration [m/s^2]	1504.9	1845.8	998.67	1048.65	818.72	1020.6
Minimum acceleration [m/s^2]	-55.26	-55.26	-483.93	-482.13	-592.72	-592.72

In table 2, there are presented speeds and accelerations of intake and exhaust valves, considering the equations from specialized literature [3]:

$$a_S = i \cdot a_t \quad (7)$$

$$w_S = i \cdot w_t \quad (8)$$

where: a_s – represents valve accelerations;
a_T – represents roller tappet accelerations;
w_s – represents valve speeds;
w_T – represents roller tappet angular speed;
i – represents the transmission ratio of the distribution system.

Table 2. Results of intake and exhaust valves kinematics.

	Convex profile		Planar profile		Concave profile	
	Intake valve	Exhaust valve	Intake valve	Exhaust valve	Intake valve	Exhaust valve
Maximum speed [m/s]	2.21	2.26	2.34	2.35	2.23	2.26
Maximum acceleration [m/s^2]	1555.01	1907.26	1031.93	1083.57	845.98	1054.58
Minimum acceleration [m/s^2]	-57.11	-57.11	-364.90	-500.05	-612.46	-612.46

Considering the results regarding the kinematics of camshaft lobes of different profiles, it was chosen to design the distribution system according to the highest lobe efficiency. Lobe efficiency can be defined as the ratio between the area below the cam lift graph and the area of a semicircle passing through the maximum cam lift height [1].

From all the three profiles of lobes, it was chosen the convex profile, due to the fact that, according to the analytical demonstration from expressions (1)-(8), this profile ensures the highest efficiency and also generates the lowest accelerations (according to table 2). As a result, the distribution system has the lowest inertia.

Relative lift height of valve, although with high influence on flow section, it varies within a narrow range, due to the fact that the upper limit is constrained by important mechanical and gas dynamics characteristics. By increasing the relative height, the h_{smax} increases, which also determines the increase of h_{cmax}, (according to formula (9)). As a consequence, it can be obtained a proportional increase of valve acceleration, with wear consequence on the cam profile.

$$h_{s\max} = i \cdot h_{c\max} \tag{9}$$

where: h_{cmax} – represents roller tappet fallower vertical lift;
h_{smax} – represents valve lift.

Because the value of the valve's relative lift is below the upper limit defined by burning chamber dimensions, the relative lift can be increased by 0.45 mm.

Therefore, the lobes of the existing distribution system are redesigned so that the lift height is higher with 0.435 mm than the initial value, because the relative height of the valve allows the increase, resulting in a final value of 12.43 mm. Camshaft lobes have a convex profile.

In table 3 there are presented comparative values of several parameters specific to the distribution system (separately for intake and exhaust), for both initial and redesigned camshaft lobes.

Table 3. Parameters of the distribution system, before and after changing camshaft lobes.

	Relative lift height of valve [mm]		Camshaft lobe efficiency	
	Initial	Redesigned	Initial	Redesigned
Intake	0.32	0.33	0.63	0.65
Exhaust	0.36	0.38	0.75	0.77

Analyzing the values presented within table 3, it can be observed that in the case of intake and exhaust valves, lobe efficiency increased by 3.1% and 2.6%, respectively. These values imply that a higher quantity of fresh air can be introduced inside engine cylinders and exhaust gases can be efficiently evacuated. If the turbocharger can introduce inside cylinders a higher quantity of air, an increased quantity of fuel can also be injected, which leads to an increase in engine power.

3. Influences on engine performances

In order to highlight the increase in performances of the analysed engine (MAN D2066 LF36), it was calculated the theoretical internal combustion engine cycle, for both the existing and modified camshaft solutions, at 1900 rev/min.

In figure 7 and figure 8 it is shown a comparative indicator diagram specific to the engine, before and after changing camshaft lobes. From the analysis of the two graphs, it can be observed that the maximum pressure in the case of lobes indicated by the manufacturer is 91.92 bar and in the case of modified lobes, the pressure reaches 96.21 bar, with the specification that it can produce premature wear of the engine. Therefore, it results in an increase of 4.66% of maximum pressure inside engine cylinders.

In order to make a comparative analysis of different engine characteristics, before and after changing camshaft lobes, first, there were determined certain parameters in order to validate the mathematical foundation. As an example, from calculus, the engine power according to the initial type of camshaft is 323.7 kW, which is similar to the value indicated by the manufacturer (324 kW) and specified at the beginning of the present paper. In both cases, calculus was made in ideal conditions (ex. Stoichiometric firing, laminar air flow etc.). The mechanical efficiency of the engine is considered 0,85.

Following the validation of results obtained in the case of the existing engine, the same parameters were determined for the engine with modified cam lobes. Hence, considering the engine power, it can be observed an increase of 3.2% (334.1 kW in the case of the modified camshaft, as opposed to 323.7 kW characteristic to the initial distribution system).

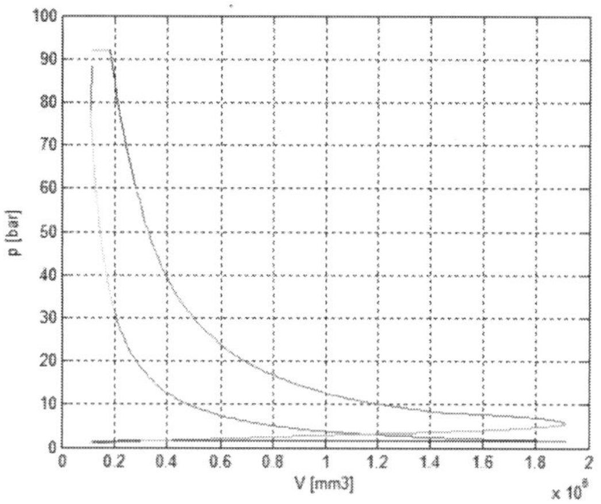

Figure 7. P-V diagram, original engine

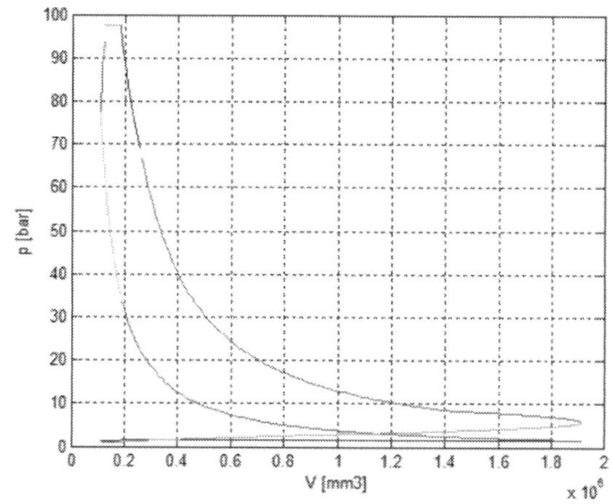

Figure 8. P-V diagram, modified engine

In table 4 there are shown results regarding engine performances, which were ascertained mathematically according to specialized literature [4]. The calculus was performed in Matlab 2019a software. Table 4 contains multiple parameters which can be submitted to comparative analysis.

Table 4. Data obtained after determining engine performances at 1900 rev/min.

Parameters	**Values characteristic to the initial distribution system**	**Values characteristic to the modified distribution system**
Indicated efficiency, η_i	0.519	0.515
Effective efficiency, η_e	0.441	0.438
Indicated pressure, p_i [bar]	17.53	18.105
Effective pressure, p_e[bar]	14.908	15.389
Effective engine power, *Pe* [kW]	**323.727**	**334.182**
Indicated engine power, P_i [kW]	380.855	393.155
Indicated fuel consumption, c_i [g/kWh]	162.168	163.472
Effective fuel consumption, c_e[g/kWh]	190.786	192.320

Following the mathematical calculus, it was also determined the average engine torque, resulting in a value of 2144 Nm in the case of the modified camshaft. Therefore, the engine torque increased by 44 Nm.

4. Conclusions

In order to improve engine performances, the distribution system was modified by increasing the lift height of valves. For that purpose, camshaft lobes have been changed regarding the height and not the profile. Convex profile with the radius of camshaft lobe head circle $r_2 = 12.5$ mm (exactly the type of profile specific to the D2066 LF 36 engine), has proven to be the most beneficial in terms of camshaft lobe efficiency.

Therefore, by increasing the lobe lift height from 12 mm to 12.43 mm, it was also increased the capability to fill in the cylinders (at ideal air flow conditions) during intake stroke from 1.02% to 1.06%. This means that filling efficiency increased by 4.1%, achieving the goal of the study.

Also, according to specialized literature, there was performed mathematical calculus in order to highlight the improvement in engine performances, generated by different camshaft lobes [4]. For example, it was determined the engine power according to the initial distribution system, obtaining 323.72 kW, which is close to the value indicated by the manufacturer, namely 324 kW. Hence, it was observed an error of 0.008%, meaning that the results were relevant for conducting a subsequent analysis. Therefore, further, it was determined the value of engine power specific to the modified lobes and it resulted in a value of 334.18 kW, which implies an increase of 10.45 kW.

Furthermore, from calculus, it results in an increase in the amount of fresh air and fuel introduced inside engine cylinders, during the intake stroke. The efficiency in this matter increased by 4.16%. The calculus was performed in Matlab 2019a software, for both initial and modified camshaft lobes.

Acknowledgements

„This work is supported by the project *ANTREPRENORDOC,* in the framework of Human Resources Development Operational Programme 2014-2020, financed from the European Social Fund under the contract number 36355/23.05.2019 HRD OP /380/6/13 – SMIS Code: 123847."

References

[1] Manea C, Năstase M 1978 *Construcţia şi calculul motoarelor de tracţiune, Instalaţii auxiliare,* Ed. (Bucharest: Ed. Academia Militară);

[2] Grunwald B 1980 *Teoria, Calculul şi Construcţia Motoarelor pentru Autovehicule Rutiere* (Bucharest Ed. didactică şi pedagogică);

[3] Gaiginschi R, Zetreanu Gh 1997 *Motoare cu ardere internă . Calcul şi construcţie* (Bucharest, Ed. Shakti);

[4] Stratulat M 1962 *Procesele energetice din motoare* (Bucharest, Ed. Academia Militară).

Study of the wear parameters of the physic-chemical and exploitation indicators of the engine oil when it is supplied with biofuel

I Besleaga and M Gadibadi

State Agrarian University of Moldova, Chisinau, 42 Mircesti Street, Republic of Moldova

i.besleaga@uasm.md, m.gadibadi@uasm.md

Abstract. The research seeks to provide the most effective techniques and methods that were used in the Laboratories of the Faculty of "Agricultural Engineering and Auto Transportation" (Department of Auto Transportation and Tractors Engineering). It provides the results of the investigation concerning on assessment of the change of physical-chemical and exploitation properties of petroleum products during engine operation and the results for testing internal combustion engines.

1. Introduction

Internal combustion engines have a large number of joints, cylinders, bearings, bolts, camshafts, rods, valves, etc., which cannot be maintained in good working order without permanent and safe lubrication. The operation of moving parts is greatly influenced by the nature and quality of the lubricant used, which must adhere well to the two moving surfaces, because reducing engine wear and tear is achieved by creating a pellicle of strong and continuous lubricant between the surfaces of parts motion. The lubricant pellicle consists of the central area in which the flow is laminar and the friction takes place between the lubricant particles; then follow the two outer areas, where the absorbed boundary layers are formed that have thicknesses of 100–200 ° A equivalent to two, three lengths of molecule [1].

The required lubricating layer, which separates the friction surfaces, will be provided when the pressure developed in the layer will be higher than the specific request of the bearing. Therefore, at low rotational speeds of the shaft and low values of the viscosity of the lubricating oil will create the conditions for fluid friction.

As a result of transformations of the fluid friction coefficient, the required viscosity of the lubricating oil can be determined. In order to ensure fluid friction, the minimum thickness of the lubricating layer being predetermined, or vice versa, the thickness of the oil layer can be controlled when using oil with a certain viscosity [2].

The operation of the moving parts is greatly influenced by the nature and quality of the used oil, which must adhere well to the two moving surfaces, and the oil pellicle must be continuous and not interrupted due to the pressure exerted by the weight of the parts. Therefore, when choosing the oil, the physic-chemical properties must be taken into account, because, depending on them, the best oil must be used at each lubrication place [3].

The quality of the engine oil is appreciated not only up to the early stages of being used in the engine, but also this ability to maintain quality and during its operation. It follows that the best oil is that which essentially changes its composition during engine operation. Therefore, the quality properties of the oil in action were assessed and analysed for this engine powered by various varieties of fuel [4].

2. Materials and Methods

The tests were carried out in the Laboratories for testing internal combustion engines and for researching the physical-chemical and exploitation properties of petroleum products of the Faculty of "Agricultural Engineering and Auto Transportation". In order to achieve the main goal, the methodology of experimental research presented in figure 1 was followed. The Stand KI-13638 GOSNITI was used to maintain the engine operating speeds, being intended for the MAI test and composed of: engine installation device, load generator, liquid rheostat, engine power system, control panel, electrical distribution panel and complete with measuring devices and control devices [5].

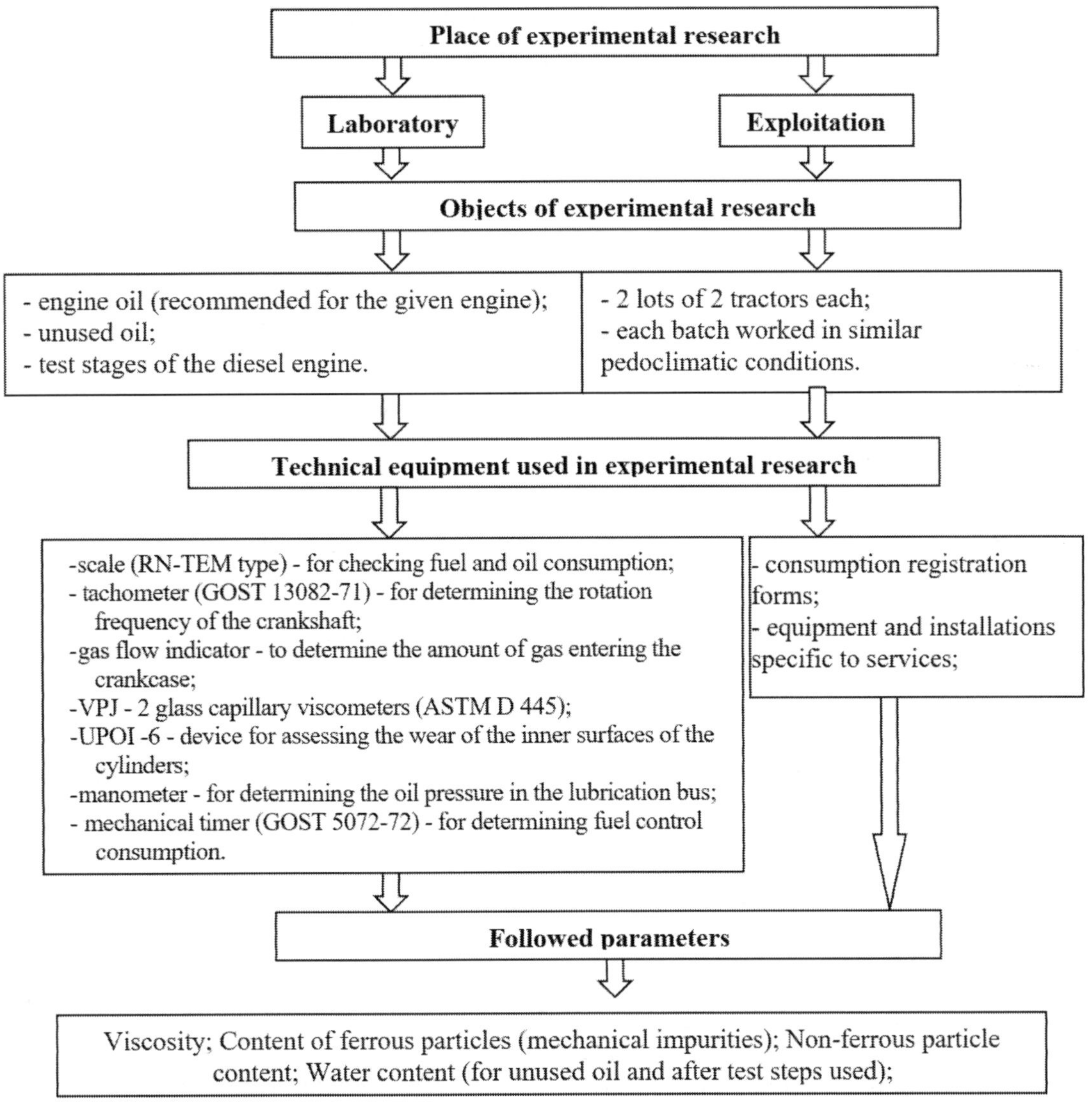

Figure 1. The general methodology of experimental research followed in the study.

3. Results and discussions

In the laboratory conditions, the physic-chemical and operating indicators of the engine oil were analysed in order to choose the optimal method of operation of the compression ignition engine powered by different varieties of fuel with the aim of increasing the operating period and reducing the degree of wear.

The following indicators have been determined for the tests: engine oil (recommended for the engine (see table 1) and oil in action in the crankcase (stand and in-service tests):

- kinematic viscosity according to ASTM D 445, using VPJ - 2 glass capillary viscometers (ASTM D 445);
- alkalinity (total basicity figure) according to GOST 11362-76, using the laboratory PH-meter;
- ash content according to GOST 12417-73, ASTM D 874;
- water content in accordance with GOST 2477-85, ASTM 95, using GOST 1594-89;
- the content of mechanical impurities according to GOST 20684-75, ASTM D 2273

Table 1: Quality characteristics of M10 Г2 engine oil (SAE10W / 40; API CE)

Characteristics	Package of additives according to standard VNIINP-360-6%; PMS-5,8% Article I. VNIINP-354-0,8%; PMS-200A-0,003%
Kinematic viscosity at 100°C, mm^2/s	10,99
Viscosity index	95
Absolute density, g/cm^3	0,895
Alkalinity (total basicity figure), mg KOH / 1g engine oil	4,65
Ash content, %	1,52
Content of mechanical impurities,%	0,019
Inflammation temperature, °C;	236
Freezing temperature, °C	-25

To evaluate the experimental results of the engine and especially of the parts in action or studied their types of wear:

- Mechanics - takes place as a result of mechanical actions.
- Corrosive-mechanical - takes place under the action of an aggressive environment on friction surfaces; such media are acids and alkalis, which already act chemically on the surfaces of the friction parts, thus causing corrosive wear. In this case, the products of the chemical action (in the form of oxides) come off during friction or are washed by the lubricating oil.

According to this classification, special attention has been paid to the following varieties of wear:

- **Abrasive** - mechanical wear of the material following the actions of cutting or scratching from solid particles in a free or fixed state.
- **Fatigue wear** - is characterized by mechanical wear following the destruction of fatigue at repeated deformation of the micro volumes of the surface layer.
- **Wear by locking (gripping)** - occurs after the deep gripping of the material, its transport from one surface to another and the action of irregularities on the adjusted surface.
- **Oxidation** - corrosive-mechanical wear, in which the main influence is exerted by the chemical reaction of the material with oxygen or another oxidant in the environment.

In the case of in-service tests of agricultural tractors, a number of factors appear that contribute to a more rigorous control over the wear of engine cylinder liners and piston rings, as a consequence of abrasive action (road or soil dust), fuel combustion products, which have an aggressive environment, as well as the molecular action following the operation of the parts in conditions of high temperatures, high pressures and oxygen in the atmosphere.

Simultaneously with the laboratory, the stand and operation tests, phenomena take place that lead to the worsening of the functional parameters of the energetic and ecological performances that contribute to the total loss of the functioning capacity. These changes in the technical condition of the engine mechanism systems are due to its wear. For the most important torques and responsible for the operation of the unit, the wear is determined directly during the operation, thus appreciating the technical condition of the contacted parts.

Depending on the experimental conditions for performing the measurements, the purpose and means of measurement and the method of execution, the discontinuous method was used to determine the wear, which involves disassembling the parts. The application of this method allows the determination of the state of the motor mechanism by using the method of micrometry, the method of fingerprints, the method of weighing, which allow the direct assessment of the wear of the parts, but not the analysis of wear in dynamics.

The micrometry method is the most common and consists in determining the actual initial and final dimensions of worn parts by measuring them with the micrometer, gear caliper, indicator, various gauges, optical and electronic devices, as well as comparators can be used (figure 2) for the direct determination of the effective linear quantities of the absolute wear of the cylinder liners.

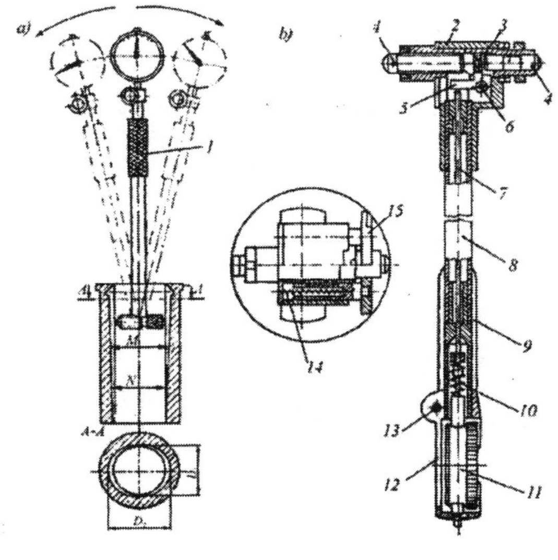

Figure 2. Schematic for measuring the bores of the cylinder liners (a) and the construction of the inner comparator (b): 1- interchangeable probe; 2 - measuring head; 3 - ball; 4 - mobile palpate; 5 - lever; 6 - lever joint; 7- rod; 8-extension; 9 - operating sleeve; 10.14 - arc-recoil; 11- indicator; 12- housing; 13 - screw; 15 - cantering blades of the comparator in the measured bore.

Absolute linear wear u is determined by:

- for diameters u = Df - Di; (1)
- for bores u = Di – Df (2)

where: Di, Df - respectively, the initial and final dimensions (after wear) of the parts.

This method does not provide the accuracy needed to measure insignificant (small) wear, as it is simple to perform.

Another category for assessing the wear of parts is the method of weighing, applying to the assessment of segments and bearings. This consists in determining the summary (global) wear proportional to the mass lost of the metal on the surface of the part and establishing the difference m_u between the initial mass of the part m_i, and its mass after a period of operation m_f, in g:

$$m_u = m_i - m_f \tag{3}$$

The average magnitude of wear h_u is determined by the relation:

$$h_u = \frac{m_u}{A_h \cdot \rho}, \tag{4}$$

where: m_u - used metal mass, g;
A_h - nominal contact surface area, mm^2
p - density of used material, g / mm^3.

The intensity of the mass wear $\mathbf{I_m}$ is determined, in *g/mm,* with the relation:

$$I_m = \frac{m_u}{A_h \cdot L_f}, \tag{5}$$

where: L_f - length of the friction path, mm.

The wear of the complete connecting rod bearings (after detaching their mass) when operating the engine on rapeseed oil (cycle 2) is 87.5 mg or 75.0% higher than the operation of the engine on diesel (cycle 1).

When operating the B 20 fuel engine (cycle 3), the wear of the complete connecting rod bearings is practically equal to the wear of the bearings when supplying the MAC with diesel fuel (cycle 1). The cylinder liner was also measured after 100 engines of the D-241 L engine with various types of fuels.

The method is recommended for determining the wear of small parts in the laboratory, in experimental research works.

This method is not effective in cases where the modification of the dimensions of the parts is carried out both by detachment from the metal and by plastic deformations; if the parts need to be thoroughly washed and dried before weighing, in particular for parts with porous surfaces which retain the lubricant, erroneous results being obtained. Moreover, the method does not allow the determination of local linear wear, determining only the average size of wear on the surface.

Therefore, the fingerprint method is used, which consists in measuring the wear on the depth of some cavities performed on the surface subjected to the friction tests. The imprints are made with the help of a diamond-tipped pyramid in the shape of a square base (figure 3, a) or with a rhombic base (figure 3, b), as well as with a rotating pyramidal knife, with alternating movement, with diamond, the shape of which is shown in figure 3.

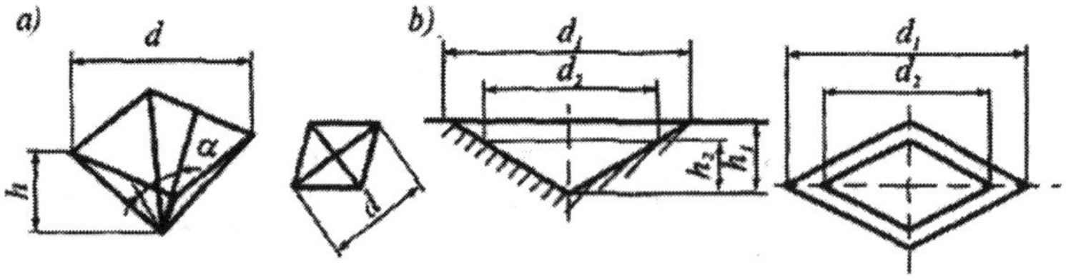

Figure 3. Imprint made with diamond base with a square base (a) and imprint made with rhombic prism border (b)

The wear of the surface is determined in accordance with the relations in Table 2 by the measurements made with the microscope, of the diagonals of the imprint executed with the square or rhombic pyramid, as well as of the lengths of the imprints made with a rotating diamond.

Engine cylinder wears analysis. The inner diameter of the cylinder is measured in 4 directions and in 3 planes. According to the measurement sheets, it is found that the cylinder wear occurs during 100 operating motors on a stand equal to 0.010 mm. This wear in the first 100 engines is not so great as to affect the proper operation of the engine. The permissible size until the cylinder liner is repaired is 110.220 mm, which represents the wear limit. Wear is achieved in about 3000 operating engines.

Analysis of the effects on the connecting rod. Based on the measurement data of the inner diameter of the connecting rod bearing (in three directions and two planes) during 100 engines of engine operation on the fuel mixture: diesel and 20% biofuel wear is not so great as to affect the good operation of the connecting rod, because the connecting rod has a wear limit of 0.11 mm.

Regarding the influence of the use of the fuel mixture: diesel and 20% biofuel on the technical condition of the engine, the partial results show that this fuel mixture does not affect the technical condition of the engine mechanism. Also, it could be found that: there were no significant wears of the bolt, connecting rod, spindle, crank assembly; at 100 engines the deposits on the piston head, the upper end of the cylinder, valves, segments were insignificant.

Table 2. Calculation of wear ratios according to fingerprint dimensions

Nr.	The process of making the imprint	Expression calculation of wear	The significance of the notations used
1	With a three-chain, negative-angled diamond pyramid knife: • on flat surfaces ; • on cylindrique surfaces.	$u = 0{,}125 \cdot (l_1^2 - l_2^2)/r$ $u = 0{,}125 \cdot (l_1^2 - l_2^2) \cdot \left(\frac{l}{r} \pm \frac{l}{R}\right)$ where we take: (+) for convex surfaces and (-) - for concave	l_1, l_2 – initial and final imprint lengths; r – the radius described by the tips of the knife; R – radius of the wearable surface.

In order to assess the wear of the inner surfaces of the cylinders with the help of the UPOI-6 device, some wells were cut in two sections (figure 4) in the middle region of the first compression segment in the position of the piston in PMS (figure 5, a; b, and 6 a; b).

Oil samples (100…150g) were taken from the engine crankcase (after every 20 engines of engine operation) to determine its physic-chemical indicators in the laboratory.

After each test cycle, the amount of oil in the crankcase is assessed, and the complete engine is disassembled and the following parameters are determined:

- calamine and varnish deposits (in points) on engine parts according to GOST 11637-65;
- model of segment wear, g (after the loss of mass);
- model of bearing wear, g (after losing their mass);
- wear pattern of the engine cylinders (according to the wear holes), mkm;
- the condition of the surfaces (mirror) of the cylinders (with the microscope of the UPOI-6 device).

The wear of the engine cylinder surface was calculated according to the relation shown in table 2.

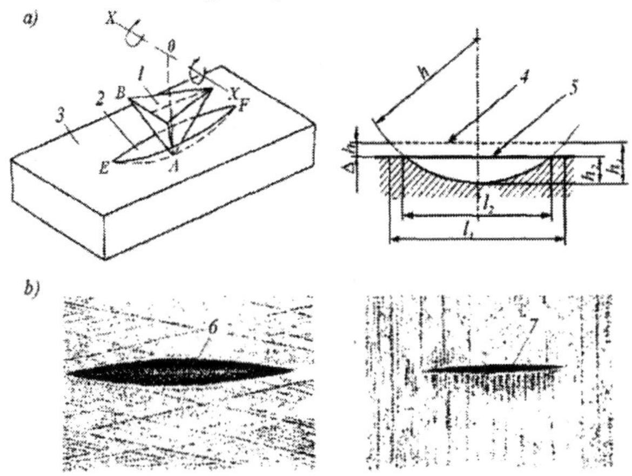

Figure 4. The scheme of execution of the imprints (a) with a rotating pyramidal knife on a flat surface and the shape of the imprint (b) on the surface of the part until and after wear: 1 - pyramidal knife; 2 - alveolus (fingerprint); 3 - the surface of the part, 4 - the level of the alveolus until wear; 5 - level after wear; 6 - the shape of the imprint until to wear; 7 - the shape of the imprint after wear.

The wells can be made on flat, cylindrical inner and outer surfaces (shaft type) with this device.

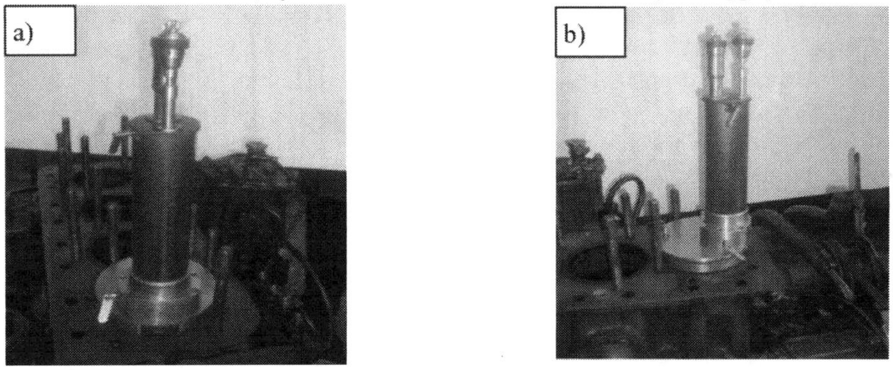

Figure 5. Execution of the imprint with the UPOI-6 device:
a) – longitudinal view; *b)* – transverse view.

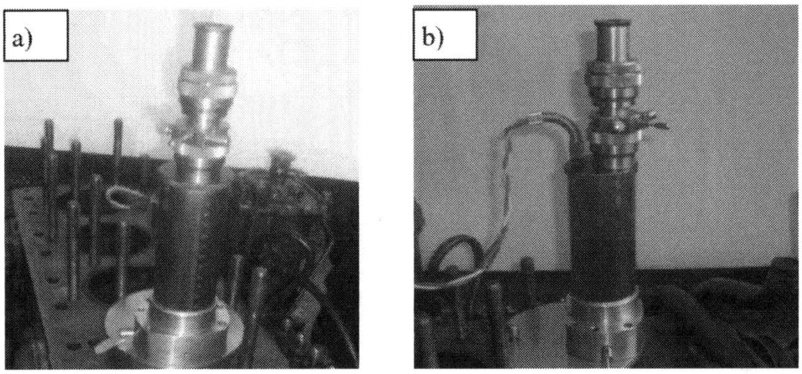

Figure 6. The assessment of the wear of the inner surfaces of the cylinders with the help of the UPOI-6 device: *a)* – longitudinal view; *b)* – transverse view.

4. Conclusions

The study of the physic-chemical and operating indicators of the engine oil when supplying it with biofuel allows to make the following conclusions:

1. The supply of B20 fuel engines ensures the maintenance of the alkalinity of the engine oil in action at a higher level, than in the case of fueling of the engines with petroleum fuel;
2. The use of B20 fuel represents a new way to save fuel of oil origin and cover the need for diesel;
3. The tests carried out at the stand demonstrate that the engine powered by a mixture of diesel and 20% biofuel can operate flawlessly to the limit of wear which is more than satisfactory.
4. At the application level, it is necessary to carry out some long-term experimental tests, which should be the subject of more complex studies under operating conditions.

References

[1] BURNETE, N.; (coord.). *Motoare Diesel şi biocombustibili pentru transportul urban.* Cluj-Napoca, Ed. Mediamira, 2008. – 1054 p. ISBN 978-973-713-217-8.

[2] LACUSTA, I., et all. *Materiale de exploatare pentru automobile.* Ch: Centrul Ed. al UASM, 2013. – 327 p. ISBN 978-9975-64-018-3.

[3] BEŞLEAGĂ, Ig. *Folosirea biocombustibilului pentru alimentarea motoarelor diesel.* Ch.: Centrul Ed. al UASM, 2011. – 111 p. ISBN 978-9975-56-023-8.

[4] Lăcustă I., Beşleagă Ig., Antoci E. Uleiurile sintetice pentru autovehicule In: *Sisteme de transport şi logistică: materialele conferinţei internaţionale*, ediţia IV-a. Chişinău 2011, p. 82-87, 0,3 c.a. ISBN 978-9975-4448-0-4.

[5] Lăcustă I., Beşleagă Ig. Încercări la stand a motorului cu aprindere prin comprimare alimentat cu ulei de rapiţă. In: *materialele Simpozionului ştiinţifico-practic internaţional „Realizări şi perspective în inginerie agrară şi transport auto"* dedicat aniversării a 65 de ani de la fondarea Facultăţii de Inginerie Agrară şi Transport Auto, Chişinău 2015 (12-13 noiembrie 2015), p.311-315, 0,31 c.a. ISBN 978-9975-64-276-7.

Electric transmission for hybrid vehicle

O I Petrov

Department of Transport, Technical University of Moldova, Chisinau, 9/8 Student Street, MD 2045, Republic of Moldova

oleg.petrov@tran.utm.md

Abstract. A technical solution is proposed for the electrical transmission of the torque of the internal combustion engine to the wheels (propeller) of a hybrid vehicle to set it in motion. The purpose of the technical solution is to reduce energy losses in the electrical power transmission device and reduce the cost of its implementation. The expected economic effect in the implementation of the proposed technical solution consists of a reduction in power losses during electric power transmission up to 10% of the engine power, and the cost of its implementation will reduce the cost of the vehicle by up to € 2 000. The proposed device is technically feasible using commercially available devices and can be used in the automotive industry, while being able to recoup the costs of implementation, that is, it meets the criterion "industrially applicable".

1. Introduction

Hybrid is a vehicle that uses the energy of several sources to rotate the wheels (propeller), including the battery. Hybrid vehicles such as HEV (hybrid electric vehicle) and PHEV (plug-in hybrid electric vehicle) have been around for many years. A lot of HEVs can be seen on the roads today as they are fairly common. However, their presence extends far beyond the world's roads. HEVs can be seen on rails, on and under the seas, and in the air. According to the International Energy Agency (IEA) Global EV Outlook 2019, the number of electric and hybrid vehicles exceeded 5.1 million in 2018, an increase of two million from 2017. The need for ever-increasing efficiency and emission reductions continues to drive the growth of the HEV market sector, as well as forcing sophisticated technologies to continually improve to support the growing demands on HEV systems. Thus, improving HEV in terms of integrated systems is critical not only in the automotive industry, but in all transport sectors [1]. High cost, limited range and long charging times are major concerns for battery-powered vehicles. Hybrid electric vehicles (HEVs), which use both internal combustion engine and battery power to propel a vehicle and do not need a mains connection to charge the battery, are preferable to a pure electric vehicle (EV) in terms of cost and range. HEV fuel consumption can be significantly reduced compared to conventional gasoline vehicles by using the engine in its most economical mode.

In hybrid vehicles, energy is transferred from sources to the wheels through various converters. Static rectifiers at voltages of hundreds of volts have a high efficiency - 0.98 ... 0.99, but the efficiency of inverters depends on the load and usually does not exceed 90% [2]. During the operation of these systems, each energy conversion has an efficiency of less than 100%, and, therefore, energy losses occur throughout the entire process. Therefore, there is a need for technologies, systems and methods that reduce energy loss through direct electrical connections between components and minimize component size.

Based on the foregoing, the object of the present technical solution is to reduce the energy loss in the electric transmission device of the hybrid vehicle and to reduce the cost of its implementation.

2. Hybrid Vehicle Electric Transmission

In prior art, for example, a sequential hybrid vehicle transmission is a transmission in which two power supplies power one electrical power unit (electric motor) that drives the vehicle. Most often, for hybrid vehicles, a series circuit is used (see figure 1). The first energy source is the fuel tank (1), the energy converter (power plant) is an internal combustion engine (2) connected to an electric generator (3). The generator output is connected to the DC power bus through a controlled electronic converter (4) (rectifier). The second energy source is a high-voltage storage battery (5) connected to the DC power bus using a controlled bidirectional power electronic converter (6) (DC / DC converter).

The traction motor (8) receives energy from the power bus through a controlled electronic frequency converter (7). The traction motor (8) is connected to the main gear (9), it can be used both as a motor and as a generator [3]. This transmission may require a battery charger to charge the batteries using a wall outlet.

The disadvantage of the typical scheme is that the power of the internal combustion engine is converted several times. First, with the help of a generator into alternating current, then with the help of a rectifier into direct current, and then with the help of an inverter again into alternating current required to control the power of the traction motor (8) by changing the frequency and magnitude of the current. Each transformation increases the cost of the electrical transmission scheme and is accompanied by energy losses.

3. Description of the proposed device for electrical transmission

In the proposed device, figure 2 uses two rather than three electronic converters. As in the well-known scheme, the first energy source is a fuel tank (1), the energy converter (power plant) is an internal combustion engine (2) connected to an electric generator (3) through a reduction gear (11). The electric generator (3) is used as a generator and as an engine for starting an internal combustion engine. The electric generator (3) in the proposed technical solution has a three-phase excitation winding on the rotor, which is connected to a controlled electronic frequency converter (10).

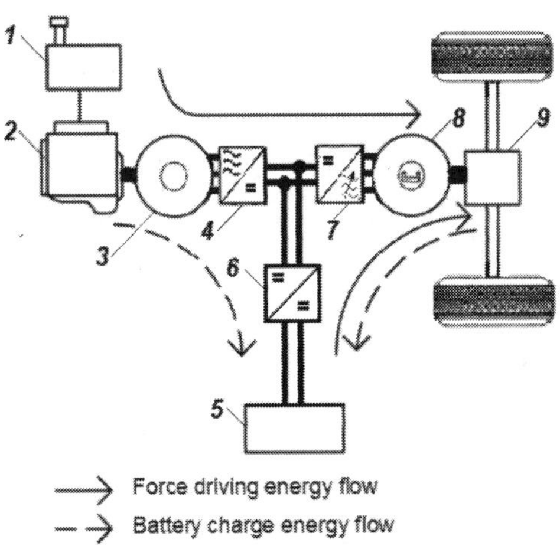

Figure 1. Series Hybrid Electric Vehicle Powertrain.

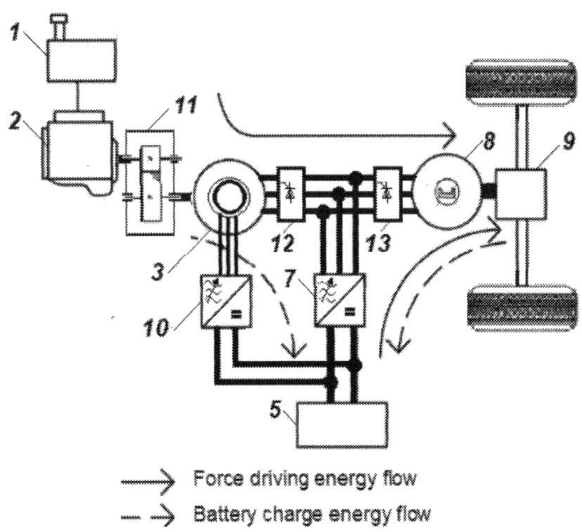

Figure 2. New Series Hybrid Electric Vehicle Powertrain.

The generator output (3) is connected to the AC power bus through a controlled electronic switch (12). The second source of energy is a high-voltage storage battery (5) connected to the AC power bus using a bidirectional controlled electronic frequency converter (7). The traction motor (8) is connected to the AC power bus through a controlled electronic switch (13). The traction motor (8) is mechanically connected to the main gear (9), it can also be used as a motor and as a generator [4].

One of the advantages of the proposed scheme is that the internal combustion engine is operated at constant speed in the optimal economical mode. The torque is transmitted to the electric generator (3) through a reduction gear (11) to generate a current with a frequency required by the traction motor (8) to ensure wheel rpm when starting off the vehicle. Further, at the constant speed of the electric generator (3), the frequency of the excitation current is increased using a controlled electronic frequency converter (10) and the electric generator (3) produces a current of the frequency that is necessary for the traction motor (8) to ensure the movement of the vehicle.

The main advantage of the proposed technical solution is that two energy converters are used, and not three, as in conventional circuits. Moreover, their total power is more than three times less than in the usual scheme. Since to power the excitation winding, an inverter (10) with a power of less than 5% of the power of an electric generator (3) [5] is used, and the second inverter (7), as a rule, is two times less power than in conventional circuits, since it's used to convert traction current from a high-voltage battery (5). Because in conventional circuits, the inverter (7) is used to convert current both from an electric generator (3) and from a high-voltage battery (5).

In the proposed technical solution, an electronic switch (12) is used to disconnect an electric generator (3) from the AC power bus for the time of regeneration of braking energy or electric traction on a high-voltage storage battery (5), as well as an electronic switch (13) to disconnect a traction electric motor (8) from the power bus for the period of starting the internal combustion engine (2).

At the same time, the proposed scheme creates certain implementation difficulties associated with the fact that a direct increase in the frequency of the rotor current will cause fluctuations in the amplitude of the supply current of the traction motor. The fact is that in conventional power transmission schemes, direct current is used to power the rotor winding ***EW***, therefore, the magnetic flux acting on the stator winding in one pass of the rotor pole under the stator winding will induce an ***EMF*** in the stator winding, which will cause half the period t of the sinusoidal current I_{SW1} to flow in it. (see figure 3a), since the unchanging magnetic flux of the rotor pole changes the force of action on the stator winding (flux linkage with the winding) in the process of passing under the winding.

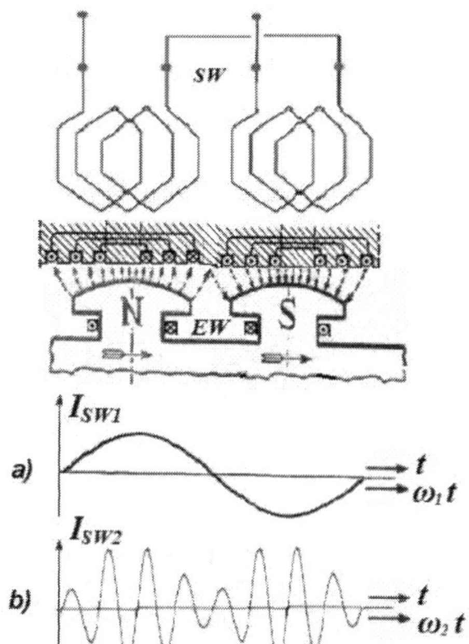

Figure 3. The current in the generator at a constant frequency of its excitation current.

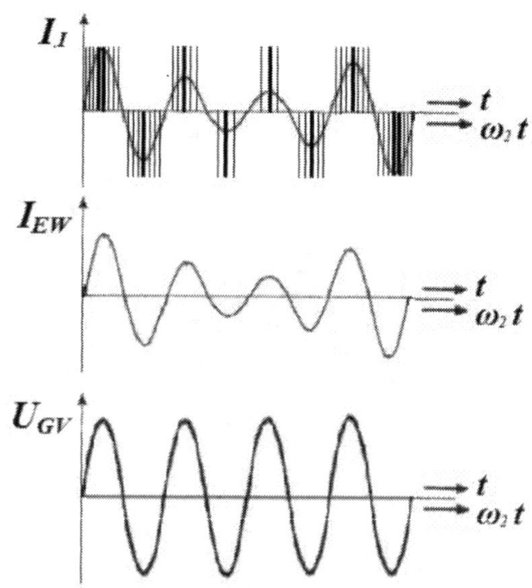

Figure 4. Excitation current and EMF of the generator.

In the proposed technical solution, the rotor magnetic flux changes sinusoidally, therefore, in one pass of the rotor pole, the force of the magnetic flux on the stator winding also changes sinusoidally, but the adhesion of the magnetic flux to the stator winding also changes, therefore, in the process of passing under the winding, the stator winding will be induced ***EMF***, which will cause the flow of sinusoidal current t with varying amplitude (see figure 3b).

The dependence of the magnitude of the ***EMF*** induced in the stator winding on the number of turns of the stator winding w and the rate of change of the magnetic flux B is expressed by the well-known formula:

$$e = - w\, dB/dt \tag{1}$$

This ***EMF*** will cause I_{SW2} to flow (see figure 3b) in the stator of the electric generator (3), which is fed into the stator winding of the electric motor (8) without conversion. The variable amplitude of the current in the stator winding of the electric motor (8) will lead to unwanted vibrations in the traction motor (8). To eliminate this disadvantage, it is proposed to supply the generator (3) with a rotor position sensor. The transmission control unit will control the electronic frequency converter (10) and generate current pulses I_I in the excitation winding of the rotor (see figure 4), smoothing of which, in the inductance of the winding, will give a sinusoidal current I_{EW} with variable amplitude, therefore the rotor magnetic flux will be more powerful in amplitude at the beginning of the passage of the rotor pole under the stator winding ***SW***, then the control unit will decrease the amplitude of the current in magnitude as it approaches the middle of the winding, and increase as the pole leaves under the stator winding ***SW***. The transmission control unit, based on the data of the generator rotor position sensor, will calculate not only the duration of the current pulses in the excitation winding ***EW*** to generate a sinusoidal current, but also the duration of the sinusoidal period in accordance with the current frequency required to control the traction motor speed, so that the required number of current periods with the required amplitude and phase caused a rotating magnetic flux in the field winding on the rotor ***EW*** and induced in the stator winding ***SW*** of the electric generator (2) corresponding to the U_{GV} sinusoidal ***EMF*** (see figure 4).

4. Operation of power transmission on a hybrid vehicle

Let us determine the frequency of the supply current of the traction motor (9) when converting the power of the high-voltage storage battery (5) for traction, for example, 55 kWh. With a constant number of pole pairs ***p*** of the electric motor (9), the angular velocity of the stator magnetic field ω_0 can be changed by changing the frequency f1 of the supply voltage (current):

$$\omega_0 = 2\pi f_1 / p \tag{2}$$

Let us express the angular velocity of rotation of the stator magnetic field ω_0 by the number of rotor turnovers ***n***:

$$n = f / p \tag{3}$$

If, take the maximum rotor speed ***n*** of an electric generator (3) with three pairs of poles equal to 10,000 min^{-1} (167 sec^{-1}), then, after reducing the speed of the main transmission gear (i_{TG} = 5:1), it is possible to ensure the rotation of the wheels 33,4 turnovers per second, and a vehicle with 15 inch wheel centres (195/60 R15) should be given a speed of 64.5 m/s or 232 km/h. In this case, the maximum frequency of the current f_{max} of the power supply of the traction motor (8) will be:

$$f_{Max} = n \times p$$
$$f_{Max} = 167 \times 3 = 500\ (Hz) \tag{4}$$

The characteristics of the frequency converter of the ALTIVAR Process ATV950D55N4 type (55 kW, 3 phases, 380 V) [7] will ensure the conversion of direct current of a high-voltage storage battery (5) into alternating current with a frequency of f_{max} = 500 (Hz).

When using the power of an internal combustion engine (2) for traction, for example, with a rated engine power of 55 kW at a speed of 3000 min^{-1}, using a reduction gear (11) with a gear ratio of

$\dot{i}_{RG}$ = 10:1, we reduce the speed by 10 times and rotate the rotor of the generator (3) at a speed of 300 min-1, which is n = 5 sec^{-1}. To control the excitation current, we use a frequency converter (inverter) from Schneider Electric ALTIVAR Process ATV950U15N4 (1.5 kW, 3 phases, 380 V) [7]. If with this inverter (10), supply an excitation current with a frequency of 1 Hz into the windings with three pairs of poles of the rotor of the electric generator (3), then, taking into account the rotation of the rotor of the electric generator (3) at a speed of 300 min^{-1}, (n = 5 sec^{-1}), in the three-phase stator windings of the electric generator (3), a current with a frequency of f_{min} = 15 Hz will be generated (see formula 4). When the stator winding of the electric motor (8) is supplied with a current of this frequency, its rotor will rotate at a speed:

$$n_{rrs} = f_{min} / p$$
$$n_{rrs} = 15/3 = 5\ sec\ \text{-}1 = 300\ (min\ \text{-}1) \quad (5)$$

and after reducing the speed of rotation of the rotor of the electric motor (8) by the main gear (9) of the transmission ($\dot{i}_{TG}$ = 5:1), the wheel of the vehicle (195/60 R15) will rotate at a speed of 1 sec^{-1} (6.95 km/h), which sufficient for the starting mode of the vehicle.

If the inverter (10) applies an excitation current with a frequency of 100 Hz to the rotor windings with three pairs of poles of an electric generator (3), then, taking into account the rotor speed, a current with a frequency of 500 Hz will be generated in the stator windings (formula 4), which will provide the speed of movement vehicle equal to 232 km/h.

Altivar Process ATV950 frequency converters are used as an example, since they are designed to control three-phase asynchronous and synchronous electric motors with power from 0.75 to 800 kW. They are used in process plants of various industries and have excellent application control algorithms and communication capabilities. Synchronous motor control profile: permanent magnet motor - drive output frequency: 0.1...599 Hz. These converters use the process of converting AC to DC and then DC to AC with specified voltage and frequency parameters. The rectifier is redundant for the control circuit of a hybrid vehicle, but converters without rectifiers are custom made, so it is not possible to set a price. The economic calculation presented below is of an approximate comparative nature.

It should also be noted that in each specific case of application of the proposed technical solution, for each type of vehicle, the gear ratio of the reduction gear (11) must be selected by calculation and the ratio of the main gear (9) of the vehicle transmission must be taken into account. The control range of the considered converters from Schneider Electric will provide the required transmission control modes.

5. Economic effect when using the proposed device

The proposed device for electrical transmission allows you to reduce the cost of the inverter (7) AC and reduce energy losses when converting the power flow to control the traction motor (8). It is known that the energy consumption for powering the excitation winding of an alternator is 0.2-0.6% of the electrical power generated by the generator [5, 6]. Frequency converters, manufactured at the current level, usually have a high rated load and efficiency, which usually reaches 90% [2]. To assess the cost reduction when using the proposed proposal, one can compare the costs of models of frequency converters for controlling a three-phase motor of the well-known company Schneider Electric [7].

According to the conventional hybrid vehicle control scheme, with an internal combustion engine power of 55 kW and a high-voltage battery of 55 kWh, a frequency converter (inverter) of the ALTIVAR Process ATV950C11N4F type (110 kW, 3 phases, 380 V) can be used to power the traction motor (8) at a cost of € 8 285.

When using the proposed technical solution, the traction motor (8) can be powered from a high-voltage storage battery by a less powerful frequency converter (inverter) ALTIVAR Process ATV950D55N4 (55 kW, 3 phases, 380 V) at a cost of € 3 485.

To control the excitation of an electric generator (3), you can use ALTIVAR Process ATV950U15N4 (1.5 kW, 3 phases, 380 V), its cost is € 595.

Therefore, after deducting the cost of a reduction gear and electronic switches, savings can be up to € 2 000.

As a result: taking into account the efficiency of the inverter, it should be expected that when implementing the proposed technical solution, a decrease in power losses during the electric transmission of energy from an internal combustion engine to the driving wheels (propeller) of a vehicle can be up to 10% of the engine power, and its costs implementation will reduce the cost of the vehicle by up to € 2 000.

6. Conclusions

The technical solution of an electric transmission for a hybrid vehicle considered in the article is technically feasible using commercially available devices and can be used in the automotive industry, while being able to recoup the costs of implementation, that is, it meets the criterion "industrially applicable".

References

[1] Chau K. T., (2015) Electric Vehicles Machines and Drives Design, Analisis and Application John Wiley & Sons Singapore Pte. Ltd.,400 p.

[2] Don Knowles, (2013) Understand Efficiency Ratings Before Chosing An AC-DC Supply, Electronic Design magazin, Endeavor Business Media, LLC, USA, https://www.electronicdesign.com/power-management/article/21795830/understand-efficiency-ratings-before-choosing-an-acdc-supply

[3] Handbook of automotive power electronics and motor drives / edited by Ali Emadi. Published in 2005 by CRC Press Taylor & Francis Group, 689 p.

[4] Petrov Oleg, „ Electric transmission for hybrid vehicle", Patent application No. 2203, filed on May 21, 2021, by the Agency for the Protection of Property Rights and the Republic of Moldova (AGEPI).

[5] Richard Marenbach, Dieter Nelles, Christian Tuttas. Elektrische Energietechnik, Springer Vieweg Fachmedien, Wiesbaden, 2013, 490 p.

[6] Soloviev I.I., Avtomaticheskie reguliatory sinchronnych generatorov, Pod red. N.I. Ovcharenco. - M.: Energoizdat, 1981. 248 p.

[7] Website Rosprivod company, https://rosprivod.com/catalog

Photovoltaic system for E-Smart electric vehicle

Radu Tarulescu, Stelian Tarulescu, Cristian Leahu, Marius Olaru

Transilvania University of Brasov, Brasov, Romania

radu.tarulescu@unitbv.ro, s.tarulescu@unitbv.ro, leahu.cristian@unitbv.ro, mariusolarson@gmail.com

Abstract. The vehicles relied on fossil fuels are rapidly being replaced by electric and plug-in hybrid vehicles. But these types of vehicles are still faced with the problem of energy availability. The abundance of solar radiation and its use as the power source in electric vehicles is a necessary condition for environmental pollution limitation. In this study, the authors present photovoltaic systems used as an electricity supply for E-Smart electric vehicles. E-Smart is an electric vehicle obtained through conversion, of a Smart ForTwo City vehicle, from the internal combustion propulsion system to a system that uses a three-phase asynchronous motor supplied from a pack of 32 batteries of LiFePO4 type.

1. Introduction

The interest for vehicles with electric propulsion systems is very high lately, which leads to the need for greater energy autonomy. The sustainable solution, from our point of view, for this type of vehicle, is the use of photovoltaic systems in order to supplement electricity for batteries.

Consequently, we are proposing, following the analysis presented in this paper, the implementation of a photovoltaic system on the E-Smart electric vehicle.

Figure. 1. The electric E-Smart vehicle used for analysis.

E-Smart is an electric vehicle, obtained by converting a Smart ForTwo City Coupe manufactured in 1999 and fueled by gasoline, with an electric propulsion system. The old gasoline engine of 33kW power was replaced with an electric three-phase asynchronous motor of 22 kW power, model N50D2. The engine control unit (ECU) of the old gasoline engine was replaced with an electronic control unit E-Car-Box based on a Curtis 1238 controller [1]. All of the commands from the vehicle's driver are processed in this controller and given to the electric motor.

Figure 2. The battery pack from E-Smart vehicle with 32 LiFePO4 cells.

To supply with energy, the electric motor has been adopted a solution using LiFePO4 (Lithium Iron Phosphate) rechargeable batteries, SP-LFP-66AHA model, with a nominal capacity of 66Ah, max charging current of 132A (optimal 22A) and max discharge current of 198A (optimal 22A) [2]. There was placed 32 batteries in the pack, in serial connection.

2. Establishing the usable surface of the E-Smart vehicle

The E-Smart vehicle's roof it's not a perfect square and it has parabolic shapes in the front of the car where the roof connects with the windshield and also on the sides where the doors meet the roof. To create a solar panel, the usable space must be rectangular, in this case, the measurements were taken so a rectangular shape can be drawn out from the roof. In figure 3 can be seen the total roof's area which we consider to be suitable for the implementation of a photovoltaic system.

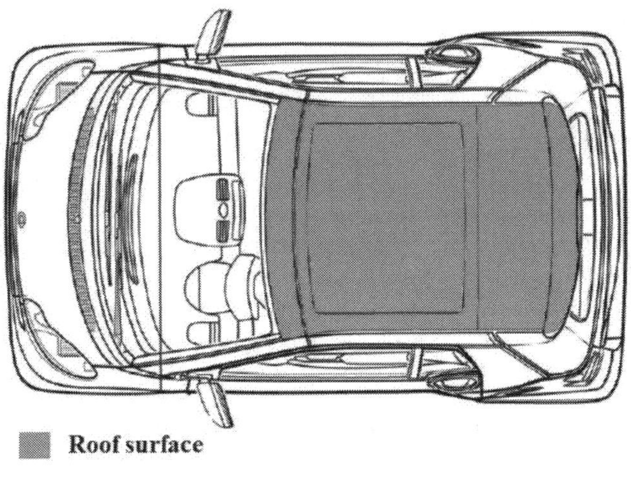

Figure 3. The total roof surface of the E-Smart vehicle.

Following the measurements, we determined that the usable area of the roof is:

$$S = W \cdot L = 0.925 \cdot 1.14 = 1.0545\ [\text{m}^2], \tag{1}$$

where: S is the surface, W is the width and L is the length of the roof.

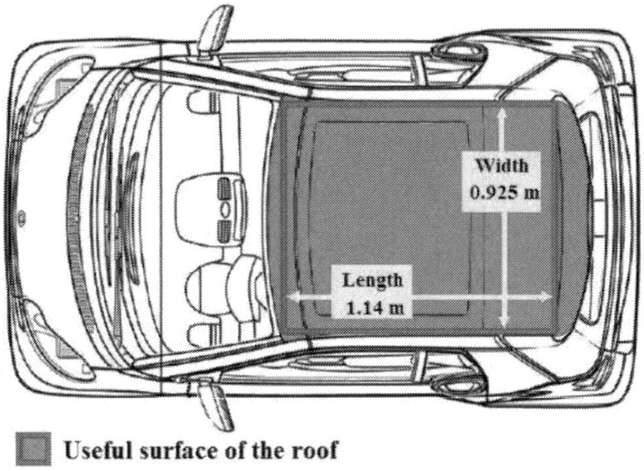

Figure 4. The total roof surface of the E-Smart vehicle.

In figure 4 is showed the usable surface, in a red rectangle. This surface, as it can be seen, does not include the spoiler due to the trunk's door that has to be opened upwards. The usable surface further limits the number of solar cells that can be arranged and implemented on the roof. Also, due to the parabolic shape that was not considered in the measurements, the solar panel measurements can differ slightly.

3. Choosing the solar cell with the highest efficiency in ratio with surface and design the panel

There are many companies on the market that are researching solar cells with high efficiency. Table 1 contains the top 10 companies that have the highest efficiency cells [3].

Table 1. Top ten models of photovoltaic cells with high efficiency

Manufacturer	Model	Rated efficiency
SunPower X series	SPR-X22-370	22.80%
REC Solar	REC Alpha Series 380AA	21.70%
LG Solar	High Ne-ON R Module	21.70%
Canadian Solar	HiDM-345MS	20.46%
Panasonic HIT panels	N340 HIT	20.30%
Upsolar	UP-M400M	20.30%
Jinko Solar	Cheetah Perc HC 60M-V	20.16%
Seraphim	SR P410-E01A-HV	20.16%
Silfab	SLA-HC 320Wp	20.00%
Solaria	Power XT 360R-PD	19.90%

In this application, due to the really small surface of the vehicle roof, it's rational to go for the highest efficiency solar cell. In this case, the chosen solar cell is one made from the SunPower Company, the model C60. This model is a reasonable option because it's flexible slightly and by placing more next to each other it can take the curvature of the roof. Also, the efficiency of 22.80% it's the highest efficiency for a cell on the market. There are some solar panels that have higher efficiencies but the companies sell the entire panel and do not sell just the cells.

The dimensions of the cell are 125 by 125mm, diagonal 160mm and the thickness of 165μm. Also, this cell provides the most power compared to the surface of the cell, one cell having a nominal power of 3.42W, voltage around 0.58V and current around 5,9A [4].

The layout of the solar cells has to form a big square panel. Knowing the dimensions, we can easily calculate how many solar panels fit into the usable surface. The surface of one solar cell is:

$$S_{cell} = 0.125 \cdot 0.125 = 0.015625 \ [\mathrm{m}^2], \tag{2}$$

And the number of cells that can fit into the usable surface is:

$$Cells = \frac{RoofSurface}{S_{cell}} = \frac{1.0545}{0.015625} = 67.488 \approx 67 \tag{3}$$

So, the number of cells that can fit it's 67. To determine the layout, the number of cells on the width and on the length of the surface it's important to calculate. So, in this case, dividing the width of the surface with the length of the cell will get us the number of cells that will fit on the width of the surface. Also in between every cell, it's necessary for a small gap, so in this case, a tolerance of 1 mm is used.

$$Cells_{width} = \frac{RoofWidth}{CellDimension + 6 \cdot Tolerance} \frac{0.925}{0.125 + 6 \cdot 0.001} = 7.06 \tag{4}$$

On the length of the vehicle's roof, a row of 9 cells can be used. In this case, the layout of the solar panel will be formed of 9 rows with 7 cells, having a total of 63 cells.

The optimal way to connect the solar cells into a panel is to have parallel and series connections as well. It's important that the series connections are made with the same number of cells so they have the same voltage and the same current. In our case, the series connection will contain 21 solar cells. This number is the result of the 3 rows with 7 cells on every row. As mentioned earlier that the series connection must have the same number of cells, so the 3 rows of series connection it's the result of dividing the 9 cells on the length of the car by the lowest common denominator, which is 3. This way 3 series connections are created like in figure 5.

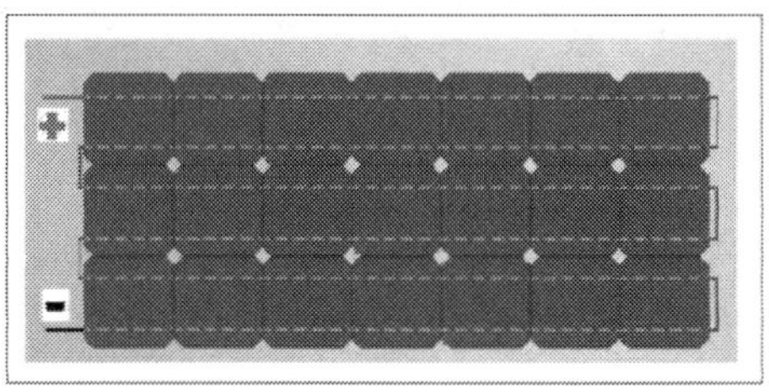

a)

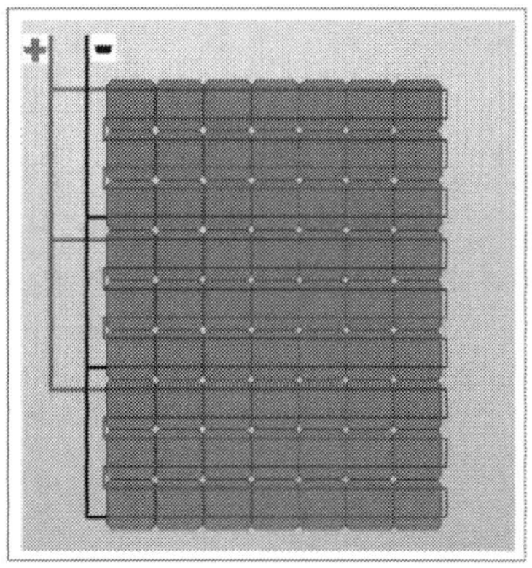

b)

Figure 5. a) The 21 solar cells connected in series. b) Those 3 obtained panels connected in parallel

4. Designing the solar panel for the E-Smart vehicle

The common solar panels are made by placing the cells on a straight plane. In our application, the solar panel will have a parabolic shape in order to take the curvatures of the car and maximize aerodynamic.

The 3D implementation (with Autodesk Inventor software) helps us in finding the optimal way of fixing the solar system on the roof and also to determine and analyze the strength of the roof and its aerodynamics.

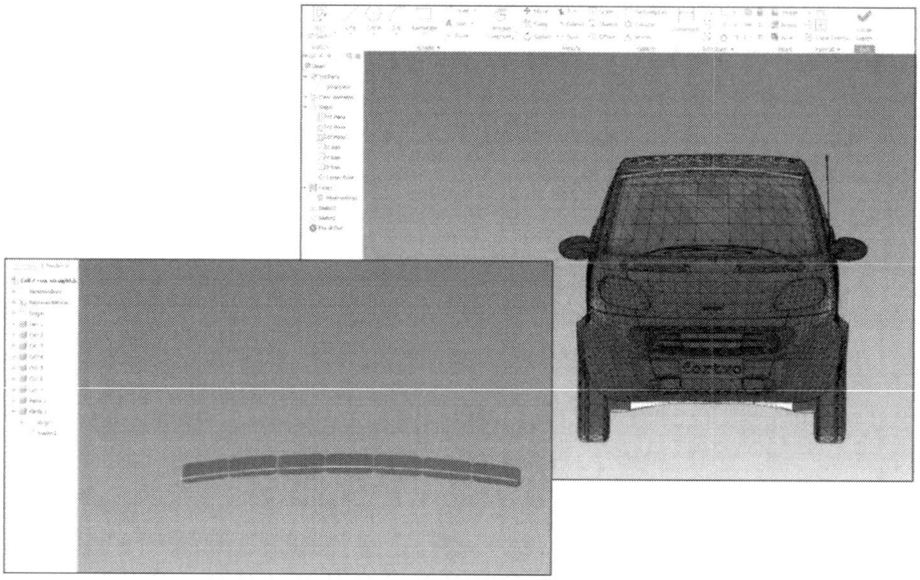

Figure 6. Designing the 3D model of the implemented solar panel

Implementing a 3D model, in this case the solar panel to an existing vehicle it can be difficult and requires a good perception of space. First, the shape of the roof was drawn out from two perspectives creating a surface that matches the roof of the car. A sketch that represents the frontal view of the roof it's considered as a simple line on which the solar cells are placed one by one so a row is formed. This row formed of 7 cells takes the curvature of the roof.

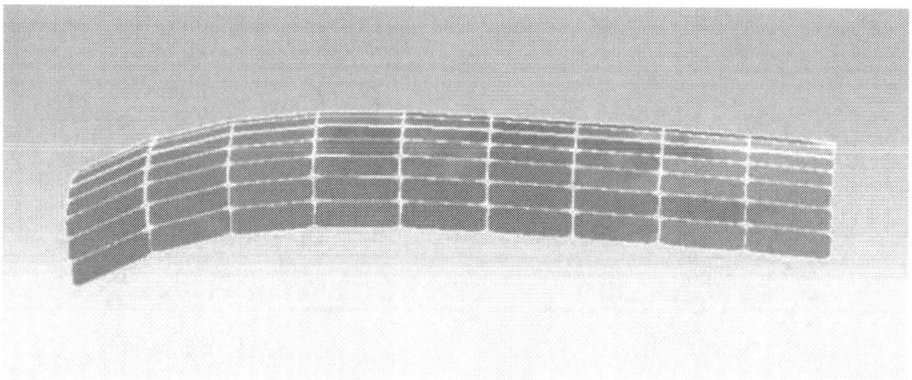

Figure 7. The 3D view of the designed solar panel

By placing the rows next to each other, row by row, the shape of the entire roof it's achieved. Finally, the shape of the solar array is consistent from the front to the back of the vehicle's roof.
Conventional solar panels are made by laminating the solar cells in between multiple thin films. The part that is facing the sun it's usually a tempered glass and underneath a thin ethylene-vinyl acetate copolymer foam (EVA) sheet. On the back of the panel, there is an EVA sheet glued with an adhesive to back substrate [5].

In our application, due to the shape of the solar panel, the tempered glass must have the shape of the roof.

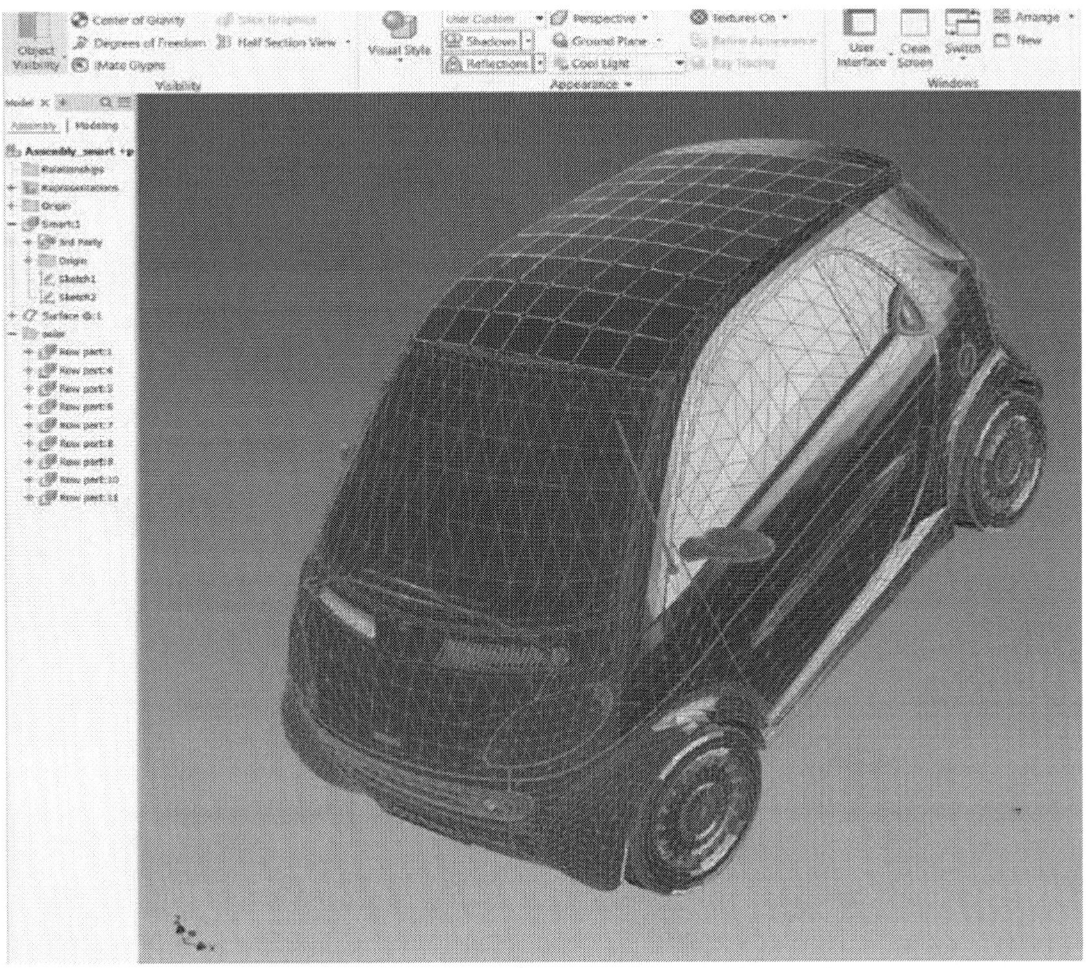

Figure 8. The solar panel placed on the vehicle roof

Another approach in implementing the solar panel on the roof is by using transparent epoxy resin which can be poured over the solar cells when are in place [6]. This method gives the flexibility to work directly on the car which is an advantage when it comes to the design of the system. The big disadvantage when using epoxy resin in this type of application is that because of excessive sunlight exposure, the epoxy resin can yellow in time. The epoxy resin can be poured in place taking the desired shape. Another disadvantage is the lack of solar cell reparability. After the resin is set in gets really hard and if there is a need for cell replacement the process of removing the hardened resin without damaging the other cells will be impossible.

The most reliable and light material that is compatible for our application is polycarbonate. Polycarbonate is a strong material used in different types of applications including automotive and aircraft components. Polycarbonate is a plastic material. At a temperature of 250 Celsius the polycarbonate sheet can be shaped to the desired form and after slowly cooling down it wardens and remains in that form. The deformation or plastic reaction can be done once for a polycarbonate sheet [7]. After this process, the sheet will gather new physical properties. The tensile strength of the material will increase and the radiation resistance will be better. These physical properties are important especially for this particular application as the solar panel that it's placed on the car it's exposed to objects that can hit and damage the solar panel. The radiation resistance it's also an important factor since the efficiency of the cells are highly correlated to the transparency of the material that is above the solar panel [8].

The 3D model needs to imitate as close as possible a real-life system and for this, all the components had to be created in the 3D Autodesk Inventor software. The solar cells are laminated between two sheets of polycarbonate. The polycarbonate was created in the 3D Autodesk Inventor software, giving the following physical properties: density, tensile strength, coefficient of friction, Poisson's ratio, Young's modulus to gather more precise measurements during simulation.

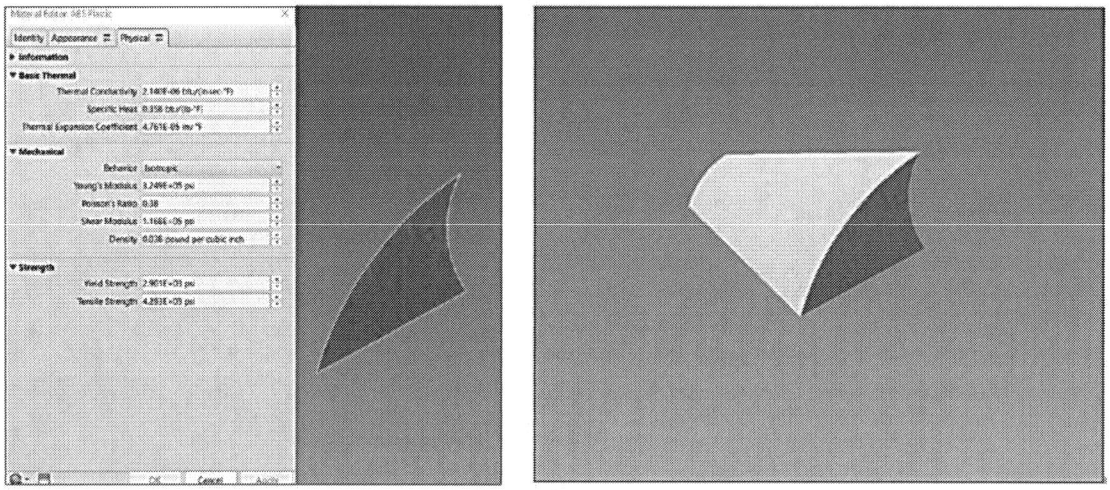

Figure 9. The spoiler used in front of the solar panel

Another element was created in 3D Autodesk Inventor software in order to modify the air flow from the front of the car where the panel will meet the roof of the vehicle. This component will act as a spoiler so the air cannot get under the panel and lift it. This component can be made from different types of materials, and in this project the materials from fiber glass, plastic to carbon fiber. In this project, the material used was Acrylonitrile Butadiene Styrene (ABS) plastic. This material was chosen due to the high impact resistance and the ability to create the component with a 3D printer. All the physical properties were manually introduced in the Material editor.

5. Conclusions

In the automotive industry, the use of photovoltaic systems it's a new technology, therefore the proposed solar system has room for more analyses and improvements [9, 10]. The study was conducted for the E-Smart electric vehicle, which is used in research in our department. Equipping this car with the proposed photovoltaic system will increase the autonomy, using renewable energy, with just a little increase of the total mass. The future tests with the electric vehicle equipped with this photovoltaic system will take place in some regions of Romania, but especially in Brasov County.

Photovoltaic systems are constantly being developed in Romania, being used more and more often for the production of clean energy. The largest solar park in Romania, with 332,000 operational photovoltaic units, is located in Ucea de Sus, Brasov County and has an installed capacity of 82 MW, covering a land surface of 200 ha [10].

The photovoltaic potential in Romania is presented in figure 10. The map [11] is developed by Global Solar Atlas, which is published by the World Bank Group, funded by ESMAP, and prepared by Solargis. It is noted that Romania has a high energy potential and any application with photovoltaic systems is eligible.

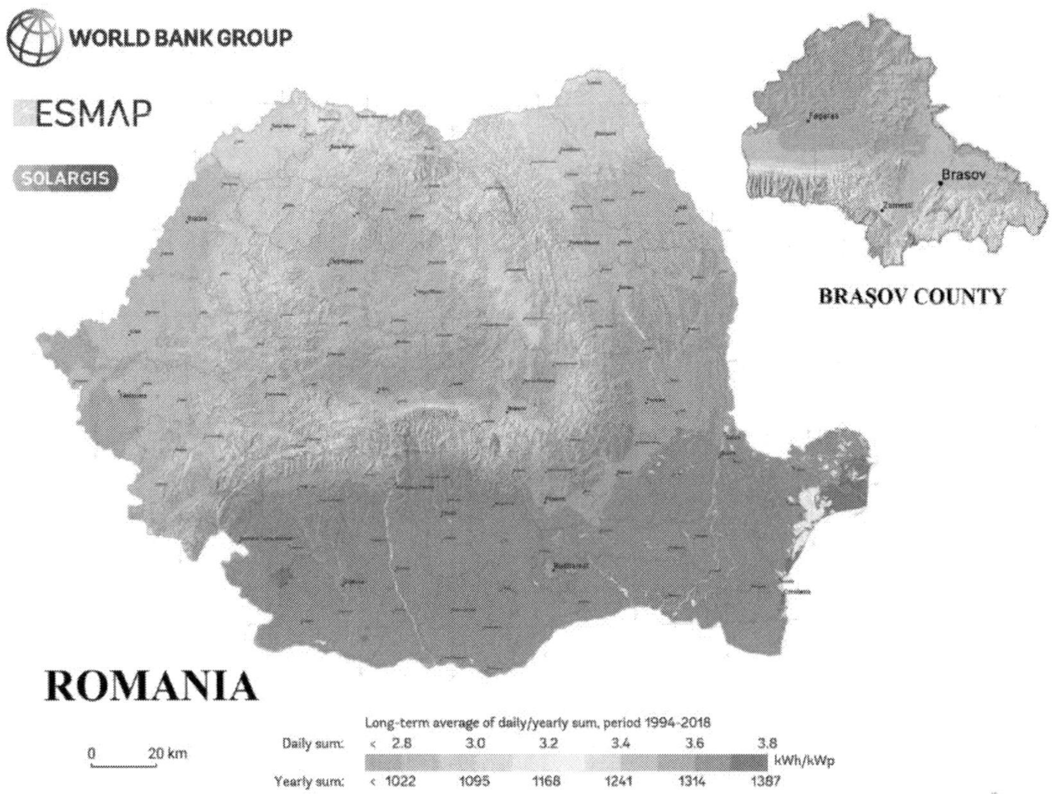

Figure 10. The map of photovoltaic potential in Romania with a detail of Brasov County [11]

The proposed photovoltaic system was calculated and designed taking into account the characteristics and parameters of the accessible photovoltaic cells on the market. The 3D model of the photovoltaic system was made with the Autodesk Inventor software. For the design process, we took into account the dimensions of the vehicle's roof, its usable area and the aerodynamic elements that must be added in order for the implementation of this system to be eligible. We also took into account the properties of all the materials used, density, tensile strength, coefficient of friction, Poisson's ratio, Young's modulus, to gather more precise measurements during simulation.
The studies will continue with an analysis of the behavior of the photovoltaic system attached to the vehicle, in terms of aerodynamics.

References

[1] Ţârulescu R., Ţârulescu S., Electronic control systems of E-Smart vehicle, International Congress of Automotive and Transport Engineering, Springer International Publishing Switzerland, pp: 509-516, (2017).
[2] Ţârulescu R., Ţârulescu S., Battery management system of E-Smart vehicle, International Congress of Automotive and Transport Engineering, Springer International Publishing Switzerland, pp: 517-524, (2017).
[3] Ewa Golisz, K. K., Efficiency analysis of photovoltaic systems. Warshaw Univerity of Life Science Publisher, (2020).
[4] SunPower data sheet, Document #001-66352 Rev** A4_en, sunpowercorp.com, (2010).
[5] Arno, S., Klaus, J., Olindo, I., Miro, Z., Solar Energy - The physics and engineering of photovoltaic conversion, technologies and systems. UIT Cambridge Publisher, Edition 1, (2016).

[6] Reinders, A., Verlinden, P., Van Sark, W., Freundlich, A., Photovoltaic solar energy: from fundamentals to applications. Wiley Publisher, (2017).
[7] Tasdemir, M., Properties of acrylonitrile–butadiene–styrene/polycarbonate blends with styrene–butadiene–styrene block copolymer, Journal of Applied Polymer Science 93(6), pp: 2521 – 2527, (2004).
[8] Abdelhamid. M., Pilla. S., Singh, R., Haque, I., Filipi, Z., A comprehensive optimized model for on-board solar photovoltaic system for plug-in electric vehicles: energy and economic impacts. International Journal of Energy Research 40, Wiley Online Library, pp: 1489-1508, (2016).
[9] Dinakaran, C., Padmavathi, T., Performance evaluation on electric vehicle by solar photovoltaic system, International Journal of Scientific and Technical Advancements, Volume 6, Issue 4, pp. 39-44, (2020).
[10] Năstase, G., Şerban, A., Dragomir, G., Brezeanu, A.I., Bucur. I., Photovoltaic development in Romania. Reviewing what has been done, Renewable and Sustainable Energy Reviews, Volume 94, pp. 523-535, (2018).
[11] ***https://solargis.com/maps-and-gis-data/download/romania

IOP Conf. Series: Materials Science and Engineering 1220 (2022) 012010 doi:10.1088/1757-899X/1220/1/012010

The analysis of a hybrid turbocharger main parameters applied to an internal combustion engine

Rareş-Lucian Chiriac, Ovidiu Andrei Condrea and Anghel Chiru

Transilvania University of Braşov, Str. Politehnicii Nr. 1, 500024 Brasov, Romania

achiru@unitbv.ro

Abstract. Internal combustion engines are constantly developing. Currently, a solution in full development is a hybrid car that has an internal combustion propulsion combined with an electrical motor. The electrical motor is powered by a battery which in turn is connected to other electricity suppliers for recharging. Therefore, the aim is to recover the thermal energy from the car, through various solutions, in order to transform it into electricity. Electricity is for the car's battery as well as for its consumers. After an analysis of the energy balance of the internal combustion engine, it can be seen that the engine energy still has great potential to be transformed into electrical energy through various solutions. Electrical energy can be produced even from braking or other mechanical devices that transform mechanical energy into electrical energy. For example, a co generator of energy hybrid electric turbocharger aims to recover exhaust gas energy and to capitalize it. The turbocharger is connected through a shaft with a generator. Between the compressor wheel and the generator is a speed reducer with a gear ratio of 1:10. This article aims to present the analysis based on the results of the experimental and simulation research of the hybrid turbocharger. The tests were performed using a diesel internal combustion engine with 4 cylinders at a capacity of 1.9 liters. The simulations will be performed using AMESim software developed by Siemens. The main parameters, which are highlighted, are: pressure ratio of the turbocharger, the rotation speed of the turbocharger and the power from the experimental results in relation to time and engine power at 100% load.

1. Introduction

Davies and the authors have shown that hybrid turbocharging is the key technology for hybridized power trains and engines [1]. Sailer and authors have shown that through simulation it was proven that the performance of exhaust gas used by turbochargers is very high [2]. Naundorf and the authors sustain that the exhaust gas from internal combustion engines has a great potential to be used for the energy system of the engine [3]. Gabriel and the authors present that many turbocharger developers are trying to develop and offer improvements for turbocharger engines applied to spark ignition engines and also compression ignition engines. They are proposing solutions like: variable turbine geometry, hybrid turbochargers, multi-stage turbocharger systems [4]. Getzlaff, Heim and authors have shown also that the main goal is to reduce fuel consumption and have more power for the engine [5, 6]. Rode and the authors have the opinion that the new electric turbocharger concepts were highly efficient for the internal combustion engines [7]. Some of the developed turbochargers have also high flow rates and high pressure ratios [8]. Hunziker and authors have proved that spark ignition engines and also compression ignition engines have been developed to increase performance, reduce pollution all in relation to costs [9]. For this paper, a hybrid turbocharger co generator of electrical power has been developed to prove that the turbocharger has a great potential to be improved.

Figure 1. The real construction of the hybrid turbocharger.

The main component parts of the turbocharger prototype, such as: turbocharger, speed ratio and electric generator presented directly as components, but also as simulation parts of the AMESim software.

Table 1. The main component parts of the turbocharger prototype

nr. crt.	Main component part	Component part in AMESim software	Designation
1.	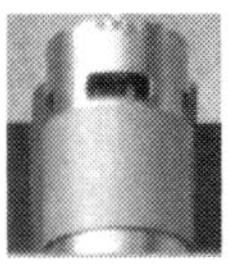	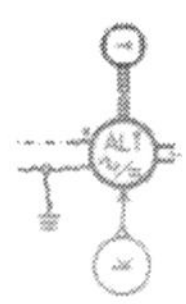	electrical generator
2.			gear ratio/ 1:10
3.	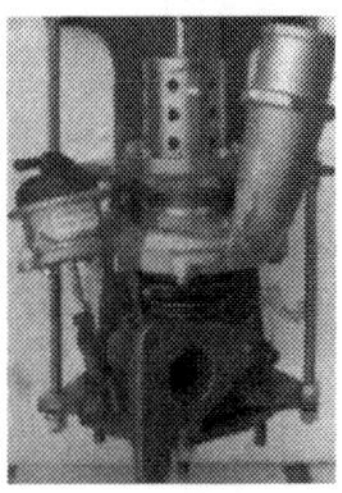	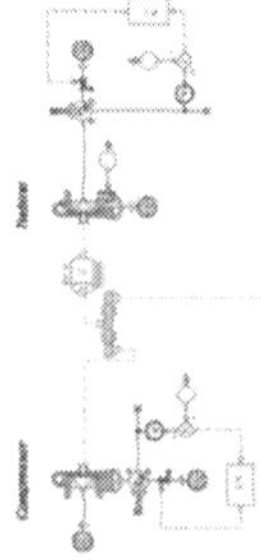	Garrett turbocharger

2. Objective

The general objective of this experimental work is to demonstrate that the hybrid turbocharger has a very good performance improvement. For that, the simulation and validation results of the hybrid turbocharger prototype must also be presented.

Figure 2 illustrates the overview of the hybrid turbocharger with the main components of the simulation system.

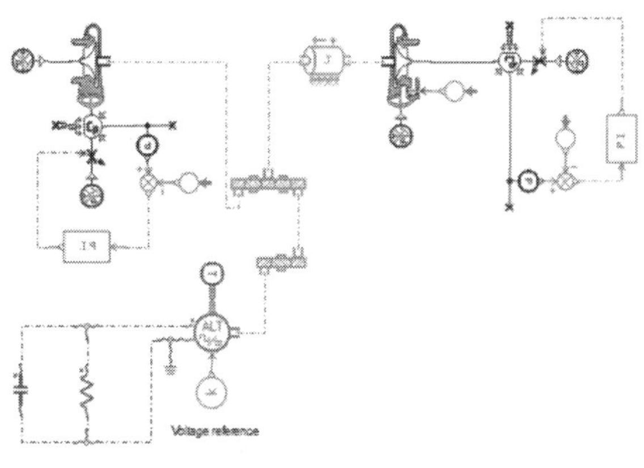

Figure 2. Hybrid turbocharger components with AMESim software.

Figure 3 illustrates the components of the engine with internal combustion diesel with 4 cylinders of 1.9 liters supercharged equipped with a hybrid turbocharger which is modelled by AMESim software.

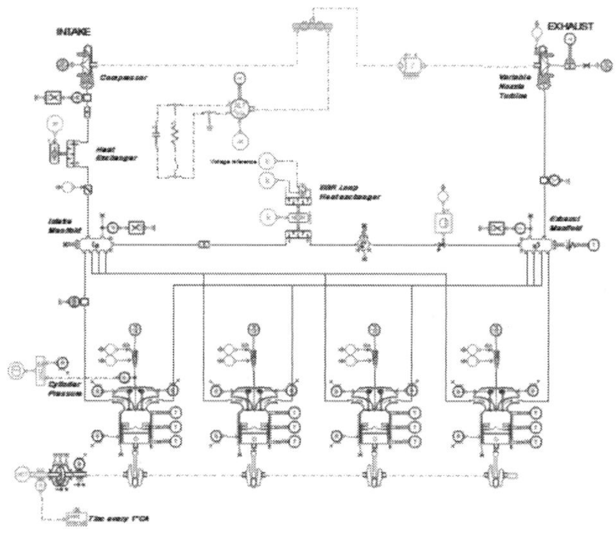

Figure 3. Diesel engine 1.9 liter cylindrical diesel internal combustion engine equipped with a hybrid turbocharger.

3. Methodology

The basic equations used for the mathematical simulation of the compressor, turbine and alternator can be divided into two categories namely for the mechanical part and for the electrical part.

The basic equations used for the mathematical simulation of the compressor, turbine and alternator can be divided into two categories namely for the mechanical part and for the electrical part.

3.1. Equations for the mechanical part

To take into account the real engine operating conditions, the rotary speed and the mass flow rate values are corrected or reduced.

IOP Conf. Series: Materials Science and Engineering 1220 (2022) 012010 doi:10.1088/1757-899X/1220/1/012010

3.1.1. Compressor equation, speed/ flow rate.

The main compressor equation, speed/ flow rate are:

$$N_{cor} = \frac{N_{act}}{\sqrt{\frac{T_{up}}{T_{st}}}} \tag{1}$$

$$N_{red} = \frac{N_{act}}{\sqrt{T_{up}}} \tag{2}$$

$$\dot{m}_{cor} = \dot{m}_{red} \cdot \frac{\sqrt{\frac{T_{up}}{T_{st}}}}{\frac{P_{up}}{P_{st}}} \tag{3}$$

$$\dot{m}_{cor} = \dot{m}_{red} \cdot \frac{\sqrt{T_{up}}}{P_{up}} \tag{4}$$

$$PR = \frac{P_{out}}{P_{up}} \bar{A}o = \sqrt{\frac{T_{out}}{T_{up}} \left(\frac{1 + \frac{\gamma - 1}{2} M_{out}^2}{1 + \frac{\gamma - 1}{2} M_{up}^2} \right)} = \sqrt{\left[1 + \frac{PR^{\frac{\gamma - 1}{\gamma}} - 1}{\eta} \right]} \tag{5}$$

3.1.2. Turbine equations, speed/ flow rate.

The main turbine equations, speed/ flow rate are:

$$N_{red} = \frac{N_{act}}{\sqrt{\frac{T_{up}}{T_{st}}}} \tag{6}$$

$$N_{red} = \frac{N_{act}}{\sqrt{T_{up}}} \tag{7}$$

$$\dot{m}_{act} = \dot{m}_{red} \cdot \frac{\sqrt{\frac{T_{up}}{T_{st}}}}{\frac{P_{up}}{P_{st}}} \tag{8}$$

$$\dot{m}_{cor} = \dot{m}_{red} \cdot \frac{\sqrt{T_{up}}}{P_{up}} \tag{9}$$

$$P_{mec} = \omega_{turb} \cdot t_{turb} \tag{10}$$

3.2. Equations for the electrical part

The main alternator parameters are: rotary velocity in [rot/min], temperature in [degC], current in [A], torque in [Nm], efficiency in [null] (i.e. no unit), losses in [W] (see Figure 4).

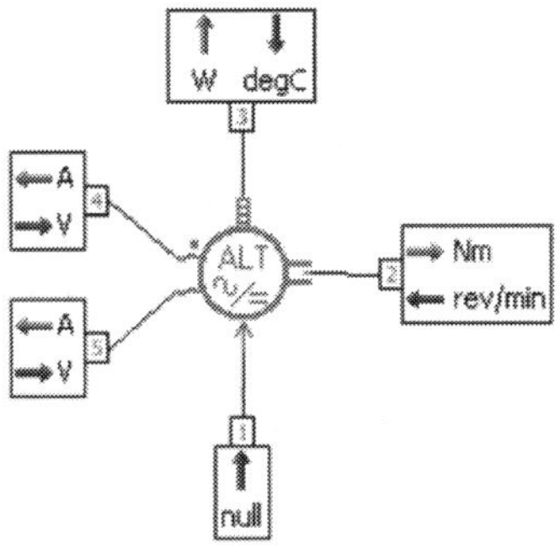

Figure 4. Generator symbol and main parameters in AMESim Software.

$$P_e = |U \cdot I| \tag{11}$$

$$I = \frac{U}{R} \tag{12}$$

[10]

where:

N_{cor} – speed corrected, [m/s];
N_{act} – speed actual, [m/s];
N_{red} – speed reduced, [m/s];
T_{up} – upstream temperature, [K];
T_{st} – standard temperature, [K];
P_{up} – upstream pressure, [kPa];
P_{st} – standard pressure, [kPa];
$\dot{m}_{cor}$ – mass flow rate corrected, [-];
$\dot{m}_{red}$ – mass flow rate reduced, [-];
PR – compressor pressure ratio, [-];
A_O – compressor sound speed ratio, [-];
P_{up} – upstreamabsolute total pressure at port 3, [bar];
P_{out} – downstream absolute total pressure at port 1, [bar];
T_{up} – upsteam temperature at port 3, [K];
T_{out} – downstream temperature at port 1, [K];
γ – specific heat ratio upstream gas mixture, [-];
η – the isentropic efficiency, [-];
P_{mec} – mechanical power, [W];
ω_{turb} – turbine rotary speed, [rad/s];
t_{turb} – turbine torque, [Nm];
P – electrical power, [W];
U – voltage, [V];
I – current, [A];
R – resistance, [Ω];
rot/min – rotations per minute;
W – Watt.

4. Results

The results will show the electrical parameters resulting from the hybrid turbocharger namely: the current intensity, the voltage and the power of the generator. The energy is produced by the hybrid turbocharger at a constant speed of 170000 rotations per minute, with a resulting pressure of approximate 1 bar.

Figure 5 shows the resulting electrical power of the electrical generator in relation to time (experimental values). The experimental range value is from 0 to 110 Watt. And in figure 6 it is shown the resulting electrical power from the AMESim simulation in relation to time. The simulation range values are from 0 to 120 Watt.

IOP Conf. Series: Materials Science and Engineering 1220 (2022) 012010 doi:10.1088/1757-899X/1220/1/012010

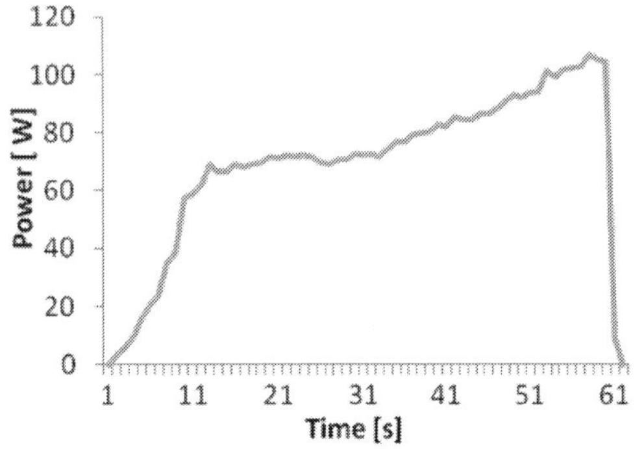

Figure 5. The generator electrical power from the experimental values in relation with time

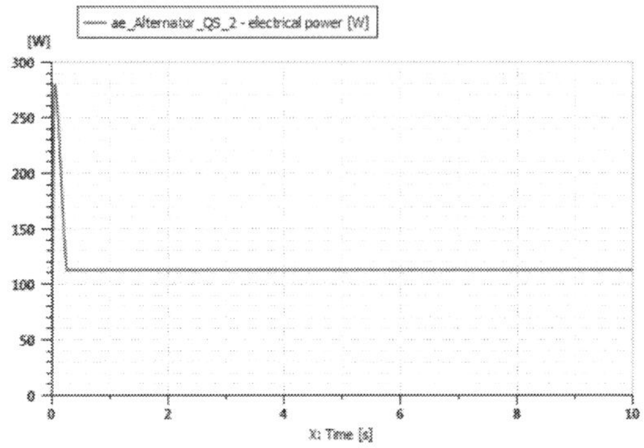

Figure 6. The resulting generator electrical power from the AMESim simulation in relation with time

Figure 7 shows the compressor outlet pressure in relation to time (experimental values). The experimental values range is from 0 to 1.55 bar. Therefore figure 8 indicates the compressor outlet pressure simulated with AMESim in relation to time with the simulation value range from 0 to 1.65 bar.

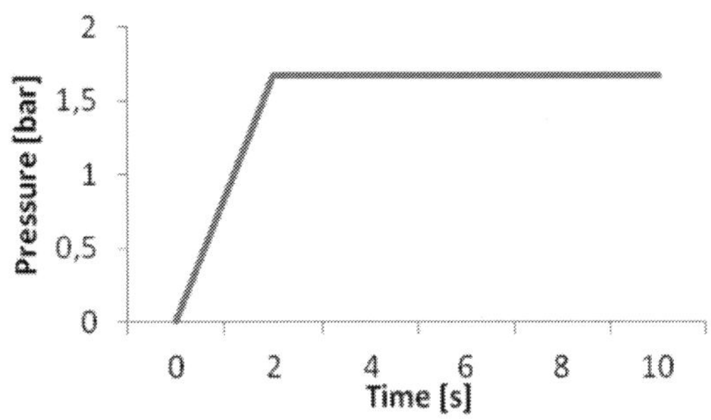

Figure 7. The compressor outlet pressure from the experimental results in relation to time

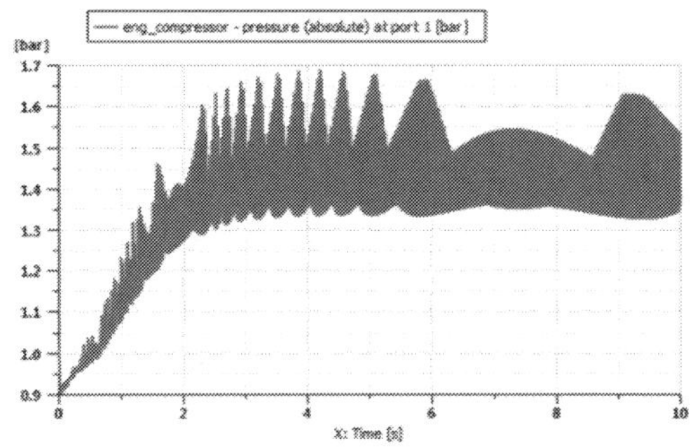

Figure 8. The compressor outlet pressure simulated with AMESim

Figure 9 illustrates the pressure ratio, PR from the experimental values in relation to time. The experimental research value range is from 0 to 1. And figure 10 illustrates the pressure ratio, PR simulated with AMESim in relation to time with the simulation value range from 0 to 1.

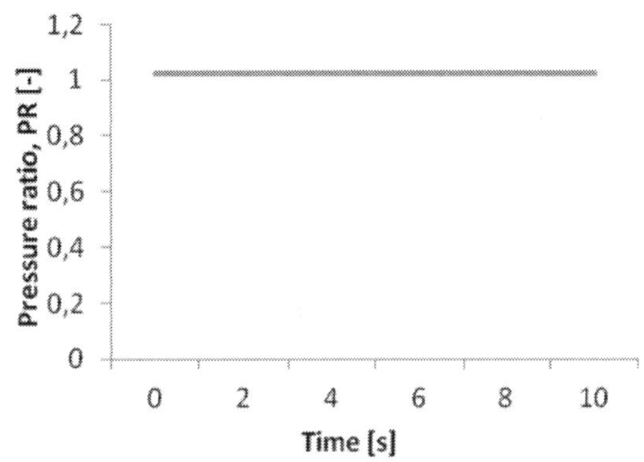

Figure 9. The pressure ratio, PR (Pressure Ratio) from the experimental results in relation to time

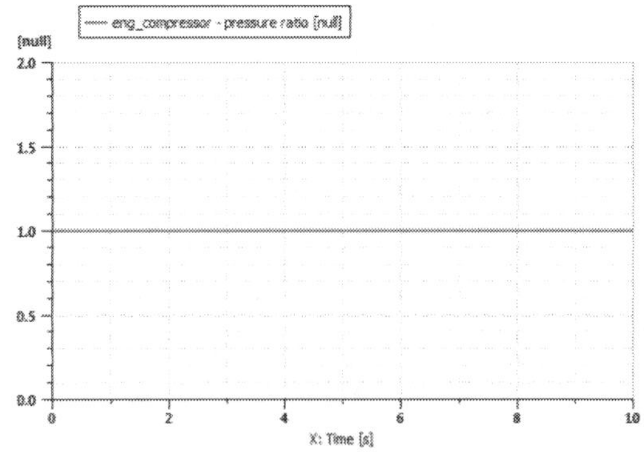

Figure 10. The pressure ratio, PR simulated with AMESim in relation to time

Figure 11 shows the experimental values between the rated hybrid turbocharger power of maximum 23 kW and the generator power produced of maximum 0.025 kW at 100% load.

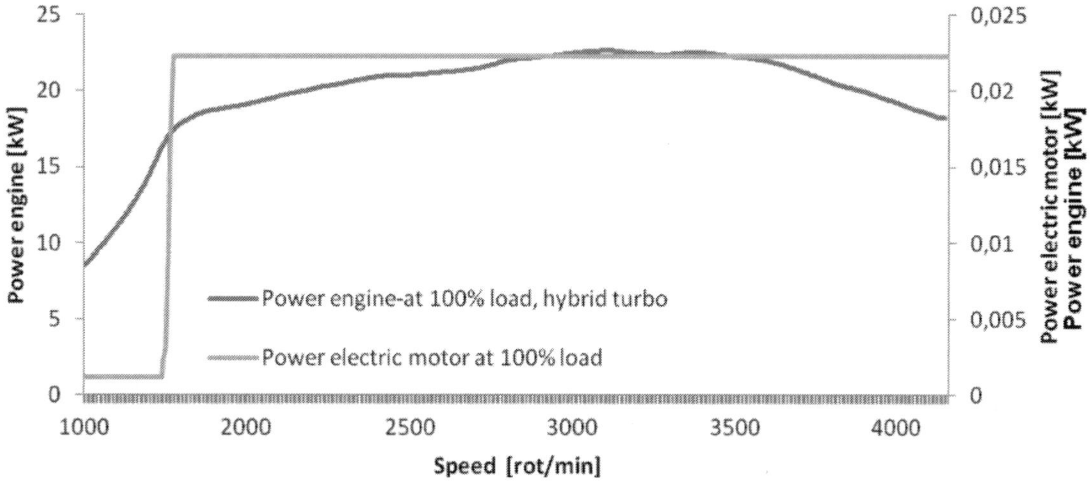

Figure 11. Comparison between the rated power of hybrid turbocharger engine and the electrical generator power produced at 100% load

Figure 12 shows the result of the simulation with the AMESim software of the nominal power of the 1.9 liter diesel engine equipped with a hybrid turbocharger of maximum 23 kW at 100% load.

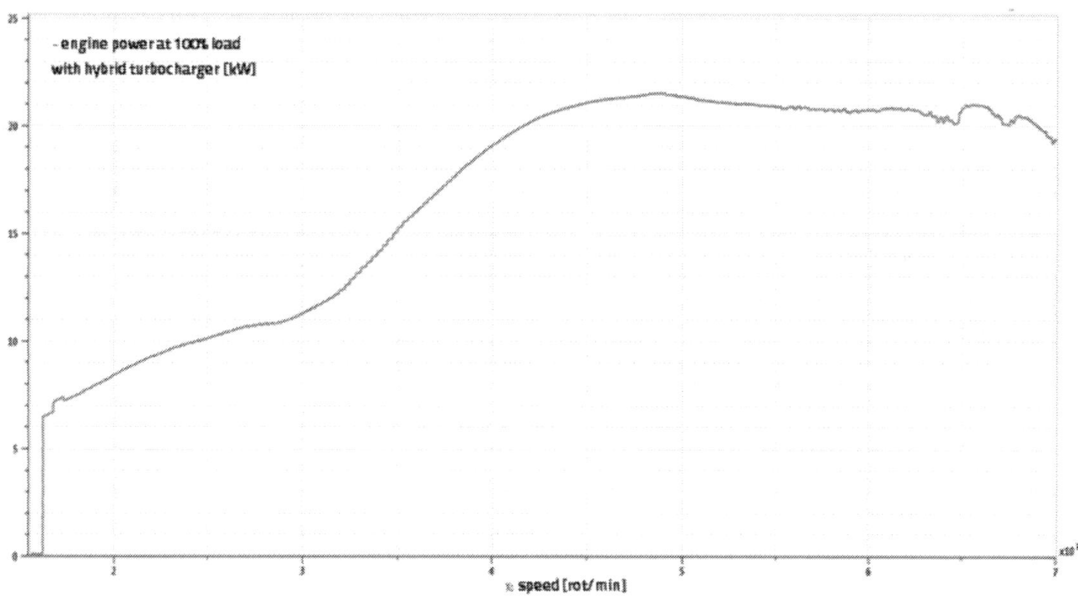

Figure 12. Rated engine power at 100% load simulated with AMESim software of Diesel 1.9 engine equipped with hybrid turbocharger

It can be mentioned that in figure 12 the diagram is the result of the AMESim software simulation which has used the speed rage from 0 to 7000 rot/min. Figure 11 shows the experimental values of the rated power of the hybrid turbocharger engine which was performed at a rage from 0 to 5000 rot/min.

5. Conclusions

It was demonstrated that the hybrid turbocharger has a great potential for development and to obtain green ecologic electricity for the vehicle. It is able to generate electrical power for the consumers and also to compress air for the turbocharging process of the engine. The electrical energy can be used for consumption and also can be stored in batteries [11].

The main advantages of the new hybrid turbocharger co generator of electrical energy are: the electrical energy and storage in batteries, also to redirect to peripheral consumers and to compress air for the engine [12, 13].

Acknowledgement

I would like to thank for the collaboration with the company Garrett by Honeywell - manufacturers of turbochargers, the Siemens Company for the AMESim software, Turbocar Serv SRL for the experimental works and the "Transylvania" University of Braşov.

References

[1] Davies, P., et al., Electric Turbocharging - Key Technology for Hybridized Powertrains, MTZ Worldw 80, 2019, https://doi.org/10.1007/s38313-019-0096-y
[2] Sailer, T., et al., Simulation of the Compress or Performance of Exhaust Gas Turbochargers. MTZ Worldw 72, 2011, https://doi.org/10.1365/s38313-011-0004-6
[3] Naundorf, D., et al., Turbocharger test stand with a hot gas generator for high-performance supercharging systems. MTZ Worldw 69, 2008, https://doi.org/10.1007/BF03226916
[4] Gabriel, H., et al., The turbocharger with variable turbine geometry for gasoline engines. MTZ Worldw 68, 2007, https://doi.org/10.1007/BF03226804
[5] Getzlaff, U., et al., Simulation of the thermo characteristics of a water-cooled turbocharger. MTZ Worldw 71, 2010, https://doi.org/10.1007/BF03227041
[6] Heim, K.: Existing and future demands on the turbocharging of modern large two-stroke diesel engines, 8th Supercharging conference, 1–2 October 2002, Dresden, Germany
[7] Rode, M., et al., Electric Turbocharger Concept for Highly Efficient Internal Combustion Engines. MTZ Worldw 80, 2019, https://doi.org/10.1007/s38313-019-0056-6
[8] Hunziker, R., et al., A new series of small turbochargers for high flow rates and high pressure ratios. MTZ Worldw 63, 2002, https://doi.org/10.1007/BF03227546
[9] Mithun D., An experimental study on the combustion, performane and emission characteristics of a diesel engine, fuelled with diesel-castor oil biodiesel blends, Renewable Energy, 2017
[10] LMS AMESim, Ameshelp Library, https://www.siemens.com/plm/support
[11] Galindo, J: Strategies for improving the mode transition in a sequential parallel turbocharged automotive diesel engine, International Journal of Automotive Technology 2009
[12] Perrot, N.: Experimental Study of Centrifugal Compressor Speed Lines Extrapolation for Automotive Turbochargers, SAE Paper, 2017
[13] Chiriac, R.: New constructive solutions for hybrid turbochargers- as electrical energy generator for increasing the green supply of the vehicule, International Conference for Doctoral Students, "Transilvania" University, Brasov, 2017

IOP Conf. Series: Materials Science and Engineering 1220 (2022) 012011 doi:10.1088/1757-899X/1220/1/012011

Modeling and simulation of an electric propulsion system equipped with Fuel Cell

A A Ancuța and G Danciu

Department of Automotive Engineering, Faculty of Transport, University „Politehnica" of Bucharest

alex.ancuta96@yahoo.com

Abstract. The paper proposes the modelling of an electric propulsion system that uses as the main energy source, a Fuel Cell. The model is based on the structure of Toyota MIRAI, the first electric vehicle equipped with Fuel Cell developed in series in 2015. The model is developed under MATLAB Simulink software and adapts existing fuel cell and vehicle blocks to the real vehicle case. The simulation results contain the time evolution of some important parameters for both fuel cell and vehicle. Conclusions analyse these results and suggest future possible pathways of study.

1. Introduction

The sharp evolution in recent years regarding stricter pollution rules, especially related to the amount of carbon dioxide (CO_2) has led to an acceleration of manufacturers in the development of less polluting propulsion systems or not at all to equip vehicles road. As for the electric propulsion system powered by energy from a traction battery, especially Li-ION or Li-Polymer type, the pollution is not even eliminated, but there is only a passing trend from the urban environment, into theenvironment rural, by the fact that the process of manufacturing and possibly recycling of batteries isan extremely complicated and important producer of CO_2. This has again become a topic of controversy and perhaps even a disadvantage in terms of the growing commercialization of electric vehicles, so that new alternative solutions are being sought for on-board sources of electricity storage. In addition to supercapacitor batteries, a variant with good efficiency, but much more expensive, the use of hydrogen (H_2) and its physical and chemical properties that position this element in the area of extremely important efficiency advantages is increasingly being considered. its high electricity production correlated with minimal pollution.

For this purpose, H_2 stored on board a vehicle in cylinders with special mechanical and thermal properties, resistant to very high pressures (approx. 700 bar) is used as fuel in the Fuel Cell to produce electricity that will supply an external load, respectively the propulsion system. This component, in turn, is also characterized by a number of disadvantages in terms of controlling the voltage generatedat the output of the Fuel Cell, which determines the need to use a power converter at the output, which maintains the voltage at the optimal value to be able to p ower the connected load; also the storage of hydrogen on board the vehicle is another problem generated by the extremely high storage pressure inthe tanks and finally, the global infrastructure to supply these tanks.

In the following, we will analyze several situations of using a fuel cell and we will draw a series of conclusions based on the experimental results obtained by simulation.

2. Theoretical aspects about the Fuel Cell

Fuel Cell is an enclosure inside which hydrogen stored in a special tank is introduced, on the one hand, and oxygen from the atmosphere on the other, and through a series of chemical reactions a certain amount of electricity is produced and can power an electric propulsion system and a series of non-polluting residual components: water (H_2O) used for additional cooling of the system and heat, which is discharged into the atmosphere by specialized systems. In conclusion, the chemical energy produced by them is converted into chemical energy used to power the electric vehicles that produce the traction of the vehicle by means of chemical reactions [1, 2, 7].

From a constructive point of view, the fuel cell consists of two electrodes (catalysts for PEM FC), one positive: cathode and the negative: the anode and an electrolyte (a membrane permeable to H_2) which has the role of developing chemical reactions together with a catalyst that allows the decomposition of hydrogen and oxygen molecules. There are obviously several types of fuel cells, but basically, the construction and operation are the same [1].

2.1. Principle of operation

The principle of operation is based on three chemical reactions that result in the electricity used in the propulsion system. Figure 1 shows schematically the operating principle of a fuel cell:

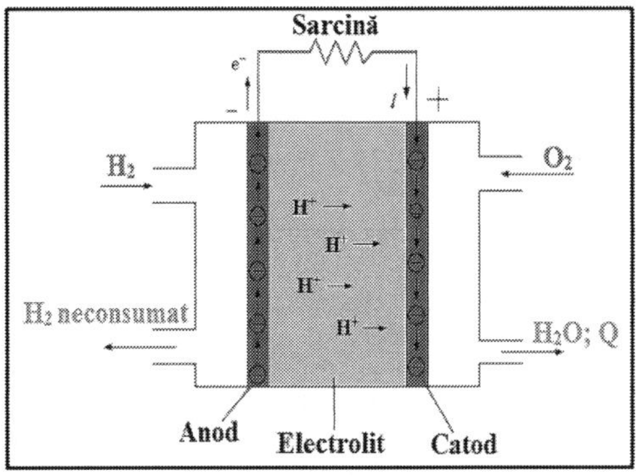

Figure 1. Principle of operation of a fuel cell [1].

The three chemical reactions that show the operation of the fuel cell are as follows, in the order in which they occur [1]:

$$2H_2 \rightarrow 4H^+ + 4e^- \quad (1)$$

Reaction 1 consists of the oxidation reaction of the fuel, which involves the breaking of the hydrogen molecule once it comes into contact with the existing electrons on the anode. Following the reaction, hydrogen electrons migrate through an external circuit generating electricity. Hydrogen protons (H +) migrate across the membrane to the cathode.

The second reaction, expression 2, is the reduction reaction that occurs in the cathode area where the oxygen introduced from the atmosphere interacts with the electrons existing on the cathode andcations occur:

$$O_2 + 4e^- \rightarrow 2O^{-2} \quad (2)$$

The last reaction consists in the interaction of the protons that crossed the permeable membrane and the oxygen anions existing in the cathode area:

$$4H^+ + 2O^{-2} \rightarrow 2\,H_2O \quad (3)$$

Following the three reactions presented above, electricity is produced with a transformation efficiency of over 70% and without generating polluting emissions harmful to the environment. Although there are many significant voltage losses in the economy and efficiency of a fuel cell, the evolution and development of technologies in recent years in this direction has led to an improvement in fuel cell performance and thus a reduction in all forms of losses due to chemical reactions. Thus, a typical fuel cell operates with average cell voltages in the range of 0.65 V 0.7 V; even at currents of about 1 A / cm2, which offers an efficiency of approx. 50%, a value that surpasses the most powerful compression ignition engine that can achieve an efficiency of up to 40-45%, but under certain conditions and operating regimes. We can conclude that the efficiency of the fuel cell is higher the lower the currents, as opposed to heat engines where at low power the efficiency is extremely low.

2.2. Architecture of an electric propulsion system equipped with a fuel cell

Figure 2. Toyota Mirai propulsion system architecture [10]: 1 – Li-ION battery; 2 – Tank of H_2; 3 – Fuel Cell and DC/DC Converter; 4 – PermanentMagnet Synchronous Motor; 5 – Power Electronics.

Figure 2 shows a constructive structure of the propulsion system that equips the Toyota Mirai car. It consists of two hydrogen storage tanks on board the vehicle, an electrochemical battery with NiMH technology in the original version, later replaced with Li-ION technology, due to the advantage of higher specific energy and higher efficiency in the latter sources. As for component number 3, it is the Fuel Cell assembly and the 650 V DC / DC power converter for controlling the supply voltage of the Fuel Cell supply system. In older versions, the converter was a separate component from the Fuel Cell,but with the optimization of the system, both functionally and in terms of achieving an optimal space from an ergonomic point of view, these components became a common block. The propulsion system consists of a three-phase, synchronous AC electric motor with permanent magnets, which is controlled by an electronic power unit [8, 9].

3. Making models in Simulink and simulating the Fuel Cell in different situations

To make the simulation models, different blocks already existing in the SIMSCAPE library of the used program were used and based on some modeling examples, the system to be presented was set up. The model was developed in many steps as follows. The first step was to setting Fuel Cell parameters: the obtained model is characterized by a maximum power of 114 kW (impulse regime) and 85 kW, continuous power. The rated operating point is characterized by a rated current of 250 A and 340 V, and the maximum point of 356.25 A and 320 V. The internal resistance of the battery is 0.21164 Ω, and the rated voltage of a cell is 1.1649 V. Operating temperature of the battery is 65°C, and the nominal percentage of use for the two elements: 95.46% H_2 and 45.22% O_2. The supply pressure with H_2 varies between 1.5…2.5 bar, and for our system we chose 1.5 bar according to the information in the technical data sheets of the Mirai car, while for O_2 the value is 1 bar. After parameterizing the Fuel Cell, the control block also gives us the voltage characteristics and operating points of interest, as we can see in figure 3. You can immediately see the extremely important influence of the battery current on the supply voltage offered by it and directly the proportionality between current and power offered by Fuel Cell in the two modes of interest (continuous and impulse):

IOP Conf. Series: Materials Science and Engineering 1220 (2022) 012011 doi:10.1088/1757-899X/1220/1/012011

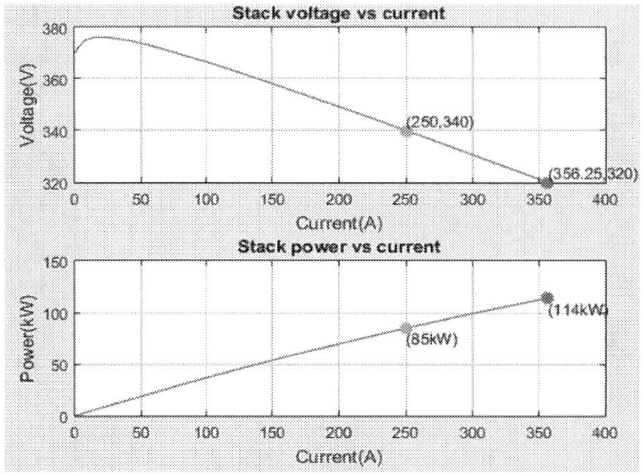

Figure 3. Parametrization of the Fuel Cell.

With the parameterization of Fuel Cell, to finalize the model for simulation we adopted a flow regulator modeled based on the calculation equation of H_2 flow according to a series of characteristic parameters, regulator connected to the battery input and output current, so to perform a closed loop operation. The flow is calculated according to the next relationship [3, 4]:

$$V_{H_2} = \frac{6000 \cdot R \cdot T \cdot N \cdot i_{FC}}{z \cdot F \cdot U_{fH_2} \cdot x\%} \tag{4}$$

Where: R – is Boltzman's constant (8.3145 J/mol·K); T – is the operating temperature of the battery (K); N – is the number of cells in the Fuel Cell; i_{FC} – is the current flowed by the Fuel Cell; z – is the number of electrons migrating as a result of chemical reactions; F – is the Faraday constant (96485 A·s/mol); U_{fH_2} – percentage of hydrogen utilization and %x hydrogen composition.

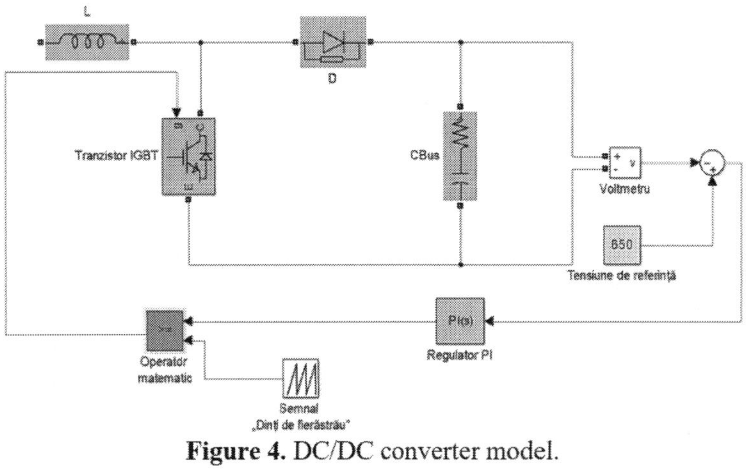

Figure 4. DC/DC converter model.

To complete the model for simulation, we made a650 V DC/DC converter designed to keep the voltage at the output of the Fuel Cell at the nominal value, so that it can supply an external load. This converter is based on a PI(s) type regulator (Proportional-Integral), a "sawtooth" type signal generator and a reference constant of 650. Regarding the electronic components, inside we find a transistor of IGBT power, a diode, a coil at the input to the converter and a capacitor. The parameters of these components are extremely important in terms of the output signal of the converter and its correct operation. The electrical sizes of the coil and capacitor were set according to the nominal parameters of the battery, and the constants of the PI controller (s), as follows: KP = 0.00005 and KI = 0, and component D (derivative) was taken out of the equation so as to we avoid as much as possible instabilities in the considered system being an incipient model in terms of determining the performance of a Fuel Cell and the fundamental influence of the power converter [5, 6].

3.1. Simulation of the Fuel Cell when is connected to a constant load

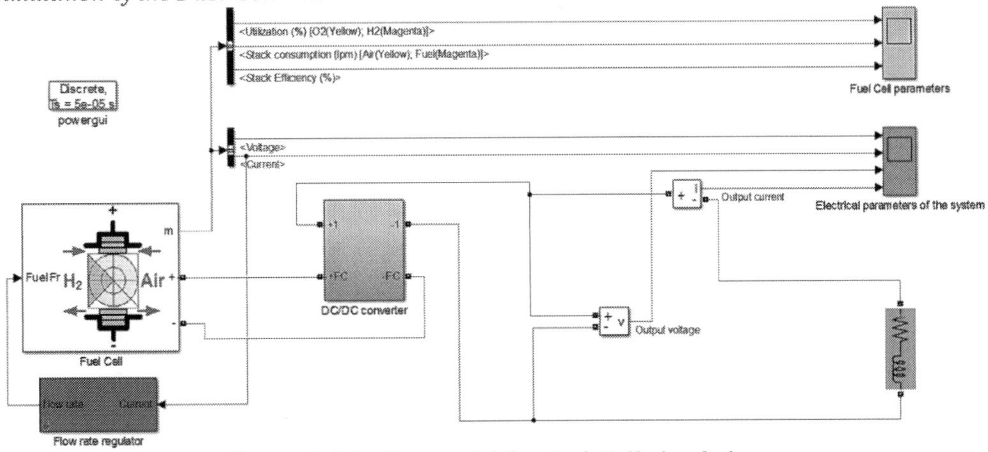

Figure 5. The first model for Fuel Cell simulation.

3.1.1. Simulation results.

The simulation was performed using a sampling time of 10 seconds, a discrete variable and the external load had a resistance of 1.36 ohm, calculated according to the parameters of the fuel cell. We obtained the following experimental results:

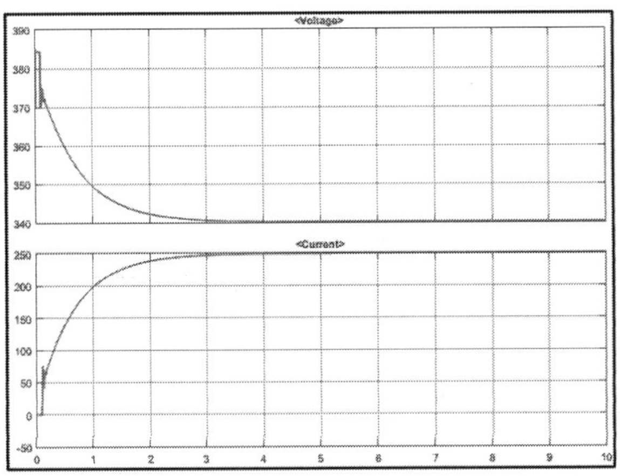

Figure 6. Fuel Cell voltage and current variation.

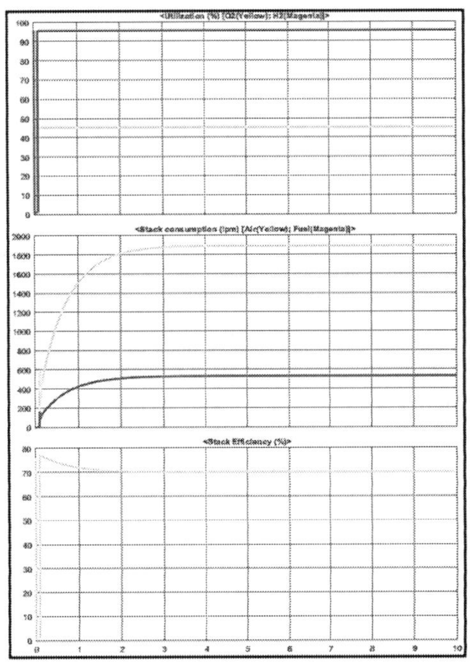

Figure 7. Utilization(%), Stack consumption (lpm) and Stack Efficiency (%).

3.1.2. Conclusions.

In this case, we noticed several aspects related to the operation of the system, as well:

- When the Fuel Cell system has been connected to a constant load, we can observe an instability of the curves of variation of the nominal voltage of the battery and respectively the current at the output of it, we can see this in figure 6; instability caused by the PI controller (s) used to control the transistor inside the converter and its response time, as well as by the internal resistance of the fuel cell, which negatively affects the performance of the Fuel Cell;

- After the first 3 seconds, the characteristic sizes of the Fuel Cell stabilize, the voltage reachingthe nominal value of 340 V, and the current reaches the value of 250 A;
- Fuel Cell efficiency is directly proportional to the fuel and air flow used by Fuel Cell, so whenthese values stabilize and reach maximum values, Fuel Cell efficiency tends to remain almost constant around the maximum value entered as the input date.

3.2. Simulation of the Fuel Cell connected to a constant load together with the Li-ION battery.

To complete the model of the electric propulsion system, we connected the model shown in figure 5 together with the Li-ION battery parameterized according to the information provided by Toyota [3], and also used an existing mask in the SIMSCAPE library.

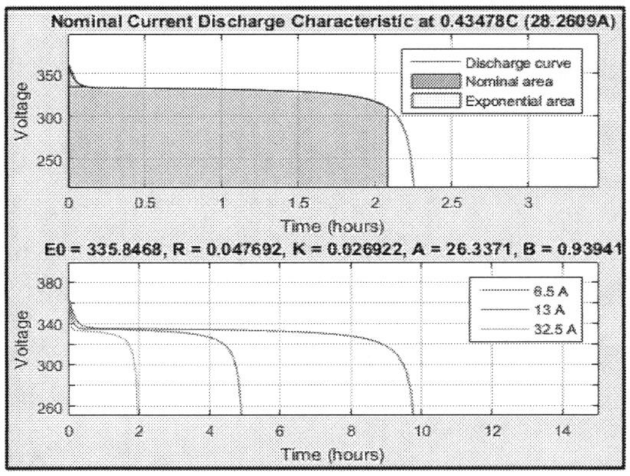

Figure 8. Li-ION battery discharge characteristic.

Similar to the Fuel Cell, the control block automatically generates in the case of the battery the voltage variation curves in relation to the battery discharge time (2 hours and 25 minutes) and the voltage variation for different values of the discharge current (6.5 A; 13 A; 32.5 A). By connecting the battery model, the system in the following figure was obtained, the simulation conditions being the same as in the previous case, using a discrete variable and the solver "ode23tb (stiff / TR-BDF2)" from Simulink library, for a sampling time of 5 seconds, on half compared to the Fuel Cell simulation alone. The following image shows the model for this simulation case:

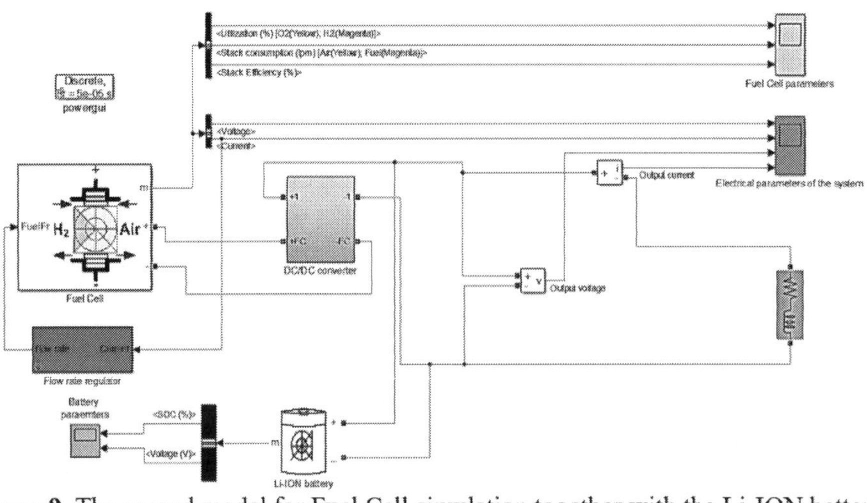

Figure 9. The second model for Fuel Cell simulation together with the Li-ION battery.

3.2.1. Simulation results.
In this case, in addition to the characteristics of the fuel cell, we also analyzed the parameters of the Li-ION battery:

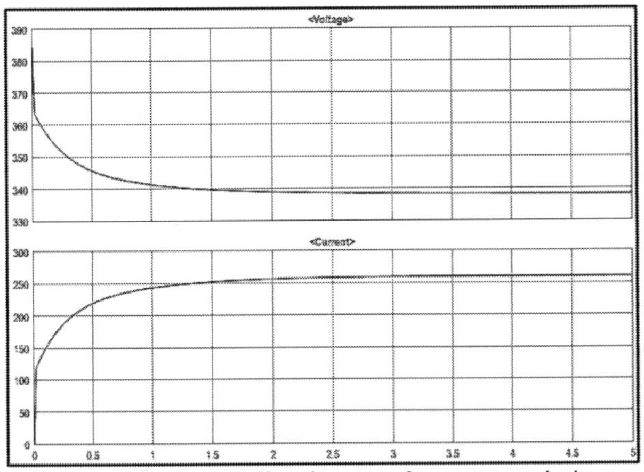

Figure 10. Fuel Cell voltage and current variation.

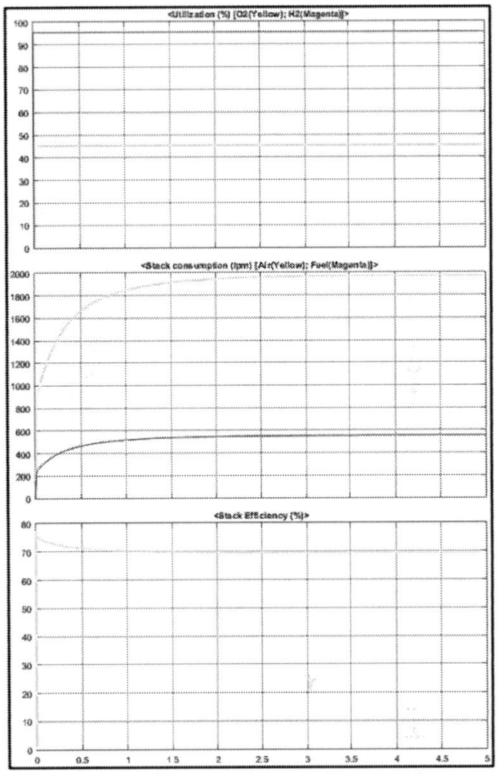

Figure 11. Utilization(%), Stack consumption (lpm) and Stack Efficiency (%).

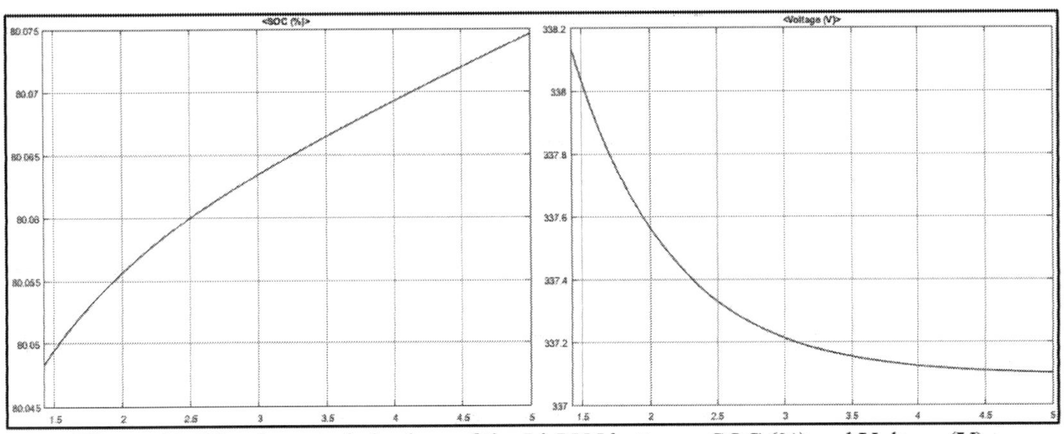

Figure 12. The characteristic sizes of the Li-ION battery – SOC (%) and Voltage (V).

3.2.2. Conclusions. This time, connecting the battery produces a faster stabilization of the system, but also compromises the value of the nominal voltage of the battery, tending to use a quantity of energy produced by it to change its state of charge in an increasing direction, which we can see it in figure 12 where an increase of 0.075% over a period of 5 seconds is easily noticeable.

3.3. Model of the electric propulsion system and its simulation

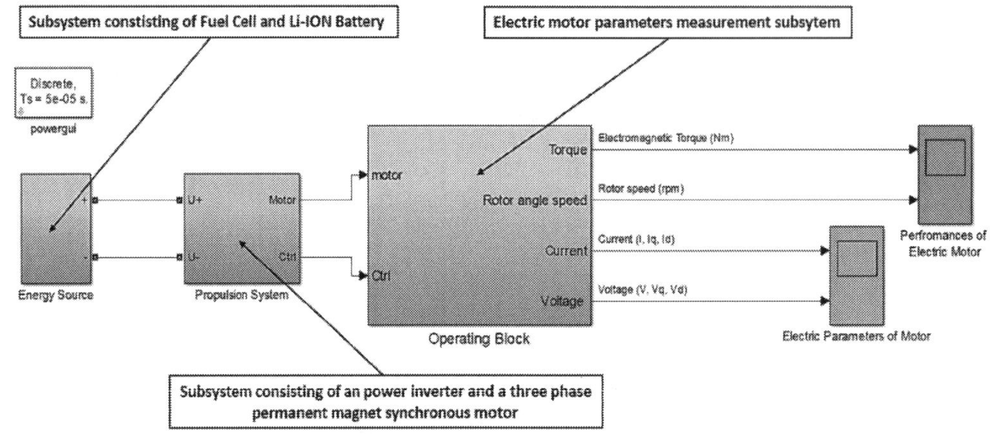

Figure 13. Electric propulsion system model.

Finally, the model of the propulsion system was made, as can be seen in figure 13, using a permanent magnet synchronous motor, an "IGBT-DIODE" inverter and an electric motor speed control module. based on two blocks taken from the SIMSCAPE library, namely: "SpeedController" and "VECTController". The torque requirement was modeled based on the theoretical characteristic of an alternating current electric motor and was correlated with the reference value in impulse mode of the motor on the Mirai car, of 335 Nm. The simulation was performed using the same discrete variable, the same solver, but a shorter sampling time, which would better present the variation of the parameters obtained from the simulations. Figure 14 shows the subsystem called "Propulsion System":

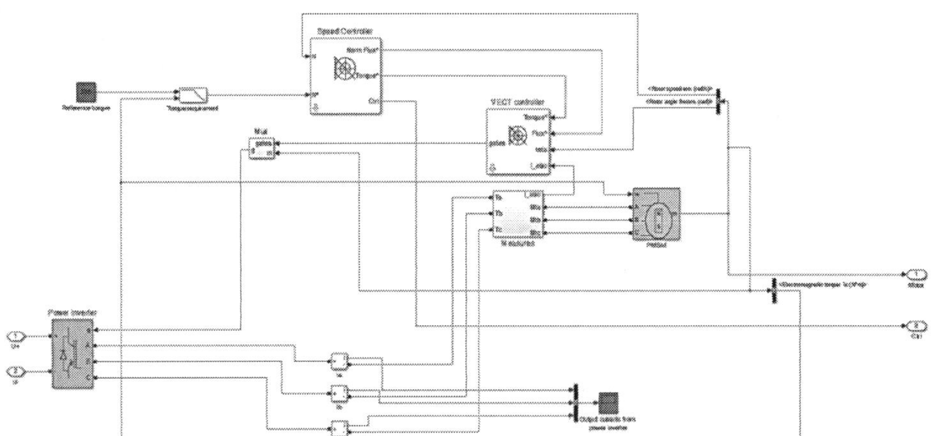

Figure 14. "Propulsion System" architecture.

As for the subsystem called "Power Source", it consists of the model shown in figure 9, Fuel Cell together with the Li-ION battery. The "Operating block" subsystem is an existing mask in the SIMSCAPE library, which allows the measurement and stabilization of the parameters of the electric motor. The actual simulation of the propulsion system was performed in an acceleration mode of the electric motor in impulse mode, until the maximum speed and the stabilization of the generated electromagnetic torque, so that we can observe how the Fuel Cell supplies the system with electricity without the need for the traction battery, which aims only to store energy, for example that obtained by the process of braking with recovery or to support at certain times Fuel Cell.

3.3.1. Simulation results.

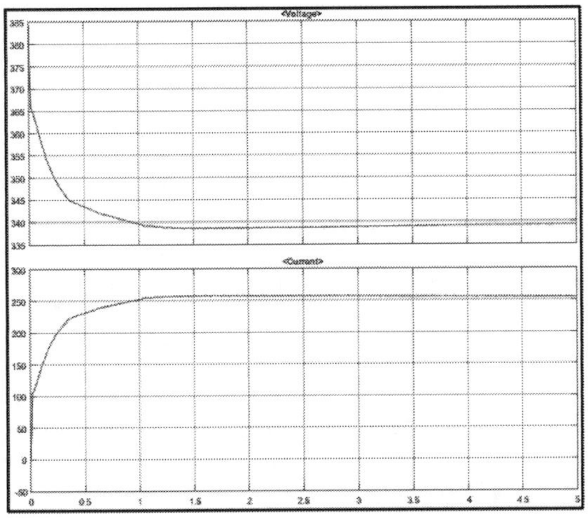

Figure 15. Fuel Cell voltage and current variation.

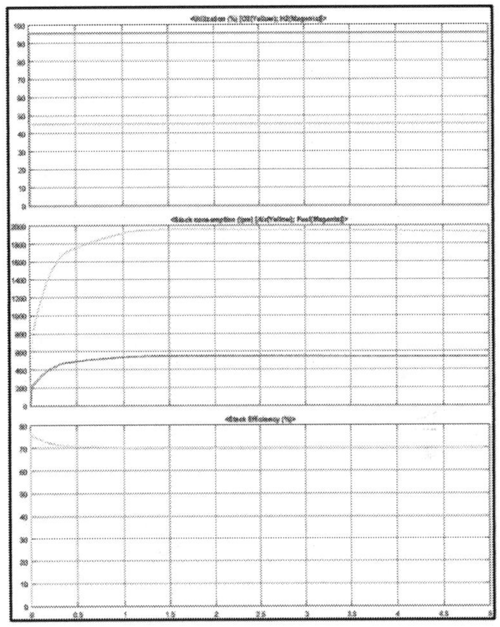

Figure 16. Utilization(%), Stack consumption (lpm) and Stack Efficiency (%).

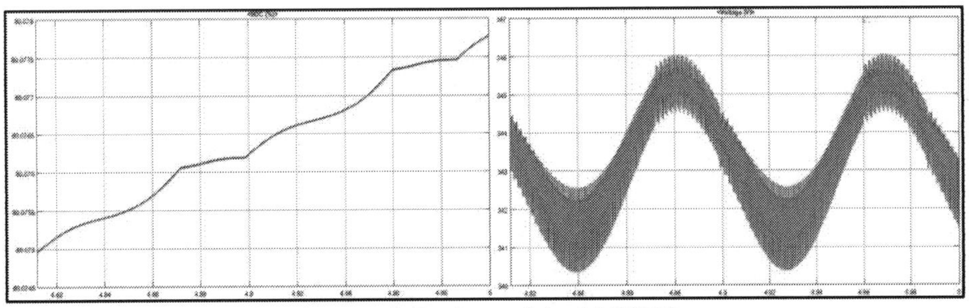

Figure 17. The characteristic sizes of the Li-ION battery – SOC (%) and Voltage (V).

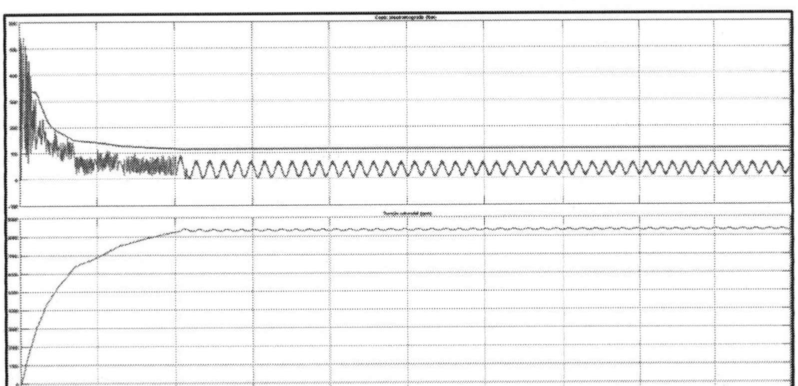

Figure 18. Electromagnetic torque (real value – red color/reference value – purple) and motor speed variation for an electric motor.

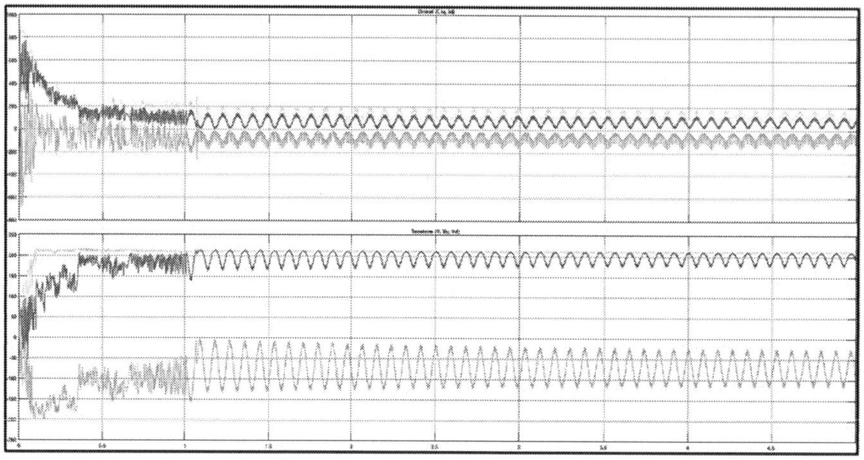

Figure 19. Current and voltage for an electric motor.

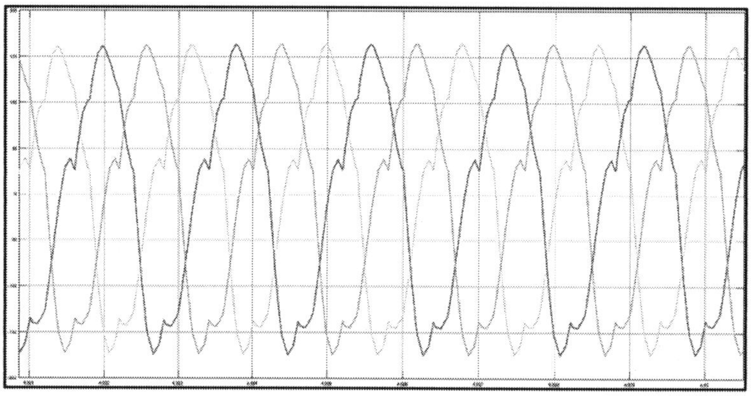

Figure 20. Output currents variation from the power inverter.

4. Conclusions

- When connecting the propulsion system, there is a strong instability in the converter control mode which shows that the IGBT transistor is more malfunctioning, thus concluding the problems caused by the error introduced in the system by the PI controller (s), aspects that can be observed in the variation of the characteristic parameters of the electric motor;
- State of charge of the battery changes by 0.003%, which over a longer sampling period would result in an approximately constant curve of around 81%, which demonstrates that the Fuel Cell is able to provide enough power to power the system. propulsion;
- Variation of the electromagnetic torque produced by the electric motor is visibly affected by the unstable operation of the power converter, thus a relatively large relative error occurs between the torque requirement introduced as input signal and the variation of the output torque;
- Improving the control mode of the converter and fine tuning the characteristic electrical quantities (coil inductance L = 500 μH and capacitor capacity C = 7500 μF [3], values used for this simulation) will determine a stabilization of the whole system and the operation of the electric machine as much as possible. close to the values imposed theoretically;

Overall, the modeled system concretely shows the variation of the parameters of a Fuel Cell in different operating situations and proposes different system optimization ideas that can be achieved by improving the input signals that control both the power converter and the propulsion system to be later on a dynamic model of the Toyota Mirai car.

Acknowledgements
I would like to thank first of all to Professor Grigore Danciu for all the support provided in conducting this research, for the advice and ideas he offered me and for the entire collaboration so far and also for colleagues in the department who supported and guided me in my professional training.

References

[1] Mehrdad, Ehsani; Yimin, Gao; Ali, Emadi – *Modern Electric, Hybrid and Fuel Cell Vehicles – Fundaments, Theory and Design* – Second Edition; Texas A&M University College Station, Texas, U.S.A
[2] Larminie, J. Lowry, J. – *Electric Vehicle Technology Explained* – Oxford Brookes University, Oxford, U.K
[3] Ahmad Yulianto, Simic, M. Taylor, D. Trivailor, Pavel – *Modelling of full electric and hybrid electric fuel cells buses* – International Conference on Knowledge Based and Intelligent Information and Engineering Systems, KES2017, 6-8 September 2017, Marseille, France
[4] Abert Cerdan Codina, - *System Level modelling of fuel cell driven electric vehicles* – Master thesis in Eelectric Engineering, CHALMERS UNIVERSITY OF TECHNOLOGY, Gothenburg, Sweden 2017
[5] Kavyapriya, S. Krishna Kumar R. - *Modeling and Simulation of DC-DC Converters for Fuel Cell System* - International Journal of Engineering and Advanced Technology (IJEAT) ISSN: 2249 – 8958, Volume-9 Issue-3, February, 2020
[6] Sandeep Gupta, Swati Singh, Lini Mathew, Shimi S. L. - *Comparative Study of Dc To Dc Converters Via Simpowerelectronics Fuel Cell Stack* - IOSR Journal of Engineering (IOSRJEN) www.iosrjen.org ISSN (e): 2250-3021, ISSN (p): 2278-8719 Vol. 04, Issue 01 (January. 2014)
[7] Bertini, L. - *Modeling and Optimization of a Fuel Cell Hybrid System* - Master thesis KTH Royal Institute of Technology, Division of Applied Electrochemistry
[8] *Medium And Heavy Duty Fuel Cell Electric Vehicles* – California Environmental Protection Agency, AIR RESOURCES BOARD, November 2015
[9] *Toyota Mirai FCV_Posters_LR_tcm-11-564265* – technical data sheet of the vehicle, 2020
[10] ***- https://seekingalpha.com/article/3965730-are-hydrogen-cars-threat-to-electric-vehicle

IOP Conf. Series: Materials Science and Engineering 1220 (2022) 012012 doi:10.1088/1757-899X/1220/1/012012

Study of the influence on gas emissions by using hydrogen in a diesel engine

Cosmin Constantin Suciu[1], Sorin Vlad Igreț[2] and Ioana Ionel[1]

[1]PolitehnicaUniversity Timisoara, Faculty of Mechanical Engineering, Timisoara, Romania
[2]Aurel Vlaicu University, Faculty of Engineering, Arad, Romania

ioana.ionel@upt.ro

Abstract. This research studies the possibility of the reduction of the emission values emitted by a diesel engine, by adding hydrogen to the fuel mixture. Due to the higher pollution restrictions implemented globally on the internal combustion engines, for new but also older vehicles, suitable solutions must be found, even for diesel engines, towards major incriminations concerning pollution are attested. The water electrolytic conversion for hydrogen production is achieved by using a special device, in symbioses with the internal combustion engine. The hydrogen quantity, as resulted from the process and fuelled into the engine, represents 10% of the total volume of the intake capacity. Being directly correlated to the stoichiometric ratio, the intake of hydrogen is finally controlled; therefore the total calorific power of the fuel mixture is increasing. The calorific value of Diesel is 44.8 MJ/kg, while that of hydrogen is 141.8 MJ/kg. Since the addition of hydrogen represents 10% of the total intake capacity, the addition of total calorific value of the mixture fuelled raises by approximate 30%. All the measurements have been done in a lab using a certified and approved gas analyser.

1. Scope of the research

During the last years, automobile constructors are globally confronting with more severe pollution restrictions. The global answer to this necessity goes towards electrification. Still, vehicles based on internal combustion engines (ICE) are in use and the owners are presently not capable to change towards electric cars, especially due to the high prices, even some states are supporting a part of it, and also due to the few charging points, that must be very much developed, assuring the continuity of the transport. One indicates also that some of the vehicle manufacturers and owners tend to keep the internal combustion engine, detrimental to electrical propulsion because the production of batteries implies an extremely polluting process, not mentioning the energy consumption, that is necessary during the production, as it is mainly based on fossil fuels and C is emitted [1, 2, 3]. Also, it is worth mentioning the high prices of the batteries, the large volume, as well as difficulties in reusing the waste batteries, after their use.

In this context, the present automobile engine manufacturing industry is investigating the possibility of emission reduction, and turn towards greenhouse effects technologies; recently in discussion the consideration of utilizing unconventional synthetic fuels [4, 5] is mentioned, as a potential solution.

Hydrogen can be an alternative fuel source, that can be obtained, at low costs and with a calorific power superior to all of the conventional fuels [6, 7, 8].

2. Measurements and results

For the achieved measurements a vehicle, equipped with a diesel engine on which it was mounted an electrolytic converter, was used.

Table 1 presents the engine specifications.

Table 1. Engine specifications.

Characteristic	Value
Number of cylinders	4
Cylinder's layout	Straight/ inline
Cylinder bore [mm]	81.0
Cylinder stroke [mm]	95.5
Compression ratio	18.5
Number of valves	8
Valve's position	SOHC
Engine code	BMP
Fuel type	Diesel
Engine's position	Front, Transversal
Engine's displacement [cm^3]	1,968
Intake	VGT + intercooler
Max power output [kW]	103 @ 4,000 rpm
Max torque output [N·m]	320 @ 1,800 rpm
Transmission layout	Front wheel drive
Gearbox	Manual
Number of gears	6
Fuel tank volume [l]	70

A gas analyser was used for the measuring of the polluting exhaust gas. There were two measurements, one at the engine's idle speed and the other at an engine speed of 2000 rotations per minute (rpm).

Using the electrical power supply of the vehicle, one dissociates water into hydrogen and oxygen, practically producing gas from water. This gas is a mixture between oxygen and hydrogen, also called Oxy-hydrogen gas (HHO gas) [9, 10].

In [11] it is indicated that, in the near future, hydrogen is expected to become a significant fuel that will largely contribute to the quality of atmospheric air. The arguments brought by the authors [11] are based on the fact that it is the most widespread one on the earth and as H_2 it can be obtained from a number of sources both renewable and nonrenewable by various processes.

The electrolytic converter device, presented in figure 1, is characterized by positive, negative and neutral plates, a water tank and a bubble filter and hoses and electrical wiring connectors. Potassium hydroxide is used in a proportion of 5-15% in the water tank further on being used as an electrolyte in the main convertor. The water tank is connected by hose to the electrolytic convertor, which is powered by the electrical supply of the vehicle's engine, using a voltage between 12-14V controlled by a switch. This system, after being powered, begins to produce HHO gas which is filtered in the bubble filter before being introduced in the engine's intake system. The bubble filter works by liquid filtering the HHO with the purpose of stopping the small gas particle and vapours. The reason for this is that the HHO gas reaches 50 degrees Celsius and also vapour may occur. The whole system requires an electrical current between 2 and 20 Amp, being able to produce between 2 and 4 litres of HHO per minute.

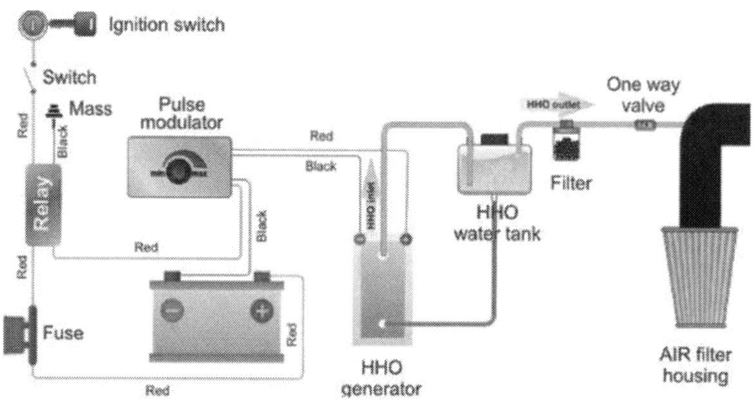

Figure 1. HHO system – installation scheme [12].

The calorific value of hydrogen is 3 times higher than diesel fuel. The resulting hydrogen from the process represents 10% of the total volume of intake capacity.

For these measurements, Capelec 3200 gas analyser was used. Initially, the measurements have been done at the engine's idle speed on only diesel fuel, values being presented in table 2. For the second set of measurements, the engine was exploited in the exact same environment and conditions but with the electrolytic convertor enabled. Values obtained are in table 3.

Table 2. Measurement results using only diesel fuel.

No.	Engine speed [rpm]	CO [%vol]	CO_2 [%vol]	HC [ppm]	O_2 [%vol]	Temperature [°C]
1	820	0.061	3.91	45	14.3	85
2	2000	0.084	4.22	37	14.1	85

Table 3. Measurement results using diesel and HHC as fuel.

No.	Engine speed [rpm]	CO [% vol]	CO_2 [%vol]	HC [ppm]	O_2 [%vol]	Temperature [°C]
1	820	0.020	2.56	14	17.02	85
2	2000	0.026	2.58	15	16.89	85

Comparing the two series of measurements, the experimental results demonstrate that the HC si CO concentrations drop in value by 30%, the CO_2concentration drops in value by over 50%, and the concentration of O_2 raises by 19%. Further research, especially concerning the particles' emission must follow.

3. Conclusion

This system presented in the article (ICE) and mounted auxiliary on a vehicle proofs that it is capable of reducing emissions of an engine. Initially, the vehicle was categorized in Euro 3 pollution class; finally, it could fit into the Euro 6 pollution class.

Due to the fact, that hydrogen is reachable anywhere in the world and quite easy to be obtained; this could be a retrofit solution for all ICE, especially diesel engines. One presented that the electrolytic convertor is capable of producing 1800 litres of HHO gas from 1 litre of water. By upgrading this system, we could produce more HHO gas and replace the amount of conventional fuel. These results conclude about more effective power of the engine, and less consumption and emissions' concentrations. Furthermore, the retrofit of the ICE for using only HHO gas as fuel might be fully possible.

The results proof around 3 times more effective power for theengine, cheaper fuel and really low emissions, even toward 0. We consider this is a good alternative for electrical power units for the automotive industry regarding that the production process of Li-ion batteries is extremely pollutant and harmful for the environment.

If this system would be better perfected, one could higher the HHO produces dosage, totally replacing the conventional fuel.

Acknowledgement
The research is part of the PhD program of the first author at the Politehnica University of Timisoara. The authors also acknowledge the Romanian Automotive Register for the substantial support.

References

[1] HS Das and MM Rahman and S Li and CW Tan 2020*Electric vehicles standards charging infrastructure and impact on grid integration: A technological review Renewable and Sustainable Energy Reviews*Volume 120
[2] Jacob W. Ward and Jeremy J. Michalek and Constantine Samaras 2021 *Environmental Science & Technology* p55
[3] A Mayer and H Burtscher and S Loretzand M Kasper and J Czerwinski 2018 *High air pollution in vehicle cabins due to traffic nanoparticle emission exposure and a solution for in-use vehicles* Vol. 421 issue 3
[4] Peters, R. and Decker M. and Eggemann L. et al. 2020 *Thermodynamic and ecological preselection of synthetic fuel intermediates from biogas at farm sites* (Energy, Sustainability and Society volume 10) article 4
[5] Jerzy Szczygieł and Marek Kułażyński2020 *Thermodynamic limitations of synthetic fuel production using carbon dioxide: A cleaner methanol-to-gasoline process* (Journal of Cleaner Production Volume 276)
[6] Y Haseli 2018 *Maximum conversion efficiency of hydrogen fuel cells* (International Journal of Hydrogen Energy Volume 43 Issue 18)pp 9015-9021
[7] Jungwoo Shin and Won-Sik Hwang and Hyundo Choi 2019 *Can hydrogen fuel vehicles be a sustainable alternative on vehicle market?: Comparison of electric and hydrogen fuel cell vehicles* (Technological Forecasting and Social Change Volume 143) pp 239-248
[8] Canan Acar and Ibrahim Dincer 2020 *The potential role of hydrogen as a sustainable transportation fuel to combat global warming* (International Journal of Hydrogen Energy Volume 45 Issue 5) pp 3396-3406
[9] Balaji Subramanian and Saleel Ismail 2018 *Production and use of HHO gas in IC engines* (International Journal of Hydrogen Energy Volume 43 Issue 14) pp 7140-7154
[10] Balaji Subramanian, Venugopal Thangavel 2020 *Analysis of onsite HHO gas generation system* (International Journal of Hydrogen Energy Volume 45 Issue 28) pp 14218-14231
[11] Christos M. Kalamaras and Angelos M. Efstathiou 2013 *Hydrogen Production Technologies. Current State and Future Developments*, Conference paper, special issue Power Options for the Eastern Mediterranean Region, https://doi.org/10.1155/2013/690627 pp1-9
[12] *** https://images.app.goo.gl/2QdheyWmAwZeJXpC8

Theoretical estimation of the drawbar pull for two-axle automotive vehicles

I Preda

Transilvania University of Brasov, Brasov, Romania

pion@unitbv.ro

Abstract. The drawbar pull, defined as the amount of force available to an automotive vehicle at the drawbar for pulling or accelerating a load, represents an important characteristic used to define its traction performances. Because the experiments necessary to determine that force are relatively difficult and expensive to realize, a theoretical assessment of that characteristic may be extremely useful to identify the traction potential of the studied vehicle. This paper presents an algorithm able to approximate the realizable drawbar force of an automotive vehicle when it moves on different types of ground. For the computation, the ground characteristics may be adopted from the literature, while the necessary input data of the vehicle is normally given by the manufacturers or can be relatively easily obtained by measurements. The algorithm can be very useful in the primary stages of vehicle design. The paper also presents some influences of the constructive data and driving conditions on the drawbar force magnitude.

Terminology

At the moment, in the vehicle dynamics specialty literature there isn't a consensus in calculus relations or even in many very important definitions. A good step in this respect was realized by SAE which standardized more definitions in the field of vehicle dynamics. But in many other countries, like the European ones, generally these standards are not adopted and respected. Also, many differences appear between the terms used by different specialists, which shows a lack of harmony in the literature.

Further in this article, many remarks must be done for a correct understanding of the notions and equations which will be presented in this article. Most of the next definitions and notations were adopted accordingly with [1] and [2].

- *velocity* indicates a vector (with magnitude and orientation); *speed* is the magnitude of velocity.
- *towing mode* – the condition of a vehicle when moving, pulling or pushing a trailer, or exerting the necessary working force over an attached implement (agricultural or industrial equipment).
- *sinkage* – the deformation of the supporting surface from its original, undisturbed surface (before the traffic), measured normal to the direction of travel [2];
- *multi-pass effect* – the supplementary compression of soil and wheel sinkage increase caused by the multiple pass of wheels on the same patch of terrain [2].
- *free rolling wheel*, corresponds to the case when no torque is applied to the wheel; the wheel is pushed by a force generated at the hub so that to sustain a constant speed movement [1], [3], [4]; sometime called *driven wheel* or *pulled wheel* [5].
- *braking wheel*, a wheel at which braking torque is applied (trying to stop its rotation); the wheel is able to generate braking force at the hub.

- *drive wheel* (*driving wheel*) represents a wheel at which propelling torque (trying to increase its rotation speed) is applied; the wheel is able to generate pulling force (tractive net force) at the hub.
- *pure rolling wheel*, a wheel at which a (small) torque is applied to sustain constant speed movement (to overcome the rolling resistance) [4]; the net tractive force, generated at the hub, is zero, i.e. no traction or braking force appears; a particular case is represented by a *straight pure rolling wheel*, which is a pure-rolling wheel moving in a straight line at zero camber angle and zero side-slip angle.
- *neutral rolling wheel*, a wheel at which both a small torque and a small pushing force at the hub are applied in order to sustain a constant speed movement [6], [5].
- R_r – *total (overall) rolling resistance*, accordingly to the test standard ISO 28580, the "loss of energy (or energy consumed) per unit of distance travelled" [7]; the sum of internal and external rolling resistances.
- R_{ri} – *internal rolling resistance*; for a wheel, the resistance to movement provided by the energy dissipation in the hysteresis of tire material; for a vehicle, the resistance to movement provided by the internal friction and the energy losses in all the traction elements (wheels or tracks) [2].
- R_{re} – *external rolling resistance*, the resistance to movement of a vehicle or a traction element provided by the surface on- and through-which it moves, due to the bulldozing effect, soil compacting, lateral friction between tire and soil; the component of shear stress on the contact area opposite to the direction of travel [2].
- $f = R_r / Z$ – *rolling resistance coefficient.*
- *Rez* – *motion resistance*, any force imposing resistance to the motion of a vehicle or assembly (rolling, grade and aerodynamic resistances); at the element level, it is composed of rolling resistance forces only [8].
- *R/W* – *motion resistance coefficient*, motion resistance divided by the weight of the vehicle (or assembly or element) [8].
- *DP* – *drawbar pull*, the amount of sustained towing force a self-propelled vehicle (or element or assembly) can produce on a given ground [8]; it is a function of the ground type and speed.
- *D/W* – *drawbar pull coefficient* (*normalized drawbar pull*); for a vehicle, is the drawbar pull divided by vehicle weight *W*; for a wheel or axle, is the drawbar pull divided by the normal ground reaction (tire load) *Z* [8].
- *NRMM II* – NATO Reference Mobility Model.
- α – *slope angle* of the ground in the direction of travel.
- θ – (*side*) *slip angle* of the wheel.
- γ – *camber angle* (*inclination angle*) of the wheel.
- ω – (*current*) *angular-speed* of the wheel.
- ω_{ref} – *reference angular-speed* of the wheel.
- ω_0 – *reference angular-speed of the wheel in the case of pure-rolling* (in SAE and European conventions).
- $s = \omega - \omega_0$ – *longitudinal slip speed*, the difference between the angular speed of the driven or braked tire and the angular speed of the straight free-rolling tire [1]; both angular speeds are measured at the same linear speed at the wheel center in the longitudinal direction of the wheel; a positive value results from driving torque.
- $SR = (\omega - \omega_0)/\omega_0$ – *longitudinal slip ratio*, the ratio of the longitudinal slip speed to the angular speed of the straight free-rolling tire; it is used to observe changes in wheel forward progress as slip occurs; this definition is adopted by SAE [1], [9] and generates positive *SR* values in the case of traction (total spinning means $SR=+\infty$) and negative SR values for braking (wheel locking means $SR=-1$).
- $\kappa = (\omega - \omega_0)/\omega$ – *longitudinal slip coefficient*; also called *slip number* [10]; it is used to observe changes in wheel forward progress as slip occurs; it is used mostly in Europe to indicate positive slip (during traction).
- σ – unspecified *longitudinal slip* (*SR* or κ).

- $TR = (v_{ref}-v)/v_{ref}$ – *travel reduction* (*speed reduction* [11]), is used to observe changes in vehicle forward progress as slip occurs [12].
- V_w – *travel speed* of the wheel center.
- V – *forward speed of travel* of the wheel center (actual tire linear speed) on the wheel longitudinal direction ($V_x = V_w \cdot \cos\alpha$).
- r_u – *unloaded radius* [2] (*free radius* [3], [1], [10], *design radius* [5]), the radius of the unloaded tire inflated to normal recommended pressure.
- r_l – *loaded radius*, the distance from the wheel rotation axis to the lowest point of the contact area (the point with the largest soil deformation, perpendicular to the direction of travel, with respect to the undisturbed surface) [2].
- r_s – *static loaded radius*, the loaded radius of a stationary tire inflated to normal recommended pressure [1], [10].
- $r_d = M_w/F_w$ – *torque radius* (*dynamic radius*), the ratio of the input torque over the gross tractive effort [2]; on hard grounds, is equal with the loaded radius r_l; is used to link forces and moments applied to a wheel [3], [9], [13], [14];
- $r_e = V/\omega$ – *effective* (*current*) *rolling radius*, the ratio of the linear speed of the wheel center in the wheel longitudinal direction to the current angular speed [1], [2].
- $r_k = V/\omega_{ref}$ – *rolling radius* [2], named also *kinematic radius* [15], is *the reference rolling radius* and is defined as the ratio of the linear speed V of the wheel center in the wheel longitudinal direction to the reference angular speed ω_{ref}; this radius (which can't be measured directly) is calculated from the kinematic parameters (V and ω_{ref}) of a wheel at zero slip angle and zero camber angle (i.e. the wheel is not skidding laterally and its plane is perpendicular to the ground) [4], [16]; for a pure rolling wheel, as is considered in this work, $\omega_{ref} = \omega_0$; rolling radius is a special case of effective rolling radius under a zero slip condition and is used to link the net tractive force (longitudinal grip force) X with the kinematics of the wheel rim.
- Z – *normal force* (*tire load* [10], [5]), the component acting perpendicular to the ground plane of the tire force (total wheel-ground interaction-force).
- F_μ – *grip force* of a wheel, the component of the tire force (wheel-ground total interaction-force) that acts on the ground plane [17], [18].
- X – *longitudinal force*; component of the grip force that acts on the longitudinal direction of a wheel.
- $F_t = X \cdot \cos\theta + Y \cdot \sin\theta$ – *tractive force*, the component of the grip force vector in the direction of travel of the center of tire contact [1]; tractive force is equal to longitudinal force times cosine of slip angle plus lateral force times sine of slip angle; in the case of null slip angle ($\theta = 0$), $F_t = X$; negative tractive force is named braking force.
- ξ – *normalized traction or braking force* (the force divided by the peak force; $-1 \le \xi \le 1$).
- *rolling characteristic of the wheel*, the graphical representation of the functions $X = X(\sigma)$ or $\xi = \xi(\sigma)$ vs. slip σ.
- M_w – *wheel moment* (*wheel torque*), the external torsional moment (torque) applied to the wheel from the vehicle about the spin axis [2]; positive values correspond with traction and negative with braking.
- F_w – *gross tractive force* of a wheel [2] (also referenced as *circumferential force* [13] or *force at the wheel* [3]; it is a theoretical force that can't be measured directly.
- $V_t = \omega \cdot r_k$ – tire *rolling speed* (or *peripheral speed* or *theoretical linear speed*).

1. Introduction

Drawbar pull (*DP*) represents the amount of sustained towing force a self-propelled vehicle (or element or assembly) can produce on a given ground [8]. The drawbar pull tests or simulations may be performed for a full automotive vehicle, an "element" (which means single wheel or axle) or even a combination of more automotive vehicles (road train).

DP tests or simulations can be extremely useful for different categories of ground vehicles, like agricultural or industrial (construction) tractors [6, 11, 19]; military vehicles[8, 20–22]; road tractor-trailer combinations (road trains) [23], or even exploratory vehicles for the research of Moon and Mars [12].

The primary objective when conducting drawbar pull tests on different vehicles is to determine the maximum allowable towing force at different speeds for different drive train configurations and vehicle layouts and loads.

Specific standards exist for *DP* test procedures in the case of earthmoving machinery (*ISO* – International Organization for Standardization, 1983 [6]) or military vehicles (*NATO AVTP 03-60* [24] and *US Army TOP 2-2-604* [25]).

Most *DP* procedures consider the vehicle traveling in a straight line with constant speed and all the wheels working at zero camber and slip angles; also the wheels of the same axle will encounter the same terrain conditions and behave identically on the right and left sides. But in practice, these hypotheses are almost impossible to be fulfilled, as presented in [18].

Vehicle *DP* depends mainly on the quality of interaction between the ground tires and surface, speed, vehicle load, layout and drive train configuration.

Many parameters associated with the *DP* may present interest for the use of a vehicle: maximum drawbar pull achievable on certain terrain, maximum drawbar pull achievable at an allowable slip (so as not to break the roots of the plants), maximum driving speed, maximum traction efficiency, the fuel quantity to travel a certain distance in imposed conditions.

2. Drawbar pull of a wheel

Drawbar pull evaluations for a full vehicle may be or not preceded by *DP* evaluations of a single wheel. Even not mandatory, the test or simulations conducted for a single wheel will help to a better understanding of the more complicated generation of drawbar pull of a complete vehicle, because the tires are the only links of the wheeled vehicle with the road; thus, theirs dynamic (kinetic) properties, i.e. the forces and moments acting on the deformed pneumatic tire, produce extremely important effects on the vehicle behaviours under all driving situations (traction, braking and steering).

The drawbar pull of a wheel or axle quantifies its ability to sustain the movement of a terrestrial vehicle. Due to the importance of these elementary components in the overall performance of the full vehicle, detailed knowledge of the influencing parameters is necessary. Such analysis will be presented further.

2.1. Wheel model and motion equations in steady-state conditions

Due to the hypothesis of a straight line movement, a planar model can be adopted for the wheel, as shown in figure 1. At this moment, it is not necessary to consider the wheel moving on the longitudinally inclined ground, because the influence of slope angle α over the *DP* will be considered in the next section of the article.

Both cases of the wheel interaction with hard (undeformable) and soft (deformable) grounds are presented. If on hard ground it is considered that only the tire is deformed, figure 1.*b*, on soft soils deformations appear on both tire and ground, figure 1.*a*. In the front region of the contact it is possible to appear the height increase of the soil surface (the tire pushes the soil forward, the so-called *bulldozing effect*), while, on the rear, the terrain presents normally an elastic rebound after the passage of the wheel.

The ground properties (road or soil) can mainly vary in two ways: rigidness and slipperiness (compression and shear). On hard grounds, these two properties aren't solely depending on the road, because rigidness should be judged also relative to tire inflation pressure and slipperiness with the tire tread pattern, including whether the tire has cleats or not. On soft ground, the ground is deformed so that the wheel continuously rolls in a *local uphill slope* [26], which is a simple explanation of higher rolling resistance on soft soil vs. on road. Higher inflation pressure will lead to more deformation of the ground, which means a steeper angle of that apparent slope. Also, the soil response changes to the tire repetitive loading. This soil behaviour is known as the *multi-pass effect* [2, 27].

To apply Newton's second law, all exterior forces and moments must be considered at the wheel interaction with the environment (gravity, air, ground and body of the vehicle).

The wheel aerodynamic interactions will be neglected because, excepting some special cases, these forces are small if compared with the other forces. Also, because stationary movement is considered (no wheel bounce), the wheel weight will be included in the force Q_w, figure 1 (the movement equations of the wheel for the case of transitory movement may be found in [28] or [18], where the inertial properties of the wheel are considered).

The interaction with the ground contains two force components: the normal force Z and the longitudinal force X (which is the longitudinal component of the grip force F_μ). In the considered hypotheses of null slip angle and camber angle, it results $X = F_\mu$ (all the grip force is used for traction).

The vehicle acts against the wheel with the load from axle Q_w, the reaction force parallel to the ground plane DP (the drawbar pull of the wheel) and the torsion moment M_w. The distances from the wheel center to the forces X and Z are respectively the torque radius (dynamic radius) r_d and the longitudinal offset e of the normal force to the wheel center, named *moment arm of the tire load* [5].

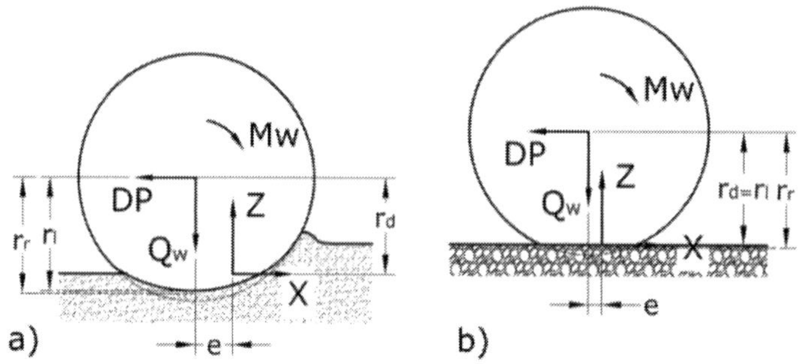

Figure 1. Planar model of the wheel: a) – on deformable ground; b) – on undeformable ground.

In the case of constant speed and constant forces Q_w and DP (but only in this situation [18]), all the accelerations of the wheel are zero, and so the movement equations for the three degrees of freedom of the wheel are:

$$\begin{gathered} DP = X \\ Z = Q_w \\ X \cdot r_d = M_w - e \cdot Z \end{gathered} \qquad (1)$$

Here, positive values of the forces correspond with their orientations existing in the figure.

As can be seen, the magnitudes of the drawbar pull DP of the wheel and net tractive force X (a longitudinal component of the grip force) are equal.

2.2. Net tractive force (longitudinal grip force) of a wheel

The numerous experimental tests realized during years indicated that the tractive force X produced by a tire is mainly a function of six parameters [4]:

$$X = X(Z, \mu, \sigma, \sigma_y, \gamma, \varphi) \qquad (2)$$

where: Z is the normal force to the ground acting on the tire (the tire load), μ – grip coefficient, σ – longitudinal slip, σ_y – side (lateral) slip, γ – camber angle and φ – spin slip (due to steering).

For the drawbar pull tests or simulations, normally null values are considered for lateral slip, camber angle and spin slip ($\sigma_y \approx 0$, $\gamma \approx 0$, $\varphi \approx 0$) and, in these conditions, the tractive force may be considered as depending on only three parameters [18]:

$$X = X(Z, \mu, \sigma) = \mu \cdot \xi(\sigma) \cdot Z \qquad (3)$$

Even relative motions exist between rim and tire belt in contact patch [5], because it is fairly inexpensive to accurately measure the rim rotational speed, it remains reasonable to compute the slip and traction force considering the rotation of the rim as the rotation of the full wheel [16]. That means, if the correctness of the previous two equations is accepted, the kinematics of the wheel rim defines completely the forces in the contact patch [4].

These last two equations show that the tractive force X isn't influenced in a direct way by the angular and linear speeds of the wheel, but suffer a major influence on their ratio ω/V. Also, X is not depending explicitly on the engine (motor), transmission, driveline and brakes, but only on the „history" of vehicle movement [18] (the vehicle instantaneous spatial position, through Z, μ and shape of the function ξ, respectively through the instantaneous values of angular and linear speeds ω and V, appearing in the definitions of the slip ratio or slip coefficient, detailed further).

Tractive force is also named *thrust* or *tractive effort*.

2.3. Wheel slip

Measuring or theoretically estimating the parameters which define the tractive or braking force X (grip coefficient μ, longitudinal slip σ and shape of the function ξ) required a large scientific effort through the years. The structure design of the tire and elasto-plastic properties of the material (rubber) are of key influence on the slip of the tires [16].

Today, different slip definitions are used and sustained (defended [9]) in the entire world.

Generally adopted (inclusively by SAE J670 standard [1]), the *longitudinal slip s* is defined as:

$$s = \omega - \omega_{ref} \tag{4}$$

where: ω is a certain angular speed of the wheel, and ω_{ref} is the reference angular speed. That means the slip s is positive for a drive wheel and negative for a braking wheel. But, there are more ways to define the reference angular speed ω_{ref} of a wheel, from which the most used are: *pure-rolling* (torque applied to sustain a constant speed movement, i.e. to overcome the rolling resistance; the net tractive force, generated at the hub, is zero, $DP=X=0$, i.e. no traction or braking force appears), *free-rolling* (no moment applied) or *neutral-rolling* (torque applied to overcome the internal rolling resistance and pushing force at the hub to overcome external rolling resistance).

Starting from that definition of the slip, SAE J670 further defines the wheel *slip ratio* as:

$$SR = \frac{s}{\omega_0} = \frac{\omega - \omega_0}{\omega_0} = \frac{\omega}{\omega_0} - 1 = \frac{\omega \cdot r_k}{V} - 1 \tag{5}$$

where: ω_0 is the reference angular speed considered for pure-rolling (a case which corresponds, as largely accepted, with minimal shear displacement in the contact patch).

The values obtained with that SAE definition of the slip ratio are positive for traction and negative for braking. The extreme cases are for *total wheel spinning* (*spinning on the spot*) during traction ($V_x=0$ and SR=$+\infty$) and *wheel locking* during braking ($\omega=0$ and $SR=-1$).

The negative value of the slip ratio is also known as *practical longitudinal slip*.

But dealing with infinite values is not so comfortable (mainly in the case of computer programs). Also, if the longitudinal grip force is represented versus SAE definition of slip, the plots present different shapes for traction and braking (the plots are non-symmetric against the origin of axes), figure 2, left side. Due to that, in Europe it is used generally the SAE definition of slip only for the case of braking, while another definition is adopted for the case of drive wheel, referred sometimes as *slip coefficient* [10]:

$$\kappa = 1 - \frac{V}{\omega \cdot r_k} \qquad \text{if} \qquad V < \omega \cdot r_k \tag{6}$$

That means $\kappa=1$ for total wheel spinning (when $V=0$).

The negative value of the slip coefficient is also named *theoretical longitudinal slip*.

Looking at the definitions from equations 5 and 6, changing from slip ratio to slip coefficient (and vice-versa) is easy: $SR=\kappa/(1-\kappa)$.

The combined definition generally adopted in Europe, similar with SAE definition for the case of braking, equation (5), and having the expression from equation (6) for the case of driving, can be written shortly

$$\sigma = \frac{\omega \cdot r_k - V}{max(\omega \cdot r_k, V)} \tag{7}$$

and has the advantage of anti-symmetry for the shapes in quadrant I and III, figure 2, right side.

The last definition, σ, which appeared already in equations (2) and (3), will be used further to indicate the relative slip of the tire.

2.4. *Grip coefficient*

The grip coefficient μ is defined as the ratio between the peak value of the tractive force and the corresponding vertical load

$$\mu = max(|X|)/Z \tag{8}$$

That maximal value is obtained at slip coefficients σ_μ which present a very large spread.

The grip coefficient μ (and its corresponding slip σ_μ, as well) depends very much on the types of ground and tire being in interaction. For example, a value μ=1 is typical for car tires on asphalt roads, but values up to 10 times smaller may be registered on ice or wet snow.

Some typical values are: snow and ice (μ=0.1-0.25 at σ_μ=0.05) [28]; sand, at a pressure of 140…200 kPa (μ=0.29 at σ_μ=0.1-0.15) [29]; wet loam (μ=0.3-0.4 at σ_μ=0.65); stubble (μ=0.7 at σ_μ=0.65); dirt or soil (dry hard terrain) (μ=0.6-0.9 at σ_μ=0.5-0.6); wet asphalt road (μ=0.6-0.8 at σ_μ=0.1-0.2); dry asphalt road (μ=0.9-1.2 at σ_μ=0.1-0.2) [28].

Experiments show that X grows less than proportionally with respect to the vertical load Z, which leads to the observation that the grip coefficient μ decreases almost linearly with the increase of load (for example, normal car tires present approximately 15% reduction of μ when the load is double as the nominal one) [4].

Also, the grip coefficient generally decreases if the vehicle speed increases [10].

2.5. *Rolling radius of the wheel*

Now, a new useful magnitude r_k, appearing in the equation (5), can be introduced: the *rolling radius* (*kinematic radius*) of the tire. This length is fully defined by the equation:

$$r_k = V_0 / \omega_0, \tag{9}$$

where: ω_0 and V_0 are respectively the *reference angular speed* and *reference linear speed* of the tire in its longitudinal direction, speeds considered to obtain pure rolling (zero slip condition). Experiments made in field or on drum or flat-track test machines may be used to obtain its value with this equation.

Representing the *effective rolling radius of the straight pure rolling wheel* (a fictitious rigid substitute wheel), r_k cannot be directly measured and is in fact the distance from the wheel axis to the instantaneous rotation center of the wheel.

On hard grounds, the instantaneous rotation center of the wheel is positioned under the ground plane, which means that r_k is only a little bigger than the loaded radius r_l. For that reason, on hard grounds, the rolling radius r_k is often replaced by the loaded radius r_l, but on soft soils that approximation isn't effective anymore due to the unknown wheel sinkage and tire deformation.

The magnitude of the rolling radius is affected by different parameters, for example, increases with the inflation pressure and decreases with the tire load; however, the sensitivity of r_k to the inflation is larger than its sensitivity to the load [16]. Also, the ground condition (hard or soft) and vehicle speed modify the rolling radius. Similar influences also suffer the loaded radius r_l. Even so, in the literature, the tire rolling radius is most often assumed constant for general computations (the use of the value provided by the tire manufacturer is generally valid for a large number of applications).

The rolling radius is used to link the drive or braking force X to the kinematics of a wheel or vehicle (respectively by slip σ and travel reduction TR).

2.6. Shape of wheel rolling characteristic

Traction slip ratio curves (graphical representations of the function $\xi(\sigma)$) can be obtained on road or on terrain by recording simultaneously the angular speed of the powered test wheel and its drawbar pull (against a dynamometer truck). Braking slip ratio curves can be obtained with a trailer where the wheel is progressively braked. Both characteristics can also be obtained based on test bench measurements. This kind of grip force plots vs. slip are named *rolling characteristic of the wheel.*

Up to the peak of the plot, figure 2, the forces depend heavily on the elastic properties of the tire tread and carcass [9]. After the peak, the forces depend on a variety of factors such as tread composition, road texture, surface moisture, speed, tire temperature, etc.

Associating a null longitudinal grip force with a null slip ratio (the so-called *zero condition*), in the SAE and European definitions of slip it is considered that the rolling radius r_k corresponds to a pure rolling wheel. But, because this condition is difficult to obtain during tests, in the experimental attainment of the wheel rolling characteristic ($X=X(\sigma)$ or $\xi=\xi(\sigma)$), generally only conditions for a free wheel are fulfilled [4]. To obtain the true characteristic, subsequent corrections are made.

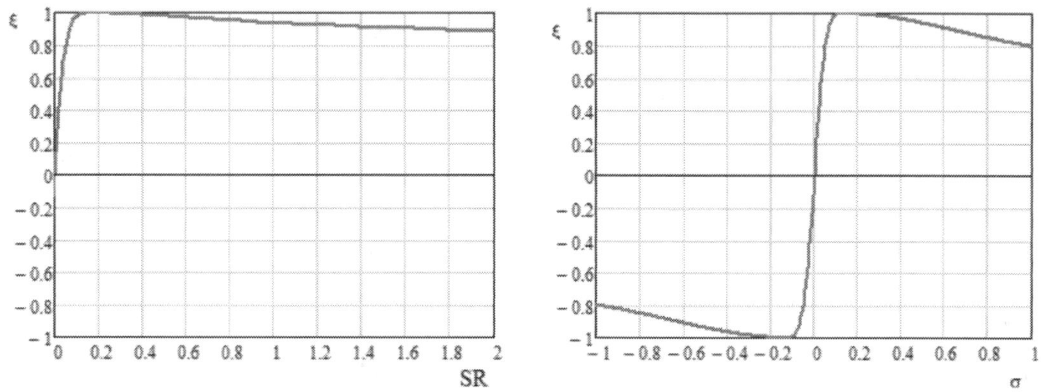

Figure 2. The rolling characteristic of the tire (normalized traction force ξ vs. relative slip): left – vs. slip ratio *SR* (SAE definition); right – vs. slip coefficient σ (European definition)).

Both SAE and European slip definitions give zero longitudinal grip force when the wheel operates at the zero slip condition [12], figure 2. This is the case of pure rolling. In the case of free rolling, the function ξ will return a negative value equal with $-f/\mu$ (as will be shown later, the negative traction force corresponding to that situation is named rolling resistance force, $R_r=f\cdot Z$).

There are numerous mathematical expressions used to describe the functions $\xi(\sigma)$ of the normalized traction or braking force [11, 30], from which maybe the best known is the Magic Formula [31].

In this work, simple polynomials of third degree were used to describe the shape presented in figure 2. To obtain the coefficients of the polynomials, only three values must be specified: the slope of the plot in origin, the slip value σ_μ corresponding to the maximal grip and the ratio of grip for extreme slip (at $\sigma=1$) and maximal grip (at $\sigma=\sigma_\mu$). One polynomial function is used for small slip σ (between 0 and σ_μ) and the other for large values.

In all these formulae, the longitudinal slip stiffness (which is proportional with the slope of each curve at zero slip) and the slip value σ_μ are quite relevant for the shapes of these $\xi(\sigma)$ functions. Normally, the maximum traction forces are obtained for hard grounds and normal car tires at $\sigma_\mu\approx0.1\ldots0.2$ [4], while on soft soils and M&S tires at $\sigma_\mu\approx0.1\ldots0.65$ [28].

When intended to obtain the shape of the normalized traction or braking force, some problems appear in the case of tires with widely spaced lugs, because the assessment of the rolling radius becomes difficult, in both experimental [6] or theoretical [32] approaches.

An aspect with high practical importance is that, on soft soils, the increase of the slip leads to the increase of the wheel *sinkage* [29]. Due to that phenomenon, the rolling resistance will increase. If on wet terrains or river fords the moderate sinkage normally produces an increase of the net traction force (the tire digs in the soil up to a harder surface, with better grip), permitting the vehicle to move further, on some soft soils, such as sand or mud, the amplification of sinkage can put the vehicle in the impossibility to move further. In such situations, it is recommended [21] that the driver to stop maneuvering (not getting worse the situation and avoiding the vehicle to remain suspended), but instead to trail its vehicle (helped by another vehicle or using its own capstan, if it exists).

2.7. Rolling resistance force

Dividing equation (1.*c*) by r_d it results:

$$X = M_w/r_d - (e/r_d)\cdot Z = F_w - f\cdot Z = F_w - R_r = DP \qquad (10)$$

where: $f=e/r_d$ is named *rolling resistance coefficient*, $R_r=f\cdot Z$ is the rolling resistance and $F_w=M_w/r_d$ is the *gross tractive force*.

This equation gives an easy way to experimentally obtain the magnitude of the rolling resistance (at least for the situation of the drawbar pull procedure). Indeed, making $M_w=0$ (and $F_w=0$), i.e. the condition of free rolling, it results $R_r=f\cdot Z=-DP$, that means the rolling resistance is the force parallel with the initial ground surface which, applied to the wheel, maintains constant rolling speed (remember the hypothesis of stationary movement). The sign minus (–) indicates that in this case the *DP* force has changed its normal orientation.

Of course, the rolling resistance exists also when the driving speed is not constant and, moreover, it is influenced by the driving conditions. At the moment, a definition for the *rolling resistance* isn't generally accepted and is balancing between the concepts of "force" and "energy loss". Although the term "rolling resistance" (i.e. force) has prevailed, the *loss* concept has been adopted into the test standard ISO 28580, which defines the rolling resistance of a tire as the "loss of energy (or energy consumed) per unit of distance travelled" [7] (in fact, energy per distance is also force). That definition may suggest at least two aspects: the rolling resistance depends on the wheel slip (because the travelled distance changes with slip); the energy consumed can appear not only on the tire, but also on the terrain and in other interactions. This third component, named also *parasitic losses*, represents the losses (aerodynamic or appearing in wheel bearings or suspension shock absorbers or bushings) which are intertwined with the rolling resistance and is unclear if must be considered or not as being a part of the rolling resistance.

It is usually considered that the total rolling resistance has two parts: the *internal rolling resistance* (or *tire rolling resistance*) $R_{ri}=f_i\cdot Z$, caused mainly by the hysteresis associated with the deformation of the tire, and *external rolling resistance* (or *soil rolling resistance*) $R_{re}=f_e\cdot Z$, caused by the soil deformation (bulldozing, compaction, tire-soil friction on the tire sides) [2, 28]. Of course, on hard grounds R_{re} may be considered zero, because the road deflections are negligible if compared with the one of the tire.

The hysteresis of the tire material causes non-symmetrical normal stress distribution along the contact patch and the tire load Z is shifted forward with the moment arm of the tire load (the distance e in figure 1).

There is often the hypothesis (not necessarily very intuitive) that, if the same tire inflation pressure is kept, the internal rolling resistance has the same magnitude on hard and soft grounds (f_i remains the same). Also, it is experimentally observed and theoretically proved that powered wheels present slightly increased rolling resistance if compared with unpowered (free-rolling) wheels.

The net traction force and the rolling resistance force can be estimated theoretically, using appropriate models and assumptions for tire and soil properties, integrating the elementary compression and shear forces [27, 32, 33]. Many scientific works present experimental values [28] or different theoretical or empirical formulae [7, 20, 21] to estimate the internal and external rolling resistance coefficients f_i and f_e. Concluding, the rolling resistance can be measured in experiments, but also can be estimated starting from values indicated in the literature from different ground and tire types.

2.8. Gross tractive force

Rewriting the equation (10) as:

$$F_w = M_w/r_d = X + R_r = DP + R_r \tag{11}$$

it gives the possibility to compute the gross tractive force F_w.

Largely adopted [2, 3, 9, 10] and basically correct, the definition of the *gross tractive force*, "the wheel input torque divided by the torque radius", isn't very useful when it tries to compute its magnitude for deformable soil condition, because it is very difficult to estimate the torque radius r_d, mainly in experiments.

Another and more useful definition (especially for driving on soft grounds) can be obtained indirectly from the definition of net tractive force existing in [2]: "the net tractive force plus the total motion resistance", which also appears clearly from the equation (11).

Even in the common perception, the torque radius r_d is necessary to be known in order to compute the gross tractive force, in reality, things can be exactly the opposite (mainly in field tests).

It also must be noted that, accordingly to the laws of dynamics stated by Newton, the gross tractive force IS NOT an exterior force for the wheel (or for the full vehicle), but a conventional quantity, which can't be measured directly.

For many years, in the specialty literature, the equations (2) or (3) are generally replaced with the equation (10) in almost any driving situations, stationary and transitory. That substitution ($X=M_w/r_d-R_r$) probably happens because the wheel traction force is so APPARENTLY instantaneously connected with the drive train performances through the torque M_w. Or, in other words, it tries to link the drawbar pull magnitude with the wheel moment and so with the driver's actions. But, in fact, there is no direct connection between the gross tractive force $F_w=M_w/r_d$ and the grip force component X (which, in fact, depends on the "history" of wheel movement), as there is no such link between the driver's commands and the vehicle response!

That means the equation (10) contains a trap for its users: it is true ONLY for stationary movement or, vice-versa, it is incorrect for transitory regime [18]. Indeed, from equation (10) it results the apparent possibility to obtain an extremely large component X of the grip force (respectively a drawbar pull DP) if it applies sufficient torque M_w at the wheel. Evidently, this affirmation is false, because the magnitude of force X is limited by the characteristics of the wheel-ground interaction, as presented earlier. Unfortunately, this mistake frequently appears in the vehicle dynamics literature. Omitting to specify the necessity of steady-state conditions (constant speeds ω and V), equations (10) or (11) are used to replace X with F_w-R_r in the wheel models, which can be used then incorrectly in the case of transitory movement conditions.

Even having many understanding problems, the gross tractive force F_w is adopted to be represented in the vehicle theoretical *traction characteristic*, largely used in the evaluation of vehicle dynamic performances. In that characteristic, the loss of speed due to the slip is not taken into account.

The gross tractive force F_w is also referred to as *rim pull*, *theoretical pull*, or *design drawbar pull* [2] or as *force at the wheel* [3] (word to word translation from Romanian). Some authors are using only the ratio M_w/r_d, as in [14], avoiding to name specifically that fictitious force.

Considering again only stationary conditions, it can be concluded that, on any type of ground, the gross tractive force F_w of a wheel can be computed with equation (11) as the sum $X+R_r$ of the longitudinal grip force and rolling resistance force. It must be noted that for a full vehicle, the grade resistance R_g (and possibly aerodynamic drag R_a) must be considered together with R_r.

2.9. Torque radius

The *torque radius* r_d is defined as the ratio of the wheel torque over the gross tractive effort [2]; another name frequently found in the literature is *dynamic radius* [3, 13, 28]; also referred to as the *kinetic rolling radius* [15]. All three words, torque, dynamic or kinetic, indicate that this radius link forces and moments applied to a wheel.

To have a visual representation of that radius, r_d is the distance from the wheel spin axis up to the action line of the ground reaction parallel with the unperturbed ground surface (the support line of the traction force), figure 1. On hard ground, the dynamic radius of a rolling wheel with zero camber angle is the shortest distance from the wheel spin axis up to the center of the contact patch.

On hard ground, the torque radius r_d is measurable, which means also that, in the same conditions, the gross tractive force F_w can be computed as M_w/r_d. But measuring r_d is not so easy, because during rolling it modifies due to many influences.

While the general perception induced by the literature is that knowing the torque radius r_d is necessary to compute the gross tractive force, in fact, things are exactly the opposite: r_d can be computed dividing the gross tractive force to the wheel torque.

There are no indications that, in the case of deformable soils, the torque radius can be directly measured, because it is hard to define or observe where the support of the tractive force is, figure 1. But, the torque radius r_d must be smaller than the loaded radius r_l, which, in turn, may be measured (at least approximately) by placing distance sensors inside the tire.

A simplifying hypothesis often used is that the torque radius r_d remains constant while the vehicle moves (for example, to compute the theoretical traction characteristic). But this assumption is far from correct [18]: the change of the tire load, the action of a variable (driving or braking) torque, the ground unevenness, and the modifications of rotation speeds lead all to the important changes of the torque radius. Also, r_d is affected by the inflation pressure (which depends in turn on the tire temperature).

Although the off-road vehicle tires are lugged, the influence of lugs (with periodic modifications of the tire radii and traction force) will not be considered explicitly here.

2.10. Drawbar pull as net force

In [27] and in [8], the next sentences can be found: "For a traction element, the net tractive effort is the same as the drawbar pull." and "Drawbar pull is a net force equal to the (gross) tractive force reduced by the motion resistance." Both statements are true, as results also from equations (1.*a*) and (10), but only for stationary conditions [18] and, for the first one, only if the ground is horizontal too. Also, the second affirmation tries to indicate that between the driver's actions and the drawbar pull a direct connection exists, which isn't always true, as already presented.

In this paper, as in [6], the term "axle" is used to describe a pair of connected wheels at the same location along the length of the vehicle and with identical driving conditions. In these hypotheses, it is obvious that the discussed forces and torques of the axle will have doubled magnitudes if compared with the ones of a single wheel. However, for some kinds of vehicles and work types, the hypothesis of "identical driving conditions" can't be accepted. For example, a tractor that plows will have different interactions with the ground in the left and right sides (at least grip coefficients and tire loads) or a gravel-leveling grader will experience a lateral push (leading at least to the diminish of the maximal drawbar pull).

Although such conclusions are generally accepted in the literature for a traction element (wheel or axle), an overview of the main quantities was necessary to be done. The cause for the previous presentation is that a lot of names and definitions (some unclear or contradictory) are used in the specialty literature (confusing the readers). Very necessary and representing excellent scientific contributions, the few existent standards ([1, 2]) may be further improved. Even more, problems appear when another language but English is used to present aspects of vehicle dynamics or different translations are necessary.

2.11. Order of assessments to obtain the experimental drawbar pull

Summarizing the information shown until now, the necessary steps to measure the drawbar pull of a wheel are:

- with no drawbar pull, it applies a torque at wheel in order to produce constant reference wheel speed V_0 (the wheel is in pure rolling situation) and it measures also the corresponding angular speed, which is the reference angular speed ω_0;

- in the same conditions, the rolling radius r_k can be computed with equation (9);
- modifying Z, ω and V and measuring the corresponding X forces, the slip values σ and the curves $X(\sigma)$ are obtained;
- from the curves $X(\sigma)$, obtained at certain Z values, one retain the maximal values X_{max}, with which will be computed the grip coefficient μ;
- using the curves $X(\sigma)$ and the value μ, the curves $\xi=\xi(\sigma)$ of the normalized traction force are obtained;
- on soft ground, pulling the wheel with constant speed and measuring the needed force, it obtains the total rolling resistance R_r; repeating this procedure on hard ground, it obtains the internal rolling resistance R_{ri}; then, it computes the external rolling resistance $R_{re}=R_r-R_{ri}$ (on hard ground $R_{re}=0$); measuring also the tire load Z, the rolling resistance coefficients will be computed;
- the gross tractive force is obtained by adding the net tractive force X and the total rolling resistance R_r;
- the torque radius r_d may be obtained by measuring the input torque at the wheel M_w and dividing it by the gross tractive force F_w; also, on hard grounds, r_d can be measured.

The values μ and the curves $\xi=\xi(\sigma)$ obtained for different ground types will be used to obtain by computing the predicted figures of the drawbar pull.

3. Drawbar pull of the full vehicle

The drawbar pull of a vehicle quantifies its ability to tow a trailer or an implement in well-defined conditions. The knowledge of the *DP* performance of any vehicle may present large practical interest, because usage optimizations or maximal performances can be achieved. For tractors, *DP* tests are typically done at low slip, in the field with unprepared, but hard terrain [6]. In the last period, electronic management systems were developed and implemented to optimize the tractor traction. They can inform the driver about optimal ballasting, driving speed, tire inflation pressures, etc. [5].

For any vehicle type, one important purpose of measuring together slip and drawbar pull on different grounds is to determine the slip at which maximum drawbar pull efficiency is obtained. The magnitude of *DP* at this optimum slip is generally the best indicator of vehicle economic performance. But, in some special cases, also knowing the maximum *DP* may be extremely appreciated.

For military applications, empirical slip relationships are now used to predict drawbar pull, slip and maximal speed of a vehicle in towing mode. The relationships are considered to be indicators of the slip and drawbar pull performances that would occur on representative soil strengths [8]. The predictions are based on motion-resistance coefficient relationships, also empirically derived from field tests with all-wheel drive vehicles or from single-wheel laboratory tests. The purpose is to predict the rolling resistance force opposing the motion of a vehicle during interaction with a given terrain strength. Normally, the aerodynamic interactions of the vehicle are not considered [8], which means that only the rolling and grade resistances are taken into account.

3.1. Travel reduction

The slip associated with a full vehicle is quantified by the so-called *travel reduction TR*, which is used to observe changes in vehicle forward progress as slip occurs. It is defined by the equation:

$$TR = \frac{V_{ref}-V}{V_{ref}} = 1 - \frac{V}{V_{ref}} = 1 - \frac{V}{\omega_{wref} \cdot r_k} \tag{12}$$

The *reference condition* (also named *zero-condition*) is not based on a specific physical phenomenon; rather it is selected to be a simple and repeatable experimental condition [12]: driving the vehicle in a straight line, with a certain constant speed and at zero *DP* load, either on hard ground or on soft terrain. To obtain the rolling radius r_k, the linear and angular speeds V_{ref} and ω_{wref} are measured simultaneously. Then, using equation 12 and making $TR=0$, r_k is computed ($r_k=V_{ref}/\omega_{wref}$).

3.2. Vehicle model and motion equations in steady-state conditions

Due to the hypotheses considered in this work (straight line motion; left and right sides of the vehicle encounter the same terrain conditions and perform the same), the model of the vehicle can be represented in two dimensions, figure 3.

The forces acting on one axle will be considered two-times bigger than the ones of a single wheel (it considers two identically behaving wheels on each side). As in the already presented case of a single wheel, to identify the exterior forces acting on the vehicle, all the interactions with the environment (gravity, air, ground and the towed trailer or implement) must be considered.

As has been already said, the wheel aerodynamic interactions will be neglected in this study (a reason is that in towing mode, the driving speed is normally small or medium).

Also, because stationary movement is considered, forward, bounce or pitch accelerations will not appear (but, the transitory movement, studied for example in [18], will lead to different results as the ones presented here).

In figure 3, the vehicle is presented moving on deformable soil, a situation more complex as on-road driving, but the next equations will be applicable in any kind of ground. The initial level of the ground (the unperturbed soil plane) is making the slope angle α with the horizontal.

The forces acting on the vehicle are: the weight W; the tire loads (normal to the *plane of unperturbed ground*) Z_1 and Z_2; the longitudinal front and rear grip forces (the net tractive forces) X_1 and X_2; the components of the force in the drawbar, DP and $DP\cdot\tan\gamma$ (the drawbar makes the angle γ with unperturbed ground plane; γ is considered positive if the force $DP\cdot\tan\gamma$ acts downwards).

The center of gravity is positioned at the distance a on the direction of travel with respect to the front axle and at the distance h_g to a *reference plane* passing through the application points of the forces X_1 and X_2 (on soft soils, due to the supplementary sinkage of the rear wheels, known as *multi-pass effect*, this plane will be a little inclined with respect to the ground plane; but here, that inclination will not be taken into account – the two planes are considered parallel).

The hook coupling (towing hitch) is positioned at the distance h_{DP} vs. the same reference plane and, in the longitudinal direction, at the distance d_{DP} to the rear of the vehicle with respect to the rear axle.

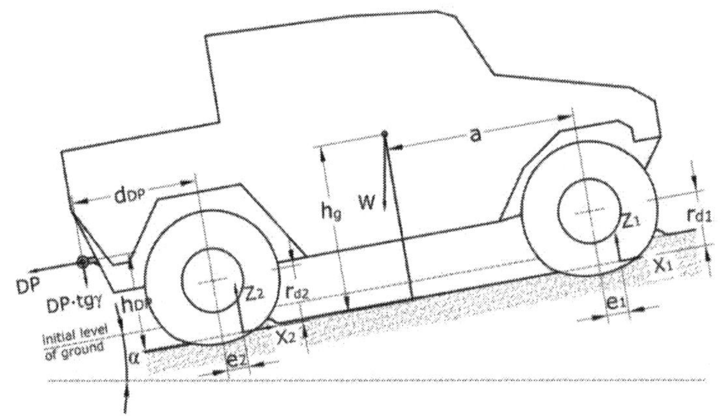

Figure 3. External forces acting on the self-propelled vehicle.

Now, using the model presented in figure 3 and applying Newton's second law (for the translation in travel direction, translation in a direction perpendicular to the unperturbed ground plane and rotation passing through the centers of the rear tire contact patches), the movement equations for the three degrees of freedom becomes:

$$\begin{gathered} X_1 + X_2 - W \sin\alpha - DP = 0 \\ Z_1 + Z_2 - W \cos\alpha - DP \tan\gamma = 0 \\ (L-a)W \cos\alpha - h_g W \sin\alpha - (L+e_1)Z_1 - e_2 Z_2 - h_{DP}\, DP - d_{DP}\, DP \tan\gamma = 0 \end{gathered} \tag{13}$$

Remembering that

$$X_1 = \mu_1\xi_1 Z_1 \quad and \quad X_2 = \mu_2\xi_2 Z_2$$
$$e_1 Z_1 = f_1 r_{d1} Z_1 = M_{r1} \quad and \quad e_2 Z_2 = f_2 r_{d2} Z_2 = M_{r2} \tag{14}$$

and introducing equation 14 in equation 13, the next equations system is obtained:

$$\mu_1\xi_1 Z_1 + \mu_2\xi_2 Z_2 = W \sin\alpha + DP$$
$$Z_2 = W\cos\alpha + DP\tan\gamma - Z_1 \tag{15}$$
$$L\,Z_1 = (L-a)W\cos\alpha - h_g W\sin\alpha - f_1 r_{d1} Z_1 - f_2 r_{d2} Z_2 - h_{DP}\,DP - d_{DP}\,DP\tan\gamma$$

On soft grounds, due to the multi-pass effect, different values may be taken for the grip coefficients (μ_1 and μ_2) or rolling resistance coefficients (f_1 and f_2) – normally $\mu_2>\mu_1$ and $f_2<f_1$.

Four quantities (Z_1, Z_2, ξ_1 and ξ_2) must be obtained, but the system (15) has only three equations, which means a supplementary equation to be found and added. However, the last two equations of that system permit to obtain the normal ground reactions (the tire loads):

$$Z_1 = \frac{W\left[(L-a-f_2 r_{d2})\cos\alpha - h_g \sin\alpha - DP[h_{DP} + (d_{DP} + f_2 r_{d2})\tan\gamma]\right]}{L + f_1 r_{d1} - f_2 r_{d2}}$$
$$Z_2 = \frac{W\left[(a+f_1 r_{d1})\cos\alpha + h_g \sin\alpha + DP[h_{DP} + (d_{DP} + f_1 r_{d1})\tan\gamma]\right]}{L + f_1 r_{d1} - f_2 r_{d2}} \tag{16}$$

These equations show that normal ground reactions Z_1 and Z_2 are sensitive to the position of the center of gravity (a/L and h_g/L), slop angle α, drawbar angle γ, drawbar pull DP and, in a smaller amount, to the total rolling resistance coefficients (f_1 and f_2) and torque radii (r_{d1} and r_{d2}).

To exemplify different influences affecting DP, it was considered a vehicle with dimensions almost proportional with the ones of the model in figure 3 and also having the center of gravity equally distanced from the front and rear axles ($a=L/2$). A concrete or asphalt road (with a big grip coefficient $\mu_1=\mu_2=1$, achievable at a slip $\sigma=0.12$) was adopted.

In figure 4, the changes of the normal ground-reactions vs. drawbar pull may be observed for two ground slopes ($\alpha=0$ and $\alpha=10^o$). As can be seen, large changes of the forces Z_1 and Z_2 appear if the drawbar pull and climbing angle are modified.

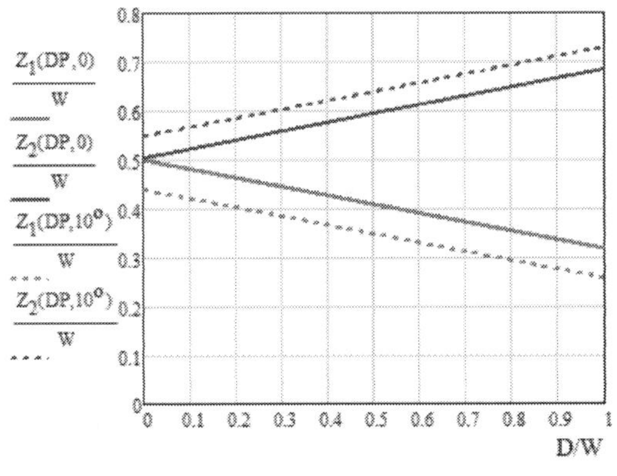

Figure 4. Normal ground-reactions change vs. drawbar pull: $\alpha=0$ – solid lines, $\alpha=10^o$ – dashed lines.

Continuing the assessment of the four unknowns mentioned previously, to obtain the normalized traction forces ξ_1 and ξ_2 it is necessary to use the first equation of the system (15), at which to add another particular equation that depends on the driving solution adopted for the propulsion system. Further, the particularities of more propulsion solutions will be analyzed.

3.3. Front and rear wheel-drive vehicles

In the case of front wheel-drive (FWD) vehicles

$$\mu_2\xi_2 = -f_2 \tag{17}$$

which, together with the first equation of the system (15), leads to the value

$$\xi_1 = \frac{DP + W \sin\alpha + f_2 Z_2}{\mu_1 Z_1} \tag{18}$$

Similarly, in the case of rear wheel-drive (RWD) vehicles

$$\mu_1 \xi_1 = -f_1 \tag{19}$$

which leads to:

$$\xi_2 = \frac{DP + W \sin\alpha + f_1 Z_1}{\mu_2 Z_2} \tag{20}$$

Replacing the normalized traction forces with their maximal values, ξ_1=1 and ξ_2=1, the transcendent equations (18) and (20) permit to obtain the maximum drawbar pull DP_{max} for vehicles with a single drive-axle. In the case of rear wheel-drive and big drawbar force, the danger that the front wheels leave the ground (Z_1=0) may exist, depending on the ground inclination-angle α, vehicle relative dimensions a/L and h_g/L and position of the towing hitch (d_{DP} and h_{DP}). Moreover, at low tire-load Z_1, the vehicle steering becomes difficult or almost impossible.

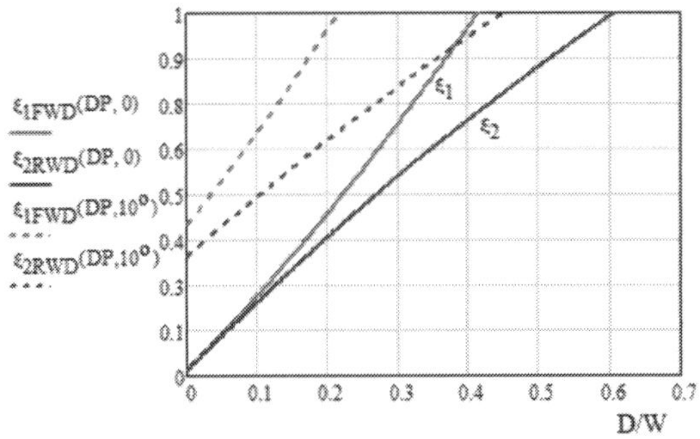

Figure 5. Example of normalized traction force ξ vs. normalized drawbar pull D/W in the case of single drive-axle vehicle and good quality road: α=0 – solid lines, α=10° – dashed lines.

The normalized traction force ξ necessary to obtain a certain normalized drawbar pull D/W is represented in figure 5. The same good quality ground (μ=1) was considered. Red and blue lines correspond respectively to FWD and RWD vehicles, while solid and dashed lines to horizontal and 10° inclined grounds. As can be observed from the figure, about 40% of the maximal grip force is necessary to climb the slope α=10° in the case of no drawbar load (DP=0). For the RWD vehicle, the maximum achievable D/W is about 0.6 on horizontal ground and reduces at about 0.45 on 10° slopes. The figures are worse for the FWD vehicle (0.42, respectively 0.22).

3.4. Vehicles with rigid connection between front and rear wheels

In the case of all wheel-drive vehicles, the mechanical distribution of the torque between the axles can be realized either using a differential or a rigid connection.

To obtain good traction performances on soft terrain, off-road vehicles are generally equipped with a transfer-case with synchronous output shafts, i.e. allowing *rigid connection* between the wheels of front and rear axles (more precisely a constant ratio k_ω between the mean angular-speeds ω_1 and ω_2 of the front and rear wheels).

Defining the *coefficient of driveline kinematics* as

$$k_\omega = \frac{\omega_1}{\omega_2} = \frac{i_{t2}}{i_{t1}} = const. \tag{21}$$

(with i_{t1} and i_{t2} the overall transmission ratios to the axles wheels), it results that for the considered type of vehicles (with no center differential or with locked differential) the next condition is true:

$$\frac{V}{(1-\sigma_1)\, r_{k1}} = k_\omega \frac{V}{(1-\sigma_2)\, r_{k2}} \quad or \quad 1-\sigma_1 = (1-\sigma_2)\frac{r_{k2}}{k_\omega\, r_{k1}} \quad or \quad \sigma_1 = 1 - \frac{(1-\sigma_2)}{k_{\omega r}} \tag{22}$$

The overall *coefficient of drivetrain kinematic non-concordance*, given by the ratio

$$k_{\omega r} = k_{\omega} \frac{r_{k1}}{r_{k2}} \tag{23}$$

proves to have a very important influence over the *DP* magnitude of these kind of vehicles. Ideally, its value should be equal with 1.

Now, combining the equation (22) with the first equation in the system (15), it is possible to calculate *DP*, because ξ_1 and ξ_2 are functions of the slip values σ_1 and σ_2. Indeed, for a certain *DP*, the transformed first equation in the system (15):

$$\mu_1 \xi_1(\sigma_1(\sigma_2)) Z_1 + \mu_2 \xi_2(\sigma_2) Z_2 = W \sin\alpha + DP \tag{24}$$

permits to find the numerical solution σ_2, because this is the only unknown. Then σ_1 is derived from equation (22) and, further, ξ_1 and ξ_2.

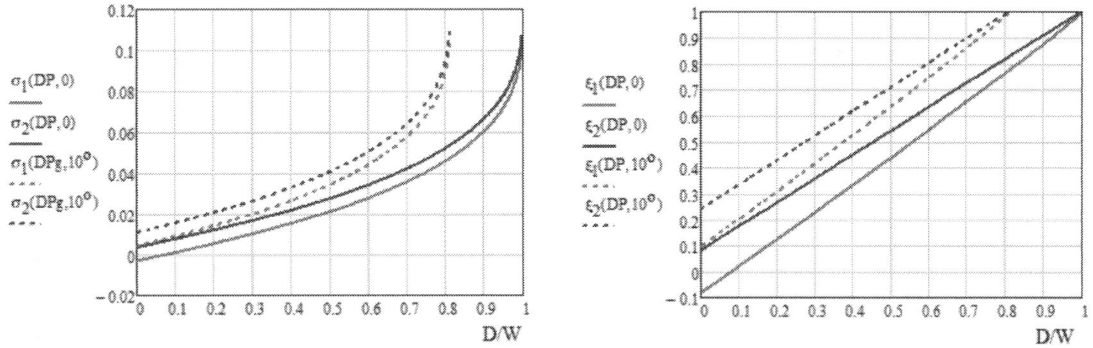

Figure 6. Example of necessary slip σ and normalized traction force ξ vs. normalized drawbar pull *D/W* in the case of good quality road: α=0 – solid lines, α=10° – dashed lines.

In figure 6 some performances of a hypothetical 4x4 vehicle (with no center-differential or with locked differential) are presented. The same good quality ground is considered. The slip values σ_1 and σ_2 necessary to generate certain normalized drawbar pull *D/W* are presented on the left-side of the figure, while on the right-side are shown the corresponding normalized traction forces ξ_1 and ξ_2. In this example, the next values, appearing in equation (22), were considered: r_{k2}/r_{k1}=1.0067 and k_{ω}=1, leading to $k_{\omega r}$=0.993. The tires' rolling radii were considered constant during driving. It observes that, for horizontal ground, σ_1 and ξ_1 present negative values and low *D/W* (as well, this phenomenon is observed experimentally [21, 34]). It also can be seen that, at drawbar pull near to maximum, the slip values tend to become equal for the front and rear wheels, $\sigma_1 = \sigma_2 = \sigma_\mu$. Another observation: the 10° slope leads to a maximal drawbar pull smaller with about 20% as for horizontal ground.

3.5. Vehicles with center differential

Another possibility to mechanically distribute the torque between axles consists in using a center (inter-axle) differential, which may be of an open or limited-slip type.

Depending on the working conditions, a differential (of any type) can behave as a locked one (with equal speeds of its output shafts) or can have unlocked functioning, permitting different speeds of the output shafts simultaneously with torque distribution in a certain (known) proportion. Due to that, the study of the differential influence over the drawbar pull is more complicated than the cases already presented. The following study approach of differential locked and unlocked functioning is based somehow on the work [35].

The behavior of any differential is governed by three equations. The first, known as the Willis equation, gives a condition for its kinematics

$$\omega_{d1} + i_p \cdot \omega_{d2} = (1 + i_p) \cdot \omega_{dc} \tag{25}$$

Here the indices $_{dc}$, $_{d1}$ and $_{d2}$ indicate respectively the differential case and the output shafts to the front and rear axles, while $-i_p$ is the *internal transmission ratio* (from the shaft $_{d1}$ to the shaft $_{d2}$ with the case $_{dc}$ fixed).

In traction mode, the dynamics of the differential obeys to the next two equations:

$$M_{d1} + M_{d2} = M_{dc} \tag{26}$$

$$\frac{M_{d2}}{M_{d1}} = \frac{M_{d20}+M_{df}}{M_{d10}-M_{df}} = \lambda \qquad (\lambda_{nf} \le \lambda \le \lambda_{pf}) \tag{27}$$

where: λ is usually called the *locking coefficient* of the differential and λ_{pf} and λ_{nf} are its extreme values.

If it considers no friction exists in the differential, then $\lambda=i_p$ and the previous equation gives the *theoretical distribution of the torque* ($M_{d20}=i_p \cdot M_{d10}$) But, in reality, the always existent friction, through the *friction moment* M_{df},

$$M_{df} = M_{dc}\frac{\lambda-i_p}{(1+\lambda)(1+i_p)} \tag{28}$$

will alter the values of the output torques M_{d2} and M_{d1} (one increases while the other decreases with the same amount M_{df}). It can be noted also that the friction moment may be either positive or negative.

Even designed to produce less friction, a so-called *open differential* still generates locking coefficients λ differing from i_p. For example, such a symmetric differential (i_p=1), typically presents values as λ_{pf}=1.07 (for the maximal positive friction moment M_{dfmax}) and λ_{nf}=0.935 (for the extreme negative friction moment $M_{dfmin}=-M_{dfmax}$). More distanced values are associated with *limited-slip* (also named *self-locking* [3, 36]) differentials, at which the friction is important. Two examples for *torque-sensitive limited-slip differentials*: a symmetric one may present the values λ_{pf}=4 and λ_{nf}=0.25, and another, asymmetric and having i_p=2, may have λ_{pf}=8 and λ_{nf}=0.8. These mentioned types of limited-slip differentials (including *torsen*) present the property that, when $\omega_{d1} \neq \omega_{d2}$, the ratio of the output moments M_{d2}/M_{d1} remains practically constant (equal with λ_{pf} or λ_{nf}).

With these remarks, it is now possible to indicate how can be computed the drawbar pull *DP* for a motor vehicle having an open- or torque-sensitive limited-slip center-differential.

At the beginning, it considers the *locked functioning* of the differential, i.e. the vehicle having a rigid connection between the wheels of the front and rear axles (situation studied in the previous subsection). Now, knowing the values ξ_1 and ξ_2 corresponding to a certain *DP* value, firstly compute the traction forces X_1 and X_2 with the equation (3), then the wheel moments M_{w1} and M_{w2} with equation (11) and finally the output moments of the center differential:

$$M_{d1} = M_{dc}\frac{1}{1+\lambda} = \frac{M_{w1}}{i_{dw1}\eta_{dw1}} \qquad M_{d2} = M_{dc}\frac{\lambda}{1+\lambda} = \frac{M_{w2}}{i_{dw2}\eta_{dw2}} \tag{29}$$

where: i_{dw1}, i_{dw2} and η_{dw1}, η_{dw2} are respectively the transmission ratios and mechanical efficiency from the differential to the front and rear wheels.

If a fictive value λ_c, computed with equation (27), is in the interval [λ_{nf}, λ_{pf}], the differential is locked ($\omega_{d1}=\omega_{d2}$) and the values ξ, X, M_w and M_d are the true ones.

If the condition $\lambda_{nf}<\lambda_c<\lambda_{pf}$ is not fulfilled, the differential has *unlocked functioning* and the output speeds are different ($\omega_{d1} \neq \omega_{d2}$). Consequently, the computation must continue.

If $\lambda_c>\lambda_{pf}$, then $\omega_{d1}>\omega_{d2}$ and the value λ_{pf} will replace λ_c in equation (29). If, contrarily, $\lambda_c<\lambda_{nf}$, then $\omega_{d1}<\omega_{d2}$ and the value λ_{nf} will be retained.

Now, replacing λ with λ_{pf} or λ_{nf}, as appropriate, and replacing equations (11), (3) and (29) in the first equation (15), the input torque of the differential M_{dc} can be expressed as function of *DP*:

$$M_{dc} = \frac{1+\lambda}{\frac{i_{dw1}\eta_{dw1}}{r_{d1}}+\lambda\frac{i_{dw2}\eta_{dw2}}{r_{d2}}}(DP + W\sin\alpha + f_1 Z_1 + f_2 Z_2) \tag{30}$$

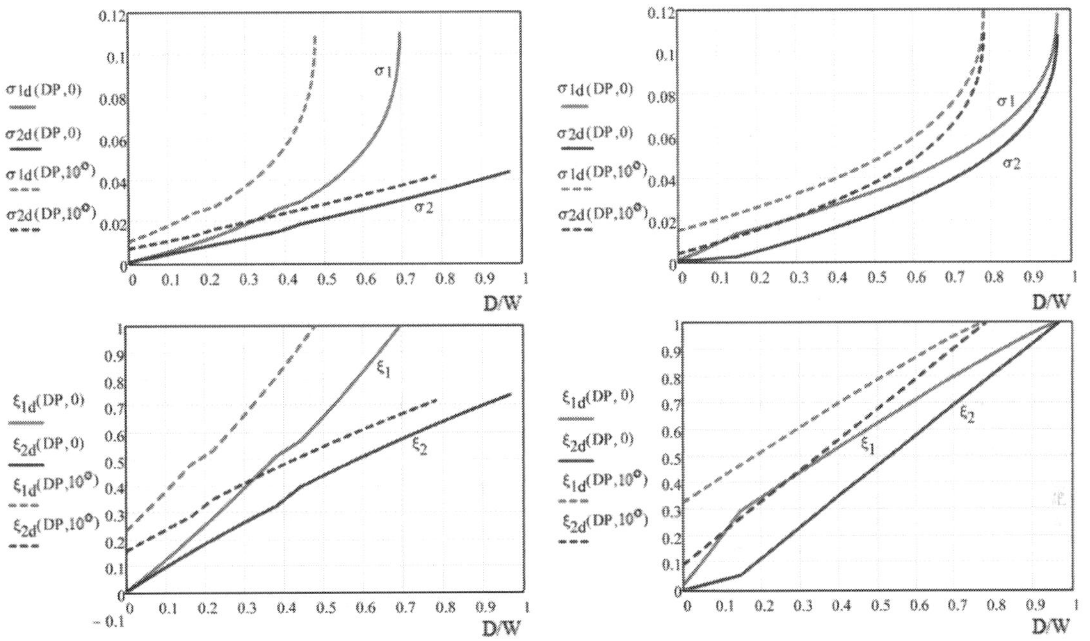

Figure 7. Example of necessary slip σ and normalized traction force ξ vs. normalized drawbar pull D/W in the case of good quality road and centre differential vehicle: left side – open differential, right side – torsen differential; α=0 – solid lines, α=10° – dashed lines.

Further, using in reversed order the mentioned equations for a certain DP value, all the quantities presenting interest can be derived: M_d, M_w, X, and finally ξ_1, ξ_2 and σ_1, σ_2.

The drawbar pull performances of a presumptive vehicle equipped with two types of center differential are presented in figure 7. The plots on the left side appertain to the open type (λ_{pf}=1.07) and the others to a torsen (torque-sensitive limited-slip) type (λ_{pf}=4).

The open differential remains locked for $0.38<D/W<0.441$ on horizontal ground, while on 10° slope for $0.165<D/W<0.218$ (for small D/W values, $\omega_{d1}>\omega_{d2}$ and for large D/W values, $\omega_{d1}<\omega_{d2}$).

In turn, the considered torsen differential remains locked for $0.147<D/W$ (up to its maximal value) in the case of horizontal ground, while on 10° slope it remains locked for all D/W interval. These "locked-unlocked" functioning changes are visible on the figure by the inflexion of the curves.

The analysis of the figure permit to ascertain that the torsen differential provides better traction performances as the open differential (maximal drawbar pull D/W_{max} equal to 0.96 vs. 0.7 on horizontal ground, respectively 0.78 vs. 0.48 on 10° slope), values which are comparable with the ones corresponding to rigid connection of the front and rear wheels.

Another observation: the slips σ, normalized traction forces ξ (and, of course, the wheel moments M_w) will not present anymore negative values: the "fighting" of the front and rear wheels (indicated by the opposite values of tires-ground grip) is transferred to the differential parts.

3.6. Relations between linear speed and angular speed

Knowing the relation between the linear speed V of the vehicle and the angular speed of a well-defined shaft of the driveline is mandatory if it wants to assess the maximal vehicle speed at certain drawbar pull. Also, this relation can be used to compute the travel reduction TR, that is necessary to obtain the tractive efficiency (that will be presented in the next subsection of the article).

The case of a motor vehicle with center differential.

Using the Willis equation (25), it is possible to compute the ratio of vehicle actual driving-speed V and angular speed ω_{dc} of the differential case (the input shaft of the differential):

$$\frac{V}{\omega_{dc}} = \frac{1+i_p}{\frac{i_{dw1}}{(1-\sigma_1)r_{r1}} + i_p \frac{i_{dw2}}{(1-\sigma_2)r_{r2}}} \tag{31}$$

Putting in the previous equation the slip values σ_{10} and σ_{20} that corresponds to constant speed driving with no drawbar pull, the reference vehicle speed V_{ref} is found and then, writing the ratio of the actual and reference speeds, it obtains the travel reduction *TR* for a vehicle with center differential:

$$1 - TR = \frac{V}{V_{ref}} = \frac{\frac{i_{dw1}}{(1-\sigma_{10})r_{r10}} + i_p \frac{i_{dw2}}{(1-\sigma_{20})r_{r20}}}{\frac{i_{dw1}}{(1-\sigma_1)r_{r1}} + i_p \frac{i_{dw2}}{(1-\sigma_2)r_{r2}}} \tag{32}$$

In figure 8 it is shown the influence of the drawbar pull over the travel reduction for the example vehicle equipped with the open differential. It can be observed that the climbing of a 10° slope with no drawbar pull leads to a speed reduction of about 1% and also that a value D/W=0.4 will determine a speed reduction of about 2% in the case of horizontal ground and 4% on the 10° slope. On both road types, the maximal drawbar pull obtains with a speed loss of about 7%.

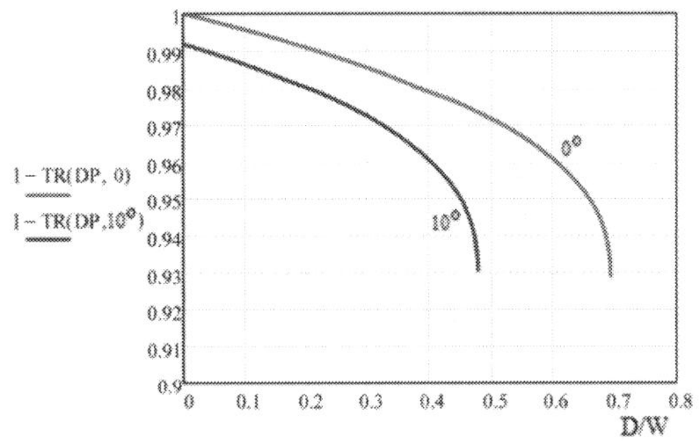

Figure 8. Influence of the normalized drawbar pull *D/W* over the travel reduction *TR* in the case of a vehicle with open center differential, on good quality road (horizontal and 10° inclined).

The previous equation shows once again that the speed loss should be taken into account when the traction performances of the vehicles are computed [18, 26], as was mentioned earlier.

The case of a motor vehicle with rigid connection between the front and rear wheels.

Adopting a similar approach but considering an equal speed ω_{to} for both output shafts of the transfer case, it results

$$\frac{V}{\omega_{to}} = \frac{r_{r1}}{i_{tw1}}(1-\sigma_1) = \frac{r_{r2}}{i_{tw2}}(1-\sigma_2) \tag{33}$$

$$1 - TR = \frac{V}{V_{ref}} = \frac{r_{r10}}{r_{r1}}\frac{(1-\sigma_{10})}{(1-\sigma_2)} = \frac{r_{r20}}{r_{r2}}\frac{(1-\sigma_{20})}{(1-\sigma_2)} \tag{34}$$

The last equation giving the travel reduction will be also used for 2WD vehicles, considering the ratio with σ_1 for the case of FWD and the ratio with σ_2 for the case of RWD.

To obtain the maximal vehicle speed at a certain drawbar pull it is necessary to know the torque vs. angular speed characteristic at the output of the transmission. Then the value of the transmission output-shaft torque, corresponding to the considered *DP*, must be computed. That is possible because:

- in the case of 2WD vehicles, the normalized traction force ξ is known and the grip force X and wheel torque M_w can be computed with the equations (3) and (11), respectively;
- in the case of vehicles with a rigid connection between front and rear wheels, the wheel torques M_{w1} and M_{w2} can be obtained similarly as for 2WD vehicles and, summing these two torques, results the torque on the output shaft of the transfer case;
- in the case of vehicles with centre-differential, the input torque (at the case) of the differential is given by equation (30).

Further, taking from the torque-speed characteristic at the output of the transmission the corresponding value of the angular speed, the speeds ω_{dc} in equation (31) and ω_{to} in equation (33) will be computed, that allows finding, with the same equations, the actual vehicle speed V.

3.7. Tractive efficiency

The *tractive efficiency* (*drawbar efficiency*) η_t is a unitless value representing the efficiency of transferring power to move an external load; it is defined as the ratio of the useful power to the consumed power.

A first equation is used to indicate the *global tractive efficiency*, i.e. the ratio of *DP* force and speed *V* to the power P_e of vehicle engine [6, 11, 19]:

$$\eta_t = \frac{DP \cdot V}{P_e} \tag{35}$$

while another to specify a local value: the *efficiency of the wheels to generate tractive power*

$$\eta_w = \frac{DP \cdot V}{P_{w1}+P_{w2}} = \frac{DP \cdot V}{\frac{V}{1-TR}(F_{w1}+F_{w2})} = \frac{DP \cdot (1-TR)}{\left(\frac{M_{w1}}{r_{d1}}+\frac{M_{w2}}{r_{d2}}\right)} = \frac{DP \cdot (1-TR)}{X_1+R_{r1}+X_2+R_{r2}+W \sin\alpha} = \frac{DP \cdot (1-TR)}{(\mu_1\xi_1+f_1)Z_1+(\mu_2\xi_2+f_2)Z_2+W \sin\alpha} \tag{36}$$

where: P_{w1} and P_{w2} represent the wheel power of the axles (ξ_2=0 for FWD vehicles and ξ_1=0 for RWD).

It is important to note that η_w (indicated by the previous equation) depends both on rolling radii r_r (by the travel reduction *TR* and normalized net traction forces ξ_1 and ξ_2) and loaded radii r_l (by the tire loads Z_1 and Z_2).

Because

$$P_e = \frac{P_{w1}+P_{w2}}{\eta_{dl}} \tag{37}$$

it results the relationship between the global efficiency η_t and wheels efficiency η_w:

$$\eta_t = \eta_{dl} \cdot \eta_w \tag{38}$$

where: η_{dl} represents the driveline overall efficiency (from the engine to the wheels).

In the case of vehicles with center differential (subsection 3.5), if the angular speeds of the output shafts aren't equal ($\omega_{d1} \neq \omega_{d2}$), the driveline efficiency η_{dl} must also consider the power lost by slippage in the differential, $|M_{df}\,(\omega_{d1}-\omega_{d2})|$.

Tractive efficiency will drop to zero when no drawbar load is applied, as well as when there is enough drawbar load to create zero forward velocity (to stop the vehicle). Different vehicle types (on wheels or on tracks) and having different configurations (4x2 or 4x4) present normally different tractive efficiency characteristics [37], being more or less suitable for a certain activity.

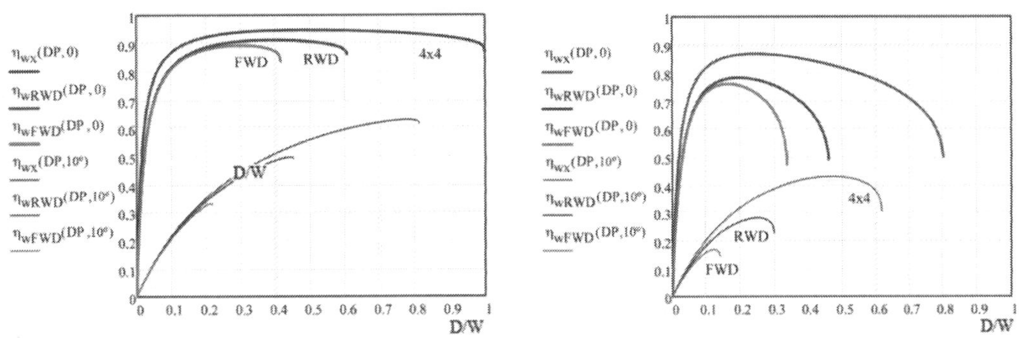

Figure 9. Traction efficiency of the wheels vs. normalized drawbar pull *D/W* in the case of a FWD, RWD and 4x4 vehicles, on asphalt (left side) and dry dirt ground (horizontal and 10° inclined).

Figure 9 presents examples of computed tractive efficiency of the wheels obtained with the data of the previously considered vehicle, in the case of three drivetrain configurations (FWD, RWD and 4x4 without center differential). Two types of grounds (asphalt and dry dirt with μ=0.8 at σ_μ=0.55), with two longitudinal inclinations, were considered. As can be seen, the best towing performances are realized by vehicles with 4x4 wheel formula. Also, the efficiency is better for the case of asphalt ground, both for maximal value and optimum performance spread (the curves present a plateau).

There is typically an optimum *DP* load that maximizes the tractive efficiency. For this reason, characteristics as the ones presented in figure 9, are used to optimize towing by adjusting vehicle configuration or drawbar load in order to operate at peak tractive efficiency.

3.8. Influence of slope angle and towing hook position

As was seen for the previous figures, the ground slope angle has a major influence over drawbar pull performances. This not only increases the total resistance (through the term $W \cdot \sin\alpha$) and reduces the total tires load, responsible for traction force generation (through the quantity $W \cdot (1-\cos\alpha)$), but also by worsening the ratios of wheel torques and tire loads.

With respect to the purpose of the vehicle, the towing hitch can have different positions: most of them are in the rear-side (for agricultural tractors, military vehicles or road tractors for full-trailers – figure 3), other in the front (as bulldozers or snow plough) or some in the middle, between axles (close to ground, as per graders, or at a considerable elevation, as per road tractors for semi-trailers). In this last case of towing truck-semitrailer combinations, the big height of the hitch (the fifth wheel) determines important load transfer between axles [23], eventually appearing the lift from the ground of the front-wheels, mainly on steep climbing angles. In such situations, vehicle control may become difficult or even impossible.

The angle γ made by the drawbar with the ground plane may also play a significant role in the towing performances. Normal values for trailing a vehicle attached at the rear side are in the range - $10^{\circ}<\gamma<10^{\circ}$. In the case of tractors carrying implements, the angle γ may also change as a function of *DP* value [19]. Furthermore, in the case of tractor-semitrailer combinations (where no physical drawbar is used), the angle γ will be also affected by the *DP* value, starting with 90° on horizontal ground and at zero towing effort, and reducing continuously while *DP* value increases [23].

3.9. Influence of wheels radii and transmission ratios

As already presented, rolling, torque and loaded radii (r_k, r_d and r_l) of the wheels suffer different influences during vehicle utilization. The present study permits to take account of some radii changes during vehicle driving, from which the tire loads and the wheel torques may have a big influence over the vehicle towing performances.

As indicated in equation (16), the tire loads Z_1 and Z_2 are not affected directly by the torques applied at the wheels (M_{w1}, M_{w2}) and suffer minor influences from the torque (dynamic) radii (r_{d1}, r_{d2}).

Even the deflection of the tire (its deformation along the z-axis) shows a declining increase with rising wheel load, that deflection vs. load characteristic is near linear [5], which indicate an almost constant stiffness coefficient for the variation of the torque radius r_d with tire load Z. A similar almost linear behavior presents r_d vs. applied wheel torque M_w [4]. The increase of both tire load and wheel torque leads to a smaller torque radius. Also, it is expected that the rolling (kinematic) radius r_k to suffer similar modifications when variable tire load and wheel torque are applied. These changes of the rolling radii (more precisely the modification of their ratio r_{k1}/r_{k2}) can have important influences over the slip, net traction forces and drawbar pull, mainly for all-wheel drive vehicles, but also for 2WD vehicles [11]. On the other hand, the changes of torque radii r_d will affect the stress of the driveline shafts.

With these clarifications, it is possible to use variable torque radii in this algorithm (replacing a constant radius with a function $r(Z,M_w)$) by considering the instantaneous tire loads and wheel torques as appears in equations (16) and (11). That is also easy because the hypothesis of linear influences is justified by experiments, mainly in the field of variation of the tire radii in normal conditions. Examples of taking into account the change of the wheel radii may be found in [11, 35].

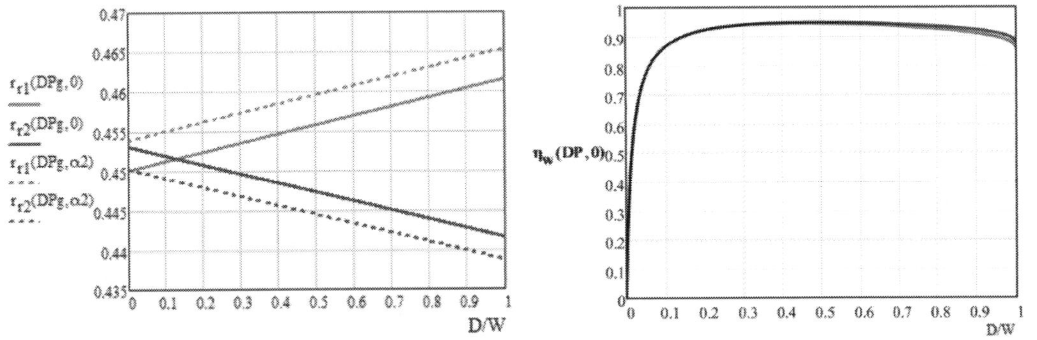

Figure 10. Change of wheels rolling radii r_k vs. tire loads Z on horizontal and 10° inclined ground (left side) and influence of the change over the traction efficiency (right side).

In the mentioned hypotheses, the variable rolling radius can be computed with the equation

$$r_k = \left(r_u - \frac{Z}{k_{ta}}\right)\left(1 - \frac{M_w}{k_M}\right) \qquad with \qquad k_{ta} = \frac{Z_s}{r_u - r_{ks}} \tag{39}$$

where: r_u is the unloaded radius of the tire, r_{ks} is the rolling radius that correspond to the static load Z_s on the axle, k_{ta} is the overall stiffness of all axle tires, Z is the actual load on the axle tires and k_M is a coefficient of the wheel radii resistance to the applied torque.

On the left side of figure 10, it is presented the computed change of the rolling (kinematic) radius r_k vs. tire loads Z, considering no sensibility to the wheel moment (infinite k_M value). On the right side of the same figure, it can be seen that the change of the radii affects the traction efficiency of the wheels – near the maximal *DP* value, on dry asphalt, the efficiency reduces by about 2%.

Together with the rolling radii ratio r_{k1}/r_{k2}, a similar important influence over the drawbar pull has the ratio i_{dw1}/i_{dw2} of the speed ratios from the torque splitting device (transfer case, center differential) to the front and rear wheels. In the case of the vehicles having front and rear wheels with different dimensions, the speed ratios may not be perfectly correlated with r_{k1}/r_{k2}, leading to some kinematic non-concordances that will affect the drawbar pull performances.

4. Conclusions

In this paper, a method for the theoretical assessment of the drawbar pull for two-axle vehicles was presented. Due to the lack of consensus in the literature, at the beginning, the necessary terms and definitions were largely and critically presented.

The performances for both single traction element (wheel or axle) and full vehicle were studied. Also, the computation particularities for single drive-axle and all-wheel drive vehicles were indicated. The emphasis was on the on-road vehicles. Most deformable soil challenges are simplified to the changes in grip and rolling resistance coefficients.

In contrast to the normal approach, from the engine and transmission data to the towing force, in this work the computation was inverted, starting from a needed drawbar pull value to the corresponding torque and angular speed in different points of the driveline.

The vehicle model permits analysis of the influences of a large number of constructive or driving parameters. For the computation, the ground characteristics may be adopted from the literature, while the necessary input data of the vehicle is normally given by the manufacturers or can be obtained by measurements.

Between the vehicle characteristics, the positions of center of gravity and towing hitch, the inclination of the drawbar, the wheels dimensions and the transmission ratios have important influences on the drawbar pull.

Ground characteristics, as grip and rolling resistance coefficients, can also be changed easily in the model. Another very important aspect, the slope angle influence, was also evaluated.

Because the experiments are costly and difficult to realize and reproduce, the theoretical assessment of the drawbar force may be extremely useful to identify the traction potential of a certain vehicle in steady-state condition, mainly for functional optimizations of an existent vehicle or in the primary stages of the vehicle design.

Also, because different grip and rolling resistance values may be adopted for the wheels of each axle, the presented algorithm gives the possibility to study the influence of the multi-pass effect over the magnitude of the drawbar pull, if such input data exists.

Moreover, based on the slip estimated to occur when a vehicle is at a certain towing magnitude, it is possible to predict the level of soil damage appearing in this situation.

Due to the possibility to estimate the modifications of the tires torque radii as result to the tire load and wheel torque changes, the presented algorithm may be adapted for vehicles with more than two axles (or even to tracked vehicles), because suspension compression at each axle, body pitch and height of the center of gravity can be computed vs. current driving condition.

References

[1] SAE J670e, *Vehicle Dynamics Terminology* (revised 2008-01-24), 1976, Society of Automotive Engineer, Inc.

[2] He R, Sandu C, Mousavi H, Shenvi M, Braun K, Kruger R and Els S, 2020, *Updated Standards of the International Society for Terrain-Vehicle Systems*, Journal of Terramechanics, vol. 91, pp. 185-231

[3] Untaru M, Poţincu G, Stoicescu A, Pereş G and Tabacu I, 1981, *Dinamica autovehiculelor pe roţi*, Bucharest: Editura Didactică şi Pedagogică

[4] Guiggiani M, 2014, *The Science of Vehicle Dynamics – Handling, Braking, and Ride of Road and Race Cars*, London: Springer

[5] Kutzbach H D, Bürger A and S. Böttinger, 2019, *Rolling radii and moment arm of the wheel load for pneumatic tyres*, Journal of Terramechanics, vol. 82, pp. 13-21

[6] Schreiber M and Kutzbach H, 2007, *Comparison of different zero-slip definitions and a proposal to standardize tire traction performance*, Journal of Terramechanics, vol. 44, pp. 75-79

[7] Ydrefors L, Hjort M, Kharrazi S, Jerrelind J and Stensson Trigell A, 2021, *Rolling resistance and its relation to operating conditions: A literature review*, Proceedings of the Institution of Mechanical Engineers, Part D: Journal of Automobile Engineering, vol. 235, no. 12, pp. 2931-2948

[8] Priddy J, 1995, *Stochastic Vehicle Mobility Forecasts Using the NATO Reference Mobility Model. Report 3. Database Development for Statistical Analysis of the NRMM II Cross-Country Traction Empirical Relationships*, US Army Corps of Engineers

[9] Milliken W F and Milliken D L, 1997, *Race car vehicle dynamics*, SAE

[10] Gillespie T D, 1992, *Fundamentals of Vehicle Dynamics*, Warrendale, USA: SAE

[11] Nastasoiu M and Ispas N, 2017, *Determining the theoretical drawbar performance characteristics of tractors, considering the variation of the wheels radii during tractor operation*, in CAR2017 Conference, Pitesti, pp. 89-96

[12] Creager C, Asnani V, Oravec H and Woodward A, 2017, *Drawbar Pull (DP) Procedures for Off-Road Vehicle Testing*, NASA

[13] Vantsevich V, Barz D, Kubler J and Schumacher A, 2005, *Tire Longitudinal Elasticity and Effective Rolling Radii: Experimental Method and Data*, SAE Technical Papers, vol. Kistler, no. 2005-01-1823, 04/11

[14] Mitschke M and Wallentowitz H, 2014, *Dynamik der Kraftfahrzeuge*, 5 Auflage, Berlin: Springer

[15] Hamersma H A, Botha T R and Els S, 2014, *Kinetic vs. kinematic roll radius on rough roads*, in 18th International Conference of the ISTVS, September 22–25, 2014., Seoul, Korea, pp. 1–6

[16] Wei Y, Oertel C, Li X and Yu L, 2017, *A theoretical model for the tread slip and the effective rolling radius of the tyres in free rolling*, Proceedings of the Institution of Mechanical Engineers, Part D: Journal of Automobile Engineering, vol. 231, no. 11, pp. 1461-1470

[17] *The tyre grip*, 2001, Société de Technologie Michelin

[18] Preda I, 2020, *Ecuatia generala de miscare – nimic mai simplu!? (The general equation of motion - nothing simpler!?)*, Ingineria Automobilului, no. 57, pp. 9-15
[19] Niţescu G, Năstăsoiu S and Popescu S, 1974, *Tractoare*, Bucureşti: Editura Didactica si Pedagogica
[20] Ciobotaru T, 2009, *Semi-Empiric Algorithm for Assessment of the Vehicle Mobility*, Leonardo Electronic Journal of Practices and Technologies, no. 15, July-December 2009, pp. 19-30
[21] Roşca P, 2019, *Studiu privind transpunerea procedurilor de testare specifice vehiculelor militare în mediul virtual*, Ph.D Thesis, Academia Tehnica Militara, Bucuresti
[22] Wasfy T M and Jayakumar P, 2021, *Next-generation NATO reference mobility model complex terramechanics – Part 2: Requirements and prototype*, Journal of Terramechanics, vol. 96, pp. 59-79
[23] Preda I, 2005, *Aspects regarding the wheel loads of tractor-semitrailer road train*, in CAR2005 Conference, Pitesti
[24] NATO Military Agency for Standardization, 1994, *AVTP 03-60 Drawbar Pull and Resistance to Motion on Hard Surface*
[25] US Army Yuma Test Center, 2007, *Test Operations Procedure (TOP) 2-2-604 Drawbar Pull.*
[26] Jacobson B, 2016, *Vehicle Dynamics*. Compendium for course MMF062, Chalmers University of Technology
[27] Wong J Y, *Theory of Ground Vehicles*, 2001, Third ed., Ottawa, Wiley
[28] Kobelski A, Osinenko P and Streif S, 2020, *A method of online traction parameter identification and mapping*, IFAC-PapersOnLine, vol. 53, no. 2, pp. 13933-13938
[29] Naranjo S D, Sandu C, Taheri S and Taheri S, 2014, *Experimental testing of an off-road instrumented tire on soft soil*, Journal of Terramechanics, vol. 56, pp. 119-137
[30] Preda I and Ciolan G, 1997, *Modelarea interactiunii dintre roata si sol*, in CAR1997 Conference, Pitesti
[31] Pacejka H B, 2006, *Tyre and Vehicle Dynamics*, Second ed
[32] Roşca R, Rakosi E and Manolache G, 2004, *Wheel Traction Prediction - A Comparison Between Models and Experimental Data*
[33] Sandu C, Taheri S, Taheri S and Gorsich D, 2019, *Hybrid Soft Soil Tire Model (HSSTM). Part II: Tire-terrain interaction*, Journal of Terramechanics, vol. 86, pp. 15-29
[34] Truţă M, Marinescu M and Vînturiş V, 2016, *Influence of the drawbar pull over the power flux within the automotive transmissions*, IOP Conference Series: Materials Science and Engineering, vol. 147, pp. 012131
[35] Preda I, Pereş G, Untaru M and Ciolan G, 1988, *Quelques aspects de l'utilisation des modelés mathématiques a l'étude des sollicitations dans la transmission des voitures*, in CONAT1988 Conference, Brasov, Romania, pp. A 229-237
[36] Gadola M, Chindamo D and Lenzo B, 2021, *Revisiting the Mechanical Limited-Slip Differential for High-Performance and Race Car Applications*, Engineering Letters, vol. 29, no. 3, pp. 824-839.
[37] Năstăsoiu M and Ispas N, 2019, *Potential Performance Characteristics for Different Types of Tractors: Two-Wheel-Drive Tractor, Four-Wheel-Drive Tractors and Crawler Tractors*, Proceedings of the 4th International Congress of Automotive and Transport Engineering (AMMA 2018). Proceedings in Automotive Engineering, pp. 521-526, Springer, Cham

IOP Conf. Series: Materials Science and Engineering 1220 (2022) 012014 doi:10.1088/1757-899X/1220/1/012014

Experimental test bench with diesel engine for the study of multi fuels usage

S Izvorean and V Stoica

Politehnica University Timisoara, Faculty of Mechanical Engineering, 1 Mihai Viteazu Blvd., Timisoara Romania

stefan_izvorean2008@yahoo.com ; virgil.stoica@upt.ro

Abstract. The diesel engine has entered a shadow cone because of emissions of nitrogen oxides and particles. However, the engine offers superior torque and power and low consumption, therefore research is needed to obtain an engine with diesel performance and spark ignition engine emissions. In this paper, we will present an experimental stand with D115 diesel engine with a common rail injection system equipped in parallel with a single point injection system in the intake manifold. The purpose of the dual supply is to obtain a homogeneous combustible air mixture in the combustion chamber with a coefficient of excess air below the detonation limit and which can be ignited by pilot injection of diesel.

The experimental test bench is presented in detail regarding both the hardware and the software part (developed in the laboratory) together with the calibration curves of the force and temperature sensors

In this paper, in order to highlight the performance of the stand, standard experimental results described in the literature will be presented.

Although there is a wide range of experimental tests in the field of internal combustion engines from the additional injection of water, alcohol, bioethanol and so on, this paper aims to provide additional data in order to understand the phenomena of combustion and pollutant emissions. By injecting alcohol into the intake manifold of the diesel engine, the aim is to ensure a poor but homogeneous mixture of fuel in the combustion chamber, which will lead to a reduction of polluting emissions while maintaining the performance of the diesel engine. It also shows the conversion of the fuel supply system of the engine, of Romanian production D115, from a classic injection system to one with a common rail. In addition to the mechanical part, the algorithms developed for engine and stand control are presented.

By the presented stand and preliminary experimental results, we show the potential of the developed experimental test bench.

1. Introduction

The diesel engine production for cars is considered to be ended due to high NOx and particulate matter (PM) produced by the combustion process, but because of its efficiency, it will probably remain the only option for agricultural, industrial, and heavy transport application. We need to find a solution that can help to reduce the emission and increase the efficiency of the engine. A possibility to improve the diesel engine emissions as pointed out by C. Stan [1] would be the ethanol injection.

Z. Stelmasiak [2] presents some results obtained in the case of diesel and spark ignition engines with dual injection systems like gasoline and alcohol or diesel and alcohol. The results clearly show an increase in efficiency when running with alcohol at high loads and also an increase of CO emissions,

NOx reduction at part loads and an increase at high loads. In the case of diesel engines, it was observed an increase in efficiency with about 3-6% when fed with alcohol. Also, a high CO and smoke reduction was observed and an increase of NOx. The author presents only some experimental results without optimization.

H. Tse et al. [3] present the results obtained on a diesel engine with a classic injection pump fueled with diesel-biodiesel-ethanol blends. Since the injection timing is fixed, the results are not going to be concluded. The cetane number of alcohol is low therefore the ignition delay will be high and the heat release will be pushed towards the exhaust phase, and the engine efficiency will decrease. With all of that, the particulate matter and NOx decrease when using alcohol. Similar experiments have been performed also by Y. Noorollahi et al. [4] on Lombardini one cylinder diesel engine. The results show that the mixture proportion influences the engine performances, but again since the engine was tared with diesel fuel we may not expect to work better with different types of fuels without any previous optimization concerning at least the injection timing. Still, the reduction of carbon monoxide and the increase of power and specific fuel consumption are observed. Experiments with ethanol-blended diesel were performed with a 4-cylinder engine with fixed injection timing by B.Q. He et al. [5], and the results show the effect of ethanol on the emissions for different loads. Mainly a decrease of NOx and particulate matter for high loads and an increase of the CO and hydrocarbons was observed.

V.P. Ferreira et al [6] show the effects of the ethanol injection within the intake manifold on one cylinder diesel engine. The ethanol injection is controlled by a microcontroller and is synchronized with the diesel injection. The results clearly show a reduction of the NOx when injecting ethanol and at the same time an increase in the CO emissions. The tests were done for one load (using as a brake an electric generator) therefore; the evolution with load is not pointed out. In addition, because an engine equipped with a mechanical injection system was used the diesel injection timing was kept fixed and it is expected that the heat release to be delayed as mentioned by the authors, and as a consequence a lower temperature in the cylinder and lower NOx emissions. The energy efficiency is decreasing when ethanol is injected, and as the authors pointed out, this can be due to the low cetane number of the ethanol and a burning delay. A better understating of the results would have been provided, if the test were done, also, for different diesel injection timings.

In the review of S.A. Shahir et al. [7], a number of somewhat contradictory results are highlighted. For example, some authors achieve a reduction in nitrogen oxides of carbon monoxide or carbon dioxide when mixtures of diesel and alcohol are used, while other researchers report increases in these parameters. We suppose that this is happening because the emissions are not correlated with the engine construction and injection characteristics and timings, and because some burning phenomena and emission formation are not completely understood.

Within the present paper, an experimental test bench equipped with a compression ignition engine is presented. The used engine is actually a D115 diesel engine equipped with a common rail injection system with a LabView based control that allows the operator to adjust any parameter or to implement any control algorithm. With the presented stand, it is possible to overcome some of the drawbacks of ethanol usage [8]. So the injection parameters can be adjusted to overcome the increase of the ignition delay due to the low cetane number of the ethanol, and it eliminates the risk of phase separation of the diesel ethanol blend by using a separate injection system for alcohol.

2. Experimental test bench

2.1. Mechanical component

The D115 engine is an inline 3 cylinder 4 stokes natural aspirated compression ignition engine, having a displacement of 2340 cm^3. It was produced by UTB (Tractor Production Plant in Brasov, Romania) under a Fiat license. Originally, the engine had a mechanical injection pump and injectors. It was capable to produce a power of 33 kW at 2400 RPM and a torque of 150 Nm at 1500 RPM.

More flexibility concerning the adjustment of injection parameters (injection pressure, injection timing, fuel dose and so on), was ensured by equipping the engine with a common rail injection system figure 1 and 2. The installation of the system required a few modifications to the cylinder head to ensure a proper fit of the solenoid injectors 2. The high-pressure pump 25 was mounted through an adaption flange instead of the auxiliary oil pump as can be seen in figure 2. The diesel fuel is brought to the high pressure pump by a low pressure electric pump 12.

It is important and necessary, for the engine electronic control, to track accurately the crankshaft position. Because the engine was not originally equipped with such a sensor, it was decided to mount one on the shaft of the old injection pump (position 5 in figures 1 and 2).

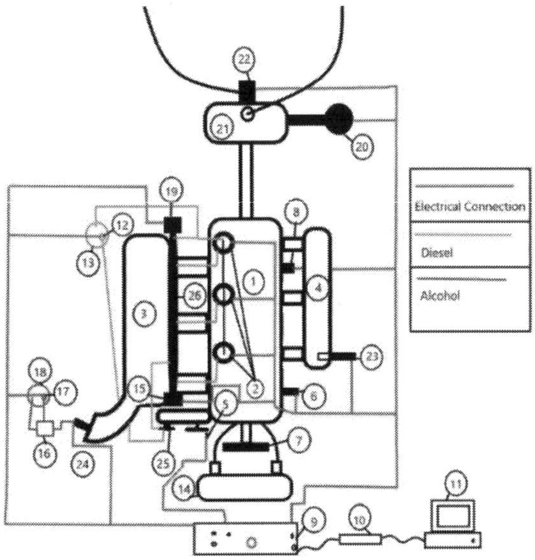

Figure 1. Schematic representation of the experimental test bench: 1 – Engine D115, 2 – Diesel Injectors, 3 – Intake Manifold, 4 – Exhaust Manifold, 5 – Crankshaft Position Sensor, 6 – Temperature Sensor, 7 – Mechanical Fan For Water Cooling, 8 – Oil Pressure Switch, 9 – Signal Conditioning Box/ Control Box, 10 – NI Compact Rio, 11 – Computer, 12 – Diesel Low Pressure Pump, 13 – Diesel Tank, 14 – Radiator, 15 – Common Rail Pressure Sensor, 16 – Alcohol Pressure Regulator, 17 – Alcohol Pump, 18 – Alcohol Tank, 19 – Diesel Pressure Regulator, 20 – Engine Load Sensor/ Force Sensor, 21 – Water Brake, 22 – Water Brake Load Control Valve, 23 – Exhaust Temperature Sensor, 24 – Alcohol Injector, 25 – Diesel High Pressure Pump, 26 – Common Rail.

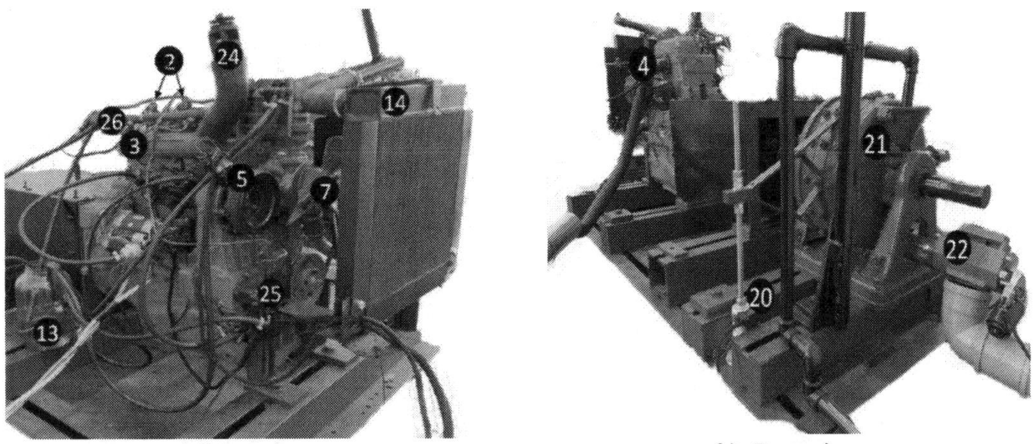

a) Front view b) Rear view

Figure 2. The experimental test bench.

This solution was adopted due to its simplicity. The injection pump shaft makes a complete rotation at two rotations of the crankshaft, so for position tracking is necessary only one sensor. This solution was possible since the movement is transmitted from the crankshaft to the injection pump through timing gears, so there is no phase shift. The used sensor is of Hall type with a digital output and works together with a toothed wheel in a 60-2 configuration. The engine was also equipped with thermocouples and a thermistor for monitoring the fluids (air, fuel, cooling liquid and oil) temperatures. A calibration curve for a thermistor based sensor temperature is presented in figure 3 b.

The hydraulic brake 21, used for ensuring different engine loads, is equipped with a force sensor 20 for measuring the torque and power, and an electronically remote-controlled valve 22 for controlling the engine load. The force sensor is based on two 50kg load cells connected to an amplifier circuit and a 24-bit analog to digital converter. Both, the sensor and the valve bodies are based on 3D printing technology and the electronics is Arduino based [9, 10]. The calibration curve of the force sensor is presented in figure 3 a.

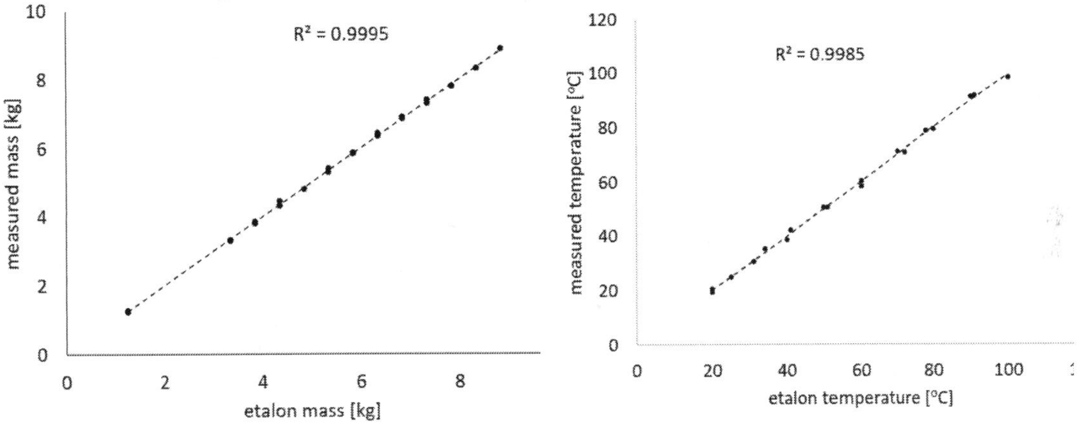

a) Force sensor b) temperature sensors

Figure 3. Calibration curves.

The main target of the test bench is to study the influence of independently and separately injection of alcohol. Therefore, the engine was supplementarily equipped with an alcohol indirect injection system. It is composed of a low-pressure pump 17 a pressure regulator 16 and a gasoline solenoid injector 24. A single injector that injects the alcohol at the entrance of the intake manifold was used. A similar solution was used within the laboratory for studying the influence of water injection on spark ignition engine emissions performances [11].

2.2. Software component

The new injection systems are controlled by a compact-RIO real time control system provided by National Instruments. Its programing is done using LabView, a graphical programing language of the same company.

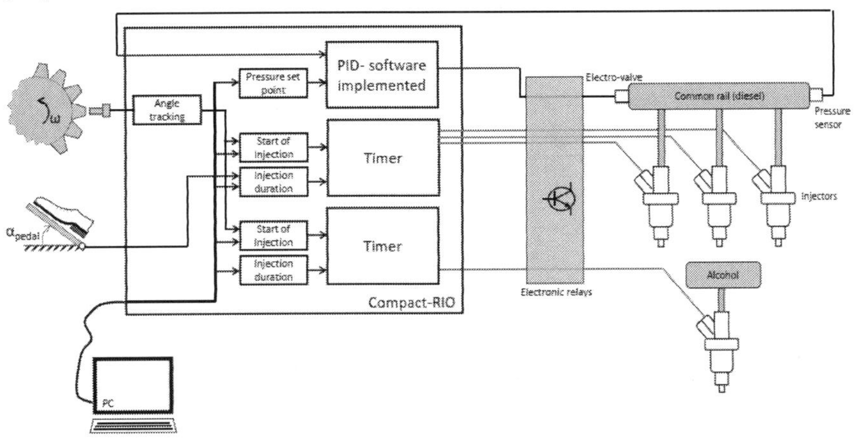

Figure 4. Schematic of the engine control system.

The developed software for the engine monitoring and control and for the brake control was foreseen with a graphical user interface figure 5. This makes it much easier to track and record in real time the functional parameters of the engine. In addition, certain parameters such as injection advance, injection duration, injection pressure, engine load, and so on can be changed from the computer keyboard.

Figure 5. Experimental test bench graphical user interface.

For the engine control, some programs and algorithms in-house developed were used [12]. The Angle tracking module converts the signal from the sensor into an angle in the range of -90 to 630 °CA so the crankshaft position is always known. The conversion is done precisely from 12 to 12 degrees, and in between, the position is obtained by extrapolation taking into account the crankshaft radial speed. As long as the position is known, it is straightforward to command the injectors using software-implemented timers. The injection is triggered by the start of injection, the value imposed by the operator, and it will last a certain amount of time (injection duration). For the diesel injection, the duration is controlled manually from a potentiometer (acceleration pedal). For alcohol injection, the operator via the PC keyboard or mouse imposes its duration. The pressure within the common rail is very sensitive to changes in the engine speed. So a PID controller was used in order to maintain the injection pressure at a certain level imposed by the operator, no matter how the engine speed changes. A schematic presentation of the control system is depicted in figure 4.

3. Experimental results

After all the above-mentioned modifications brought to the engine, the first step was to accommodate the new engine behavior and control. For the beginning, some preliminary experiments were performed in order to see the engine characteristics and to compare the results with the ones presented using the manual and literature. In figure 6 the engine power is presented as a function of the start of injection for different injection pressures and durations.

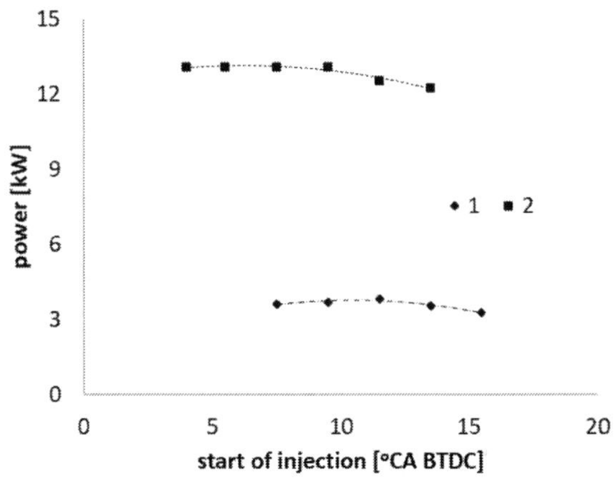

Figure 6. The influence of the injection timing on the engine power – speed 1300 rpm: 1 – injection pressure 300 bar, injection time 1.76 ms, 2 – injection pressure 650 bar, injection time 1.81 ms.

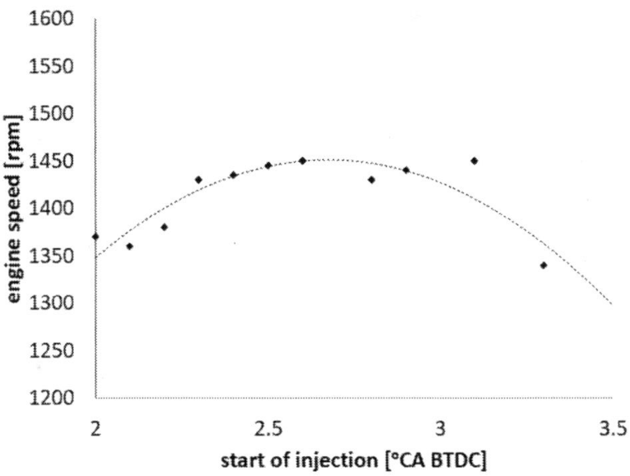

Figure 7. Injection characteristics at idle speed: injection pressure 1100 bar, injection duration 0.7 ms.

As can be seen in the first test, the injection pressure is rather small 300 bar, and we obtain an optimum start of injection somewhere around 11 °CA BTDC. In the second test with higher pressure and slightly higher injection duration, the power increases significantly, and the optimum start of injection is about 6-7 °CA BTDC. In the case of the classical injection system, the injection advance should increase with a load, but in our test, the injection pressure was increased, which probably means a better fuel pulverization and a reduced fuel-air mixture preparation, which lead to a reduced injection advance.

A similar explanation can be provided also in the case of the third experiment presented in figure 7, where can be seen a very small advance of injection but correlated with a high injection pressure and a very low duration (when injection pressure is increasing the injection duration should decree for keeping the same amount of injected fuel).

4. Conclusions

Within the present paper, the D115 diesel engine conversion from a classical injection system to a common rail injection one was presented together with some preliminary tests. The engine and the experimental bench, as a whole, worked as expected. Preliminary tests show that it is now possible to change all the injection parameters from the computer keyboard. In this way in the future, it is possible to ensure the modification of all injection parameters in a wide range. As for the auxiliary injection system in the intake manifold, there were only cold-tests. Alcohol injection experiments have not yet been performed. In the future, we intend to make a characterization of the operation of the engine with the new injection system and to continue with the in-depth study of the effects of alcohol injection.

References

[1] Stan C 2018 *Automobilul, aerul, apa, tuica si Dieselul de Cluj* Bucuresti
[2] Stelmasiac Z 2014 *Application of alcohols to dual – fuel feeding the spark-ignition and self-ignition engines* Polish Maritime Research vol. 21 no. 3(83) pp. 86-94
[3] Tse H, Leung C and Cheung C 2015 *Investigation on the combustion characteristics and particulate emissions from a diesel engine fueled with diesel-biodiesel-ethanol blends* Energy vol. 83 pp. 343-350
[4] Noorollahi Y, Azadbakht M and Ghobadian B 2018 *The effect of different diesterol (diesel biodiesel ethanol) blends on small air-cooled diesel engine performance and its exhaust gases* Energy vol. 142 pp. 196-200
[5] He B Q, Shuai S J, Wang J X and He H 2003 *The effect of ethanol blended diesel fuels on emissions from a diesel engine* Atmospheric Environment vol. 37 pp. 4965-4971
[6] Ferreira V P, Martins J, Torres E A, Pepe I M and De Souza J R 2013 *Performance and emissions analysis of additional ethanol injection on a diesel engine powered with A blend of diesel-biodiesel* Energy for Sustainable Development vol. 17 pp. 649-657
[7] Shahir S, Masjuki H, Kalam M, Imran A and Ashraful A 2015 *Performance and emission assessment to diesel–biodiesel–ethanol/bioethanol blends as a fuel in diesel engines: A review*, Renewable and Sustainable Energy Reviews vol. 48 pp. 62-78
[8] Burnete N V and Barabas I 2015 *Diesel-ethanol blends and their use in diesel engines* Romanian Journal of Automotive Engineering vol. 21 no. 3
[9] Izvorean S, Nechita A, Roteliuc C, Nica C, Cioabla A and Stoica V 2018 *Possible achievements in component manufacturing and their applications in motor vehicles for experimental stands by FDM 3D printing* Nonconventional Technologies Review vol. 22 no. 2
[10] Izvorean S 2018 *Modernizarea standului cu motor D-115* diploma project Timisoara
[11] Borborean A, Stoica V and Lelea D 2019 *Test bench for studying the effects of water injection inside an internal combustion engine* International Conference on Applied Sciences 9–11 May 2018 Banja Luka Bosnia and Herzegovina
[12] Stoica V, Borborean A, Ciocan A and Manciu C 2018 *Minimal algorithm for running an internal combustion engine* International Conference on Applied Sciences (ICAS2017) 10–12 May 2017 Hunedoara Romania

Effectiveness of the measures for CO_2 emission reduction in real world

C Constantin[1], M Bățăuș[2], C Rențea[2] and G Frățilă[2]

[1]University Politehnica of Bucharest, Faculty of Transports, 313 Spl. Independentei, Bucharest, Romania
[2]University Politehnica of Bucharest. Automotive Engineering Department. 313 Spl. Independentei, Bucharest, Romania

rentea.cristian@yahoo.com

Abstract. To reduce the CO_2 emission level for the new Corporate Average Fuel Economy regulations, a series of measures are implemented by the manufacturers. Besides the use of hybrid and electric vehicles, efficiencies packages are frequently applied to improve the CO_2 emissions of conventional vehicles. Some examples of CO_2 reduction measures are: Start&Stop system, eco tires, powertrain thermal management and aerodynamic kits. Usually, they are optimized for the WLTC (Worldwide Harmonized Light Vehicles Test Cycle). The aim of this paper is to assess the influence of these packages in real-world driving cycles. The study is done by means of modelling and simulation, using an industrial modelling and simulation environment (LMS Imagine.Lab AMESim). Two passenger cars from different classes, a compact type and a SUV are used.

1. Introduction

The new CAFE (Corporate Average Fuel Economy) regulations shift the paradigm in the automotive industry towards a more global approach. The car manufacturers need to find the correct balance between the costs and the CO_2 emissions for all the fleet. This type of regulation is promoted by the EU (European Union), by the USA and by other countries worldwide, more notably by China, India and Japan.

European Union legislation sets targets for fleet average CO_2 emissions per kilometer for every manufacturer. The target value for 2021 is 95 g/km and by 2030 the CAFE regulations involve supplementary 30% reductions in CO_2 emissions from this target. The EU Green Deal is expected to impose a supplementary reduction of 10% (to 43-48 g/km) in 2030 [1].

In the EU (European Union) a single gram per kilometer over the average limit is taxed with €95 for every vehicle and the penalties can rise to hundreds of millions of euros for a car manufacturer. This has boosted the implementation of a series of measures for the vehicle and powertrain through which CO_2 emissions can be reduced.

A very efficient but costly method of CO_2 emission reduction is the electrification of the powertrain. Nevertheless, it is expected that by the year 2030 to still have a proportion of 30% of the vehicles running with only an internal combustion engine and the rest to be EV (Electric Vehicles) in a share of 30% and HEV (Hybrid Electric Vehicles) in various configurations for the rest of 40% [2]. The conventional powertrains are expected to be used on the entry-level models in C-segment, MPV (Multi-Purpose Vehicle) or SUV (Sport Utility Vehicle) car classes.

Efficiency packages are frequently applied to improve the CO_2 emissions of conventional vehicles. Some relevant examples considered by the manufacturers are for: engine friction reduction, thermal management improvement, efficient electric energy production and use, vehicle mass reduction, vehicle drag and rolling resistance decrease. The influence of these measures was analyzed in [3] and they demonstrate a good efficiency in WLTC.

This paper investigates the efficiency of CO_2 emission reduction measures for two entry-level passenger cars from different classes: a compact SUV and a compact car (C segment medium car). Both are equipped with the same 1 liter turbocharged gasoline engine and a 5 gears MT (Manual Transmission). The study is done for the WLTC (Worldwide Harmonized Light Vehicles Test Cycle) and CADC (Common Artemis Driving Cycles) using a validated detailed model presented in [4].

2. Model description

LMS Imagine.Lab AMESim disposes of dedicated libraries of components that allow complex studies in most of the technical domains including fuel consumption or CO_2 and emissions studies [5].

To attain the purpose of this study, the model must have the following capabilities:

- Follow an imposed velocity cycle;
- Take account of the engine and gearbox thermal regime;
- Allow for engine stop&start strategies.

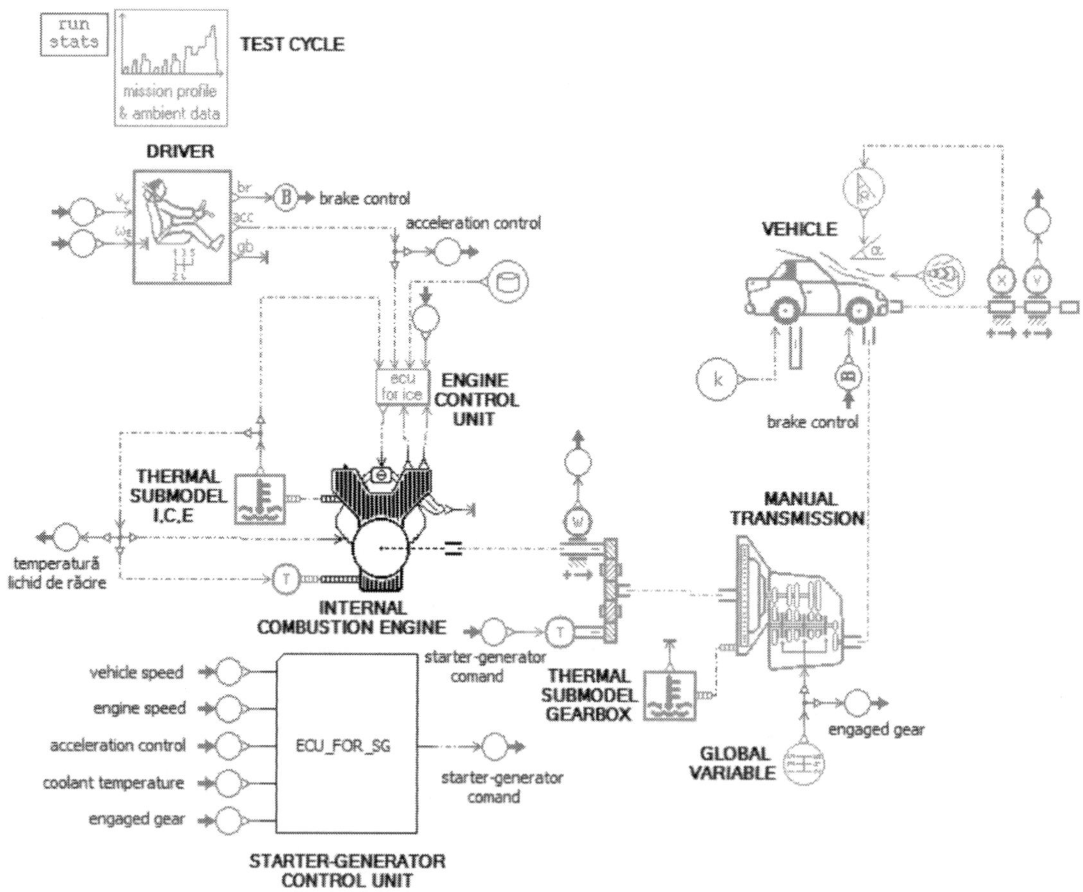

Figure 1. The vehicle and powertrain model for CO_2 emissions computation.

The model developed for CO_2 emissions estimation for an imposed test cycle is based on IFP Drive library but also contains components (submodels) from Mechanical and Signal and Control libraries. It comprises the following main submodels: driver (DRVDRVA00A), mission profile and ambient data (DRVMP2A), internal combustion engine (DRVICE01E), engine control unit (DRVVECU1A), manual transmission with included clutch (DRVMGC01), 1-D vehicle (DRVVEH4A), two thermal submodels (DRVCOO0A), a dedicated supercomponent for the starter-generator (ECU_FOR_SG) as shown in figure 1.

The two thermal submodels (DRVCOO0A) used for the internal combustion engine and for the gearbox are calibrated to compute accurately the components temperatures [4]. The dedicated component ECU_FOR_SG is responsible to generate the torque of the starter-generator and apply the Start&Stop strategy. It is developed using Signal and Control library and encapsulated in a supercomponent for the ease of use.

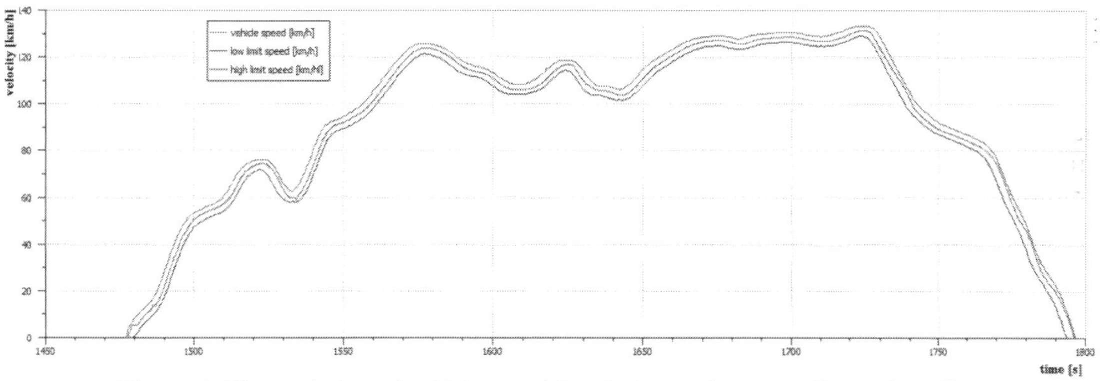

Figure 2. The variation of vehicle speed in relation to the required speed profile.

The model has a total number of 12 state variables and is fast enough to allow for parametric studies. It is a causal model and therefore it is compulsory to check whether the vehicle velocity stays within the speed limits for each simulation. In figure 2 it is presented an example of correct vehicle velocity variation within the limits imposed for phase 4 of the WLTC (determined according to the test regulations [6]).

3. Solutions for CO_2 emission reduction

In [3] several low-cost solutions for CO_2 emission reduction were proposed considering the nature of the studied vehicles (entry-level with conventional powertrains): improved alternator, Start&Stop system, engine rolling bearing, reduced vehicle mass, low resistance tires, aerodynamic improvement, increase of the number of gears.

The alternators are offered in two efficiency ranges: the standard ones have a maximum efficiency rating somewhere in the 55–60% range and the high-efficiency models are generally in the 68–75% range [7].

A Start&Stop system will reduce the engine idle losses but there are inhibitors that reduce the system benefits. The inhibitors are needed to maintain safe use, drivability and comfort.

For the internal combustion engine, there are still resources to minimize the frictional losses. An efficient measure that will also be of interest for mild hybrids is to use rolling bearings instead of plain bearings for the crankshaft. The replacement of the first crankshaft plain bearing with a rolling bearing will reduce the FMEP (Friction Mean Effective Pressure) with 0.06 to 0.14 bar across the entire engine performance map for a 1 liter supercharged gasoline engine [8].

Vehicle mass reduction is a practical measure to reduce the energy needed for vehicle propulsion. A realistic 65 kg mass reduction can be obtained by the elimination of the spare wheel and the use of plastic materials for the rear door (hatch door) construction.

Rolling resistance contributes with approximately 20% to the vehicle fuel consumption. The new eco (low rolling resistance) tires can have a life duration as good as conventional tires with similar traction performances.

The vehicle air resistance can be influenced by the frontal area and the drag coefficient. In this paper, the drag coefficient is reduced for both vehicles by 0.02. This improvement is possible by optimization of vehicles underbody and front air intake [9].

Because the reference models have a 5 gears MT this will be replaced with a 6 gears MT.

The study was done for WLTC and the effectiveness of the proposed improvements was demonstrated. Different cycles will impose different engine operating points in terms of torque-speed distribution and frequency of occurrence. WLTC is a great improvement over NEDC (New European Driving Cycle) in terms of vehicle real usage representativeness but it is still less dynamic than a real driving vehicle velocity profile.

The CADC (Common Artemis Driving Cycles) was developed within the European Artemis (Assessment and Reliability of Transport Emission Models and Inventory Systems) project and is based on a statistical analysis of a large database of European real-world driving patterns. It includes three driving schedules: Urban, Rural road and Motorway [10]. Figure 3 pictures the operating points of the SUV engine for NEDC (in red), WLTC (in green) and CADC (in blue).

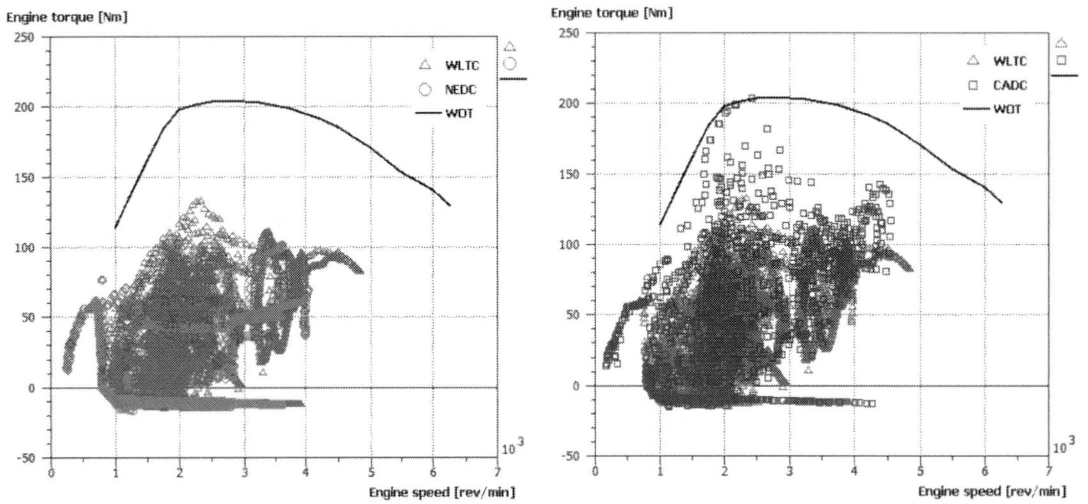

Figure 3. Engine's operating points for NDEC, WLTC and CADC (WOT: wide open throttle).

The value and composition of the traction force are influenced by the vehicle velocity and acceleration levels. It is of interest to study if the efficiency of these measures is maintained for a more realistic usage of the vehicles.

4. Results

The analysis of the CO_2 emissions during each phase of WLTC for the two original configurations confirm that the mean vehicle velocity and the VSF (Vehicle Speed Fluctuation) have a great influence and that is dependent on the vehicle type (table 1).

The parameter VSF (Vehicle Speed Fluctuation) was defined in [11] and can be calculated with relation (1) for each test cycle.

$$\mathrm{VSF} = \frac{\sigma}{\mathrm{m}} \cdot 100\ [\%] \quad (1)$$

where: σ – standard deviation of vehicle speed; m – average vehicle speed.

Table 1. CO_2 emissions for each phase of WLTC and the entire cycle.

No.	Phase	Reference SUV	Reference C-class vehicle	Average velocity	VSF
		CO_2 emission [g/km]		km/h	%
1	Low speed	225.4	220.4	25.3	51.5
2	Medium speed	149.7	138.8	44.5	39.3
3	High speed	148.5	133.6	60.7	43.2
4	Extra High speed	198.9	173.6	94	35.9
5	Total	176.9	160.4	53.3	63.8

Figure 4 presents the CO_2 cumulative emissions in g/km for each phase of WLTC. It can be observed that the difference between the two vehicle types is increasing with the average speed.

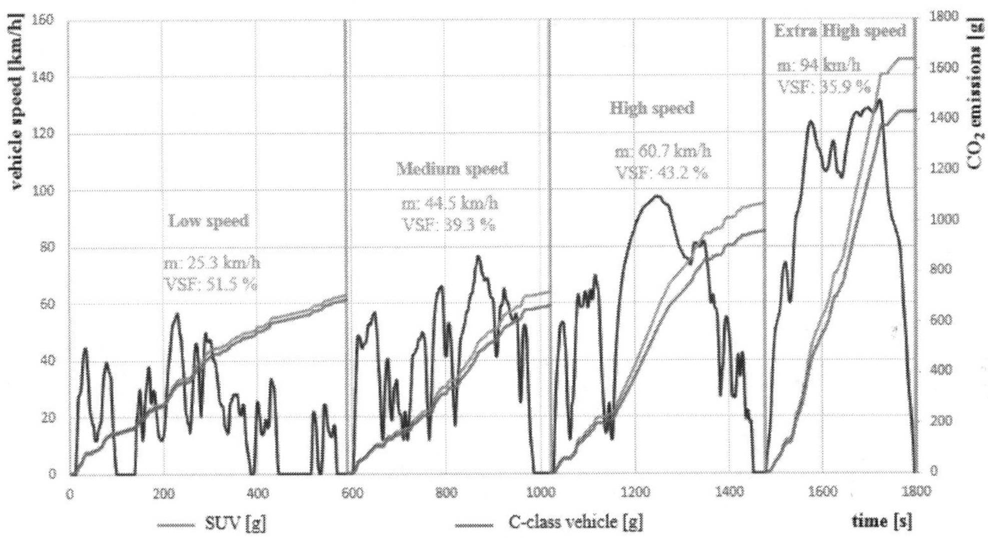

Figure 4. CO_2 cumulative emissions at each phase of WLTC.

For WLTC and CADC, the simulations start with the original configurations and then the measures are implemented step by step. Table 2 presents the CO_2 emissions in g/km for both original (reference) vehicles and with all improvements applied (optimized).

Table 2. CO_2 emissions of reference and optimized vehicles for WLTC and CADC.

No.	Phase	Reference SUV	Optimized SUV	Reference C-class vehicle	Optimized C-class vehicle
		CO_2 emission [g/km]			
1	WLTC	176.9	152.6	160.4	134.2
2	CADC	195.6	167.2	172.6	143.7

When comparing the two vehicles, there is a consistent difference regardless of the optimization and test cycle between 17 to 23 g/km in absolute value (9.3 to 14 %) to the advantage of C-class vehicle.

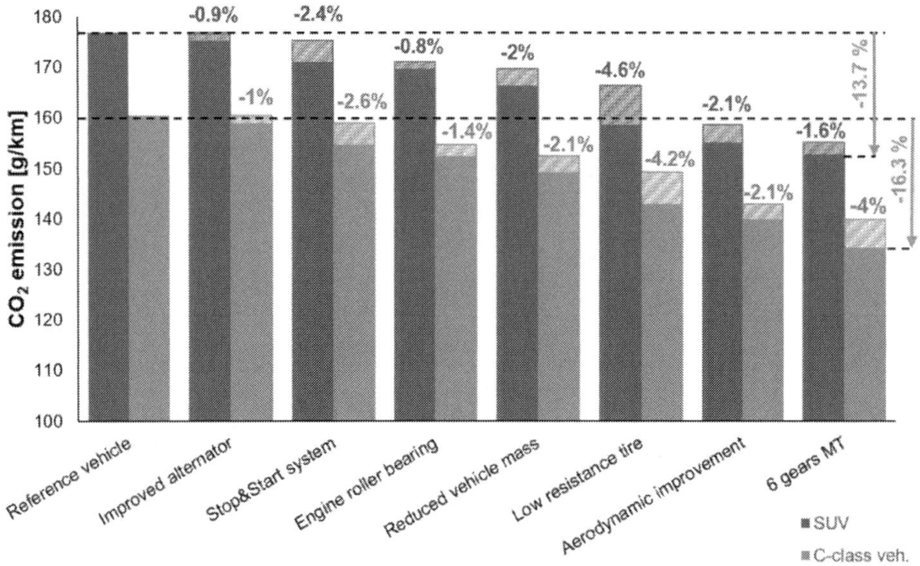

Figure 5. CO_2 emissions reduction in WLTC.

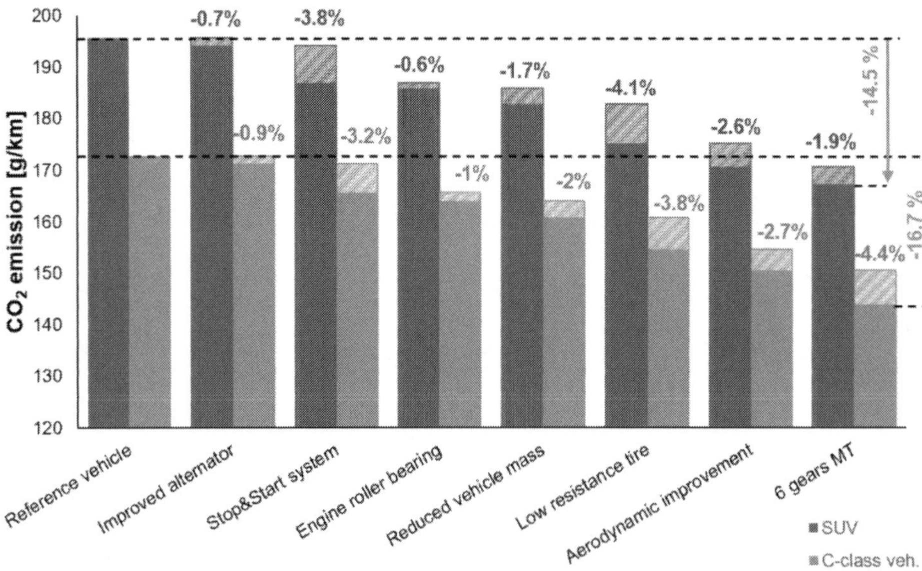

Figure 6. CO_2 emissions reduction in CADC.

For the CADC, there is an increase of CO_2 emission of about 10% for the SUV and 7% for the C-class vehicle. This can be explained by the increased average speed (60.5 km/h) and VSF (73.7 km/h). In figures 5 and 6, the impact of each successive improvement is detailed for the two vehicles and every cycle.

The important decrease of CO_2 emissions was obtained for the SUV and C-class vehicles equally for WLTC (13.7 %, respectively 16.3 %) and CADC (14.5 %, respectively 16.7 %).

To evaluate correctly the impact of the measures, a simulation of vehicle dynamic performance was also done. For the SUV it was determined an increase of 3.5% of maximum velocity and a decrease of 8.9% of the acceleration time from 0 to 100 km/h. For the C-class car, it was determined a small increase of 0.2% of maximum velocity and a decrease of 6.1% of the acceleration time from 0 to 100 km/h.

IOP Conf. Series: Materials Science and Engineering 1220 (2022) 012015 doi:10.1088/1757-899X/1220/1/012015

5. Conclusions

The conventional powertrain will still equip a consistent percent of cars by the year 2030 and CO_2 emission measurement will become more stringent for such vehicles especially for entry-level ones.

Accurate modelling and simulation of the propulsion systems are very important to allow an early prediction of CO_2 emissions for different vehicle configurations.

It was demonstrated that using available technologies, an important reduction of 13.7% to 16.7% for the CO_2 emission may be obtained regardless of vehicle type and adopted test cycle.

The influence of different measures varies between test cycles, but the correct mix will maintain a comparable relative reduction.

The studied measures have a different impact on the two vehicles; the most obvious is the use of a 6 gear MT that has an impact of less than 2% on SUV and over 4% on C-class car.

Low rolling friction tires can have a big impact regardless of the vehicle type and test cycle. However, the efficacy of this method can be maintained during the full vehicle life if it is imposed on the owners by regulation.

The dynamic performances of the vehicles are also improved, from which the acceleration time from rest to 100 km/h is significantly decreased.

References

[1] Taylor J and Bassett M 2021 *Dedicated Hybrid Engines*. Mahle Powertrain Limited

[2] Gutzmer P, Wagner U and Zink M 2018 *Mobility for Tomorrow*, Schaeffler Symposium 2018

[3] Constantin C, Bățăuș M, Rențea C and Frățilă G. 2021 *The use of modelling and simulation to ensure CO_2 emissions regulations*, 25th edition of IManEE 2021 International Conference

[4] Rențea C and Frățilă G. 2021 *The modelling and simulation of propulsion systems to estimate CO_2 emissions* (Ingineria Automobilului, nr. 58)

[5] Siemens Industry Software NV – LMS Imagine.Lab Amesim 2015 *IFP Drive Library 14* (User's guide)

[6] *Commission Regulation (EU) 2017/1151 of 1 June 2017* supplementing Regulation (EC) No 715/2007 of the European Parliament and of the Council on type-approval of motor vehicles with respect to emissions from light passenger and commercial vehicles (Euro 5 and Euro 6) and on access to vehicle repair and maintenance information, amending Directive 2007/46/EC of the European Parliament and of the Council, Commission Regulation (EC) No 692/2008 and Commission Regulation (EU) No 1230/2012 and repealing Commission Regulation (EC) No 692/2008

[7] Bradfield M 2008 *Improving Alternator Efficiency Measurably Reduces Fuel Costs,* White-Paper

[8] Schlerege F 2018 *Breakthrough of Rollerized Crank Shafts* Schaeffler Symposium 2018

[9] Yuan Z and Wang Y 2017 *Effect of underbody structure on aerodynamic drag and optimization*, Journal of Measurements in Engineering, Vol. 5, Issue 3, 2017, p. 194-204. https://doi.org/10.21595/jme.2017.19210

[10] André M 2004 *The ARTEMIS European driving cycles for measuring car pollutant emissions,* Sci Total Environ. 2004 Dec 1; 334-335:73-84. doi: 10.1016/j.scitotenv.2004.04.070

[11] Oota Y 2013 *The advantage of CVT in the real word*, 12th International CTI Symposium (Berlin: Car Training Institute)

A comparative study of calculated and experimental indicated diagrams of a S.I. Engine

V Blaga, T Mitran, A Rus, H Beleş and G Dragomir

Mechanical Engineering and Automotive Department, University of Oradea, 1 University Street, Bihor County, Romania

tudor_mitran@yahoo.com

Abstract. The functioning of the engine in a closed loop with lambda sensor and catalyzer makes the excess air factor to be maintained as close as possible to the value $\lambda=1$. Based on an own model, the authors realized an analytical calculation of the indicated diagram. The indicated diagram was also raised on an experimental stand in p-V coordinates, and the real indicated diagram in p-α coordinates. The equations for the thermal processes that take place in the engine have been determined and, after that, these were introduced in a computer. Using these equations, the state parameters in the characteristic points of the engine cycle were calculated.

1. Introduction

The theoretical and experimental research was done in order to obtain information that can be used to improve the characteristics of a S.I. engine.

The functioning regime of the engine is defined by the rotational speed and by the load, but it assumes the knowledge, in addition to these, of the thermal regime (the parts temperature, the cooling fluid temperature, the intake air temperature, the exhaust gas temperature, altitude corrections and so).

The appreciation of the mixture's quality is made by the air excess factor (λ).

The block scheme of the general model used for the modelling of the gasoline injection systems is presented in figure 1 [1].

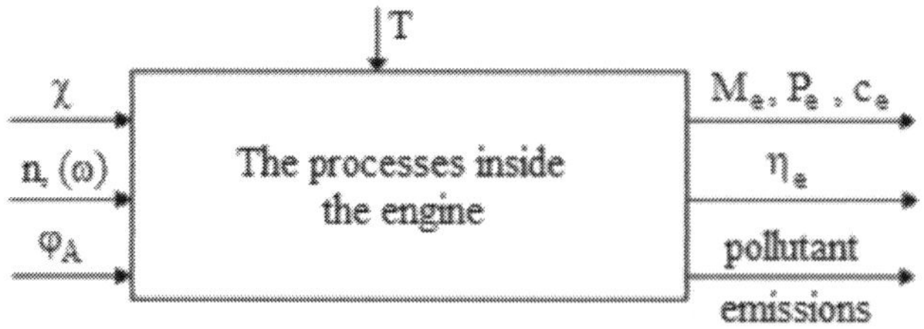

Figure 1. The general model of a S.I. engine.

This includes a system to correlate the quantity of the intake air with the quantity of the fuel injected in one engine cycle. The input parameters of the system are: the load (χ), the rotational speed (n), the angular velocity (ω) and the throttle position (φ_A), and the output parameters are: the effective torque (M_e), the effective power (P_e) and the effective fuel consumption (c_e). The processes inside the engine are influenced by the thermal regime of the engine (T).

Following the studies performed on different injection systems, one can observe that each one, no matter if it's a mechanical one or an electrical one, have their principal scheme based on the functioning of the ICE. The mechanical systems have at the base of the principal scheme two circuits: the fuel circuit, the air circuit and the appropriate corrections. The electronic systems include, in addition, the electrical circuit and the ignition circuit. The main task of the injection equipment is the correlation between the quantity of the intake air and the quantity of the fuel injected per cycle, so that the optimum air excess ratio will result for each functioning regime. This can be realized through the correlation between the quantities of fuel and air and a series of the engine's functional parameters (the pressure in the intake manifold, the throttle position and the engine's rotational speed).

The quantity of the intake air can be expressed depending on the listed functional parameters. When expressing the quantity of fuel depending on these parameters, one must take into account the injection pressure, the flow section through the injector and the injection duration.

Most often it is preferred that the duration of the electromagnetic injector's opening to depend on the depression inside the intake manifold. This is because the quantity of fuel injected per cycle is correlated with the quantity of the intake air per cycle which, as it is known from the mechanical injection equipment depends on the depression inside the intake manifold. In this case, the influence of rotational speed is smaller. The correction for the rotational speed is going to be done at the functioning regimes that require such corrections [2] and [3].

2. The proposed model for the indicated diagram

The scheme of the fuel supply system used for the calculus is presented in figure 2.

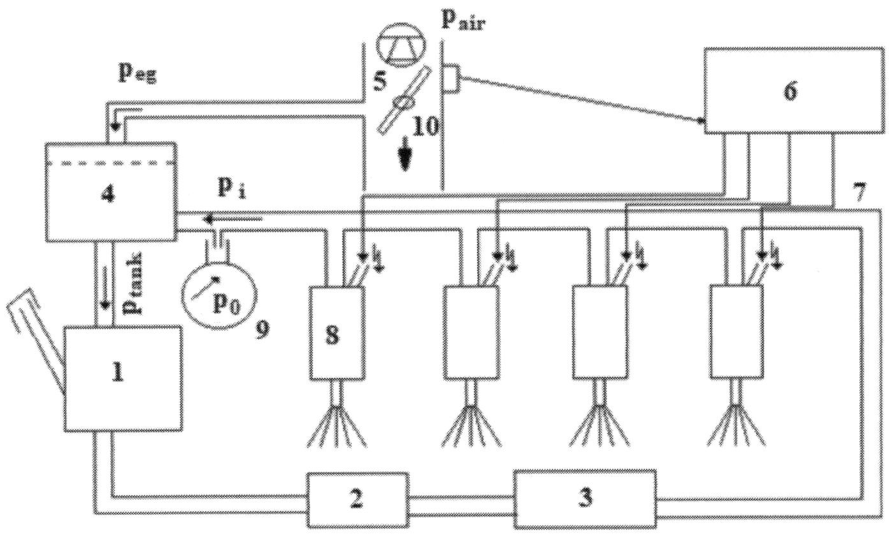

Figure 2. The fuel supply system: 1 – fuel tank; 2 – electric pump; 3 – fuel filter; 4 – pressure regulator; 5 – flowmeter to measure the intake air quantity; 6 – computer; 7 – injection manifold; 8 – electromagnetic injectors; 9 – valve to measure the pressure in the system; 10 – throttle; p_{air} – the environmental pressure; p_{eg} – the pressure of the exhaust gases; p_i – the injection pressure; p_{tank} – the pressure in the fuel tank.

At the electronic injection systems, the dependence between the quantity of the fuel injected per cycle at each functioning regime and the time for the injector's opening is established on the test stand, with command units manually operated, under the criteria of the minimum effective specific fuel consumption, the maximum effective torque and minimum pollutant emissions. After that, the data are memorized in the injection equipment's computer, tabular or under the form of variation curves of the injector's opening time depending on the rotational speed, having as a variable parameter either the depression in the intake manifold or the throttle position. That's why the pressure transducer must have a high precision [3] and [4].

The adjustment of the quantity of fuel injected per cycle for various functioning regimes is made depending on the quantity of intake air, using an air flowmeter provided with a transducer which transmits to the computer information regarding the air flow.

The air flowmeter with paddle and transducer element is replaced by Karman-Vortex type air flowmeter which functions based on ultrasonic waves, which improves the engine's performances. A study for the adoption of the air excess factor (λ) was made. The regulator pressure was determined. Also, the electromagnetic injector modelling was made and the relation of the injection duration was calculated [2].

The modelling of the SI engine cycle proposed by the authors is realized by running a computer program.

The engine cycle proposed by the authors for the analytical calculus is presented in figure 3.

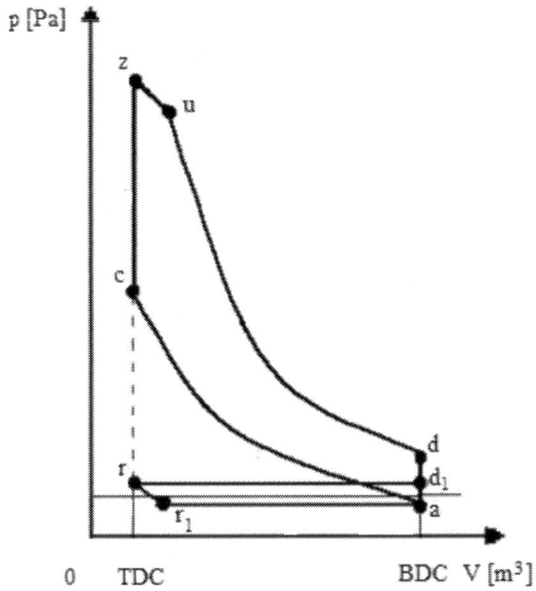

Figure 3. The engine cycle for the analytical calculus [1].

The engine cycle begins in point (r), when the piston is in the TDC, at the beginning of the intake stroke. It is considered that in this point the exhausted valve closes instantly and the intake valve opens instantly at the maximum value. The cycle is based on some assumptions [5] and [6].

The connection between the exhaust valve and the intake valve is made through the isentropic expansion ($r - r_1$). After the opening of the intake valve, the pressure drops from the value corresponding to the pressure of the residual burned gases (p_r) to the value ($p_{r1}=p_a$), equal to the intake pressure (considered to remain constant during the intake stroke). This drop of the pressure appears due to the losses at the fresh charge flow through the orifice controlled by the intake valve. Being a normally aspired engine, the pressure during the intake stroke (p_a) is lower than the environmental pressure (p_0) with a value (Δp_a), equal to the total pressure losses inside the intake manifold.

The compression is considered to be a polytropic pressure with constant exponent (n_c). It is considered that $n_c<k_c$, where (k_c) is the adiabatic exponent of the compression process. It starts in BDC at the end of the intake stroke (point a) and ends at the end of the compression stroke (point c).

The burning process takes place in two phases:

- the isochoric burning (c-z);
- the afterburning (z-u); this is considered to be a polytropic process of exponent $n_u<1$.

The assimilation of the afterburning with a polytropic process of subunit exponent is based on experimental data [7] that shows, that the final temperature of the burning (T_u) is higher than the temperature at the end of the isochoric burning (T_z) because of the heat released by burning of the last quantities of fuel. So, by combining the afterburning of a subunit exponent and the actual expansion of an exponent $n_{d1}>k_d$, one can obtain the real expansion (z-d), characterized by a polytropic exponent $1<n_d<k_d$.

The expansion process (u-d) is considered to be a polytropic one, of constant exponent $n_d>k_d$. Where (k_d) is the adiabatic exponent of the expansion process.

The exhaust process includes two phases:

- the free exhaust, when the burned gases exit the cylinder because of the difference between the superior pressure inside the burning chamber and the pressure inside the exhaust manifold. The pressure suddenly drops from the value (p_d) to the value (p_{d1});
- the forced exhaust takes place during the exhaust stroke, and it is considered to be at constant pressure $p_r=p_{d1}$); the pressure (p_r) is higher than the environmental pressure with a value (Δp_r) equal to the total pressure losses in the exhaust manifold.

The initial data for the computation are the following: $D=77\cdot 10^{-3}$ m – the cylinder bore; $S=83{,}6\cdot 10^{-3}$ m the piston stroke; $V_s=0.389\cdot 10^{-3}$ m^3 – the swept volume; R=290 J/kg – the working fluid constant; $Q_i=44\cdot 10^6$ J/kg – the net calorific value; $L_0=15$ kg air/kg fuel – the stoichiometric air requirement; $T_0=293$ K – the environmental temperature; $p_0=1\cdot 10^5$ Pa – the environmental pressure; $\lambda=1$ – the excess air factor; $\eta_{ar}=0{,}9$ – the burning process efficiency; $\xi_0=0{,}8$ – the heat release coefficient; $\xi_{ga}=2$ the gas-dynamic resistance factor of the intake manifold; $\rho_0=1{,}177$ kg/m^3 – the intake air density; $d_0=0{,}42\cdot D$ [m] – the inner diameter of the intake manifold at the valve port; n=500 ... 5500 rot min (with a variation from 100 to 100 rot/min) – the crankshaft rotational speed; $W_{pm}=S\cdot n/30$ [m/s] – the mean piston speed; $n_u=0{,}9$ – the polytropic coefficient of the afterburning process; $\eta_p=0{,}96$ – the plenitude coefficient of the engine cycle; $p_r=1{,}13\cdot 10^5$ Pa – the residual exhaust gas pressure (the pressure in point r).

The main notations used at the modelling are: $\alpha=p_z/p_c$ – the pressure raise ratio in the isochoric phase of the burning process; $\delta=V_u/V_z$ – the volume raise ratio in the afterburning process; $\varepsilon_1=V_u/V_z$ – the volume raise ratio during the expansion process; $\psi_0=\Delta p_a/p_0$ – the relative pressure drop coefficient during the intake process; $\psi_1=\Delta p_r/p_r$ – the relative pressure drop coefficient during the exhaust process; $\psi=p_a/p_r$ – the global coefficient of the pressure losses; $\varphi_r=T_{d1}/T_r$ – the ratio between the temperature at the end of the forced exhaust and the temperature at the end of the free exhaust.

Below is presented the algorithm for the calculation of the state parameters of the working gas in the characteristic points of the engine cycle: point (a) – the end of the intake stroke; point (c) – the end of the compression process; point (z) – the end of the isochoric burning process; point (u) – the end of the afterburning process, point (d) – the end of the expansion process; point (d_1) – the end of the free exhaust process.

The logical scheme for the analytical calculation is presented in figure 4.

The computation program is structured on 10 procedures and functions. In addition to the values of the constants declared at the beginning of the program, one also considers as initial data, assumed known arbitrarily chosen from statistic data of the S.I. engine cycle: $T_{a0}=322$ K, $T_{z0}=2530$ K, $T_{u0}=2660$ K, $k_{c0}=1{,}3$, $k_{v0}=1{,}3$, $k_{u0}=1{,}2$, $k_{d0}=1{,}3$ and $k_{e0}=1{,}3$. Without these, it is not possible to calculate, generally speaking, all the other parameters that characterize the engine cycle. Thus, this initial data will play the role of parameters, the variables being explicitly the temperature between these thermal processes evolve. These temperatures are closely depending on the adiabatic coefficients which, in fact, are stabilized by the completion for several times of the engine cycle until these coefficients become constant, with an error of 0,000009. The decision for the exit from the cycle for a particular rotational speed is given by the decreasing of the constant error of the intake temperature (T_a) under the value 1,5 K.

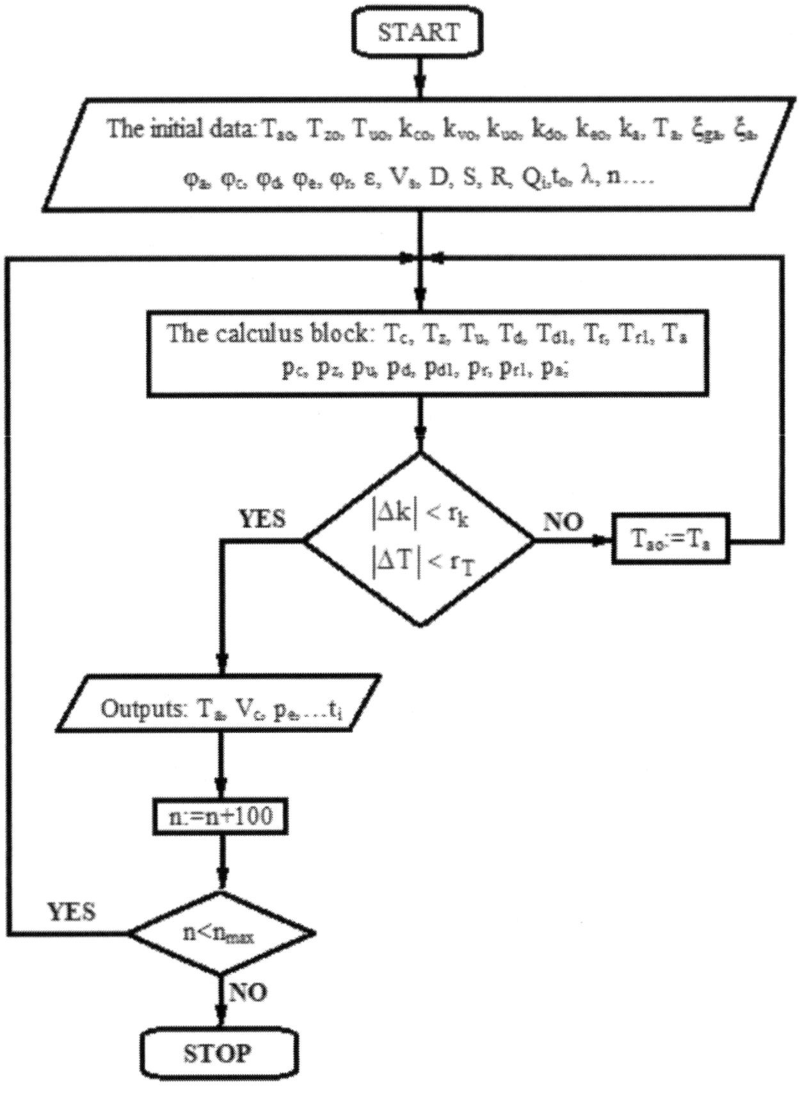

Figure 4. The logical scheme for the analytical calculus.

The computational algorithm consists of the following steps:

1) The temperatures T_{a0}, T_{z0}, T_{u0} and the adiabatic exponents k_{c0}, k_{v0}, k_{u0}, k_{d0}, k_{e0} are taken as initial parameters.
2) The engine cycle is completed, having as results new values for the initial parameters from point 1.
3) The differences between the old values and the new values (ΔT and Δk) are calculated.
4) If the differences (ΔT and Δk) are smaller than the imposed errors, the new calculated values are valid.
5) If the differences (ΔT and Δk) are greater than the cycle is completed once more, by recalculating the values of the temperatures and of the adiabatic exponents. This step continues until the differences drop under the imposed value for the errors.

One must notice that the program generally computes the parameters in the characteristic points of the engine cycle with the restriction of the energetical efficiency of the burning process (η_{ar}) and of the heat utilization coefficient in the isochoric burning phase (ξ_0).

The pressure at the intake valve port [1] and [5]:

$$p_{ga} = p_0 - k_1 \cdot (1+\xi_{ga}) \cdot n^2 \cdot \left(\frac{V_s}{d_0^2}\right)^2 \; [Pa]; \; k_1 = \frac{1}{1+(1+\xi_{ga}) \cdot \frac{k_a}{2} \cdot \left(\frac{4}{30 \cdot \pi}\right)^2 \cdot \left(\frac{V_s}{D^2}\right)^2 \cdot \left(\frac{n}{a_{sa}}\right)^2} \quad ; (1)$$

where: a_{sa} [m^2] – the effective flow area through the orifice controlled by the intake valve; k_a=1,4 – the adiabatic exponent of the intake process.

2.1. The calculus of the state parameters at the end of the intake stroke

The pressure at the end of the intake stroke [1] and [8]:

$$p_a = p_{ga} - k_1 \cdot n^2 \cdot \frac{V_s}{d_0^4} \cdot \left[(1+\xi_{ga}) \cdot \left(\frac{d_0}{D}\right)^4 - 1\right] [Pa] \quad ; \quad (2)$$

The temperature at the end of the intake stroke:

$$T_a = T_0 \cdot \frac{\varepsilon}{\varepsilon - 1} \cdot \frac{1-\psi_0}{\eta_V} \cdot \left\{1 - \frac{\varphi_r}{\varepsilon} \cdot \left[\frac{(\varepsilon/\delta)^{n_d} \cdot \delta^{n_u}}{\psi \cdot \alpha \cdot \varepsilon}\right]^{1/k_e}\right\} [K] \quad ; \quad (3)$$

where: - ε – the compression ratio;

$$\eta_V = (1-\psi_0) \cdot \frac{\varepsilon}{\varepsilon - 1} \cdot \left[1 - \left(1 + \frac{1}{n_a} \cdot \frac{k_a - k_e}{k_e - 1}\right) \cdot \frac{1}{\varepsilon} \cdot \frac{1}{\psi^{1/k_e}}\right]; \quad (4)$$

The polytropic exponent of the intake process:

$$n_a = 1 + \frac{k_a - 1}{\varphi_a} \quad ; \quad (5)$$

where: φ_a=1,05 – a coefficient that takes into consideration the heat exchange between the cylinder's walls and the working fluid during the intake process.

The adiabatic exponent of the exhaust process:

$$k_e = 1{,}268 - \frac{0{,}035}{\lambda} + \frac{126}{T_d + T_\eta} \quad ; \quad (6)$$

$$\delta = \left[1 + \frac{\varepsilon - 1}{\varepsilon} \cdot \frac{1 - n_u}{p_0 \cdot (1-\psi_0)} \cdot \frac{k_u - 1}{k_u - n_u} \cdot \frac{q_u}{\alpha}\right]^{1/((1-n_u))} \quad (7)$$

where: $q_u = Q_u/V_s$ [J/m^3] – the unitary heat input that characterizes the afterburning.

The polytropic exponent of the compression process (n_c) can be express as a function of the adiabatic exponent of the compression process (k_c) and of the coefficient that takes into consideration the heat exchange between the working fluid and the cylinder's walls during this process (Φ_c=1,06) [1] and [6]:

$$n_c = 1 + \frac{k_c - 1}{\varphi_c}; \; k_c = 1{,}438 - 0{,}525 \cdot 10^{-4} \cdot (T_a + T_c) \quad (8)$$

where: k_c – the adiabatic exponent of the compression process; T_c [K] – the temperature at the end of the compression process.

The polytropic exponent of the expansion process:

$$n_d = 1 + \frac{k_d - 1}{\varphi_d}; \; k_d = 1{,}268 + \frac{0{,}035}{\lambda} + \frac{126}{T_u + T_d} \quad \text{[1] and [9]} \quad (9)$$

where: φ_d=0,92 [10] – a coefficient that takes into consideration the heat exchange between the cylinder's walls and the working fluid during the expansion process; T_u [K] – the temperature at the end of the isochoric burning phase; T_d [K] – the temperature at the end of the expansion process.

$$\alpha = 1 + \frac{\xi_0 \cdot q_{cb} \cdot (\varepsilon - 1) \cdot (k_v - 1)}{p_c} \tag{10}$$

where: $\xi_0=Q_v/Q_{cb}$ – the utilization coefficient in the isochoric burning phase;
Q_v [J] – the heat released in the isochoric burning phase;
Q_{cb} [J] – the total heat released by the burning of the fuel inside the cylinder;
$q_{cb}=Q_{cb}/V_s$ [J/m^3] – the cylinder's thermal input;
k_V – the mean adiabatic exponent, that characterizes the temperature interval $T_c - T_z$;

The volume at the end of the intake process is equal to the maximum volume of the burning chamber: $V_a=V_c+V_s$ [m^3]; $V_c=V_s/(\varepsilon-1)$ [m^3].

2.2. The calculus of the state parameters at the end of the compression process
The pressure and the temperature at the end of the compression process:

$$p_c = p_a \cdot \varepsilon^{n_c} \left[Pa\right]; T_c = T_a \cdot \varepsilon^{n_c - 1}\left[K\right] \tag{11}$$

The volume at the end of the compression is equal to the minimum volume of the burning chamber V_c.

2.3. The calculus of the state parameters at the end of the isochoric burning
Because of the isochoric process, the pressure and temperature at the end of the burning are:

$$p_z = \alpha \cdot p_c \left[Pa\right]; T_z = \alpha \cdot T_c \left[K\right] \tag{12}$$

The volume at the end of the isochoric burning phase is equal to the minimum volume of the burning chamber (V_c).

2.4. The calculus of the state parameters at the end of the afterburning
The temperature, the pressure and the volume at the end of the afterburning:

$$T_u = T_z \cdot \delta^{1-n_u}\left[K\right]; p_u = p_z \cdot \delta^{n_u}\left[Pa\right]; V_u = V_c \cdot \delta\left[m^3\right] \tag{13}$$

2.5. The calculus of the state parameters at the end of the expansion process
The temperature and the pressure at the end of the expansion process:

$$T_d = T_u \cdot (\delta / \varepsilon)^{n_d - 1}\left[K\right]; p_d = p_u \cdot (\delta / \varepsilon)^{n_d}\left[Pa\right] \tag{14}$$

The volume at the end of the expansion process is equal to the maximum volume of the burning chamber (V_a).

2.6. The calculus of the state parameters at the end of the free exhaust process
The temperature at the end of free exhaust process:

$$T_{d_1} = T_d \cdot \left[\frac{(\varepsilon / \delta)^{n_d} \cdot \delta^{n_u}}{\psi \cdot \alpha \cdot \varepsilon^{n_c}}\right]\left[K\right] \tag{15}$$

The pressure at the end of the free exhaust process is (p_r). The volume at the end of the free exhaust process is equal to the maximum volume of the burning chamber (V_a).

2.7. The calculus of the state parameters at the end of the forced exhaust process
The temperature at the end of forced exhaust process:

$$T_r = T_{d_1} / \varphi_r \left[K\right] \tag{16}$$

The pressure at the end of the forced exhaust process is (p_r). The volume at the end of the forced exhaust process is equal to the minimum volume of the burning chamber (V_c).

2.8. The calculus of the state parameters in the point (r_1)

The temperature in the point (r_1):

$$T_{r_1} = T_r \cdot \psi^{k_e - 1/k_e} \left[K\right] \tag{17}$$

The pressure in the point (r_1) is (p_a).

Figure 5. Part of the calculation program listing.

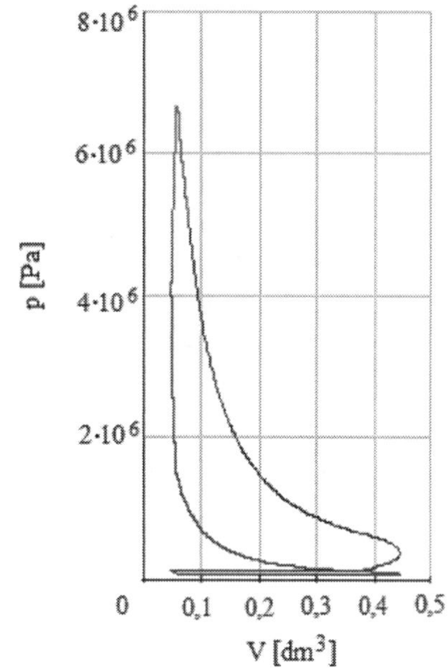

Figure 6. The calculated indicated diagram.

The computation program was realized in Mathcad and part of it is shown in figure 5. The calculated indicated diagram, obtained with this computer program, is presented in figure 6.

3. The real indicated diagram

The general configuration of a test bed for the measurement of the parameters of a S.I. engine consists, mainly, of:

- the foundation with a pate and frame on which the engine is fixed;
- the braking device, that can be hydraulic or electric;
- the fuel supply system and the devices that can measure the fuel consumption;
- the engine's starting system;
- the exhaust gas system;
- the system for the oil consumption;
- the system for the air and of the cooling liquid flow measurement;
- the devices for the measurement of the pressures and of the temperatures;

The test bed on which the experimental data were acquired is provided with an electric machine with eddy currents of a W130 Shenk type.

The data acquisition system sums 32 measurement lines to measure the pressure, the temperature, the rotational speed, the torque, the environmental parameters and so. The data acquisition and processing are automatically made by using a processes computer that is provided with a set of programs necessary for the execution of the set of especially programs to emit the commands necessary to execute the probs included in the testing procedures and for the automatic processing of the data recorded during the testing.

The transducers and the experimental devices are of:

- a resistive type – their verification was made with testers to verify the resistance depending on the cooling liquid temperature and on the intake air temperature;
- an inductive type, in which a magnetic field created by a permanent magnet has a good magnetic conductibility. The variation of the magnetic field induces in the electromagnetic coil an electric current of voltage (V), that is directed through a cable to the electronic command unit.

The setup of the test bed is presented in figure 7.

Figure 7. The test bed used to obtain the indicated diagram.

The parameters that were measured in order to raise the real indicated diagram are presented in table 1.

Table 1. The parameters that were measured in order to raise the real indicated diagram.

Parameter	Symbol and measurement unit
The rotational speed	n (rot/min)
The hourly fuel consumption	C_f (kg/h)
The air consumption	C_{air} (kg/h)
The intake air temperature	T_{air} (K)
The cooling liquid temperature	T_{cl} (K)
The exhaust gas temperature	T_e (K)
The fuel temperature	T_f (K)
The environmental pressure	p_0 (Pa)
The pressure in the exhaust manifold	p_e (Pa)
The pressure in the intake manifold	p_{im} (pa)
The pressure drop in the air filter	Δp_{af} (Pa)
The ignition advance	β (deg)
The pressure inside the cylinder	p_{cyl} (Pa)

The pressure inside the cylinder was measured using a pressure sensor for combustion analysis AVL GH 14 DK. Its characteristics are presented in figure 8.

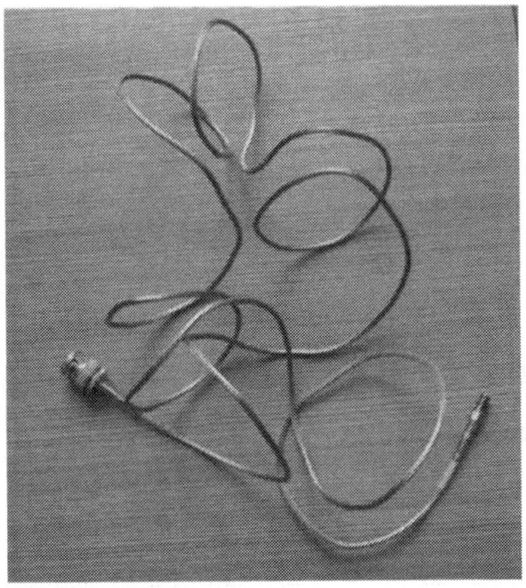

Specifications				
Measuring range			0...300 bar	
Overload			350 bar	
Lifetime	≥		10^8	load cycles
Sensitivity			19 pC/bar	nominal
Linearity	≤	±	0.3%	FSO
Natural frequency	~		170 kHz	
Acceleration sensitivity	≤		0.0005 bar/g	axial
Shock resistance	≥		2000 g	
Insulation resistance	≥		10^{13} Ω	at 20 °C
Capacitance			7.5 pF	
Operating temperature range			-40 ... 400 °C	
Thermal sensitivity change	≤		2 %	20 ... 400 °C
	≤	±	0.5 %	250 ± 100 °C
Load change drift			1.5 mbar/ms	max. gradient
Cyclic temperature drift *	≤	±	0.7 bar	
Thermo shock error **				
Δp	≤	±	0.4 bar	
Δp_{mi}	≤	±	2 %	
Δp_{max}	≤	±	1.5 %	
Thread diameter			M5x0.5	front sealed
Cable connection			M4x0.35	negative
Weight			2.2 grams	without cable
Mounting torque			1.5 Nm	

*) at 7 bar IMEP and 1300 rpm, diesel
**) at 9 bar IMEP and 1500 rpm, gasoline

Figure 8. The pressure sensor used for tests and its characteristics.

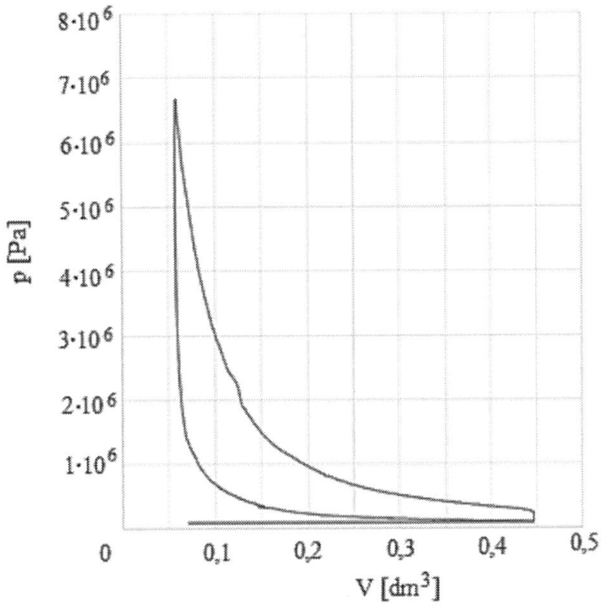

Figure 9. The real diagram (obtained on the test bench).

The real diagram, obtained on the test bench, is presented in figure 9.

4. Conclusions

The modelling of the SI engine cycle proposed by the authors consists in the presentation of the initial data of the computing program, the calculation and the correlation between the parameters of the engine for the realization of computing programs. The proposed model is a helping cycle to simulate the gasoline injection. The computer simulation permits the determination of the theoretical technical-economical parameters: the mechanical work corresponding to the rounded diagram, the indicated mean pressure, the indicated efficiency and the indicated specific fuel consumption.

After the calculation of the mechanical and the pumping losses, one can calculate the engine's effective technical-economical parameters. Using the model proposed in this paper, one can calculate the fuel mass injected per cycle, depending on the rotational speed at total load, the fuel flow through the electronic injector, reaching the purpose pursued, the determination of the injection duration depending on the rotational speed, at total load.

The change of the environmental state conditions affects the intake process. In this way, changes occur in the development of the burning process, because the state conditions of the initial mixture are modified. As a result, the volumetric efficiency, the air excess factor, the indicated and the effective efficiency, that is all the factors that decide the level of the power and of the specific fuel consumption will suffer deviations from their optimal values. In order to establish optimal values for the measurements, one must consider the conditions in which these measurements are made and the net calorific value of the fuel.

In table 2 is presented a comparison between the values of the calculated pressures in the characteristic points of the engine cycle.

Table 2. The comparison between the calculated and the real pressures in the characteristic points of the engine cycle.

	The calculated pressure (Pa)	The measured pressure (Pa)
p_a	87000	87000
p_c	1944000	1999000
p_z	666200	6664000
p_u	499700	5072000
p_d	376500	375000
p_r	112000	122000

It can be observed that the differences between the two data sets are in the interval 0-8,9 %.

The calculated value of the pressure at the end of the intake process is equal to the measured one.

The pressure measured at the end of the compression stroke is 2.8% greater than the one calculated. This difference is due to the fact that, in the proposed model, one considers that the burning begins at the end of the compression stroke (the piston in TDC). In reality, the triggering of the electric spark takes place in advance regarding the TDC, and the pressure raise is greater than that in the cycle without burning. However, the pressure difference in this point is small, because in this part (to the end of the compression stroke) only small quantities of fuel are burning and, because of that, no important heat quantity is released.

Regarding the difference between the calculated and the measured value of the pressure at the end of the isochoric burning, this is of only 0,03%. The measured value is still greater than the calculated one.

At the end of the afterburning, the measured is 1.5% greater than the calculated one. This can be explained by the fact that, because the assumption that the burning process begins in TDC, in the theoretical cycle the quantity of fuel burned in the isochoric phase is greater than the real one. So, in the afterburning, in the theoretical cycle remains a smaller quantity of fuel to be burned.

Conversely, at the end of the expansion stroke, the measured value of the pressure is smaller than the calculated one. This is because, in the theoretical cycle, one considers that the exhaust vale opens at the end of the expansion stroke (BDC). In reality, the exhaust vale opens before the pistons reach the TDC.

The greatest difference is registered in the case of the exhaust pressure (8.9%). Probably this difference appears because the pressure is not measured in this region, but is estimated based on the minimum, the maximum and the mean pressures in the characteristic points of the cycle.

The values of the calculated temperatures in the characteristic points of the engine cycle were compared to the ones from statistical data. It was observed that the calculated values are inside the recommended intervals.

For a better comparison between the calculated and the experimental data, the two diagrams were overlapped in figure 10.

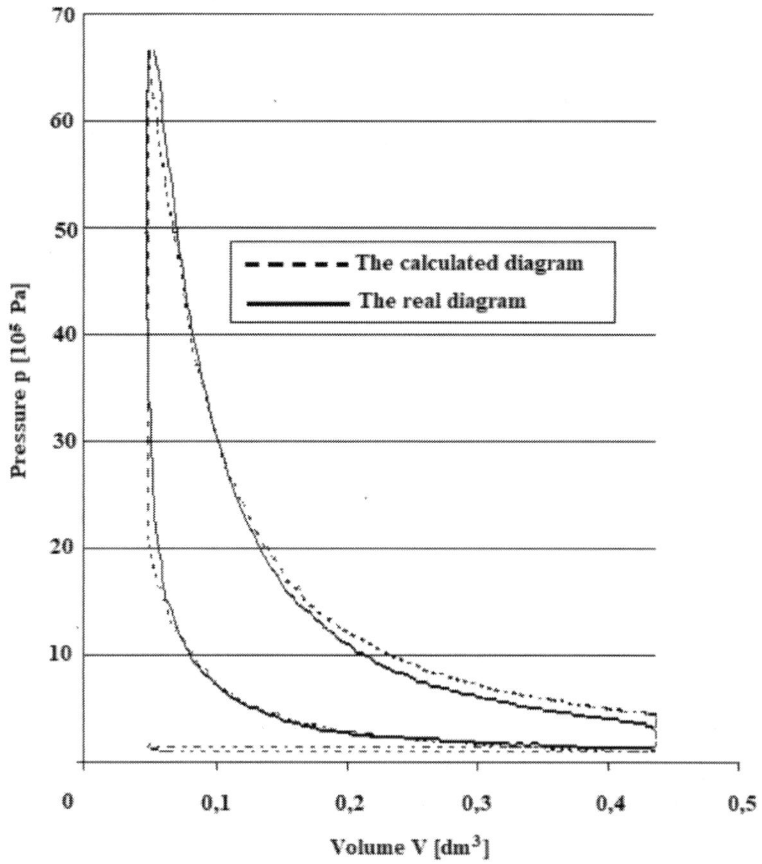

Figure 10. The overlap of the calculated and the measured indicated diagram.

In figure 10 it can be observed that, at the end of the compression stroke, the difference between the calculated and the measured values is the maximum (maximum 18,75%), but the area between the two curves is very small. As it was mentioned before, this difference is due to the fact that in the theoretical cycle it is considered that the burning process begins at the end of the compression stroke, in TDC.

A relatively big difference (maximum 4,5 %) is registered also at the beginning of the expansion stroke, corresponding to the afterburning in the theoretical cycle. This difference can be explained by the fact that the estimated burned fuel quantity in the theoretical cycle is greater than the real one, also because of the assumption that the burning process begins in the TDC. So, the quantity of the fuel that is estimated to be burned in the afterburning in the theoretical cycle is smaller than the real one, and the calculated pressures are smaller than the real ones.

There are some differences between the calculated and the measured pressures in the second part of the expansion stroke, where the calculated values are bigger than the measured ones (around 3%). The error can be lowered by changing the exponent of the expansion process.

The practical application of this calculus involves, on one hand, the perfectioning of the physical-mathematical model, so that it will be as close as possible to the real development of the gasoline injection process.

This means the reduction of the theoretical assumption and the calibration of some of the computation parameters, based on the experimental values.

A theoretical model to compute with sufficient precision the thermal processes that take place in a SI engine can be useful for the development of new propulsion systems for road vehicles.

Many think that the transition to the full electric vehicles is a hybrid propulsion system. A variant is the inline hybrid propulsion system. In this case, the thermal engine should provide the energy to power the traction electric motor. The main advantage is the fact that the thermal engine can function at a single regime. This means that the adjustments that are necessary for the functioning at different regimes are no longer necessary.

If it is possible the development of a model for the study of the ICE functioning at low or medium loads, it will be possible to determine the parameters that ensure minimum fuel consumption and polluting emissions at a specific functioning regime.

So, the model proposed in this paper, which proved to have a very good precision (and that can be improved) can be used to study the functioning of SI engines in different conditions.

The program was used for the study of other normally aspirated SI engines with gasoline injection. The authors intend to use this model for turbocharged engines also.

References

[1] Blaga V 2005 *Engines with Gasoline Injection* (Oradea: University of Oradea Publishing House)

[2] Delanette M 1989 *Les Motors a Injection* (Edition Tehniques pour L'automobile et L'industrie)

[3] Grunwald B 1980 *The Theory, the Calculus and the Constuction of the Engines for Road Vehicles* (Buchareŕst: Didactic and Pedagogical Publishing House)

[4] Mitran T, Blaga V and Moca S 2017 *The Calculation Algorithm for the Determination of the Intake Stroke for GDI Engines* (Brasov: International Congress of Automotive and Transport Engineering) pp 311-318

[5] Blaga V 2000 *Contributions to the Modelling of the Gasoline Injection at the SI Engines* Cluj-Napoca: PhD Thesys)

[6] Blaga V, Beles H, Dragomir G, Vlad I, Vlad M and Timar M 2009 *The Theoretical Contribution Modelling Gasoline Injection at Spark Lightening* (Viena: DAAM International Scientific Book) pp 43-58

[7] Radcenko V 1977 *Criteria for the Optimization of the Thermal Processes* (Bucharest: Technical Publishing House)

[8] Blaga V 2013 *Processes and Characteristics of the Internal Combustion Engines* (Oradea: University of Oradea Publishing House)

[9] Blaga V, Bara L, Bara C, Bara V and Domuta C 2008 *The Experimental Stand of Engines with Gasoline Injection* (Viena: DAAM International Scientific Book) pp 109-110

IOP Conf. Series: Materials Science and Engineering 1220 (2022) 012017 doi:10.1088/1757-899X/1220/1/012017

Aspects of electric transmission implementation on a battle tank

C O Ilie[1], L Barothi[1], M Marinescu[1] and I T Giurgiu[2]

[1]Military Technical Academy, 39th-49th George Coşbuc Blvd. Bucharest, Romania
[2]Training School for Communications, Information Technology and Cyber Defense, 3rd-5th Vasile Milea Blvd, Sibiu, Romania

constantin.ilie@mta.ro, laszlo.barothi@mta.ro, marin.marinescu@mta.ro, giurgiutiberiu@yahoo.com

Abstract. In the paper, a high performance electric transmission designed for heavy military vehicles is presented. The power source is a 1300 hp gas turbine and a 200 hp supercharged diesel engine. Both engines drive a synchronous generator. The electric current supplied by the generator is rectified. Each sprocket is mounted on the output shaft of a gearbox driven by an asynchronous electric motor. An Insulated Gate Bipolar Transistors (IGBTs) inverter control AC motors in a separate way to achieve dynamic performance. Diesel engine alone is used for driving on paved roads with speeds of up to 15 km/h with low fuel consumption. Gas turbine and diesel engine are automatically coupled to the generator to reach the maximum torque and dynamic performances. The tank can thus reach a maximum speed of 70 km/h.

1. Introduction

A tank may be efficient the longer it survives on the battlefield. It must be extremely mobile and noiseless. Also, it must have low fuel consumption, reduced maintenance costs and flexibility in traction system layout. As Brian R. Zahn [1] indicated in the strategically report about the future combat system, "reducing the vehicle's size, camouflaging its silhouette, attenuating engine and drive train noise, and reducing exposure time have become compulsory design factors for future combat vehicles". Shorter exposure time is achieved by being agile on the battlefield, moving faster in different directions. A high-performance transmission and a high power to weight ratio (specific power) should provide military vehicles a good behaviour. Propulsion efficiency is strongly affected by the ground-track interaction and the increased vehicle lifecycle cost associated with increased power. Also, it is by far the most influential parameter of all vehicles. It must deliver continuously to each track the exact amount of torque to be transformed in traction force. As well, the more power the transmission can transfer, the lower its efficiency [2]. Current hydrodynamic power transmissions consume a significant amount of energy and require a powerful cooling system that needs 14-15 percent of engine power. Considering new develops of electric motors and power electronics a new direction of development opens that of electric transmissions [3-4]. On the other hand, electrical systems for weaponry and protection such as high-power electromagnetic railguns were developed. The electromagnetic impulse, obtained by discharging a condenser battery over a coil's circuit (a rail type) may propel projectile to hypersonic velocities. The electrical power to fire a round may reach 10 MJ. This power is delivered by an electric power generator, driven by an internal combustion engine, the same power plant of the transmission. Initially, electromagnetic railguns were developed for the Navy Forces. Yet, due to their important advantages such are the hypersonic firing velocity high range and hitting power, the low mass of the shell, as well as rapid rearming, adapting them for armored vehicles became of interest.

At this very moment, there is no combat vehicle with electric transmission, although electrification is the general trend. Numerous studies have already been developed as well as analyses for electrical and hybrid power plant layouts. Thus, [5] presents the modeling of a series-hybrid transmission for a heavy, armored tracked vehicle. Paper [6] shows, as can be seen in figure 1, a system design and details of mobility testing of a Tracked Combat Hybrid Electric Vehicle (T-HEV), while paper [7] analyses the possibility to use an electromechanical transmission on a tracked combat vehicle. Also, several countries start Hybrid Electrical Vehicles (HEV) development programs such as Future Rapid Effects Systems (FRES) in the UK, Spliterskyddad EnhetsPlatform (SEP) in Sweden, Future Combat System (FCS) and Joint Light Tactical Vehicle (JLTV) in the US [8-10].

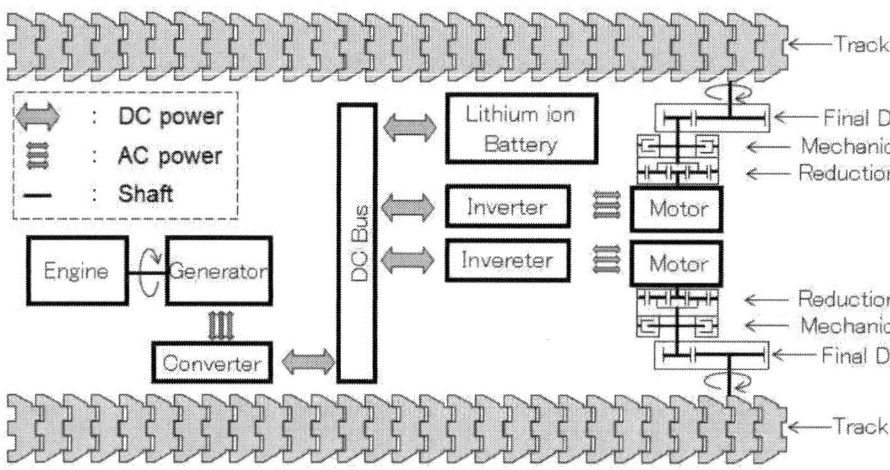

Figure 1. Structure of component arrangement of T-HEV [7].

Given the general trend of vehicles' electrification, including those for military use, this paper proposes a direction in the process of modernizing the Romanian tank TR-85M1. This is one with a total weight of 50 t powered by an 860 hp (640 kW) diesel engine and hydrodynamic power transmission. It reaches a top speed of 60 km/h [11]. The new structure replaces the engine, gearbox and transmission and retains the armored housing, armament, fire control system and protection systems.

2. General Issues

The approach is aimed to increase engine power, improve powertrain efficiency and vehicle maneuverability. The general architecture of the vehicle propulsion system is presented in figure 1. A gas turbine, diesel engine, electric generator and electronic power module are in the back of the vehicle. On the front side are the sprocket wheels and the electric machines designed to be easily mounted/dismounted. The central part is dedicated to the crew and the other specific systems. This solution ensures a maximum grip force with a good use of body volume. The driver controls the vehicle (accelerating, steering and braking) by operating the electric machines. Even if the electric machines ensure a large amount of regenerative braking, mechanical brakes are also necessary. These are mounted on the shafts of the sprockets.

3. Power Module

The power module consists of a generator driven by a gas turbine and a turbo-diesel engine. The gas turbine chosen is TURBOMECA TURMO IIIC4, a model for military applications. As it is presented in [12], "the TURMO IV C is a free power turbine turboshaft engine. The gas generator consists of a single-stage axial and single-stage centrifugal compressor driven by a two-stage axial turbine, and an annular combustion chamber with radial fuel injection by means of a rotating injection wheel. The free power turbine consists of a single axial stage. It is coupled directly to the output shaft". Some of its technical characteristics are presented in table 1 [12-14].

Table 1. TURBOMECA TURMO III C4 specifications [12-14]

Length (mm)	2 184
Height (mm)	719
Width (mm)	637
Rated shaft power - Intermediate Contingency (unlimited)	1 100 kW/33350 rpm
Rated shaft power - Maximum Contingency (2½ minutes)	1 160 kW/33800 rpm
Rated shaft power - Take Off (5 minutes)	1 115 kW/33450 rpm
Rated shaft power - Maximum Continuous	960 kW/32500 rpm

The diesel engine is a turbocharged, electronically controlled one, able to develop 150 kW (200 hp). It drives the input shaft of an intermediary gearbox while the output shaft drives the inductor of the generator. The gas turbine drives the generator rotor through a special gear box [15]. The inductor is located outside and the rotor moves in the opposite direction. Thus, depending on the needs of the mission, the powertrain can be set for a fuel-efficient operation or for maximum dynamic performance. The generator is cooled with a Roots compressor with variable speed. When necessary, the air absorbed by the gas turbine will first pass through the generator to cool it. It is used as an additional cooling device that has an important contribution to the peak power of the generator. In special cases, it takes in fresh air from the gas turbine inlet.

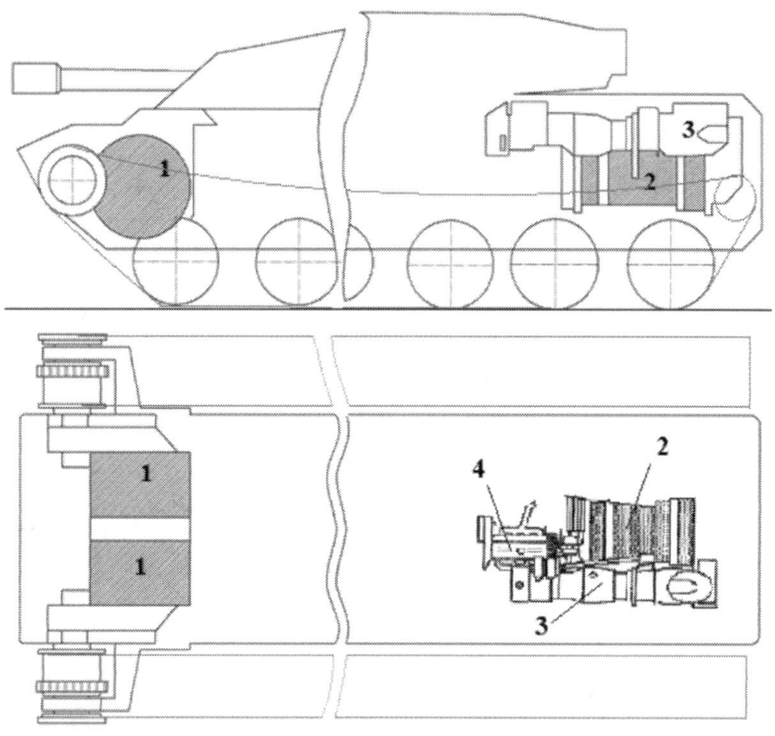

Figure 2. The general architecture of the vehicle propulsion system:
1 – AC traction motor; 2 – electric generator; 3 – gas turbine; 4 – turbo-diesel engine.

4. Traction Module

The traction module consists of two induction motors, with regular lubrication and housing cooling. Each motor has a rated power of 880 kW and withstands overloads 3 times higher for 8 minutes and 10 times higher for 30 seconds. The traction motors have been mounted so as to make an important frontal space shield with a thickness of 15-30 cm. In case of breakage of the front armor, the strong mechanical resistance of induction motors achieves the protection of the crew compartment. Also, for the same purpose, traction modules are filled with a special liquid used to cool the traction motors. Both motors have ring windings for pole switch variation. The outer part of the windings is cooled directly by liquid. Each engine transmits torque to the sprocket via a traction gearbox. Figure 3 shows the coupling of the motor with the sprocket wheel reducer by an elastic transmission [16]. It is patented by ASEA and is used for electric locomotives. Each traction motor is fixed to the armored case of the vehicle via a set of supporting arms. The shaft of the rotor 2 is hollow and it rests on the bearings in the cage 4 of stator 3. At one of the ends, it fits an internal toothed crown that engages pinion 6 of the torsion shaft 7. At the other end, the torsion shaft fits an elastic coupling that connects the engine to gear 11. The elastic coupling is made of the disc 8 (rigid on the torsion shaft), the rubber bearings and hub 9, which is fixed to the gear's shaft 10. Al this assembly plays the role of a torque damper, also allowing the engine to perform angular displacements up to 12 degrees and linear ones up to 50 mm. Sprockets are designed with symmetrical bearings for greater impact resistance. The external bearing support and the traction gearbox have a common design for a good lateral protection of the crew compartment.

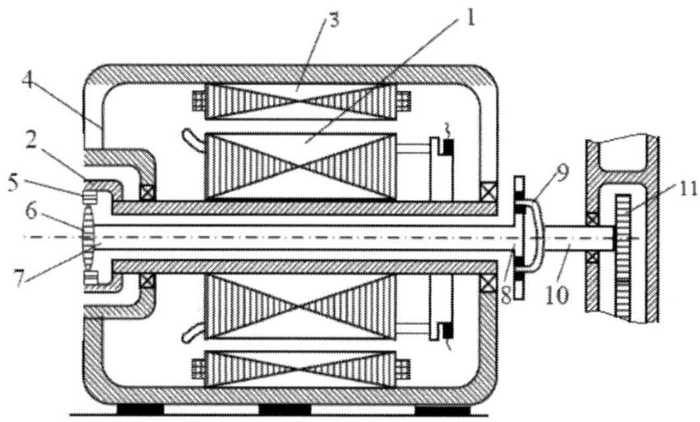

Figure 3. The coupling of the motor with the sprocket wheel reducer by an elastic transmission: 1 – rotor; 2 – shaft of the rotor; 3 – stator; 4 – cage of stator; 5 – inner teeth; 6 – pinion; 7 – torsion shaft; 8 – disc; 9 – hub; 10 – gear shaft; 11 – pinion

5. Power electronics

In figure 2 it is presented the main circuit diagram. The generator supplies electric power via a rectifier and four Insulated Gate Bipolar Transistors (IGBTs) inverters, two for each traction motor. This solution was adopted for switching pole variation. Thus, with a minimum number of poles, both inverters of each motor are in synchronous mode functioning. To obtain the maximum number of poles, one inverter of each motor changes the phases of the free components currents. A special controller module was designed for braking and cornering. The electronic power modules are located under the gas turbine and cooled with fresh air intended for its operation. Table 2 presents characteristics of traction motors [17]. The main brake is electric. In this situation, the asynchronous machines are both switching to a generator mode. Decreasing the vehicle's velocity on high speed motion in a straight line is achieved by dissipating energy within the brake by the means of electric resistance. Its efficiency is superior to the mechanical one, with drums or discs due to the higher heat dissipation capacities. When low speeds are involved, the electric power may be sent within the electric net of the vehicle to recharge batteries, power a railgun or other specific systems. In cornering conditions, the electric machine from the inside of the steering starts braking, turning itself into a generator while supplementary powering the electric machine from the outside of the steering.

Table 2. Traction motors specifications.

Design	12 Pole/ 6 Pole Asynchronous
Rating	650 kW Continuous
Nominal Design	1000 rpm/102.5 Hz
Power Factor	Cos φ = 0,85
Max. Frequency	192 Hz at 40 km/h
Max. Speed	3200 rpm at 161 Hz
Cooling Method	Liquid, Flow 30 l/min.

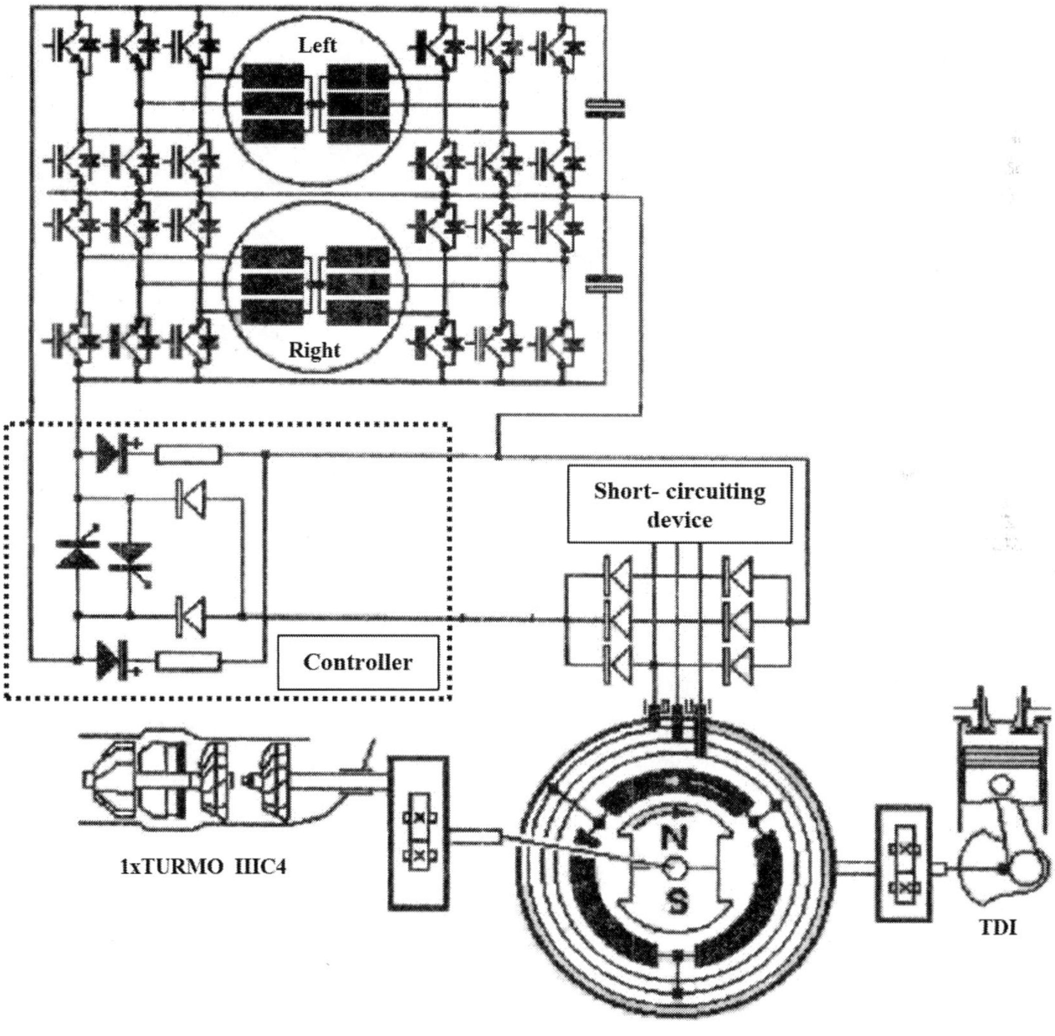

Figure 4. Power electronics circuit diagram

Thus, an important amount of power is recovered (otherwise, it would have been lost as heat). To further decrease the steering radius, motors are instructed to spin in different directions by differently controlling their phases. Both motors' control is electronic and permanently keeps the whole system's working mode at the grip's limit (electric ABS). The mechanical (conventional) brakes will operate only either the electric motors fail or when working as parking brakes. If necessary, it can be mounted at the rear on the shaft of the tension wheel.

6. General performance overview

As already mentioned, the gas turbine drives the rotor of the generator, and the diesel engine its stator, in the opposite direction to the rotor. In this way, a rotation speed for the generator equal to the sum of the other two is obtained. This solution reduces the volume and mass of the generator for the same delivered power.

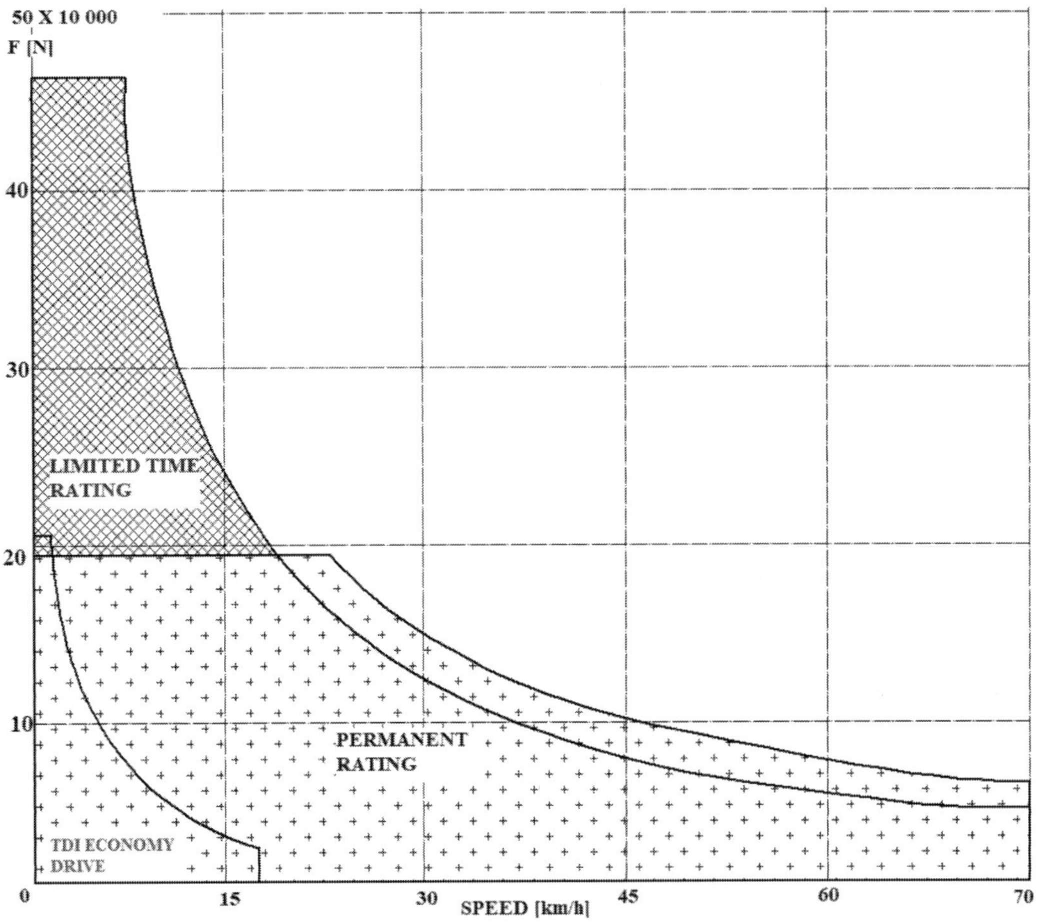

Figure 5. Vehicle general performance overview

The reducers arranged between the engines (diesel and gas turbine) and the generator elements (rotor and stator) have the role of equalizing the engine torques entering the generator. Only the speeds differ. Thus, when traveling on flat ground with speeds of up to 25 km / h, the diesel engine is sufficient. When the power demand increases, the turbine also starts operating. Figure 5 shows the traction characteristics of a 50 t vehicle obtained by simulation. The transmission is a continuous type offering a characteristic close to the ideal one. Top speed is limited by the tracks at about 70 km/h [18]. If future development of the tracks design would allow, the top speed could be around 80 km/h. Under normal operating conditions, the power is supplied by the gas turbine. When the gas turbine is running, the vehicle can reach the maximum speed of 70 km/h on the flat terrain with a concrete surface. The tank can climb a maximum slope of 32 degrees with a top speed of 7 km/h. There are situations when the vehicle has to move in the column or at a relatively low speed or to perform different maneuvers in the car park. In these cases, only the turbo-diesel engine (150 kW) is used. It can transfer 90 kW to the sprockets wheels and the vehicle can reach a top speed of 15 km/h on the flat terrain with a concrete surface and 800 km autonomy. The difference represents losses and necessary power for other systems and equipment that are vital for the tank.

7. Conclusions

This work is a starting point in the process of modernizing the TR-85M1 tank and transforming it into an armored hybrid combat vehicle. An energy unit consisting of a diesel engine and a gas turbine and an electric transmission with an alternating current generator, two traction motors and an electronic power module were proposed. At the same time, a simulation of the dynamic performances of the TR-85M1 tank equipped with the presented powertrain was performed. The following conclusions can be emphasized:

- A doubling of the power of the powertrain was obtained, which leads to the increase of the tank mobility and survivability in the battlefield;
- By arranging the traction motors in the front, an improvement of the protection in case of frontal impact and dynamic performances is obtained.
- AC electric transmission technologies offer the possibility to obtain a convenient combination of maximum advantages of different motors;
- AC electric transmission technologies could be a very attractive alternative in the construction of a high performance tank;
- The results of this study can be the basis for the future development of TR-85M1 and other armored vehicles electric transmissions.

On the other hand, electrification raises a number of difficulties such as:

- Designing of control algorithms that must be implemented in the electronic power circuit and optimize the operation of the powertrain, a laborious and expensive task;
- Designing an effective ventilation system of the energy room.

References

[1] Zahn B R, 2000, *The Future Combat System: Minimizing Risk While Maximizing Capability* USAWC Strategy Research Project pp.3

[2] Dix D M, 1981, *Propulsion System Technology for Military Land Vehicles* Institute for Defense Analyses and Technology Division pp. S-2

[3] Cantemir C-G, 2006, *Concept design of a new generation military vehicle* Proc. SPIE 6201 Sensors and Comand, Control, Communications, and Intelligence (C3I) Technologies for Homeland Defense V Vol.6201 620113

[4] Ilijevski Ž, 2017, *A Hybrid-Electric Drive Concept For High Speed Tracked Vehicles https://bib.irb.hr/datoteka/257455.ILIJEVSKI-ISTVS-HU-Final.pdf*

[5] Rondinelli E, Velardocchia M, Galvano E, 2012, *Electro-mechanical transmission modeling for series-hybrid traked tanks* International Journal of Heavy Vehicle Systems vol.19 n 03, pp.256-280

[6] Hidetaka T, Takeshi Y, Koki J, 2018, *Development of Tracked Combat Hybrid-Electric Vehicle https://www.dst.defence.gov.au/sites/default/files/basic_pages/documents/ICSILP18Thu1430_Taira_et_al-Tracked_Hybrid-Electric_Combat_vehicle.pdf*

[7] Glebov V V, Klimov V F, Volosnikov S A, 2017, *Assessment of the Pssibility to Use Electromechanical Transmission in Combat Tracked Platforms* Mechanic, Material Science & Engineering

[8] Kramer D, Parker G, 2011 *Curent state of military hybrid vehicle development* Int. J. Electric and Hybrid Vehicles Vol.3 No 4 pp.369-387

[9] Dalsjo P, 2008 *Hybrid electric propulsion for military vehicles-Overview and status of the technology* FFI-raport 2008, Norwegian Defense Research Establishment

[10] Sivakumar P, Reginald R, Venkatesan G, Visvanath H, Selvathai T, 2017 *Configuration Study of*

Hybrid Electric Power Pack for Tracked Combat Vehicles Defense Science J. , Vol 67, No 4, pp.354-359
[11] Alexa O, 2015, *Contributions to the study of the influence of the inertia masses on the longitudinal dynamics of military tracked vehicles PhD Thesis* Military Technical Academy Bucharest
[12] Marinescu M, Lespezeanu I, Radu V, Ilie C O, Alexa O, 2018, *Drag Phenomena Within a Torque Converter Driven Automotive Transmission – A Turbulent Flow Approach* Proceedings of the 4th International Congress of Automotive and Transport Engineering (AMMA 2018) pp 512-520
[13] European Aviation Safety Agency (EASA), No. E.074, Safran Helicopter Engines, 2016 *TYPE-CERTIFICATE DATA SHEET FOR TURMO IV SERIES ENGINE,* https://www.easa.europa.eu/sites/default/files/dfu/E.074_issue%2002_20160108_1.0.pdf
[14] Turbomeca Safran Group , 2006, *The Turmo powers the SA321 Super Frelon and SA 330,* https://web.archive.org/web/20061122214952/http://www.turbomeca.com/public/turbomeca_v2/html/en/produits/sous_famille_home.php?sfid=512&mid=615
[15] Goulos I, Giannakakis P, Pachidis V, Pilidis L, 2013 *Mission Performance Simulation of Integrated Helicopter-Engine System Using an Aeroelastic Rotor Model* Journal of Engineering for Gas Turbines and Power Vol.135 / 091201
[16] Cantemir C G, *Planetry gear box. Patent RO-92868*
[17] Popovici O, 2008, *Tractiune electrica* Editura Universitatii Oradea, pp. 128-129
[18] Gorianu M, 1977 *Mecanica vehiculelor rapide pe șenile* Academia Militară, București

Mathematical simulation of the economic-energy parameters of the engine 4DC-125/110 supplied with mixed fuel based on thermal calculation

P Voleac and A Dichii

State Agrarian University of Moldova, Chisinau, 42 Mircesti Street, Republic of Moldova

p.voleac@uasm.md

Abstract. The article presents a mathematical model of thermodynamic calculation of compression ignition engines. The 4DC-125/110 engine was chosen as an example, and on the basis of the mathematical model, its energy and economic performances were simulated supplying various types of fuel to identify the best operation results.

1. Introduction

Lately, there has been registered an increase in fuel prices especially for gasoline and diesel and this requires the identification of alternative solutions for powering contemporary engines. One source would be the use of compressed natural gas as fuel which can improve the ecological situation. In the total mass of pollutants from the operation of motor vehicles, its emissions are relatively low and they usually do not exceed 4% of total emissions from pollution sources. Anyway emissions from cars cause intense local pollution significantly endangering not only drivers but also people, animals, plants and soil [1].

The successive processes that occur in compression ignition engines are complex gas-dynamic and thermodynamic processes accompanied by rapid exchanges of heat, mass and mechanical work with the external environment; these also occur physical and chemical phenomena that determine the qualitative and quantitative changes of the working fluid (the formation of the fuel mixture between air and fuel and its combustion over time) [2].

The thermal calculation of the compression ignition engine allows the determination of the comparative parameters of the work cycle processes, of energy and economy indices, gas pressure in the engine cylinders during its operation on both diesel and biofuel [2].

The purpose of the work is to perform a thermal calculation for the diesel engine model 4DC-125/110, powered by different types of fuels. As a result, this calculation will provide the estimation and comparison of the economic-energy parameters of the respective engine.

2. Material and method

As a basis for a mathematical model of thermal calculation, the method described in the paper "Methodological indications of thermal calculation of ICE (internal combustion engines)" was accepted [3].

This method can also be applied when replacing the traditional fuel with another type of fuel (biofuel, partly with methane gas), and this article aims to present the calculation method with the results of estimation and comparison for the 4DC-125/110 engine powered by a mixture of different types of fuel.

3. Results and discussions

The following is the thermal calculation performed according to the chosen method, the engine having the following characteristics presented in table 1. A series of necessary preliminary values will be used from the existing information in the specialty literature, engine characteristics and on the experimental basis.

Table 1. Initial parameters for the calculation.

	Parameters	Type of fuel			
		Diesel fuel	**Diesel fuel+CH_4 (80%+20%)**	**Diesel fuel+ Biofuel from rapeseed oil (80%+20%)**	**Biofuel from rapeseed oil + CH_4 (80%+20%)**
1	Rotations, n, min^{-1}	2200	2200	2200	2200
2	The number of cylinders, i	4	4	4	4
3	Compression ratio, ε	16	16	16	16 [2]
4	Cylindrical capacity, V_l	4,75	4,75	4,75	4,75
5	Atmospheric pressure, P_0, MPa	0,102	0,102	0,102	0,102
6	Atmospheric temperature, T_0, ^0K	293	293	293	293
7	Preheating the mixture, ΔT, ^0K	15	15	15	15 [2]
8	Pressure losses in the intake manifold, ΔP_a, MPa	0,02	0,02	0,02	0,02[6]
9	Exhaust temperature, T_r, ^0K	850	850	900 [2]	850
10	The degree of pressure increase, λ	1,4	1,42	1,3 [2]	1,42 [6]
11	Excess air coefficient, α	1,5	1,44	1,25 [2]	1,44 [6]
12	Heat utilization coefficient, ζ	0,82	0,82	0,82	0,82[6]
13	The fullness coefficient of the indicate diagram, ν	0,94	0,94	0,94	0,94 [6]
14	Lower calorific value, Q_d, MJ/kg	42,70[6]	43,38[6]	41,68[6]	39,31[6]

The lower calorific value of the fuels will be:

- Diesel fuel, Q_d =42,7 MJ/kg [5,6];
- Biofuel from rapeseed oil, Q_d =43,24 MJ/kg [4];
- industrial methane, Q_d =46,1MJ/kg [5].

The parameters of the gas change process.

Primarily, the pressure is determined at the end of the admission process –P_a, and in the case of ordinary engines it can be obtained from the relation:

$$P_a = P_0 - \Delta P_a, MPa; \tag{1}$$

Exhaust gas pressure (at the end of the evacuation) – Pr, can be obtained from the equality:

$$P_r = P_0\left[1 + (0,55 \cdot 10^{-4} \cdot n)\right], MPa. \tag{2}$$

The exhaust gas coefficient is estimated according to the formula:

$$\gamma_r = \frac{T_0 + \Delta T \cdot P_r}{T_r \cdot \varepsilon \cdot P_a - P_r} \tag{3}$$

The temperature at the end of the intake is determined by the relationship:

$$T_a = \frac{T_0 + \Delta T + \gamma_r T_r}{1+\gamma_r}, {}^0K \quad (4)$$

Compression process parameters. The compression process in compression ignition engines is carried out polytropically with the exponent n_c=constant [2]. Its value can be determined by the expression of Petrov V. A. [3].

$$n_c = 1,41 - \frac{100}{n} \quad (5)$$

The pressure at the end of compression is determined by the relationship:

$$P_c = P_a \cdot \varepsilon^{nc} MPa. \quad (6)$$

The temperature at the end of compression will be equal:

$$T_c = T_a \cdot \varepsilon^{n_c - 1}, {}^0K \quad (7)$$

Combustion process parameters. Knowing the mass composition of the air of 77% nitrogen and 23% oxygen, the theoretical minimum amount of air required for complete combustion is obtained:

$$L_{\min} = \frac{O_{2\min}}{0,23} = \frac{1}{0,23}\left(\frac{8}{3}C + 8H_2 - O_2\right), \frac{kg.air}{kg.fuel} \quad (8)$$

or in kmol: $$L_0 = \frac{L\min}{29} = \frac{1}{6,67}\left(\frac{8}{3}C + 8H_2 - O_2\right), \frac{kg.air}{kg.fuel} \quad (9)$$

The amount of exhaust gas is determined by the expression:

$$M_r = \gamma_r \cdot \alpha \cdot L_0, \frac{kg.air}{kg.fuel} \quad (10)$$

The amount of exhaust gas products per kilogram of fuel if $\alpha>1$.

$$M = \alpha \cdot L_0 + \frac{H_2}{4} + \frac{O_2}{32}, \frac{kg.air}{kg.fuel} \quad (11)$$

The quantity of products before burning and after burning will be:

Before burning: $$M_c = \alpha \cdot L_0 (1+\gamma_r), \frac{kmol.air}{kg.fuel} \quad (12)$$

After burning: $$M_z = M + M_r, \frac{kmol.air}{kg.fuel} \quad (13)$$

The molecular coefficient of variation is determined by the relationship:

$$\mu = \frac{M_z}{M_c} \quad (14)$$

The specific molar heat of the fresh fluid – C_{v1}, for the temperature range (500...3000 ^{0}K) is obtained by the expression:

$$C_{v1} = A + B \cdot T_c = 20,16 + 0,00173 \cdot T_c \quad (15)$$

otherwise:

$$(C_{v1} + 8,315 \cdot \lambda) + \frac{Q_d \cdot \xi}{M_c} = S_1 \quad (16)$$

$$A^{11} = \frac{0,92}{\alpha} + 28,315 \quad (17)$$

$$B^{1} = \left(\frac{13,8}{\alpha} + 15,5\right) \cdot 10^{-4} \quad (18)$$

The temperature at the end of combustion is determined from the expression:

$$T_z = \frac{-\mu A^{11} \pm \sqrt{(\mu A^{11})^2 + 4\mu B^1 \cdot S_1}}{2\mu B^1}, {}^0K \quad (19)$$

Gas pressure at the end of the combustion process:

$$P_z = \lambda \cdot P_c, MPa \quad (20)$$

The degree of preliminary expansion is determined by the relationship:

$$\rho = \frac{\mu}{\lambda} \cdot \frac{T_z}{T_c} \quad (21)$$

Expansion process parameters. Admitting that the expansion process aims to release the energy of the engine fluid to the piston, it can be considered to flow with a polytrophic exponent n_d= constant [3].

The polytropic exponent of expansion is determined from the expression:

$$n_d = 1,22 + \frac{130}{n} \quad (22)$$

The degree of expansion is determined from the expression:

$$\delta = \frac{\varepsilon}{\rho} \quad (23)$$

The pressure at the end of expansion is calculated from the relationship:

$$P_b = P_z \frac{1}{\delta^{n_d}}, MPa \quad (24)$$

The temperature at the end of expansion is calculated from the relationship:

$$T_b = T_z \frac{1}{\delta^{n_d - 1}}, {}^0K \quad (25)$$

The average pressure of the theoretical cycle will be:

$$P_i' = \frac{P_c}{\varepsilon - 1}\left[(\rho - 1)\cdot\lambda_p + \frac{\lambda_p \cdot \rho}{n_d - 1}\left(1 - \frac{1}{\delta^{n_d - 1}}\right) - \frac{1}{n_c - 1}\left(1 - \frac{1}{\varepsilon^{n_c - 1}}\right)\right], MPa \quad (26)$$

The average indicated pressure is determined from the expression:

$$P_i = P_i' \cdot \nu, MPa \quad (27)$$

The indicated efficiency is determined from:

$$\eta_i = P_i \frac{8,315 \cdot M_c \cdot T_c(\varepsilon - 1)}{P_c \cdot Q_d} \quad (28)$$

The average piston speed is calculated from the expression:

$$w_{pm} = \frac{2 \cdot S}{60 / n} = \frac{S \cdot n}{30}, m \cdot s^{-1} \quad (29)$$

The pressure of mechanical losses is obtained from the relation:

$$P_m = 0,089 + 0,0118 \cdot w_{pm}, MPa \quad (30)$$

The efficiency that reflects the mechanical losses is obtained from the formula:

$$\eta_m = \frac{P_e}{P_i} = \frac{P_i - P_m}{P_i} = 1 - \frac{P_m}{P_i} \quad (31)$$

The average effective pressure results from the relationship:

$$P_e = P_i - P_m, MPa \quad (32)$$

The engine efficiency is obtained from the expression:

$$\eta_e = \eta_i \cdot \eta_m \quad (33)$$

The specific efficiency fuel consumption is determined from the expression:

$$g_e = \frac{10^3}{\eta_e \cdot Q_d}, \frac{g}{MJ} \quad (34)$$

The effective power of the engine is determined by the relationship:

$$N_e = \frac{P_e \cdot V_l \cdot n}{120}, kW \quad (35)$$

The torque is obtained from the relationship:

$$M_e = \frac{9554 \cdot N_e}{n}, N \cdot m \quad (36)$$

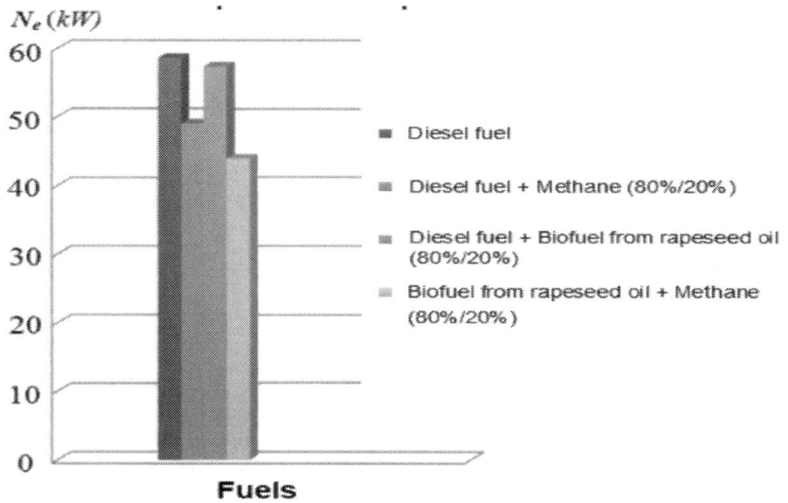

Figure 1. Modification of the effective power of the engine depending on the qualitative and quantitative composition of the used fuel.

Table 2. Energy and economic parameters which are obtained.

	Parameters	Type of fuel			
		Diesel fuel	**Diesel fuel +CH_4 (80%+20%)**	**Diesel fuel + Biofuel from rapeseed oil (80%+20%)**	**Biofuel from rapeseed oil + CH_4 (80%+20%)**
1	Indicated efficiency, %	40,7	43,4	0,0	45,5
2	The pressure of mechanical losses, MPa	0,19	0,19	0,19	0,19

	Parameters	Type of fuel			
		Diesel fuel	**Diesel fuel +CH_4 (80%+20%)**	**Diesel fuel + Biofuel from rapeseed oil (80%+20%)**	**Biofuel from rapeseed oil + CH_4 (80%+20%)**
3	The average effective pressure, MPa	0,67	0,56	0,65	0,5
4	The engine efficiency, %	31,5	32,2	30,8	32,7
5	The specific efficiency fuel consumption, g/MJ	74	72,6	76	81,1
6	The effective power, kW	58,9	49,0	57,4	43,9
7	The torque, N·m	255,1	212,6	249,1	190,5

4. Conclusions

Based on the mathematical model of thermal calculation for the chosen engine and the simulation of its supply with different fuel mixtures, among which the traditional one persists, we can conclude the following:

- The effective power decreases by 16.6% (49 kW) compared to the case of substituting a massive amount of 20% diesel with methane; respectively there is a decrease of 2.38% (57.4 kW) in the case of replacement of diesel with 20% biofuel, and for the last case we have a decrease of 25% (43.9 kW) if the used mixture will consist of biofuel and methane in a mass proportion of 80% and 20% respectively;
- The engine efficiency reaches the minimum value of 30.8% (for the case of the mixture of diesel and biofuel), and the maximum of 32.7% is registered in the case of using the mixture of biofuel and methane in the respective proportion.

Drawing a parallel between the effective power and the economic performance of the engine, which are reflected by the specific actual fuel consumption, it is clear that the most economical case is obtained when the engine works on a mixture of diesel and methane, meaning a thermodynamic gas-diesel cycle, obtaining a decrease in power of 16.6%.

References

[1] Voleac P. Particularităţile funcţionale ale MAI alimentat cu policarburanţi (gazos-lichid). Chişinău, UASM.

[2] Beşleagă, Ig. Particularităţile calcului termic al motorului diesel alimentat cu biocombustibil. Ştiinţa Agricolă, nr. 2/2011. ISSN 1857-0003.

[3] Bazele teoretice ale MAI, Indicaţii metodice pentru îndeplinirea proiectului de an "Calculul termic al MAI", Ch, Centrul ed. al UASM, 2005, 37 p.

[4] Frunze O. Cercetări privind eficienţa economică şi ecologică a utilizării uleiurilor vegetale pentru alimentarea MAC. Braşov, 2005, 169 p.

[5] *** http://antonio-merloni.ru/teplotvornaja-sposobnost-razlichnyh-vidov-topliva.

[6] Lacusta, I. et al. Performanţele energetice ale motorului diesel alimentat cu biocombustibil. In: Agricultura Moldovei. 2009, nr 7-8, p. 26-28.

Comparative numerical evaluation of ECE-EUDC cycle performances of a hypothetical car equipped with three types of propulsion systems: conventional, series hybrid and parallel hybrid

M Ionaş, V Argeşanu and N Lontiş

Department of Mechanical Engineering, Faculty of Mechanical Engineering, Politehnica University Timişoara, Timişoara, Romania

mihai.ionas@student.upt.ro

Abstract. This publication refers to the studies which were made to a compact car model to observe the behavior of the three drivetrain versions that the car can be equipped with: the version in which the vehicle model is equipped only with a Diesel engine (classic), a plug-in hybrid version with the combustion engine placed in a series circuit (pSHEV) and a second plug-in hybrid variant, where the thermal engine is arranged in parallel (pPHEV) with the electric motor (EM). The study emphasizes the behavior of three different vehicle propulsion systems equipped on the compact car on a standardized road cycle condition defined by the Urban Driving Cycle (UDC) referred as ECE 15 cycle combined with the Extra-Urban Driving Cycle (EUDC). The scope is to set up the three different powertrain systems through numerical modeling and to generate the conclusions based on the simulations results obtained in ADvanced VehIcle SimulatOR (ADVISOR).

1. Introduction

For decades, internal combustion engine (ICE) vehicles proved to be the most reliable source of transport. In the last thirty years, due to the widespread use of thermal engines, negative factors such as environmental pollution, global climate change and the health problems caused by exhaust emissions demanded imposing restrictive policies to gradually reduce the pollution and to improve the thermal engine performance.

In recent years, major automotive manufacturers tried to gradually bring hybrid compact cars to the market and prepared a step-by-step transition to fully electric vehicles, hydrogen vehicles or fuel cell vehicles. With the continuous raise of fossil fuel prices and upcoming environmental issues, hybrid electric vehicle technology has become the best option for buyers (end users) [1].

To mitigate the effects of global warming, the hybrid electric vehicle uses a thermal engine in combination with at least one electric motor (some are acting as electric generators), therefore fuel efficiency is improved compared to conventional vehicles [2]. Thus, through the efficient management of the hybrid propulsion system, the consumer demand is satisfied by the increased power, the decreased fuel consumption and the ecological factor of the vehicle.

The existing hybrid electric propulsion system solutions are: hybrid system with a thermal engine located in series (range extender); hybrid system with a thermal engine located in parallel; series-parallel hybrid system (full hybrid); mild hybrid system; hybrid plug-in systems and high-performance hybrid systems.

Recent studies show that various types of hybrid Diesel systems were simulated. The simulations were made on vehicles like 7000 kg Diesel vehicle, Diesel hybrid electric bus and even on vehicles, which had their drivetrain converted to hybrid on their most common driving cycle. All the simulation results showed a good influence on emission reduction [3, 4, 5]. The most harmful exhaust products generated by Diesel engines are nitrogen oxides (NOx), particulate matter (PM) in high concentrations and carbon monoxide (CO) and hydrocarbons (HC) in minimal concentration [6]. Furthermore, the possibility of a more significant reduction of emissions was tested by using enriched hydrogen as an extra control system, including the classic systems represented by the diesel oxidation catalyst (DOC) – used to control CO and HC emissions, the diesel particulate filter (DPF) – used to control PM emissions and the selective catalytic reduction (SCR) – used to control NOx emissions [9].

2. The experimental model

The research consists of performing three simulations on the compact vehicle on the same standardized ECE-EUDC road cycle condition, but the propulsion system will be different; the first simulation - the propulsion system is of the classic type, the vehicle being equipped only with a compression ignition (CI) engine; the second simulation - the propulsion system is of the hybrid type, having the ICE placed in series with the EM and the third simulation - a hybrid propulsion system, having the ICE placed in parallel with the EM.

Next, the parameters corresponding to each type of propulsion proposed for testing will be selected. The ICE vehicle in the conventional system has a CI engine, which develops a maximum power of 93 kW. The CI engine was chosen the same for the SHEV, respectively PHEV and the aerodynamic drag coefficient (Cd) is identical so that the data obtained from each simulation are conclusive. The continuously variable transmission (CVT) type is an automated manual five-speed transmission with optimized gear ratios for cost, space and performance reasons.

Hybrid Electric Vehicle (HEV) systems, in addition to the ICE, have electric generators (EG), EMs and battery packs (BA). Thus, the components found simultaneously in the pSHEV and pPHEV configurations were selected as identical, so that there are no differences in performance due to the mass difference or the performance of these components.

The main input parameters used for the simulation are given by the fuel converter – 150 kW ICE for the conventional system, respectively 67 kW for the hybrid versions; the BA with the capacity of 65 kW; the EM with the maximum power of 83 kW and the EG with the maximum power of 33 kW.

The traction type for the three propulsion variants is the common front-wheel drive, so the drive wheels have the additional role of handling.

The output parameters which will be analyzed and compared are: the total fuel usage [l/100km], the emissions report [g/km], the discharge of BA, the acceleration tests [s] (0-50 km/h; 0-100 km/h; 80-120 km/h) and the weight of the vehicles [kg].

Once the parameters for the three propulsion systems have been selected the next step namely is the selection of the single standardized cycle based on which the three configurations will be tested, the choice of the number of test cycles to be subjected to each powertrain variant and the definition of the initial temperature conditions of the ICE during the simulations. The chosen test cycle to perform the simulations is ECE_EUDC, also referred as the MVEG_A cycle, used as an approval test for emissions and fuel consumption of the EU light vehicles type according to EEC Directive 90 / C81 / 01. The whole cycle comprises four continuous ECE 15 segments followed by an EUDC segment shown in figure 1.

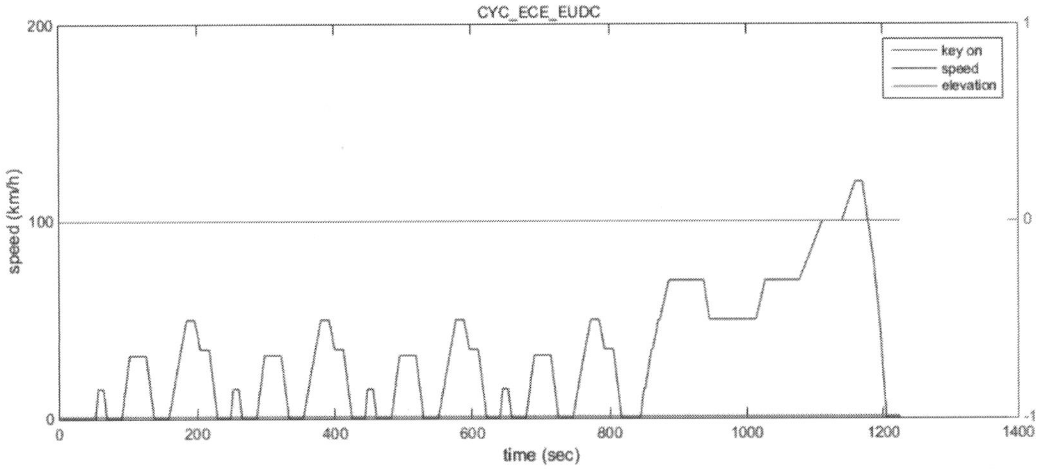

Figure 1. ECE-EUDC road driving cycle.

The ECE_EUDC cycle has a total time of 1180 s, an ECE segment lasts 195 s and the EUDC segment lasts 400 s. The constant speeds during the ECE_EUDC cycle are: 0 km/h, 18.35 km/h 25.93 km/h, 50 km/h for the ECE sections, respectively 0 km/h, 62.59 km/h, 69.36 km/h, 120 km/h for the EUDC portion. For each propulsion system, the simulation of a single ECE_EUDC test cycle with the compression ignition engine running in optimal thermal conditions.

The initial loading conditions for each vehicle model were considered to have four persons on board, each having a weight of 80 kg and a total of 50 kg of luggage. The last step is running the simulation for the three different models, preceded by the analysis of the results obtained from the numerical simulation and to conclude with the results comparison.

3. Classic system – Compression ignition engine

The most common engine used for automobiles is the internal combustion piston engine. The engine is composed of several main subsystems, which work together for the optimal operation of the assembly, as follows: the engine mechanism, the distribution mechanism of the air and fuel supply systems, the electrical system, the cooling system, and the lubrication system represented in figure 2.

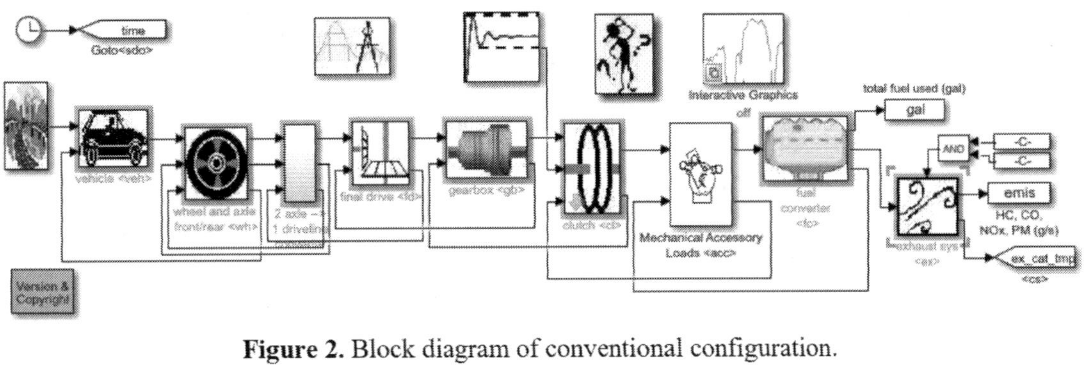

Figure 2. Block diagram of conventional configuration.

Following the simulation, the reports related to the emissions, overall ratio were obtained and represented in figure 3 and figure 4. The emissions report obtained for the classic system – HC: 0.293 g/km; CO: 1.233 g/km; NOx: 0.447 g/km; PM: 0.032 g/km. On the traveled distance of 10.9 km, the fuel consumption is of 8.4 l/100 km, with the gasoline equivalent of 9.6 l/100 km. Three sets of acceleration tests were made – 0-50 km/h: 3.6 s; 0-100 km/h: 8.4 s; 80-120 km/h: 4.7 s.

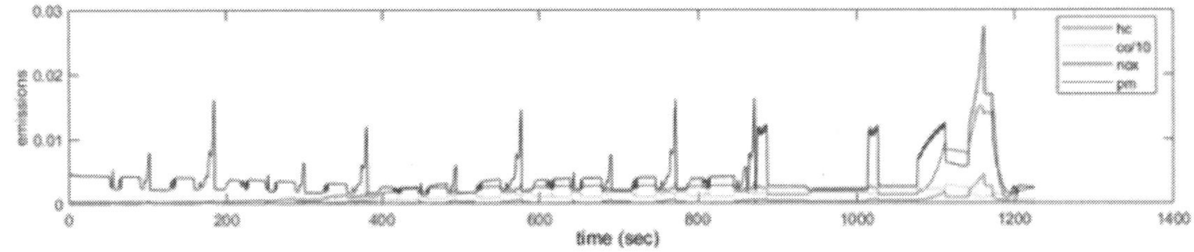

Figure 3. Emissions report for the conventional system.

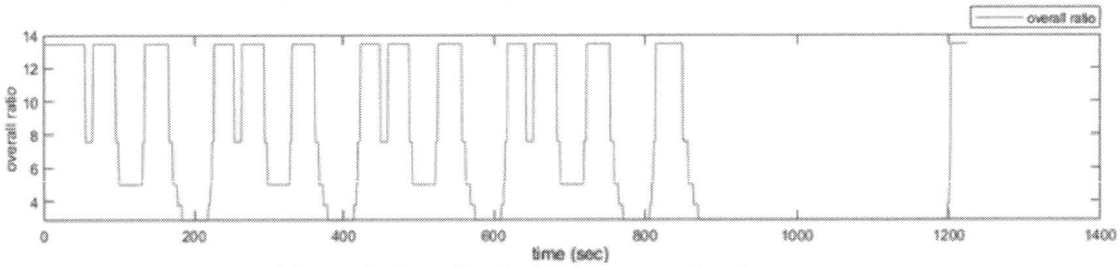

Figure 4. Overall ratio for the conventional system.

4. Hybrid system – Compression ignition engine placed in series with the electric motor

The plug-in hybrid system with the thermal engine placed in series has the propulsion provided by a single energy converter; one of the systems produces a form of energy that is used by the other system, which ultimately ensures the full propulsion of the drive wheels.

In the case of the vehicle subjected to simulation, the series powertrain system has the propulsion provided entirely by the EM, the ICE has the role of driving the electric generator that powers the EM and charges the BA as shown in figure 5.

In general, the installed BA have a higher capacity than the BA installed pPHEV, but in the case of the simulation, the BA with the same mass and the same capacity were considered. The advantage of the ICE, in this case, is that it can work exclusively in optimal mode, which minimizes fuel consumption when simulating an urban route. pSHEV allows the EM to be powered from both BAs and a secondary power source, the ICE, which can charge the batteries independently or simultaneously with the generation of propulsion power [7, 8].

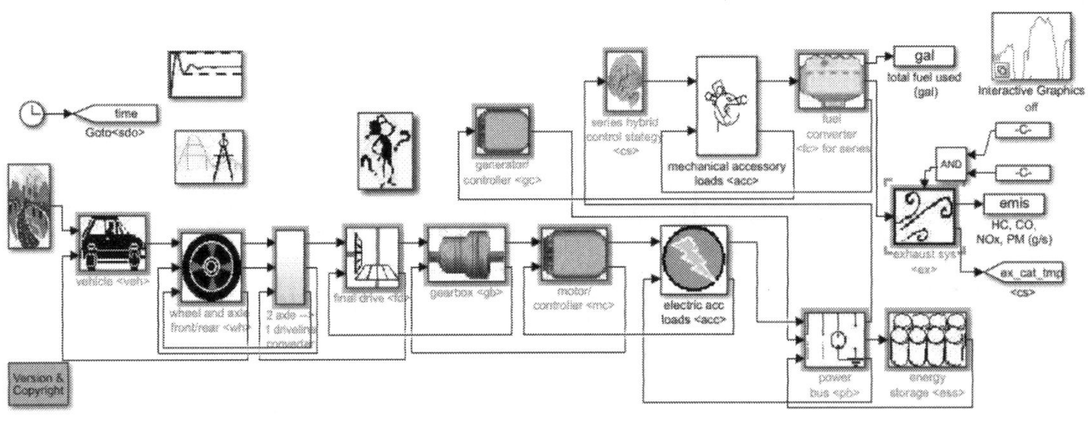

Figure 5. Block diagram of pSHEV configuration.

The pSHEV model was simulated and the results were obtained regarding to the emissions, fuel consumption, state of charge (SOC) and acceleration tests are displayed in figure 6 and figure 7. The emissions report obtained for the series hybrid system – HC: 0.012 g/km; CO: 0.018 g/km; NOx: 0.455 g/km; PM: 0.036 g/km. On the traveled distance of 10.9 km, the fuel consumption is of 2.6 l/100 km, with the gasoline equivalent of 3 l/100 km. Three sets of acceleration tests were made – 0-50 km/h: 3.8 s; 0-100 km/h: 11.5 s; 80-120 km/h: 8.9 s. The overall ratio for the pSHEV is constant.

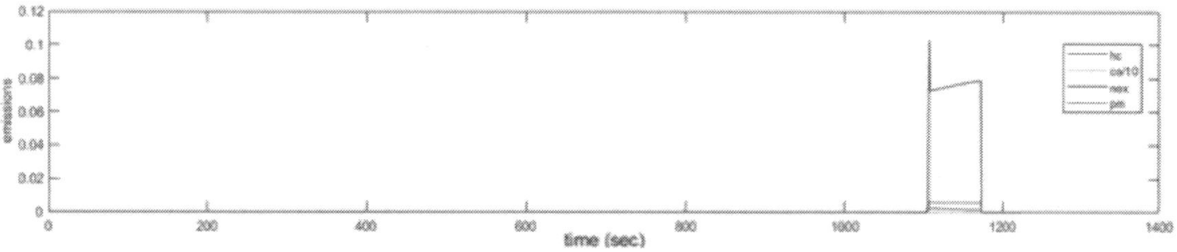

Figure 6. Emissions report for the pSHEV.

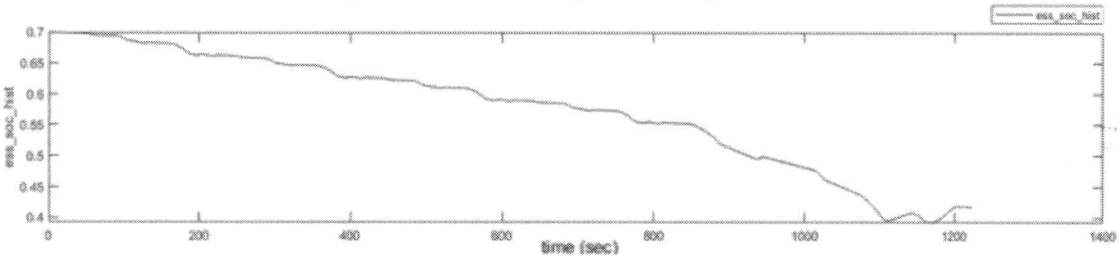

Figure 7. Discharge of BA for the pSHEV.

5. Hybrid system – Compression ignition engine placed in parallel with the electric motor

In the case of the pPHEV which is displayed in figure 8, the ICE and the EM can be connected simultaneously to the transmission, thus being able to drive the wheels simultaneously (charge-depleting mode), the ICE and the EM drive the wheels simultaneously but the ICE increases the SOC (charge-sustaining mode) and when the ICE is off and the EM is the only one running (electric-only mode) [7, 8].

The EM and the ICE are therefore located on the same axis, the role of the EM is to ensure a surplus of power for the ICE in certain transient regimes. The architecture of parallel hybrid vehicles consists of fully hybrid and semi-hybrid systems, the simulated vehicle being part of the first category.

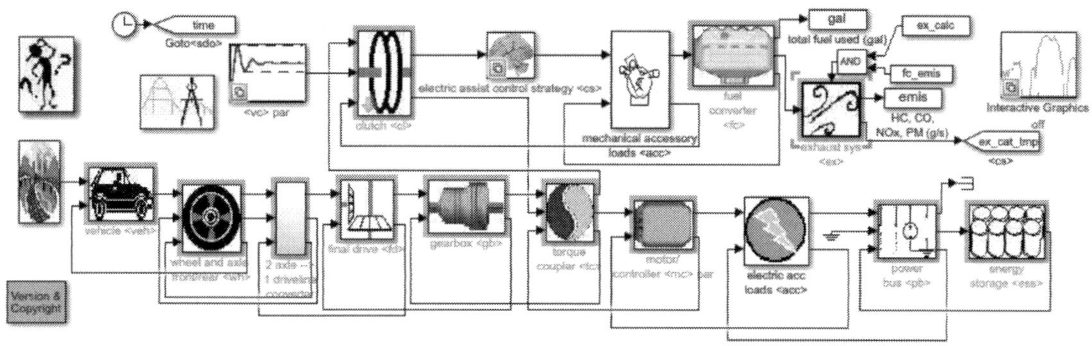

Figure 8. Block diagram of pPHEV configuration.

The pPHEV model was simulated and the results obtained regarding to the emissions, SOC and overall ratio tests are shown in figure 9 to figure 11. The emissions report obtained for the parallel hybrid system – HC: 0.137 g/km; CO: 0.719 g/km; NOx: 0.361 g/km; PM: 0.024 g/km. On the traveled distance of 10.9 km, the fuel consumption is of 6.7 l/100 km, with the gasoline equivalent of 7.6 l/100 km. Three sets of acceleration tests were made – 0-50 km/h: 3.6 s; 0-100 km/h: 7.3 s; 80-120 km/h: 3.5 s.

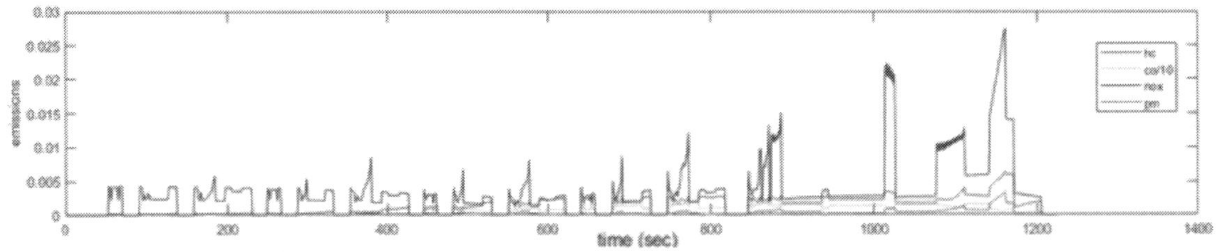

Figure 9. Emissions report for the pPHEV.

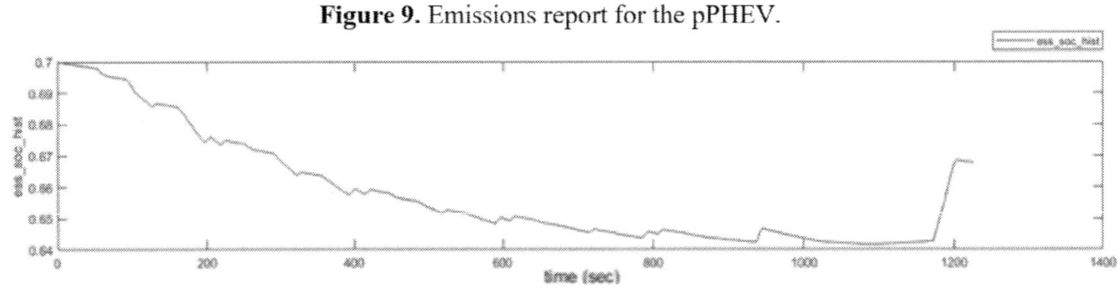

Figure 10. Discharge of BA for the pPHEV.

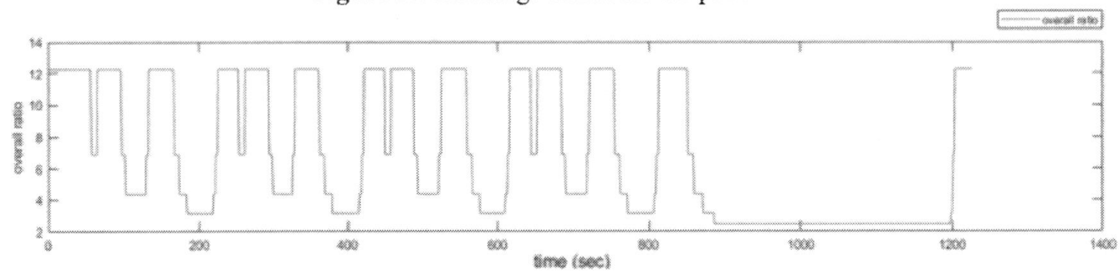

Figure 11. Overall ratio for the pPHEV.

6. Conclusions

Following the simulation, the data received were interpreted according to the total mass of the vehicle, which differs from one propulsion system to another, fuel consumption, emissions generated by each type of system, but also according to the advantages and disadvantages of each system.

The mass of the vehicle with the classic propulsion system is the smallest - 1614 kg, being followed by the pPHEV system - 1798 kg, respectively pSHEV system - 1836 kg.

The fuel consumption on the urban standardized driving cycle is the lowest for the series hybrid system: 2.6 l/100 km (petrol equivalent 3 l/100 km); the next system is the parallel hybrid: 6.7 l/100 km (petrol equivalent 7.6 l/100 km); the classic system equipped with a Diesel engine only is the least economical: 8.4 l/100 km (petrol equivalent 9.6 l/100 km).

In terms of emissions, the most environmental-friendly solution is the parallel hybrid system - NOx: 0.361 g/km, followed by the classic system – NOx: 0.447 g/km, and the serial hybrid system ranks last with NOx emissions of 0.455 g/km. If the ICE was designed with a smaller cylinder capacity for the hybrid versions, then the NOx and PM amount would decrease remarkably.

In conclusion, although the hybrid versions of the compact car are complex and more expensive than the conventional version, they remain the best alternative to propulsion systems. The plug-in technology reduces the fuel consumption and the pollutant exhaust emissions even more than a HEV, the BA charging takes less time than for a pure electrical vehicle and the consumed electricity can come from a non-renewable or a renewable source of energy. Maybe the standalone CI engine will have problems in the future to pass the next Euro regulations, but the plug-in HEV equipped with a CI engine will be a reliable powertrain solution for future years. After all, the Diesel engines have low-operating costs, they are energy efficient, have high durability and reliability in comparison to gasoline engines.

References

[1] G. Montazeri, K. Mahmoodi and A. Keskin, 2015, *An optimal energy management development for various configuration of plug-in and hybrid electric vehicle*, Journal of Central South University, vol. 22, pp. 1737-1747.

[2] I. Cho, K. Mahmoodi and J. Lee, 2020, *Characteristics of Battery SOC According to Drive Output and Battery Capacity of Parallel Hybrid Electric Vehicle*, Journal of Applied Sciences, vol. 10.

[3] S. B. Han et al., 2010, *Emissions simulation in a 7000 kg-grade diesel hybrid electric vehicle*, International Journal of Automotive Technology, vol. 11, pp. 105-110.

[4] B. Suh et al., 2012, *Simulation of a powertrain system for the diesel hybrid electric bus*, International Journal of Automotive Technology, vol. 13, pp. 701-711.

[5] H. Danial, M. I. M. Rashid, 2017, *ADVISOR Simulation and Performance Test of Split Plug-in Hybrid Electric Vehicle Conversion*, Energy Procedia, vol. 105, pp. 1408-1413.

[6] I. A. Reşitoğlu, K. Altinişik, A. Keskin, 2015, *The pollutant emissions from diesel-engine vehicles and exhaust aftertreatment systems*, Clean Technologies and Environmental Policy, vol. 17, pp. 15-27.

[7] C. Mansour, D. Clodic, 2012, *Dynamic modeling of the electro-mechanical configuration of the Toyota Hybrid System series/parallel power train*, International Journal of Automotive Technology, vol. 13, pp. 143-166.

[8] I.J. Albert, E. Kahrimanovic, A. Emadi, 2004, *Diesel sport utility vehicles with hybrid electric drive trains*, IEEE Transactions on Vehicular Technology, vol. 53, pp. 1247-1256.

[9] H. T. Arat, 2019, *Simulation of diesel hybrid electric vehicle containing hydrogen enriched CI engine*, International Journal of Hydrogen Energy, vol. 44, pp. 10139-10146.

Analytical study regarding topological optimisation of an internal combustion engine cylinder block

M Gaidur, I Pascal, E Rakosi, T Ulian, S Talif and Gh Manolache

"Gheorghe Asachi" Technical University of Iasi, Faculty of Mechanical Engineering, Romania

mihai.gaidur@gmail.com

Abstract. The continuous rush for development has led the world of automotive engineering to progress that seemed unimaginable just a few years ago. This research paper examined the impact of a topological optimisation of an internal combustion engine block. Starting with the initial conditions, a constraint and load case scenario was established, after which a FEA simulation was conducted to set the reference values. In the last part of the research paper, test results are head-to-head compared to determine and quantify the improvements. This study aims to reduce the volume of the optimised part and increase its overall rigidity. The topology optimisation process represents a good solution for "what if" scenarios. By varying optimisation constraints, the designers can quickly check if the mass reduction of the analysed part is worth the changes.

1. Introduction

The topology optimisation of engine cylinder block came as an economic necessity caused by the increasing price of the base materials. For this research paper, two scenarios were taken into consideration. The first study case simulated the engine cylinder block made from aluminium alloy AlSi8Cu3/EN AC-46200. The second study case simulated the engine cylinder block made from grey cast iron. Both simulations highlight each material type's advantages and disadvantages, alongside their proprietary design constraints, such as machinability, load responses, material prices, maintenance costs and lifetime.

The ultimate goal for this research is to determine the feasibility grade of potential topology optimisation of the aforementioned part, regardless of the chosen manufacturing material, and the economic benefit of potential topology optimisation.

2. Load case scenario

For the first simulation, where the initial results were established, the following settings had been used:

- The maximum pressure generated by combustion was established at 7 [MPa], a general pressure used in gasoline engines [4]. The pressure was applied inside the analysed cylinder;
- The engine block was constrained at its base;
- Also, for better results, a thermal load of 270 [°C] was applied inside the cylinder;
- Because of hardware resources limitation, only one cylinder of the V8 engine was simulated, with maximum load. The simulation environment can be seen in figure 1.

For the first case study, aluminium alloy AlSi8Cu3/EN AC-46200 was used. This alloy is widely used in the automotive industry, mainly due to its resilience to high loads. In addition, it is easy to cast, has high heat resistance, and has proven results in engine blocks and cylinder heads applications. This alloy has a tensile strength R_m equal to 240 [MPa] and a yield strength $R_{p0,2}$ equal to 140 [MPa], both values being obtained from pressure die-cast manufactured parts [3].

Figure 1. Representation of the used cylinder in the baseline values establishment.

Aluminium engine blocks have better thermal conductivity than their counterparts made from grey cast iron. This material allows the engine to work at a higher temperature in the combustion chamber, which increases the engine's power rating with more minor to no modifications to its existing cooling system [6].

For the second study case, grey cast iron was used. This material is widely used in industrial applications due to its high machinability. This material has a flake-like graphite structure mainly responsible for its excellent machinability. The tensile strength R_m has a value between 100 and 500 [MPa] [1].

3. Load case simulation

The Simulation Module of AUTODESK FUSION 360 was used for finite element simulations. The results were obtained via an Educational License granted by the AUTODESK group.

The simulated CAD tridimensional model can be seen in figure 2, and its meshed version is illustrated in figure 3:

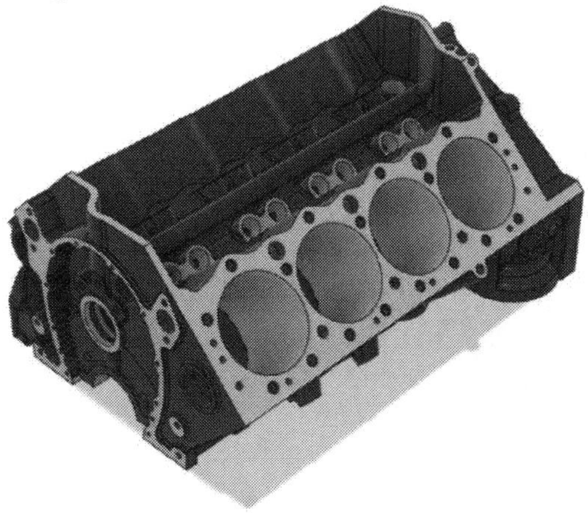

Figure 2. Representation of the used CAD model for finite element simulation

Figure 3. Meshed representation of the engine cylinder block

After all the input data was established and defined, the first study case simulation was conducted. The first case used the following base material: AlSi8Cu3. The simulation results can be seen in figure 4, figure 5, figure 6, figure 7 and figure 8:

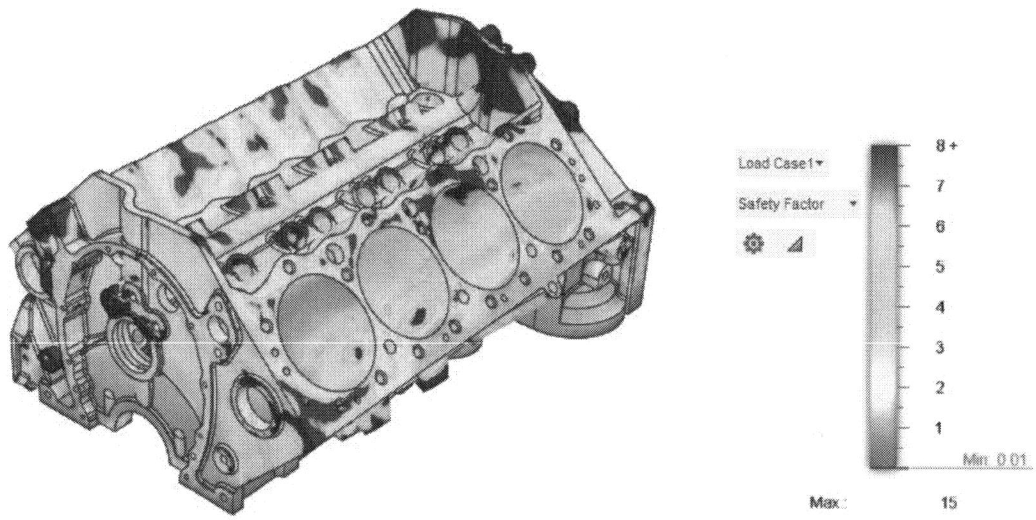

Figure 4. Safety factor.

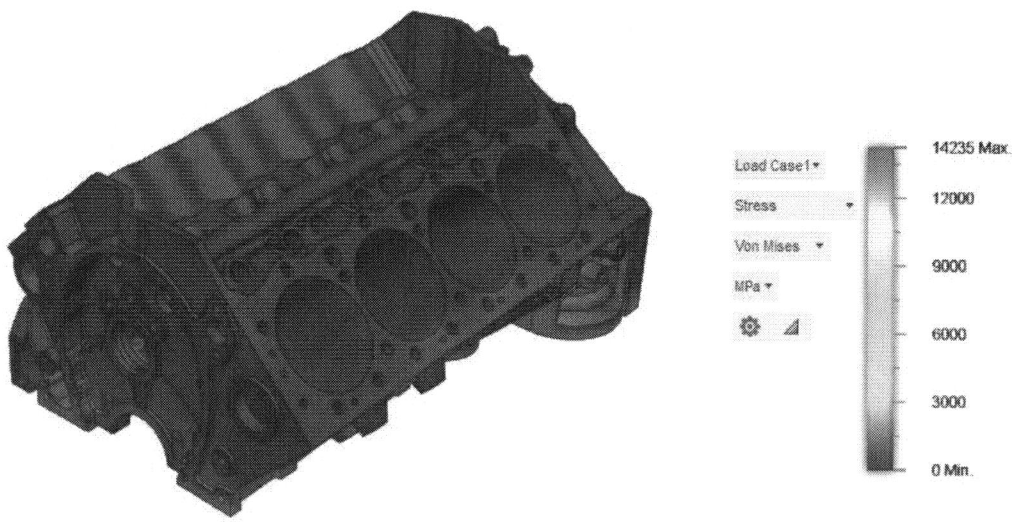

Figure 5. Von Mises stress.

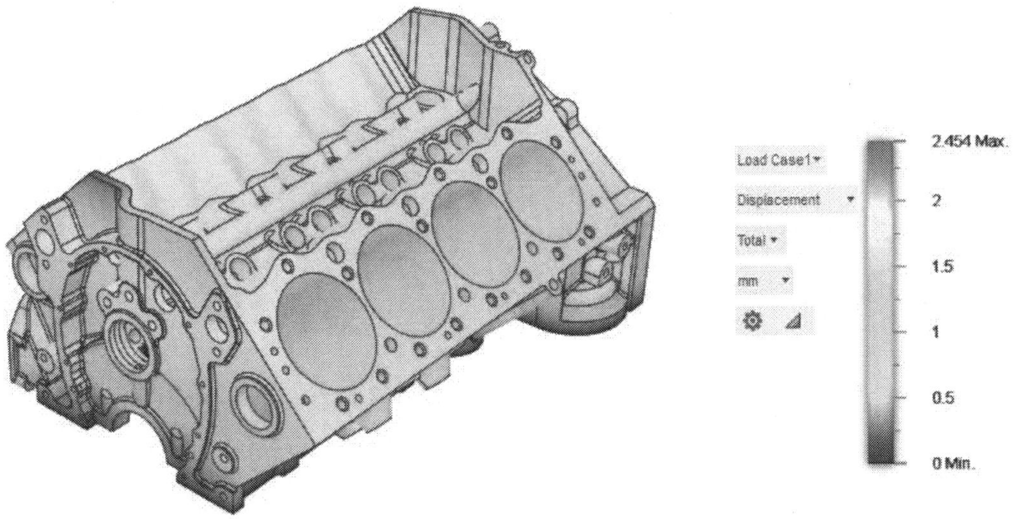

Figure 6. Displacement.

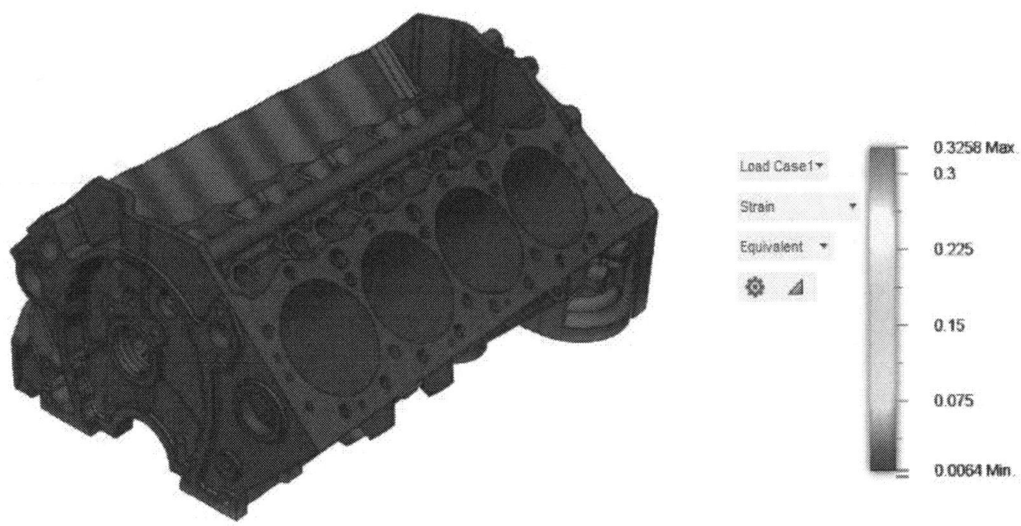

Figure 7. Equivalent strain.

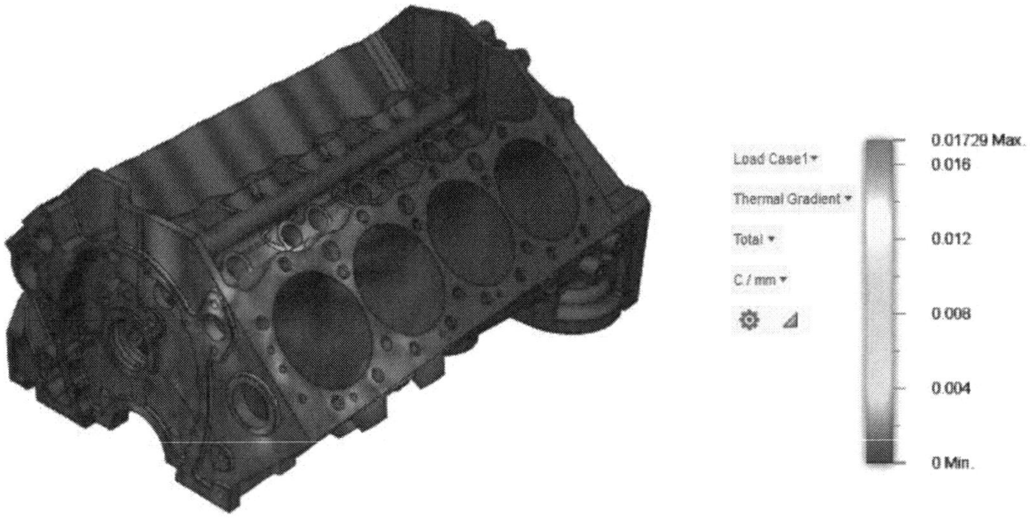

Figure 8. Thermal gradient.

For the second study case simulation, the grey cast iron was used as a base material. The simulation results can be seen in figure 9, figure 10, figure 11, figure 12 and figure 13:

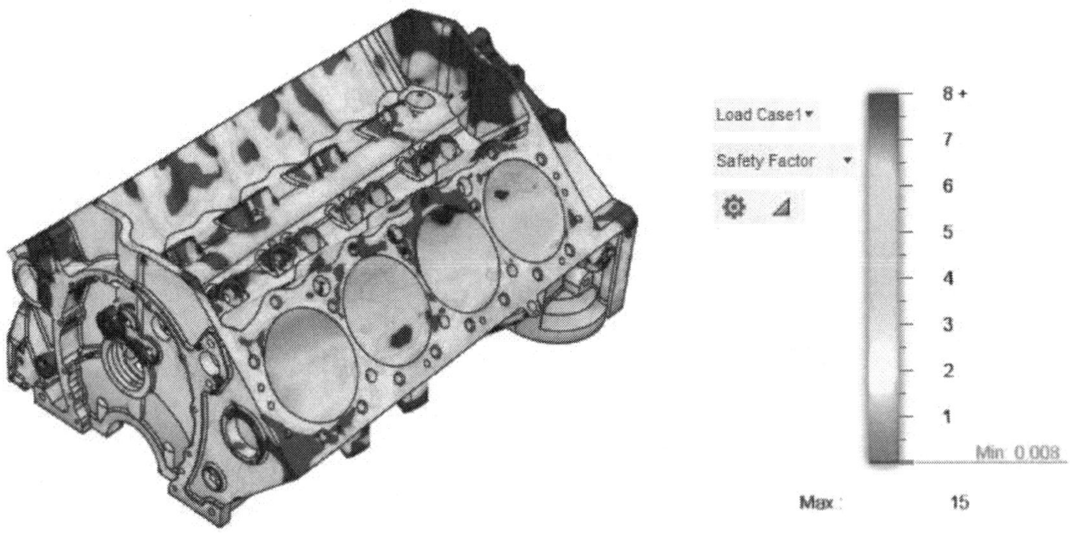

Figure 9. Safety factor.

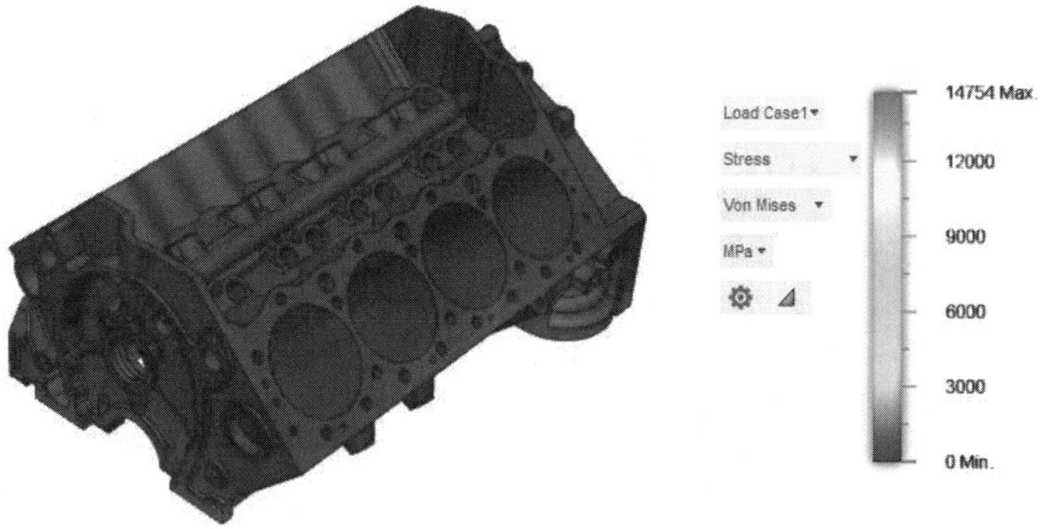

Figure 10. Von Mises stress.

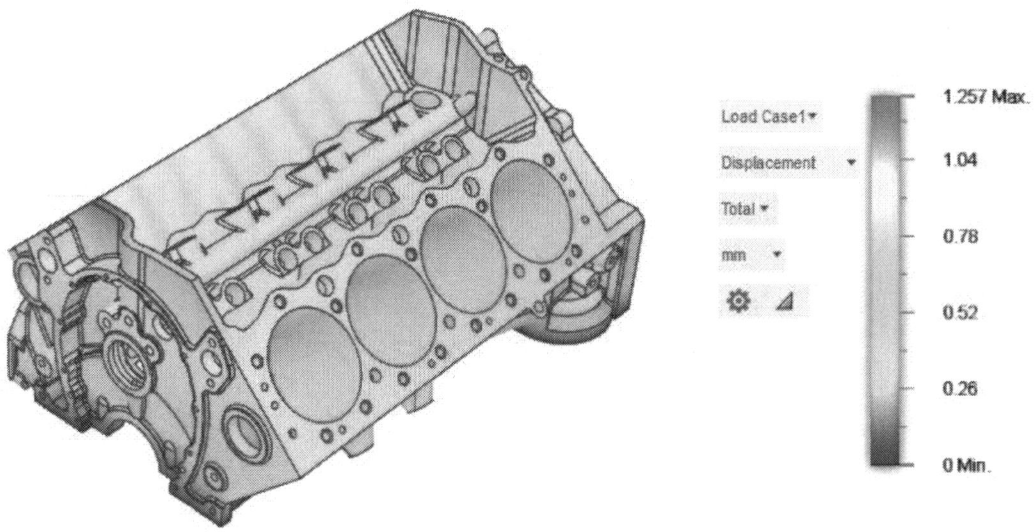

Figure 11. Displacement.

Figure 12. Equivalent strain.

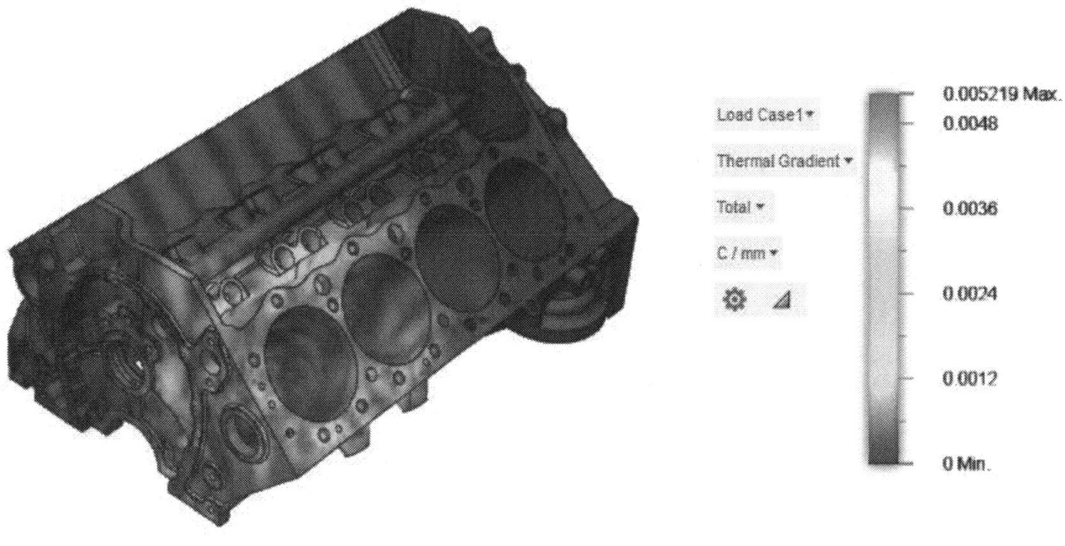

Figure 13. Thermal gradient.

Both of the simulation results are summarised in table 1:

Table 1. Simulation results.

The material used for simulation	Minimum safety factor	Maximum Von Mises stress [MPa]	Maximum displacement [mm]	Maximum equivalent strain	Thermal gradient [°K/mm]
AlSi8Cu3 alloy	0.01019	14235	2.454	0.3258	0.01729
Grey cast iron	0.008065	14754	1.257	0.244	0.005219

It can be seen that the grey cast iron engine block has a better displacement and equivalent strain load response due to the higher material rigidity. However, the engine block made from aluminium has better thermal conductivity than the grey casted iron version.

4. Topology optimisation of the engine block

The purpose of this optimisation is to reduce the engine block's overall mass while keeping its initial rigidity. With the help of specialised calculation toolsets, the optimum engine block shape was determined whilst respecting the user imposed constraints. The base of these algorithms underlies genetic algorithms, optimal criterion methods, topological derivatives, and many more mathematical modelling relationships [2, 5].

For the topological optimisation results summarisation, three simulations were compiled, for each material type, using the following constraint criteria:

a) Reduction of 1% of the initial mass while keeping 100% of the initial rigidity;
b) Reduction of 2% of the initial mass while keeping 100% of the initial rigidity;
c) Reduction of 3% of the initial mass while keeping 100% of the initial rigidity.

The selection of areas that were topology optimised was based on the part load behaviour. Therefore, certain areas that are not critical for the applied load are prone to be eliminated through topology optimisation. The topological optimisation results for each mass reduction scenario can be seen in figure 14, figure 15, figure 16:

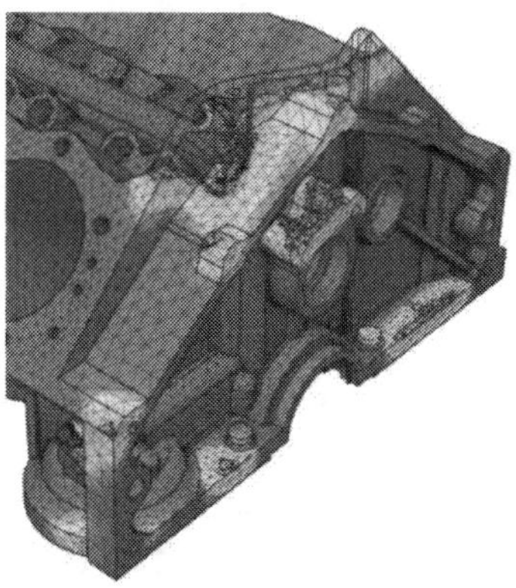

Figure 14. 1% mass reduced cylinder engine block.

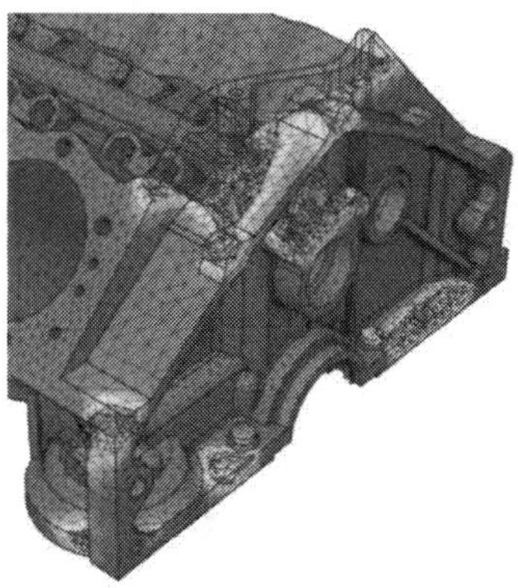

Figure 15. 2% mass reduced cylinder engine block.

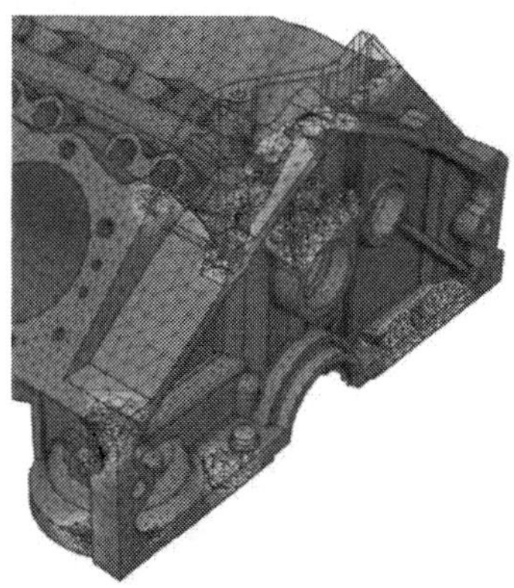

Figure 16. 3% mass reduced cylinder engine block.

The simulation results are centralised with the help of table 2:

Table 2. Simulation results.

The material used for simulation	Optimisation scenarios	Minimum safety factor	Maximum Von Mises stress [MPa]	Maximum displacement [mm]	Maximum equivalent strain	Thermal gradient [°K/mm]
AlSi8Cu3 alloy	1% mass reduction	0.01015	14285	2.477	0.3142	0.03403
	2% mass reduction	0.01	14348	2.495	0.3085	0.04561
	3% mass reduction	0.008	14569	2.516	0.3103	0.05586
Grey cast iron	1% mass reduction	0.01279	9302	1.268	0.1602	0.034
	2% mass reduction	0.013	9422	1.288	0.1688	0.04
	3% mass reduction	0.01368	9502	1.295	0.1723	0.041

The percentages variation of each relevant parameter, compared to the initial reference values, are centralised with the help of table 3:

Table 3. The percentages variation of each relevant parameter, compared to the initial reference values.

The material used for simulation	Optimisation scenarios	Minimum safety factor	Maximum Von Mises stress [MPa]	Maximum displacement [mm]	Maximum equivalent strain	Thermal gradient [°K/mm]
AlSi8Cu3 alloy	1% mass reduction	-0.393%	0.351%	0.937%	-3.560%	96.819%
	2% mass reduction	-1.865%	0.794%	1.671%	-5.310%	163.794%
	3% mass reduction	-21.492%	2.346%	2.526%	-4.758%	223.077%
Grey cast iron	1% mass reduction	58.586%	-36.953%	0.875%	-34.344%	551.466%
	2% mass reduction	61.190%	-36.139%	2.466%	-30.820%	666.430%
	3% mass reduction	69.622%	-35.597%	3.023%	-29.385%	685.591%

In Table 3, it can be observed that for the engine block that is manufactured from aluminium alloy, the minimum safety factor starts to massively decrease from -1.8% at 2% mass reduction to -21% at 3% mass reduction, as the 3% mass reduction milestone was achieved. The minimum safety factor does not represent the safety factor of the whole part. Because the simulation is of a static type, small areas with complex geometries that directly result from the topology optimisation might not withstand a constant maximum load; thus, the importance of further interpretation and risk analysis is being highlighted. The maximum Von Mises stress and the maximum displacement have slightly increased from the initial state values with a mean of 2%. It can be noticed that the thermal gradient parameter has drastically improved, with an increase of 223% at 3% mass reduction.

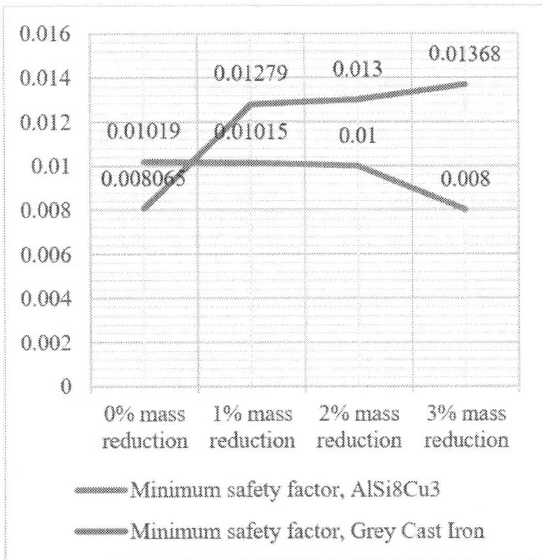

Figure 17. Variation of the minimum safety factor.

The minimum safety factor improved for the cylinder engine block manufactured from grey cast iron, with a 69% increase at 3% mass reduction. The maximum Von Mises stress decreased with a mean value of 36%, regardless of the percentage of mass reduction. The maximum displacement tends to be directly proportional to the percentage of the mass reduction of the analysed part. The maximum equivalent strain tends to reduce with a mean value of 30%, regardless of the percentage of mass reduction. The significant improvement can be seen at the thermal gradient parameter, with an increase of 685% at 3% mass reduction.

The variation of the minimum safety factor for both load case scenarios can be seen in figure 17. The minimum safety factor of aluminium alloy tends to have a descending trendline, whilst the trendline of the minimum safety factor of the grey cast iron engine block has an ascending trendline. The minimum safety factor does not represent the overall safety factor of the part. Minimum values are located in small areas where the material removal left sharp edges, which cannot withstand the applied loads. For this reason, further result interpretation and risk analysis need to be conducted.

For both load case scenarios, the variation of the Von Mises stress can be seen with the help of figure 18. The mean value of the von Mises stress tends to remain constant for the aluminium alloy. In contrast, the mean value of the Von Mises stress of the grey cast iron engine block decreases when 1% of the mass is reduced and then stabilises at a mean value of approx. 9000 [MPa]. Two trendlines were extracted with the help of this figure, which are the following:

- For aluminium alloy, the Von Mises stress vary by the following formula:

$$y = 14194x^{0.0145} \tag{1}$$

- For grey cast iron, the Von Mises stress vary by the following formula:

$$y = 13607x^{-0.323} \tag{2}$$

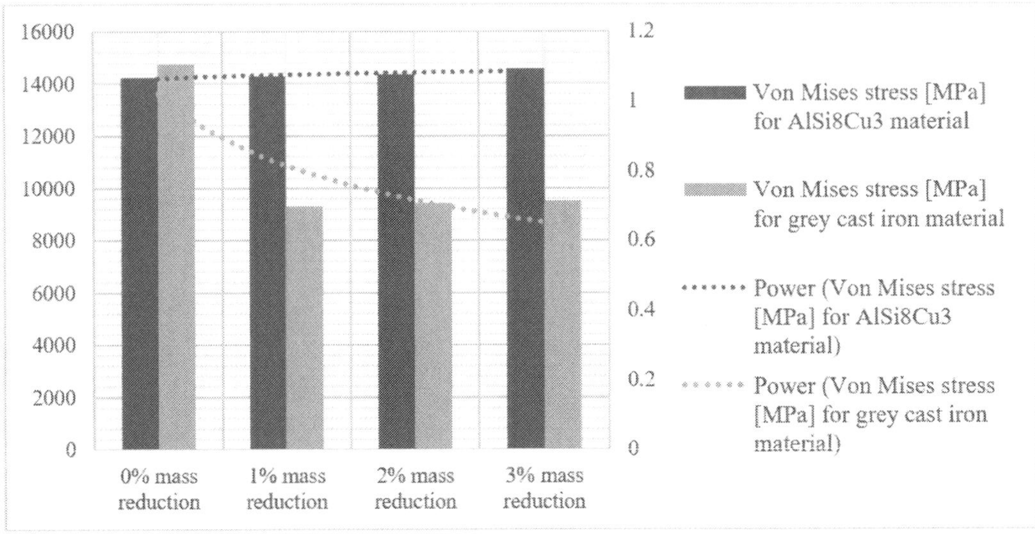

Figure 18. Von Mises stress variation for both load case scenarios.

In figure 19, variation of the maximum deflection correlated with mass reduction percentage was plotted. The variation law of maximum deflection for both load cases was approximated with the following formulas:

- For aluminium alloy, the maximum deflection varies by the following formula:

$$y = 0.0204x + 2.4345 \tag{3}$$

- For grey cast iron, the maximum deflection varies by the following formula:

$$y = 0.0134x + 1.2435 \tag{4}$$

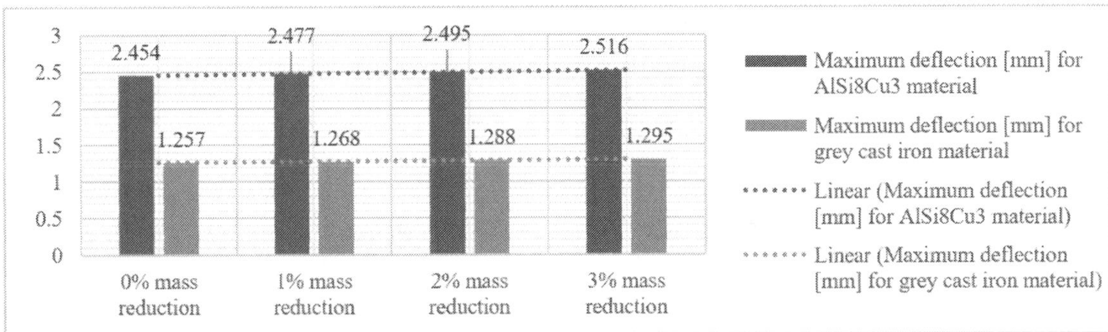

Figure 19. Maximum deflection correlated with the mass reduction percentage.

In figure 20, the variation of the maximum deflection from its reference value, correlated with its mass reduction in percentages, was plotted. As in percentages variation, two trend lines were generated with the help of figure 20. The variation laws were approximated by the following formulas:

- For aluminium alloy, the maximum deflection, in percentages, varies by the following formula:

$$y = 0.014\ln(x) + 0.087 \tag{5}$$

- For grey cast iron, the maximum deflection, in percentages, varies by the following formula:

$$y = 0.0199\ln(x) + 0.093 \tag{6}$$

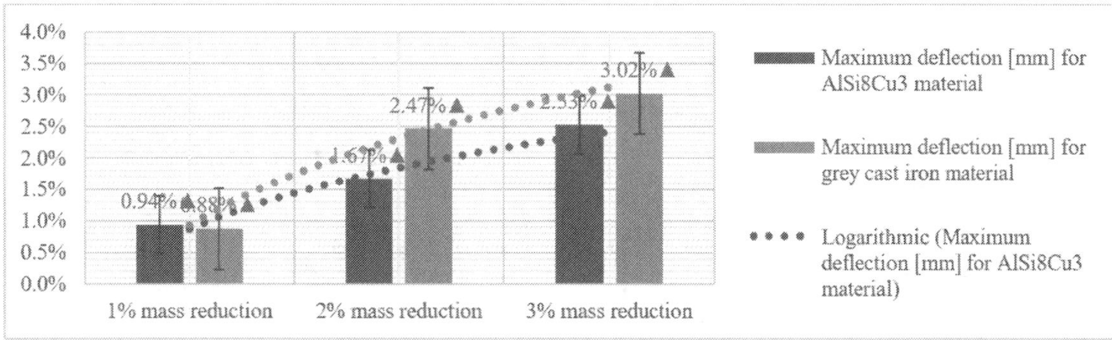

Figure 20. Variation of the maximum deflection from the reference value, in percentages, in correlation with the mass reduction percentage.

In figure 21, the variation of the maximum deformation correlated with the mass reduction percentage was plotted. The variation law of the maximum deformation for both load cases was approximated using the following formulas:

- For aluminium alloy, the maximum deformation varies by the following formula:

$$y = -0.012\ln(x) + 0.3244 \quad (7)$$

- For cast-iron grey, the maximum deformation varies by the following formula:

$$y = -0.052\ln(x) + 0.228 \quad (8)$$

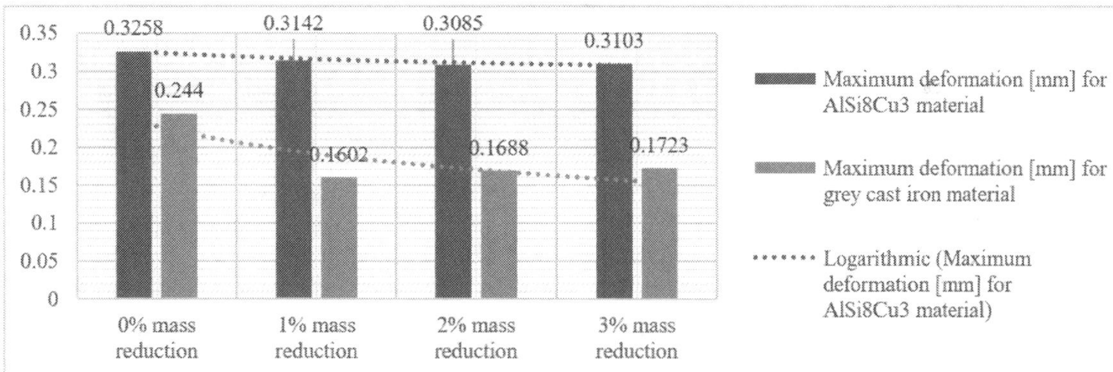

Figure 21. Maximum deformation correlated with the mass reduction percentage.

In figure 22, the variation of the maximum deformation from its reference value, correlated with its mass reduction in percentages, was plotted. As in percentages variation, two trend lines were generated with the help of figure 22. The variation laws were approximated by the following formulas:

- For aluminium alloy, the maximum deformation, in percentages, varies by the following formula:

$$y = -0.012\ln(x) - 0.038 \quad (9)$$

- For grey cast iron, the maximum deformation, in percentages, varies by the following formula:

$$y = 0.0458\ln(x) - 0.3425 \quad (10)$$

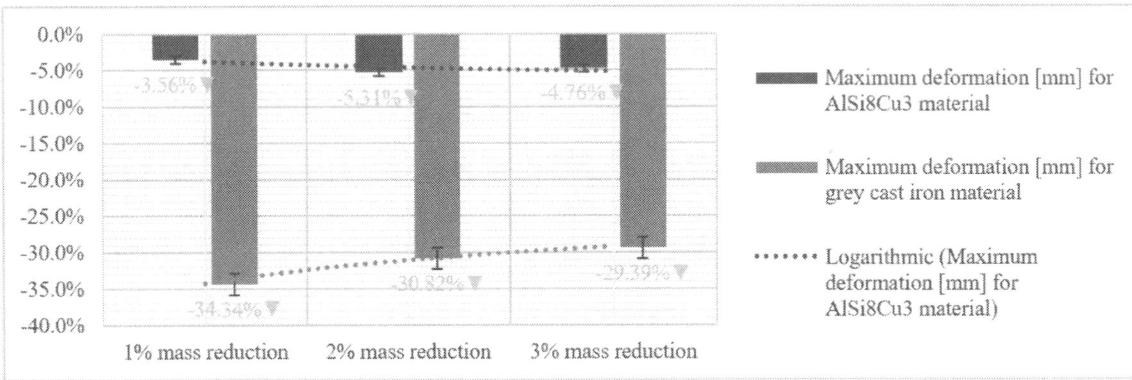

Figure 22. Variation of the maximum deformation from the reference value, in percentages, in correlation with the mass reduction percentage.

The thermal gradient evolution was plotted in figure 23:

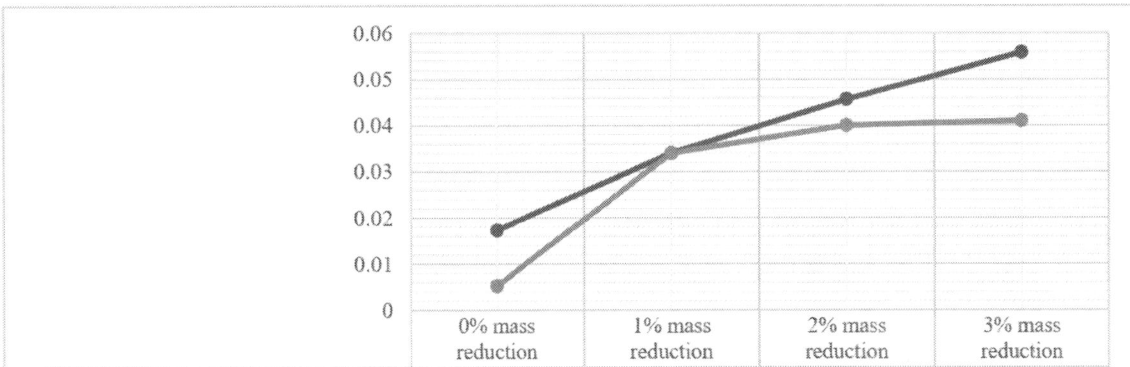

	0% mass reduction	1% mass reduction	2% mass reduction	3% mass reduction
Thermal gradient [K/mm] for AlSi8Cu3 material	0.01729	0.03403	0.04561	0.05586
Thermal gradient [K/mm] for grey cast iron material	0.005219	0.034	0.04	0.041

Figure 23. Thermal gradient variation correlated with the mass reduction percentage.

5. Conclusions

1. Topology optimisation of the cylinder engine block is plausible, but constraint conditions will apply. All the functional areas need to be precisely defined and protected against material alteration during the topology optimisation process;
2. Based on the results of the finite element analyses, we can conclude that the topological optimisation is best suited for the engine block made of aluminium alloy;
3. The product development team has to decide the feasibility grade of the topology optimisation process, which can differ from one material to another and load scenario;
4. The topology optimisation process represents a good solution for "what if" scenarios. By varying optimisation constraints, the designers can quickly check if the mass reduction of the analysed part is worth the changes.

References

[1] Evans R, *Metalworking Fluids (MWFs) for Cutting and Grinding, Fundamentals and Recent Advances*, Woodhead Publishing Series in Metals and Surface Engineering, 2021;

[2] Gaidur M, Pascal I, et al., *Analytical study regarding the topological optimisation of an internal combustion engine piston*, Materials Science and Engineering, Volume **997**, 2020;

[3] Garb M, Leitner M, et al., *Effect of elevated temperature of the fatigue strength of casted AlSi8Cu3 aluminium alloys*, Procedia Structural Integrity, Volume **7**, 2017;

[4] Grünwald B, *Theory, calculus and engine construction for road autovehicles*, 2nd edition, Bucharest, pp **341**, 1980

[5] Kober M, Kuhhorn A, et al., *Nonlinear topology optimisation of centrifugally loaded aero-engine part with newly developed optimality-criteria based algorithm*, Aerospace Science and Technology, Volume **39**, 2014;

[6] Torres R, Esparza J, et al., *Characterisation of an aluminium engine block*, International Journal of Microstructure and Materials Properties (IJMMP), Volume **1**, 2006.

IOP Conf. Series: Materials Science and Engineering 1220 (2022) 012021 doi:10.1088/1757-899X/1220/1/012021

Contributions towards improving the displacement and strain responses of an internal combustion engine piston through topology optimisation

M Gaidur, I Pascal, E Rakosi, T Ulian, Gh Manolache and S Talif

"Gheorghe Asachi" Technical University of Iasi, Faculty of Mechanical Engineering, Romania

mihai.gaidur@gmail.com

Abstract. State-of-the-art parts require state-of-the-art engineering. On this occasion, in this research paper, contributions towards improving the displacement and strain responses of an internal combustion engine piston are presented. The load case scenario and the topology constraints are established in the beginning. In the following stage, the reference values are defined with the means of an FEA simulation. Afterwards, the topology of the piston is optimised. To quantify the improvements, a comparison analysis was made in the ending part of this paper. This study aims to reduce the displacement and strain responses of an internal combustion engine piston through topology optimisation.

1. Introduction

In the field of shape optimisation and design, one of the most important contributions to the modern concept of structural optimisation can be traced back to the work of Bendsøe [1]. In their study, the authors considered an optimisation paradigm that went beyond optimising shapes alone and incorporated changes in the connectivity of the structure. In addition, they discussed a distribution method based on the use of artificial composite material with microscopic voids. Since then, topology optimisation has become the most general form of structural optimisation and has provided evidence that efficient solutions can be obtained starting from given specifications, such as the power design domain, load cases, and boundary conditions [6].

The ultimate goal for this research is to determine the feasibility grade of potential topology optimisation of the part as mentioned above, regardless of the chosen manufacturing material, and the economic benefit of possible topology optimisation.

The need for topology optimisation came as a performance increase requirement. Through topology optimisation, the mass reduction of the piston is achieved; thus, the optimal mass distribution throughout the piston's geometry will ensure improved dynamic behaviour of the piston during engine strokes.

The optimized engine piston has the following mechanical characteristics: piston bore B = 70 [mm], piston stroke S = 64 [mm], and the nominal speed n = 6000 [rpm]. The aforementioned data was used to determine the piston's forces to withstand during workload [3].

2. Load case definition

For simulation purposes, the highest load values were taken into consideration. The first and the main force that the piston has to withstand is the force generated by the pressure of expanding gases during piston stroke. This force was calculated with the following formula:

$$F_P = \frac{\pi B^2}{4}(p - p_{cart}) \tag{1}$$

B is the piston's bore, p is the gas pressure in the cylinder that for this scenario is equal to 9 [MPa], and p_{cart} is the gas pressure inside the engine bay [3]. p_{cart} usually has a value of 1 [bar], which is equal to 0.1 [MPa]. Once the input parameters are determined, the force will have the following value:

$$F_P = \frac{\pi 70^2}{4}(9 - 0.1) = 34250 \text{ [N]} \tag{2}$$

Once F_P is known, the inertia force of the masses that are in translation F_i is calculated with the following formula:

$$F_i = -m_p r \omega^2 (\cos\alpha + \Lambda \cos 2\alpha) \tag{3}$$

Where m_p is the sum of the piston masses that is equal to 0.4 [kg], r is the radius of the crank that is equal to 32 [mm], ω is the angular velocity which is 628 [rad/s], and Λ is the ratio between the crank radius and the connecting rod length. Thus, for this scenario, Λ is equal to 0.333(3). Finally, α is the rotation angle between the crank and the vertical axis. Because we will use the worst-case scenario values, the α angle was adopted at 370 °RAC, where RAC stands for "Rotation Angle Crank" [3]. Thus, F_i will have the following value:

$$F_i = -6729 \text{ [N]} \tag{4}$$

The maximum resulting force, F_{RMAX}, is equal to the sum of the forces mentioned above, which is:

$$F_{RMAX} = F_P + F_i = 34250 - 6729 = 27521 \text{ [N]} \tag{5}$$

Consequently, F_{RMAX} will be used as input for the finite element analysis.

3. Load case simulation, used to establish reference values

For reference values establishment, the simulation model had the following parameters:

a) Structural constraints.
The engine piston was constrained with the piston pin radially and axially.

b) Mechanical loads.
The force generated by the pressure of the expanding gases, F_P, has been decomposed into two mechanical loads. The maximum resulting force, F_{RMAX}, is applied on the piston head along the cylinder's normal axis and the other is the normal force. This force was applied normal to the cylinder axis.

c) Thermal loads.
A thermal load was also applied. A 249 [°C] temperature was applied on the piston head [2].

d) Material Al 7075 alloy.

This material has the following mechanical properties: the tensile strength equals 276 [MPa], and the elasticity limit is 145 [MPa]. This alloy has a strong resistance to deformation, comparable to many steel alloys, and it has high fatigue resistance, making it suitable for intensive load usage. The 7000 series alloys are mainly used in the automotive industry due to their excellent resistance and density ratio [3, 4].

The software used for this analysis is AUTODESK FUSION 360, specifically, the Shape Optimization module. The results were obtained with the help of an Educational License granted by the AUTODESK group.

IOP Conf. Series: Materials Science and Engineering 1220 (2022) 012021 doi:10.1088/1757-899X/1220/1/012021

The simulated CAD model of the piston is represented in figure 1, and its meshed model is represented in figure 2:

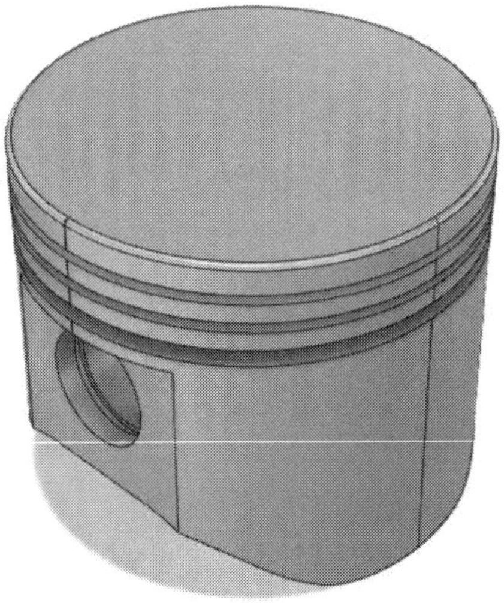

Figure 1. Representation of the used piston CAD model for finite element simulation.

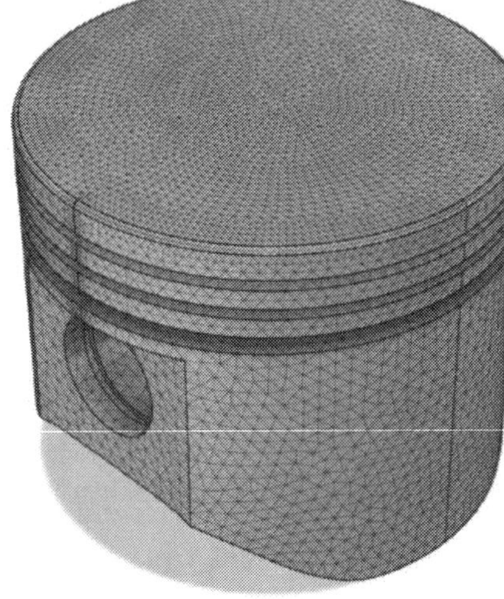

Figure 2. Representation of the meshed simulated piston geometry.

After all the previous criteria were defined, the simulation compiled the initial results, which are represented through **figure 3**, **figure 4**, **figure 5**, **figure 6** and **figure 7**:

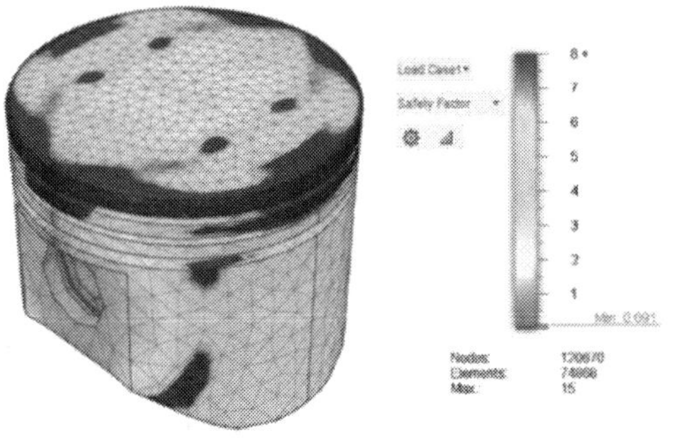

Figure 3. Resulted safety factor.

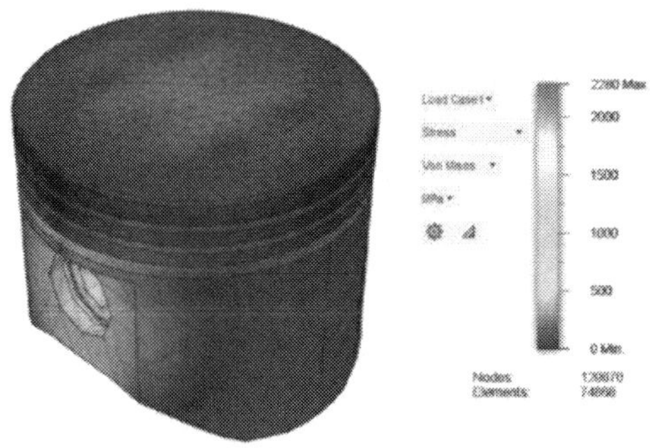

Figure 4. Resulted Von Mises stress.

The safety factor is a significant result in evaluating how suitable a design is for its intended application. The safety factor indicates if a design is likely to survive unharmed, bend, or break when subjected to the applied loads. Various criteria and considerations influence the safety factor for a given material, manufacturing process and applications [8].

Stress is defined as the force acting per unit area. It is computed from strain (elongation or compression per unit length) and the material stiffness. Finally, the solver outputs the Von Mises stress that is the effective total stress magnitude from combining the six stress tensor components. This value is always positive [8].

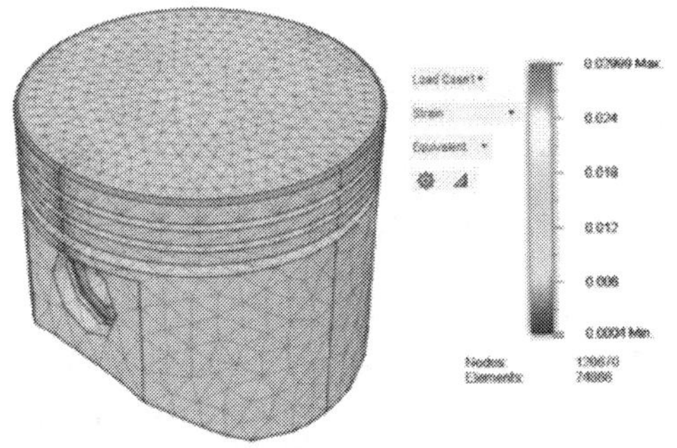

Figure 5. Resulted equivalent strain.

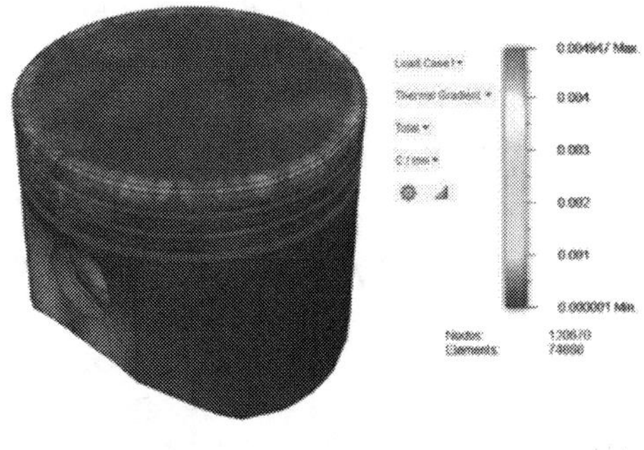

Figure 6. Resulted thermal gradient.

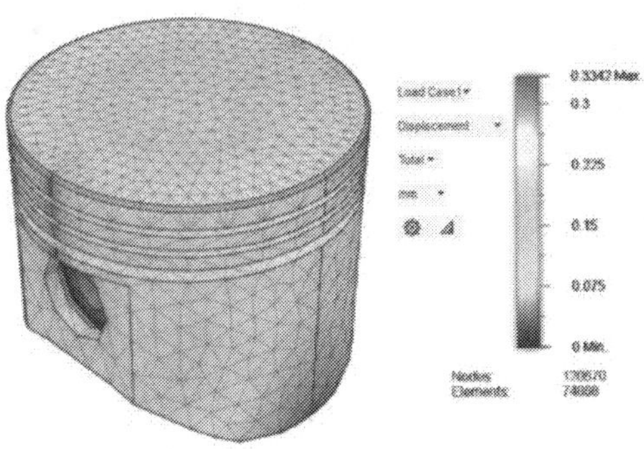

Figure 7. Resulted displacement

Strain measures the change in the length of a material along a particular direction per unit length of the material. Strain is a unitless or dimensionless quantity since the length unit appears in both the numerator and the denominator, and they cancel out. The equivalent strain represents the effective total strain magnitude from the combination of the six strain tensor components. This value is always positive [8].

The displacement result is a colour contour plot that shows how much the model has moved relative to its original position. The total displacement displays the magnitude of the displacement, regardless of its direction. This value is always positive. The magnitude is the total distance any given point has moved [8].

The thermal gradient parameter represents the dimensional quantity, expressed in [°K/mm], of units of degrees per unit of length. It describes what direction and at what pace the temperature changes around a specific location [8].

The numerical values of the aforementioned simulated parameters were summarised in table 1 [7]. By analysing the variation of these parameters, we can establish a default behaviour of the simulated part under the user-defined loads.

Table 1. The numerical values of the aforementioned simulated parameters.

Minimum safety factor	Maximum Von Mises stress [MPa]	Maximum displacement [mm]	Maximum equivalent strain	Thermal gradient [°K/mm]
0.09	2279	0.334	0.0283	0.0066180

4. Topology optimisation of the piston

This optimisation aims to improve the displacement and strain response of the piston under certain loads. Using a specific toolset of calculations, the optimal shape of the piston was determined, consequently, for user imposed constraints. The calculation relationships underlying this process aim to determine the values of the minimum stresses, the minimum deformations, the minimum displacements, the minimum mass, and the minimum volume of the part. The core of these algorithms is based on genetic algorithms, optimal criterion methods, level analysis, topological derivatives and many more. [3, 5].

For the topological optimisation results, three simulations were compiled, using the following simulation criteria:

a) Reduction of 1% of the initial mass, keeping 100% of the initial piston rigidity;
b) Reduction of 2% of the initial mass, keeping 100% of the initial piston rigidity;
c) Reduction of 3% of the initial mass, keeping 100% of the initial piston rigidity.

The results of each mass reduction scenario are represented with the help of figure 8, figure 9 and figure 10. The piston region that is in shadow represents the reduced geometry that was once part of the initial part. Thus, a more considerable percentage of mass reduction will be represented by a more extensive material "penalisation".

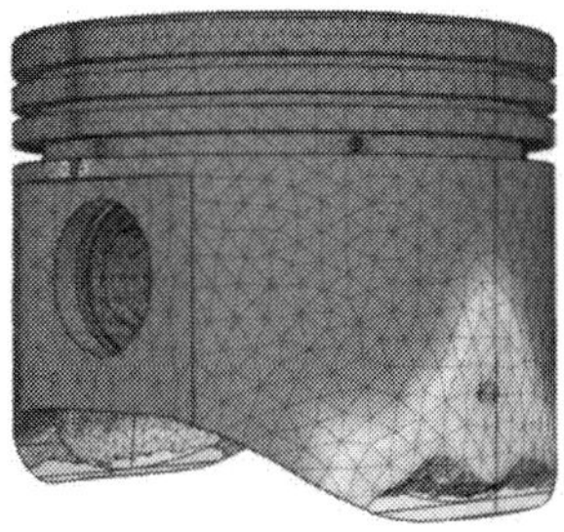

Figure 8. 1% mass reduced piston.

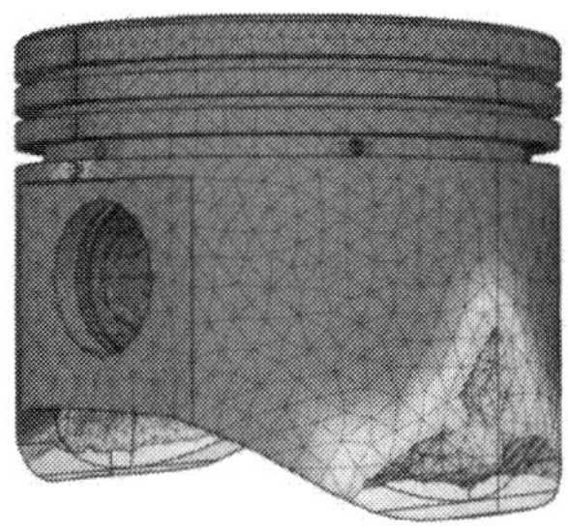

Figure 9. 2% mass reduced piston.

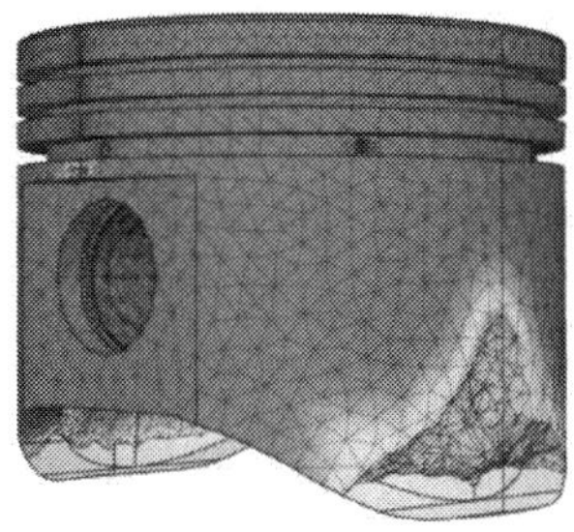

Figure 10. 3% mass reduced piston.

Once the optimised topology geometries are finished, the new geometries will be requalified with the help of the initial simulation scenario. The simulation results of the new optimised pistons are summarised in table 2:

Table 2. The simulation results of the new optimised pistons.

Optimisation scenarios	Minimum safety factor	Maximum Von Mises stress [MPa]	Maximum displacement [mm]	Maximum equivalent strain	Thermal gradient [°K/mm]
1% mass reduction	0.03	2346	0.01183	0.01504	0.0000360
2% mass reduction	0.02544	2490	0.007623	0.01993	0.0000045
3% mass reduction	0.03	2510	0.009287	0.01757	0.0000041

As for percentage values, the variation of each relevant parameter, for each optimisation scenario, compared to the initial reference values, are represented in table 3:

Table 3. The variation of each relevant parameter, for each optimisation scenario, compared to the initial reference values.

Optimisation scenarios	Minimum safety factor	Maximum Von Mises stress	Maximum displacement	Maximum equivalent strain	Thermal gradient
1% mass reduction	-66.67%▼	2.94%▲	-96.46%▼	-46.89%▼	-99.46%▼
2% mass reduction	-71.73%▼	9.26%▲	-97.72%▼	-29.63%▼	-99.93%▼
3% mass reduction	-66.67%▼	10.14%▲	-97.22%▼	-37.96%▼	-99.93%▼

The cells colouring varies by the following rules:

- For the minimum safety factor, increased values are coloured in green, and decreased values are coloured in red;
- For the maximum Von Mises stress, increased values are coloured in red and decreased values are coloured in green;
- For the maximum displacement, increased values are coloured in red and decreased values are coloured in green;
- For the maximum equivalent strain, increased values are coloured in red and decreased values are coloured in green;
- For the thermal gradient, increased values are coloured in green, and decreased values are coloured in red.

In table 3, it can be noticed that while increasing the mass reduction of the piston, the maximum displacement value decreases by a stable value of 97%. The maximum equivalent strain has more significant variations in relation to the mass reduction percentage. Although the maximum displacement and the maximum equivalent strain were optimised, these come at the cost of increasing the maximum Von Mises stress and also lowers the minimum safety factor.

The variation of minimum safety factor was plotted in figure 11. The minimum safety factor decreased when the first mass reduction was made but tended to stay constant when the mass reduction rate increased. Thus, the minimum safety factor does not represent the whole part safety factor. Furthermore, because the finite element simulation is a static type, small areas where complex geometries are located might not withstand a constant maximum load.

The variation of safety factors can be correlated with the Von Mises stress response via figure 12. The mean value tends to remain constant, even though, at 2% mass reduction, a slight decrease can be observed. Due to the part's complex geometry, the topology optimisation process's sharp edges might not withstand the applied load locally.

A polynomial trendline that fits the plotted value best was determined with the help of figure 12. The trendline, which is represented by the green dotted line, has the following formula:

$$y = 0.046x^2 - 0.0182x + 0.0437 \qquad (6)$$

R-squared, R^2 or the coefficient of determination, is a statistical measure in a regression model that determines the proportion of variance in the dependent variable that the independent variable can

IOP Conf. Series: Materials Science and Engineering 1220 (2022) 012021 doi:10.1088/1757-899X/1220/1/012021

explain. In other words, R^2 shows how well the data fit the regression model. R-squared can take on any value between 0 and 1. Thus, a trendline is most reliable when its R-squared value is at or near 1.

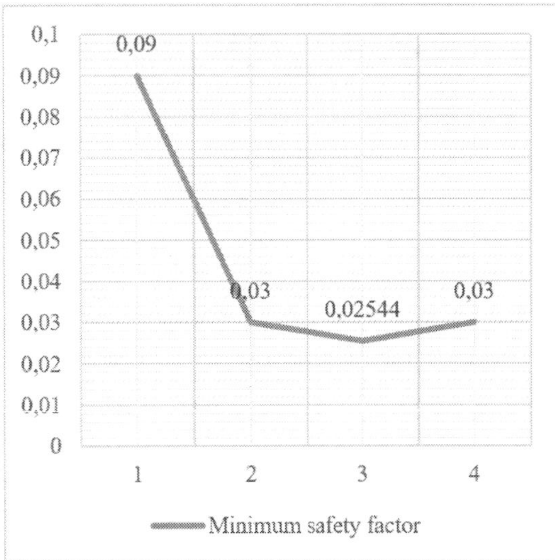

Figure 11. Variation of the minimum safety factor for topological optimized piston.

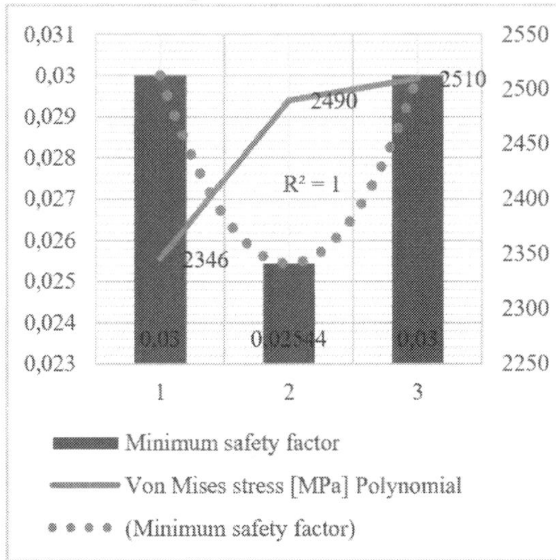

Figure 12. Von Mises stress variation correlated with the minimum safety factor and mass reduction percentage.

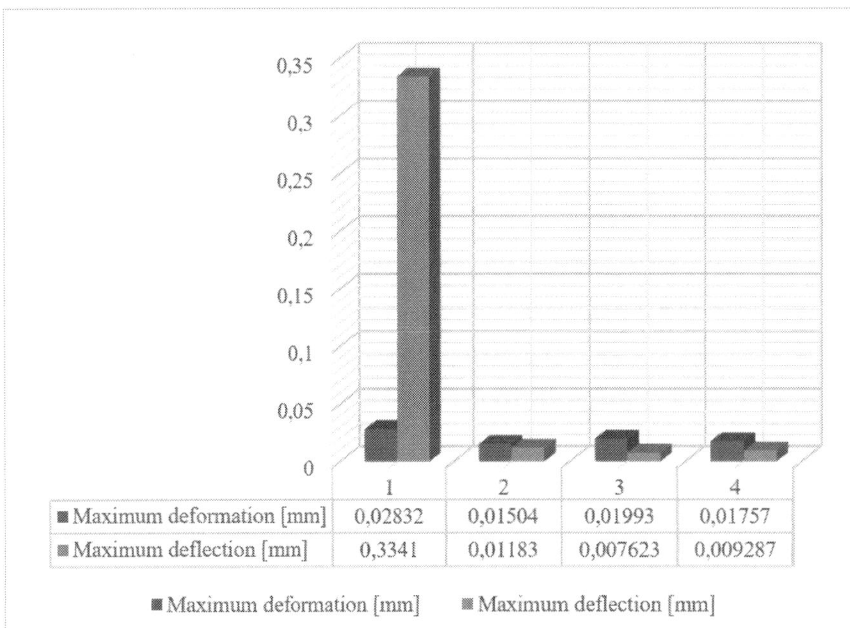

Figure 13. Deflection and deformation values, correlated with the percentage of the mass reduction of the analysed part.

The deflection and deformation values were plotted with the help of figure 13. It can be seen that via topology optimisation, the maximum deflection value decreased signifycantly, from a nominal value of 0.33 [mm] to 0.009287 [mm] at the 3% mass reduction stage, thus improving the overall load response of the optimised part.

A value change and trendline of the von Mises stress, represented with a dotted red line, correlated to the optimised part mass reduction percentage, are determined with the help of figure 14. The variation law has the following formula:

$$y = 179\ln(x) + 2264 \tag{7}$$

As in percentages variation, a trendline represented by the dotted green line and value plot can be seen in figure 15. The variation of percentages can be approximated using the following formula:

$$y = 0.0683\ln(x) + 0.0337 \quad (8)$$

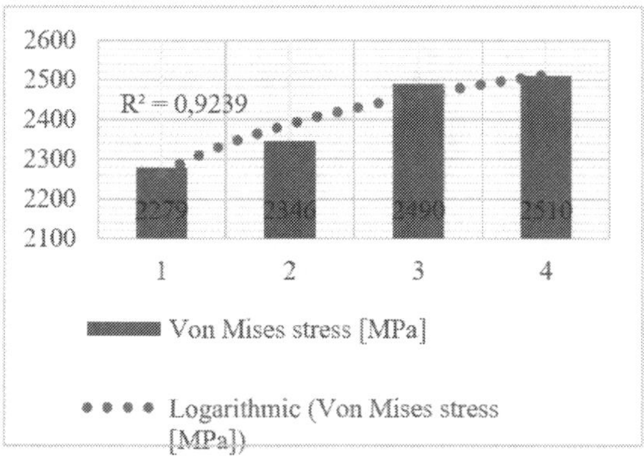

Figure 14. Von Mises stress correlated with mass reduction percentage.

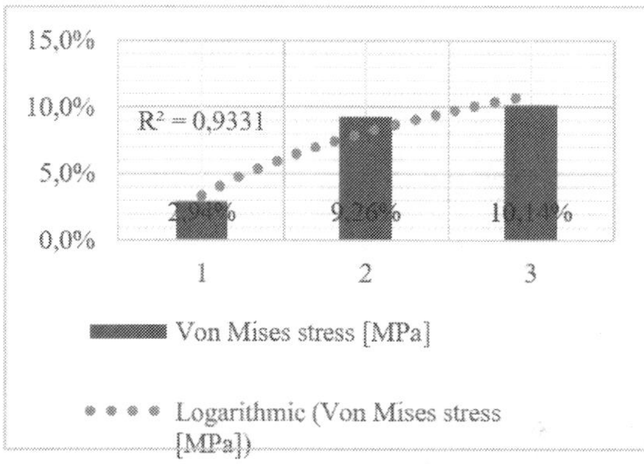

Figure 15. Von Mises stress correlated with its initial value and mass, in percentages.

The evolution and trendline of the maximum deformation correlated to the optimised part mass reduction percentage are represented with the red dotted line and shown in figure 16. The variation law, which has a polynomial profile, has the following formula:

$$y = 0.0027x^2 - 0.0164x + 0.0407 \quad (9)$$

As in percentages variation, a trendline represented with a dotted green line and value plot can be seen in figure 17. The variation law, which has a polynomial profile, has the following formula:

$$y = -0.128x^2 + 0.5567x - 0.8976 \quad (10)$$

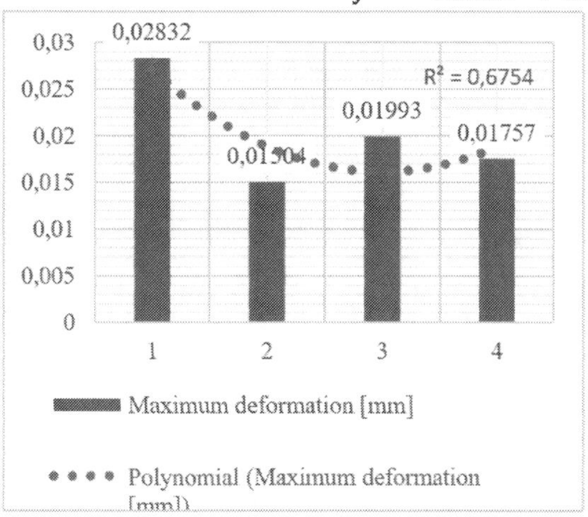

Figure 16. Maximum deformation correlated with mass reduction percentage.

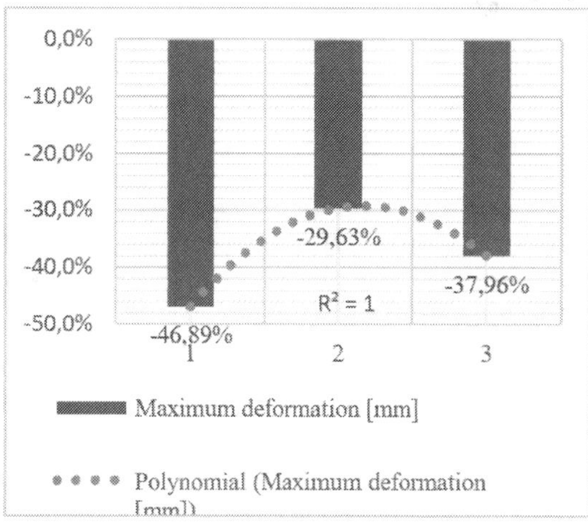

Figure 17. Variation of maximum deformation, from the reference value in correlation with mass reduction, in percentages.

The evolution and trendline of the maximum deflection correlated to the optimised part mass reduction percentage are represented with the dotted red line and shown in figure 18. The variation law can be approximated with the following formula:

$$y = 0.197x^{-2.704} \quad (11)$$

As in percentages variation, a trendline represented with the dotted green line and value plot can be seen in figure 19. The polynomial variation law has the following mathematical formula:

$$y = 0.0088x^2 - 0.039x - 0.9344 \quad (12)$$

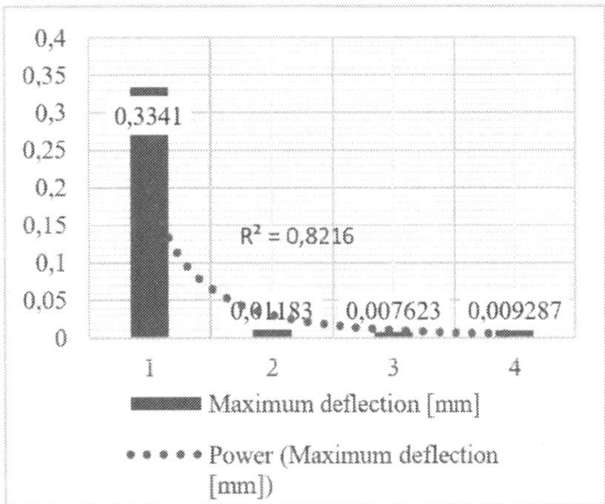

Figure 18. Maximum deflection correlated with mass reduction percentage.

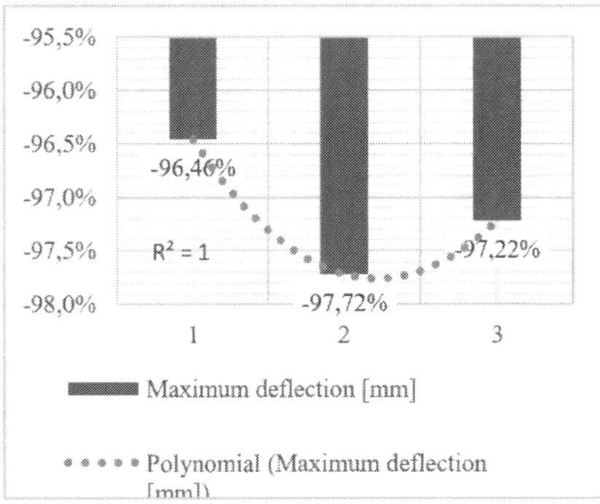

Figure 19. Variation of maximum deflection, from the reference value in correlation with mass reduction, in percentages.

A thermal gradient evolution can be seen in figure 20. During topology optimisation, no significant changes were noticed.

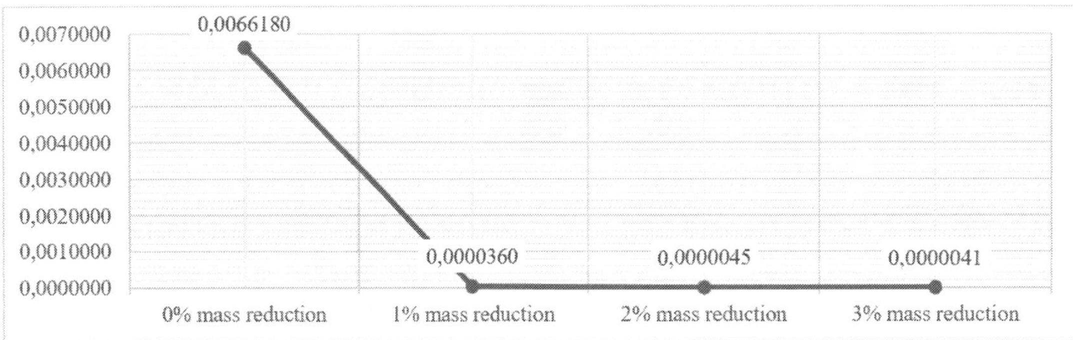

Figure 20 Thermal gradient variation correlated with mass reduction percentage.

5. Conclusions

Given the topology optimisation results, the following conclusions were made:

1. Mass reduction between 1 and 3% is feasible. Even though the maximum displacement and stress response improved, the Von Mises stress and the minimum safety factor, as performance, decreased. The performance decrease is still well between the optimum value range;
2. The economic side of the problem also has to be considered for a complete feasibility check. Because of the expensive software solutions that are commercially available right now, a topology optimisation might be suitable only for mass-produced parts;
3. The topology optimisation analysis can be used as an additional checkup tool to determine the part's response to the user imposed loads. It is up to the development team to decide the improvement factor that can be feasibly achieved while keeping the production cost at an optimum level;
4. Part's quality can be improved through topology optimisation, which means additional monetisation can be done.

References

[1] Bendsøe, M P, Sigmund, O, et al., *Topology optimization theory, methods and applications*, 2nd Edition, Springer Publishing Berlin, 2004;

[2] Berthold G, *Theory, calculus and engine construction for road autovehicles 2nd edition*, Didactic and pedagogic publishing house, Bucharest, 1980;

[3] Gaidur M, Pascal I, et al., *Analytical study regarding the topological optimisation of an internal combustion engine piston*, Materials Science and Engineering, Volume **997**, 2020;

[4] Gökhan Ö, Karaaslan A, *Properties of AA7075 aluminum alloy in aging and retrogression and reaging process*, Transactions of Nonferrous Metals Society of China, Volume **27**, 2017;

[5] Kober M, Kuhhorn A, et al., *Nonlinear topology optimisation of centrifugally loaded aero-engine part with newly developed optimality-criteria based algorithm*, Aerospace Science and Technology, Volume **39**, 2014;

[6] Pastore M, Mercuri V, et al., *Topology optimization of stress-constrained structural elements using risk-factor approach*, Computers and Structures, Volume **224**, 2019;

[7] Seabra M, Azvedo J, et al., *Selective laser melting (SLM) and topology optimisation for lighter aerospace components*, Procedia Structural Integrity XV Conference on Fracture, Volume **1**, 2016.

[8] Fusion 360 Product Documentation. (n.d.). Autodesk. Retrieved September 13, 2021, from https://help.autodesk.com/view/fusion360/ENU/?guid=GUID-296056DA-ED8C-4EBF-8E3A-2209B1DE9AEE.

Numerical investigation of a pressure wave supercharger

I Costiuc[1]**, L Costiuc**[2]

[1]Automotive and Transport Engineering Department, Faculty of Mechanical Engineering, Transilvania University of Braşov, BV 500036, Romania
[2]Mechanical Engineering Department, Faculty of Mechanical Engineering, Transilvania University of Braşov, BV 500036, Romania

costiuciuliana@unitbv.ro

Abstract. The paper aims at a numerical investigation of the evolution of velocity, pressure and temperature field along the wave rotor channels for a pressure wave supercharger. Since in literature most of the studies are made considering the working fluid as incompressible and inviscid in a 2D field, the present study is using the compressible and viscous terms in unsteady Navier-Stokes equations for fluid in 3D field. The geometry was drawn in CAD software using measurements made on a real model of the CX-93 pressure wave supercharger. The simulation was conducted using a CFD code for unsteady 3D k-ε, k-ω model approach to reproduce data such as pressures, temperature and mass flows which are usually measured in real engine pressure wave supercharging. The computational domain for uRANS was modeled as a moving rotational domain with adaptive meshing. Results such as velocity, pressure and temperature field in the rotor channels were obtained for exhaust gas inlet pressure of 0.28 MPa and 1465 K temperature at different rotational speeds. The air inlet state considered was: 0,098 MPa and 293 K. Supercharging by means of a pressure wave supercharger, in order to improve the performance of an internal combustion engine, appears to be a promising solution since the exhaust gas generates a benefice boost of intake air with significant advantages when compared to the conventional turbocharging. The numerical modelling of the complex phenomena occurring within the narrow channels might be a useful tool for improving the pressure exchange between the working fluids, either by modifying the input parameters or by optimizing the geometry of the rotor, ports or pockets.

1. Introduction

The European Union established as one of its key targets for 2020 a 20% reduction in greenhouse gas emissions compared with 1990, and a 40% reduction for 2030 [1]. Consequently, the EU has taken legislative action to restrict higher levels of pollution, especially arising from burning fossil fuels. The primary consumers of fossil fuels are the propulsion systems, being stated [2] that the road transportation sector is one of the main sources of greenhouse gas emissions, e.g. CO_2, and air pollutants. Therefore, the internal combustion engines (ICE) have become the point of interest in terms of reducing emissions, as well as efficient use of energy.

Among the measures to reduce vehicle emissions EU, the most common are: the introduction of alternative propulsion technologies - by developing the hybrid or electric cars; eco-innovation; improving the efficiency of conventional engines - by implementing advanced and specific technologies, such as technologies for the exhaust system: oxidation catalysts, cellular filters, absorbers, catalysts substances injected into the gas stream, etc. Well-conceived thermal management of the ICE can contribute in achieving the targets outlined above, by:

- energy conservation and waste heat recovery,
- improving the overall efficiency and performance,
- lowering fuel consumption and, thus,
- reducing gas emissions.

Supercharging aims to increase the torque and engine power by raising bmep while reducing the maximum engine speed. Thus, the mechanical losses of the engine, the noise and fuel consumption decrease, with the benefit of prolonging engine life [3]. Superchargers are efficient in low and medium engine speeds, but their improvement on thermal efficiency is very limited because the effective work is reduced with the amount consumed to drive the compressor. On the other hand, turbochargers have higher thermal efficiency because the compressor is driven by the energy coming from the exhaust gases, but they work properly on the engine high-speeds range and they induce increased exhaust gas pressures.

2. Pressure Wave Supercharging

2.1. Fundamentals of PWS. General Operating Principles. Design and Construction

A PWS operation is based on the pressure waves ability to transfer energy from the combustion gases to the intake air. The PWS geometry consists of a rotor (called wave rotor or cell wheel) with channels shaped parallel to the rotation axis, which are radially positioned on one or more rows (Fig.1a). The PWS, by means of shockwaves, transfer the energy inside the channels directly from the exhaust gas towards the fresh intake air, the fluids being in short direct contact.

The fluids are the high pressure hot gases (HPG) and the low pressure cold air (LPA), and their interaction induces boost, as the exhaust gases compress the intake fresh air. Within the channels the process is extremely fast, thus, the phenomenon of mixing between the two fluids is insignificant.

The cell wheel rotates inside a casing, in between two fixed end plates (figure 1a.): one connected to the intake manifold, called the "cold stator" (figure 1b.) and one connected to the exhaust gas passages, called the "warm stator" (figure 1c.). The rotor is driven by a pulley-belt drive mechanism that rotates the cell wheel with a multiple of the engine speed [3].

(a) (b) (c)

Figure 1. (a) rotor, also named cell wheel, (b) warm stator, (c) cold stator.

The acoustic and thermodynamic processes between the inlet fluids result in the outlet fluids leaving the PW supercharger: the low pressure gases (LPG) leaving the rotor towards the exhaust system and the compressed high pressure air (HPA) heading towards the cylinders (figure 2).

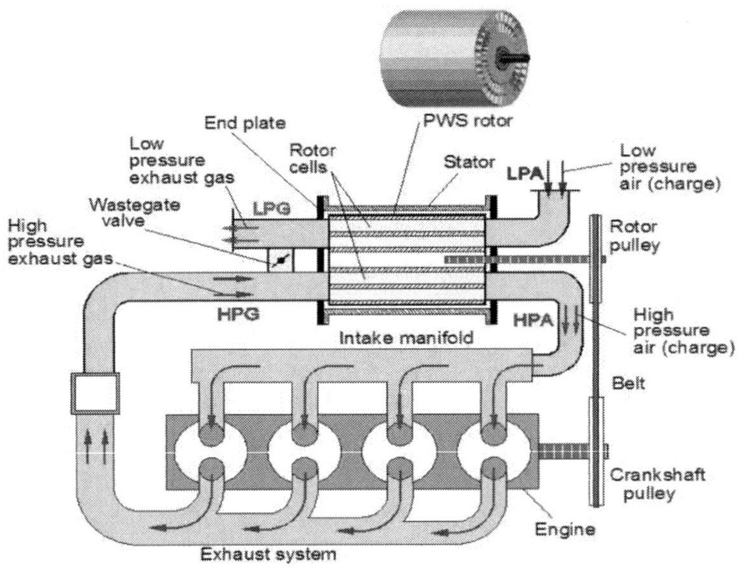

Figure 2. Four port PWS working fluids and assembly elements.

The configuration with four end plates that contain the inlet and outlet ports consists of: two ports for the exhaust gases (HPG and LPG), and two ports for the air (LPA and HPA). PWS can be designed as shown in figure 3 in two configurations: passing-through flow (TF) - when the working fluids flow in the same direction, and reverse flow (RF) - when each fluid enters and exits on the same side.

These two configurations ensure similar performance values but they are substantially different. In the 4 ports TF rotor, both fluids travel the rotor longitudinally, maintaining an average temperature relatively constant over the entire length of the rotor, allowing a self-cooling effect to the rotor. On the RF configuration, the air casing remains cold while the gases casing gets hot, being suitable for other types of applications [5]. In this paper, a RF configuration is used for the numerical modelling of the PWS operation.

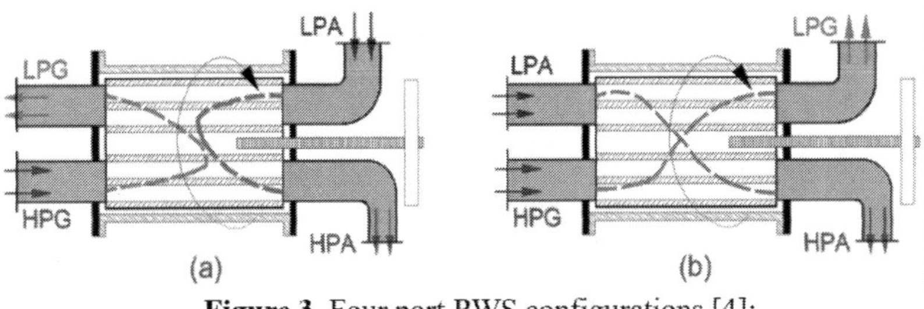

Figure 3. Four port PWS configurations [4]:
(a) The passing-through flow configuration, (b) The reverse flow configuration.

In short terms, the PWS rotor, as it rotates, allow the channel openings to be exposed alternatively to the inlet or outlet openings, therefore the fluids flow through the ports. The compression and expansion waves are initiated inside the rotor channels; the pressure waves evolving inside the channels compressing the intake charge.

In practice, the PWS design is rather difficult, concerning the number of channels, the number of rows or, for instance, breaking the symmetry of the rotor cells for reducing the noise [4-9]. Also, the leakage is of great importance as it seriously affects the performance – it has to be minimized, yet preventing the occurrence of contact, regardless of the thermal regime. A 0.5mm gap was taken into account in the present model design.

3. Pressure Wave Supercharger Modelling and Calculations

3.1. Introduction
In order to model the pressure wave compressor processes, we need to build the basic equations based on fundamental physical principles for viscous flow, choose a suitable model of the fluid and obtain mathematical equations, which properly describe the physics of the flow. In this part, the equations are be presented for unsteady, compressible and viscous three-dimensional flow in a general form. The viscous approach was chosen because the effects of viscosity, thermal conduction, and mass diffusion are important. Also, particularized models for pressure wave compressor used during the years are presented after the general model. The steps for mathematical modeling usually are:

I. State three fundamental physical principles of nature, namely:

a. Mass is conserved (i.e., mass can be neither created nor destroyed).

b. Newton's second law: force = mass × acceleration.

c. Energy is conserved; it can only change from one form to another.

II. Determine a suitable model of the fluid: the finite control volume, the infinitesimal fluid element and molecular model.

III. Apply the fundamental physical principles to the chosen model of the fluid in order to obtain mathematical equations which properly describe the flow.

3.2. The equations of compressible viscous flow
A complete set of equations presented in the differential form below are the compressible Navier-Stokes equations for an isotropic Newtonian fluid with variable properties [11, 12]. The first three equations are conservation of mass, momentum and energy, and the rest of the equations are constitutive equations to complete the model.

$$\frac{D\rho}{Dt} + \rho\nabla \cdot V = 0 \quad \text{(1 eq.)}$$
$$\rho\frac{DV}{Dt} = -\nabla p + \rho f + \nabla \cdot \tau \quad \text{(3 eqs.)}$$
$$\rho\frac{D}{Dt}\left(e + \frac{1}{2}V \cdot V\right) = \rho\dot{q}_{gen} - \nabla \cdot \dot{q}_{visc} - \nabla \cdot (pV) + \rho(f \cdot V) + \nabla \cdot (\tau \cdot V) \quad \text{(1 eq.)}$$
$$\tau = \mu(\nabla V + \nabla V^T) + \lambda(\nabla \cdot V)I \quad \text{(6 eqs.)}$$
$$\dot{q}_{visc} = \dot{q}_{cond} = -k\nabla T \quad \text{(3 eqs.)}$$
$$\dot{q}_{gen} = 0 \quad \text{(0 eq., if neglected)}$$
$$\mu = \mu(\rho, T) \quad \text{(1 eq.)}$$
$$\lambda = \lambda(\rho, T) \quad \text{(1 eq.)}$$
$$k = k(\rho, T) \quad \text{(1 eq.)}$$
$$p = p(\rho, T) \quad \text{(1 eq.)}$$
$$e = e(\rho, T) \quad \text{(1 eq.)}$$

The unknown variables of the above differential system are:

- ρ – density, scalar, 1 variable [kg m^{-3}]
- V – velocity, vector, 3 variables [m s^{-1}]
- p – pressure, scalar, 1 variable [N m^{-2}]
- e – internal energy, scalar, 1 variable [J kg^{-1}]
- T – temperature, scalar, 1 variable [K]
- τ – viscous stress, symmetric tensor, 6 variables [N m^{-2}]
- q_{cond} – heat flux vector, vector, 3 variables [W m^{-2}]
- μ – first coefficient of viscosity, scalar, 1 variable [kg m^{-1}s^{-1}]
- λ – second coefficient of viscosity, scalar, 1 variable [kg m^{-1}s^{-1}]
- k – thermal conductivity, scalar, 1 variable [W m^{-1}K^{-1}]

The volumetric body force considered in most models is given only by a gravitational field and there f = g and is the constant gravitational acceleration. The result is a differential system with 19 equations and 19 variables, which can be solved. But, this system must be and become consistent with considering the second law of thermodynamics that states:

$$\rho \frac{Ds}{Dt} \geq -\nabla \cdot \left(\frac{\dot{q}_{ht}}{T}\right)$$

and the result is:

$$-\frac{\dot{q}_{ht}}{T^2}\nabla T + \frac{1}{T}(\tau : \nabla V) \geq 0$$

The constitutive equations for thermal conductivity $k(\rho, T)$, or heat flux $\dot{q}_{ht}$, first viscosity coefficient $\mu(\rho, T)$, the second viscosity coefficient $\lambda(\rho, T)$ and then the viscous stress τ must be constructed and verified as that to respect the second law of thermodynamics.

4. Numerical model

In the present work, the PWS geometry was created in a 3-D model using ACAD software and was imported into Comsol software [13, 14] (see figure 4). The model uses baseline dimensions for Comprex CX-93 four port reverse flow pressure wave supercharger. The mesh resulted after some geometry repair operations are presented in figure 5 consisting of 2386820 tetrahedral elements only on the air domain after meshing.

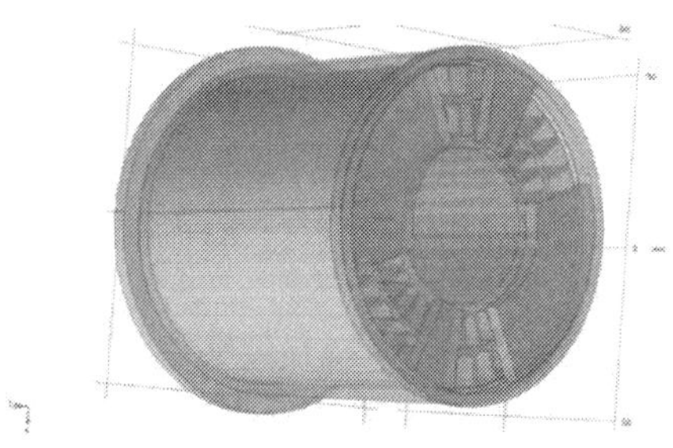

Figure 4. PWS imported geometry.

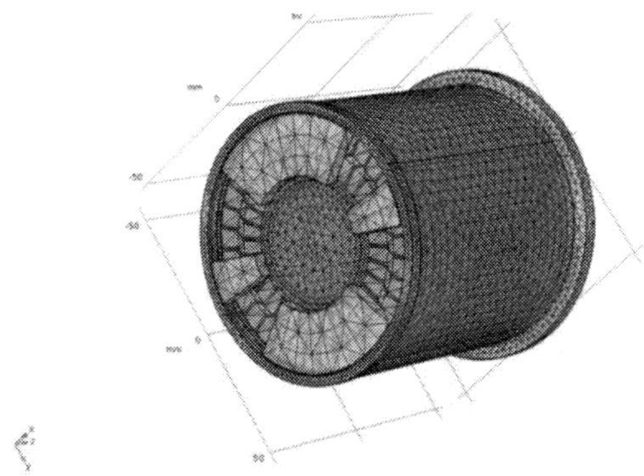

Figure 5. PWS resulted mesh.

The rotor material was considered steel and the fluid used is the air, considered a compressible gas. The specific heat, thermal conductivity and viscosity were considered as temperature dependent.

The model uses the Reynolds-averaged Navier-Stokes (RANS) equations in the air domain. The properties for the fluid are the air at atmospheric pressure and for the solid those of steel. The air properties are considered as temperature dependent. The equations for air domain, which includes equations for turbulent kinetic energy k and for dissipation ε, are used for the rotating domain model. To simulate the PWS time dependent behaviour, a rotating domain model was used. The first step was the frozen rotor study to provide an initial guess for the velocity field and pressure in the 3D domain. In the second step, the flow was treated as turbulent and time dependent using the coupling of the conservation and momentum equations with the energy equation.

The boundary conditions at inlet and outlet ports were set up as: a velocity and pressure input on the gas ports, and a pressure input on fresh air ports. To the entire rotating domain, rotor domain containing fluid, an axial motion was set up with the rotational speed *n*. The boundary conditions are presented in table 1.

Table 1. Entry data for the model

Property	Value	um
Air inlet pressure	$0.98 \cdot 10^5$	$N\ m^{-2}$
Air inlet temperature	293 K	K
Exhaust gas inlet pressure	$1.80 \cdot 10^5$	$N\ m^{-2}$
Exhaust gas inlet temperature	1465	K
Air specific heat ratio	1.4	-
Air specific gas constant	287	$J\ kg^{-1}K^{-1}$
Channel length	93	mm
Rotor diameter	93	mm
Revolutions per time	12600	rpm

5. Results and conclusion

The 3D model results are presented below. Figure 6 presents the pressure distribution in PWS channels in a contour plot, as figure 7 shows the velocity field.

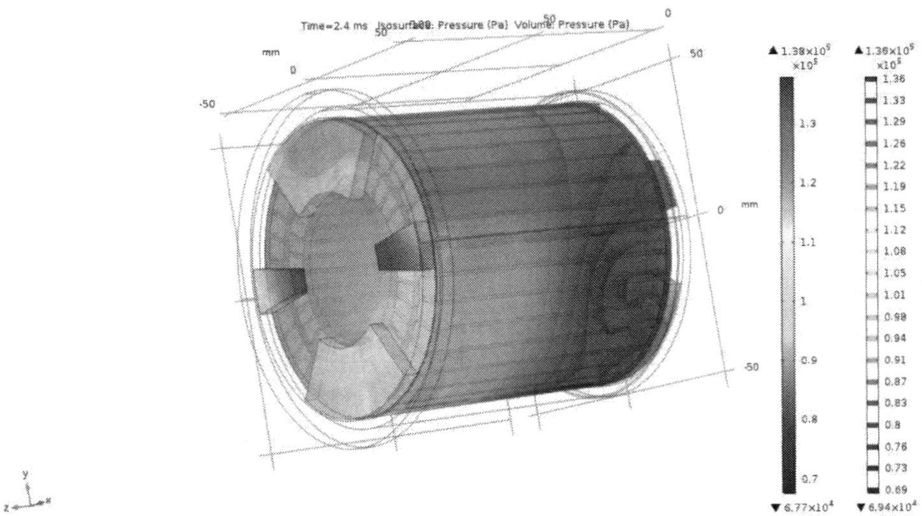

Figure 6. Pressure field @2.4ms time for 12600 rot/min.

The results obtained for the pressure on PWS air ports on the left side of figure 6 show that the compressed air at the moment of 2.4 msec. of time reaches the value of 1.36 bar on exit air ports. In figure 6 also are shown the resulting pressure field inside channels from exhaust gas inlets, the right side of the figure, to air side on the left. For the velocity field the results obtained are presented in figure 7. The PWS air ports are on the right side of figure 7 that shows that the exhaust gas enters with a maximum value of 75 m/s through inlet ports and travels to air side. Because of the rotating domain of the CX rotor, the velocity field is deflected from exhaust gas inlets to air ports. In figure 7 was shown the resulting velocity field inside channels from exhaust gas inlets, from the left side of the figure, to air side on the right side.

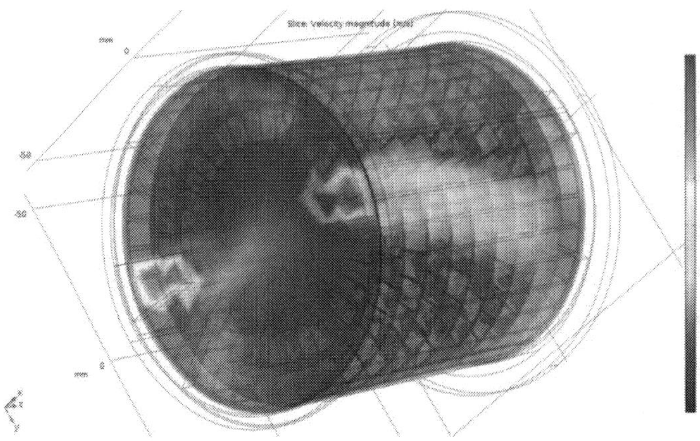

Figure 7. Velocity field @2.4ms time for 12600 rot/min.

As a conclusion, using a pressure wave supercharger to improve the performance of an internal combustion engine appears to be a promising solution since the exhaust gas generates an increase of air pressure on the aspired air, which has a beneficial effect on the internal combustion engine cycle. All theoretical and numerical results presented here encourage the idea that at microscale compression by pressure waves may be more efficient than by conventional mechanical compressors or turbochargers. The presented model can be used on other types of PWSs to enhance or optimize the engine performance or energy and ecological management.

References

[1] *** http://ec.europa.eu/clima/citizens/eu/index_ro.htm - European Comision - Climate Action

[2 *** http://www.eea.europa.eu/ EEA, 2015b, Monitoring CO_2 emissions from new passenger cars and vans in 2014. EEA Technical report No 16/2015, European Environment Agency.

[3] Heisler, H. - Advances Engine Technology, SAE International, 1995, ISBN 1560917342

[4] Fu JQ, Liu JP, Xu ZX, Ren CQ, Deng BL 2013 A combined thermodynamic cycle based on methanol dissociation for internal combustion engine exhaust heat recovery. Energy **55**, 778–86

[5] Fu J, Liu J, Wang Y, Deng B, Yang Y, Feng R, Yang J 2014 A comparative study on various turbocharging approaches based on IC engine exhaust gas energy recovery, App.Energ. **113**

[6] Spring P 2006 Modeling and Control of Pressure-Wave Supercharged Engine Systems, Diss. ETH No. 16490, Swiss Federal Institute of Technology, ETH Zurich

[7] Liu JP, Fu JQ, Ren CQ, Wang LJ, Xu ZX, Deng BL 2013 Comparison and analysis of engine exhaust gas energy recovery potential through various bottom cycles. Appl.Therm.Eng **50**

[8] Iancu F, Piechna J, Müller N 2008 Basic design scheme for wave rotors, Shock Waves **18**, Springer Verlag

[9] Frackowiak M, Iancu F, Potrzebowski A, Ackbari P, Müller N, Piechna J 2004 Numerical Simulation of Unsteady Flow Processes in Wave Rotors, Proc. of IMECE04 2004 ASME International Mechanical Engineering Congress, Anaheim, California USA

[10] Heywood J 1988 Internal Combustion Engine Fundamentals, McGraw-Hill International Editions

[11] Powers JM 2015 Lecture notes on gas dynamics. University of Notre Dame, USA

[12] Çengel YA, Boles MA 2006 Thermodynamics: An Engineering Approach, 5th edition, McGraw-Hill, 2006

[13] Costiuc I, Chiru A 2017 Thermodynamic Process Modeling in Pressure Wave Superchargers, RoJAE **23**, 83-88

[14] Comsol v.3.5a – Academic Licence

Investigations of pressure field along a channel of a pressure wave supercharger

I Costiuc[1], L Costiuc[2], A Chiru[1]

[1]Automotive and Transport Engineering Department, Faculty of Mechanical Engineering, Transilvania University of Braşov, BV 500036, Romania
[2]Mechanical Engineering Department, Faculty of Mechanical Engineering, Transilvania University of Braşov, BV 500036, Romania

Abstract. The aim of the paper is a numerical investigation of the evolution of the pressure field along the wave rotor channels of a pressure wave ICE supercharger. In the present literature, most of the studies are considering the fluids as incompressible and inviscid in a 2D field. The present study is using the compressible and viscous terms in the unsteady Lattice Boltzmann method for fluid in a 3D field. The geometry was drawn in CAD software using measurements made on a real model of the CX-93 pressure wave supercharger. The simulation was conducted using a code for native unsteady LBM approach to reproduce data such as pressures, temperature and mass flows, which are usually hard to be measured in a real pressure wave supercharger. The computational domain was modelled as a moving rotational domain with adaptive refinement. Results such as velocity, pressure and temperature field in the rotor channels were obtained for exhaust gas inlet pressure of 0.292 MPa and 721 K temperature at different rotational speeds. The air inlet state considered was: 0,096 MPa and 313 K. The simulated values obtained are similar to the reported experimental results found in the literature showing a good concordance with the model.

1. Introduction

A supercharger is a device positioned on the intake side of the engine to increase the air pressure at the inlet to the engine cylinder. By using a mechanical supercharger, the improvement of the thermal efficiency of the internal combustion engine (ICE) is limited, because part of the effective power of the ICE is consumed to drive the compressor, as shown in [1, 2]. Another conventional approach to increase the fresh air pressure is the use of the exhaust gas energy within turbines, which drive a suction air compressor (turbocharging). Turbochargers are the most widely used solutions for car manufacturers to produce useful boost, because of their advantages, such as a higher thermal efficiency due to the energy recovery of the engine's exhaust gas.

Other constructive alternatives for supercharging manage to overcome the shortcomings of the engine turbocharger, such as the turbo-lag [2]. One alternative is the pressure wave supercharger (PWS). Pressure wave aggregates, known also as wave rotors, use shock waves to transfer energy directly between fluids without using mechanical components, thus having the potential to increase efficiency [3, 5-7]. In a PWS, there is a direct interaction between high pressure and high temperature exhaust gas and low pressure and low temperature air. In short, hot gases produce a shock wave that extends through the wave rotor channels and compresses the fresh air. The quick response to engine torque for the full range of engine speeds and the intake air pressure boost are reasons to consider PWS a good ICE supercharging option for vehicles.

The present paper aim is a numerical investigation of the evolution of pressure field along the wave rotor channels for a PWS, taking into account the compressible and viscous character of the fluids when using the unsteady Lattice Boltzmann method for fluids in a 3D field [7]. The PWS was modelled in CAD software after the real geometry of a COMPREX CX-93 pressure wave supercharger.

2. Numerical model

The numerical modelling used a 3D CFD model using FEM for PWS [4]. The 3D approach showed the distribution of the velocity, pressure and temperature field in the entire modelled volume. This approach involves a consistent calculation effort, even if the fluid is considered inviscid or in a stationary state. That is due to a large number of finite elements used. In the present study, it is used a 3D numerical model using the Lattice Boltzmann Method. This method was chosen because it is implicitly non-stationary and allows native modelling of rotational motion. The implementation of the LBM method in XFlow-2013 made possible the realistic simulation of the possible conditions in the PWS channels. The 3D-PWS channels in the rotor are modelled using the basic dimensions for the CX-93 pressure wave rotor; also, the dimensions of the cold and hot side of the PWS with compression and expansion pockets are considered. The 3D geometry was created using AutoCAD as shown in figure 1 and figure 2.

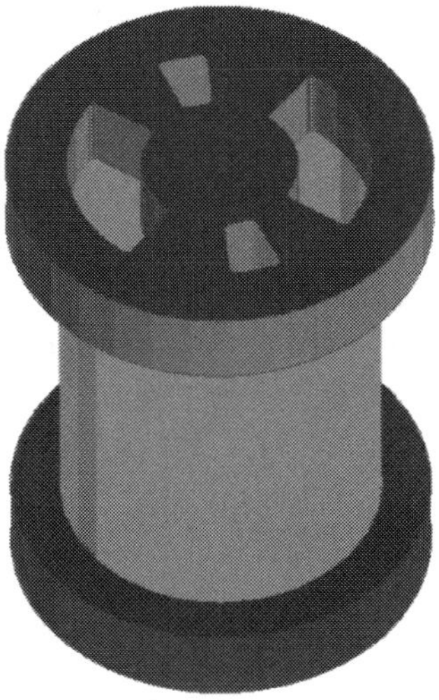

Figure 1. PWS cold air (blue) and hot gases (red) side CAD geometry.

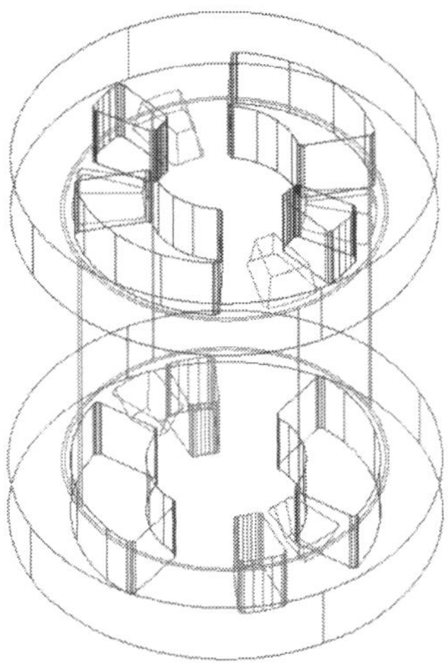

Figure 2. PWS raster geometry.

The final numerical model, after some import resulting geometry repair operations, is shown in figure 3 and figure 4 and consists of 42839 D3Q27 elements. The rotor material used in modelling was considered steel and the fluid used is air. The fluid was considered a compressible gas, and the specific heat, thermal conductivity and viscosity were considered as temperature dependent.

An implicit solver model was used, coupling the conservation and momentum equations with the energy equation and the flow was treated as turbulent and time dependent. To simulate the PWS behaviour in the rotating machinery model a fixed rotational speed of 12500min^{-1} was used. The simulation time was set to 10 seconds to investigate the PWS stable response.

The boundary conditions at inlet and outlet ports were set up as: a pressure, temperature input on the hot side of the inlet gases duct, and a pressure and temperature input at the cold side for fresh air inlet duct. The PWS channels and the ports were initially assumed to be filled with air at a reference constant pressure and temperature. The entire rotating domain containing fluid and an axial motion was set up with the rotational speed.

To investigate the pressure in channels, a plot variable line is used as it is shown in figure 3, where the rotor is hidden, and figure 4, where the hot side is hidden. This plot line setting connects the inlet of hot gases with the exit of the air and is used to visualize the pressure distribution.

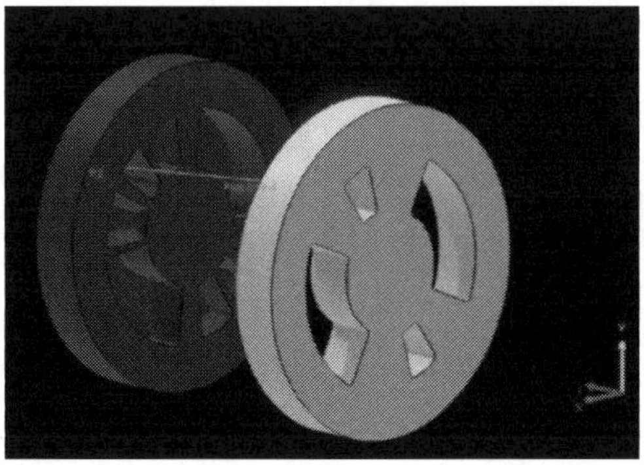

Figure 3. PWS cold air (blue) and hot gases (green) side CAD geometry.

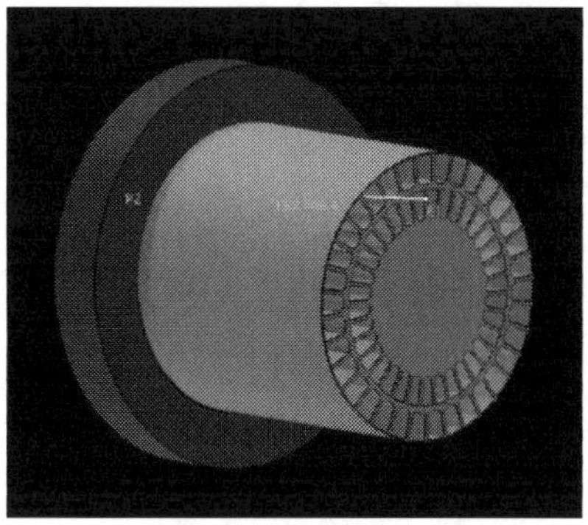

Figure 4. PWS rotor geometry.

The model was solved using single phase fluid, resolving both momentum and energy equations, with an enabled viscous term in the energy equation. The equation of state for air is used as Boussinesq state equation model. For viscosity, the Sutherland model is used. The solution used for turbulence modelling is the Large Eddy Simulation (LES) with introducing an additional viscosity, called turbulent eddy viscosity to model the subgrid turbulence. The LES scheme uses the Wall-Adapting Local Eddy viscosity model as presented in that provides a consistent local eddy-viscosity and near wall behaviour. The near wall treatment in the boundary layer uses a generalized law of the wall that takes into account the effect of adverse and favourable pressure gradients as presented in [7].

The boundary conditions used in the simulation are presented in table 1.

Table 1. Entry data for the model

Property	Value	UM
Static cold-side inlet pressure ($P_{1,0}$)	$0.96 \cdot 10^5$	Nm^{-2}
Static cold-side inlet temperature ($T_{1,0}$)	313	K
Static hot-side inlet pressure ($P_{2,0}$)	$2.92 \cdot 10^5$	Nm^{-2}
Static hot-side inlet temperature ($T_{2,0}$)	721	K
Air specific heat ratio (k_{air})	1.4	-
Air specific gas constant (R_{air})	287	$Jkg^{-1}K^{-1}$
Channel length (L)	93	mm
Rotor diameter (D)	93	mm

The software code for the LBM model imposes a stability parameter against time [8] with values between 0 and 1, with respect the feedback regarding the stability of the numerical scheme as the Courant–Friedrichs–Lewy (CFL) condition. A stability parameter near to 1 means the stability of the numerical scheme is not ensured and the simulation may diverge. The ideal value should be around 0.2 - 0.4. In this simulation, the resulted stability parameter is between 0.18 and 0.31 as is presented in figure 5, captured at $3x10^{-3}$ seconds.

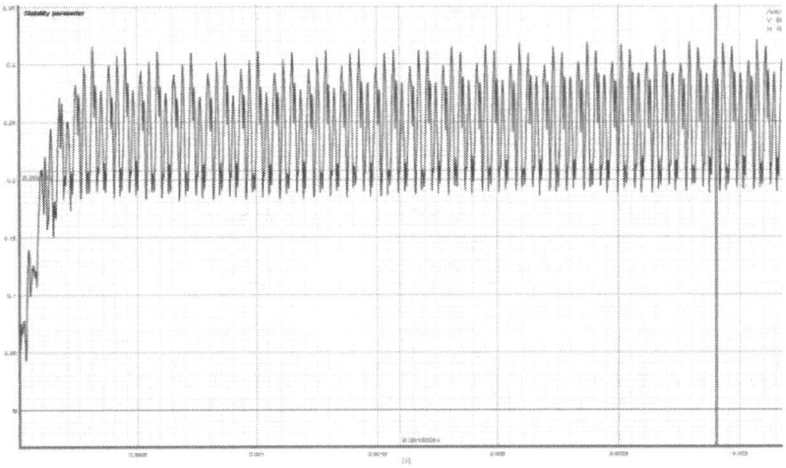

Figure 5. Stability parameter against time.

3. Results and discussions

The results obtained with the LBM model described above are presented below for the 19×10^{-3} seconds of simulation. The velocity distribution in PWS channels in the contour plot is presented in figure 6. The temperature field is presented in figure 7. The pressure field and the pressure distribution for a single PWS channel, connecting the exhaust gas inlet and the compressed air outlet, are presented in figure 8 and figure 9.

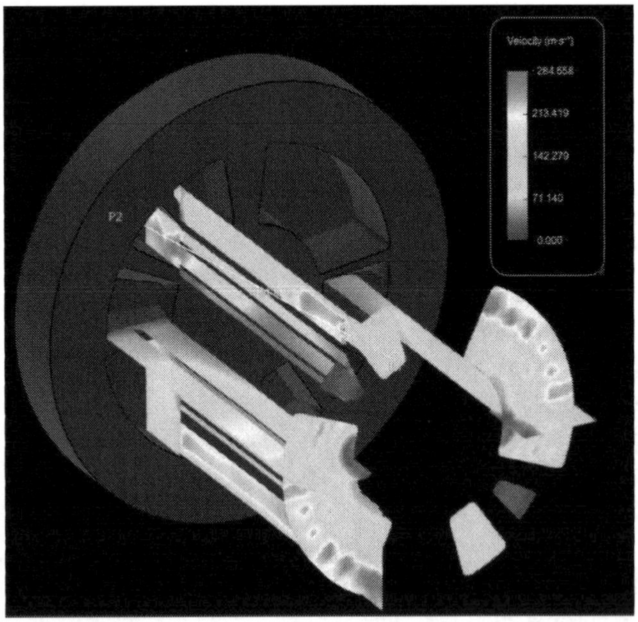

Figure 6. Velocity field for a 3D model using 3 cut-planes (XY, YZ and XZ).

From the simulation point of view, the model shows the non-uniform thermal and velocity fields. The velocity field, figure 6, shows that the variations of velocity in channels, as well in the output exhaust gas ports and output compressed ports were quite large from 70 to 284 m/s, depending on the position of the rotor channels.

IOP Conf. Series: Materials Science and Engineering 1220 (2022) 012023 doi:10.1088/1757-899X/1220/1/012023

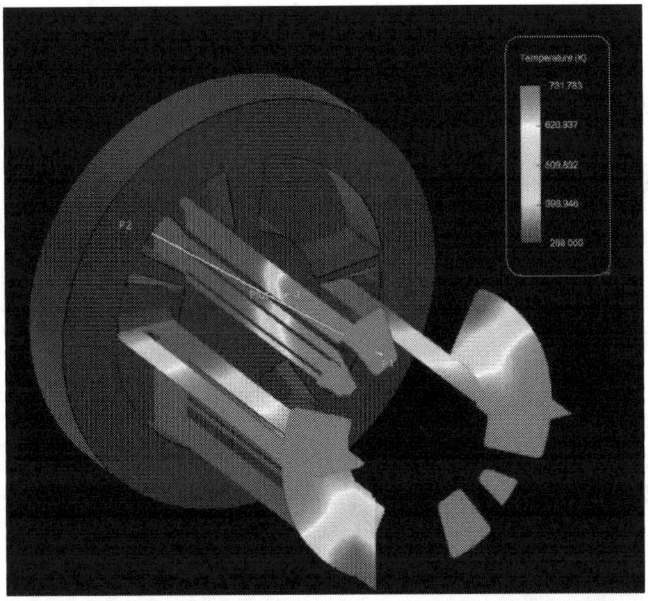

Figure 7. Temperature field for a 3D model using 3 cut-planes (XY, YZ and XZ).

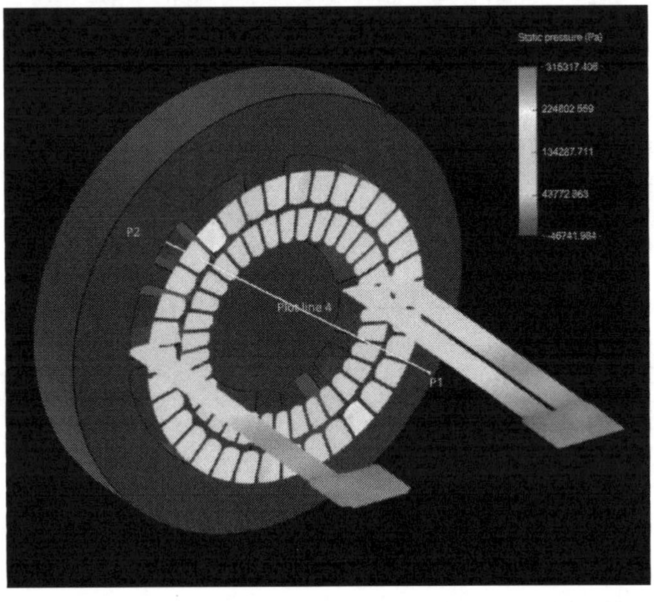

Figure 8. Pressure field for a 3D model using 3 cut-planes (XY, YZ and XZ).

The temperature field, as can be seen in figure 7, is also non-uniform along the rotor case. The main result for the temperature field is the fact that the interfaces of air-exhaust gases do not reach the cold side, and it remains almost near to the hot side (exhaust gases). The pressure field is presented in figure 8.

The pressure evolution along the channel is presented in figure 9 and shows that the compressed air is evacuated from PWS with a value of 1.47 bar, which correspond to a 1.53 compression ratio. These results are similar to the experimental values found in the numerous reported studies about CX-93 found in the literature, validating thus the present model.

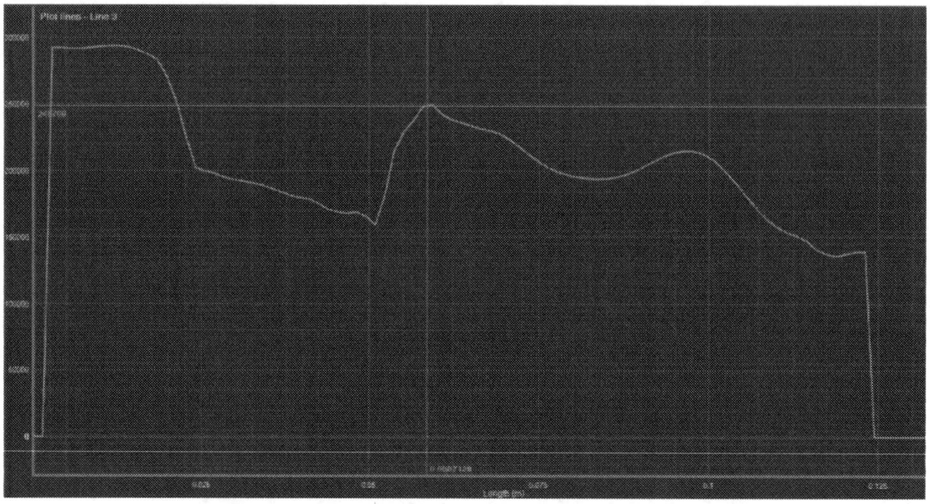

Figure 9. Pressure along a channel.

4. Conclusion

Using a wave rotor for the ICE supercharging reduces the baseline compression pressure ratio and the compression exit temperature. Furthermore, this may reduce the rotational speed which results in reduced mechanical and thermal stresses and relaxed design constraints.

The numerical results presented here encourage the idea that at microscale compression by shock waves may be more efficient than by conventional centrifugal compressors or turbo-compressors. However, the simulation results will be useful for further validation of the presented model with data considering variable rotational speed and experimental data. Also, a similar model for other different PWS types (dimensions, geometry, material, number of channels, etc) can be developed in future studies, as an instrument to improve any feature or performance of existing wave rotors.

References

[1] Spring P 2006 Modelling and Control of Pressure-Wave Supercharged Engine Systems, Diss. ETH No. 16490, Swiss Federal Institute of Technology, ETH Zurich

[2] Fu J, Liu J, Wang Y, Deng B, Yang Y, Feng R, Yang J 2014 A comparative study on various turbocharging approaches based on IC engine exhaust gas energy recovery, Appl. En.**113**

[3] Liu JP, Fu JQ, Ren CQ, Wang LJ, Xu ZX, Deng BL 2013 Comparison and analysis of engine exhaust gas energy recovery potential through various bottom cycles. Appl. Therm. Eng. **50**

[4] Costiuc I, Costiuc L 2017 Mesh based modeling of a pressure wave supercharger, Proc. Int. Conf. on Computational Mechanics and Virtual Engineering, COMEC 2017

[5] Iancu F, Piechna J, Müller N 2008 Basic design scheme for wave rotors, Shock Waves **18**, Springer Verlag

[6] Frackowiak M, Iancu F, Potrzebowski A, Ackbari P, Müller N, Piechna J 2004 Numerical Simulation of Unsteady - Flow Processes in Wave Rotors, Proc. of IMECE04 2004 ASME International Mechanical Engineering Congress, Anaheim, California USA

[7] Holman MD, Brionnaud RM, and Abiza Z 2012 Solution to industry benchmark problems with the Lattice-Boltzmann code Xflow, EU Congress on Computational Methods in Applied Sciences and Engineering (ECCOMAS 2012), J. Eberhardsteiner et.al. (eds.), Austria

[8] XFlow Tutorial Guide 2013. Next Limit, www.nextlimit.com

IOP Conf. Series: Materials Science and Engineering 1220 (2022) 012024 doi:10.1088/1757-899X/1220/1/012024

Numerical evaluation of vehicles aerodynamics in platoon using CFD simulation

I L Scurtu, M I Gheres

Department of Automotive Engineering and Transports, Technical University of Cluj-Napoca, Romania

marius.gheres@auto.utcluj.ro

Abstract. Due to increased traffic and new technologies developed to improve road safety, a new vehicle driving technique is being studied. The vehicles' platooning driving method has as objective to minimize the aerodynamic drag and therefore the fuel consumption. The aim of this study is to evaluate the distance between platoon traveling vehicles and to propose an optimal travel distance. The CFD evaluation is performed for two categories of vehicles at a given velocity and distance between them for two cases: in the first simulation scenario, a tractor-trailer is represented, followed by a SUV positioned at a given distance from the rear of the vehicle combination; in the second simulation, two tractor-trailers at the same boundary condition as for the first case are simulated. The vehicles models are made by using a CAD modelling environment, respecting the overall dimensions of an existing vehicle. The numerical evaluation of a distance between vehicle models is performed using the CFD method based on the Navier-Stokes equations averaging. The last part of the paper presents the interpretation of CDF simulation results, establishing the recommended travel distance between vehicles and conclusions of this study.

1. Introduction

The growth of the vehicles fleet requires the creation of more efficient transport infrastructure. The aerodynamic interaction of the vehicles in platoon makes possible a lower fuel consumption depending on the distance between the vehicles. For conducting this aerodynamic study several papers are studied with the purpose to determine the state of the art regarding the CFD evaluation of vehicles in platoon. The vehicles equipped with Intelligent Transport System, like distance sensor, adaptive braking system, global position system, adaptive cruise control, allow the lower distance between the vehicles than a vehicle length keeping the road safety of the vehicles in platoon [1]. An experimental investigation study of a two-truck platoon considering the inter-vehicle distance, lateral offset and yaw is performed by Törnell et al. [2], where, in experiments, were used two 1:6 scale detailed models in a wind tunnel with a moving ground. In conclusion, a short inter-vehicle distance leads to a reduction of drag coefficient, which allows a more efficient running on the road.

Iozsa et al. [3] has performed an CFD study for four truck convoy at 25 m/s air velocity in Ansys software. This study of aerodynamics consists in streamlines analysis, velocity vector analysis and pressure map for four identical trucks that follow each other at a 3 m distance. The results show that this distance between vehicles can reduce by 75% the aerodynamic resistance for the vehicles included in platoon A CFD comparative experiment is performed by Ebrahim et al. [4] where is studied the platoon aerodynamics between Ahmed model for two inter-vehicle spacing configurations, with 0.125 L and

1 L intervehicle spacing, where L is the body vehicle length. The Ahmed body represents a simplified vehicles model described originally by S.R. Ahmed in 1984. The Ahmed model geometry was modified without mounts (generic wheels) at 0.75 L, 0.5 L and 0.4 L, was based on a 25° slant angle. In conclusion, this study is capable to simulate platoon cases in a virtual wind tunnel without sacrificing aerodynamic resolution.

Gnatowska and Sosnowski [5] described the CFD analysis inter-vehicles distance effect in platoon arrangement. The vehicles models are represented by the two identical Ahmed bodies. In terms of determining the drag coefficient values, the frontal area was 0.007 m^2 and the slant angle is to 25 degrees. The inter-distance between the bodies is in the range of 0÷1.2 L. The CFD experiment is conducted at two velocities of 8 m/s and 88 m/s. Following the simulations, the complex flow fields have been visualized and it can be observed the velocity fields and pressure map generated at different regions. Simeon et al. [6] presented an aerodynamic CFD study for four vehicles in platoon, using TruckSim software for predicting the fuel consumption efficiency for each vehicle. The vehicles are inter-spaced in platoon at 30 ft (9.144 m), 50 ft (15.24 m) and 100 ft (30.48 m) by using a cruise control system on the truck. Another platoon CFD study was performed using OpenFoam software [7]. This presents the estimation error-based performance of five learning algorithms: Support vector regression, Polynomial regression, Linear regression and two different models of Neural Networks. The drag coefficient for each vehicle in two, three and four vehicle platoons are provided at different inter-vehicle distances. Aerodynamic drag analysis of autonomous electric vehicle platoons was studied by Kaluva et al [8]. The vehicles platoon was studied in urban environments for two vehicles model: a minibus and a passenger car. The platoon size is in the range of 1÷7 vehicles, running at 10 m/s velocity. The obtained results show a reduction of drag coefficient of the platoon of up to 24% at a 1 m inter-vehicle distance. Another CFD study of three platoon trucks with a sensing and control system was performed at 80 km/h velocity, with an inter-distance between trucks of 10 m. By using the vehicles platooning the fuel consumption has improved by approx. 14 % [9].

2. Work method

This research concerns a numerical investigation of the airflow velocity distribution, air pressure on the vehicle body and drag coefficient computed at 25 m/s velocity of the vehicles in platoon. The vehicles are represented by the CAD models of a tractor-trailer and a medium size SUV. These models were designed at a real overall dimension.

2.1. CAD modelling

The model of the medium SUV is designed in SolidWorks CAD environment to the real car size. This model has been used in an aerodynamic study that showed the aerodynamic influence of the bicycle rack [10].

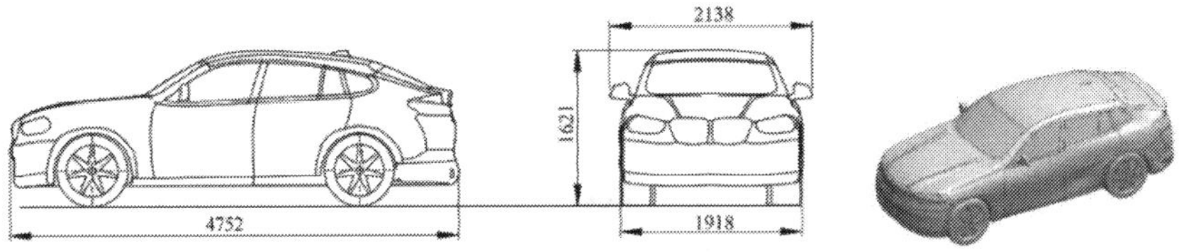

Figure 1. The CAD model and overall dimensions of the SUV.

The overall dimensions of the SUV model are presented in figure 1.

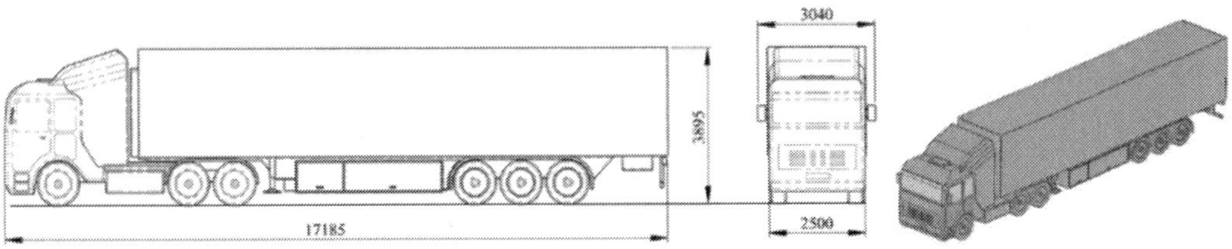

Figure 2. The CAD model and overall dimensions of the tractor-trailer.

In figure 2, the designed CAD model of the tractor-trailer is presented. This model presents most auxiliary elements of a real tractor-trailer. The frontal area of the tractor-trailer model is 9.46 m^2.

2.2. Mathematical approach

2.2.1. Force on vehicle body.
When the vehicle is driving on the road, the aerodynamic forces act on the bodywork, which generates a movement resistance and affects the stability. In this case study, the aerodynamic evaluation of the vehicles in platoon is performed by the computation of the aerodynamic coefficient and aerodynamic forces acting on the vehicles in the opposite travelling direction [11]. The expression of the aerodynamic force is given in relation 1:

$$F_x = \frac{1}{2} \cdot \rho \cdot V^2 \cdot C_D \cdot A \tag{1}$$

where: ρ is the air density, V is the air velocity, A is the frontal area of the vehicle body and CD is the aerodynamic drag coefficient.

From the expression of the drag force, it is deducted the expression of the drag coefficient presented in relation 2:

$$C_D = \frac{2 \times F_x}{\rho \times V^2 \times A} \tag{2}$$

2.2.2. Mathematical model.
The airflow around the vehicle body is characterized by the turbulent flow and the fluctuating velocity fields. The largest fluctuating velocities appear between the space from the cabin and semi-trailer and between the vehicles in case of vehicles platoon. The mathematical approach used the equations Reynolds Averaged Navier Stokes, a model used in most engineering applications that use fluid flow simulations [12]. The equations can be written in the Cartesian tensor form, presented in relations 3 and 4:

$$\frac{\partial U_i}{\partial x_i} = 0 \tag{3}$$

$$\rho = \frac{\partial}{\partial x_j}\left(U_j U_i + \overline{u_j' u_i'}\right) = -\frac{\partial P}{\partial x_i} + \frac{\partial}{\partial x_j}\left(2\mu S_{ij}\right) \tag{4}$$

where: U_i is the mean of velocity tensor, x_i is vector position, $\overline{u_j' u_i'}$ is the average of fluctuating velocity, P is the mean static pressure, μ is the molecular viscosity.

In relation 5 the strain rate expression is given:

$$S_{ij} = \frac{1}{2}\left(\frac{\partial u_i}{\partial x_j} + \frac{\partial u_j}{\partial x_i}\right) \tag{5}$$

The standard k-ε turbulence model is applied in this study and is given in equation 6:

$$\rho U_j \frac{\partial k}{\partial x_j} = \tau_{ij} \frac{\partial U_i}{\partial x_i} - \rho\varepsilon + \frac{\partial}{\partial x_j}\left[\left(\mu + \frac{\mu_T}{\sigma_k}\right)\frac{\partial \varepsilon}{\partial x_j}\right] \tag{6}$$

where: τ_{ij} is the Reynolds stress, μ_T is turbulent viscosity, σ_k is closure coefficient, k is turbulent kinetic energy and ε is the dissipation rate of turbulent energy.

The dissipation rate of turbulence kinetic energy equation is given in the next relation:

$$\rho U_j \frac{\partial \varepsilon}{\partial x_j} = C_{\varepsilon 1} \frac{\varepsilon}{k} \tau_{ij} \frac{\partial U_i}{\partial x_j} - C_{\varepsilon 1}\rho + \frac{\varepsilon^2}{k} + \frac{\partial}{\partial x_j}\left[\left(\mu + \frac{\mu_T}{\sigma_\varepsilon}\right)\frac{\partial \varepsilon}{\partial x_j}\right] \tag{7}$$

The turbulent viscosity expression is given in the relation 8:

$$\mu_T = \rho C_\mu \frac{k^2}{\varepsilon} \tag{8}$$

2.2.3. Reynolds number.
To perform the aerodynamic simulation is necessary to establish the airflow regime. This is established by calculating the Reynolds number that is an important dimensionless variable and represents the ratio between inertial and viscous forces [13]. When the number value is high, the inertial force dominates and viscous forces can be neglected; if the number value is lower, the viscous forces dominate and the inertial forces are usually neglected. The expression of the Reynolds number is presented in the relation 9:

$$R_e = \frac{\rho \cdot V \cdot L}{\mu} \tag{9}$$

To determine the flow rate around the truck and semi-trailer, the Reynolds number is calculated following the relation 1 at V=25 m/s airflow velocity, the air density is $\rho = 1.2041$ kg/m^3, corresponding to a temperature T=20°C, the length of the truck assembly is L = 17.185 m and the dynamic air viscosity $\mu = 1.802 \cdot 10^5$ kg/(m·s). To establish the external flow regime in this case, the airflow is considered over a plane surface. The error that can be highlighted between reality and simulation is given by the fact that the undisturbed air does not move in relation to the ground while in the case of simulation the air is moving in relation to the ground. If the value of the Reynolds numbers is $R_e<5x10^5$ the airflow is laminar. The transient flow regime is situated in the range of $5x10^5< R_e<10^7$, while after $R_e>10^7$ the flow is turbulent. The obtained value of the Reynolds number in the studied case is $2.87x10^7$, which indicate a turbulent airflow.

3. Boundary conditions and numerical simulation

The CFD domain is designed to allow the change of the distance between the vehicles and permits to consider the road in the simulation. For better visualisation of the airflow around the vehicle body's, the computational domain has a width of 6 m, a height of 7 m and a length of 80 m. The first vehicle is placed into the CFD domain at a distance of 8 m from the air inlet. In this study, two simulations are performed: the first CFD simulation case consists of a tractor-trailer model that has a 25 m/s velocity followed with the same speed by the SUV at 5 m, 10 m, 15 m, 20 m, 25 m and 30 m; the second simulation case is performed at the same initial conditions, but a similar tractor-trailer type is replacing the SUV. The velocity value was chosen due to the European speed limitation for the tractor-trailer vehicles. The computational domain for the first case is meshed into 93266 fluid cells, of which 16397 fluid cells are in contact with the solid surface; for the second case, the considered volume is meshed into 107005 cells, of which 25940 fluid cells are in contact with the vehicles surface. For both cases, the mesh is generated automatically at a ratio factor of the initial mesh, which means that the nearby of the fluid layer is denser, to increase the accuracy of the results.

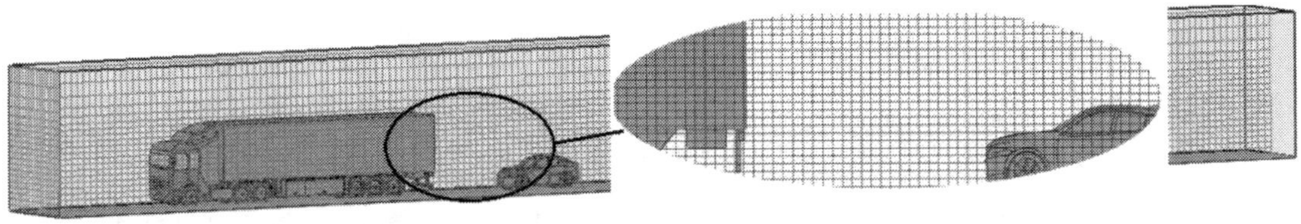

Figure 3. The CFD computational domain and mesh grid.

In figure 3, the meshed grid of the CFD domain is presented. It can be observed that the mesh density is high in the nearby area of the vehicle surface.

To have a comparative term of the results, separate tractor-trailer and SUV simulations are performed to compute the pressure on bonnet and windscreen surfaces, and also the drag coefficient. For the interpretation of these results, the surface of the bonnet and windscreen are chosen as target surfaces. All simulations are performed at a velocity of 25 m/s and with parameters values applied to the calculation of the Reynolds number. Figure 4 shows the bonnet and windscreen surface for both vehicles from which the pressure are taken.

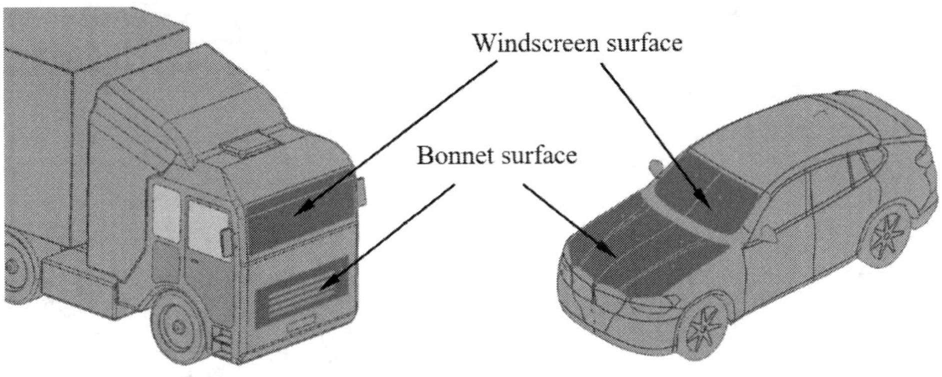

Figure 4. The reference surfaces chosen for both vehicles.

4. Results

This section presents the simulation results. The results are discussed in terms of global drag forces and, respectively, bonnet and windscreen relative pressure obtained for different varied inter-vehicle distances.

4.1. Single truck and single SUV

In this section are presented the first reference results of separated simulations for single tractor-trailer and SUV (table 1). The bonnet and windscreen surface has been chosen because their shape can influence the vehicle aerodynamics.

Table 1. Results for single truck and single SUV

	Tactor-trailer	SUV
Drag coefficient	0.728	0.433
Drag force (N)	2594.13	442.61
Relative pressure on the bonnet surface (Pa)	101633	101235
Relative pressure on the windscreen surface (Pa)	101568	101320

The values of the aerodynamics coefficient, aerodynamic force, relative pressure of the bonnet surface and windscreen are determined at an initial air speed of 25 m/s for each vehicle. Relative pressure values are related to atmospheric pressure.

4.2. Truck followed by SUV

Table 2 presents the numerical results of the truck followed by the SUV in the range of 5-30 m.

For the validation, the results for the leading tractor-trailer followed by SUV are presented in figure 5, in ascending order of the distance between the vehicles. The airflow velocity distribution is in the range of 5-30 m. Following the visual distribution of the velocity, which is presented in figure 5 and the numerical results from table 2. With increasing the distance between vehicles, the drag coefficient and drag force increase. The relative pressure on the SUV bonnet decreases with the increase of inter-distance from vehicles. The longitudinal view of the velocity streamlines distribution is represented on the middle plane of the truck body and the top view is represented on the plane situated at 1 m distance from the ground.

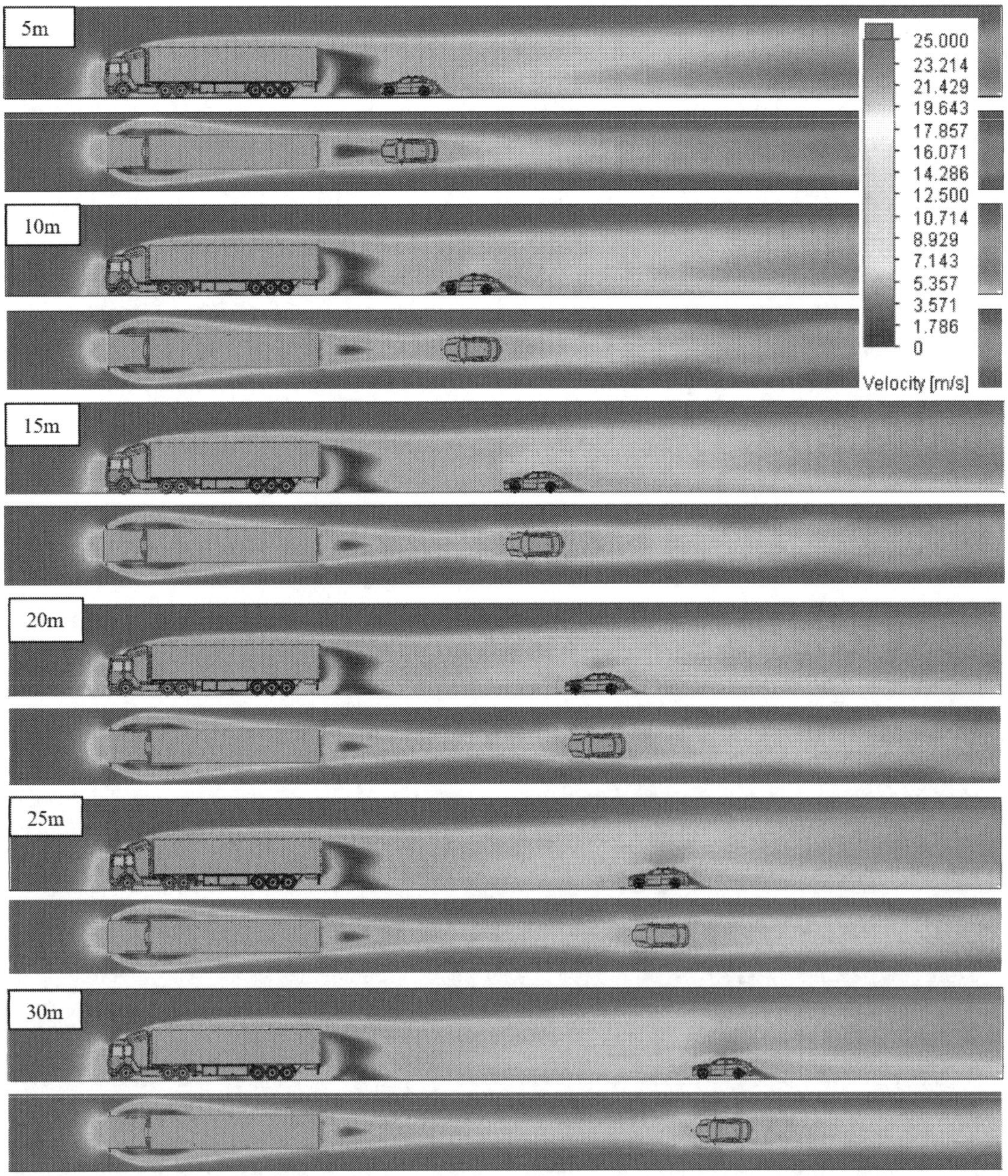

Figure 5. Airflow velocity distribution at different distances of a tractor-trailer followed by SUV.

The value of the drag coefficient and aerodynamic force corresponds to the vehicles assembly consisting of the truck followed by SUV.

Table 2. Numerical results of the truck followed by SUV.

	Distance between truck and SUV (m)					
	5	10	15	20	25	30
Drag coefficient	0.757	0.770	0.781	0.784	0.785	0.788
Drag force (N)	2697.27	2744.93	2781.56	2795.05	2793.03	2809.04
Relative pressure on the SUV bonnet surface (Pa)	101341	101303	101291	101284	101280	101278
Relative pressure on the SUV windscreen surface (Pa)	101338	101314	101316	101315	101312	101322

The relative pressure on the surface of the SUV bonnet is higher in the case of two vehicles than the relative pressure value obtained in the individual SUV simulation, decreasing with the increase in the distance between the vehicles and the pressure on the surface of the windscreen is less in the range of 10-25 m.

4.3. Tractor-trailer followed by another similar tractor-trailer
The airflow velocity distributions between the identical tractor-trailer combinations are presented in figure 6, where it can be observed the simulation results in the same boundary condition as in the previous simulations.

Table 3 presents the numerical results of the simulation between the similar two tractor-trailers at the same boundary conditions.

Table 3. Numerical results for the two tractor-trailer combinations.

	Distance between trucks (m)					
	5	10	15	20	25	30
Drag coefficient	0.950	1.041	1.080	1.119	1.126	1.133
Drag force (N)	3385.44	3708.40	3846.34	3984.71	4009.36	4032.82
Relative pressure on the bonnet surface (Pa)	101393	101441	101465	101489	101495	101500
Relative pressure on the windscreen surface (Pa)	101368	101394	101435	101453	101471	101476

In the case of driving in platoon the aerodynamic resistance value of the second truck is less than in the case of a single truck and the aerodynamic drag on the leading truck increases.

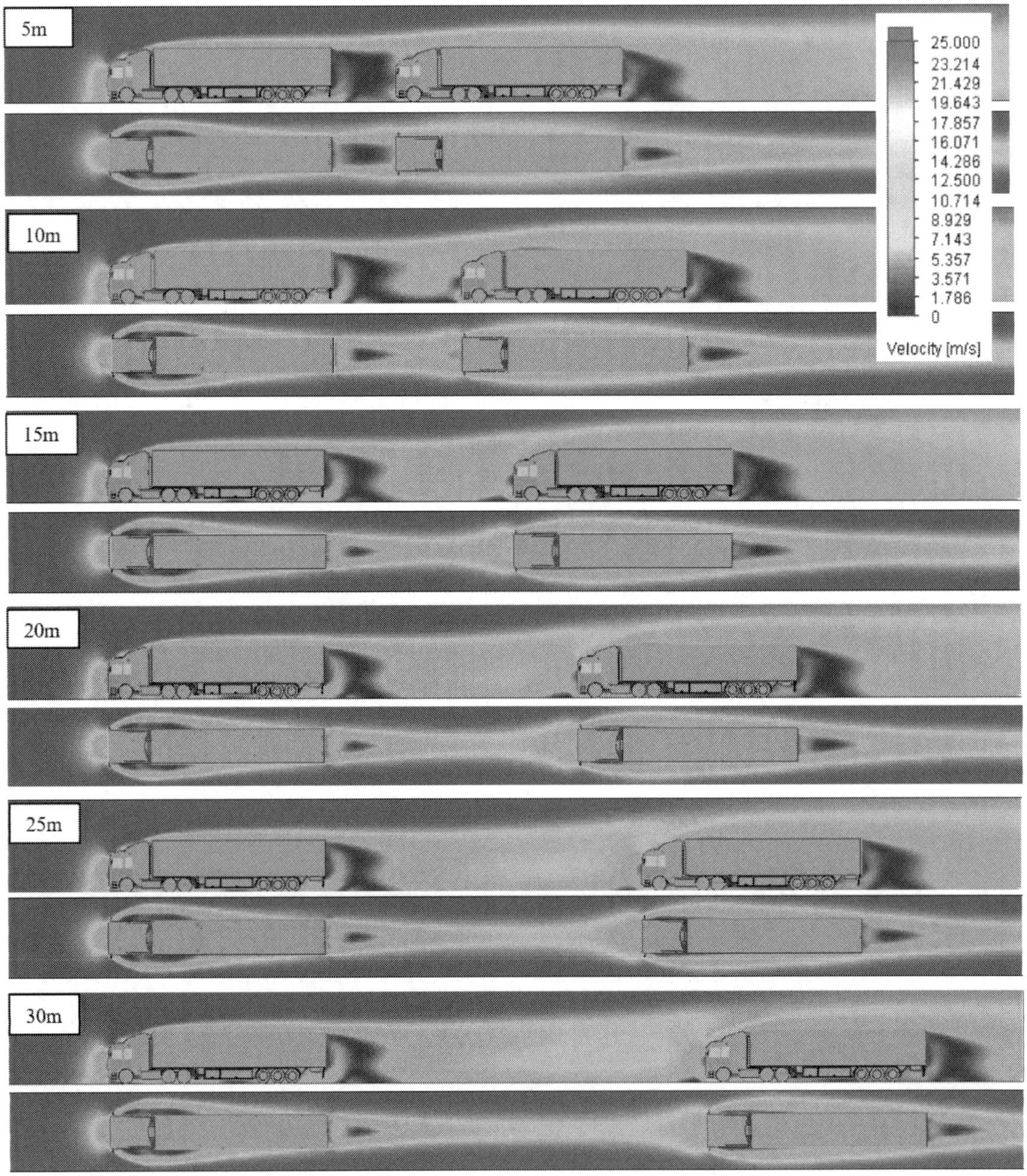

Figure 6. Airflow velocity distribution at different distances of two same tractor-trailers.

4.4. Tractor-trailer during the overtaking operations by the second vehicle

This section presents the case when the tractor-trailer is overtaking by the second vehicle: in the first case, the tractor-trailer is followed by the SUV and in the second case by the tractor-trailer. The overtaking scenario is composed by the trailer-tractor as a leading vehicle followed by the second vehicle positioned at a 5 m distance and the half part of the vehicle body is placed on the right side of the tractor-trailer.

Figure 7 presents the airflow distribution in overtaking intention in case of tractor-trailer followed by SUV, the 7a present the pressure on the vehicle surface and 7b show streamlines distribution-top view at 1 m. In figure 8, it is represented the airflow distribution in overtaking intention in the case of tractor-trailer followed by tractor-trailer, the 8a presents the pressure on the vehicle surface and 8b show streamlines distribution-top view at 1 m.

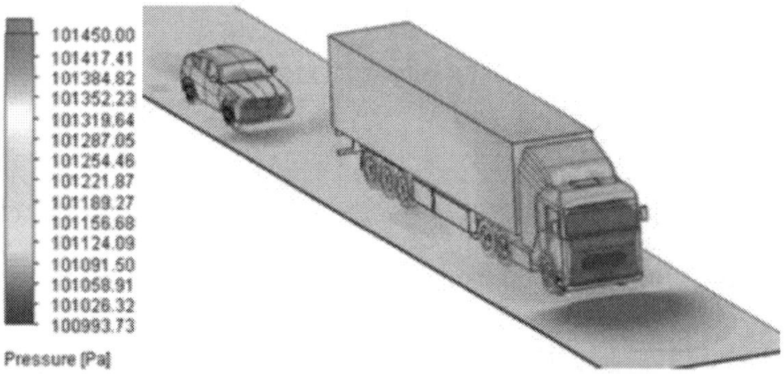

a) Pressure distribution on the vehicles surface

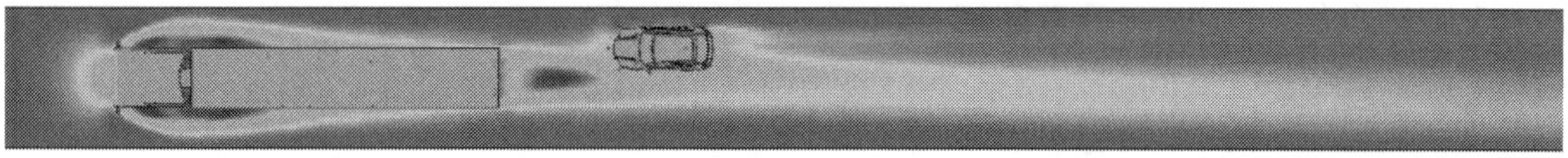

b) Streamlines distribution -top view at 1 m

Figure 7. Airflow distribution in overtaking intention.

Table 4. Numerical results for the overtaking intention case.

	Drag coefficient	Drag force (N)	Relative pressure on the bonnet surface (Pa)	Relative pressure on the windscreen surface (Pa)
Tractor-trailer followed by SUV	0.783	2787	101302	101349
Tractor-trailer followed by tractor-trailer	1.136	4047	101523	101496

a) Pressure distribution on the vehicles surface

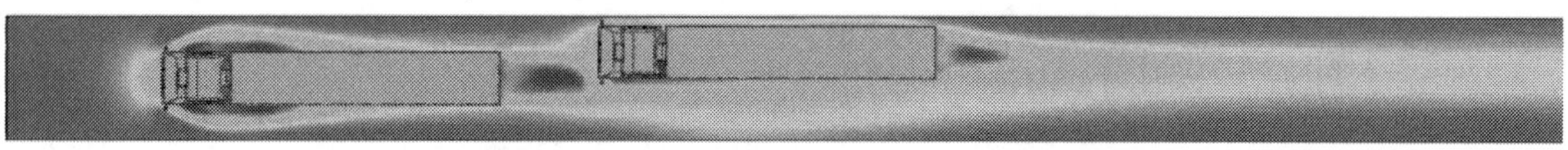

b)Streamlines distribution -top view at 1 m

Figure 8. Airflow distribution in overtaking intention

The drag coefficient and the drag force it is higher in the case of the tractor-trailer followed by a tractor-trailer in overtaking intention because the tractor-trailer has a larger front surface in comparison with SUV.

5. Conclusions

In this paper, the CFD based numerical evaluation of the aerodynamic behaviour in platoon of tractor-trailer and SUV is determined. Accurate numerical simulations are performed to describe the airflow structures around the vehicles body and to obtain the numerical values of the pressure on the bonnet and windscreen surface. For better road safety, 20 m between the vehicles can be considered as a safe choice. The movement of vehicles in convoy is recommended to be applied on the highway so as not to affect the traffic on the two-way public road. The results of this study can be used as a reference for keeping the optimal distance between vehicles through the adaptive cruise control.

References

[1] Schito, P. 2012 *Numerical and experimental investigation on vehicles in platoon* SAE International Journal of Commercial Vehicles, 5(2012-01-0175), 63-71

[2] Törnell, J., Sebben, S., & Elofsson, P. 2021 *Experimental investigation of a two-truck platoon considering inter-vehicle distance, lateral offset and yaw* Journal of Wind Engineering and Industrial Aerodynamics, 213, 104596

[3] Iozsa, D., Stan C., and Ilea L. 2017 *Study on the influence of the convoy rolling over aerodynamic resistance* IOP Conf. Series: Materials Science and Engineering 252 (2017) 012035

[4] Ebrahim, H. M., Dominy, R. G., & Leung, P. S. 2016 *Evaluation of vehicle platooning aerodynamics using bluff body wake generators and CFD* In 2016 International Conference for Students on Applied Engineering (ICSAE) (pp 218-223) IEEE

[5] Gnatowska, R., & Sosnowski, M. 2018 The influence of distance between vehicles in platoon on aerodynamic parameters. In EPJ Web of Conferences (Vol. 180, p. 02030) EDP Sciences.

[6] Siemon, M., Smith, P., Nichols, D., Bevly, D., & Heim, S. 2018 An integrated CFD and truck simulation for 4 vehicle platoons (No. 2018-01-0797). SAE Technical Paper

[7] Jaffar, F., Farid, T., Sajid, M., Ayaz, Y., & Khan, M. J. 2020 *Prediction of Drag Force on Vehicles in a Platoon Configuration Using Machine Learning* IEEE Access, 8, 201823-201834

[8] Kaluva, S. T., Pathak, A., & Ongel, A. 2020 *Aerodynamic drag analysis of autonomous electric vehicle platoons* Energies 13(15) 4028

[9] Tsugawa, S., Kato, S., & Aoki, K. 2011 *An automated truck platoon for energy saving* IEEE/RSJ international conference on intelligent robots and systems (pp. 4109-4114) IEEE

[10] Scurtu I.L. 2021 *Aerodynamic performance evaluation for a vehicle structure equipped with a bicycle rack* Automotive engineering ISSN 1842-4074 vol 58 pp 10-14

[11] Bodea S.M., Prodan C.V., Scurtu I.L. 2019 *The Aerodynamic Study of a Body Truck* In: Burnete N., Varga B. (eds) Proceedings of the 4th International Congress of Automotive and Transport Engineering (AMMA 2018). AMMA2018 2018. Proceedings in Automotive Engineering. Springer, Cham.

[12] Jurco A.N. 2021 *Study of the influence of roof luggage box on a vehicle aerodynamics* Automotive engineering ISSN 1842-4074 vol 58 pp 15-18

[13] Balcau M.C. 2021 *Aerodynamic study of a car towing a motorcycle trailer* Automotive engineering ISSN 1842-4074 vol 58 pp 22-26

Research of gear drives

A Dobreva, V Dobrev and G Mollova

University of Ruse, 7017 Ruse, Bulgaria

adobreva@uni-ruse.bg

Abstract. Worm gear drives possess important advantages such as: compact design, high power density due to a high gear ratio within a single gear stage. However, these gear drives often show high sliding speeds within the tooth contact, resulting in relative low efficiency coefficient in the meshing. Therefore, an exact calculation regarding efficiency in the meshing and load capacity is important during the beginning stage of the machine design process. Currently, no calculation method is available to analyze worm gears considering efficiency in meshing and contact strength. The objective of the research implemented was to analyze extensively the meshing efficiency and the contact stress behavior of various geometry designs of worm gears drives. Optimization research has been implemented considering the criteria mentioned. All calculations and theoretical analysis have been implemented for different center distances, different modules, gear ratios and rotational speeds of the worm shaft. Simulation results are shown and discussed. Important conclusions and recommendations are suggested.

1. Introduction

Worm gear drives are quite suitable for transmitting large gear ratios in one stage. A worm drive is a power transmission device in which a worm meshes with a worm gear to transmit power between two non-intersecting shafts that are oriented at a right angle. An important feature of these drives is the probability to achieve self-locking (for large gear ratios) when reversing the direction of the power flow.

Due to the great values of the sliding speed, heating is often the determining limit of the load capacity. On the other hand, one of the significant advantages of worm gear drives is their quiet operation capacity.

Worm drives have been widely used in various applications: noise is to be limited; space is limited; absorption of shock loading is required; minimum maintenance is necessary. The application of the worm gear drives is approximately 10 percent of all mechanical power transmissions.

Research of worm gear drives is carried out worldwide, which is described in the books and publications of Niemann and Winter [1], Hoehn et al. [2], Octrue [3, 4] and Houser [5]. Based on the scientific results achieved, a method for overall worm gear capacity estimation has been developed within the DIN 3996 standard [6].

The axes of the worm and the worm wheel intersect (most often) under 90°, figure 1.

There are three types of worm-gear drives: a) those with cylindrical worms (figure 1 a); b) globoid worm - cylindrical gear (figure 1 b); c) globoid worm - globoid wheel (figure 1 c).

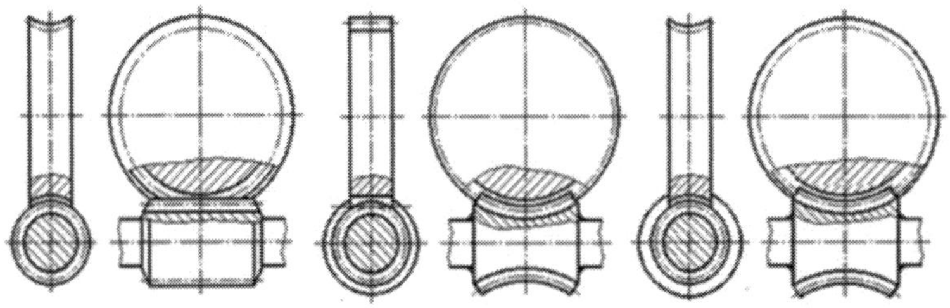

Figure 1. Types of worm gears: a) cylindrical worm - globoid wheel; b) globoid worm – helical involute gear; c) globoid worm - globoid wheel.

2. Theoretical Background

Depending on the method of generation, the following types of cylindrical worms are to be differentiated according to the German Standard DIN 3975 [7]:

- ZA – Flank form A (straight sided axial worm - type ZA): The straight shape-producing line of the cutter and the tooth generator of the worm coincide. Flank form A is obtained by setting a trapezoidal rotary cutter in such a way that its cutting edges lie in the axial section. It is produced by milling or grinding with a relevant profiled cutter or by skiving.
- ZN – Flank form N (straight sided normal worm - type ZN): The straight shape-producing line of the cutter and tooth flank generator of the worm coincide again, but do not cut the worm axis. Flank form N is obtained by setting a trapezoidal rotary cutter at axis level in such a way that its cutting faces lie in the plane inclined by the reference lead angle γ_m, thus determining the normal profile. It is produced by milling with a conical end-milling cutter or a relatively small disc-milling cutter.
- ZK – Flank form K (milled helicoid worm-type ZK): Flank form K is obtained by engagement of the worm with a rotary cutter of double cone form or by a milling cutter. Convex profiles in axial planes are achieved.
- ZI - Flank form I (involute helicoid worm-type ZI): The tooth flanks are parts of involute helicoid surfaces. They are tangential to a plane whose normal with respect to an axial section is slewed round by reference lead angle γ_m and inclined to the worm axis by the angle of generation $\alpha°$. Flank I is obtained by grinding with a flat sided grinding wheel.

The load capacity calculation methods of worm gears are described in details in DIN 3996 [6]. The most significant criterion is the contact strength calculation due to the following reason: the relative high friction coefficient and the high sliding speed between the flanks of worm and worm gear leads to high local contact pressure.

The other important criteria are: bending capacity, wear (usually appears on the tooth flanks of bronze worm wheels), scuffing and heating.

The efficiency of a worm gear drive is usually evaluated between 50 and 96 percent, depending on lubricant; speed; surface roughness; load; material pair; worm profile; worm gear size; worm number of starts and temperature [8-12].

The efficiency calculations for worm drives are standardized in AGMA 6034-B92 [13], BS 721 [14] and DIN 3996 [6]. The author's team completed successfully some research tasks in the field of energy efficiency in the area of gear drives and machine science. The results of these investigations are described in details in the following publications: [15–19]. Gear losses cause a significant proportion of the overall power loss. Friction within the meshing contact of two tooth flanks relates to the applied load of the tooth system and results in load-dependent gear losses.

3. Methods

Categories of criteria have been selected according to which an optimization study of worm gear drives is performed. The most significant group of criteria are the tribological parameters: coefficient of friction and meshing efficiency in worm gearing. The friction coefficient is calculated in accordance with [17] and [18]. The meshing efficiency is determined according to [1] and [6].

The second important group are the kinematic criteria: frequency of rotation of the input shaft, sliding speed and gear ratio. All calculations have been implemented with three values of rotational speed of the input shaft: 750 rpm, 1000 rpm and 1500 rpm.

The third main category of criteria includes the load capacity of the worm gear drive on the criterion contact strength. Strength calculations have been performed considering the contact strength criterion for worm gears of type ZI with the nominal output torque for the corresponding center distance and for the relevant rotational speed of the worm shaft.

The contact stresses obtained are to be compared with the permissible contact stress calculated according to the methods of [1] and [6]. All calculations have been implemented for different center distances, different modules, gear ratios and input shaft rotational speeds.

4. Results and Discussion

Considering the described methodology, the values of the efficiency in the worm gear meshing have been calculated and compared for different values of the gear ratio, center distance and rotational speeds of the worm. A conclusion can be deduced that as the gear ratio increases, the meshing efficiency decreases. Worm gear drives with a gear ratio between 12 and 30 are characterized by the highest efficiency values.

A significant tendency can be observed that at higher rotational speeds of the worm, the meshing efficiency of the gear is reaching higher values. The charts in figure 2 show the results obtained at center distance a = 125 mm.

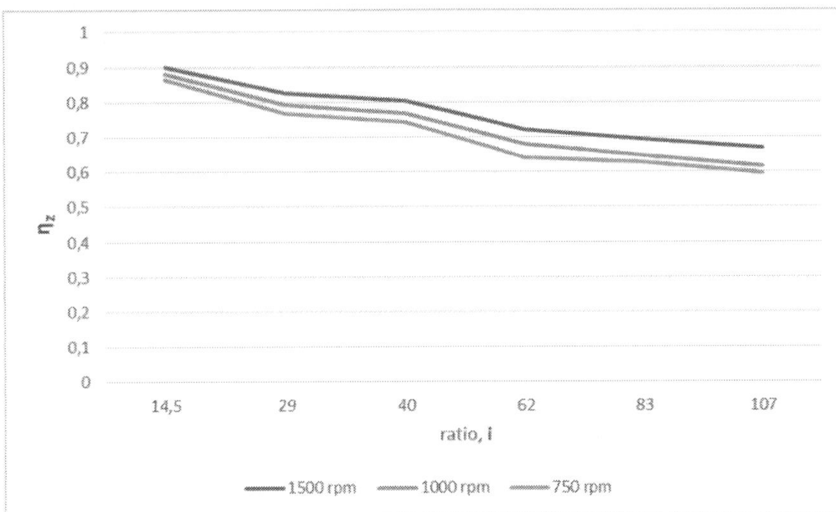

Figure 2. Efficiency coefficients in the worm gear meshing for center distance a =125 mm.

Strength calculations have been performed considering the contact strength criterion for worm gears of type ZI with the nominal output torque for the corresponding center distance and for the relevant rotational speed of the worm shaft.

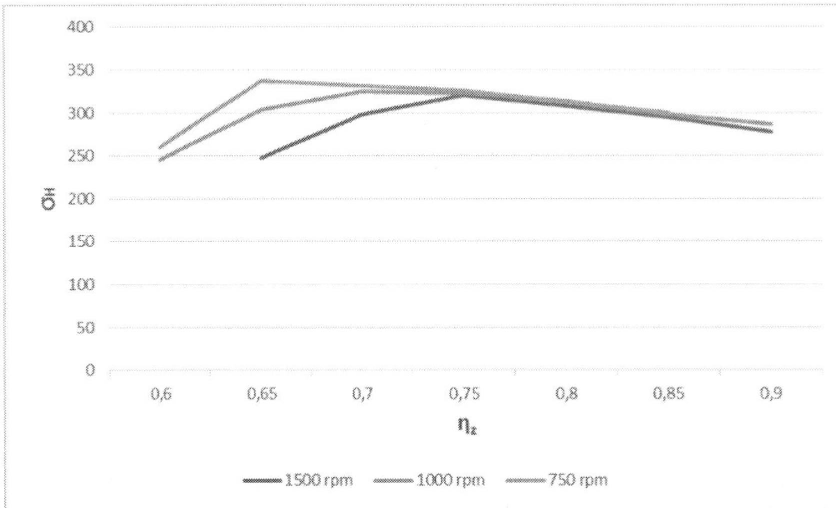

Figure 3. Efficiency coefficients and contact stresses in the worm gear meshing for center distance a =125 mm.

The contact stresses obtained are lower than the permissible contact stress calculated according to the methods of [1] and [6].

There is a positive trend that at higher rotational speeds the values of contact stresses decrease by the same values of meshing efficiency in the worm gearing. The lowest values of efficiency in the engagement - from 0.6 to 0.7 - are observed in worm gears, which are of the type: self-locking. For the same type of gears, the contact stresses have the highest possible values: about 300 to 340 MPa.

5. Conclusions

An optimization study of worm gear drives is presented. Three categories of criteria have been applied: tribological, kinematic and load capacity. Calculations and analyzes of the results have been performed.

The influence of the gear ratio and the rotational speed of the worm shaft on the meshing efficiency in the worm gearing and the variation of the values of the contact stresses has been established.

The results obtained can be used at an early stage of the design process of worm gear drives, allowing the designer to choose the optimal parameters in terms of maximum meshing efficiency in worm gear drives.

Acknowledgements

The authors' team expresses gratitude and acknowledgements for the support of the scientific fund at the University of Ruse, Bulgaria.

References

[1] Niemann, G. and H. Winter: Maschinenelemente, Springer Verlag, (1996).

[2] Hoehn, B.-R., Neupert, K. & Steingroever, K.: Wear Load Capacity and Efficiency of Worm Gears", VDl Berichte, No. 1230, pp.409-425, (1996).

[3] Octrue, M.: Relationship between wear and pitting phenomena in worm gears. American Gear Manufacturers Association, paper 97FTM9, (1997).

[4] Octrue, M.: Evolution des methodes de calcul de la capacite de charge des engrenages a r'oue et vis tangentes pression et usure", 4th World Congress on Gearing and Power Transmission, Paris, France, 1, pp. 405-417, (1999).

[5] Houser, D. R., Su, X. & Vaishya, M.: Effects of wear on the meshing contact of worm gearing, AGMA 99FTM18, (1999).

[6] DIN 3996: Calculation of Load Capacity of Cylindrical Worm Gear Pairs with Rectangular Crossing Axes, Beuth Verlag, Berlin, Germany, (2012)
[7] DIN 3975: 2/2007, Begriffe und Bestimmungs größen für Zylinder-Schnecken-getriebe mit sich rechtwinklig kreuzenden Achsen, (2007).
[8] Liou, J. J., Rakuff, S.: Development of Worm Drives. In: Power Transmission Engineering, Febr., pp. 38–43, (2018).
[9] Mautner, E. M., Sigmund, W., Stemplinger, J. P and Stahl, K: Efficiency of Worm Gearboxes," Proc. Inst. Mech. Engrs., Part C: J. Mechanical Engineering Science, 230, pp. 2952–2956, (2016).
[10] Turci, M., Ferramola, E., Bisanti, F. and Giacomozzi, G.: Worm Gear Efficiency Estimation and Optimization," AGMA Fall Technical Meeting, Detroit, USA, (2015).
[11] Muminovic, A., Repcic, N. and Zezelj, D.: The Efficiency of Worm Gears Lubricated with Oils of Mineral and Synthetic Bases," Trans. FAMENA, Vol. 37, no. 4, pp. 65–72, (2013).
[12] Murphy, W. R. and Cheng, V.M: The Effect of Lubricant Traction on Worm Efficiency, Gear Technology, Jan.-Feb., pp. 26–34, (1985).
[13] AGMA 6034-B92: Practice for Enclosed Cylindrical Worm gear Speed Reducers and Gearm otors, AGMA, Alexandria, USA, (2010).
[14] BS 721-1: Specification for Worm Gearing — Imperial Units, BSI, London, UK, (1963).
[15] Dobreva, A.: Theoretical Investigation of the Energy Efficiency of Planetary Gear Trains. Mecha nisms and Machine Science, No 13, pp 289-298, (2013).
[16] Dobreva, A.: Methods for Improving the Geometry Parameters and the Energy Efficiency of Gear Trains with Internal Meshing. VDI – Berichte, No 2199.2, pp 1291 – 1302, (2013).
[17] Dobreva, A. & Dobrev, V.: Improving the Tribological Characteristics of Heavy Loaded Gear Boxes. Proceedings of the First Balkan Conference on Tribology "Balkantrib'93", Volume 2.3, Sofia, pp 166-170, (1993).
[18] Dobreva, A. & Dobrev, V.: Innovative Methodology for Decreasing Mechanical Losses in Vehi cles. Proceedings of the 4th International Congress of Automotive and Transport Engineering (AMMA 2018), Springer Verlag, pp 234 – 242, (2018).
[19] Dobreva, A. & Stoyanov, S. Optimization Research of Gear Trains with Internal Meshing. Ruse, University Publishing Centre, pp 144, (2012).

Toward blockchain adoption for the automotive industry

F Gîrbacia , D Voinea , R Boboc , M Duguleană , C C Postelnicu

Transilvania University of Brasov, 29 Eroilor, 500036 Brasov, Romania

garbacia@unitbv.ro

Abstract. Blockchain proves to be a powerful technology that can drive forward the automotive industry. This paper describes how blockchain has been applied, which automotive areas are most interested in the technology, as well as the benefits and challenges of blockchain for the automotive industry by a systematic literature review. Blockchain offers many advantages, such as: transparency, security, high speed of operations. There are numerous areas where this technology could be applied in the automotive domain: better protection against counterfeit spare parts and materials, effective preventive maintenance, by informing the services based on the data received from the car sensors, monitoring of the cars by the rental companies, the safe storage of the data used by intelligent navigation systems, in order not to affect the safety of the passengers, autonomous transactions, triggered by the vehicle for making payments in charging stations, all with the purpose of facilitating a good and safer medium for drivers. The future vehicle will be connected both with other vehicles, but also with the existing infrastructure, so a database containing cryptographically secure information is needed, and blockchain is such a solution that can be applied in the dynamic nature of traffic.

1. Introduction

The automotive industry is facing a number of challenges related to the design, manufacturing, promotion, sale, insurance and history of vehicles and their components. The automotive sector needs to adopt novel technologies to stay ahead of the competition. Blockchain is a disruptive technology that has the potential to provide an answer to most of these problems (such as just in time delivery, ownership verification and counterfeit protection), being a reliable and transparent tool for collecting, storing, managing and using car data in a more efficient way [1]. Furthermore, artificial intelligence (AI), machine learning and blockchain are seen as three technologies that go hand-in-hand. When it comes to the automotive industry, the implementation of AI within vehicles can make driving more intuitive for users, as well as enhancing safety and security for drivers. A blockchain, which has as its basic unit the data block, is a distributed mechanism that records transactions efficiently, verifiably and permanently, in a linked structure. Thus, the complexity of the process and implicitly the costs are reduced, eliminating the intermediaries between the two parties. The user of an application using blockchain technology could track the complete history of a vehicle, without the risk that they have been falsified.

Traditional automotive companies like Volkswagen [2], BMW [3], Porsche [4], Mercedes Daimler [5], Renault [6] working on their own blockchain initiatives or testing the existing blockchain protocols IOTA or Stellar. In this paper, we discuss recent use cases that have been developed in representative automotive domains.

2. The blockchain technology

The blockchain technology works much like a digital ledger that is public, decentralized, and distributed. Transactions are kept with distributed hash-chains that form a peer-to-peer network of nodes (a node is a user). The "decentralized" part means that no one can manipulate the database. Additionally, the databases themselves aren't held by any single party, and they are also interconnected to the internet in a way that prevents data manipulation. Because the ledger is public, anyone can see and verify transactions in real time. Additionally, the ledger is secured by cryptography, with new transactions signed and hashed into the ledger.

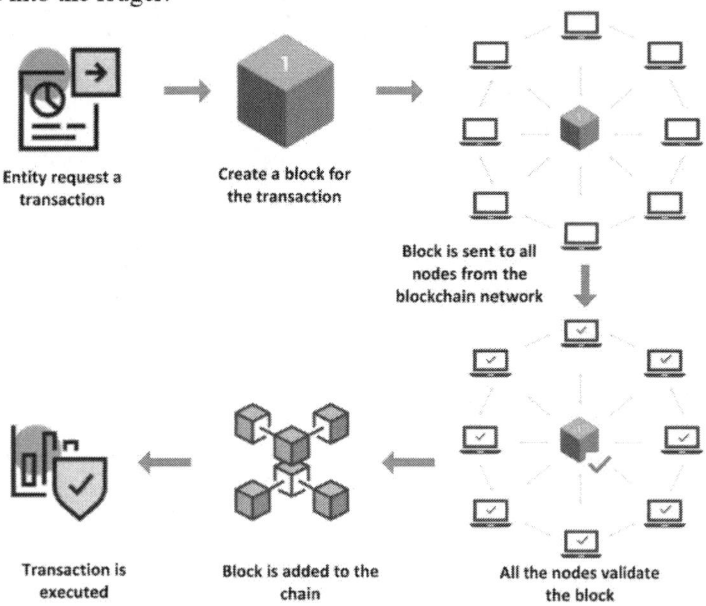

Figure 1. The blockchain technology.

Nodes of the network will execute computations and solve complex problems in order to verify a newly added transaction. Once the nodes have all verified the transaction the node that solves the problem first creates a block of transactions and generates a block header (see figure 1). This header contains a block identifier, a block height and a transaction list (consisting of hashes of the new transaction and hashes of the old transactions). Another node of the network compares this block header with their own block headers and verifies that the contents are the same. When the contents of a block header are verified by all the nodes, then the block is a valid block and will be added to the list of all the blocks.

3. Blockchain for the automotive industry: use cases

The following subsections present the different types of use cases (see figure 2).

3.1. Autonomous vehicles

Several studies target the use of blockchain technology in autonomous vehicles (AVs). One of the first approaches was related to simply storing data parameters of the current state that are received either directly from the car's operating system or from neighbouring vehicles [7]. Moving on, the proof of liability in the case of unforeseen events (accidents) was another idea explored by the research community. The problem that resides here is who is to blame in case of an accident, since the control is entirely autonomous. In [8], the authors propose a framework that can be used by stakeholders (car owners, insurance companies, automotive manufacturers) to solve this issue. Forensic data storage is also explored in [9], where authors use the Hyperledger Fabric network to prove such a system is feasible. Security problems of AVs are a concern of several studies.

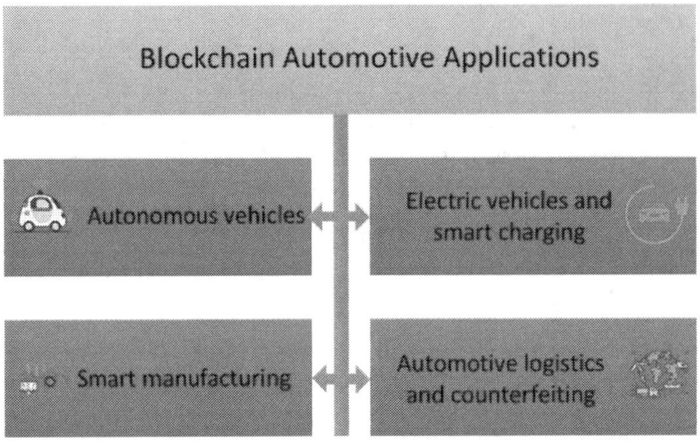

Figure 2. Automotive applications in which the blockchain was experimented.

Researchers try to capitalize on the benefits of the blockchain (immovability, transparency, decentralization and consensus) for solving attacks such as denial-of-service, GPS spoofing, timing or man-in-the-middle [10]. Mitigating security and privacy issues with blockchain is already an effervescent field of study, with experts focusing on data control, trackability, accountability, fault tolerance and scalability, to name just a few areas of research [11]. Refuelling is another use case on the list. According to the EU, all new vehicles, including the future AVs, will be electric starting with 2035 [12]. That means that refuelling with electricity should also suffer major improvements, to cope with the increasing consumer numbers. Authors from [13] use Ethereum Virtual Machine based smart contracts to organize this activity. The firmware update paradigm is another sensitive area that can use blockchain to increase its resilience [14]. Other blockchain uses refer to the management of the AVs traffic. Imagine a scenario within a crowded city. Based on a simple smart contract that is implemented on a blockchain that supports fast and inexpensive micro-transactions, some vehicles could allow others to automatically move faster, depending on a fee-based algorithm.

3.2. Electric vehicles

Today, Electric Vehicles (EVs) have sparked a lot of interest owing to improvements in battery life and reduced emissions, but there are still some challenges related to their market penetration, of which the most important is related to the charging process [15]. Due to the limited quantity of charging stations, there are issues with home charging as well as public charging. Drivers of electric vehicles must also be informed and find out where charging facilities are located in public areas. This lack of charging infrastructure which causes range anxiety [16], can be prevented by using blockchain-based applications, which allow drivers to share their private chargers with others. Such an example is presented in [17], a peer-to-peer (P2P) framework for private charging pile sharing systems. Another implementation of a smart charging system based on blockchain is proposed in [18], with the aim to reduce load congestion of the electric power distribution grid. In addition, the benefits of blockchain technology include real-time payments using cryptocurrencies and increased security [19]. As shown in [20], blockchain technology could also motivate EV users to share their energy surplus with others, improve the availability of clean energy, and reduce range anxiety. Among other opportunities offered by technology in the field of Evs, are: enhanced security of energy trading, online monitoring of EVs, managing the charging infrastructure, developing the concept of the internet of EVs, promoting the role of shared energy [21, 22].

3.3. Smart manufacturing

The new production methods and processes that are needed to manufacture the next generation of vehicles will be based on the integration of Industry 4.0 smart concepts [23, 24]. One of the difficulties of smart manufacturing is the increased number of interactions between manufacturing entities. The blockchain technology can be used to streamline the smart manufacturing process of vehicles by improving the integration of shared systems and components maintained by different companies and organizations. In this context, the blockchain technology allows improvement of security, traceability and immutability of smart manufacturing shared data [25, 27, 29].

The possibilities for blockchain as a manufacturing service are multiple. An example of a middleware solution that implements blockchain features for smart manufacturing is Man4Ware [25]. The blockchain services implemented in this platform allows improvement of security, traceability and immutability of smart manufacturing shared data. Another blockchain-based conceptual framework design for a smart manufacturing system is presented in [28]. The framework relies on the blockchain to facilitate device data transmission and manufacturing services. In [26] blockchain services are used for quality assurance by providing more transparent and secure data related to material provenance and production equipment management. Another example is presented in [30], a platform that uses the Internet of things (IoT), Machine Learning (ML) and the blockchain technology for monitoring the manufacturing system. The blockchain technology allows preventing fake data and improving the security of data transmission and costs. The proposed system was evaluated for automotive manufacturing in South Korea.

3.4. Counterfeiting prevention
The automotive industry is estimated to expand by 8%-10% in 2021, making it a key component in any nation's economy [40]. Therefore, there is an increased need for automotive manufacturers to improve their protection against supply chain counterfeiting and to raise their security standards against malicious attacks [31]. A possible solution to mitigate counterfeiting is to use permissioned blockchains such as Hyperledger Fabric by means of a distributed consensus algorithm [32].

3.5. Automotive logistics
The automotive sector is under pressure to rapidly innovate and release products that are competitive, environmentally friendly and have a reasonable price. Therefore, one important issue is resolving the complexity of supply chains and logistics [33]. Blockchains and other disruptive technologies such as IoT, Big Data and AI have a high potential of transforming supply chains and logistics by 2023 [34]. For example, tracking tools and blockchain coupled with smart contracts can facilitate payments to suppliers once goods were successfully delivered to their destination.

A proper integration of blockchain technology (BT) with other cutting-edge technologies (cloud computing, IoT, big data analytics, cybersecurity) can build trustworthiness, transparency, immutability and reliability in a supply chain network [35]. A blockchain-based logistics monitoring system was proposed, programmed and tested on Ethereum and its main scope was to allow parcel tracking [36]. Another study focused on resolving traceability issues and proposed TokenTrail, a decentralized blockchain application based on the semi-fungible ERC 1155 token [37]. Smart contracts can be used to create a network of hyperconnected logistics by leveraging blockchain and IoT [38]. Vehicle tracking and freight management could also be significantly improved by the use of BT, although a high-scale deployment is far from reality [39].

4. Conclusion
The blockchain can provide to the automotive industry services which can have a significant impact on a large number of different aspects, from the mobility of vehicles to the manufacturing and maintenance of them. Blockchain has already shown potential on improving autonomous, electric or shared vehicles through decentralized apps. Based on the presented use cases, it can be concluded that blockchain represents the future of the vehicles. It will improve the driving experiences, safety and even the possibility of sharing information between entities.

5. Acknowledgments
This work was supported by a grant of the Romanian Ministry of Research, Innovation and Digitization, CNCS/CCCDI – UEFISCDI, project number PN-III-P2-2.1-PED-2019-4366 within PNCDI III (431PED).

References
[1] Fraga-Lamas P and Fernández-Caramés T M 2019 A review on blockchain technologies for an advanced and cyber-resilient automotive industry IEEE access 7 17578
[2] Putting blockchains on the road, Volkswagen, https://www.volkswagenag.com/en/news/stories/2018/08/putting-blockchains-on-the-road.html#, Last accessed 14.07.2021
[3] How blockchain automotive solutions can help drivers, BMW, https://www.bmw.com/en/innovation/blockchain-automotive.html, Last accessed 14.07.2021
[4] Blockchain: the key technology of tomorrow, Porsche, https://newsroom.porsche.com/en/company/porsche-blockchain-technology-opportunities-digitization-16800.html, Last accessed 14.07.2021
[5] Once round the block, please! Daimeler, https://www.daimler.com/innovation/blockchain-2.html, Last accessed 14.07.2021
[6] The blockchain, transformation vector for the future of the automotive industry, Renault, https://www.renaultgroup.com/en/news-on-air/news/the-blockchain-transformation-vector-for-the-future-of-the-automotive-industry/, Last accessed 14.07.2021
[7] Narbayeva S, Bakibayev T, Abeshev K, Makarova I, Shubenkova K and Pashkevich A 2020 Blockchain technology on the way of autonomous vehicles development *Transportation Research Procedia* **44** 168
[8] Oham C, Kanhere S S, Jurdak R and Jha, S 2018 A blockchain based liability attribution framework for autonomous vehicles arXiv preprint arXiv:180205050
[9] Guo H, Li W, Nejad M and Shen C C 2020 Proof-of-event recording system for autonomous vehicles: A blockchain-based solution IEEE Access, 8, 182776
[10] Gupta R, Kumari A and Tanwar S 2021 A taxonomy of blockchain envisioned edge-as-a-connected autonomous vehicles *Trans. on Emerging Telecommunications Technologies* **32**(6) e4009
[11] Gupta R, Tanwar S, Kumar N and Tyagi S 2020 Blockchain-based security attack resilience schemes for autonomous vehicles in industry 4.0: A systematic review *Computers and Electrical Engineering* **86** 106717
[12] EU to Urge 2035 Goal to End Combustion-Engine Era in Autos, Bloomberg https://wwwbloombergcom/news/articles/2021-07-09/europe-to-propose-end-of-combustion-engine-era-in-green-overhaul, Last accessed 14072021
[13] Pedrosa A R and Pau G 2018 ChargeltUp: On blockchain-based technologies for autonomous vehicles *Proc. of the 1st Workshop on Cryptocurrencies and Blockchains for Distributed Systems* p 87
[14] Baza M, Nabil M, Lasla N, Fidan K, Mahmoud M and Abdallah M 2019 Blockchain-based firmware update scheme tailored for autonomous vehicles *Conf. IEEE Wireless Communications and Networking 2019* p1
[15] Florea B C, Taralunga D D 2020 Blockchain IoT for Smart Electric Vehicles Battery Management *Sustainability* **12**
[16] Pevec D, Babic J, Carvalho A, Ghiassi-Farrokhfal Y, Ketter W and Podobnik V 2019 Electric Vehicle Range Anxiety: An Obstacle for the Personal Transportation (R)evolution? *4th Int. Conf. on Smart and Sustainable Tech. 1*
[17] Fu Z, Dong P, Li S, Ju Y and Liu H 2021 How blockchain renovate the electric vehicle charging services in the urban area? A case study of Shanghai China *J. of Cleaner Production* **315** 128-172
[18] Okwuibe G C, Li Z, Brenner T and Langniss O 2020 A Blockchain Based Electric Vehicle Smart Charging System with Flexibility *IFAC-PapersOnLine* **53**(2) 13557

[19] Foti M, Vavalis M 2021 What blockchain can do for power grids? Blockchain: Research and Applications *Res. and Apps*. **2**(1) 100008
[20] Lasla N, Al-Ammari M, Abdallah M and Younis M 2020 Blockchain Based Trading Platform for Electric Vehicle Charging in Smart Cities *IEEE Open J. of Intellig Transp Systems* **1** 80
[21] Hasankhani A, Hakimi S M, Bisheh-Niasar M, Shafie-khah M and Asadolahi 2021 Blockchain technology in the future smart grids: A comprehensive review and frameworks H *Int. J. of Electrical Power and Energy Systems* **129** 106811
[22] Wang Q, Li R, Zhan L 2021 Blockchain technology in the energy sector: From basic research to real world applications *Computer Science Review* **39** 100362
[23] Jadoon G, Ud Din I, Almogren A and Almajed H 2020 Smart and agile manufacturing framework, a case study for automotive industry *Energies* **13**(21) 5766
[24] Alptekin B, Tunaboylu B, Zaim S and Perlo P 2020 Smart Manufacturing of Electric Vehicles *The International Symp. for Production Research* p767
[25] Mohamed, N and Al-Jaroodi J 2019 Applying blockchain in industry 4.0 applications Conf. IEEE 9th annual computing 2019 p 0852
[26] Zhang Y, Xu X, Liu A, Lu Q, Xu L and Tao F 2019 Blockchain-based trust mechanism for IoT-based smart manufacturing system *IEEE Trans. on Comp. Social Sys.* **6**(6) 1386-1394
[27] Leng J, Ye S, Zhou M, Zhao J L, Liu Q, Guo, W and Fu L 2020 Blockchain-secured smart manufacturing in industry 4.0: A survey *IEEE Trans. on Sys ,Man and Cybernetics: Systems* **51**(1), 237
[28] Lee C K, Huo Y Z, Zhang S Z and Ng K K H 2020 Design of a smart manufacturing system with the application of multi-access edge computing and blockchain technology *IEEE Access* **8** 28659
[29] Zuo Y 2020 Making smart manufacturing smarter–a survey on blockchain technology in Industry 4.0 *Enterprise Information Systems* **1** 31
[30] Shahbazi Z and Byun Y C 2021 Smart Manufacturing Real-Time Analysis Based on Blockchain and Machine Learning Approaches *Applied Sciences* **11**(8), 3535.
[31] Oham C, Michelin R A, Jurdak R, Kanhere S S and Jha S 2021 B-FERL: Blockchain based framework for securing smart vehicles *Information Proc. and Management*, **58**(1) 102426
[32] Lu D, Moreno-Sanchez P, Zeryihun A, Bajpayi S, Yin S, Feldman K and Kate A 2019 Reducing automotive counterfeiting using blockchain: benefits and challenges *IEEE Int. Conf. on Decentralized Applications and Infrastructures* (DAPPCON) 39
[33] Viriyasitavat W, Da Xu L, Bi Z, and Sapsomboon A 2020 Blockchain-based business process management (BPM) framework for service composition in industry 4.0 *J. of Intellig. Manufacturing* **31**(7) 1737
[34] Pournader M, Shi Y, Seuring S and Koh S L 2020 Blockchain applications in supply chains, transport and logistics: a systematic review of the literature *Int. Journal of Production Res.* **58**(7) 2063
[35] Kamble S S, Gunasekaran A, Subramanian N, Ghadge A, Belhadi A and Venkatesh M 2021 Blockchain technology's impact on supply chain integration and sustainable supply chain performance: evidence from the automotive industry *Annals of Operations Res.* 1
[36] Helo P and Hao Y 2019 Blockchains in operations and supply chains: A model and reference implementation *Computers and Industrial Engineering* **136** 242
[37] Kuhn M, Funk F, Zhang G and Franke J 2021 Blockchain-based application for the traceability of complex assembly structures J. of Manuf. Sys. **59** 617
[38] Betti Q, Khoury R, Hallé S and Montreuil B 2019 Improving hyperconnected logistics with blockchains and smart contracts *IT Prof* **21**(4) 25
[39] Kamble S, Gunasekaran A and Arha H 2018 Understanding the Blockchain technology adoption in supply chains-Indian context *Int. J. of Production Res.* **57**(7) 2009
[40] Global auto sales forecast, S&P,https://www.spglobal.com/ratings/en/research/articles/210511-global-auto-sales-forecasts-the-recovery-gears-up-11952032, Last accessed 14.07.2021

IOP Conf. Series: Materials Science and Engineering 1220 (2022) 012027 doi:10.1088/1757-899X/1220/1/012027

Modelling a virtual wind tunnel for vehicle aerodynamic measurements

L Ilea, D Iozsa and G Fratila

University Politehnica of Bucharest, Autovehicule Rutiere Department, 313 Splaiul Independentei St., Romania

laurentiu.ilea@yahoo.com

Abstract. The automotive makes of our days are challenged to fulfil more and more criteria to remain competitive and financially performant, considering the legislation that must be respected to be able to commercialise each model. One highly important set of rules focus on emissions and strict limits are implemented. To develop compliant products, each make must focus on aerodynamic performance so to reduce the drag of air, thus the required force to overcome for movement at high speeds. While homologation is currently done exclusively as a physical test, to be able to reduce the aerodynamic coefficient, strong virtual tools are used. These measurements performed with computation fluid dynamic software are translated into a virtual wind tunnel that must be able to give relevant and coherent results in respect to final physical validations. The paper will indicate the main types of wind tunnels, parameters to consider, error source and how results can be used. While the computational power is growing each moment, the industry is continuously adapting to obtain the most performant and appealing designs with minimal resources, this being the key of reducing the CO_2 footprint of passenger vehicles transport segment.

1. Vehicle general aerodynamics

Aerodynamics and its related phenomena were the objects of study and experiments since antiquity. With advancements regarding understandings of air behavior and flow laws, the flight was made possible and right after, the transfer toward the automotive industry started to take place.

Along with the elegance that may come with the aerodynamic body design of a vehicle, the main advantage is the improved easiness to overcome the aerodynamic resistance, force which varies with the square of the vehicle speed [1]. While dynamic performance can be highly improved for sports vehicles if the aerodynamic coefficient (Cd) is reduced, the need to reduce transport CO_2 footprint hired aerodynamic expertise in the development of high efficiency passenger vehicles [2].

Aerodynamic performance can be determined indirectly by physical tests performed on track, using methods that continuously changed along the time, from trying to isolate the vehicle from the environment to strings placed on body to identify turbulence and pressure sensors array, just to name a few, as indicated in figure 1 [3].

IOP Conf. Series: Materials Science and Engineering 1220 (2022) 012027 doi:10.1088/1757-899X/1220/1/012027

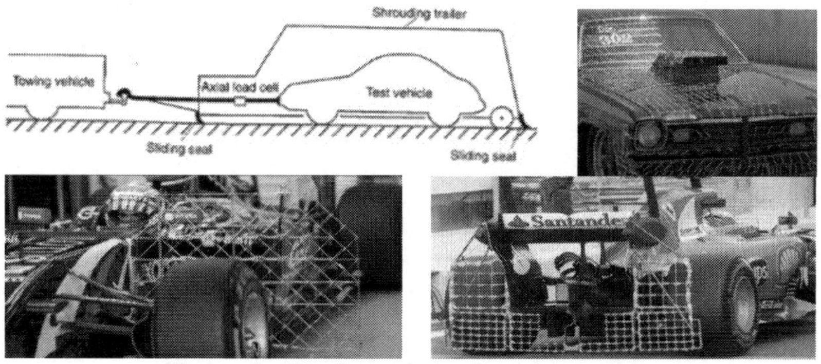

Figure 1. Aerodynamic track measurements:
(a) isolated method; (b) string method; (c) pressure sensors method.

While physical tests represent a big leap in understanding aerodynamics and what may be done to reduce this encountered resistance, thus fuel consumption and emissions, all tests performed in an outside environment are influenced by changing parameters like temperature, pressure, wind direction and guests, track friction coefficient, etcetera. These conditions represent of course the usage of the vehicle in the real world, but measurements are very difficult to be done and used in the development phase due to big variation of input parameters, iterative results that can't be compared, impossibility to compare with benchmark or competition. Currently, the deceleration method on track is used to homologate the aerodynamic performance of a model in countries using NEDC (New European Drive Cycle) legislation.

To avoid the above-mentioned external influences a new test was imagined. Starting from the premise that the vehicle must be tested in a well controlled environment, where iterations can be performed in the same conditions, the wind tunnel was designed and adopted by the majority of today vehicle makers. As it happened with tests on track, the wind tunnel had multiple development phases based on working parameters control and associated costs since this type of facility requires very big investments for high accuracy.

The wind tunnel was in its initial form, while innovative, still a rudimentary tool (Eifel wind tunnel – figure 2a) and after derived into a slotted wall wind tunnel (this method allows smaller testing rooms, thus less expensive, but results accuracy may be affected – figure 2b), open section wind tunnel (aerodynamic phenomena better simulated, but low energetic efficiency due to necessity to continuously bring air to required speed – figure 2c) and finally closed section wind tunnel (the energetic efficiency is solved by keeping the air in a closed loop, thus low speed loss – figure 2d) [4].

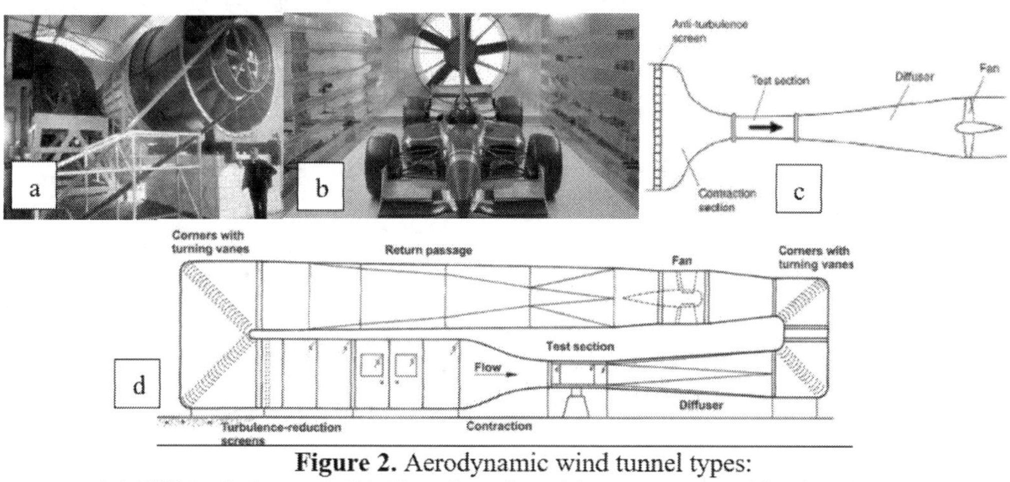

Figure 2. Aerodynamic wind tunnel types:
(a) Eiffel wind tunnel; (b) Slotted section; (c) Open section; (d) Closed section.

IOP Conf. Series: Materials Science and Engineering 1220 (2022) 012027 doi:10.1088/1757-899X/1220/1/012027

2. Automotive aerodynamic development

The development of vehicle aerodynamic performance initially started with scaled mockup tests and was finalized with validations on full size, representative designs. Since the timeline today between the birth of the project idea to a vehicle on street is as reduced as 2.5 years, the mentioned validation strategy could be considered obsolete.

To be able to meet the more ambitious Cd targets, correlated with current CO_2 rules, new tools had to be found and perfectioned so to permit early Cd design assessments and corrections. Taking advantage of the exponential growth of computing power and the existence of software for the aeronautical industry, once again the automotive world made a leap by applying computational fluid dynamic methods in its development process.

Starting from the early stages, 3D models of the vehicle are constructed by engineers and introduced in a virtual environment, simulating in this way the phenomena that could be previously observed just at the end of the project, in a real wind tunnel test. To obtain a final model coherent with Cd homologation method, a virtual wind tunnel with a closed section is used. This method allows the use of the same input environment parameters (temperature 20 degrees Celsius, fine controlled air directionality, air speed of 140 km/h), a robust benchmark between different models and the correlation between virtual tests performed in the development phase and final physical tests and homologation. It is mentioned that under WLTP (worldwide harmonized light procedure) the homologation of Cd coefficient is performed in closed section wind tunnels and are complemented with Real Drive Emissions (RDE) tests.

By applying this new logic, the performance is improved (Cd reduced drastically for current vehicles as compared with models from a few decades ago), risks can be identified in advance and corrected, the time required to finalize a project was also reduced which improved the dynamic of industry and accentuated the competition between makes. All these advantages lead to a more performant and environment friendly automotive park.

3. Virtual wind tunnel

As exposed in chapter 2, the most common and chosen method of study for the current study is the closed section wind tunnel due to its high accuracy in results and also the relative energy efficiency. One of the downsides remains the high cost of building. The modelled facility is S2A wind tunnel (Montigny-le-Bretonneux, France), as described in figure 3 [5].

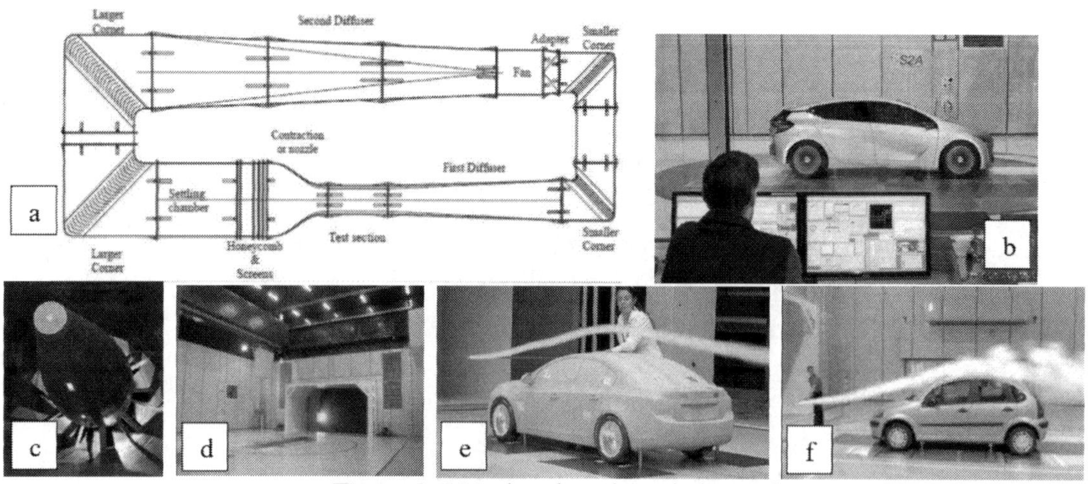

Figure 3. S2A closed section wind tunnel.
(a) Functional scheme; (b) Control room; (c) Fan; (d) Test room; (e) 2:5 scale test; (f) 1:1 scale test

This test facility was built on the latest standard and applies up to date research in aerodynamic measurements, being able to accommodate experiments for vehicles (passenger and utility), plane sections, competition vehicles and even bicycles, of both aerodynamic and acoustic nature. The air inlet has a surface of 24 m^2 with a 3/4 jet section, the air speed can be raised to 240 km/h, the floor can be rotated to allow side wind tests, the road roughness and influence is simulated with a central rotating belt. The aerodynamic torque is measured with a balance of 6 components as described in figure 4, data being processed and shown in real time for Cd coefficient, CdA (A being projected object surface), temperature, Cy (lateral aerodynamic coefficient) and Cz (vertical aerodynamic coefficient), for both front and rear axle.

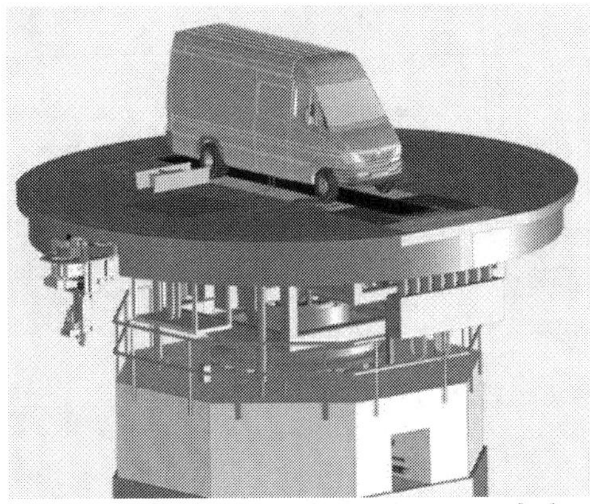

Figure 4. S2A type aerodynamic torque balance.

Considering this real-life environment, the virtual testing facility is built using the same proportions and functional devices like the belt found on the vehicle X axis. This 3D testing tool (described in figures 5(a) and 5(b)) will be immersed in a CFD software that will simulate the passage of high-speed air over the vehicle body. The air volume in the testing room is meshed from the finest dimension (cells with 2mm length) to the course near the testing walls (cells with 64mm length), the passage being made with a multiplication of 2 to facilitate the convergence process, as indicated in figure 5(c). Results are recorded in each cell and are assembled for visualization of pressure, speed, wake and numerical value by applying the Lattice Boltzmann mathematical model. It is mentioned that other mathematical models like Navier-Stokes's equations and their derivatives can be used along with a diversity of turbulence models (K-epsilon, K-omega, etcetera) chosen depending on application [7].

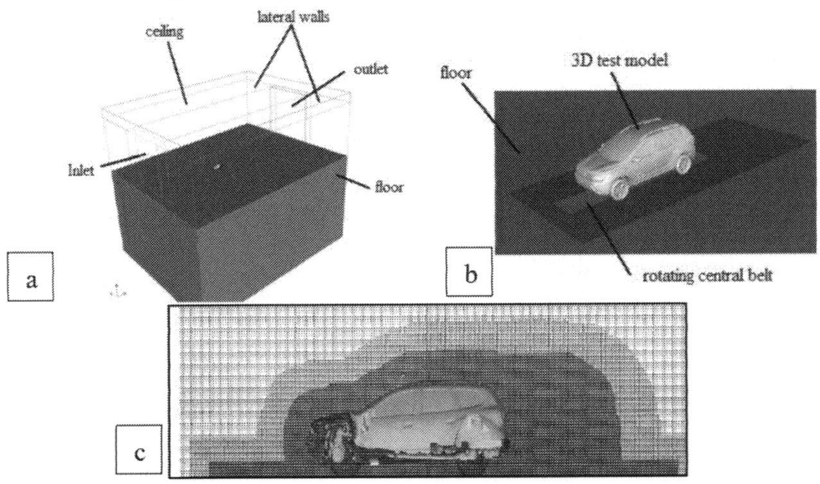

Figure 5. (a) Virtual wind tunnel; (b) Zoom on 3D test model; (c) Mesh of environment.

With boundary condition applied on walls, air flow simulated on X direction from inlet to outlet the environment can be modelled in coherence with the S2A wind tunnel. For an air speed of 140 km/h, generic results like the ones in figure 6 can be easily obtained and analyzed to evaluate the air flow, identify possible optimizations, and apply them in an iterative manner on the 3D tested vehicle.

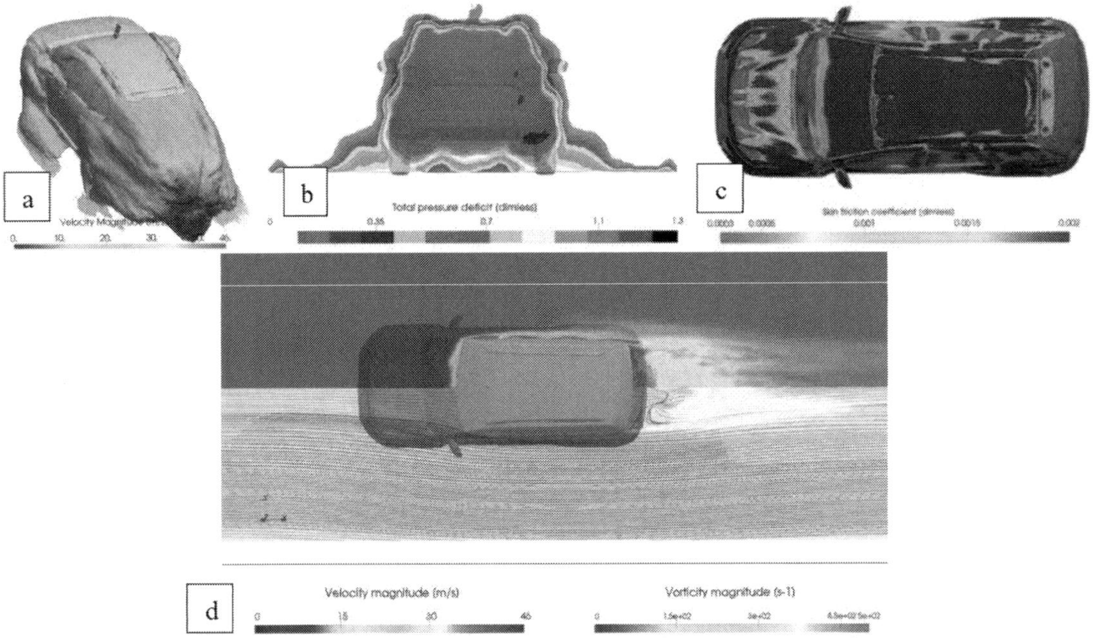

Figure 6. (a) 3D Velocity magnitude representation; (b) Total pressure deficit representation; (c) Skin friction representation; (d) Velocity magnitude.

It must be considered that while the CFD tools allow an early assessment and correction, with relative reduced resources in comparison with physical mockup tests, the application of mathematical models implies the use of chosen wake models and coefficients which finally can induce a difference between 3D results and final physical ones. If software and models are fine tunned the relative error is reduced but the process must be started over when a different body type is tested since the air flow will not show the same behavior [8]. Today, big investments in research and development, high computing power and elaborate mathematical models can offer very precise results within a tolerance of under 2%. Besides errors linked to mathematical model and results convergence (number of timesteps) [9], differences may appear also from fabrication dispersions between different vehicles of the same model, tire diversity and temperature. Nevertheless, the aerodynamic measurements process used today by most of automotive manufacturers can provide valuable information that allows engineers and designers to converge toward a performant, appealing and aerodynamic body shape, which help reduce CO_2 emissions.

4. Conclusions

The paper presents a general overview of the aerodynamic performance of the automotive industry and the progressivity from the exclusive world of motorsport where the focus was dynamic performance and downforce, toward the mass-produced passenger vehicles to increase movement efficiency by shaping the body design to slice easier thru the air. If aerodynamic resistance at high travelling speeds is easier overcome, the power required is diminished and emissions, highly constrained by authorities, are also reduced.

Automotive manufacturers had to adapt the development and testing procedures for the aerodynamic coefficient to adapt to the requirements and available CFD tools. While the real world and virtual wind tunnels are very important measurement facility that allows the faster development of improved performance vehicle, errors must be taken into consideration and treated with care to avoid project issues and even possible official test incoherencies.

The current paper is part of a study on the influence of automotive accessories on overall aerodynamic performance with regards to CO_2 emissions. The work will be continued with the detailed analysis and results impact when different mathematical models are used.

References

[1] Andreescu C *Dinamica Autovehiculelor lecture* (Transporturi Faculty, Dept. Autovehicule Rutiere, UPB, 2012-2013)
[2] Katz J, *Race Car Aerodynamics*, 1st Edition
[3] Standford Course *Jones Classical Aerodynamic Theory*
[4] Hucho W-H *Aerodynamics of Road Vehicles: From Fluid Mechanics to Vehicle Engineering* (4th edition)
[5] *** https://www.soufflerie2a.com/en/
[6] Houghton E L, Carpenter P W *Aerodynamics for engineering students* (5th edition)
[7] Nakayama Y *Introduction to Fluid Mechanics* (Tokai University, Japan)
[8] Eastlake C N *An Aerodynamicist's View of Lift, Bernoulli and Newton*
[9] Sodja J *Turbulence models in CFD* (Faculty for mathematics and physics Department of physics, University of Ljubljana)

The reliability of automotive subsystems influence on the motor vehicles fuel economy

R I Negrilă and S V Igreţ

Aurel Vlaicu University, Arad, Romania

radu.negrila@premiermotors.ro

Abstract. The reliability theory consistently regards all phases of vehicle construction and operation. The properties determining the vehicle reliability, which are also called potential objectives, are precisely formulated from early design and manufacturing phases and accomplished in the production process. The analysis of reliability from the perspective of potential properties allows us to identify certain categories of potential properties. One refers to potential properties that depend on the deterioration of operating functions. In this category, we can find the resistance to breakage and deformation, stability of operating processes, stability of the physical-mechanical properties of the construction materials and of the working environment. Secondly, the potential properties required to accomplish the required level of production and reduce the manufacturing costs, such as manufacturing technology, the corresponding volume of production, simplicity of manufacture and failure-free manufacturing. Thirdly, there are potential properties needed to maintain the reliability at a high level during service, such as developing the quality of diagnoses, technical maintenance, reducing all types of costs and increasing productivity. Here we find the technology of the diagnosis and the totality of the external expenses and the adaptability to the automation of diagnosing tools. Operating reliability is the result of the full accomplishing of the potential reliability of components. The reliability of automotives treated as a subsystem can successfully be evaluated by defining certain operating indicators and characteristics, the economic characteristic being of great interest and importance.

1. Introduction

Within the operating costs of vehicles, the cost of fuel, lubricants and other operating materials used intervenes with an average percentage of 25 - 35%, a high share having fuel consumption. Based on these considerations, this consumption is one of the most important characteristics of these vehicles.

Fuel consumption depends on the type, power and construction of the engine, as well as the construction features of the chassis and body, as well as the conditions under which the vehicle moves (load, speed, level of the resistance forces).

2. The fuel economy characteristics of the vehicle

The basic quantity by which the fuel consumption can be assessed is expressed by the specific fuel consumption, necessary to obtain a unit of power (one kilowatt) for one hour, at a certain operating regime.

This parameter is established when testing the engine on a special stand where the effective power P_e developed by the engine at the considered speed (in kW) is measured, as well as the mass of fuel, C_h, consumed within one hour (in kg), so that the relation for the determination of the specific consumption takes the form:

$$c_e = 10^3 \cdot \frac{C_h}{P_e} \quad [g / kW \cdot h] \tag{1}$$

In order to determine the dynamic qualities of a motor vehicle, it is necessary to know its maximum possibilities, for which purpose it is necessary to start from the speed characteristic to the total load of the engine. Under normal operating conditions, however, the car engine works most of the time on low loads, so on partial loads. Therefore, in order to determine the fuel economy qualities of a motor vehicle, it is necessary to consider the normal operating conditions of the engine in operation and to examine the whole complex of possible working regimes, i.e. both at total loads and at various partial tasks.

Based on the experiments performed on all engines made so far for vehicles, it results that the lowest values of specific consumption are obtained at high load (80-100%), and specific consumption increases with decreasing engine load. Both in the case of total load and in the case of partial loads, the curve of the specific consumption that presents a minimum (optimal specific consumption), for a speed between the speed corresponding to the maximum power and the one corresponding to the maximum engine torque (figure 1).

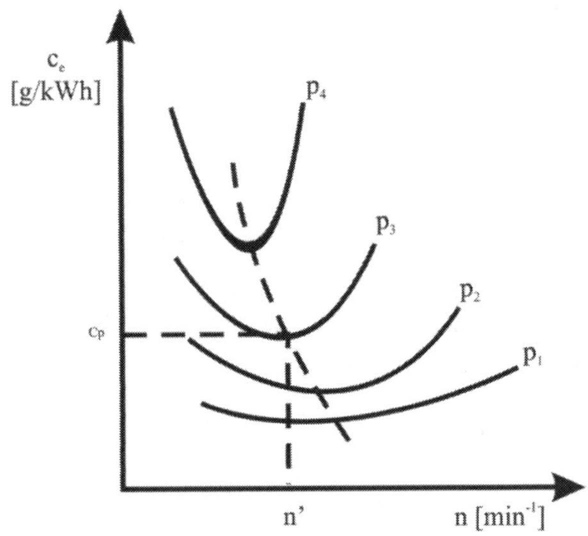

Figure 1. Variation of specific actual fuel consumption as a function of engine speed at different degrees of intake of fresh load decreasing, starting with p_1, p_2, p_3, p_4 [5].

The specific fuel consumption of the engine is a parameter that can be used to assess the economical quality of the engine, its operation in a certain regime, but it can not give an idea of the fuel consumption of the vehicle in operating conditions.

For this, it is necessary to use such chosen parameters, so that with their help it is possible to highlight both the size of consumption and the influence of traffic conditions on the profitability of the vehicle (speed, type and condition of the road), as well as certain characteristics. of the vehicle (weight, aerodynamic factor, gearbox transmission ratios, etc.)

The most used fuel consumption parameter of the vehicle is the fuel consumption per 100 km travelled, expressed in either kilogram (C_{100}^{kg}) or liter (C_{100}^{ℓ}).

Knowing the hourly fuel consumption C_h (kg/h) and the vehicle speed, v (km/h), the consumption is given by the relation:

$$C_{100}^{kg} = C_h \cdot \frac{100}{v} \quad [kg / 100\,km] \tag{2}$$

Fuel consumption per 100 km is most often expressed in liters, because in this way it corresponds to the way of expressing the capacity of fuel tanks in cars and those in gas stations.
If it is noted with $\rho\left(kg/dm^3\right)$ the density of the fuel, then the relation between the consumption per 100 km in liters and the one expressed in kilograms is:

$$C_{100}^{\ell}=\frac{C_{100}^{kg}}{\rho}=C_h\cdot\frac{100}{\rho\cdot v}\quad\left[\ell/100\,km\right] \tag{3}$$

Taking into account the relation (1), the consumption per 100 km depending on the specific consumption and the effective power developed by the engine is:

$$C_{100}^{kg}=\frac{c_e\cdot P_e}{10\cdot v}\quad\left[kg/100\,km\right] \tag{4}$$

and

$$C_{100}^{\ell}=\frac{c_e\cdot P_e}{10\cdot\rho\cdot v}\quad\left[\ell/100\,km\right] \tag{5}$$

Another convenient parameter for expressing profitability, from the point of view of fuel economy, is the number of kilometers, L, which can be covered by consuming one liter of fuel. The dependence between this parameter and C_{100}^{ℓ} consumption is as follows:

$$L_{\ell}=\frac{100}{C_{100}^{\ell}}\quad\left[km/\ell\right] \tag{6}$$

The use of the L_{ℓ} parameter, which is very little introduced in the practice of vehicle operation, is however very rational, because in operation it facilitates and makes clearer the evidence of consumption, simplifies the calculation of the distance that can be covered by the vehicle with a certain amount of fuel contained in the tank. comparing, by profitability, the various models and types of vehicles.

In order to study the fuel economy of a vehicle, it is necessary to know how to vary the consumption parameter depending on the speed and the specific resistance coefficient of the road. In addition to these factors, the specific engine sizes, which influence the specific fuel consumption, must also be taken into account. This graphically represented dependence is the consumption characteristic of the vehicle. The consumption parameter, which is represented graphically as a function of speed, for different values of the specific road resistance, is the fuel consumption per 100 km driven.

The effective power delivered by the engine during the operation of the vehicle, with a certain uniform speed, corresponds to the powers required to overcome the forward resistances, by the relation:

$$P_e=\frac{P_{rul}+P_p+P_a}{\eta_t}=\frac{P_{\psi}+P_a}{\eta_t} \tag{7}$$

where:

P_{rul} is the power at the drive wheels, corresponding to rolling resistance;
P_p is the corresponding power when climbing the slope;
P_a is the power corresponding to air resistance;
P_{ψ} is the power corresponding to road resistance; $P_{rul}=P_{\psi}+P_p$
η_t is the transmission efficiency [1].

If relations (5) and (7) are taken into account, the following relation may be written, for the C_{100}^{ℓ} consumption of the motor vehicle, when driving at a uniform speed v, the following relation:

$$C_{100}^{\ell}=\frac{c_e\cdot P_e}{10\cdot\rho\cdot v}=c_e\cdot\frac{P_{\psi}+P_a}{10\cdot\rho\cdot\eta_t\cdot v} \tag{8}$$

The specific fuel consumption can be expressed by the relation:

$$c_e=3600\cdot\frac{\rho_0\cdot\eta_v}{\lambda\cdot L_{\min}\cdot P_e}\quad\left[g/kW\cdot h\right] \tag{9}$$

Hence it can be observed in connection with relation (9) that the C_{100}^{ℓ} consumption is also affected by the technical condition of the engine and the transmission of the vehicle, which can be quantified by the corresponding quantities in the respective expressions.

Fuel consumption compared to the L distance achieved by the vehicle is also a specific consumption expressed by the amount of fuel consumed by the vehicle in liters (or kilograms) to cover a unit distance (100 km), so:

$$C_{100}^{\ell} = 100 \cdot \frac{Q}{L} \quad [\ell / 100km] \tag{10}$$

In general, for a practical application, it is proposed that L be the length of a duty cycle defined by the descriptive parameters shown in figure 2.

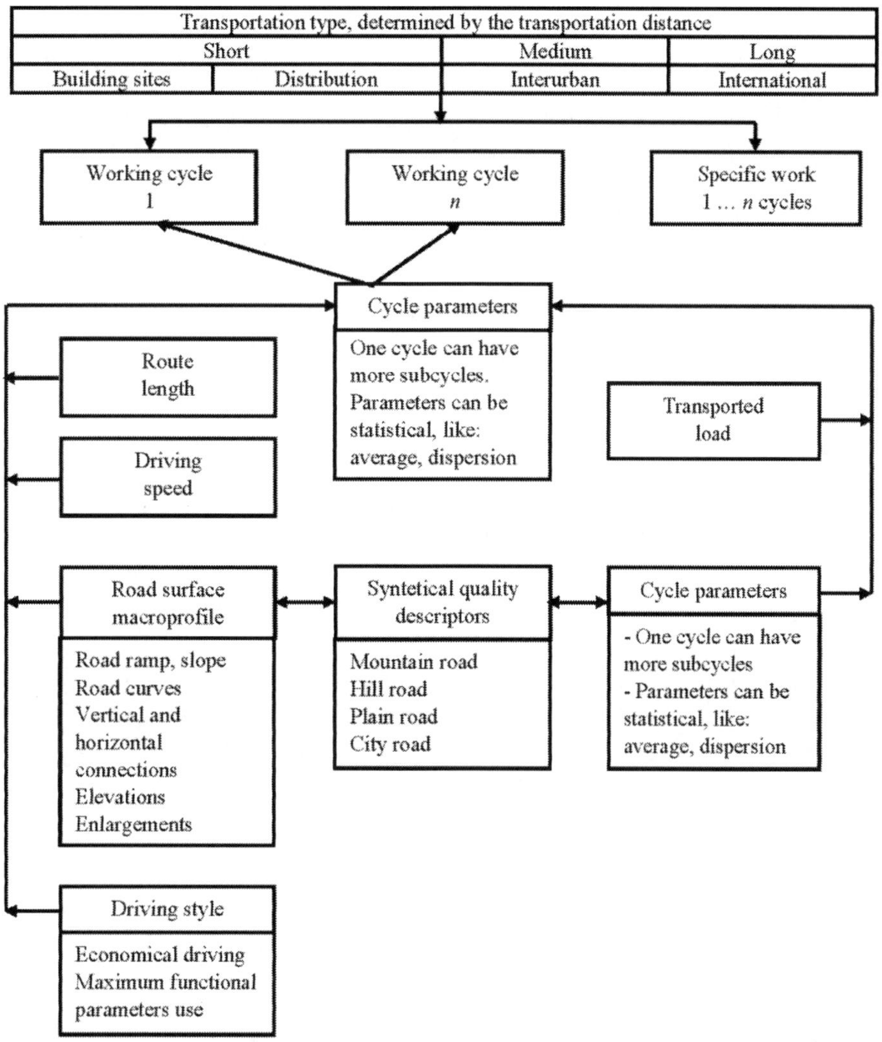

Figure 2. Synthesis of the descriptive parameters of the vehicle operation [5].

By doing so, at least in theory, a direct link can be made with the use of the vehicle. In this sense, some macrocycles can be defined for their use, which can be characterized by the type of road traffic and the average speed that can be achieved on the respective route:

- highway type traffic, with high average speeds over 55 km/h;
- intercity traffic, with an average speed of over 38 km/h;

- urban traffic, with an average speed between 16… 38 km/h;
- congested urban traffic, with an average speed below 16 km/h.

As shown in figure 2, a duty cycle can be characterized by several parameters and factors, which are defined by the length of the route, travel speed and load transported and customized in terms of transport conditions, by the macro- and microprofile of the path. driving style, driving style etc. This hypostasis can be highlighted by the diagram in figure 3 which determines the percentage of power (load) required to overcome the resistances, depending on the speed and mass transported, correlated with the road conditions.

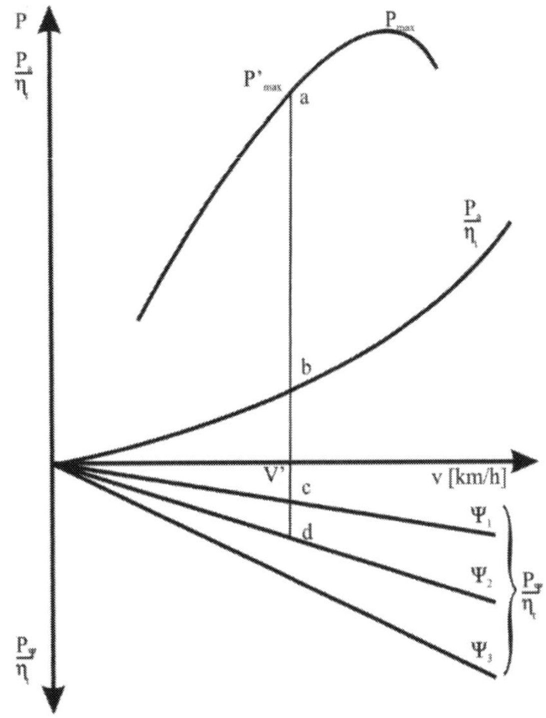

Figure 3. Diagram for determining the percentage of power required to overcome road resistance as a function of speed [5].

In quadrant 1 of the diagram is represented the variation of the effective power, P_e, in a certain gear of the gearbox and the variation of the power necessary to overcome the air resistance, P_a / η_t, calculated at the flywheel; in quadrant 4 of the diagram, the lines were drawn by origin, which represents variations in powers P_ψ / η_t depending on the speed v, also calculated at the level of the engine flywheel, for the different values of the specific resistance coefficient ψ_1, ψ_2, ψ_3 of the road.

From figure 3 it can be seen that, for a certain value of the road coefficient ψ_2 and a certain speed gear v', in the considered gear, the percentage p of power required to be developed by the engine from the power to the total load is:

$$p = \frac{\overline{bc} + \overline{cd}}{\overline{ac}} \cdot 100 [\%] \tag{11}$$

Also, from the respective diagram it can be seen that, as the speed of the vehicle increases or the road conditions worsen, the engine load will increase, which will directly influence the fuel consumption. But, during operation, the technical condition of the engine components, the transmission units and the running system also change, which leads to functional anomalies, i.e. to the diminution, including the reliability of the vehicle [3]. This decrease in reliability is reflected both indirectly, by reducing the actual engine efficiency, the filling coefficient and the transmission efficiency, and directly, by increasing the fuel consumption per 100 km travelled.

As the total transmission ratio i_t in the respective gear is known, the engine speed n', corresponding to the speed v', can be determined using the known relation:

$$n' = \frac{v' \cdot i_t}{0,377 \cdot r_r} \quad [rot/\min] \tag{12}$$

where r_r is the running radius of the vehicle wheel.

From the characteristic of the specific consumptions of the engine (figure 1) it is possible to determine the specific consumption, c_p, corresponding to the engine speed n' and the percentage p of engine loading, possibly by interpolation between the variation curves in the diagram. Also, knowing the percentage p and the power at total load P_{max}, corresponding to the engine speed n' (figure 3), we can calculate the power at low load, P_{red}, which must be delivered by the engine at any engine speed:

$$P_{red} = p \cdot P_{\max} \tag{13}$$

Then, using relation (8), we calculate the fuel consumption per 100 km for the respective operating mode:

$$C_{100}^{\ell} = c_p \cdot \frac{P_{red}}{10 \cdot \rho \cdot v} \tag{14}$$

Proceeding successively in this way, at the same ψ resistance, for different values of speed v and then for different ψ resistances, the consumption characteristic of the vehicle in the respective gear is obtained $C_{100}^{kg} = f(v,\psi)$ (figure 4).

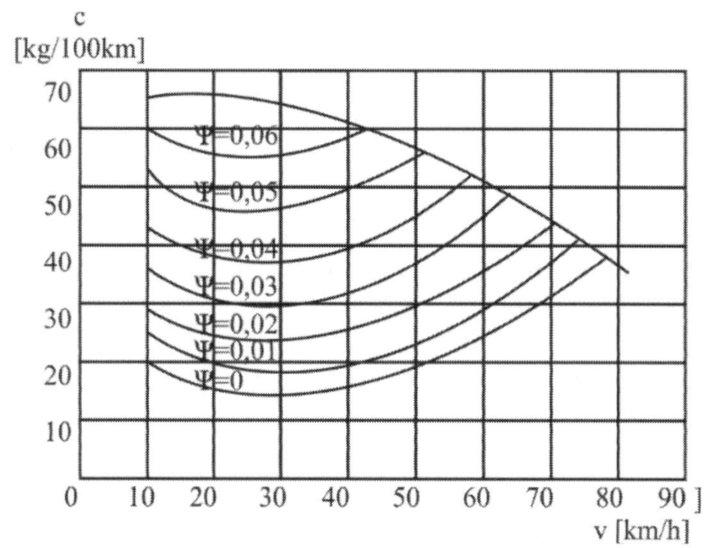

Figure 4. Consumption characteristic for a gearbox gear for different values of the coefficient of resistance ψ [5].

It should be noted that the appearance of the specific consumption curves of the engine depending on the load is very different between spark ignition (SI) and compression ignition (CI) engines.

The characteristics of the specific fuel consumption at the average operating speed of a SI and a CI engine of the same maximum power are represented in figure 5.

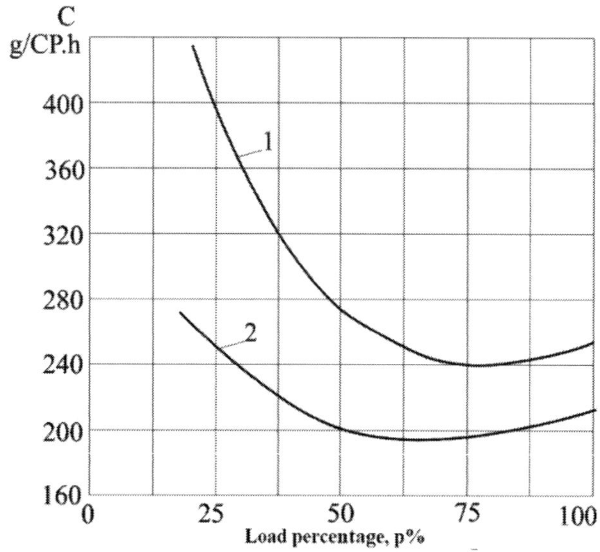

It is observed that at SI (curve 1), the specific consumption increases sharply as the load decreases, compared to CI (curve 2), where the variation of the load causes much smaller variations of the specific fuel consumption. Also, at CI, the specific consumption remains much lower over the entire range of load variation.

Figure 5. Characteristic of specific fuel consumption depending on the loading percentage *p%*: 1 – SI; 2 – CI [6].

The economic advantage - from the point of view of fuel consumption - of the use of compression-ignition (diesel) engines has led to the equipping with such engines of vehicles intended for the transport of materials or of a large number of passengers; in the last 40 years, the use of diesel engines has expanded even in small passenger cars, although the expansion of gasoline injection fuel systems has greatly improved the specific consumption curve at SI, which is today much closer to the corresponding curve of diesel engines [4].

From the shape of the characteristic consumption curves (figure 4) it can be seen that, for a certain road resistance, there is a certain optimal speed, at which the fuel consumption is minimal and which must be taken into account when driving the vehicle.

If we consider as an evaluation parameter the number of kilometers that can be covered with a liter of fuel, L_ℓ, than the relations (3) and (6) allow us to draw the consumption characteristic $L_\ell = f(v,\psi)$ starting directly from the engine characteristic, measured on the test bench, or from an existing consumption characteristic for the parameters C_{100}^{ℓ} or C_{100}^{kg}.

3. Conclusions

In the planning of regular car transports, there is a certain need to determine the fuel consumption on a certain route, which has sections with different known ψ coefficients of resistance. The problem can be conveniently solved with the help of the dynamic characteristic [2] and the consumption characteristic. Figure 6 shows the graphical solution of this problem, for a route that comprises four portions of lengths S_1, S_2, S_3, S_4, with the respective specific resistances ψ_1, ψ_2, ψ_3 and ψ_4.

The first quadrant shows the dynamic characteristic of the vehicle, calculated and constructed for the normal load with which the transport is to be made. Quadrant 4 shows the consumption characteristic for the values of the resistance coefficient of the respective road ψ_1, ψ_2, ψ_3 and ψ_4.

On the left extension of the speed axis, on a certain scale, there are represented the lengths S_1, S_2, S_3 and S_4 of the road sections of equal ψ resistance.

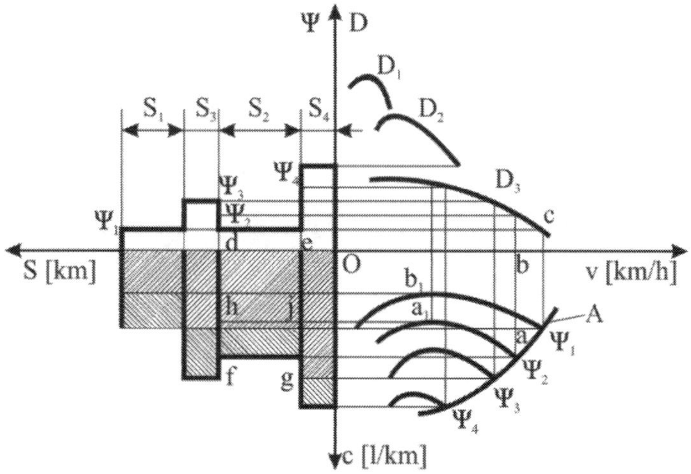

Figure 6. The complex characteristic of the vehicle, dynamical and economical [5].

The value of the coefficient of resistance ψ and the dynamic characteristic allow to determine the value of the maximum speed with which each of the sections of the route can be covered, as well as the fastest gear that can be engaged in the gearbox. The amount of fuel that would be consumed for traveling at this speed, on each of the sections of the route, is determined by the consumption characteristic and is equal to the area of the respective rectangle in quadrant 3, i.e. $S \cdot C_{100}^{\ell}$. Thus, for example, the highest speed with which it can run on the portion S_2 is the speed *Ob*, corresponding to the point a of the intersection of the consumption curve for the ψ_2 resistance with the consumption curve at full load, the sector S_2 being able to be passed in the third gear (the point *C* on dynamic characteristic D_3); the amount of fuel required to travel through sector S_2 at this speed is given - taking into account the size of the representation scales - by the area of the rectangle *defg*.

For speeds of the vehicle lower than the highest speed corresponding to the respective sector, the fuel consumption decreases. Thus, for sector S_2, the most economical speed is Ob_1, corresponding to the a_1 point, of minimum consumption; the amount of fuel required is reduced to the area of the right hatched rectangle *dehi*. As can be seen, an even greater reduction in speed leads, again, to an increase in fuel consumption and a decrease in transport productivity. In the final choice of speed for each of the sectors of the route, in addition to fuel economy, other factors are taken into account that determine the profitability of transport, including, first of all, productivity and transport security.

References

[1] Untaru M 1981 *Dinamica autovehiculelor pe roţi* (Bucureşti: Editura Didactică şi Pedagogică)

[2] Bobescu Gh 1998 *Motoare pentru Automobile şi Tractoare, Dinamică, calcul şi construcţie vol. II* (Chişinău: Editura Tehnică)

[3] Nagy T, Stănescu M A, Ţurea N, Dima D 1997 *Fiabilitatea şi terotehnica autovehiculelor Vol I* (Braşov: Editura Universităţii „Transilvania")

[4] Dietsche K-H, Klingebiel M 2007 *Automotive Handbook 7-th Edition* (73201 Plochingen, Deutschland: Robert Bosch GmbH)

[5] Negrilă R I 2005 *Studiul influenţei factorilor de exploatare asupra fiabilităţii autovehiculelor în vederea optimizării activităţii de mentenanţă* (Arad: Editura Universităţii „Aurel Vlaicu").

[6] Negrilă R I 2006 *Fiabilitatea sistemului conducător-autovehicul-drum-mediu* (Arad: Editura Universităţii „Aurel Vlaicu").

Proposal of an analytical method based on linear regression for modeling reliability by Weibull law

A Boroiu[1], V Nicolae[1], O Alssadi[2] and T Udma[2]

[1] Full Professor, PhD eng., University of Pitesti, Pitesti, Romania
[2] PhD student, University of Pitesti, Pitesti, Romania

alexandru.boroiu@upit.ro

Abstract. The paper aims to identify a sufficiently precise and easy method that will ultimately lead to a working tool easily accessible and used for modeling reliability through the most widely used mathematical law in the field of reliability - the Weibull law. An analytical method for linearizing the Weibull function is proposed in order to allow, by using the linear regression function of the Microsoft Office software, to obtain the parameters of the reliability model. The calculation program created in Microsoft Excel allows the identification of the 3 parameters of the Weibull model and, based on the Weibull model thus determined, the validation or invalidation of the reliability objectives. The application presented reveals that the proposed working tool is easy to use and the Weibull model obtained has a very good probability.

1. Introduction

Reliability modeling is done by different methods: graphical, analytical or grapho-analytical [3].

Graphical methods involve the use of special probabilistic networks (in Cartesian coordinates, but with appropriate scales for each type of distribution) on which the discrete values of one of the reliability indicators, determined on the basis of experimental data, are represented as points. With the help of the obtained graph, the values of the mathematical model parameter are determined (average m - for the exponential distribution, average m and the quadratic mean deviation σ - for the normal model, parameters γ, η, β - for the Weibull model).

Analytical methods consist in modeling reliability in a purely analytical way, by approximating the experimental distribution with a theoretical law of distribution, using the established mathematical methods in modeling theory (moment method, maximum plausibility method, least squares method). Based on these methods, numerical calculation programs are made, which allow the identification of the theoretical model of reliability with an imposed precision.

There are numerous laws of continuous distribution used in reliability: exponential, normal (Gauss-Laplace), lognormal, χ^2, Student (t), gamma, beta, Fischer – Snedecor (F), Rayleigh, Gumbel, Weibull, etc. In general, these distribution laws have been taken from the mathematical literature and have been adapted to the specifics of reliability. But there are also laws that have been discovered precisely as a result of the need to describe as accurately as possible the phenomena related to reliability, as is the case of the Weibull law.

Of these, the most used in the reliability study are the following:

- the exponential law, due to its simplicity;

- the normal law - because it is a distribution that characterizes very well the processes in nature when the number of experiences is very high;
- the Weibull law, which can be customized so as to represent with sufficient precision any type of distribution.

There are also grapho-analytical methods, combined methods in which some parameters are determined analytically and others are determined graphically.

The most convenient working method is adopted [1, 9], depending on the number of experimental data (graphical methods are applicable for a limited number of data), depending on the desired operability (graphical methods are particularly operative and provide fast an image of the allure of the distribution, the analytical ones are operative only if there are already realized calculation programs) and of the desired precision (the analytical methods obviously offer a high degree of precision).

2. Formulation of the problem

One problem faced by those who want to conduct reliability studies is the difficulty of having a dedicated algorithm or software to identify the most appropriate model for the study, because this requires either the purchase of that software or the use of some of the known graphical or grapho-analytical methods, which have limits in terms of the plausibility of the model obtained [5, 10].

As a result, the aim of this paper is to identify a sufficiently precise and easy method (supported by a calculation program that can be done using the Microsoft Office package), which would ultimately lead to a work tool that is easy to access and used to model the reliability by the most comprehensive mathematical law in reliability - the Weibull law.

3. Parameters and main reliability indicators of Weibull law

Weibull law is defined by the triparametric expression proposed for the reliability function:

$$R(t) = e^{-\left(\frac{t-\gamma}{\eta}\right)^{\beta}} \tag{1}$$

where: γ = location parameter or position parameter, is a constant that defines the start time of the variation of the reliability function R(t);

η = scale parameter, expresses the extent of distribution over the time axis.

Thus, if $(t - \gamma)$ is equal to η, R(t) calculated with relation (1) becomes:

$$R(t) = e^{-1^{\beta}} = e^{-1} = 0.368 \tag{2}$$

that is, the scale parameter represents the time, measured from the moment γ, at which 63.2% of the elements can be damaged. For this reason, this parameter expresses a characteristic operating time.

As a result, the parameters γ and η are expressed in units of time and can be graphically highlighted (figure 1).

β = shape parameter, is dimensionless and is the parameter that determines the allure of variation curves for reliability indicators - figure 2 (for the case $\gamma = 0$).

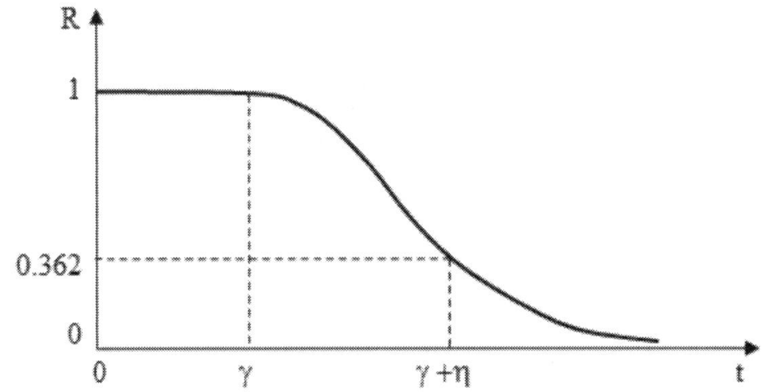

Figure. 1. Highlight the parameters γ and η on the graph R(t) for Weibull law.

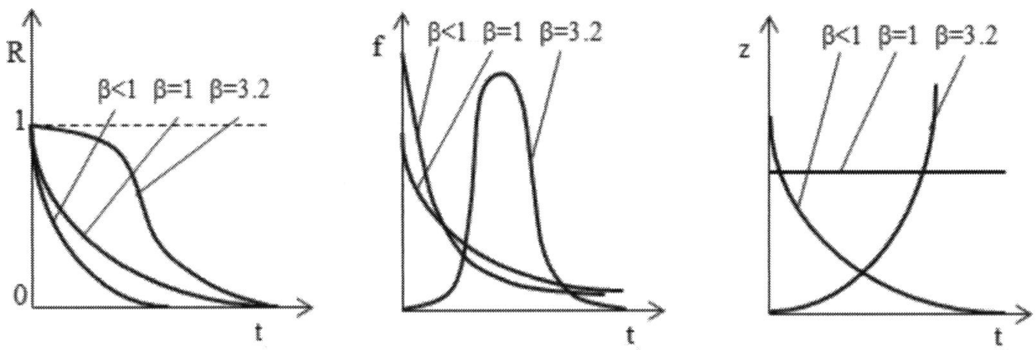

Figure. 2. Graphs of reliability indicators R(t), f(t) and z(t) for Weibull low with different values of the shape parameter β.

Figure 2 shows the graphical variations of the indicators R (t), f (t) and z (t) considering the location parameter $\gamma = 0$ (so the faults appear from the moment $t = 0$) and the scale parameter $\eta = 1$, for different values of the parameter β. If $\eta \neq 0$the shape of the distribution does not change, only the time scale changes (narrower or more extensive).

Analyzing the graph z (t) in figure 2 it is found:

- For $\beta < 1$, the failure rate z (t) is decreasing, so that the Weibull law represents the first part of the well-known "bathtub" curve, being able to characterize the failures during the running-in period;
- For $\beta = 1$, the failure rate z (t) is constant, as in the case of the exponential law, being able to characterize the behavior of the product during the duration of use;
- For $\beta > 1$, the failure rate z (t) is increasing (for the case $\beta = 3.2$ in the figure, f (t) corresponds even to the normal law), so in this case, it can characterize the wear period of the product.

Thus, the Weibull law can represent distributions of failure times in all three periods of the product's life.

In conclusion, for certain values of the parameters γ, η and β, the Weibull law can be used in the study of a large number of continuous distributions, including the two laws that have special use in reliability studies: the exponential law and the normal law.

An additional quality of the Weibull model is that it encompasses the exponential model ($\beta = 1$), normal model ($\beta = 3.2 \ldots 3.4$), Rayleigh model ($\beta = 2$) etc., in the form of particular cases, depending on the shape parameter β.

Based on the expression of the reliability function R (t), the main reliability indicators are determined:

- Non-reliability function (distribution function):

$$F(t) = 1 - R(t) = 1 - e^{-\left(\frac{t-\gamma}{\eta}\right)^{\beta}} \tag{3}$$

- Probability density of failure time:

$$f(t) = -\frac{dR}{dt} = \frac{\beta}{\eta}\left(\frac{t-\gamma}{\eta}\right)^{\beta-1} \cdot e^{-\left(\frac{t-\gamma}{\eta}\right)^{\beta}} \tag{4}$$

- Failure rate (intensity):

$$z(t) = \frac{f(t)}{R(t)} = \frac{\beta}{\eta} \cdot \left(\frac{t-\gamma}{\eta}\right)^{\beta-1} \tag{5}$$

4. Weibull model identification methods

The identification of the Weibull model that best approximates an experimental distribution can be done by analytical methods based on known special modeling methods (moment method, maximum plausibility method, least squares method). These methods are complex, but if calculation programs are made for this purpose, the problem is reduced only to entering numerical data (operating time values) to determine the three parameters that define the Weibull model (γ - location parameter, η - parameter scale, and β - shape parameter).

In the papers [7, 8] is presented a calculation program in TurboPascal made by the method of smallest squares which allows, by the successive halving of the interval, obtaining values for the three parameters with an accuracy of 10^{-8}.

Graphic methods or grapho-analytical methods (in which the indicators β and η are determined graphically, and the indicator γ is determined graphically-analytically) have a quite high efficiency when the number of data is not large [12].

Several types of probabilistic networks have been developed to estimate the Weibull model, the most widely used being the one proposed by Allan Plaît in 1962 [6]. This is based on the linearization of the Weibull distribution by two successive logarithms of the reliability function given by the relation:

$$R(t) = e^{-\left(\frac{t-\gamma}{\eta}\right)^{\beta}} \tag{6}$$

It follows successively:

$$\ln R(t) = -\left(\frac{t-\gamma}{\eta}\right)^{\beta} \tag{7}$$

$$\ln\frac{1}{R(t)} = \left(\frac{t-\gamma}{\eta}\right)^{\beta} \tag{8}$$

Logging in a second time, results:

$$\ln\ln\frac{1}{R(t)} = \ln\ln\frac{1}{1-F(t)} = \beta\ln(t-\gamma) - \beta\ln\eta \tag{9}$$

which represents a linear equation of the form:

$$Y = \beta X + c \tag{10}$$

where:

$$Y = \ln\ln\frac{1}{1-F(t)},\quad X = \ln(t-\gamma)\ \text{ şi }\ c = -\beta\ln\eta \tag{11}$$

It follows that in the probabilistic network having on the ordinate the Y scale and on the abscissa the X scale, a distribution of statistical data defined by the Weibull law of parameters γ, η and β will be aligned after a line, also called "Weibull line", having the equal inclination. with the parameter β.
To determine the scale parameter η we start from the observation:

$$F(t=\gamma+\eta) = 1 - e^{-\left(\frac{\eta}{\eta}\right)^{\beta}} = 1 - e^{-1} = 1 - 0.368 = 0.632 \tag{12}$$

which leads to the conclusion that the value $\gamma + \eta$ is identified on the time scale corresponding to the value F (t) = 0.632 on the Weibull line.

The determination of the location parameter η is performed by a grapho-analytical method, based on the line identified by the probabilistic network Allan-Plait.

5. The proposed method for identifying the Weibull model

The proposed method has as a starting point the observation that the analytical expression for the Weibull line in relation (9), which contains the shape parameter β and the scale parameter η, can be capitalized to determine the values of the parameters β and η by linear regression of the distribution of points Yi (Xi) using the "trendline" function in Microsoft Excel [4].

Thus, according to the concepts presented above, the Microsoft Excel program is used and the following steps are performed:

Step 1. Enter on a column the values of the n failure times representing the input data and then they are ordered ascending from t_1 to t_n.

Step 2. The value of the first failure time is adopted, absolutely reasonable, for the location parameter γ:

$$\gamma \equiv t_1 \tag{13}$$

Step 3. Create a second column with serial numbers for failure time, from 1 to n.

Step 4. Create a third column with the values ti – Y.

Step 5. Calculate on a fourth column the proportion of defective products, using a formula that avoids the lack of operationality or relevance for extreme values, recommended by the literature [2, 11]:

$$F_i = (i - 0.5)/n \tag{14}$$

Step 6. Calculate on a fifth column the values of the argument Xi:

$$X_i = \ln(t_i - \gamma) \tag{15}$$

Step 7. Create a last column with Yi values:

$$Y_i = \ln \ln \frac{1}{1 - F_i} \tag{16}$$

Step 8. Construct the graph with the distribution of points Yi = f (Xi). For the start time of faults t0, where the function ln (t-t0) is not defined, the value 0 is automatically assigned.

Step 9. Using the "trendline" function, the regression line of the distribution of pairs of points (Xi, Yi) is constructed, for which the analytical relation and the value of the square of the degree of correlation R are obtained.

According to relation (9), the coefficient of the variable X is the form parameter β, and the free term contains the parameters β and η:

$$c = -\beta \ln \eta \tag{17}$$

Step 10. From this last relation, knowing the value of the parameter β, the parameter η can be obtained analytically, according to the relation:

$$\eta = e^{-\frac{c}{\beta}} \tag{18}$$

As mentioned before, at time t = γ + η the proportion of defective products is 63.2%, which can be seen in column F of the table with the processed data.

Now that the Weibull model has been identified (the values of the 3 parameters are known: γ, η and β), the main reliability indicator can be estimated for any time value (relations 1, 2, 3 and 4).

It can also be verified that at a specified value of time tc (tc is a quantum of operating time) the product meets a predetermined reliability objective R_{lim}:

$$R(t_c) = e^{-\left(\frac{t_c - \gamma}{\eta}\right)^{\beta}} \geq R_{\lim} \tag{19}$$

Following the steps of the algorithm presented above, a calculation program was created in Microsoft Excel which performs, based on the input data, the identification of the 3 parameters of the Weibull model and the verification of the predetermined reliability objective.

6. Application

For a batch of 25 air conditioning systems with the same characteristics and used on the same type of vehicle, under similar conditions, the times of the first failure [months] were recorded and the data were entered in the calculation program, resulting in the values presented in Table 1 and the Weibull regression line presented in Figure 3.

Table 1. Input data and values obtained with the calculation program.

t [months]	Rank	t-t0 [months]	F	X = ln(t-t0)	Y = ln[-ln(1-F)]
t1 = 6	1	0	0.02	0.0000	-3.9019
10	2	4	0.06	1.3863	-2.7826
12	3	6	0.10	1.7918	-2.2504

t [months]	Rank	t-t0 [months]	F	X = ln(t-t0)	Y = ln[-ln(1-F)]
15	4	9	0.14	2.1972	-1.8916
17	5	11	0.18	2.3979	-1.6172
17	6	11	0.22	2.3979	-1.3925
21	7	15	0.26	2.7081	-1.2003
21	8	15	0.30	2.7081	-1.0309
28	9	22	0.34	3.0910	-0.8782
32	10	26	0.38	3.2581	-0.7381
37	11	31	0.42	3.4340	-0.6075
38	12	32	0.46	3.4657	-0.4842
39	13	33	0.50	3.4965	-0.3665
45	14	39	0.54	3.6636	-0.2529
46	15	40	0.58	3.6889	-0.1421
48	16	42	0.62	3.7377	-0.0330
54	17	48	0.66	3.8712	0.0759
62	18	56	0.70	4.0254	0.1856
62	19	56	0.74	4.0254	0.2979
64	20	58	0.78	4.0604	0.4148
64	21	58	0.82	4.0604	0.5393
73	22	67	0.86	4.2047	0.6761
81	23	75	0.90	4.3175	0.8340
89	24	83	0.94	4.4188	1.0344
92	n = 25	86	0.98	4.4543	1.3641

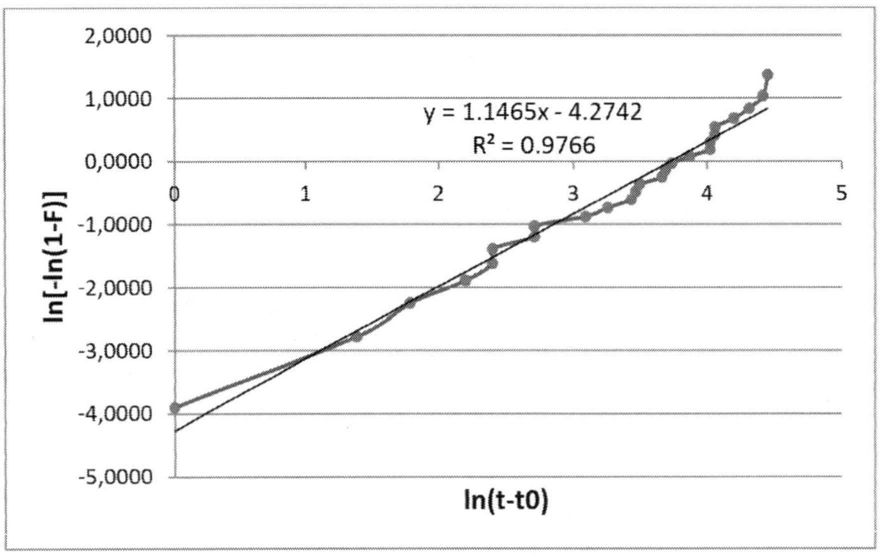

Figure 3. The regression line of the Weibull model.

We obtained, with a very high degree of confidence ($R^2 = 0.9766$), the equation of the Weibull line:

$$y = 1.1465x - 4.2742, \tag{20}$$

where the argument of the function (coefficient of the variable x) is the form parameter itself: $\beta = 1.1465$.

According to relation (13), it is considered:

$$\gamma \equiv t_1 = 6 months \tag{21}$$

The free term of the equation allows, by a direct calculation (rel. 18), the determination of the value of the parameter :

$$\eta = e^{-\frac{c}{\beta}} = e^{-\frac{-4.2742}{1.1465}} = 41.5 months \tag{22}$$

For the validation of the specified reliability objective (in the first year of operation the reliability should not fall below the value $R_{lim} = 0.95$, or the quantum of 95% should be at least 12 months), according to relation (19) it resulted:

$$R(12) = e^{-\left(\frac{12-6}{41.5}\right)^{1.1465}} = 0.897 < R_{\lim} = 0.95 \tag{23}$$

So, the proposed reliability target is not met (has been invalidated).

7. Conclusions

The proposed analytical method for linearizing the Weibull function allows the use of the linear regression function of Microsoft Office software to obtain the Weibull line equation, where the form parameter β is identified with the function argument, and the scale parameter η results from the analytical calculation in relation to free term.

The calculation program created in Microsoft Excel allows the identification of the 3 parameters of the Weibull model and, based on the Weibull model thus determined, the validation or invalidation of the reliability objectives.

The application presented also reveals that the Weibull model determined on the basis of this method has a sufficiently high degree of correlation in relation to the distribution of available data, even if they are not very numerous.

Thus, the proposed working tool leads to reliable estimates of reliability, is easy to use and offers the possibility to perform extensive reliability analyzes.

References

[1] O Alssadi, M Bâldea, M Istrate, A Boroiu 2018 Using the reliability logic schemes to forecast the vehicles systems reliability Proc. of the 5-th International Conference ADEM 2018 pp. 97-102

[2] C Andreescu 1996 Aplicaţii numerice la studiul fiabilităţii automobilelor (Bucureşti: Magie Publishing House)

[3] A Boroiu, A-A Boroiu 2019 Vehicle reliability (Pitesti: University of Piteşti Publishing House)

[4] A Boroiu, M Istrate, M Bâldea, A-A Boroiu 2016 Automotive maintenance plan improvemenet based on distribution of failures times Fiability and Durability 1 (Târgu-Jiu : "Academica Brâncuşi" Publishing House) pp. 316-320

[5] A Boroiu, M A Titu, C Oprean, A-A Boroiu 2013 Improving reliability modeling in the cases of truncated tests Proc. of the 6-th International Conference on Manufacturing Science and Education-MSE 2013 pp. 211-214
[6] A Boroiu 2013 Modelarea fiabilității autovehiculelor (Pitesti: University of Pitești Publishing House)
[7] A-A Boroiu, I Tabacu, A Ene, E Neagu, A Boroiu 2017 Computing programs for determining traffic flows from roundabouts Proc. of the 11-th International Automotive Congress "Automotive Engineering and Environment CAR 2017, IOP Conf. Series: Materials Science and Engineering 252 pp. 1-7
[8] A-A Boroiu, F Ivan, S Pârlac 2016 Modelarea zgomotului rutier (Pitesti: University of Pitești Publishing House)
[9] N Cordoş, N Filip 2000 Fiabilitatea autovehiculelor (Cluj-Napoca: Todesco Publishing House)
[10] J Grigore, A Boroiu, A-A Boroiu 2012 Proposals to improve the reliability modeling in the cases of truncated tests Fiability and Durability Supplement 1 (Târgu-Jiu : "Academica Brâncuşi" Publishing House) pp. 343-349
[11] P O'Connor 1991 Practical Reliability Engineering (New York: John Wiley&Sons Publishing House)
[12] F Palisson 1989 Détermination des paramètres du modèle de Weibull à partir de la méthode de l'actuariat Revue de statistique appliquée 37 no. 4 pp.5 – 39

New composite polymer sensor for vehicle weighing

A I Dontu, E Rakosi, P D Barsanescu and T M Ulian

Faculty of Mechanical Engineering, "Gheorghe Asachi" Technical University of Iasi, Iasi, Romania

andrei-ionut.dontu@academic.tuiasi.ro

Abstract. The work presented in this paper describes a new concept of the sensor for weighing-in-motion of vehicles. This sensor has small dimensions and can be used for weighing the vehicles. The sensor has four strain gauges, two of them are external and two are embedded into polymer. It's well known that the influence of temperature can affect the measurement process and also can change the output signal of strain gauges. The hysteresis and linearity graphics of strain gauges are compared in order to establish which one has higher values. The measurements lead to interesting results which are presented in this paper.

1. Introduction

The importance of weigh-in-motion (WIM) sensors for monitoring vehicles in traffic is now bigger. Basically, because the road transportation is more and more important and the density of vehicles is increasing the speed of traffic, weighing vehicles can no longer be achieved without the use of dedicated sensors [1]. In this context, it should be known that there is a wide range of sensors used for collecting data traffic. Thus, there are sensors for determining the speed, dimensions of vehicles, especially their length, width and height, sensors for determining weight, proximity sensors, video cameras, and so on [2]. Road transportation is responsible for producing 17% of greenhouse gases in the world [3]. WIM sensors are working in difficult conditions like strong vibrations, differences of temperature, ice, dust, and so on. The price of the sensors is influenced by the accuracy of the WIM sensors in weighing vehicles. The accuracy depends on environmental conditions but, also, on the dynamic behaviour of each vehicle (suspension, tire type, speed, driving style etc.).

The WIM sensors mounted in the road have advantages like detecting overloading trucks or detecting the vehicles that are speeding, but also disadvantages like installing them that takes time, short maintenance interval time or calibrating the sensor on each 1-2 years. The most known solutions are: bending plate sensors, load cell sensors, piezoelectrical sensors, bridge weigh-in-motion sensors, optic fibre, capacitive sensors and hydraulic WIM sensors. Although many of these systems are used, only a few are common. All of these sensors are used to determine the speed of vehicles, for weighing the cars and trucks and determine the overloading vehicles which are driving out of city, so, the sensors are for interurban routes [4].

2. Actual types of WIM sensors

The actual Weigh-in-Motion (WIM) sensors are usually made of: a pair of wired magnetic loops and a force sensor. The WIM station needs to determine the speed of vehicles, classify the vehicles, weigh the cars and trucks and provide real data traffic useful in avoiding traffic jams [5].

The WIM sensors can be classified in wide width plate sensors and strip/bar sensors. The first type of sensor consists of one bending plate which measures the force applied by the tire on the force sensor. In this case, the entire footprint of the tyre is on the force sensor. The second type of WIM sensor measures the force by integrating the output signal with time, which makes them less accurate. Despite the fact that there is a wide range of WIM sensors, only three of them are currently commercially successful - bending plate sensors, load cell sensors and piezo quartz sensors [6].

For the *bending plate sensors*, as the car passes over the sensor, the strain gauges measure the deformation of the plate by force applied by the tyre. This kind of sensor consists of two active plates made of steel mounted with two inductive loops [7]. One plate can be used for both car's wheels or two plates for each wheel of the car. For the second case, with two bending plates, the sensor is more accurate, because the load on each wheel is different and depends on how the load is distributed on the car. The bending plate sensors are less accurate than the piezoelectric ones, but under dynamic road and environmental conditions, both systems are performing with the same accuracy [7].

Load cell sensors are more precise than bending plate sensors and contain a steel box with a rigid top plate on the same level with the road, which is mounted in concrete. Now, a lot of patents are presenting new solutions of load cell sensors with different methods of weighing. In figure 1 it can be seen the elements of one patented motion weighing system [8, 9].

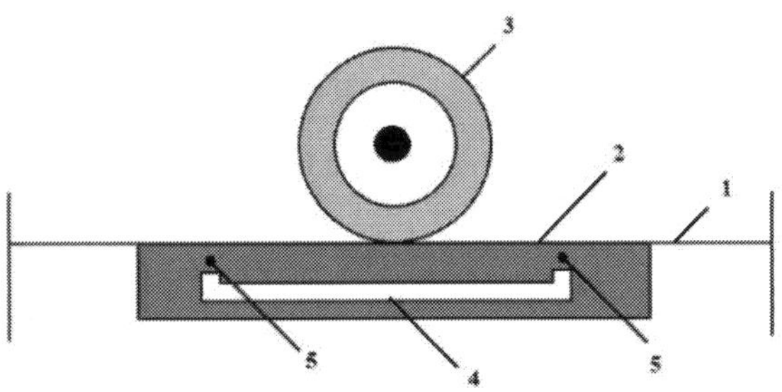

Figure 1. Weigh-in-motion system with force cells: 1 – roadway; 2 – weighing platform; 3 – wheel; 4 – sensor chamber; 5 – weighing sensors [8].

Piezo quartz sensors are made from an aluminium profile, which is mounted in the piezoelectric material (quartz). The contact between the tire and the sensor is made by an epoxy sand pad. The tire applies the force on the sensor and the piezoelectric materials give an output signal which is the function of the force applied. This kind of sensor is narrow and have the width smaller than the tire footprint [4, 5].

All these types of WIM sensors have disadvantages like: they are expensive, they are sensitive to mechanical stress, the installation is expensive and gets time consuming, the maintenance perturbs traffic for significant periods of time and some of them are smaller than tire footprint so the calibration process needs to be more often.

3. New composite polymer sensor

The new concept of polymer WIM sensor consists of a bar with a rectangular section on which strain gauges were applied. They are arranged on the exterior of the sensor but also, two of them are embedded in the composite polymer. The strain gauges were attached to an elastic element made of polymethylmethacrylate (PMMA), with chemical formula $(C_5O_2H_8)n$, also known as acrylic, acrylic glass, or plexiglass.

This sensor, with dimensions of 750 x 20 x 30 mm, is also made up of three PMMA plates, glued together. In figure 2 is presented the 3D sensor, developed using CAD software – SolidWorks.

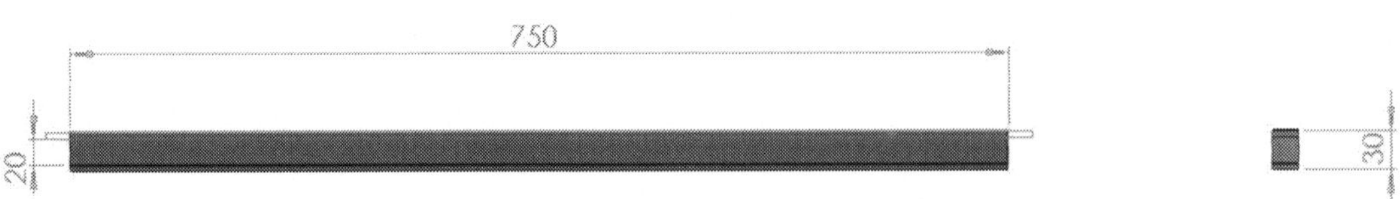

Figure 2. CAD Modeling of the new sensor

In order to physically achieve the designed model, the preliminary stages were completed, including the selection and verification by measurements of the basic material, as suggested in figure 3. Subsequently, three plates of different thicknesses were cut from a plexiglass plate and then milled. In figure 4 and figure 5, the operation of cutting the material on the machine tool is observed, and figure 6 shows the milling of the plates to bring them to the shape and dimensions established in the conception-design phase. It is mentioned that these operations of preparation of the elastic element were made inside of the Faculty of Mechanical Engineering from Iasi. The external strain gauges are made by Micro-Measurements and their resistance value is 350 Ω, meanwhile, the embedded ones are made by HBM (Hottinger Baldwin Messtechnik GmbH) with a resistance value of 120 Ω.

Figure 3. Dimension the material.

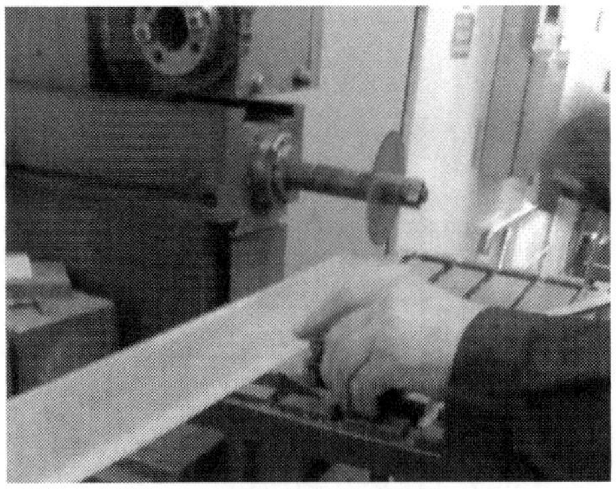

Figure 4. Material cutting.

Figure 5. Cloth cutting of the material.

Figure 6. Milling of the material for the final shape and dimensions.

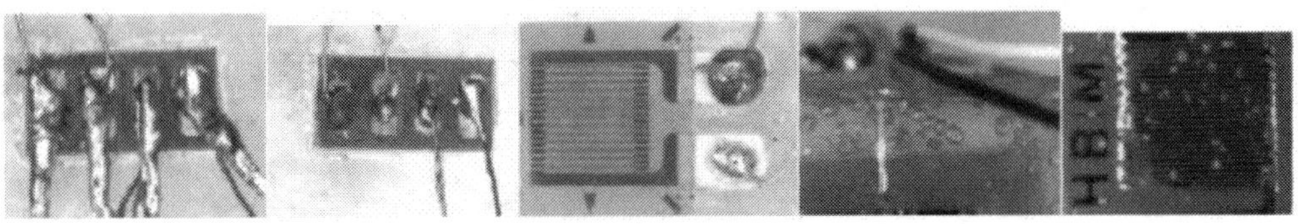

Figure 7. Strain gauges wire connections.

Preliminary verification under the microscope over the electrical connections was made to be sure the electric connections are good. Thus, in figure 7, it can be observed the embedded strain gauges inside the elastic element, as well as the connections that were made.

4. Experimental study of new WIM sensor

The requests of this new WIM sensor are to monitor the vehicles and classify them by structure and also to be cheap and to have a satisfying accuracy. To determine the measured errors that could appear, experimental tests were made in order to determine the hysteresis and the calculated deviations from nonlinearity for each strain gauge.

For this purpose, a force was applied, which produces a maximum deformation of the elastic element of the sensor of 3.17 mm. The application of the force in order to achieve this was done in eight loading stages, performing three loading cycles each.

In figure 8, it is representing the scheme of the strain gauges applied on the elastic element of the sensor.

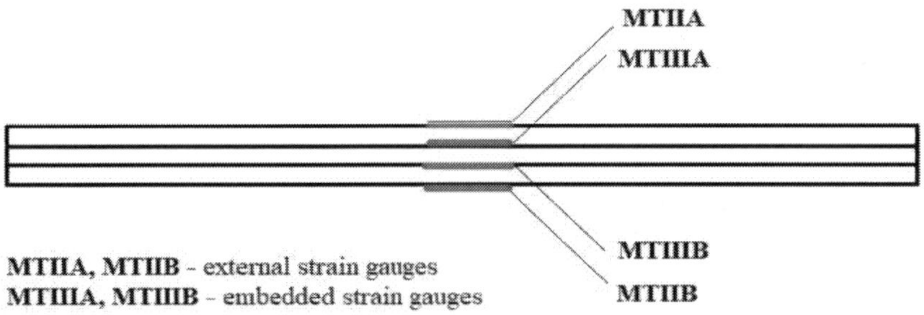

Figure 8. Strain gauges position.

The tests were made on four point bending machine presented in figure 9.

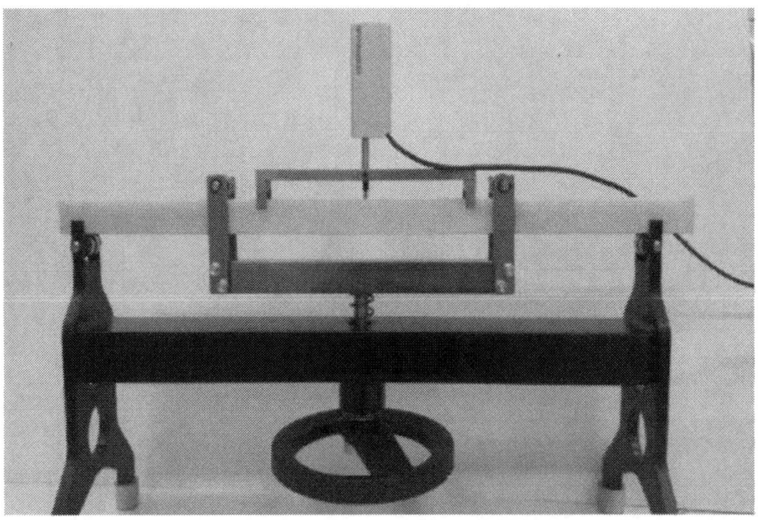

Figure 9. Four point bending machine (HTW Saar – University).

The experimental study of the hysteresis phenomenon at the level of each strain gauge was considered. Thus, figures 10-13 contain the representation of these hysteresis variations for one test.

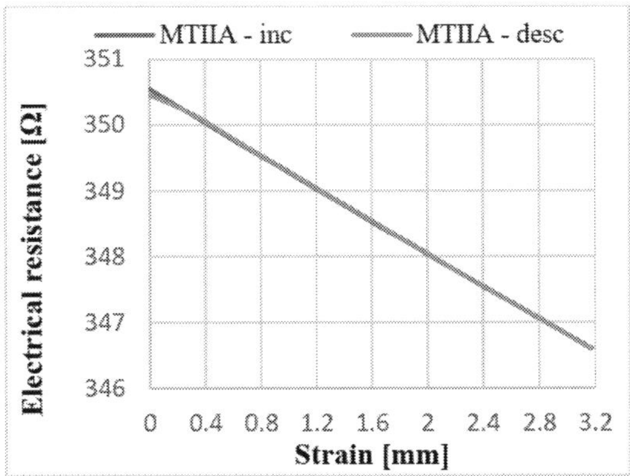

Figure 10. Hysteresis variations for strain gauge MTIIA.

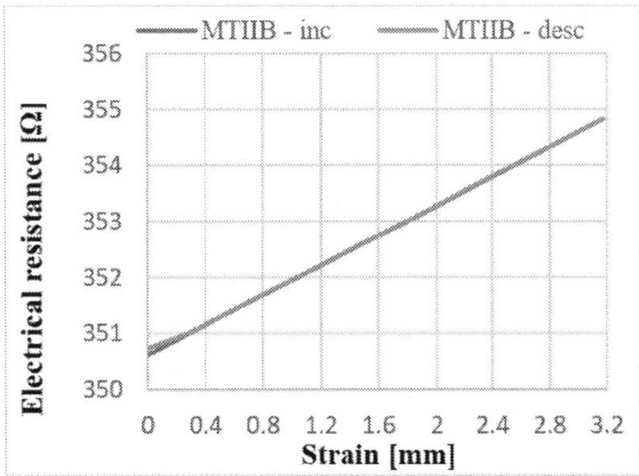

Figure 11. Hysteresis variations for strain gauge MTIIB.

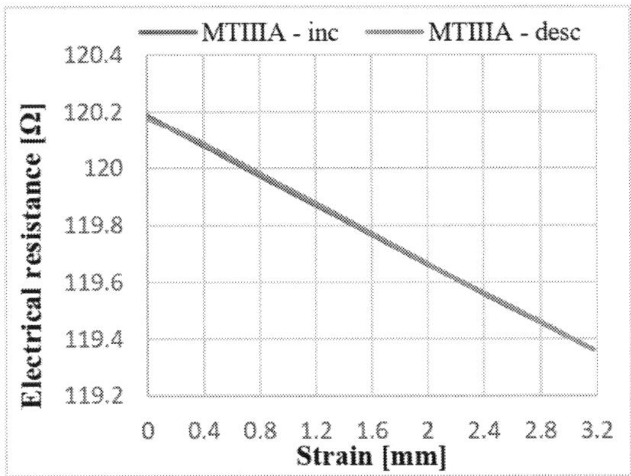

Figure 12. Hysteresis variations for strain gauge MTIIIA.

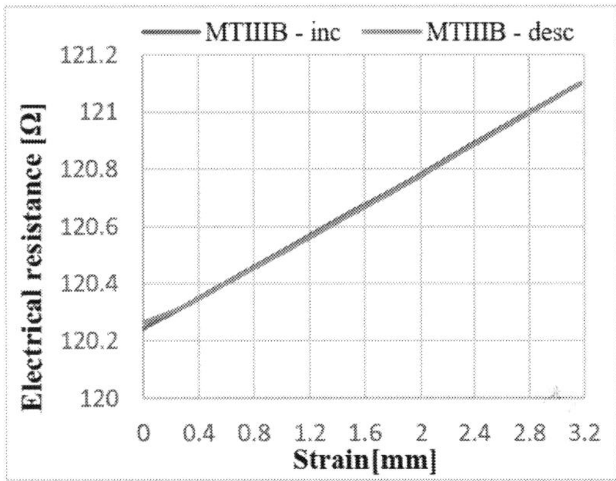

Figure 13. Hysteresis variations for strain gauge MTIIIB.

It is noted that the first and last values of the measurements were excluded because they did not show stable values. This phenomenon probably occurred due to the behaviour of the material used for the elastic element of the WIM-PMMA sensor

The analysis, in this case, highlights a very narrow range between the hysteresis curves. At the same time, it is found that the change in electrical resistance as a function of deformation is much smaller in the embedded strain gauges in the sensor as opposed to those on the outside of the sensor. The variations of the electrical resistance for the external strain gauges have values of 4 Ω, unlike those on the inside where the resistance varies by 0.8 – 0.9 Ω.

The experimental research aimed at studying the linearity of strain gauges. Also, in this case, three measurements were performed for each of the strain gauges MTIIA, MTIIB, MTIIIA, MTIIIB. Following the determinations performed, the variations of the nonlinearity deviations were represented graphically, expressed as a percentage, depending on the deformation. Figures 14-17 contain the representation of these linearity variations for one test.

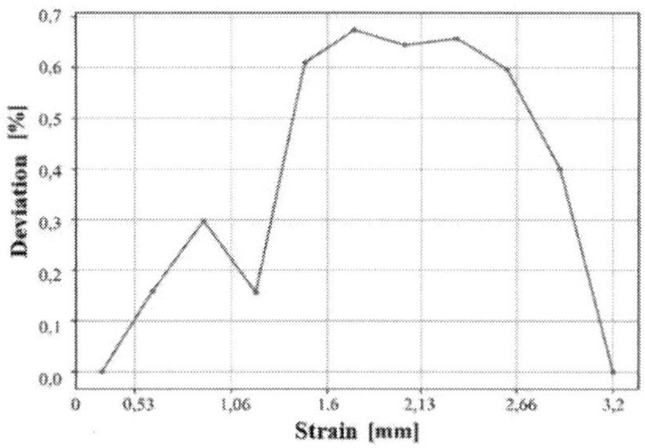

Figure 14. Linearity deviation for strain gauge MTIIA.

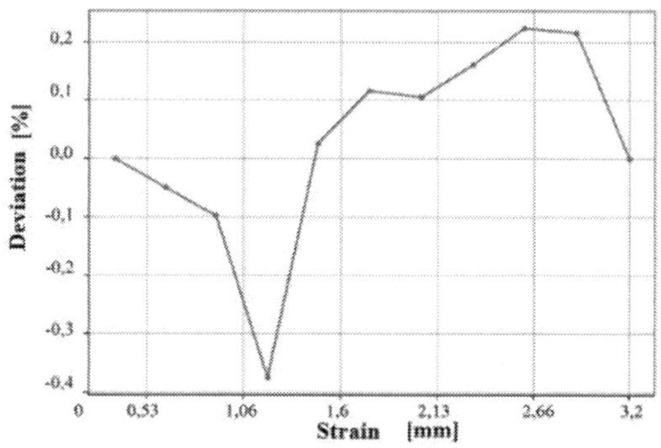

Figure 15. Linearity deviation for strain gauge MTIIB.

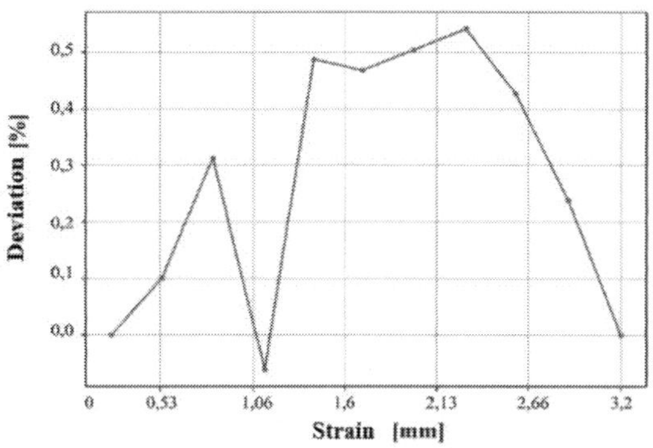

Figure 16. Linearity deviation for strain gauge MTIIIA.

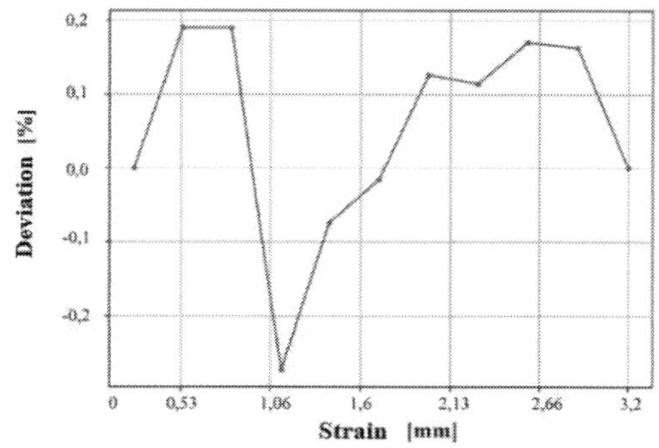

Figure 17. Linearity deviation for strain gauge MTIIIB.

To get an overview of the linearity deviation for the strain gauges, in figure 18, there are graphically represented the average values of the maximum deviations of nonlinearity for each strain gauge.

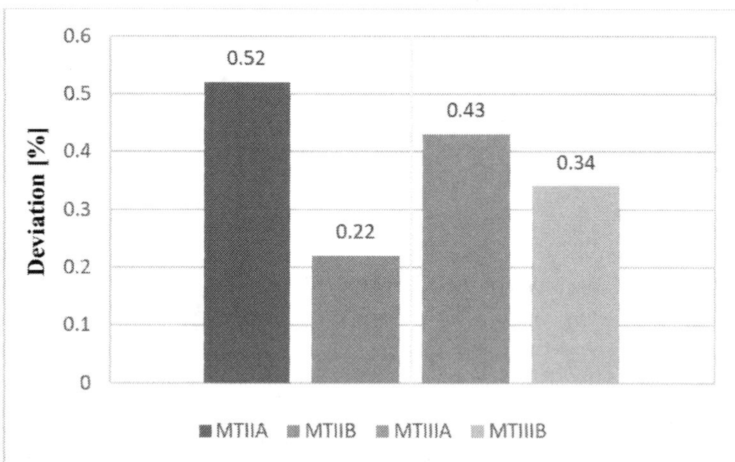

Figure 18. The average values of the maximum deviations of nonlinearity.

The measurements revealed average values of the maximum deviations of nonlinearity of 0.52%, 0.22%, 0.43% and 0.34% for strain gauges MTIIA, MTIIB, MTIIIA and MTIIIB.

It should be noticed that for all the strain gauges mounted both in the outside and inside of the sensor, the measurements are showing a deviation of less than 1%, which is a confirmation of the proper behavior of the device.

The influence of the temperature over the strain gauges was tested and the results conduct to the fact that the difference of values of the output signal at temperature variations is higher for the strain gauges on the superior part than those on the lower part (the opposite side).

5. Conclusions

The WIM technology evolves day by day and more and more countries use this technology for controlling the illegal overloading of trucks. The new sensor concept can be integrated into intelligent smart systems, using them for controlling urban traffic and avoiding traffic jams, which will conduct to fluidizing the traffic flows in the main intersections of the cities.

This new concept of WIM sensor shows:

- a very narrow area between the hysteresis curves;
- the change in electrical resistance as a function of strain is much smaller at the strain gauges embedded in the sensor as opposed to those on the outside of the sensor;
- the variations of the electrical resistance for the external strain gauges have values of 4 Ω, unlike those on the inside where the resistance varies by 0.8 – 0.9 Ω;
- it is found that the material used for the elastic element of the sensor, by its behavior, induces a phenomenon of instability for the first and last value in the measurements;
- all the strain gauges obtain a nonlinearity deviation of less than 1%, which is a confirmation of the proper functioning of the device for the success of this solution generally studied.

The studies and tests reveal a new WIM sensor that could be used for weighing vehicles or can be future developed to obtain traffic data in real time by using with other sensors. This solution can be integrated in the concrete being considered as an underground solution or in an overground device that could use to control the overloading trucks.

References

[1] Guerson, L; Van Loo, H.; Francheschi, L.; Tani, V.Z.; Valente, A.M. (2016), Development of a WIM Data Quality Management System for the Brazilian Federal Road Network. In: Proceedings of the 7th International Conference on Weigh-In-Motion (ICWIM 7), Eds. Jacob, B. et al., J. Wiley, 98-107

[2] P D Barsanescu, P Carlescu and A Stoian 2009 Senzori pentru cantarirea autovehiculelor in miscare Ed. Tehnopress, Iaşi ISBN 978-973-702-685-9

[3] A I Dontu, L Gaiginschi and P D Barsanescu 2019 Reducing the urban pollution by integrating weigh-in-motion sensors into intelligent transportation systems. State of the art and future trends *IOP Conf. Ser.: Mater. Sci. Eng.* **591** 012087 DOI: 10.1088/1757-899X/591/1/012087

[4] I Agape, A I Dontu, A Maftei, L Gaiginschi and P D Barsanescu 2019 Actual types of sensors used for weighing in motion *IOP Conf. Ser.: Mater. Sci. Eng.* **572** 012102 doi:10.1088/1757-899X/572/1/012102

[5] A I Dontu, P D Barsanescu, L Andrusca and N A Danila 2020 Weigh-in-motion sensors and traffic monitoring systems – Sate of the art and development trends *IOP Conf. Ser.: Mater. Sci. Eng.* **997** 012113 doi:10.1088/1757-899X/997/1/012113

[6] P. Burnos and J. Gajda 2016 Thermal Property Analysis of Axle Load Sensors for Weighing Vehicles in Weigh-in-Motion System *Sensors* **16** 2143

[7] Donald Jalmar Clark A comparison of bending plate and piezoelectric weigh-in-motion systems *master degree thesis*, Montana State University – Bozeman, 2002

[8] Patent US2015075297A1 – 2015 (S.U.A.): D Cornu and A Hofmann Hollow profile for a weight-in-motion sensor (Holder: Kistler Holding AG, Winterthur-Elveţia)

[9] Patent WO2013139284A1 - 2013: Z Yingjie, L Yuan and Z Tao Device for weighing vehicles in motion (Holder: Pantian Technology Development Co.)

Patent UA66460A -2003 (Ukraine): B G Yakovych Device for weighing vehicles in motion

New concept of car door reinforcement

I Pascal, M Gaidur, E Rakosi, T Ulian, Gh Manolache and S Talif

Mechanical and Automotive Engineering Department, "Gheorghe Asachi" Technical University of Iași, Romania

ionut.pascal@yahoo.com

Abstract. The purpose of this study is to make a series of finite element analyses and to show if the new concept of car door reinforcement brings any improvements. For this reason, was made a comparison between the finite element analyses of the actual concept of car door reinforcement and of the new proposal. The conditions for analyses were chosen as to reflect the reality of the situation. The analyses are of a dynamic type. In this way, we took into consideration the same material type for the car door and different material types and dimensions for the reinforcements. With the help of this study, the authors want to prove the improvements that the new proposal of car door reinforcement can bring into the automotive industry.

1. Introduction

One of the most important objectives of the design and manufacturing process of vehicles is the passive safety. This increasingly vehement requirement directs designers to new body architectures, but also to new materials from which they are made [1, 3].

In terms of the passive safety concept adapted to the body, it must be remembered that all constructive solutions and materials used aim to absorb the impact force resulting in a collision through structures adjacent to the passenger compartment and prevent deformation or penetration of solid bodies from the body skeleton into the passenger compartment [2, 5]. Usually, the body consists of three compartments: the engine, the passenger and the luggage. It is desirable that the impact be absorbed as much as possible by the engine compartment and luggage compartment, leaving the passenger compartment untouched as much as possible.

Some vehicles are equipped with a system that detects the danger of a side impact and causes the prone part to rise so that the impact takes place in the sill area, thus increasing the degree of protection offered by the vehicle [3, 5]. The main disadvantage of this system is the fact that it can be mounted only on vehicles that come equipped with air suspension or electronically controlled which limits its use.

The criteria for assessing the protection capacity of a body are represented by a score obtained after performing a series of well-defined and internationally recognized tests. These tests to which the bodies of road vehicles are subjected vary from one continent to another, so for Europe, we have EURO NCAP (side impact) and ECE R95 (1998-side impact) [6, 7, 11].

The evolution of technology lately allows the tests provided by these standards in the virtual environment. The impact analysis, being performed with the help of the computer, allows the manufacturer to obtain the results in a shorter time if we take into account the time of assembly of the vehicle and also to implement solutions to improve its strength structure in a considerably shorter time.

If it is desired to use finite element analysis for testing road vehicle bodies, another potential advantage is the cost of performing such a test, which in this case could be significantly lower [4].

2. The theoretical basis of study

Through this study, the authors conducted a series of finite element analyses to highlight the possible improvements brought by a new reinforcement concept for road vehicle doors.

The first step in carrying out this study is the 3D design of the door, in this sense we chose as a landmark the DACIA car, the Solenza model. The 3D model is presented using the image in Figure 1.

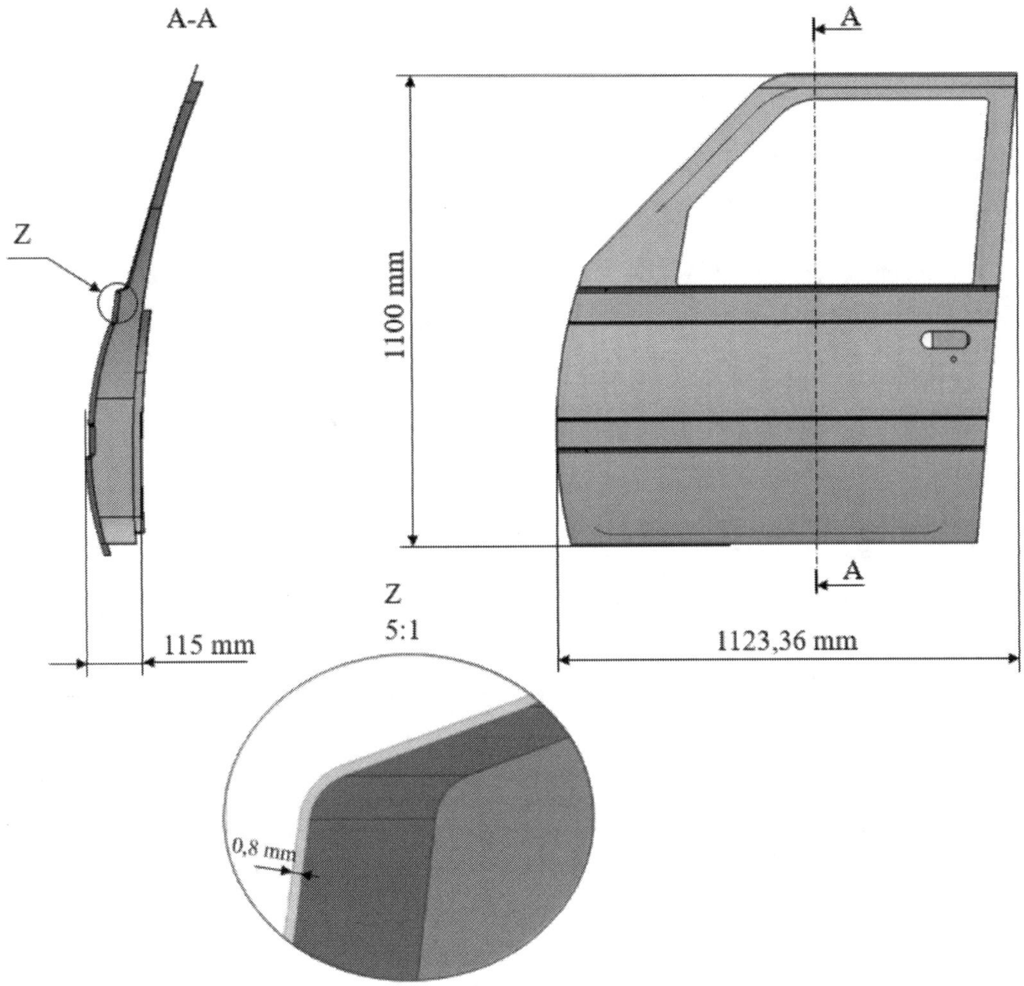

Figure 1. Car door model.

In addition to the overall dimensions of the door model, in figure 1 we can see with the help of the Z detail the thickness of the door plate, which is 0.8 [mm].

The scenario according to which the finite element analyzes are carried out is in accordance with the simulated situation within the tests carried out by EURO NCAP or ECE R95. Figure 2 describes the situation transposed in the ANSYS program.

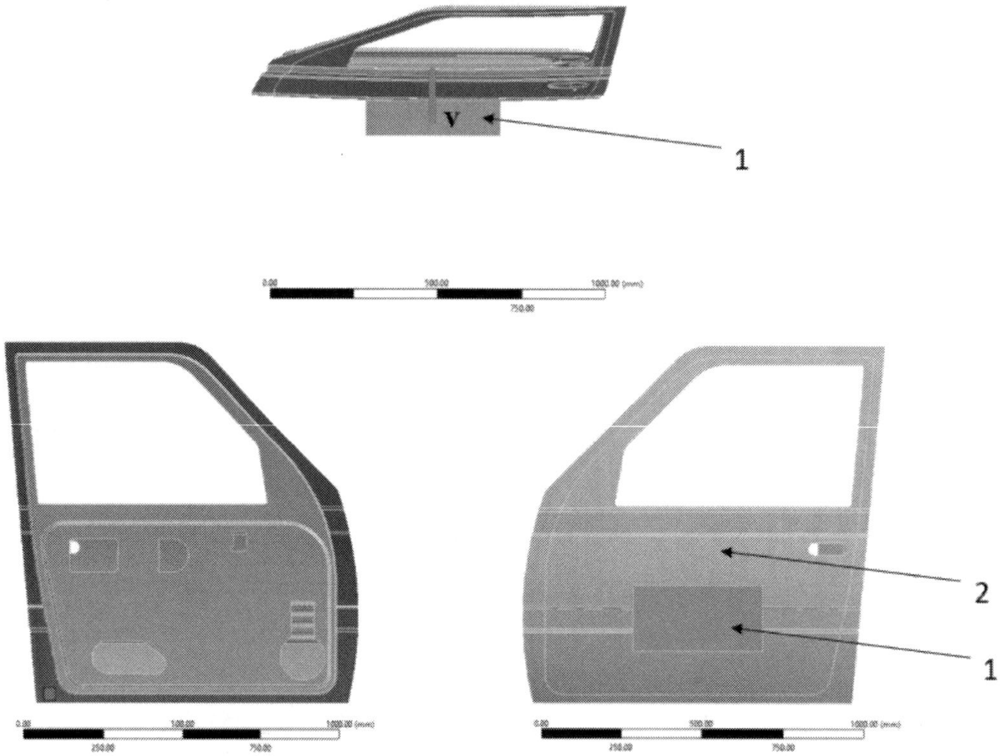

Figure 2. The initial conditions for the analysis.

The purple areas on the door (2) are the fixed areas, in reality, these are the areas with which the door comes in contact with the car body. The movable barrier (1), whose mass is equal to 2000 [kg], moves with a speed v, as the red arrow shows in figure 2 and will hit the door (2) [8, 10]. Basically, these data are converted in ANSYS and after that, the mesh is created. The existing reinforcement model and the proposed new reinforcement model can be observed in figure 3 [12].

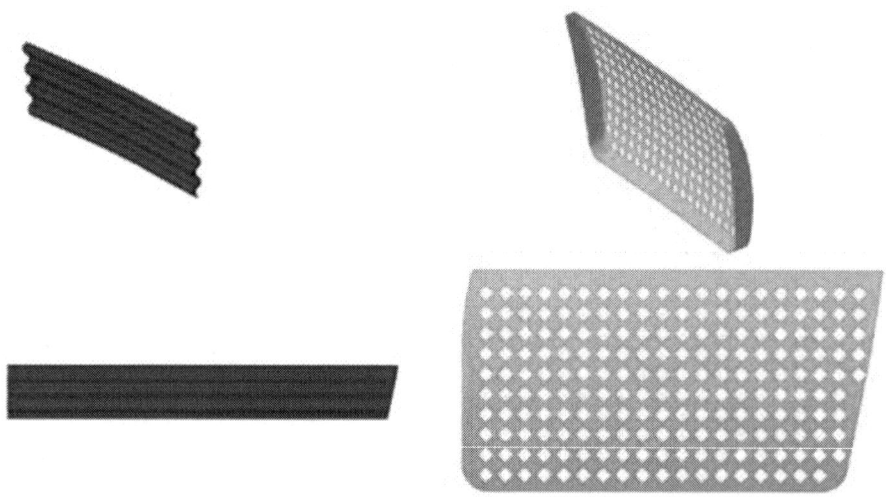

a. Existing concept b. New concept
Figure 3. Reinforcement concepts.

For the existing reinforcement model, we considered ultra-high strength steel with a wall thickness of 1.6 [mm], and for the new proposal, we decided to take into consideration an aluminium alloy with a material thickness of 5 [mm] [13]. The material for the car door is galvanized steel. The mechanical properties used in performing finite element analyses are mentioned in the bellow table.

Table 1. Mechanical properties.

Parameters / Material	Young's modulus [MPa]	Mass density [Kg/m^3]	Poisson ratio
Car door	$193 \cdot 10^3$	7930	0.3
Existing concept	$210 \cdot 10^3$	7890	0.3
New concept	$71.7 \cdot 10^3$	2810	0.3

3. The finite element analysis. Results and interpretations

Both for the door with the existing reinforcement model and for the door with the new reinforcement concept, a set of 3 finite element analyzes was performed in which we varied the speed of the mobile barrier. The values we assigned to the moving speed of the mobile barrier are: 50 [Km/h], 70 [Km/h] and 90 [Km/h] [9]. What was followed through these finite element analyzes is the total deformation on the direction of impact. Comparing the obtained results, we want to highlight if the new reinforcement concept has any contribution in terms of vehicle occupant protection.

During the presentation of the obtained results, images with sections A-A and B-B are inserted for each case. The sectioning paths are shown in figure 4.

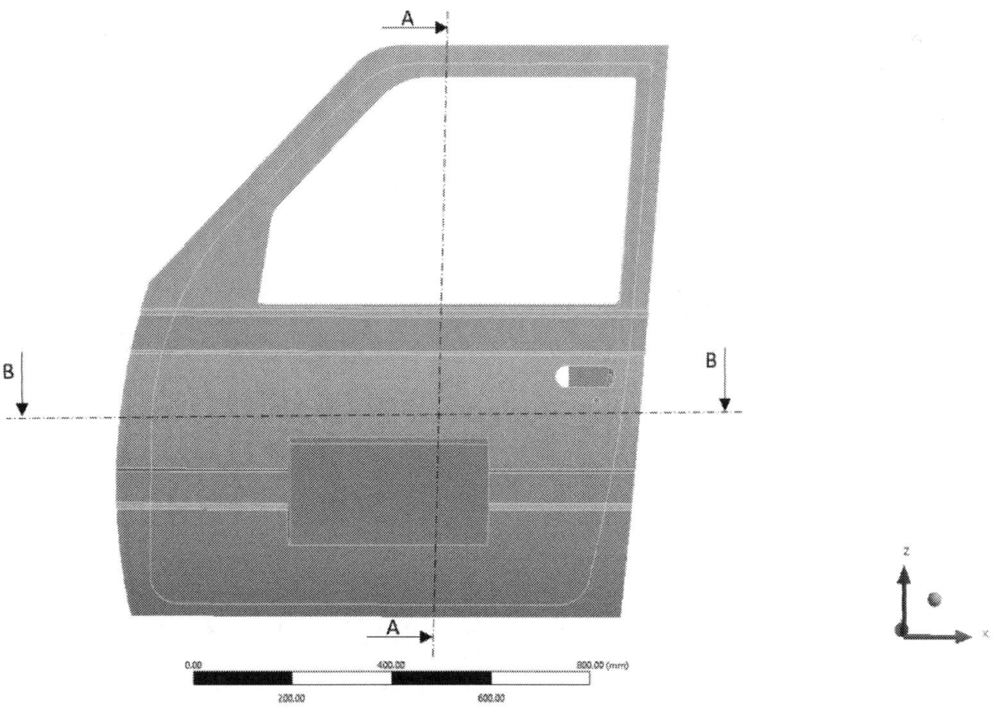

Figure 4. Section path.

The first set of finite element analyzes that we performed was the one in which the speed of the mobile barrier is equal to 50 [Km/h], and the results are presented using the figures below.

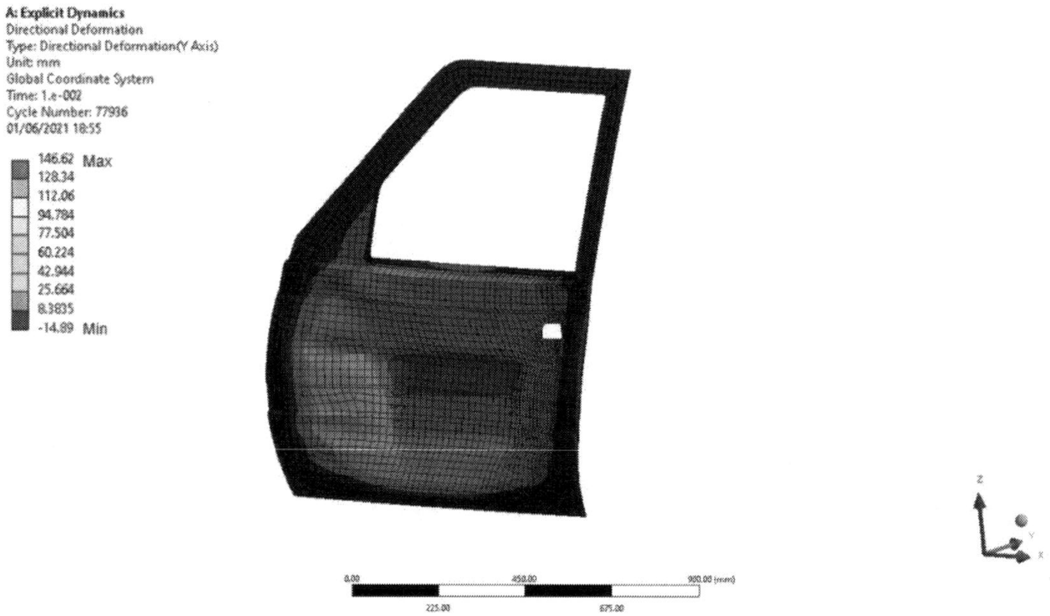

Figure 5. Total deformation on the y-axis for the car door with existing reinforcement concept v=50 [Km/h]

Analyzing the scale of total deformation on the y direction, we observe values between 146.62 [mm] and 8.38 [mm]. The red color is representative of the upper limit of the scale and marks the areas where the deformation has reached its maximum value. At the opposite pole, we meet the color light blue, this being significant for the areas where the deformation reaches the minimum limits on the scale.

From figure 5 we can conclude that the outer face of the door, the one that first comes in contact with the movable barrier, contains the largest deformations.

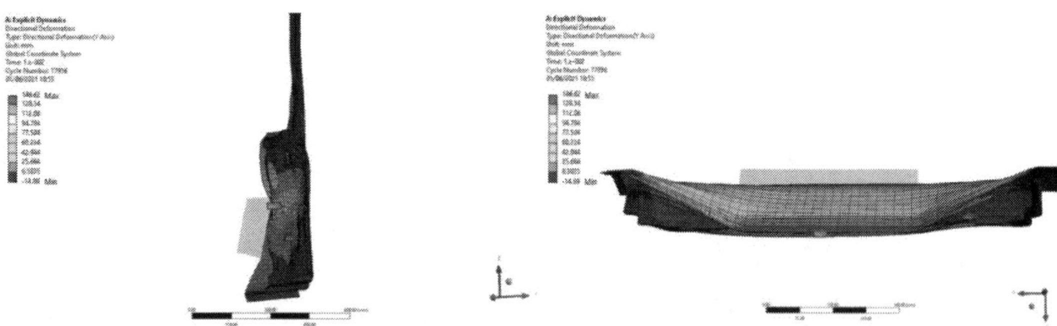

a. A-A section b. B-B section

Figure 6. Section views for the car door with existing reinforcement concept.

With the help of sections A-A and B-B we can observe the way in which the existing reinforcement is deformed, but also what values of the deformation we meet on the inner face of the door, the one that is closer to the occupant of the vehicle. According to the scale of values, we see in figure 6 that on the inner face of the door we find portions in which the deformation reaches values between 25 [mm] and 42 [mm].

Figure 7. Total deformation on the y-axis for the car door with new reinforcement concept v=50 [Km/h].

In the case of the door with the proposed new reinforcement concept, we notice that the upper limit of the deformation scale has a lower value compared to the previously presented case in which the door contained an existing reinforcement concept. In the situation presented in figure 7, on the face of the exterior door, we observe deformations of maximum 89 [mm].

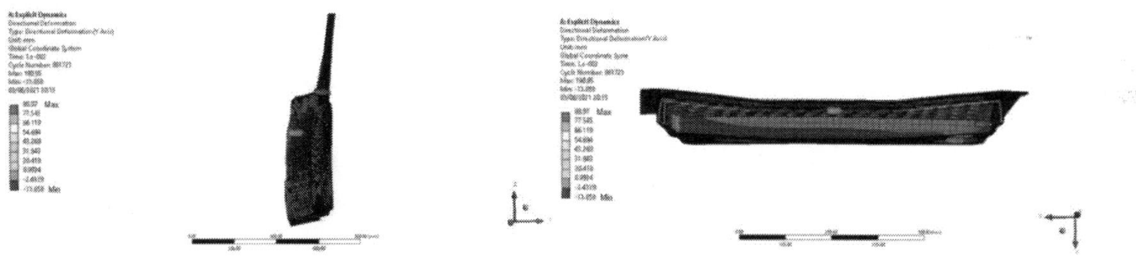

a. A-A section b. B-B section

Figure 8. Section views for the car door with a new reinforcement concept.

Sections A-A and B-B, from figure 8, highlight in this case the deformations appeared in reinforcement and on the inner face of the door. On the reinforcement surface, we find deformations of 89 [mm], while the inner face of the door does not seem to be affected in case of impact with a speed of 50 [Km/h].

The next set of finite element analyzes performed similar conditions, the only change made was to increase the speed at which the mobile barrier moves from 50 [Km/h] to 70 [Km/h].

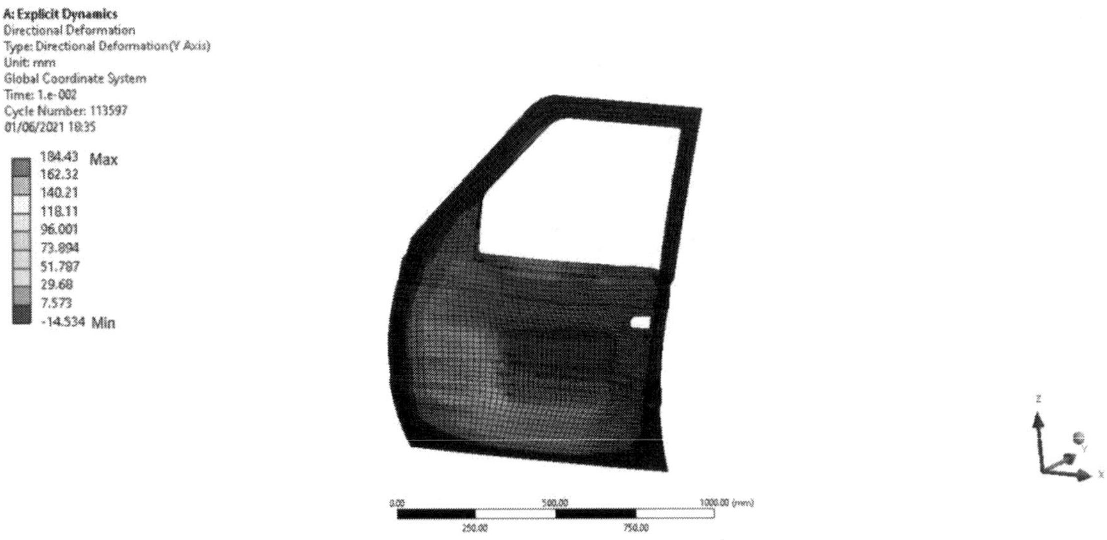

Figure 9. Total deformation on the y-axis for the car door with existing reinforcement concept v=70 [Km/h].

As expected, with the increase of the mobile barrier movement speed, the upper limit of the deformation scale will also increase. Analysing the deformation scale in figure 9 we notice that the upper limit is marked by the value 184.43 [mm], the red color representing this value on the door surface.

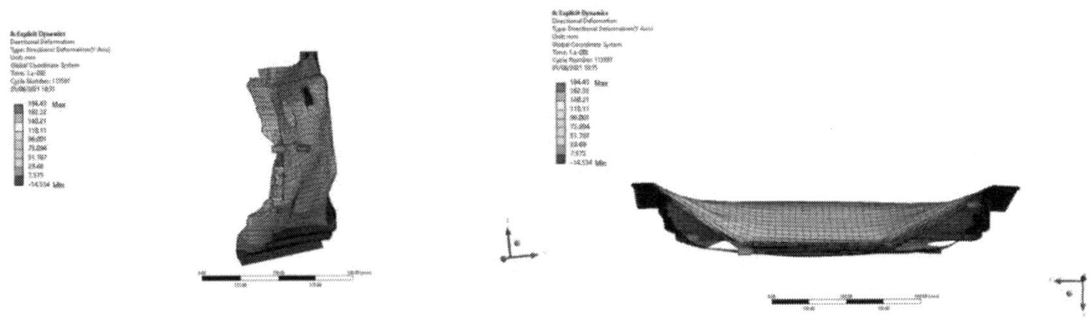

a. A-A section b. B-B section

Figure 10. Section views for the car door with existing reinforcement concept.

The sections in figure 10 provide information on the deformation values found on the reinforcement and on the inner face of the door. Compared to the first case in which the speed of movement of the mobile barrier was 50 [Km/h], in this case, the deformations on the inner face of the door have values of up to 74 [mm].

Also in the case of the door reinforced with the new concept we have increased the speed of movement of the mobile barrier, the results can be observed with the help of the images below.

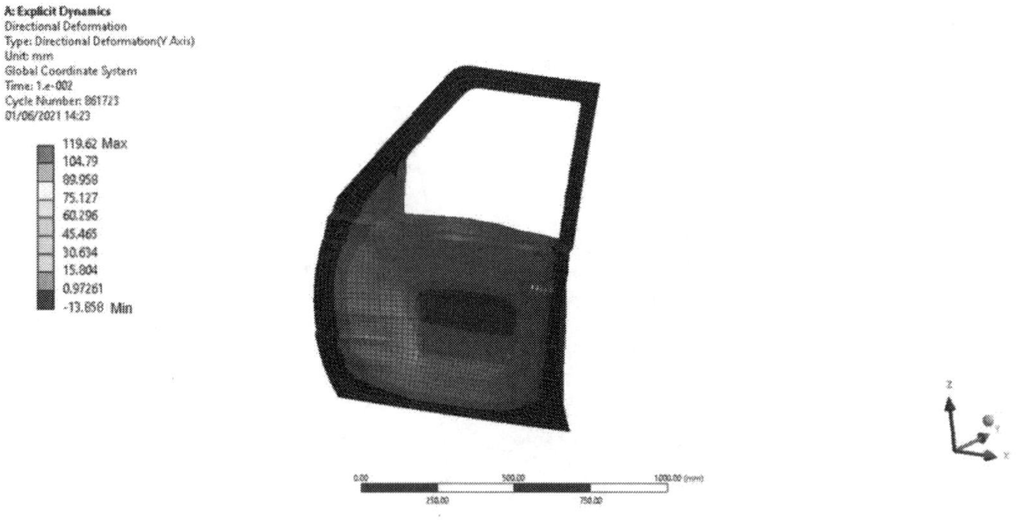

Figure 11. Total deformation on the y-axis for the car door with a new reinforcement concept v=70 [Km/h].

The deformation values, according to the scale shown in figure 11, are between 119.62 [mm] and 1 [mm]. As in previous cases, the red color is representative of the maximum value of the deformation.

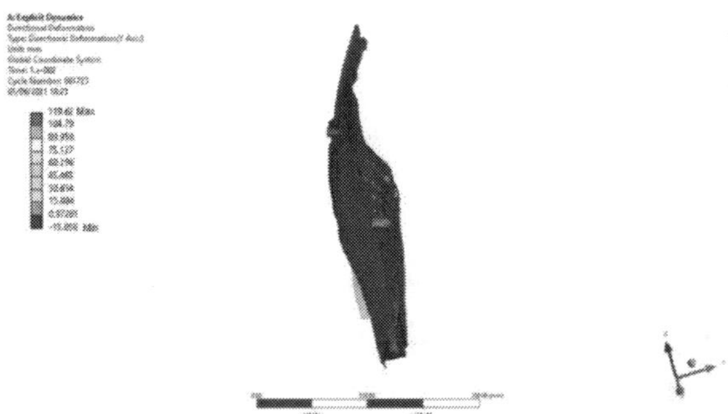

a. A-A section

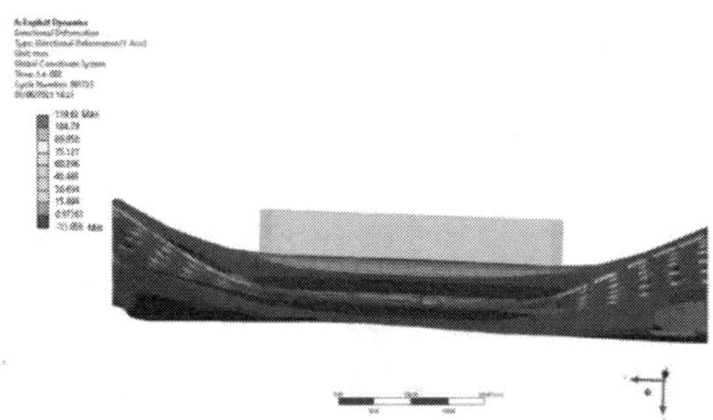

b. B-B section

Figure 12. Section views for the car door with a new reinforcement concept.

With the help of the sections from figure 12 we can observe that once the speed of movement of the mobile barrier is increased from 50 [Km/h] to 70 [Km/h], deformations also appear on the inner surface of the door. The deformations encountered here have values between 15 [mm] and 30 [mm].

The last set of finite element analyses was performed with a moving barrier speed equal to 90 [Km/h].

IOP Conf. Series: Materials Science and Engineering 1220 (2022) 012031 doi:10.1088/1757-899X/1220/1/012031

Figure 13. Total deformation on the y-axis for the car door with existing reinforcement concept v=90 [Km/h].

According to the deformation scale shown in figure 13, the red color represents the value of 231.23 [mm] that can be found on the surface of the door and the reinforcement.

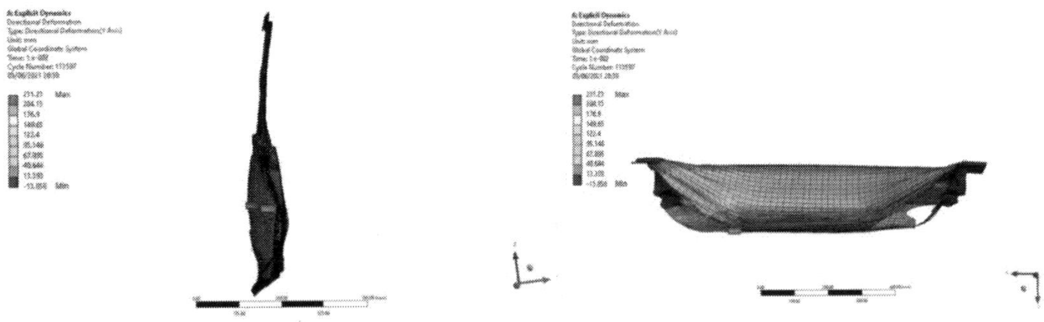

a. A-A section b. B-B section

Figure 14. Section views for the car door with existing reinforcement concept.

Analysing the sections presented in figure 14, we notice that in this case, the inner face of the door includes areas where the deformation value reaches 122.4 [mm]. Compared to the first set of finite element analyzes in which the speed of movement of the mobile barrier was 50 [Km/h], in this case, the deformations on the inner face of the door are 3 times higher, this means that the danger that the occupant from the vehicle to be larger test wounds.

In the case of the reinforced door with the new solution, when the speed of movement of the mobile barrier is 90 [Km/h], the results are presented through the following figures.

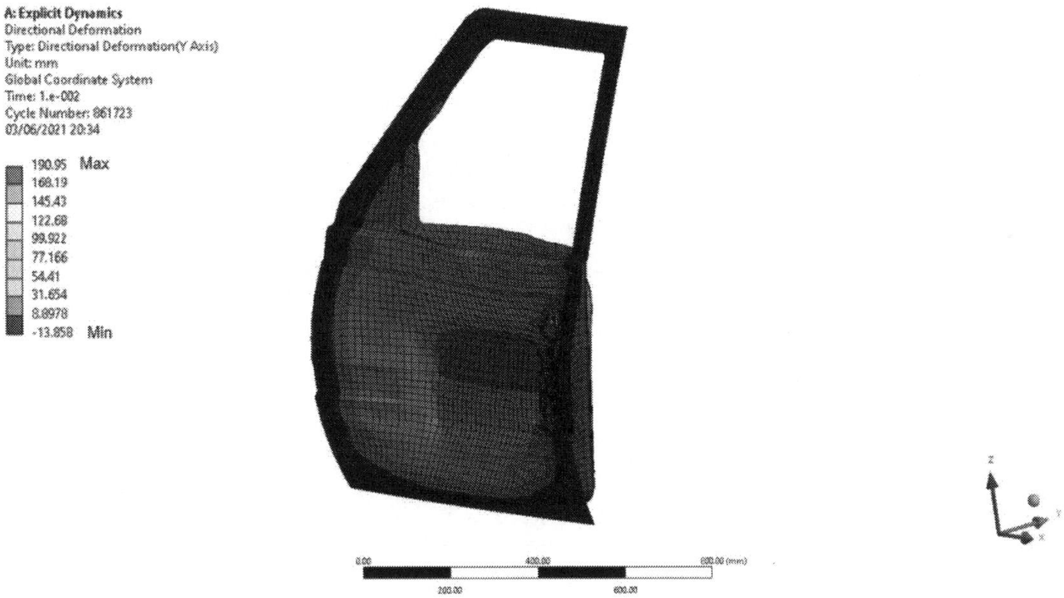

Figure 15. Total deformation on the y-axis for the car door with new reinforcement concept v=90 [Km/h].

The maximum value of the deformation can be identified with the help of the scale shown in figure 15. The areas where we find this type of deformation are highlighted by the red color.

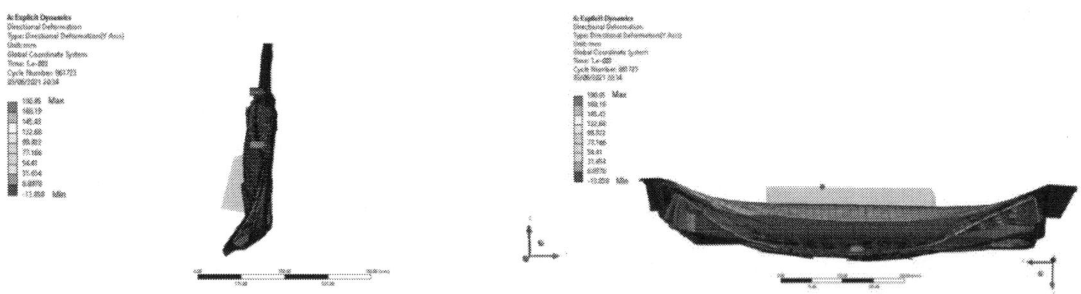

a. A-A section b. B-B section

Figure 16. Section views for the car door with a new reinforcement concept.

4. Conclusion

Before proceeding to the comparison of the obtained results and to draw the final conclusions, some clarifications must be made regarding the analysed situations. The analysed door model does not contain the window support assembly, the closing mechanism or the plastic face that is mounted towards the passenger compartment. All these components would have reduced the values of deformations in both cases.

Comparing the results obtained after the finite element analysis, we can conclude that in the case of the reinforced door with the new proposed concept, the obtained results confer better protection on the occupants of the vehicle in case of a side impact.

If the mobile barrier moves at a speed of 50 [Km/h], we can conclude that in the case of the reinforced door with the new proposed concept the total deformation is about 40% lower compared to the reinforced door with the existing model. This means that the distance that the mobile barrier travels from the moment it collides with the door face until it stops, is less in the case of reinforcement with the new concept, so the risk of the vehicle occupant being injured is approximately 40% less.

Likewise, in the case of the other two situations, in which the mobile barrier moves with 70 [Km/h], respectively 90 [Km/h], we noticed such a difference between the two reinforcement concepts. In the scenario in which the barrier moves by 70 [Km/h], the new concept deforms by 35% less, while in case the barrier moves by 90 [Km/h], the new concept deforms by about 18% more little bit.

In all the analysed situations, the new reinforcement concept proposed to be deformed less compared to the already existing ones, but the space that this concept occupies compared to the already existing one attracts a disadvantage on this concept. Another possible disadvantage is the price of obtaining such a concept. These two identified disadvantages can be combated by the fact that the new concept offers a sporty safety for the occupants of the vehicle and by the fact that the manufacturers can reduce the thickness of the material used to make the door from 0.8 [mm] to 0.6 [mm] or even 0.4 [mm].

References

[1] Radu Gaiginschi, Radu Drosescu, Adrian Sachelarie s.a – "*Siguranța Circulației Rutiere vol II* " , Ed.Tehnică
[2] Guangyao Zhao – „*Whole Car Side Impact Mode and Response Evaluation*"
[3] Radu Gaiginschi – „*Reconstrucția și expertiza accidentelor rutiere*",Ed.Tehnică
[4] Ștefan I.Maksay, Diana A.Bistran –„*Introducere în metoda elementelor finite*",Ed.Cemi Iași 2008
[5] Mariana Domnica Stanciu, Ovidiu Mihai Terciu, Ioan Curtu – "*Compozite Lignocelulozice Aplicații in industria automobilelor*",Ed. Universității Transilvania din Brașov, 2014
[6] Mark Halford Warner – "*Development of Pole Impact Testing at Multiple Vehicle Side Locations As Applied To The Ford Taurus Structural Platform*".
[7] *** https://fr.wikipedia.org/wiki/Euro_NCAP
[8] Markus Bröckerhoff – "*Structural Improvements for the Lateral Pole Impact*", ATZ 07-08I2007 Volume 109.
[9] Florentina Rotaru (Parsachiv) – "*Fenomene de degradare la impactul mecanic al structurilor compozite de tip sandwich*", teză de doctorat – seria 16: Inginerie mecanică nr.42, Galați 2018.
[10] Ling, Liang & Dhanasekar, Manicka & Thambiratnam, David & Sun – "*Lateral impact derailment mechanisms, simulation and analysis*" - International Journal of Impact Engineering. 94. 36-49.
[11] Hoegner, C, Gajewski, M, and Zippel – "*Simulation of the lateral pole-impact on a crash simulation system*". Germany: N. p., 2001. Web.
[12] Deb A, Srinivas KC – "*Development of a new lumped-parameter model for vehicle side-impact safety simulation*". Proceedings of the Institution of Mechanical Engineers, Part D: Journal of Automobile Engineering. 2008;222(10):1793-1811.
[13] Patberg L, Philipps M, Dittmann R. - "*Fibre-reinforced composites in the car side structure*". Proceedings of the Institution of Mechanical Engineers, Part D: Journal of Automobile Engineering. 1999;213(5):417-423.

Bearings fault detection by means of frequency analysis

R M Stoica[1], D Voicu[1], R Vilau[1], L Barothi[1] and L Deleanu[2]

[1]Department of Military Vehicles and Transportation, Military Technical Academy "Ferdinand I", 39-49 George Cosbuc blvd., Bucharest, Romania
[2]Department of Mechanical Engineering, "Dunărea de Jos" University of Galati, 47 Domneasca Str., Galati, Romania

ramona.stoica@mta.ro

Abstract. The present paper aims to analyse the possibility to prevent machine failure due to bearing faults, with the use of frequency analysis. The experimental research consisted in mounting a simple mechanical transmission, integrating an electric motor, a rigid coupling, a driven shaft, two pillow block bearings and a uniaxial accelerometer (connected to a data acquisition board), which was tested at two constant motor speeds. Follow-up, with the use of recorded data, it was plotted the envelope spectrum of bearing functioning, based on BPFO, BPFI, FTF and BSF harmonics, which were later used to determine on which element of the bearing is the defect located, or if the bearing lacks proper greasing.

1. Introduction

Bearings are machine elements used for the purpose of reducing friction between moving parts of an assembly. The extensive use of bearings in all sorts of applications has led to numerous designs and dimensions, being manufactured of different materials.

The importance of proper functioning when comes to bearings consists in ensuring a long working period of time. Therefore, it is not enough to choose the right bearing according to the application, but also to ensure correct mounting within the assembly.

To that end, lately, there have been developed methods to analyse and prevent faulty functioning of bearing, thus avoiding the damage of machine elements. According to (Kansanaho, 2019) [4], vibration measurement is the most forehand detection method, which can prevent bearing failure.

For that purpose, the present paper aims to analyse the functioning of a bearing and identify the characteristic elements such as amplitude, velocity or displacement, which could indicate functioning problems. From all types of faults encountered at bearings, whether there are localized or distributed defects, the study if focused on the former, meaning cracks or pits on bearing surfaces [4, 9].

The present paper aims to apply the technique of envelope analysis, detailed by Kim S et al [8] in their paper, in order to analyse the functioning of rolling elements within mechanical transmissions.

2. Theoretical bases

Therefore, it is considered a deep groove ball bearing type 6205, with the characteristic parameters detailed in Table 1 [1, 5, 10]:

Table 1. Characteristic parameters of the radial single row ball bearing.

Type of characteristic	Symbol	Value	Unit
Diameter of inner race	d	25	mm
Diameter of outer race	D	52	mm
Width	B	15	mm
Diameter of ball	d_b	7.94	mm
Pitch diameter	p_D	39.04	mm
Contact angle between the ball and the raceway	ϕ	0	°

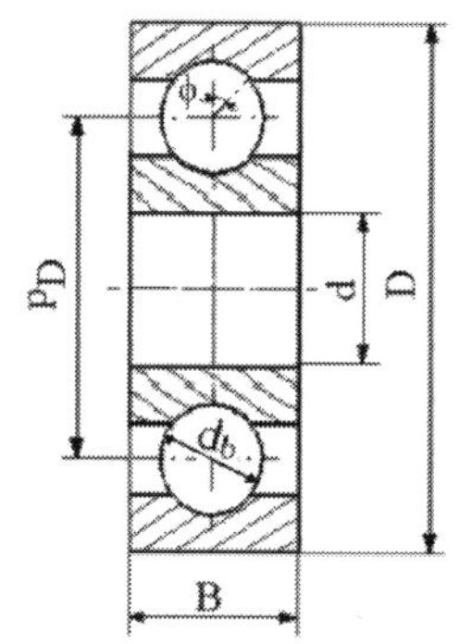

Figure 1. Characteristic parameters of the analysed radial single row ball bearing [7].

In order to prevent failure of a bearing, one can use frequency analysis to detect faults on the surfaces of the bearing's components (inner or outer race, rolling elements, cage etc.). Such an analysis can point out the existence of one or multiple cracks on the bearing surface, which are visible on the graphs every time one component passes over the defect.

Also, the position of the fault on one of the bearing components can be determined based on the envelope spectrum, which uses the Fast Fourier Transform [2, 3].

Knowing the geometrical details of a bearing, each time an element passes over a crack, it can be determined the repetition rate of the impact (vibration frequencies) as follows:

$$BPFO = \frac{n_b \cdot n}{2}\left(1 - \frac{d_b}{p_D} \cdot \cos\phi\right) \qquad [\text{Hz}] \tag{1}$$

where: $BPFO$ – the ball pass frequency specific to the outer race;
n_b – the number of balls specific to the bearing, which equals 9 for a 6205 deep groove ball bearing;
n – the shaft speed, converted from rev/min to Hz.

$$BPFI = \frac{n_b \cdot n}{2}\left(1 + \frac{d_b}{p_D} \cdot \cos\phi\right) \qquad [\text{Hz}] \tag{2}$$

where: $BPFI$ – the ball pass frequency specific to the inner race.

$$FTF = \frac{n}{2}\left(1 - \frac{d_b}{p_D} \cdot \cos\phi\right) \qquad [\text{Hz}] \tag{3}$$

where: FTF – the fundamental train frequency (cage speed, if any).

$$BSF = \frac{p_D \cdot n}{2 \cdot d_b}\left[1 - \left(\frac{d_b}{p_D} \cdot \cos\phi\right)^2\right] \qquad [\mathrm{Hz}] \tag{4}$$

where: BSF – the ball spin frequency.

3. Experimental research

Considering the formulae (1) - (4), the bearing parameters mentioned before and the testing speed of 300 rev/min, the calculated bearing defect frequencies are: $BPFO = 71.69$ Hz, $BPFI = 108.3$ Hz, $FTF = 7.96$ Hz and $BSF = 39.16$ Hz.

Also, the formulae (1) - (4) are consistent with a no slip functioning of the bearing elements, which is improbable, therefore, the results obtained can vary up to 1-2% from the calculated results.

For the experimental research, it was used a simple mechanical transmission, to avoid unwanted sources of vibrations, consisting in an electric motor, a rigid coupling used to transmit the torque from the driving shaft of the motor to a driven shaft and two pillow blocks, each containing a deep groove ball bearing type 6205. The mechanical transmission is depicted in figure 2.

Figure 2. Mechanical transmission setup.

To measure the vibrations during bearing functioning, as well as the bearing speed, there were used a uniaxial accelerometer type SV 160.01 (figure 3a) and a non contact laser tachometer (figure 3b). The acceleration sensor has the following specifications: the frequency domain between 0.5-14000 Hz, resonance frequency – 25 kHz, limit shock acceleration – 5000 g, transversal sensitivity - < 5%, maximum temperature range between -55...+120/max 150°C, connector type M12.

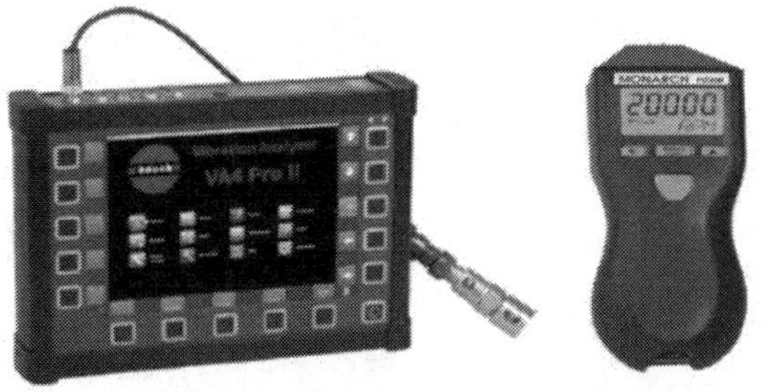

a) accelerometer SV 160.01 b) laser tachometer

Figure 3. Mechanical transmission setup.

The experimental research consisted in performing three tests with the deep groove ball bearing, in order to detect any crack or defect on bearing elements (on the inner or the outer race, or on the rolling elements). Each test was performed at constant motor speeds of 300 rev/min (10 Hz). Also, the three tests resulted by positioning the accelerometer in three mutually perpendicular directions, according to standard ISO 10816-1 (the accelerometer positions are depicted in figure 2) [6].

Therefore, during the first test, the accelerometer was positioned in the radial plane, normal to the horizontal-transverse direction, during the second test, the accelerometer was positioned in the radial plane, normal to vertical-transverse direction and during the third one, it was performed an axial measurement, normal to shaft's longitudinal axis (according to ISO 10816-1) [6]. Also, it was avoided the speed variation during tests, which would have implied to remove shaft speed variation by performing a pre-processing step. Each test lasted 60 seconds.

4. Results

Considering that the bearing is sealed and it is mounted out of the reach of the human eye to allow inspection, or the crack is merely not on the visible parts, the envelope spectrum analysis will be used to determine where is the defect on bearing elements, if such exists.

To that end, first, it is performed a time-domain analysis, by plotting the RMS (root mean square) values of three fundamental quantities (e.g. displacement, amplitude and velocity), considering the accelerometer mounted in the radial plane, normal to the horizontal-transverse direction (figure 4). Second, it was performed the frequency-domain analysis, determining the RMS velocity spectrum and follow-up, the amplitude demodulation, which is used to extract the bearing frequencies from the vibration spectrum of the other mechanical transmission constituents (figure 5).

According to standard ISO 10816 regarding vibration severity, the value of vibration velocity depicted in figure 4, namely 0.624 mm/s, implies that the bearing encloses within the Good category. Yet again, from the 4-spectrum graph, it can be observed that the highest value of velocity corresponds to a frequency of 20 Hz, which is not equal to any calculated bearing defect frequencies.

Also, from figure 5, it can be observed that the highest values of RMS velocity and acceleration were recorded at 10 Hz, 20 Hz, 50 Hz, 70 HZ, 155 Hz, 179 Hz and 186 Hz. Comparing these values with the bearing defect frequencies, it can be concluded that there is no indication for faults on bearing elements.

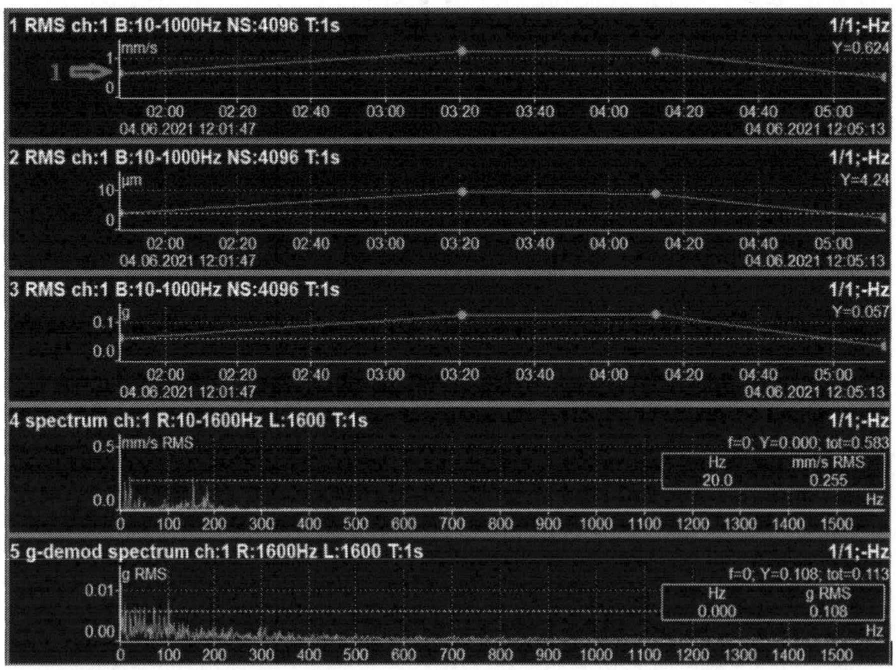

Figure 4. RMS values for the first measurement of the accelerometer.

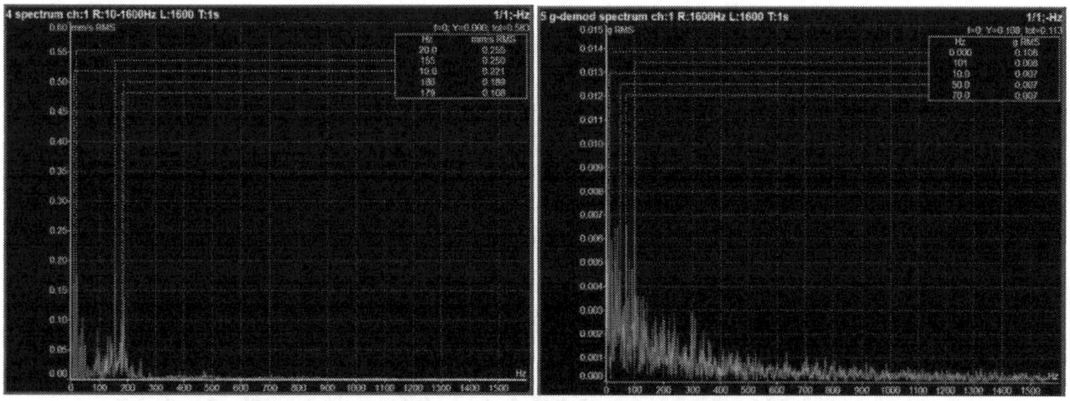

Figure 5. RMS velocity spectrum and amplitude demodulation from the first accelerometer position.

Similar to figure 4, in figure 6 and figure 7 there are shown the RMS values for the second and the third measurement position of the accelerometer. Both graphs show that the velocity values indicate the proper functioning of the bearing (according to ISO 10816). Also, the spectrum analysis does not indicate the characteristic signature of fault frequencies on bearing elements.

Figure 6 depicts the maximum RMS velocity value obtained for a frequency of 20 Hz, and maximum RMS acceleration for 0 Hz. The same results are obtained in the case of the third measurement (shown in figure 7). These frequency values, non consistent with any bearing defects, may be generated by the shaft rotation or other mechanical elements.

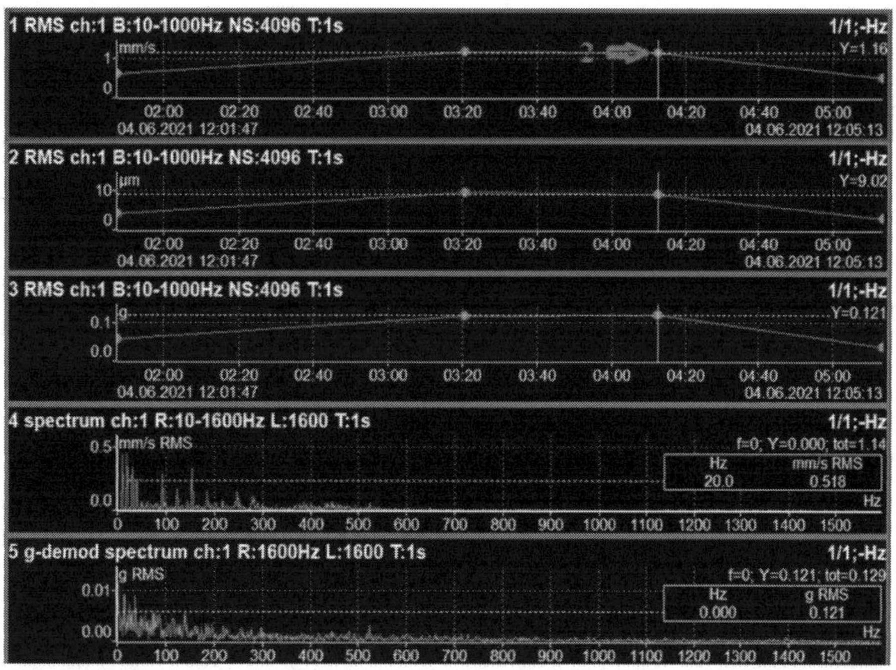

Figure 6. RMS values for the second measurement of the accelerometer.

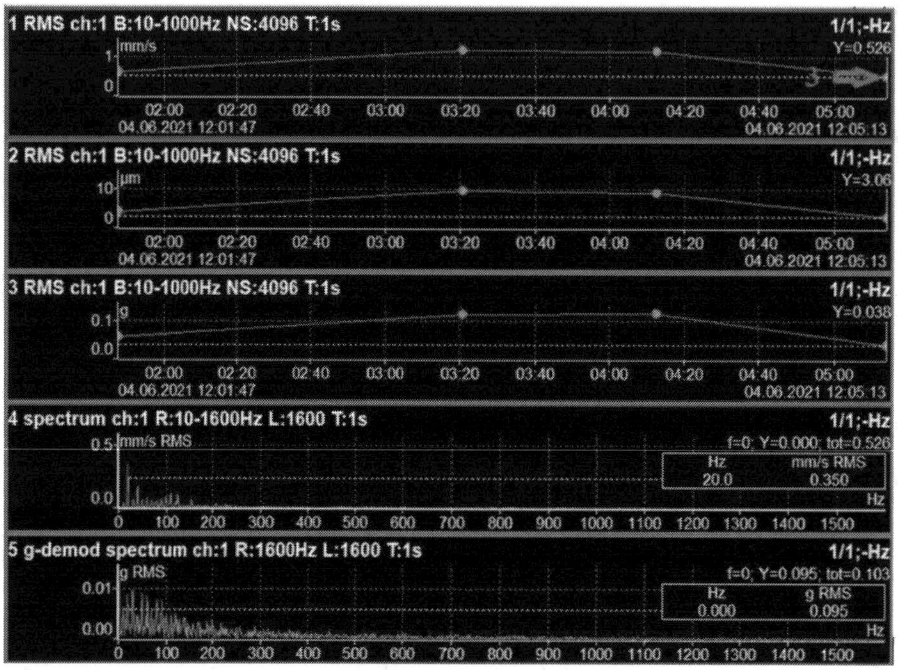

Figure 7. RMS values for the third measurement of the accelerometer.

5. Conclusions

It is important to diagnose bearings functioning in order to prevent machinery failure, which implies high costs for remedial maintenance. That is why it results in the need to perform different analyses to determine defects of bearings.

In the early stages of deterioration, the frequency analysis of bearings or the values of acceleration and velocity amplitude are irrelevant. Once the deterioration level increases, the spectrum analysis can be used to depict bearing defect frequencies such as BPFO, BPFI, FTF and BSF.

From the graphs, it can be concluded that the frequency analysis did not indicate any faults on bearing elements. Also, the values of vibration velocities for all three measuring points on the pillow block bearing did not suggest any defects.

In addition, by comparing bearing fault frequencies values with figures 4-7, the envelope analysis revealed that the bearings do not show periodic impulse in the time domain or frequency domain.

Acknowledgement

This work is supported by the project ANTREPRENORDOC, in the framework of Human Resources Development Operational Programme 2014-2020, financed from the European Social Fund under the contract number 36355/23.05.2019 HRD OP /380/6/13 – SMIS Code: 123847.

References

[1] Fowler P H and Frank Manders *Estimation of bearing contact angle in-situ by X-ray kinematography* Space and Technology Group

[2] Harris T A and Kotzalas 2006 M N *Advanced Concepts of Bearing Technology: Rolling Bearing Analysis* Fifth Edition ISBN: 9780429123344 DOI:10.1201/9781420006582

[3] Hans K H *Envelope analysis for diagnostic of local faults in rolling elements bearings-Application Note* Bruel&Kjaer Sound and Vibration measurement Denmark

[4] Kansanaho J 2019 *Data-driven methods for diagnostics of Rolling Element Bearings* JYU Dissertations 151 University of Jyväskylä ISBN: 978-951-39-7936-2 (PDF)

[5] Smith W A and Randall R B 2015 *Rolling Element Bearing Diagnostics Using the Case Western Reserve University Data: A Benchmark Study* Mechanical Systems and Signal Processing DOI:10.1016/j.ymssp.2015.04.021

[6] Standard ISO 10861-1 Mechanical vibration – Evaluation of machine vibration by measurements on non-rotating parts – Part 1: General guidelines, First edition 1995-12-15

[7] Stoica R M Voicu and D Vilau R 2020 *Desen tehnic în ingineria autovehiculelor* ISBN 978973640310-1

[8] Kim S, An D and Choi J H 2020 *Diagnostics 101: A Tutorial for Fault Diagnostics of Rolling Element Bearing Using Envelope Analysis in MATLAB* MDPI Appl. Sci. 2020, 10, 7302; doi:10.3390/app10207302

[9] Ştefănescu I Tomescu (Deleanu) L and Bucşă M 1993 *Ungerea Lagărelor cu rulmenţi* Sesiunea ştiinţifică "45 de ani de Învăţământ Superior la Galaţi"

[10] *** https://uk.mathworks.com/help/predmaint/ug/Rolling-Element-Bearing-Fault-Diagnosis.html

Researching the influence of various factors on the superficial layer materials used in car body construction and protection

Oleg Frunze

Department of Transport, Technical University of Moldova, Str. Studentilor 9/8, Chisinau, MD-2012, Republic of Moldova

frunze.oleg@gmail.com

Abstract. To carry out the restoration of the car body, in addition to modern equipment for body repair and painting, it is necessary the use of high-quality consumables for body repair and painting, which are selected for each brand of car, strict following the technology of repair car body, painting work and qualified specialists in body repair and painting, as well as knowledge of the factors affecting the degradation and condition of the body and its elements to improve the quality of final results. In average value, the thickness of the protective layer varies from manufacturer to manufacturer in the range of values 80-170 μm. The use of different materials for the manufacture of vehicle components and materials for protection against external factors of the body is a major interest for the vehicle industry and is the subject of this article.

1. Introduction

Factors affecting the safety of the car body: mileage, service life and seasonality, storage conditions, climate and care. They are all interconnected. But if such as the climate do not depend on us, then the seasonality, storage conditions and care are in one way or another up to us, and therefore, they allow us to influence the durability of the car.

With seasonality is simple. If you want to use a car longer - give up driving at an unfavorable time of the year.

The storage conditions mean ambient temperature, humidity, precipitation and other influences on the car, but care - washing the car body and engine compartment, treating the paint coating, cleaning the interior.

It seems that everything is clear with the storage and maintenance conditions: the better when the car is protected from precipitation, sun and wind, and when the car is washed more often. If we remember that with an annual mileage of 15,000 kilometers, it is in motion for only about 300 hours that means less than an hour a day! And the rest of the time it stands: in a warm or cold garage, under a tent, in an open area with or without a cover storage place.

The benefits of a warm garage don't need to be described. But disadvantages: the main one is the continuation of the advantages. The higher the all-year temperature, the worse. With its growth, the ageing of protective coatings and the process of metal corrosion, if damaged, are more intense. Externally, a dry car is not at all like that. In the overwhelming majority of cases, ventilation in such garages is insufficient to dry out all hard-to-reach car body parts in a short time. If you arrive at your garage during the rain or after a wash, it's safe to say that the interior surfaces of the car body parts will stay wet for days and weeks.

The car especially suffers when stored in such a garage in the winter after returning from the street. Due to the sharp temperature difference, both the outer and inner surfaces of the body are covered with dew, which falls out of the relatively humid warm air of the garage. The amount of condensed moisture directly depends on the temperature difference and can be very significant. In some cases, damp areas can be observed on the floor along the perimeter of the body after a few minutes, as if the car had been placed in the garage immediately after washing. Outside, it dries up in a few hours. And the water that has not flowed out of the hollow body elements remains there until the next trip. In addition, the smaller the area occupied by the wet area inside the body, the more intense the corrosion process. Only a very time-consuming, carefully thought-out anti-corrosion treatment, for various reasons inaccessible to most drivers, can neutralize the negative consequences of storing a car in a warm garage. If you do not regularly take special measures, the car body gets through corrosion damage already at the beginning of the third year of operation and serves four, maximum five years.

For these reasons, it is better not to leave a warm garage without extreme need in winter.

In this sense, a cold garage is even somewhat preferable. However, in the off-season, the processes proceed quite noticeably, since when the car enters a garage, the engine and body give off heat to the surrounding small volume of air and the garage becomes warm for some time. Until the car cools in it, a corrosive process will take place on wet surfaces. And therefore, in a cold garage, a car sometimes degrades in five years [6].

Of all possible storage locations, it is preferable to keep the car under a ventilated covered storage shed. Besides the fact that it is protected from precipitation there, this type of parking has the greatest advantages. A wet car dries out here relatively quickly, and in the cold season its heated parts cool down very quickly to below-zero temperatures, at which corrosion stops.

An open parking lot, especially in winter, looks terrible: cars covered in snow, cold and windy. But this is a very well place for long-term car storage. As well as under covered storage shed, in a parking lot, the car cools down quickly and is well ventilated. The paintwork and most of the exterior parts are designed to withstand long-term exposure to solar radiation and natural precipitation. The main disadvantage of an open parking lot is that the car is not protected from dew deposition on the body, containing harmful substances from the atmosphere, as well as freezing on glass and horizontal panels of snow and ice. The exterior plastic parts also suffer from the degradation of the exterior appearance to detachment and deformation. After heavy or prolonged rains, water may appear inside the car, which means that you need to constantly monitor the condition of the floor under the rugs and in other places, find and eliminate microleaks, the number of which increases with the age of the car.

We can also see cars covered with various tent covers in open parking. This seemingly obvious protection measure has more disadvantages than advantages. The latter is that the car cover tent protects the body from precipitation, freezing of snow and ice. However, after some time, the car cover tent may be the cause of the appearance of indelible colored spots on the outer coating. This happens because the car cover tent itself, after prolonged use and repeated humidification, is impregnated with chemicals deposited from the atmosphere. The concentration of these substances in the moisture impregnating the fabric of the tent when it gets wet changes and as the water evaporates, it necessarily reaches such a value that even automobile paint cannot resist this evil solution. The whole question is only how long the car cover tent stays in this state. If it is long enough, the texture of the tent is printed on the most prominent places of the car body and this pattern can only be removed mechanically or by repainting.

In windy weather, the car cover tent scratches the paint and can cause blistering of the paint, which is especially readily formed in summer. Imagine a thunderstorm has passed on a hot summer day. Naturally, the car cover tent and the car under it got wet, then the sun reappear. Soon, the temperature under a wet and, especially, a dark case, at 100% humidity, rises by several tens of degrees. In these conditions, moisture penetrates to the metal through any smallest damage to the coating, which is always on the surfaces of the body. A localized area of corrosion occurs between the paint layers and the metal, and since the products of this process take up more space than the components from which they were formed, the body coating bubble up [5].

A particular interest is the processes associated with the drop of various kinds of chemicals from the atmosphere, which affect the durability of the paint coating and the car body as a whole.

Most often, the red rash makes itself felt. Having found small rusty spots and drips on the paintwork of the car body, many believe that "local corrosion" of the car body, associated with the poor quality of the metal, has begun to break through the paint. These fears are completely unfounded.

An attentive observer will notice that the rash is located only on the horizontal surfaces of the car body.

The rash is caused by metal particles falling out of the atmosphere. Having settled on the surface of the body, they begin to corrode and stick tightly to the body when dew or light rain falls. The rash usually disappears after a few good washes. If not, you can wash it off with a 5% solution of oxalic acid.

Corrosion of the outer seams of the car often appears. After some time of operation, the mastic that seals the outer side of the joints of body parts cracks, and corrosion products protrude from there. This is most often seen in white cars, where the leaking rusty moisture is more noticeable. The reason is winter driving on uneven roads, not cleared of ice. Due to the mutual movement of car body parts, cold mastic cracks appear. This in itself is harmless. But salt solution gets into the cracks, causing corrosion. You can deal with this by smearing the seams with a heated special compound and then wiping the surface with a rag.

Also, some studies have shown that the change in color was caused by exposure to nitric acid from the atmosphere with acid rain or fog. As the moisture evaporated, its concentration changed to such an extent that it burned through the synthetic paint.

Under the influence of sulfurous precipitation, especially the red pigment of the paint loses its color, and the blue, which is more resistant to similar influences, appears. The color change occurs to a depth of ten microns, and it could be removed mechanically. Their appearance in the future will mean a deterioration of the ecological situation in terms of sulfur emissions into the atmosphere.

The order and frequency of maintenance are of great importance for the safety of the car.

There is an opinion that the car should be washed before putting it in the garage. But during washing, water penetrates between the glass and the seals into the cavity of the doors and other joints of the body. In addition, in almost any, especially a used car, due to the loss of elasticity of the seals during washing, accompanied by a large influx of water, leaks and microleaks occur, which do not occur during driving and in the heaviest rain. No matter how dirty the car is during the trip, the hollow body cavities, which necessarily have drainage holes, are blown through by air pressure. In a garage, under a cover shed or in a parking lot, such a car dries out faster than after washing, since the bulk of the water is removed from it during the movement. In the garage, to speed up drying, it is advisable to put it with the door glasses down. From the above, it is clear that it is necessary to wash the car before the trip, and the longer the trip, the better.

However, not everything is clear here either. If the car is used all year round, is kept warm and does not have time to dry out or freeze to freezing temperatures for the next trip, it is better to wash it before putting it in the garage in order to remove dirt that retards the drying of individual cavities and parts, and in winter - salt adhering to snow.

The development of the automotive industry demands increased requirements for quality and safety, demand formation and use of new forms and materials. Carbon fiber materials most fully meet modern requirements, as they have a number of unique characteristics and demonstrate the best cost value [1].

Composite materials for cars have noticeably pushed the usual metal on the market. And not only steel, but also aluminum alloys, which until recently were considered the best in all aspects. Currently, composite materials are used to create almost any car assembly. Even concept cars are produced, the body of which consists entirely of composites.

Composite materials and products based on continuous fibers and reinforcing fabrics are widely used for the production of external car parts. Most often they do:

- Load-bearing structures - of doors and seats, protective elements of the bottom.
- Fastening elements for bumpers and radiators.
- Decorative items - interior decorative panels, external decorative panels.
- Trunk lids, body panels, brake discs, body elements, thermal and sound insulation.

Increasingly, the bodies of many types of cars (including heavy trucks) are entirely made of lightweight, durable and inexpensive carbon fiber reinforced plastics.

Composite materials for the automotive industry are primarily carbon fiber products. It has been used in the automotive industry for many years, and its use is growing every year. The most important advantage of carbon fiber is its light weight and high strength. Carbon-fiber-reinforced polymers (CFRP) is 5 times lighter than steel and 1.8 times lighter than aluminum. The use of composites in the automotive industry can reduce vehicle weight by 20-25%. This significantly increases the efficiency of the engine and reduces fuel consumption.

Carbon fibers are made from synthetic and natural polymer-based fibers. Depending on the processing mode and feedstock, materials of different structures and properties are obtained. This is the main advantage of composite materials. They can be created with initially specified properties for a specific task.

Carbon is 12.5 times more durable than steel. When we say "carbon", we remember, of course, the hoods of tuning cars. Today there is not a single body part that is not made of carbon fiber. Not only hoods are made from it, but also fenders, bumpers, doors and roofs. The weight savings are obvious. The average weight gain when replacing the hood with a carbon fiber one is 8 kg. However, for many, the main thing will be the fact that carbon parts look very stylish on almost any car.

Car carbon fiber is widely used in racing clothing. These are carbon helmets, boots with carbon inserts, gloves, suits, back protectors, etc. etc. Such equipment not only looks good, but also increases safety and reduces the weight of the suit (very important for a helmet). Carbon is especially popular with motorcyclists. The most advanced bikers dress themselves in carbon from head to toe.

The development of technology in the automotive industry is primarily associated with the development of motorsport. Observing technical progress in the development and application of composite materials, we can confidently say that in the near future there will be production cars with a fully composite body and many components and assemblies [2].

2. Methodology

Today, there are many methods for identifying the condition of the car body, which make it possible to find out not only about the damage and abnormal condition of the paintwork, but also about the areas where metal deformations have been fixed, to consider welds and any additional body reinforcements.

Car body diagnostics consists of the following steps:

- it is necessary to determine the VIN-cod of the body and compare it with the manufacturer's sample;
- second, is need to determine the actual age of the body (the exact date of the car's release) and changes in the VIN-cod of the car body;
- at the third stage, a step-by-step check of the body parameters in problem areas.

If the thickness of the paint-and-lacquer coating is scanned repeatedly and everywhere (this is possible with the help of ferromagnetic materials, aluminum and brass), then it is possible to determine the zones of corrosive destruction of the integrity of the body. This makes it possible to identify the following parameters:

1. The thickness of the paint;
2. Places of partial or complete repainting;
3. Renovates in the paintwork;
4. Deformation in the load-bearing elements of the body frame;
5. Cracks, corrosion;
6. Traces of restoration after repair.

It is now possible to carry out such an examination, since there are many modern high-tech and mobile equipment.

After collecting the data, it is necessary to analyze them by statistical methods [3].

For example, below, table 1 shows the thicknesses of the coating surfaces for different vehicle manufacturers.

Table 1. Thickness of coating for different manufacturers [7].

	Manufacturer	Min. μm	Max. μm	Avr. μm	St.dev. μm
1	Audi	90	147	118.5	9.50
2	BMW	89	165	127.0	12.67
3	Chery	105	120	112.5	2.50
4	Chevrolet	75	150	112.5	12.50
5	Citroen	75	130	102.5	9.17
6	Daewoo	95	115	105.0	3.33
7	Fiat	111	130	120.5	3.17
8	Ford	119	160	139.5	6.83
9	GAZ	80	105	92.5	4.17
10	Honda	87	145	116.0	9.67
11	Hyundai	70	130	100.0	10.00
12	Infiniti	114	147	130.5	5.50
13	Lada	60	100	80.0	6.67
14	Lexus	125	150	137.5	4.17
15	Mazda	85	130	107.5	7.50
16	Mercedes Benz	90	250	170.0	26.67
17	Mitsubishi	53	125	89.0	12.00
18	Nissan	75	150	112.5	12.50
19	Opel	110	157	133.5	7.83
20	Peugeot	58	115	86.5	9.50
21	Renault	55	140	97.5	14.17
22	Skoda	99	125	112.0	4.33
23	SsangYong	100	110	105.0	1.67
24	Subaru	110	140	125.0	5.00
25	Suzuki	77	115	96.0	6.33
26	Toyota	77	130	103.5	8.83
27	Volkswagen	70	105	87.5	5.83
28	Volvo	95	130	112.5	5.83
	Average, μm	**87.46**	**136.29**	**111.88**	**8.14**
	Standard deviation, μm	**19.84**	**28.37**	**19.41**	

3. Results and discussions

Today, there are many methods for identifying the condition of the car body, which make it possible to find out not only about the damage and abnormal condition of the paintwork, but also about the areas where metal deformations have been fixed, to consider welds and any additional body reinforcements. Thickness for different parts of the car is shown in table 2.

Table 2. Thickness for different parts of the car.

	Part of body	Min. μm	Max. μm	Avr. μm	St.dev. μm
1	Hood	128	142	118.5	2.33
2	Wing	125	132	127.0	1.17
3	Roof	119	135	112.5	2.67
4	Front door	108	129	112.5	3.50
5	Back door	128	136	102.5	1.33
	Average, μm	**121.6**	**134.8**	**128.2**	**2.20**
	Standard deviation, μm	**3.33**	**2.17**	**2.75**	

When comparing the reference and actual values of the coating layer thickness, small discrepancies of tens of microns are normal, which is associated with uneven material deposition. If the difference is hundreds of microns, then we can say that in this place the body was painted again, and an even greater increase in discrepancies indicates large-scale repair work, which includes straightening and a set of subsequent measures. A decrease in the layer thickness by several tens of microns indicates polishing with the use of abrasive materials.

The primer surfacer provides a measure of corrosion protection. The usual dip priming system is [4]:

- Anti-corrosive dipping primer: 12-18μm;
- Primer surfacer: 40-50 μm;
- Finish: 45-55 μm.

4. Conclusions

The thickness of the protective layer can have a wide range of formation, especially in the second-hand vehicles market, from almost zero layer to values that exceed the maximum level provided by special measurement devices. It was noticed that the producers choose different fields depending on the possibilities they have and the market demands. In average value, the thickness of the protective layer varies from manufacturer to manufacturer in the range of values 80-170 μm. Typically, the factory paint layer on steel auto parts is no more than 200-250 microns. Readings above 250 microns indicate re-painting. Readings up to 300 microns are possible if a cosmetic defect is painted over, for example, a scratch from a key. Readings closer to 1000 microns indicate that there is still filler under the paint. The body element with such readings definitely received a strong deformation in an accident and was putty during repairs. If the work is done poorly, the paint on the putty may crack and fall off over time. More than 1000 microns is a sign of serious body repair, which means participation in a serious accident. It is better not to buy a car with a coating layer of more than 1000 microns. The maximum value is 2000 μm which the thickness gauge can display. If the layer is thicker, the device will show nothing. This means that there is a lot of putty at this point or another material is used. Carbon fiber materials most fully meet modern requirements, as they have a number of unique characteristics and demonstrate the best cost value. The body car is and will remain the basic and most important component for the car industry, so it is important to find effective design solutions and new materials to cover the demanding needs of environmental protection and the customer needs.

References

[1] Crolla D. A. (2009) Automotive Engineering - Powertrain, Chassis System and Vehicle Body, *(Butterworth-Heinemann Ed.), Elsevier Inc.*, pp.593-596

[2] Danner M., Mauer F. (1991) Die moderne Unfallreparatur, *Verlagsgesellschaft Eurotax AG*, pp.176-181

[3] Frunze O., Stratulat S., (2004) Fiabilitatea si mentenenta in transport, *Evrica, Chisinau*, pp.10-12

[4] Lambourne R., Srivens T., (1999) Paint and surface coatings, Woodhead Publishing, p.411-415

[5] Morello L., Rossini L. R., Pia G., Tonoli A. (2011) The Automotive Body, Volume I: Components Design, *Springer Dordrecht Heidelberg*, pp.135-137

[6] Talbert R., (2008), Paint technology handbook, CRC Press, Taylor & Francis Group, pp.89-90

[7] Standard paint thickness of vehicles by car brand, visited at 14.06.2021, https://etari.de/en/standartlackdicke-lackdickentabelle

Overview on globally applied used engine oil recycling technologies

S A Rațiu[1], G O Tirian[1], N L Mihon[2], and M D Armioni[1]

[1]Politehnica University of Timişoara, Faculty of Engineering Hunedoara, Department of Engineering and Management, 5 Revoluției Street, 331128 Hunedoara, Romania
[2]Politehnica University of Timişoara, Mechanical Faculty, Department of Road Vehicles/Mechanical Machines, Equipment and Transportation, Mihai Viteazu Street 1, 300222, Timişoara, Romania.

sorin.ratiu@fih.upt.ro

Abstract. Due to the growing importance of environmental issues in modern society, the global economy has been compelled to refocus further towards eco-friendly, low-pollution technologies based on waste recycling. Furthermore, these technologies not only offer economic value to products at the end of their life cycle, but they have the ability to address the issue of inadequate waste storage in landfills. At the same time, given the growing trends in the number of vehicles, the demand for lubricants is also increasing, which involves the generation of larger and larger amounts of used engine oil. Being included in the category of hazardous waste, it requires management based on sustainable principles, and recycling is recommended as a preferred method of treatment, as it still has a high economic value. This article aims to present some of the technologies that have been effectively implemented on a global scale for the recycling of used engine oil, as well as a series of advantages that these processes present, both ecologically and from an economic perspective.

1. Introduction

Engine oils, whether derived from petroleum products or synthesized, are used for a wide range of applications, including lubrication, heat and power transfer, protection of engine parts, cleaning the components, and so on. These oils become contaminated and/or degraded depending on the application and operating environment and must be therefore discarded, leading to the formation of used engine oils. These waste oils are classified as hazardous waste and fall into Basel Convention subgroups Y8 and Y9 [1].

Used engine oil is considered a threat to both the environment and human health, because of its high level of dangerous contaminants. Of these, the most known are:

- Polycyclic aromatic hydrocarbons (benzanthracene, benzopyrene, naphthalene, etc.): they appear as a result of incomplete combustion of the fuel and present a high carcinogenic risk [2];
- Polychlorinated biphenyls: these are organic compounds that have been used in the past as additives for lubricants. They are particularly dangerous due to their carcinogenic risk and high toxicity [3];
- Heavy metals (Fe, Cr, Ni, Pb, Cu, Zn): results from the wear process of the engine's metal components [4].

The main focus of this article is to present a comprehensive overview of the current engine oil recycling methods in use around the world, including a description of their technological flow and the enterprises that apply them. In addition, the benefits of employing these technologies will be discussed from both an ecological and an economic standpoint.

2. Literature review

The increase in attention to the recycling of used engine oils is mainly due to environmental problems that have arisen with increasing pollution levels and the need to conserve natural resources. Also, the recent technological development and the accentuation of the recovery of waste have led to the emergence of technologies that facilitate recycling processes. In some countries with a high degree of technological innovation, up to 50% of used engine oils come from recycled oils. However, in the European Union, for example, only 13% of the amount of base oils required on the market is produced from used engine oil that has undergone a refining process [5].

The oldest way of treating used engine oil, being at the same time the method that generated the best results until the development of advanced recycling technologies [18], is the treatment with acid and clay. The process is based on treating the used oil with different acids (sulfuric, acetic, formic) to remove pollutants and then using different types of clay (such as bentonite) to neutralize the resulting product [6]. Activated carbon can also be used for neutralization [7]. The process is not currently encouraged, however, due to the low degree of sustainability (the resulting acidic sludge is considered a hazardous waste and without ecological potential for recovery).

Vacuum distillation and clay treatment is another process that allows the removal of impurities from used engine oil to transform it into base oil. In the first stage, that of atmospheric distillation, the separation of water and light compounds takes place. Subsequently, vacuum distillation (TDA column – Thermal de-asphalting) and fractionation are carried out: at this stage, the organo-metallic compounds and asphalt minerals are separated and three quantities of used oil are fractionated. TCT (thermal clay treatment) follows: this is when the characteristics of the three fractions of waste oil separated in the previous phase are improved. The last step is pressure filtration [6].

Vacuum distillation and hydrogenation is a technology in which clay has been replaced by hydrogenation, for reasons of efficiency and sustainability. The oil is dehydrated at atmospheric pressure, in order to remove light hydrocarbons and water. Vacuum distillation is then applied at a temperature of 250 °C, using a conventional vacuum tube or a thin film evaporator to separate the diesel fuel. Materials such as residue, heavy metals, additive degradation products, etc. a flow of capitalization of asphalt products follows. The last step is the hydrogenation of the distilled product, to remove toxic nitrogen and sulfur compounds, as well as other oxides present in the chemical composition of the used oil. The spent catalyst resulting from the hydrogenation process is disposed of outside the company. Also during this stage, the odors from the oxidation of some chemical elements are eliminated and the color of the final product is improved [8]. During the vacuum treatment process, the hydrogenated oil is subsequently separated into fractions used for engine, industrial or hydraulic oils. The residue resulting from the vacuum distillation process is used for the production of bitumen and other asphalt products [9].

One method that involves the use of chemical solvents to recycle used engine oil is solvent extraction. It consists in dissolving the aromatic components that affect the properties of the oils, but keeping the desired components, such as saturated hydrocarbons [10]. Before treatment, the used oil is filtered to remove solids. Then the oil is treated with solvent, mixing a predetermined amount of oil and solvent for one hour, then letting the solution settle for 24 hours. The oil separated from the sludge is transferred to a vaporizer to facilitate the separation of the solvents from the recycled oil. Subsequently, the oil is mixed with active alumina to remove dark color and specific odors [11]. Solvents that can be used to treat waste oils include 1-butanol, methanol, ethanol, propane, toluene, methyl ethyl ketone, acetone, and so on [12].

A number of methods and techniques for recycling used engine oils have been developed, applied differently in various parts of the world. These technologies depend mainly on the nature of the base oil (mineral or synthetic) as well as the nature and amount of contaminants.

3. Recycling methods and technologies applied on a global scale for used engine oil

Recycling used engine oil is not only a benefit for the environment but also a way to reduce pollution. This process also brings with it strong economic benefits and financial gains for companies that have developed or implemented recycling processes and technologies for this type of waste. For this reason, the recycling of used engine oil is practiced, through various processes and techniques, in factories around the world, belonging to already well-known companies or to some that are in full development. Next, seven examples of such currently applied technologies on a global scale will be presented.

3.1. HyLube technology – Puraglobe

Direct Contact Hydrogenation (DCH) is the technical principle of HyLube technology. The procedure (as shown in figure 1) consists of several phases, the first of which is combining used engine oil with heated hydrogen. Following that, the mixture is transferred into a separating flash, where some of the impurities (such as metals, solid particles, and a trace of hydrocarbons) are separated. The oil is transferred from the separator to a catalytic reactor, which removes the residual soluble metals, and then to the hydroprocessing (HPS) stage (both hot and cold), which ensures the reactions that lead to the refining of used oil. Desulfurization, dechlorination, oxygenate conversion, denitrification, aromatic saturation, and mild hydrocracking are the most common of these reactions [9]. The last stage is the separation of water and gases from the oil fraction, which is then collected and transferred to the additive section to be improved and reintroduced into the automotive circuit.

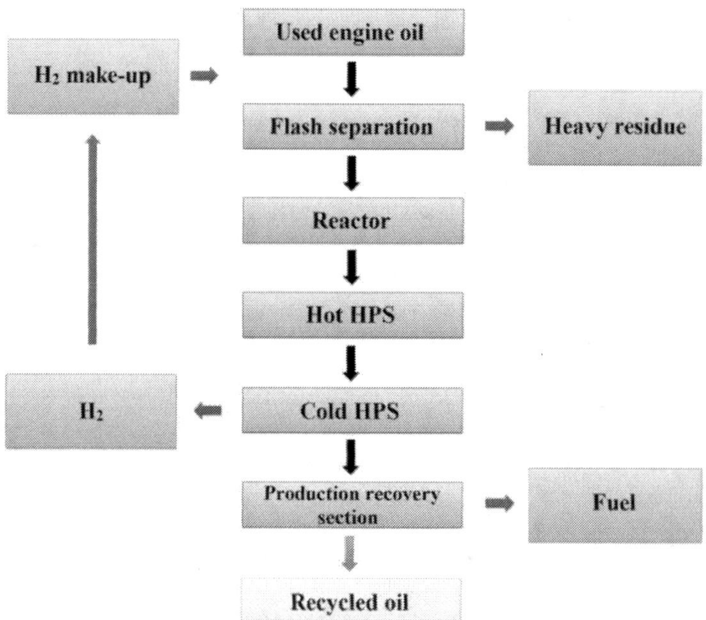

Figure 1. Technological flow chart of the HyLube process [9].

The HyLube technology is the first used engine oil recycling method in which the lubricant is treated in a pressurized hydrogen environment without any pretreatment. The technique is implemented by the corporation PURAGLOBE Holding GmbH, which is headquartered in the Chemical and Industrial Park Zeitz in Germany [13].

3.2. Revivoil technology – Itelyum Regeneration (Viscolube)

The Revivoil technology (figure 2) is based on two main processes: thermal de-asphalting (TDA) and hydrofinishing (HDF).

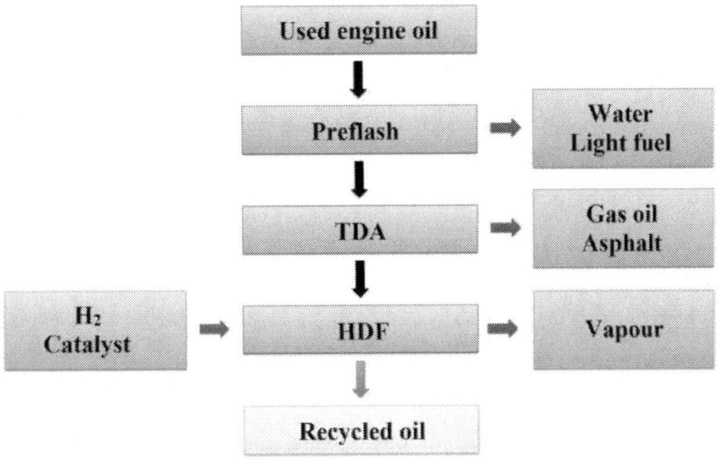

Figure 2. Technological flow chart of the Revivoil process [14].

Filtered spent oil from storage tanks is pre-heated and blended with additives to reduce fouling. The mixture is heated to 140 °C before being transferred to a preflash column. The blend of water and light hydrocarbons is then condensed with the separation of components after being removed from the top of the separation column.

The dehydrated oil from the preflash unit is sent to the Thermal De-Asphalting unit (TDA), where it is split from the compounds that can cause fouling in an intermediate tank; the oil is heated to 350°C and fractionated in the TDA distillation column under vacuum (15 Torr), with separation of the diesel fraction from the top of the column and the asphalt fraction from the bottom of the column. The cuts are stripped and cooled in the stripping column. Oil cuts following TDA are finished by contact with hydrogen over the catalyst, which permits the elimination of metals and metalloids, organic acids, sulfur- and nitrogen-containing compounds, and enhancing the color and thermal stability of the oil [14].

This technique is used in Viscolube factories across Europe. Viscolube, founded in 1963, manufactures lubricating base oil whose characteristics are said to be superior to those of virgin base oils already on the market. Viscolube is present in Italy with two production facilities and has the capacity to treat more than 190 000 tons of waste oil per year, thus producing more than 110 000 t/y of re-refined base oil and 20 000 tons/year of bitumen [15].

3.3. Mohawk technology – Southern Oil

This technology (depicted in figure 3) consists of two main methods: vacuum distillation and hydrofinishing. In the first step, used oil is chemically processed to prevent impurities from precipitating and causing corrosion and fouling of the equipment. Pre-treated used oil is then distilled for the first time to separate water and light hydrocarbons. Water is filtered before being sent to a waste water treatment station. Light hydrocarbons are utilized as a fuel in plants or delivered as a product.

High vacuum distillation of free-of-water oil in a thin film evaporator separates diesel fuel, which may be utilized in the facility or sold as fuel; heavy particles such as residue, metals, additive degradation products, and so on are transferred to a heavy asphalt flux stream [9].

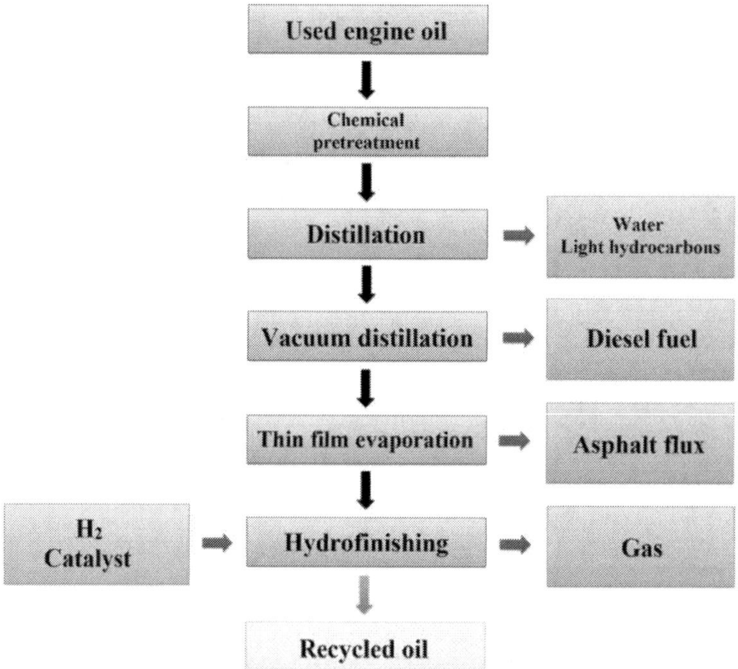

Figure 3. Technological flow chart of the Mohawk process [16].

One of the enterprises that have implemented this method of treating used engine oil is Southern Oil. The goal of this company, as defined by them, is keeping lube oil in productive use, delivering improved oil security through avoided imports and a lower carbon footprint compared to the production of imported crude. The process recovers 99% of the lube oil component of the waste oil, creating a near perpetual cycle of use and re-use of a diminishing and finite resource. Southern Oil has the capacity to re-refine 38% of Australia's annual waste / used lube oil production through its Wagga Wagga plant in New South Wales and Northern Oil Refinery near Gladstone in central Queensland [16].

3.4. Safety Kleen technology – Clean Harbors Inc.

The essential stages of the Safety Kleen process, as seen in figure 4, are thin film evaporation and hydrogenation. Following atmospheric dehydration, the elimination of light ends takes place, a stage that improves thermal stability, color and odor while reducing polynuclear aromatics and removing higher-boiling halogenates and polar compounds.

The waste oils are distilled again under a higher vacuum, and the lubricating oil base stock/fractions are recovered. The residues generated as a byproduct of the process can be utilized in industrial applications such as asphalt extenders. Vacuum distillation debris is utilized as an asphalt extender or as fuel in blast furnaces with pollution controls. The used catalyst from the hydro treatment phase can be regenerated. The remaining oils are treated with hydrogen to remove sulphur, nitrogen, chlorine, heavy metals and other impurities. This step also corrects any issues with odor, color and corrosion performance [9].

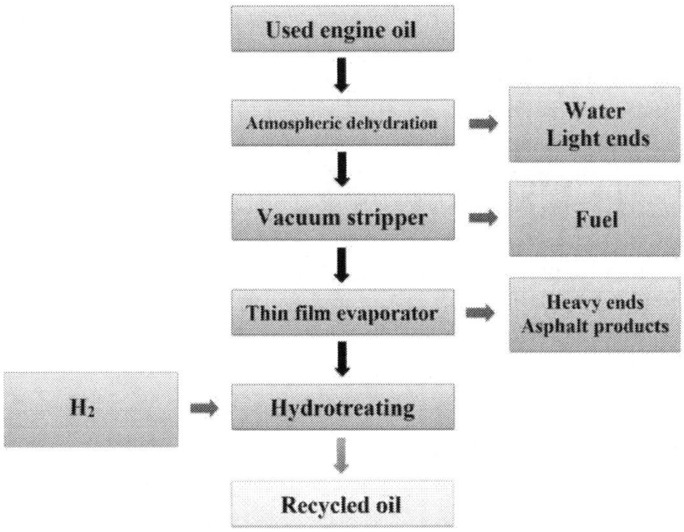

Figure 4. Technological flow chart of the Safety Kleen process [9].

Safety Kleen is North America's largest re-refiner of used oil. Along with its parent company, Clean Harbors, it offers a full range of used oil products collection and recycling services, among which the recycling of used engine oil is an important branch. Used engine oil recycling plants owned by Clean Harbors Inc., which use Kleen Safety technology, can be found throughout the United States as well as in Canada [17].

3.5. Bechtel MP extraction – Bechtel

Bechtel technology (shown in Figure 5) is based on the solvent extraction process, which uses chemicals (in this case, n-methyl-2-pyrolidone) to ensure optimal treatment of contaminated oil and recovery of the base oil.

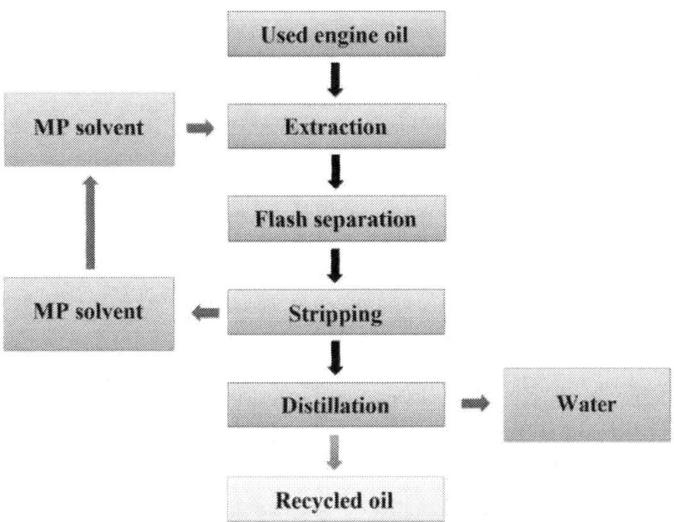

Figure 5. Technological flow chart of the Bechtel MP extraction process [18].

N-methyl-2-pyrolidone (MP) was selected because of its ability to specifically eliminate unwanted aromatic (PCBs) and hetero-components (oxygen-, nitrogen-, sulphur-containing compounds). In the extraction tower, used oil and solvent come into contact with each other at the right pressure and temperature. The raffinate and the majority of the solvent are taken from the column's top and directed to a solvent recovery section. The extract section is taken from the tower's bottom and sent to the solvent recovery section. MP is then removed from the bottom fraction (extract oil) in the solvent recovery section by multiple-effect evaporation at varying pressures, followed by flash separation and steam stripping under vacuum [9].

Bechtel's activity in recycling used engine oil takes place in the United States. It has been found that this process improves the viscosity index (VI), thermal stability, oxygen stability, and color of the base oil resulting from recycling. Also, the solvent used is recirculated and refreshed, which adds sustainability to the technological process [19].

3.6. STP technology – Afriquia SMDC

STP technology (figure 6) recycles used engine oil into base oil by vacuum distillation and chemical treatment (for API group I oils) or hydrofinishing (for API group II oils). Initially, the used oil is preheated and passed through a drum flash to remove water and light hydrocarbons. The product thus treated is subjected to the stripping process, thus eliminating the rest of the gas oil. The oil is then high-vacuum distilled in a thin-film evaporator to isolate heavy impurities such as metals, heavy polymers, carbon, and dust that deposit heavy residue at the bottom of the evaporator.

The final fractionation of vacuum distillate enhances product quality and generates treated gas oil as well as two lubricant fractions, a light cut (SN-150) and a heavy cut (SN-500). The hydrofinishing process improves the quality of the treated oil [14].

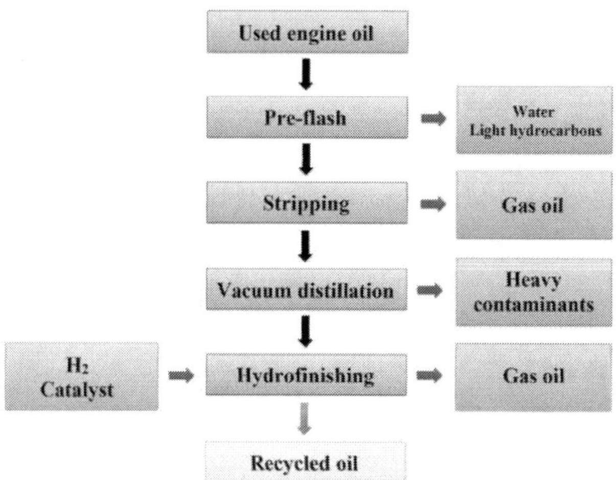

Figure 6. Technological flow chart of the STP process [14].

One of the companies that applies STP technology (the hydrofinishing variant) for used oil recycling is Afriquia SMDC (with its subsidiary Afrilub), located in Morocco. With a production of over 40,000 tons of recycled oil per year and a strategic partnership with Chevron, the company offers the possibility of recycling for waste oils from the northwest region of Africa [20].

3.7. Lwart technology – Lwart Lubrificantes

The process of recycling used engine oil applied by Lwart Lubrificantes (illustrated in Figure 7) involves as main methods of treatment TDA (Thermal De-Asphalting) and solvent extraction.

Used engine oil is initially subjected to the process of atmospheric dehydration, thus eliminating the additional amount of water. Subsequently, flash distillation is applied, where the removal of light hydrocarbons takes place. The thermal de-asphalting process removes heavy contaminants (polycyclic aromatic hydrocarbons, heavy metals), and the solvent treatment achieves an additional purification, as well as the neutralization of the finished product [21].

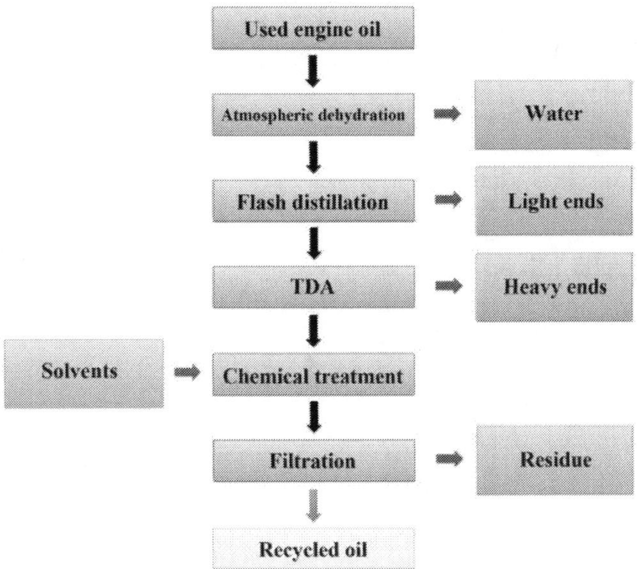

Figure 7. Technological flow chart of the Lwart process [21].

Lwart Lubrificantes is a company based in Brazil, which has 15 collection centers and a recycling plant for used engine oil. The recycling plant processes approx. 140,000 tons of used oil/year, thus avoiding its burning as fuel, as well as the import of base oil [22].

4. Advantages of implementing the presented recycling technologies

The process of recycling used engine oil has a number of advantages common to all technologies applied worldwide. Depending on the specifics of the field to which the analysis of these advantages refers, they can be both ecological and economic. In this form, they can be found in Table 1.

Table 1. Advantages of used engine oil recycling technologies.

Ecological advantages	Economic benefits
Decreasing the amount of used engine oil that reaches the environment, thus reducing pollution	Promoting the circular economy and a sustainable future
Reducing the concentrations of heavy contaminants discharged into the environment	Reducing the demand for virgin oils and their production costs
Lowering the risk of pollution in the hydrographic network	Reducing the expenses of disposing of used engine oil through alternative means
Toxic emissions that would be released into the atmosphere as a result of waste oil incineration activities are eliminated	Construction of recycling facilities that have the potential to employ a significant number of people
Pollution related to the exploitation of petroleum products is being reduced	Increasing a region's economic worth through creating used engine oil recycling businesses

5. Conclusions

Due to the high content of contaminants, many of which (heavy metals, acids, PAHs, PCBs) have a major negative impact on the environment, used engine oil is classified as hazardous waste. Improper storage or disposal can lead to soil and groundwater pollution, an increased risk of animal death and, in essence, to the degradation of ecosystems. Used engine oil also poses risks to human health, as some of the pollutants (such as PAHs and PCBs) have been classified as carcinogenic. Other types of adverse health effects (immunological, reproductive, fetotoxic and genotoxic) have also been reported, and oil vapors are toxic, affecting the respiratory and digestive systems [23]. Thus, a special management strategy is required for this type of waste. From an ecological point of view, as well as for economic reasons, the recycling of used engine oil is encouraged as the first option for its management.

This article exemplifies a number of seven such used engine oil recycling technologies currently applied in different parts of the world. Both economically and ecologically, they are an example of responsible management of used oils, avoiding their discharge into the environment, thus contributing to the protection of the environment and the development of a more sustainable society. The technologies presented (as well as other methods of recycling used engine oil) are in a permanent process of improvement, to maximize the amount of recycled oil, as well as the efficiency of the recycling process.

Annexes

List of abbreviations:

Abbreviation	Meaning
DCH	Direct Contact Hydrogenation
HDF	Hydrofinishing
HPS	Hydroprocessing
MP	N-methyl-2-pyrolidone
PAH	Polycyclic Aromatic Hydrocarbons
PCB	Polychlorinated Biphenyls
STP	Scientifically Treated Petroleum
TCT	Thermal Clay Treatment
TDA	Thermal De-Asphalting
VI	Viscosity Index

References

[1] Basel Convention on the Control of Transboundary Movements on Hazardous Wastes and Their Disposal 1995 *Technical guidelines on used oil re-refining or other re-uses of previously used oil*, Geneva, Switzerland

[2] *** https://www.umweltbundesamt.de/en/publikationen/polycyclic-aromatic-hydrocarbons

[3] U.S. Department of Health and Human Services 2014 *Case Studies in Environmental Medicine: Polychlorinated Biphenyls (PCBs) Toxicity, Agency for Toxic Substances and Disease Registry*, WB 2460

[4] *** https://www.gov.scot/binaries/content/documents/govscot/publications/advice-and-guidance/2015/07/planning-and-waste-management-advice/documents/planning-waste-management-advice-pdf/planning-waste-management-advice-pdf/govscot%3Adocument/Planning%2Band%2Bwaste%2Bmanagement%2Badvice.pdf

[5] GEIR (Groupement Européen de l'Industrie de la Régénération) 2016 *Waste Framework Directive revision: European waste oil re-refining industry position*, Brussels, Belgium

[6] *** Regional Activity Centre for Cleaner Production (RAC/CP) Mediterranean Action Plan 2000 *Possibilities for the recycling and reuse of used oils*, Barcelona, Spain

[7] Ahamad M T, ChadraSekhar B P, Mohan P N, Joshi K S, Sree T D R 2015 Recycling and Analysis of Spent Engine Oil, *International Journal of Scientific & Engineering Research*, 6(11), pp 711-715

[8] Akilimali F C 2017 *Feasibility study of recycling used lubricating oil*, The University of Dodoma, India
[9] United Nations Environment Programme Division of Technology, Industry and Economics International Environmental Technology Centre 2012 *Compendium of Recycling and Destruction Technologies for Waste Oils*, Osaka, Japan
[10] Rincon J, Canizares P 2005 Waste oil recycling using mixtures of polar solvents, *Industrial & Engineering Chemistry Research*, 44. 10.1021/ie0580452, pp 7854-7859
[11] Osman D I, Attia S I, Taman A. R. 2018 Recycling of used engine oil by different solvent, *Egyptian Journal of Petroleum*, 27, pp 221-225;
[12] Boadu K O, Joel O F, Essumang D K, Evbuomwan B O 2019 A Review of Methods for Removal of Contaminants in Used Lubricating Oil, *Chemical Science International Journal*, 26(4), pp 1-11;
[13] *** https://www.puraglobe.com/company/
[14] Dalla Giovanna F, Khlebinskaia O, Lodolo A, Miertus S 2003 *Compendium of Used Oil Regeneration Technologies*, International Centre For Science And High Technology, Trieste
[15] *** https://www.itelyum-regeneration.com/en/revivoil/
[16] *** https://www.sor.com.au/what-we-do/waste-oil-refining-process/
[17] *** https://www.cleanharbors.com/services/technical-services/recycling-services/used-oil-and-oil-products-recycling
[18] Audibert F 2011 *Waste Engine Oils: Rerefining and Energy Recovery*, Elsevier, Amsterdam, The Netherlands
[19] *** https://www.bechtel.com/services/chemicals/bhts/oil-processing/
[20] *** http://www.afrilub.ma/
[21] *** https://www.youtube.com/watch?v=bF5xDnjFcg4
[22] *** https://www.lwart.com.br/coleta-de-oleo-lubrificante-usado/
[23] Nixon H, Saphores J D 2002 *Used Oil Policies to Protect the Environment: An Overview of Canadian Experiences* 10.1061/40630(255)11

IOP Conf. Series: Materials Science and Engineering 1220 (2022) 012035 doi:10.1088/1757-899X/1220/1/012035

Evaluation of temperature sensors used in automotive applications type NTC and PTC

Dragos Tutunea, Dumitru Ilie, Laurentiu Racila, Oana Otat and Ionut Geonea

Vehicles, Transport and Industrial Engineering, Faculty of Mechanics, Craiova, Romania

dragostutunea@yahoo.com

Abstract. The coolant temperature sensor is an important tool for measuring the temperature of the engine improving performance and life time. Using this information's the engine control module control different parameters achieving optimal engine control. Thermistor-based temperature sensors have been the choice for a variety of automotive applications enhancing the performance of car systems. In this paper, an experimental stand was designed and realized in the laboratory to evaluate the accuracy of two thermistors NTC and PTC used to measure the engine coolant temperature. Based on the variation of the temperature characteristics curves function of resistance will be determined.

1. Introduction

The utilization of thermistors has increased over the past years and at present are used in a variety of applications. Sensors play a critical role in the collection of information regarding the functionality of the engine. Despite their use in various domains (biochemical, material sciences, energy accounting, process control) they have not been characterized with regard to their thermal stability [1]. When an engine is starting, the combustion chambers have a lower temperature and the electronic control unit ECU needs to have information about the coolant temperature. ECU use this information to adjust the fuel injection in the combustion chamber [2].

Thermistors are usually used as engine temperature sensors based on a thermistor, which is a variable resistor influenced by the variation of the temperature. The ECT (Engine Coolant temperature sensor) and IAT (Intake Air temperature sensor) are the same using thermistors to measure the temperature. There are 2 types of sensors PTC-positive temperature coefficient thermistors used in electric current control and NTC-negative temperature coefficient thermistors used in temperature sensing [3-5].

If in the past only ceramic materials were used to produce thermistors by 2003 Si and Ge high temperature NTC thermistors were used with high performances. The range of temperature is relatively small compared to other sensors but there are precise within that range [6]. The disadvantages of both thermistors are their nonlinear temperature dependence of resistivity and the deterioration of the sensor characteristics due to the aging and degradation. The R(T) dependencies vary in function of the technology applied during construction and material properties. The Steinhart-Hart equation is used to describe temperature-resistance T(R) dependencies in steps with intervals of 1°C.

$$\frac{1}{T}=A_0+A_1\cdot R_T+A_3\cdot(\ln R_T)^3+A_5\cdot(\ln R_T)^5 \quad (1)$$

where: RT is the thermistor resistance at the temperature T; T is the temperature in K; A0, A1, A2, A3, A4 and A5 are the coefficients [7].

To calculate the coefficients (equation 1), it is sufficient to measure the resistance of the thermistor at three variable temperature points. The purpose of this study is to evaluate NTC and PTC temperature characteristics on an experimental stand to find temperature-resistance dependencies.

2. Material and methods

The experimental stand was designed and built in the laboratory of Elements of Mechatronics of the Faculty of Mechanics of Craiova. The scheme of the principle of the stand is shown in figure.1. The stand can be used to determine the variation of resistance for NTC and PTC thermistors function of temperature (figure 2).

The stand uses the following equipment:

- Electrical resistance of 2 kW with three steps of adjustments
- Recirculation pump type Lukon 25-40 180;
- Digital thermometer type TM-902C;
- 2 digital multimeters type DT-832;

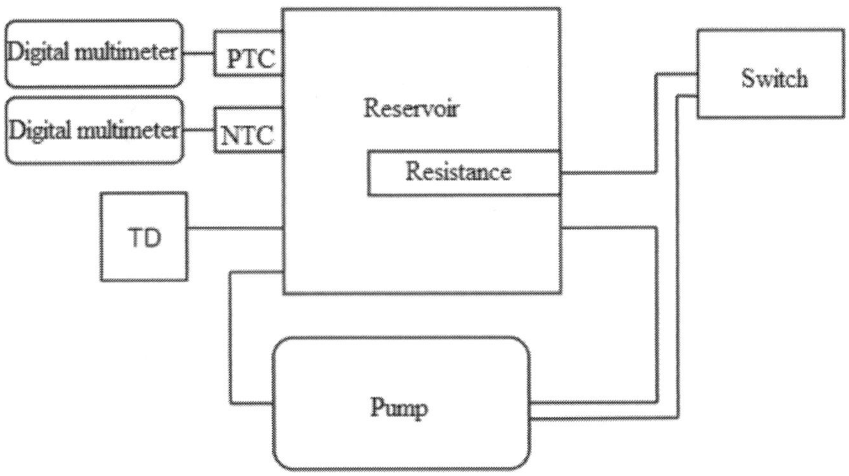

Figure 1. Diagram of the stand for testing NTC and PTC thermistors.

Figure 2. Front view of the experimental stand.

IOP Conf. Series: Materials Science and Engineering 1220 (2022) 012035 doi:10.1088/1757-899X/1220/1/012035

To obtain experimental results we will proceed as follows:

- the reservoir is filled with water;
- the pump is turned on to recirculate the water in the reservoir;
- the pump has three manually selectable speed stages, is selected the first speed;
- the electric resistance is turned on to heat the water;
- the electric resistance has three manually selectable heating stages, is selected the third stage;
- the resistance of PTC and NTC sensors is measured with 2 digital multimeters and is correlated with the temperature indicated by the digital thermometer mounted on the reservoir;
- the maxim heating value of the stand is 95 °C near the boiling point of the water. Heating with higher temperatures can damage the pipes;
- the measurements are realized in 55-70 minutes.

3. Results and discussion

The tests start with the turn on of the pump and electrical resistance and an initial temperature of 28 °C. The water is heated until 85°C and the resistance is measured at intervals of 1˚C. Three tests were performed maintaining the same condition and the average values were taken into considerations. The response of resistance vs. temperature for temperature sensors is highly nonlinear and is almost exponential. (Table 1).

Table 1. Experimental data.

T [˚C]	PTC [Ω]	NTC [Ω]
28	2.02	1.82
29	2.02	1.75
30	2.03	1.7
31	2.04	1.64
32	2.05	1.57
33	2.06	1.51
34	2.07	1.46
35	2.08	1.4
36	2.09	1.36
37	2.1	1.31
38	2.11	1.26
39	2.12	1.21
40	2.13	1.16
41	2.14	1.13
42	2.15	1.09
43	2.16	1.06
44	2.17	1.03
45	2.18	0.98
46	2.19	0.95
47	2.2	0.92
48	2.21	0.88
49	2.22	0.85
50	2.23	0.83
51	2.24	0.81

T [°C]	PTC [Ω]	NTC [Ω]
52	2.25	0.79
53	2.26	0.76
54	2.27	0.74
55	2.28	0.71
56	2.29	0.69
57	2.3	0.67
58	2.31	0.65
59	2.32	0.63
60	2.34	0.6
61	2.35	0.59
62	2.36	0.57
63	2.37	0.55
64	2.38	0.54
65	2.39	0.53
66	2.4	0.52
67	2.41	0.49
68	2.43	0.48
69	2.44	0.46
70	2.45	0.45
71	2.47	0.43
72	2.48	0.42
73	2.49	0.41
74	2.5	0.4
75	2.51	0.39
76	2.52	0.38
77	2.54	0.37
78	2.55	0.35
79	2.56	0.35
80	2.57	0.33
81	2.58	0.32
82	2.59	0.31
83	2.6	0.31
84	2.62	0.3
85	2.63	0.29

Based on experimental data, a logarithmic approximation for the curve of variation resistance-temperature for both thermistors is presented in figure 3 and 4.

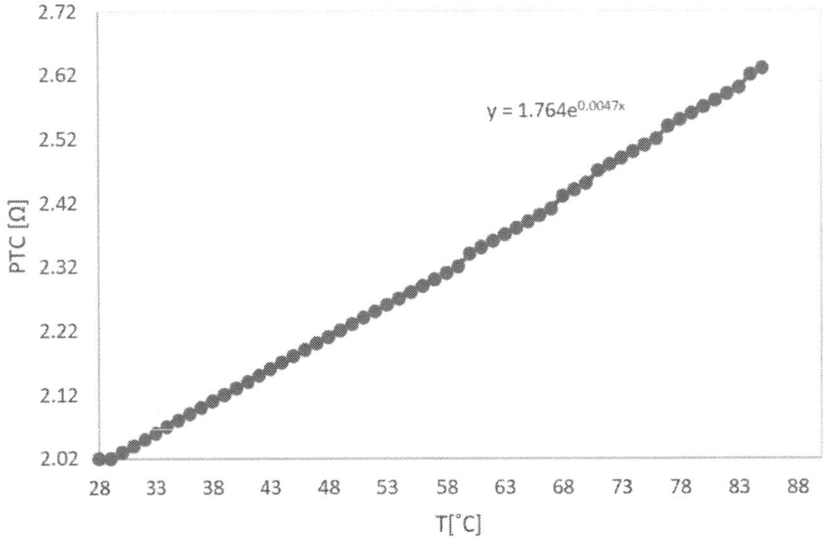

Figure 3. Variation of resistance function of temperature for PTC thermistors.

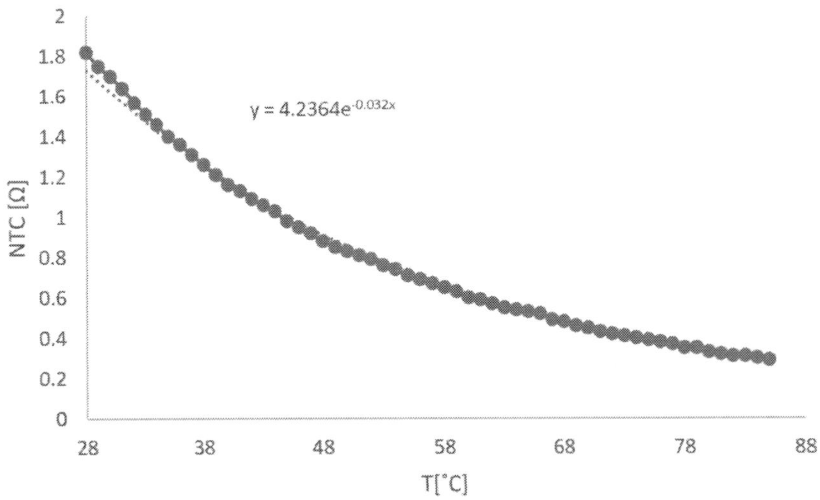

Figure 4. Variation of resistance function of temperature for NTC thermistors.

4. Conclusion

In this study, an experimental stand was conceived to analyze the resistance temperature dependencies for temperature sensors used in the automotive industry. The proposed stand allows to test the NTC and PTC sensors in a temperature range between 0 to 95 ˚C. The stand uses a pump to recirculate the heated water to have the same temperature on both thermistors. Several tests were performed and the average value was taken into consideration for measurement. The results show a logarithmic expression for both thermistors. This stand can be used in further experiments to test different thermistors in a wide range of temperatures.

References

[1] Wood SD, Mangum BW, Filliben JJ, Tillett SB, *Journal Of Research of the National Bureau of Standards* Volume B3, No.3, May-June 1978.

[2] Sugiarto T, Putra DS, Purwanto W, Analysis on the role of engine coolant temperature sensor in gasoline engine, ISSN 2528-2611, e-ISSN 2528-2700 Vol.2, No.2, December 2017, Page.145-152.

[3] Jagtap S, Rane S, Gosavi S, Amalnerkar D, Preparation, characterization and electrical properties of spinel-type environment friendly thick film NTC thermistors, *Journal of the European Ceramic Society*, Volume 28, Issue 13, 2008, pp. 2501-2507.

[4] Liu X, Luo Y, Li X, Electrical properties of BaTiO3-based NTC ceramics doped by BaBiO3 and Y2O3, *Journal of Alloys and Compounds*, Volume 459, Issues 1-2, 2008, pp. 45-50.

[5] Zubair MA, Leach C, The influence of cooling rate and SiO2 additions on the grain boundary structure of Mn-doped PTC thermistors, *Journal of the European Ceramic Society*, Volume 28, Issue 9, 2008, pp. 1845-1855.

[6] Lee WY, Kim TM, Kim MJ, Ko YW, Kim JD, User-friendly Calibration Tool for Temperature Measurements of PCR Devices with NTC Thermistors, *International Journal of Control and Automation* Vol. 8, No. 7 (2015), pp. 13-24

[7] Dumcius A, Gailius D, Kuzas P, Stability of Negative Resistance Coefficient Thermistors for Long-term Temperature, *Elektronika Ir Elektrotechnika*, ISSN 1392–1215, VOL. 20, NO. 6, 2014.

IOP Conf. Series: Materials Science and Engineering 1220 (2022) 012036 doi:10.1088/1757-899X/1220/1/012036

Experimental investigation on the alternator charging capacity for automotive applications

Dragos Tutunea, Dumitru Ilie, Laurentiu Racila and Cristina Rotea

Vehicles, Transport and Industrial Engineering, Faculty of Mechanics, Craiova, Romania

dragostutunea@yahoo.com

Abstract. This research paper investigates the use of an automotive alternator on an experimental stand to evaluate the capacity to load the battery and produce electricity for car systems. The efficiency was calculated by measuring the electric power supplied to the battery for different regimes of speed. Various aspects regarding the efficiency of energy conversion by using an alternator were also discussed. The stand uses an electric motor with various transmission ratios to rotate the alternator. The speed was recorded with a digital tachometer and the energy consumption of the electric motor with a digital wattmeter. The results obtained showed that the increase in alternator speed generated more electricity and improved the load of the battery.

1. Introduction

In recent years, automotive producers have been looking to improve vehicle performance by increasing component performance and power density. Such a system is the on-board electrical power charging system, which is usually composed of an alternator, a battery, and a power converter. It is obvious that the electrical systems of the vehicle are pushed to their limits and now various components are replaced by modern systems that are electrically assisted. Furthermore, new legislation increased car functionalities, and customer demands have pushed automakers to improve and increase electrical power in their vehicles. For the last 60 years, the claw pole alternator (or Lundell alternator) has been used to generate power for automotive applications [1]. The principal advantages of Lundell are the simplicity of construction and low cost. His performance is influenced by alternator speed and load conditions [2]. Modern alternators must produce electricity for motors for cooling fans, heated windows, seats, and catalysts, electric actuators for power steering, various engine sensors and alarm systems. If at the beginning of 1990, typical consumption of 500 W was sufficient, now 3-5 kW is considered normal. The latest vehicles produced are equipped with several electrical devices such as a rear camera, a DVD player, a USB port for phone charging, LED lights, etc. [3]. The output of normal alternators ranges between 40 and 100 amps. In most vehicles, the charging system of the alternator will produce a voltage between 13.5 and 14.4 volts. The charging system must produce a higher voltage than the car battery voltage to overcome its internal resistance. The output of the alternator depends on the engine speed to generate electricity. For example, the ratio speed between the engine and the alternator is 1:1 for a racing car, 1:2 for a drag car, and for a street car, 1:3 [4]. Researchers have been working for years to develop higher power alternators with improved performance, such as: the claw pole alternator where the DC-excited winding is in the stator or the doubly excited brushless claw pole alternator [5, 6]. Another direction of research is the use of rare earth, high-power permanent magnets (PM), which have higher efficiency and power density. These types of alternators improve the output quality and dynamic performance [7, 8]. The objective of this study was to evaluate the output energy delivered by an alternator mounted on an experimental stand in the laboratory.

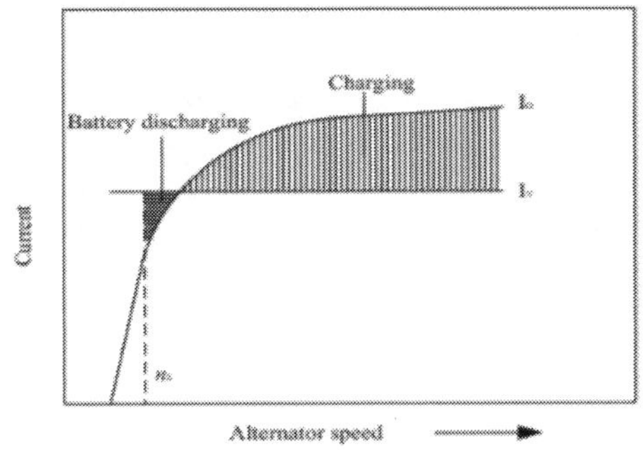

Figure 1. Predicted alternator output power versus output voltage [3].

2. Material and methods

The experimental stand was designed and built in the laboratory of Elements of Mechatronics of the Faculty of Mechanics of Craiova. The scheme of the principle of the stand is shown in Figure 2. The stand can be used to verify the output energy delivered by the alternator function of the speed of rotation simulating the real condition of usage in a vehicle. The performance and the output power of the optimal claw pole alternator over the entire speed range from the low speed of 200 rpm to the top speed of 2200 rpm are presented.

The stand uses the following equipment:

- Bosch alternator 0 123 310 048;
- Electric motor Einhell Hauswasserwerk HW 40;
- Bosch battery S4 008;
- Digital multimeter KDM-83;
- Digital wattmeter PZEM-021;
- Voltage variator x26;
- Digital tachometer AXIO MET AX-2901.

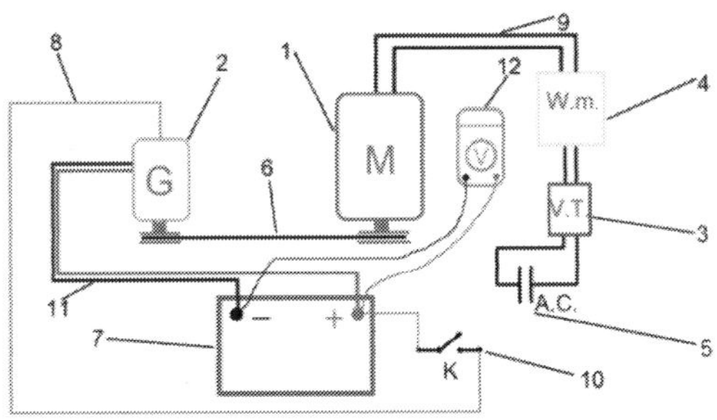

Figure 2. Diagram of the stand for testing alternators:
1 – Electric motor; 2 – Alternator; 3 – Voltage variator; 4 – Digital wattmeter; 5 – AC current source (220 V); 6 – Transmission belt; 7 – Battery; 8 – Power supply cable alternator; 9 – Power supply cable alternator electric motor; 10 – Excitation switch; 11 – Battery power cables; 12 – Digital multimeter

To obtain the experimental data, proceed as follows:

- The electric motor is powered by a 220 AC source;
- The speed of the alternator is measured with a digital tachometer. The alternator is actioned by a drive belt from the electric motor. The motor speed is changed by means of a voltage variator;
- The charging voltage supplied by the alternator to the battery is measured with a digital multimeter;
- The electric parameters of the electric motor are measured with a digital wattmeter.

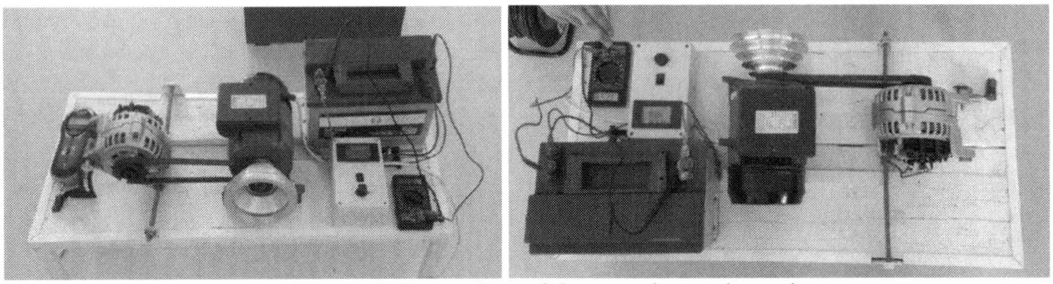

Figure 3. Front view of the experimental stand
(battery, alternator, electric motor, digital wattmeter, multimeter).

3. Results and discussion

The alternator output depends on various factors such as battery temperature and condition, wire size, engine RPM, alternator temperature, and engine horsepower. From the data shown in table 1, it is shown that there is a speed ratio between the engine and the alternator of 1:3. It means the speed of the alternator is three times the speed of the electric motor. Different vehicles have various alternator-engine speed ratios, depending on the power needed for their electrical systems. Each car has a speed ratio that is set so that the alternator can produce more than the car requires. A fully charged battery has a voltage of 13.7 V. From the table, it is seen that after 1200 rpm, the alternator is recharging the battery. The alternator will supply all the current demand for the electrical systems while at the same time charging the battery continuously (Fig.4).

Table 1. The dependence between speed of alternator and voltage.

Nr. Crt.	Speed of alternator [rpm]	Output voltage [V]
1.	204	12.21
2.	467	12.33
3.	780	12.43
4.	832	12.57
5.	881	12.61
6.	967	12.64
7.	973	12.65
8.	983	12.69
9.	1004	12.73
10.	1053	12.78
11.	1136	12.89
12.	1154	13.07
13.	1184	13.23
14.	1187	13.61
15.	1210	14.01
16.	1264	14.12
17.	1345	14.15
18.	1545	14.17
19.	2109	14.20
20.	2112	14.21
21.	2132	14.26
22.	2133	14.31
23.	2135	14.33
24.	2140	14.34
25.	2143	14.35

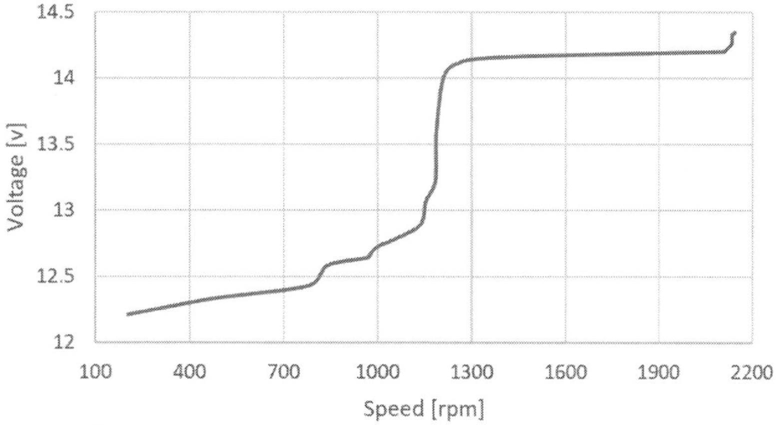

Figure 4. Variation of voltage vs speed of the electric motor.

4. Conclusion

In this study, an experiment was done to find the relation between electric motor speed and alternator. The results show that the alternator produces output energy even at low speeds and its efficiency is reached at higher values of around 1200 rpm. These experiments were done to get a better understanding and better knowledge of vehicle charging systems. This stand can be used in further experiments to study temperature influence, mechanical parameters, friction force, maximum torque values and electromechanical aspects.

References

[1] Brahimi N, Tahi S, Boudissa E, Bounekhla M, Optimal Rotor Design of Claw-Pole Alternator for Performance Improving, Adv Syst Sci Appl 2019, 02, pp.120-133.

[2] Ivankovic R, Cros J, et al., power electronic solutions to improve the performance of lundell automotive alternators, New advances in vehicular technology and automotive engineering, 2012.

[3] Mazlan RK, Dan RM, Zakaria MZ, Hamid AHA, Experimental study on the effect of alternator speed to the car charging system, MATEC Web of Conferences 90, 01076, 2017.

[4] Giri NK, Automobile Mechanical, New Delhi, Khanna Publisher, pp. 376, 2009.

[5] Dajaku G, Lehner B, Dajaku Xh, Pretzer A, Gerling D,Hybrid Excited Claw Pole Rotor for High Power Density Automotive Alternators. Electrical Machines (ICEM), XXII International Conference of the IEEE, Lausanne, Switzerland, 2016.

[6] Rebhi R, Ibala A, Masmoudi A, On the Modeling of a Doubly-Excited Brushless Claw Pole Alternator: Application to a Solar-Wind-Wave Hybrid Energy Conversion System, Eighth International Conference and Exhibition on Ecological Vehicles and Renewable Energies, 2013.

[7] Wang Q, Niu S, Yang L, Design Optimization and Comparative Study of Novel Dual-PM Excited Machines, IEEE Trans. Ind. Electron. 2017, 64, pp. 9924–9933.

[8] Pop AC. Gyselinck JJ, Pinto DE, Vintiloiu I, Optimization of Low-Power Brushless PM-Machines for Automotive Applications with Focus on High-Volume Mass Production. IEEE Trans. Ind. Electron. 2017, 64, pp. 9767–9775.

IOP Conf. Series: Materials Science and Engineering 1220 (2022) 012037 doi:10.1088/1757-899X/1220/1/012037

Engine oil analysis to evaluate the degree of its wear during the period of operation of the vehicle

C Zaharia, R Niculescu, M Năstase, A Clenci, V Iorga-Simăn

University of Pitesti, Faculty of Mechanics and Technology, Department Automobiles and Transport, 1, Tg. din Vale Street, Pitești, 110040, Argeș, Romania

catalin.zaharia@upit.ro

Abstract. The lubricating oil fulfills several purposes, such as: reducing friction processes, protecting the parts of the engine mechanism from wear, preventing the occurrence of corrosion phenomena, taking over and transporting or transferring impurities to the oil filter, lubricating the moving parts as well as their cooling.

The lubrication fluid of the internal combustion engine has a significant contribution on its service life and involves its proper functioning during operation. Changing the engine oil before reaching the manufacturer specified mileage and/or time corresponding to the planned preventive maintenance becomes an imperative action sometimes, but with negative effects on the operating costs of the vehicle.

Prolonged operation in different conditions could lead to the deterioration of the lubricating oil, respectively of its properties and quality. Thus, in order for the user to benefit from maximum engine performance, it is necessary to change the engine oil at well-defined time/running intervals. So, it is very important to know when the technical overhaul of the engine will be carried out because the late change of the engine oil can affect the parts of the engine mechanism and implicitly, the performance of the vehicle or the impossibility of driving. On the other hand, premature change of engine oil increases the costs, affects the environment considering the recycling process and resource depletion.

The authors of this paper used samples of SAE 5W30 engine oil and analyzed the following parameters: viscosity, density, flash point, contamination with fuel and solids, degree of oxidation, dispersion power. The information on the evolution during the period of operation (considered post-warranty period) of these parameters is the basis for the elaboration of engine maintenance strategies considering the following three goals: the reliability, the operating costs of the vehicle and environmental issues.

1. Introduction

It is well known that the oil present in the lubrication system fulfills the following roles: reduces friction in couplings (bearings, piston - segment - cylinder, etc.) by lubricating contact surfaces, washes and transports impurities or metal debris caused by the wear process; also, cools the hot surfaces of the engine (piston, valves, cylinder, etc.), contributes to the transfer of heat to the outside and, through the formed film, realize the sealing of the piston-segment-cylinder coupling, opposing resistance to gas leaks toward the lower crankcase.

Driving style and operating conditions can generate long periods of operation of internal combustion engine (i.c.e.) at high speeds and high temperatures which causes damage to the oil, respectively its properties and, consequently, its quality. This whole process can lead to the wear and tear of the engine mechanism parts. In the process of operating of i.c.e., the lubrication system is subjected to mechanical - chemical processes that contribute to changing the technical condition of its parts and subassemblies. Among them are: the physicochemical properties of the oil, the thermal balance of the engine, the oil mass in the lower crankcase, clogging of the oil filters and modification of the pressure in the lubrication system.

The change in the properties of the oil can be based on its contamination (impurity), caused by impurities from the outside environment, metal debris due to wear and tear, oil particles due to changes in its composition, and dilution with cooling liquid or fuel [1]. During engine operation, the amount of engine oil in the lubrication circuit varies as follows: it increases with the contamination of the oil with a certain amount of coolant and/or fuel or decreases due to combustion or loss on the seals.

Thus, for operation in normal technical parameters, the quality and quantity of engine oil must be appropriate and are considered to be extremely important factors in the period between two technical revisions imposed on the lubrication system by the car manufacturer.

It is known that in order to benefit from the warranty period of a new car or to use a car for as long as possible, the technical inspections must be carried out according to the manufacturer's instructions, considering the number of months of use or the number of kilometers traveled. Currently, the time period for which a car engine is subject to technical overhaul has become generalized to 1 year, and the maximum mileage varies between 20,000 and 30,000 km, depending on the year of manufacturing of the car. If the vehicle is older than 5 years, certain car manufacturers in Romania recommend performing a technical overhaul of the engine at each 15,000 km.

If from the point of view of the quantity of engine oil there are simple levers of control and supervision in the period between two technical revisions, however, related to the quality of the engine oil there is no concrete control based on generalized technical solutions and cost efficient.

However, manufacturers of automotive equipment have recently come up with some relatively simple technical solutions, but the implementation process must take into account the harsh working conditions of electrical and environmental installations of cars and, in particular, design aspects, which involves obtaining excellent electromagnetic immunity and emission performance [2] and hence significant application costs.

At the beginning, oil conditions sensors have not identified any specific variable and were capacitance or die-electric constant instruments. They are, firstly, signalizing change when either water, oxidation or acidity changes appear and were not sensitive to other key parameters. These have been developed primarily for the automotive sector as a means of improving the engine management protocol to alert the driver of the most appropriate servicing interval [3].

The need for the monitoring of changes in real time, in order to be able to manage any significant risk, and the potential for cost improvement over the life of the system, are the key factors behind the move towards oil sensors. The management of modern technology demands engineering staff to identify and manage risks quickly. Nowadays the modern equipment helps the engineer to do this.

The latest sensors are able to detect not only the degradation of the overall quality of the oil, but also to estimate the percentage of soot, the base number, the relative humidity, the additive depletion, etc., the oil life while the i.c.e. is running. Although these sensors cannot replace the results of laboratory tests, they can provide the information needed to make preventive maintenance decisions before damage occurs. New sensor technologies use electrochemical impedance spectroscopy (EIS), which covers a wide range of oil properties and allows the sensitivity to most failure modes [4].

Harish Shinde et al. in [5] mention that a capacitive sensor may be used in order to quantify the quality of the oil used, analyzing the changes in the capacitance of the engine oil. There is also being developed sensors which operation is based on the detection of free radicals produced by the degradation of engine oil [6]. Harish Shinde et al. in [7] also mention that changes in engine oil viscosity were analyzed using a fluid property sensor, the results obtained were compared with the results from a standard laboratory setting and they showed a very good correlation.

Very important for an engineer is the capability of the sensor to operate accurately in order to program in a specific response. Sensor capability can be based on:

- oil condition parameters (changes in the main properties: viscosity, reserve alkalinity, oxidation, nitration, additive levels);
- oil contaminant parameters (externally generated: water, fuel, soot, abrasives and acids);
- system wear debris parameters (internally generated: from bearings, shafts, gears, liners, rings, seals).

Alarms based upon rates of change, in addition to upper and lower limits, would ideally be considered for each individual system. Any specific sensor might respond to changes in data within each of the above groups, however, each signal must be separately identified [3].

Some cars manufacturers designed programs that are managed by the injection computer of the i.c.e., which, depending on the driving style, respectively the number of rotations of the crankshaft in a known period of time and the operating conditions, commands the activation of the warning light "service", informing the user that it is compulsory to carry out an unplanned technical overhaul, in order to protect the car engine.

Clearly, in the last century, oil analysis has been the backbone of reliability programs. From the best-known instruments used by laboratories to more modern, miniaturized table equipment, this technology has transformed the way and speed in which car maintenance decisions are made. Thus, oil quality monitoring is a benefit for optimizing oil exchange [8].

Late replacement of used engine oil can affect engine parts with the cost of repairs, which are often considerable. Therefore, the recommendation to change the engine oil prematurely based on the analysis of its physicochemical properties certainly contributes greatly to reducing the maintenance costs of the car and its use without losing part or all of its operating capacity. According to the literature, the adoption of a maintenance system based to a greater extent on preventive interventions involves lower costs than the maintenance system that contains corrective interventions.

On the other hand, the premature change of suitable engine oil also influences the environment and, at the same time, can be considered an inefficient action if the resource depletion process is taken into account [9]. Thus, when discussing oil as a valuable resource, every effort should be made to recycle it whenever possible. Several effective ways are known to bring used oil back into use; for example, some oils may be cleaned by filtration or reconditioning and reused [10]. The recycling process is the easiest way to benefit from degraded oil. Whether it is the i.c.e., gearbox or hydraulic systems, all oils reach the end of use and must be replaced at some point. Thus, economic operators have the obligation to collect waste oils separately and to hand over the entire amount of waste oil to economic companies authorized to carry out recycling or disposal activities [11, 12].

In this paper, the authors aim to develop a methodology that can issue preventive recommendations on internal combustion engine maintenance based on assessing the quality of engine oil by determining its physicochemical properties through a set of laboratory analyzes. Thus, the following will be determined: viscosity, flash point, dilution with fuel, density and an evaluation of the dispersion power, of the degree of oxidation and contamination with liquids and carbonaceous substances will be performed by the oil stain method.

2. Characterization of the physicochemical properties of the engine oil

In order to be able to develop a method for evaluating the used engine oil and issuing recommendations regarding the planned preventive maintenance, the following parameters are briefly reviewed: kinematic viscosity, flash point, density, fuel dilution/content and oil stain. The oil stain will be analyzed according to the presence of sediments, the degree of oxidation and the presence of water.

Viscosity is a measure of a fluid's resistance to flow at a given temperature. In the case of lubricants, it is commonly perceived as the "thickness", of the oil [13]. Used oil analysis laboratories and lubricant manufacturers rely on viscosity measurements as a key parameter in determining the condition of used oil, or specification of new oil. For example, changes in the viscosity of used oils can be an indication of lubricant degradation (increase/decrease in viscosity) or oil dilution (decrease in viscosity).

The kinematic viscosity (kV) is commonly referred to as just 'viscosity' and is determined by measuring the time a fluid needs to flow through a defined capillary at a defined temperature. The viscosity is typically expressed as a kV@40 or kV@100 test result, where 40 or 100 denote the test temperature used (°C). The viscosity of lubricating oil is defined at 100°C according to the SAE J300 Industry Classification System and is expressed in mm^2/s or cSt (centistokes). In practice, this parameter is also measured at 40 °C and many used oil analysis laboratories report the viscosity measurement at 40 °C, 100 °C or both. Kinematic viscosity is the absolute or dynamic viscosity in (Pa · s) divided by the fluid density.

Flash point (FP) measures the tendency of a sample to form a flammable mixture with air under controlled laboratory conditions. This is the lowest temperature, at which vapors from the oil can be ignited. Flash point data are also used in shipping and safety regulations to define "flammable" and "combustible" materials. Flash point data can indicate the possible presence of highly volatile and flammable material in a relatively non volatile or non flammable material, such as the contamination of lubricating oils by small amounts of diesel fuel or gasoline. Change of flash point is mainly influenced by fuel contamination and to a certain extent by oil degradation [14]. There are different industry flash point test methods including the "open cup" and "closed cup" type methods. It must be noted that due to evaporation losses the result with the closed cup method is generally lower than the result with the open cup method [3].

Density (d) is an important property of petroleum products being part of product specifications. Materials are usually bought and sold on that basis or if on a volume basis then converted to mass basis via density measurements.

For the determination of oil degradation, it is used oil analysis on filter paper, known as the "oil stain method." This analysis enables assessment of capacity detergent - dispersant, the degree of contamination of the oil with particles of carbon and chemical degradation by oxidation engine oil increasing [15].

3. Methodology and experimental settings

The working methodology is based on collecting engine oil samples and their detailed analysis in the laboratory; at the completion of the laboratory analysis, based on the results, will be issued recommendations on the maintenance to be applied (figure 1).

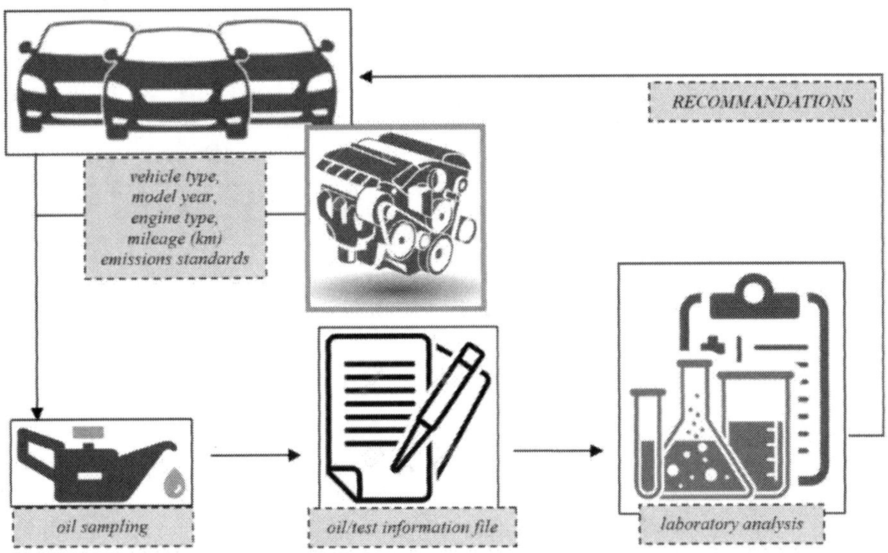

Figure 1. The structure of oil sample steps.

The study was carried out on oil samples taken from i.c.e. with the year of manufacture between 2019 and 2020, with the following characteristics:

- engine oil type: oil used by the car manufacturer, SAE 5W30;
- engine type: diesel engine, with DPF;
- cylinder capacity: 1500 cm^3;

- mileage: 1769 - 82245 km;
- emissions standards: EURO 6;
- operating conditions: real different conditions; non standardized conditions according to the use of a test bench or tests on a special circuit.

Viscosity - equipment description. The equipment used in this article is called U-VIsc and is a fully automatic instrument that utilizes glass capillary viscometers to obtain values for the kinematic viscosity of Newtonian fluids. It is considered a good tool for mineral and synthetic oils (fresh and used) and for any fluid within the viscosity range of the instrument [15].

Method description. The standard test method for kinematic viscosity of transparent and opaque liquids D 445 is under the jurisdiction of ASTM Committee D-2 on Petroleum Products and Lubricants. The time is measured for a fixed volume of liquid to flow under gravity through the capillary of a calibrated viscometer under a reproducible driving head and at a closely controlled and known temperature. The kinematic viscosity is the product of the measured flow time and the calibration constant of the viscometer [15]. The time it takes for the sample to flow through a section of the tube is proportional to viscosity as follows:

$$kV = C \cdot t \tag{1}$$

where: kV – kinematic viscosity, (mm^2/s); C – viscometer constant, (mm^2/s^2); t – flow time, (s).

In order to correct this kinetic energy effect, the result should be corrected. Following Hagenbach and Couette theory [15], the result should be corrected as follows:

$$kV = C \cdot t - E / t^2 \tag{2}$$

The kinetic energy factor E ($mm^2 \cdot s$), though is not a constant, can be derived from the tube's dimensions (capillary length and diameter).

Flash point and dilution - equipment description. Eraflash apparatus produced by Eralytics, Austria was used also in this study in order to find the oil flash point and dilution degree. The measurement with this instrument is directly under the latest and safest standards ASTM D6450 & D7094. It uses the Continuous Closed Cup method for finding the Flash Point (CCCFP).

Method description. The method used is ASTM D7094. It is considered an extremely versatile flash point method that can be used for accurate flash point measurement and also, to obtain the influence to the flash point of highly flammable fractions in samples like gasoline in diesel or diesel in oil or gas in oil. When the start temperature is reached the sample cup is lifted. Then the sample cup is lifted and the measurement takes place. The electric arc ignites in the pre-programmed interval steps. The study was conducted with an interval of 1 second. The flash point was detected when the pressure after ignition is 20 kPa or higher at the level of the measurement chamber.
The method also permits to determine the degree of dilution of the oil with diesel fuel. For this, it was necessary to determine the reference curve by controlled dilution of fresh oil with diesel.

Density - equipment description. The equipment is Density Meters DS7800 made by A. KRÜSS OPTRONIC – Germany [15].

Method description. The Density Meters DS7800 operates according to the frequency oscillator principle and covers the determination of the density or relative density of crude oils with high vapor pressures in order to prevent vapor loss during the transfer of the sample to the density analyzer. A small volume of the liquid sample is introduced into an oscillating sample tube and the change in oscillating frequency caused by the change in the mass of the tube is used in conjunction with calibration data to determine the density of the sample [15].

Oil stain analysis - method description. The method is based on viewing the stain that a drop of oil leaves on a paper filter. Usually, a stain of used oil consists of three zones (fig. 6). A small quantity of used oil sample and the corresponding fresh oil is dropped onto a sheet of special filter paper and allowed to spread. The size dispersion is calculated with the formula below [15, 16]:

$$SD = (d/D) \cdot 100 \tag{3}$$

where: d – diameter of central zone; D – diameter of 2nd circle.

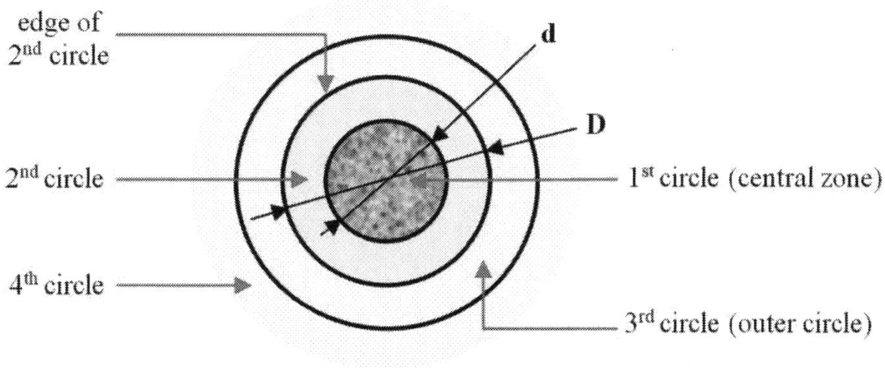

Figure 2. Oil stain structure.

The central zone indicates the presence of sediments, such as dust, wear particles, carbon, insoluble matter and other contaminants. Depending on the age and condition of the oil, the focal point will vary from light gray, in clean oil to black, in degraded oil. A clean engine, usually, shows easy-to-identify circles.

The second circle provides information about the condition of the oil. The range of results is from light gray to black. If the engine and oil are in good condition the second circle appears lighter than the center area. If the engine is in good condition but the oil is degraded, the second circle will appear gray to black. If the oil is in poor condition, the central area can extend over the second circle and over its edges.

This area, called the diffusion zone, indicates the dispersing power of the oil. If the oil has lost its dispersing property, the carbon particles can no longer diffuse through the pores of the filter paper and concentrate in the central area. The size of the dispersion, SD, indicates the ability of the oil to hold insoluble particles in suspension. In other words, when the oil loses its dispersing capacity the insoluble solid particles coagulate, forming large particles that obviously damage by abrasion the surfaces of the moving parts.

The edge of circle 2 indicates the possibility of water content. If water is present, the test medium will immediately show clear peaks on the edges. When the water content is high the results will be visible in a few minutes. The test differentiates between two types of water: from condensation or coolant (with antifreeze/glycol). Both types of water form peaks at the edge of the circle. The water in the coolant with excess antifreeze/glycol will turn yellow after about 30 minutes.

The outer circle, also called the transverse zone, consists of the oil separated from the carbon particles and shows the degree of oxidation of the oil by its color which can vary from yellow to brown.

When there is fuel in the oil, a transparent yellow ring (4th circle) will form in the exterior area of the outer circle. The ring will appear immediately after the droplet enters the test medium. A narrow ring of light is normal. If the ring widens after a few hours, the results are moderate to poor.

4. Results and discussions

For the analysis, engine oil was taken from a number of 14 cars with the technical characteristics already mentioned in chapter 3. The samples of used oil were analyzed from the point of view of the physicochemical properties of the oil, determining values for the following parameters:

- flash point, FP (°C);
- diesel content, DC (%);
- kinematic viscosity at 40°C, kV (mm^2/s);
- density at 15° C, d, (g/cm^3), table 1.

The study also considered the realization of the oil stain that provides information about oxidation processes, contamination with water or glycol, the presence of sediments and the dispersion power. The results are presented in table 1.

Table 1. The results of oil samples test.

Sample no.	Tested oil turnover / total car turnover (km)	MY*	FP (° C)	DC (%)	kV (mm²/s)	d (g/cm³)	SD	Oil stain: Sediment	Oil stain: Oxidation	Water / glycol
New oil, SAE 5W30 (DPF)			196,8	0	68,79	0,8545				
1	1769 / 1769	2020	196,7	0,1	38,41	0,8407	very good	normal level	no	no
2	3046 / 3046	2019	197,8	0	40,26	0,8426	good	normal level	no	no
3	7432 / 7432	2020	189,8	0,6	39,92	0,8463	good	normal level	no	no
4	11046 / 11046	2020	184,8	0,8	38,18	0,8494	good	normal level	no	no
5	14600 / 26544	2019	191,8	0,5	65,71	0,8645	very good	high level	high level	no
6	15200 / 34025	2019	191,7	0,5	64,46	0,8658	very good	high level	high level	no
7	18407 / 18407	2020	182,8	0,9	37,74	0,8559	very good	high level	high level	yes
8	19200 / 59964	2020	183,8	0,8	60,36	0,8712	good	high level	high level	yes
9	20100 / 82245	2019	185,7	0,8	59,55	0,8667	very good	high level	high level	yes
10	25891 / 25891	2020	175,7	1,2	42,28	0,8705	very good	high level	high level	yes
11	26800 / 52219	2019	191,8	0,5	64,16	0,8717	good	high level	high level	yes
12	29458 / 29458	2020	183,7	0,8	51,03	0,8656	very good	high level	high level	yes
13	29980 / 29980	2020	184,9	0,8	43,07	0,8707	very good	high level	high level	no
14	30254 / 30254	2020	179,9	1	37,81	0,8620	Very good	high level	high level	yes

* MY: Model Year

Regarding the flash point, the reference value for the oil approved by the manufacturer, SAE 5W30, for engines with DPF, is 196.8 ^{0}C. From the analysis of the obtained results, it is observed that the higher the car's turnover, the lower the flash point; the causes of this phenomenon can be attributed to the contamination of the oil with water/fuel or to the level of degradation of the engine oil. If we analyze the values obtained for the diesel content, they demonstrate that the oil samples at which the flash point decreased had a diesel content between 0.1 - 1.2%. An explanation for the decrease in flash point is the fact that in the case of samples taken from engines with significantly higher turnover (18407 - 30254 km, except 29980 km) is noted the presence of water and glycols in the analyzed oil.

Oil oxidation is the process by which oxygen reacts with hydrocarbon molecules to form insoluble carbonaceous residues and resins. This process is influenced by temperature and by the degree of oil contamination. Checking the oxidation status of the oil for the samples taken from engines with small turnover (1769 km, 3046 km, 7432 km and 11046 km), it is observed that they do not show oxidation process; the oil samples, taken from the following turnovers: 14600 km to 30254 km, show a high level of oxidation and a high degree of sediment; their diesel content is between 0.5 - 1.2%. Oxidation also influences the density of the oil [17]. As the oxidation progresses, the density of the oil increases, evolution observed in analyzed samples no. 5 - 14, which have increased oxidation level, so implicitly increased density.

The value of 68.79 mm^2/s is the reference value for the kinematic viscosity (kV, table 1) of the new oil (SAE 5W30, DPF). Table 1 shows that all samples of oil analyzed show a decrease compared to that value, the largest decrease reaching the percentage of 45%. And in this case, the causes are related to the unwanted process of oil contamination (dilution with diesel, the appearance of insoluble, soot, water) and its oxidation. The water present in the engine oil is a potential pollutant and a major engine hazard, even in small amounts. Unfortunately, with any internal combustion engine, traces of water in its oil seem inevitable. Normally, for removal, the causes will be sought both in the direction of the internal leaks of the engine and by interpreting the causes of the appearance of condensation.

Another way in which the lubricant could lose its viscosity is by breaking the oil molecules due to high operating temperatures. Oils can also contain additives that improve the viscosity of the oil. During use, the additives may decompose and break, causing the oil viscosity to decrease.

Decreasing the kinematic viscosity is more harmful than increasing it. By decreasing compared to the reference, the thickness of the oil film will be reduced, with the occurrence of the risk of blocking the bearings [18]. The main cause of the decrease in viscosity is the dilution of the oil with light fuel, the addition between two technical revisions with oil with lower viscosity or its contamination with cleaning liquids.

The results for the size of the dispersion, SD, are good and very good, and that demonstrates the ability of the oil to hold insoluble particles in suspension.

Regarding the identification of the causes that led to the modification of the physicochemical properties of the analyzed oil, the following can be mentioned:

- defective injectors, excessive idling and running with frequent starts - stops tend to increase the amount of fuel that can pass near the piston rings to the oil pan;
- modern diesel engines are equipped with a diesel particulate filter (DPF), which is periodically subjected to regeneration processes, by an additional injection of diesel into the engine cylinders [19], in some cases or by external dosing of fuel, in other cases; the fuel is injected into the combustion chamber, but the combustion is not deliberately carried out to push this fuel to the exhaust system, where it will be used to burn the ash and particles deposited in the DPF; during this process, it is possible that due to the tolerances between the piston and the engine block, a certain amount of fuel may mix with the oil from the lower crankcase of the engine,
- also the fuel pumps are lubricated with oil, which could be another source of diluting the fuel in oil; as mentioned above, the dilution of the fuel in the oil results in a decrease in the viscosity of the oil, which could lead to metal-to-metal contact between the components, thus increasing wear; the corrosivity of the oil also increases.

The parameters that were taken into account for the study have undesirable changes during the turnover between two technical revisions imposed by the car manufacturer. Their changes are not major and here it can be remembered that the samples were taken from car engines relatively new (years of manufacture: 2019 – 2020), without significant wear. We consider that the situation would be different if oil samples were taken and analyzed from older engines, with higher total turnover and therefore, with a higher degree of wear.

The graphs are not conclusive in this kind of approach and for the issuance of maintenance recommendations, we have to look at the whole analysis chain for a single test, considering that the vehicles did not work in the same conditions and therefore cannot be compared. In future projects, we will make analyzes on vehicle fleets with the possibility of clear/detailed control of the operating conditions between oil exchanges.

5. Conclusions

The structure of the maintenance activity has changed continuously, finding the development of the activity of strict supervision of the technical condition of the vehicles, with direct consequences regarding the volume of works to correct the technical condition, which is continuously reduced. Thus, the major importance of a modern preventive maintenance system that requires real-time data and immediate feedback is emphasized, so that the reliability engineers can make rapid maintenance decisions. Also, such studies based on the analysis of the oil used for lubrication can identify much earlier the wear degree and potential defects that may occur in the operation of an i.c.e. and remedy before they become imminent and resource consuming of any kind.

Monitoring the condition of the oil in order to issue preventive maintenance plans for a new or old engine can certainly lead to a decrease in the cost of ownership of car owners and not only.

On the other hand, when it comes to car fleets, one of their major interests is to reduce maintenance and repair costs. For this segment, the authors of this paper can propose the present method by periodically analyzing engine oils between two technical revisions, knowing clearly: the oil used (the same for the entire fleet), the number of km traveled with the oil analyzed, the total number of km of the car, year of manufacture of the car, amount of oil completed / type of oil completed and even driving style.

Based on the results obtained, the degree of engine wear can be assessed, and preventive maintenance recommendations can be developed in the period between engine technical overhauls to efficiently manage costs for the entire car fleet.

Author Contributions

The authors have an equal contribution to the work presented in this article.

References

[1] Rammohan A 2016 *Engine's lubrication oil degradation reasons and detection methods: A review* (Journal of Chemical and Pharmaceutical Sciences, Volume 9 Issue 4)
[2] D'Abramo P Maccioni A Pasetti G Tinfena F 2017 *Advanced Sensor Solutions for Automotive Applications* (Wideband Continuous-time ΣΔ ADCs, Automotive Electronics and Power Management)
[3] CIMAC 2011 *The International Council on Combustion Engines, Used engine oil analysis – user interpretation guide*, No. 3
[4] Stephen S 2019 *Advantages of Using Real-Time, In-Line Oil Sensors* (Poseidon Systems, Machinery Lubrication)
[5] Harish Shinde Anand Bewoor 2018 *Capacitive sensor for engine oil deterioration measurement* (AIP Conference Proceedings)
[6] Fan Cheng Takayuki Shibata Yoshifumi Aoki Hiroshi Hirata 2019 *Development of an Oil Degradation Sensor Based on Detection of Free Radicals* (JSAE/SAE Powertrains, Fuels and Lubricants)
[7] Harish Shinde Anand K Bewoor 2020 *Viscosity Assessment for Quantifying Engine Oil Deterioration Using New Cost-effective and Handy Fluid Property Sensor* (International Journal of Advanced Science and Technology, Vol. 29, No. 7)
[8] Stephen S 2019 Advantages of Using Real-Time, In-Line Oil Sensors (Poseidon Systems, Machinery Lubrication)
[9] Paul Farless 2021 *The DOs and DONT's of Used Oil* (Poseidon Systems, Machinery Lubrication)
[10] Michael C. Brown 2015 *Why You Should Regenerate, Reclaim or Recycle Used Oil* (Poseidon Systems, Machinery Lubrication)
[11] DIRECTIVE 2008/98/EC OF THE EUROPEAN PARLIAMENT AND OF THE COUNCIL 2008 *on waste and repealing certain Directives*
[12] HOTARARE nr. 235 2007 privind gestionarea uleiurilor uzate (MONITORUL OFICIAL nr. 199 / 2007
[13] Isuru Premaratne A J M Dayawantha 2017 *Vehicle service date prediction system using lubricant viscosity degradation pattern* (Jaipur, India: International Conference on Computer, Communications and Electronics)
[14] Dana Kareem Hameed Kameran Ali 2020 *Analysis of 0W-20 Totachi Brand Oil to Determine the Rate of Oil Deterioration* (UKH Journal of Science and Engineering, Volume 4, Number 2)
[15] Zaharia C Niculescu R Iorga Ducu C Clenci A Aron B 2016 *Diagnosing the operation of a locomotive diesel engine based on the analysis of used oil in the period between two technical revisions* (Brașov, Romania: International Congress of Automotive and Transport Engineering, 319-327, Editor Springer)
[16] Landowski B Baran M 2019 *Analysis of selected results of engine oil tests* (MATEC Web of Conferences, 18th International Conference Diagnostics of Machines and Vehicles)
[17] Noria Corporation 2021 *Density's Role in Lubricant Performance* (Poseidon Systems, Machinery Lubrication)
[18] Story R 2019 *Why an Oil's Viscosity Drops* (Poseidon Systems, Machinery Lubrication)
[19] Tingjun H Teng H Xuwei L Bin C 2015 *Impact of Fuel Injection on Dilution of Engine Crankcase Oil for Turbocharged Gasoline Direct-Injection* (SAE 2015 World Congress & Exhibition)

Results concerning manufacturing of tubular car components using hydroforming and hydroperforation processes

D I Poiana, I Ionel, R M Popa and I Bordeaşu

Politehnica University of Timisoara, Faculty of Mechanical Engineering, Piata Victoriei no. 2, RO-300006, Timisoara, Romania

ioana.ionel@upt.ro

Abstract. The purpose of the paper is to highlight the advantages that the automotive industry gains by using hydroforming and hydroperforation processes in automobile manufacturing. In recent years, progress has been made concerning the equipment and tools needed to carry out the hydroforming and hydroperforation processes. As these machines and tools have been perfected and produced on a large scale, becoming cheaper and more cost-effective were main scopes of the development, increasing the mass production, supporting the automotive industry, as well, is another goal. The improvement of the tools used and the equipment determined that these processes have to be extended for a quite large number of components of the vehicle, in the production of large series. A comparative analysis and advantages of the technology are presented regarding the impact on the environment of these processing, hydroforming, and hydroperforation processes, in relation to other traditional processing processes.

1. Characteristics of hydroforming and hydroperforation processes for the tubular parts

1.1. Principle scheme of the hydroforming and hydroperforation processes

The hydroforming and hydroperforation processes are based on the use of a special die, which exerts deformation forces on the half-finished product, with the help of hydraulically operated pistons and punches, for the deformation of a tubular half-finished product using a hydraulic pressure inside this tubular half-finished product.

The term hydro-piercing appears in the specialized literature for hydroperforation [1, 2] as well.

In general, when applying the hydroforming process, certain pre-forming operations of the half-finished products subjected to hydroforming are performed, also with the same hydroforming equipment.

Figure 1 shows the principal elements and components of de tool for hydroforming and hydro perforating process for a tubular work piece: upper closing tool; axial punch; work piece (tubular); lower tool; the forces that appear F_{close} and F_{axial}.

IOP Conf. Series: Materials Science and Engineering 1220 (2022) 012038 doi:10.1088/1757-899X/1220/1/012038

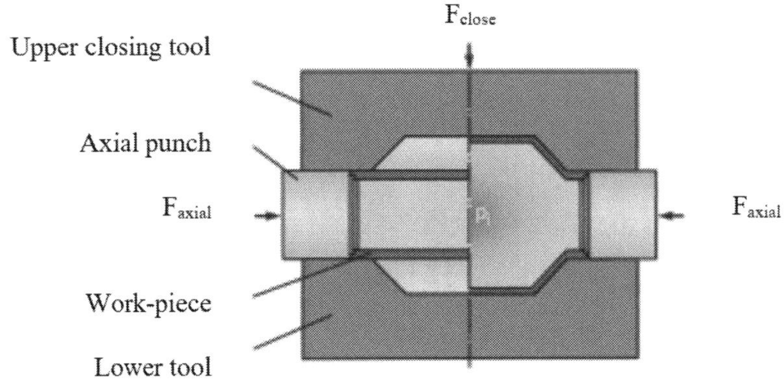

Figure 1. Principle of hydroforming, according to [3].

Figure 2 shows four ways in width after hydroforming the tubular part, holes can be made in the tubular part through the hydroperforation process. These four situations are: a) with the complete inward detachment of the slug inside the piece; b) with the slug remaining attached to the wall of the piece on the inside piece; c) with the formation of an inward flange of the wall; d) with the detachment of the de slug in a niche, towards the outside de piece.

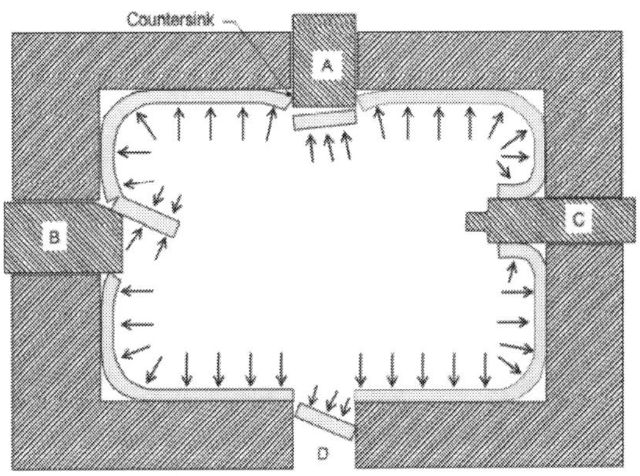

Figure 2. Cross-section schematic illustrating inward hydroperforation with uncoupled slug (A), inward hydroperforation with attached slug (B), hydroperforation with flanged land for self-threading screws (C), outward hydroperforation with uncoupled slug (D) [1].

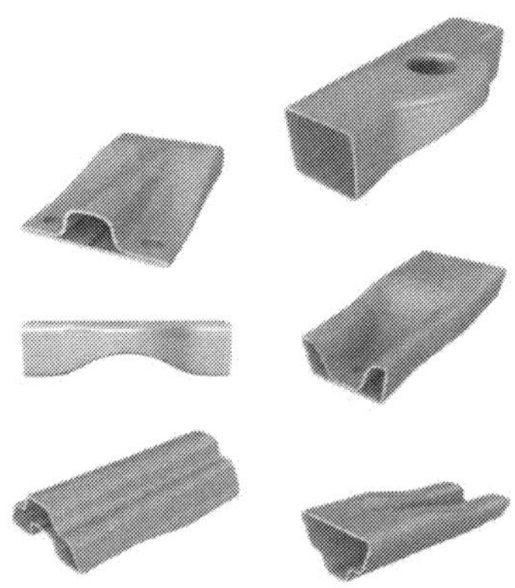

Figure 3. Different cross-sections, according to [1].

In most cases, before the actual hydroforming, a pre-formation of the tubesoccur, especially of the cross sections, occurs. From the point of view of the sections that can be made, in figure 3, there are presented different types of sections of the executed parts.

1.2. Steps taken by hydroforming of tubular half-finished parts

In general, the hydroforming process goes through several specific stages:

a) Extruded profile; b) Cutting; c) Bending; d) Hydroforming; e) Machining; f) Cleaning; g) Heat treatment; h) Surface treatment; i) Final component.

Figure 4 shows the technological stages of hydroforming on a tubular part: 1 – introduction of the half-finished part in the die; 2 – closing the die and sealing the die; 3 – introduction of the liquid under pressure inside the tubular half-finished product; 4 – actuation of the axial pistons; 5 – forming the finished part by pressing with axial force F_{axial} ; 6 -removing the finished part.

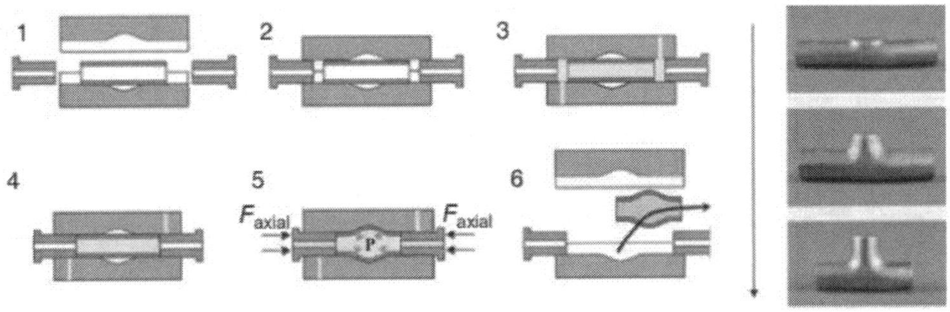

Figure 4. Steps in a typical hydroforming process shown with a small tubular part [1].

Figure 5 shows schematically, in this section, obtained by hydroforming the tubular part of figure 4, by deformation of the die.

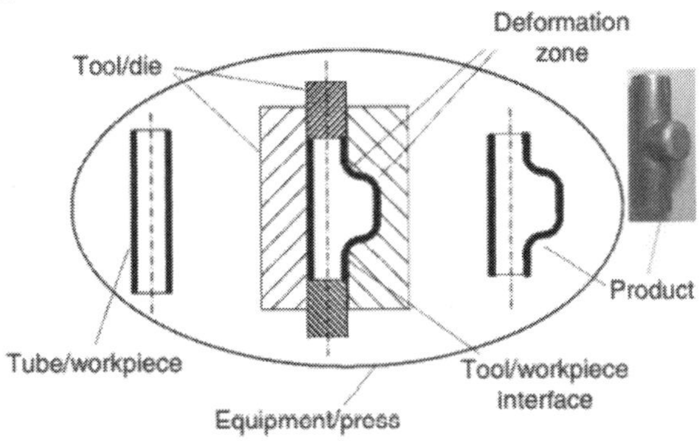

Figure 5. Achieving the formation of a part by deforming a tubular half-finished product [1].

2. Car components obtained by hydroforming and hydroperforation

2.1. Types of tubular parts in the body and chassis of the car

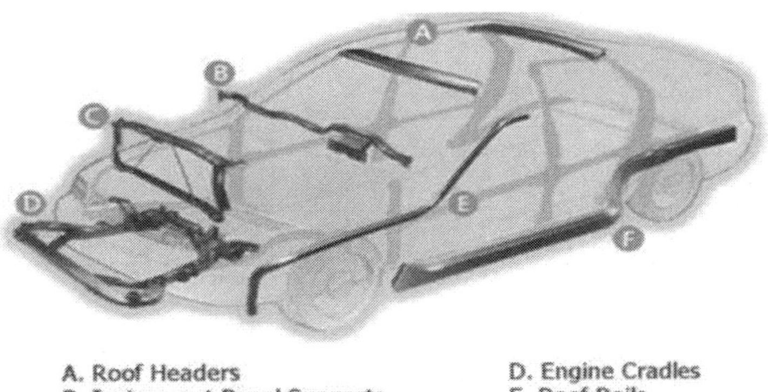

Figure 6. Tubular parts positioned at the car body, [1].

By hydroforming, certain components of the car can be made, starting from tubular half-finished products made of pipes. Some characteristic types of car components, shown in figures 6, 7, 8, 9 and

10 [1], are: roof header (figure 6, A); instrument panel supports (figure 6, B); radiator supports (figure 6, C); engine cradle (figure 6, D); roof rail (figure 6, E), frame rail (figure 6, F); rollover bar protection systems (figure 3, b); steel camshafts (figure 3, a), frame rails; rear axle frames; radiator frames; suspension cross members; drive shafts; gear shafts; exhaust parts; engine tubes; pressure tubes.

Figures 7 and 8 show the way how the body is assembled with certain components obtained by hydroforming and hydroperforation. Figures 9 and 10 present other examples from the automotive manufacturing industry.

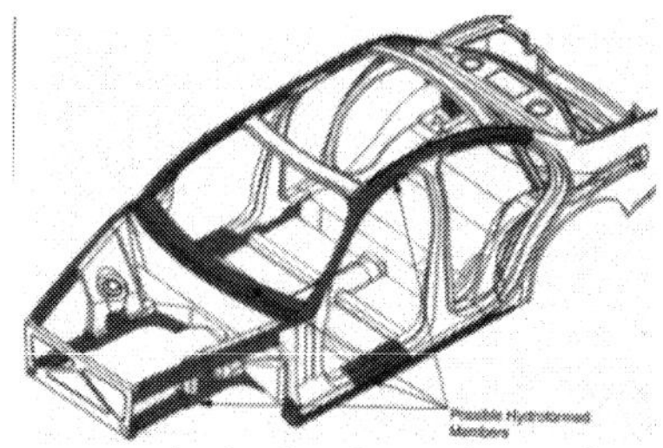

Figure 7. The hydro formed members are integrated into a body structure [2].

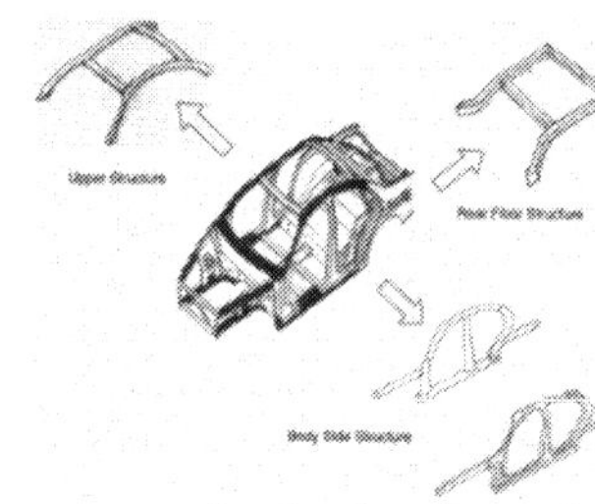

Figure 8. The hydro formed body structure of figure 7, shown as assemblies according to [2].

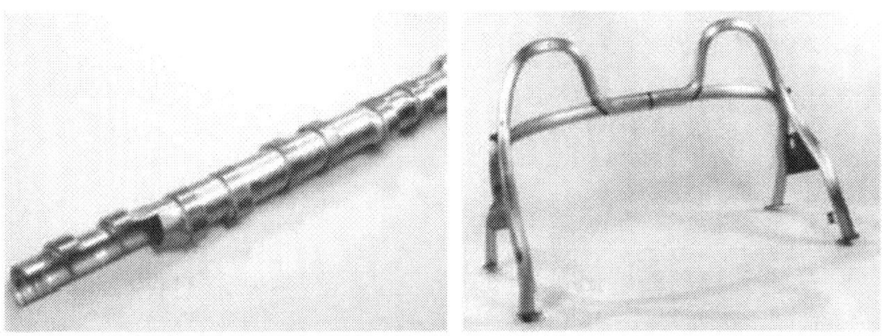

a b

Figure 9. Hydroforming (tubular) parts for various automotive applications: a – hydro formed steel camshaft (BMW 3.0 L DOHCI-6 Ward's 2005) offers 28 to 50% mass savings; b - 2-piece roll bar for a convertible car, according to [1].

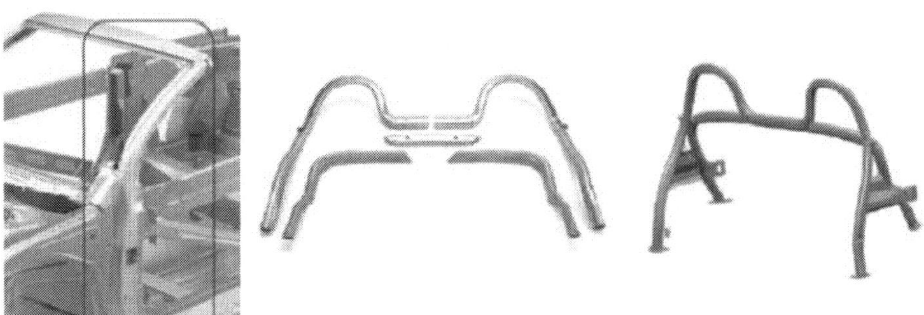

a b

Figure 10. Car components: a – hydro formed pillar in Volvo C 70); b - rollover bar protection system in Porsche Boxter, according to [1].

2.2. Engine cradle and structure manufactured by hydroforming and hydroperforation. Stages of the process

Figure 11 shows the transformations undergone by a pipe-type half-finished product to achieve an engine support, with hydroforming and hydro perforation performed at the end of the hydroforming. The technological stages are: 1 – half-finished initially from the pipe; 2 – bending the pipe; 3 – preforming the piece of the sections along the piece; 4 -making the finished piece by making holes in the formed piece.

Figure 12 shows the hydro formed structure of the car.

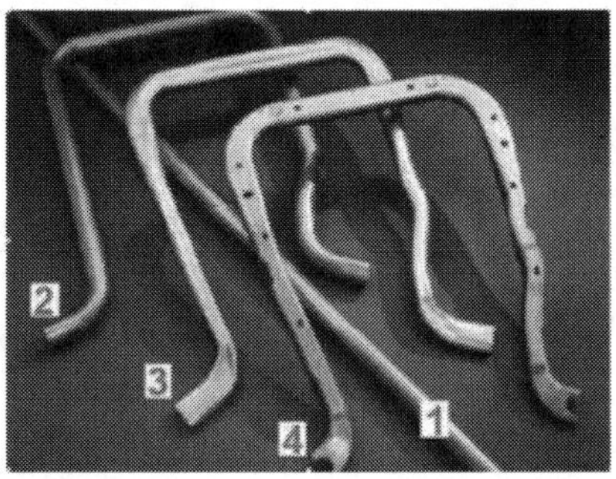

Figure 11. Hydro-formed engine cradle: the initial straight tubular blank (1), bent (2), preformed (3), hydro-formed and hydro-perforated (4), according to [1].

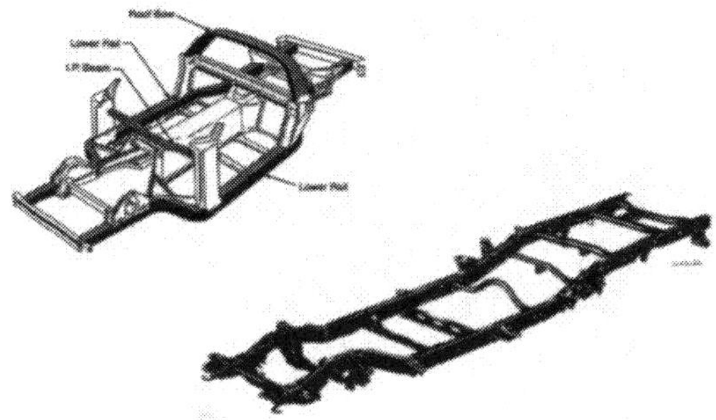

Figure 12. Hydro-formed structure in production (Courtesy General Motors Corp.) according to [2].

Figure 13 shows the hydroforming tooling for an engine cradle (figure 13, a) and a hydroforming tooling, for a side rail of an SUV chassis (figure 13, b).

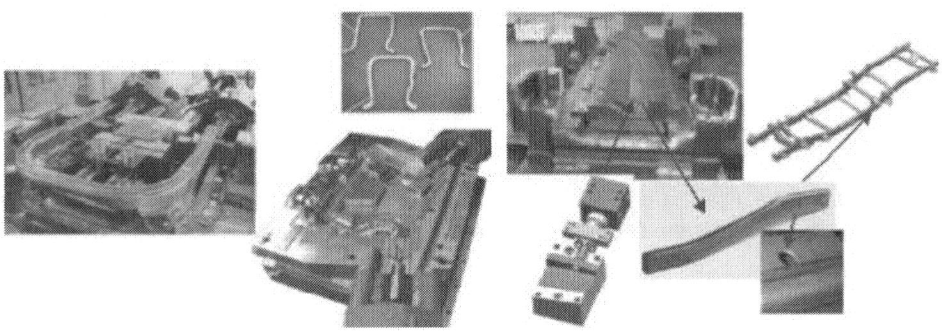

a b

Figure 13. a – Hydroforming tooling for an engine cradle (courtesy of Siempelkamp Pressen Systeme – SPS); b – hydroforming tooling for a side rail of an SUV chassis. Detail shows the close-app of a hole piercing tooling subcomponent, according to [1].

2.3. Advantages of using hydroforming in the automotive industry

Many processing processes are used in the manufacture of automobiles. One of the modern processing processes developed in the automotive industry is hydroforming together with hydroperforation. Hydroforming is used to obtain components of the car with relatively complex shapes, replacing in many cases certain subassemblies with a single part obtained by hydroforming. Hydroperforation is used to make holes in the parts obtained by hydroforming. In general, hydroperforation is used at the end of hydroforming operations [1].

The execution of the holes in the parts obtained by hydroforming can be done by other modern procedures, such as laser perforation [1, 2] and thermal erosion drilling [4, 5, 6]. The hydroforming and hydroperforation processes have many advantages, but also some disadvantages. Compared to traditional processing processes, however, hydroforming and hydroperforation have important advantages for the automotive industry.

The advantages of hydroforming and hydroperforation for the automotive industry are:

a) the use of hydroforming has allowed to significantly reduce the weight of half-finished products for certain components of the car body and chassis, as different types of thin-walled pipes can be used as half-finished products, without the need for further processing of the thickness of the pipe wall;
b) the number of components of a part that was previously obtained by the assembly is reduced, and by hydroforming a piece with a complex shape can be obtained following the modeling by hydroforming;
c) lower costs of the components and assemblies of the car component are achieved by reducing the number of operations and using smaller quantities of materials;
d) increased strength and rigidity of the finished parts obtained by hydroforming is obtained;
e) the costs of the necessary equipment are reduced:
f) high dimensional stability of the obtained finished parts is obtained;
g) holes with small positioning tolerances and good dimensional accuracy are obtained;
h) use of light materials, magnesium and aluminum alloys, contributes to reducing the weight of the car;
i) the possibility of automating hydroforming production lines by using process computers.

The main disadvantage of the hydroforming process is the cost of the equipment. In recent years, by increasing the demand for this equipment and producing it in large quantities in the automotive industry, considerable reductions in these costs were performed [1].

3. Research on the impact of the hydroforming and hydroperforation process on the environment, compared to other processes

3.1. Determining factors that produce negative effects on the environment and on the worker

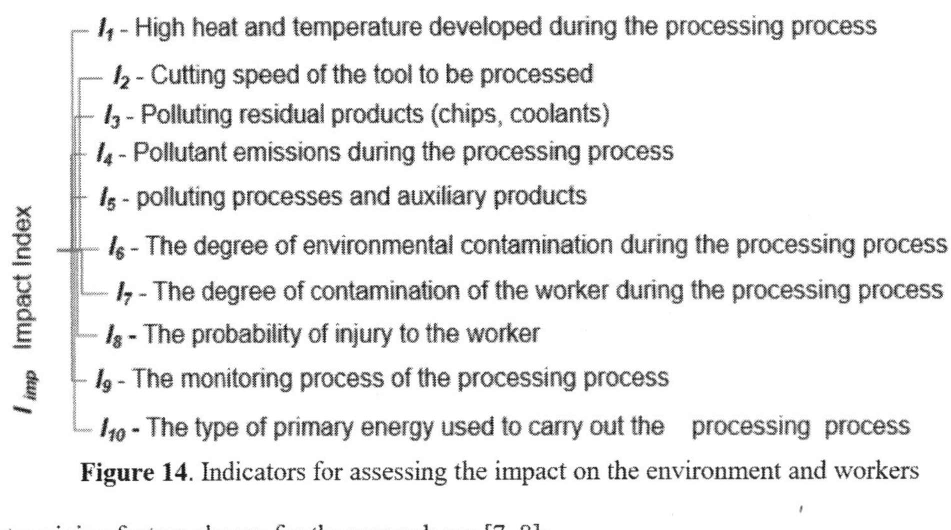

Figure 14. Indicators for assessing the impact on the environment and workers

The determining factors chosen for the research are [7, 8]:

1) high heat and temperature developed during the processing process,
2) cutting speed of the tool to be processed,
3) polluting residual products (chips, coolants),
4) pollutant emissions during the processing process,
5) polluting processes and auxiliary products,
6) the degree of environmental contamination during the processing process,

IOP Conf. Series: Materials Science and Engineering 1220 (2022) 012038 doi:10.1088/1757-899X/1220/1/012038

7) the degree of contamination of the worker during the processing process,
8) the probability of injury to the worker during the processing process,
9) the monitoring process of the processing process,
10) the type of primary energy used to carry out the processing.

For each determining factor, an indicator was associated from I_1 to I_{10}. Each indicator can take discrete values between 0 and 5. Figure 14 shows the ten indicators. To obtain the value of the impact index denoted I_{imp}, the values are assigned to the ten indicators, following the analysis of the considered processing process. The sum of the values of the indicators is calculated:

$$I_s = I_1 + I_2 + I_3 + I_4 + I_5 + I_6 + I_7 + I_8 + I_9 + I_{10} \tag{1}$$

where: I_s can take values in the interval $0 \leq I_s \leq 50$.

The impact index I_s is calculated by dividing the sum of the indicators by 10.

$$I_{imp} = \frac{I_s}{10} \tag{2}$$

where: I_{imp} takes values between the following interval: $0 \leq I_{imp} \leq 5$.

The ten indicators taken into account when determining the impact index I_{imp} with the corresponding discrete values, from the best standard 0, without pollution, to the worst standard 5, which is very polluting and/or very dangerous, are further calculated. The values that are assigned based on the analysis of the processing process, which is established based on the algorithm, are presented in table 1.

Table 1. The possible values of the indicators and the calculation algorithm of the impact index I_{imp}.

	Indicators / possible values (0 ... 5)
I_1	heat and temperature developed during the processing process / 0 = 0-50°C, 1 = 51-100°C, 2 = 101-200°C, 3 = 201-300°C, 5 = over 301°C
I_2	cutting tool speed / 0 = very high over 1001 m / min, 1 = high 701-1000 m / min, 2 = medium 401-700 m / min, 3 = low 101-400 m / min, 5 = very low 0-100 m / min
I_3	polluting residual products (chips, coolants) / 0 = without residual products, 1 = residual products with very low pollution, 2 = residual products with medium pollution, 3 = residual products with very high pollution, 5 = hazardous waste products
I_4	pollutant emissions during the processing process / 0 = zero emissions, 1 = with very low pollution, 2 = medium pollution emissions, 3 = very pollution emissions, 5 = hazardous emissions
I_5	polluting processes and auxiliary products / 0 = non-polluting, 1 = very low degree of pollution, 2 = medium degree of pollution, 3 = high degree of pollution, 5 = very high degree of pollution
I_6	degree of environmental contamination during the processing process / 0 = no contamination, 1 = low contamination, 2 = medium contamination, 3 = high contamination, 5 = dangerous contamination
I_7	the degree of contamination of the worker during the processing process / 0 = no contamination, 1 = small contamination, 2 = medium contamination, 3 = high contamination, 5 = dangerous contamination
I_8	the probability of injury to the worker during the processing process /

	Indicators / possible values (0 ... 5)
	0 = not dangerous, 1 = small dangerous, 2 = medium dangerous, 5 = very dangerous
I_9	processing process monitoring system / 0 = with complex sensors and automation, 1 = with complex sensors and without automation, 2 = with simple sensors, 3 = monitoring without sensors, 5 = without monitoring
I_{10}	the type of primary energy used in the processing / 0 = renewable, 1 = low carbon footprint, 2 = medium carbon footprint, 3 = high carbon footprint, 5 = very high carbon footprint
I_s	$I_s = I_1 + I_2 + I_3 + I_4 + I_5 + I_6 + I_7 + I_8 + I_9 + I_{10}$
I_{imp}	Impact index $\boldsymbol{I_{imp} = I_s / 10}$

3.2. Calculation of the impact index for the hydroforming process
The analysis of the hydroforming process is materialized by the following results:

1) high heat and temperature developed during the processing process
   - for hydroforming temperature is lower to 50°C, I_1= 0,
2) cutting speed of the tool to be processed
   - is very high, over 1001 m / min, I_2= 0,
3) polluting residual products (chips, coolants)
   - is the residual product with very low pollution, I_3= 1,
4) pollutant emissions during the processing process
   - is zero emissions, I_4= 0,
5) polluting processes and auxiliary products
   - is a very low degree of pollution of the working fluid, I_5= 1,
6) the degree of environmental contamination during the processing process
   - no contamination, I_6= 0,
7) the degree of contamination of the worker during the processing process
   - small contamination of the working fluid, I_7= 1,
8) the probability of injury to the worker during the processing process
   - not dangerous, I_8= 0,
9) the monitoring process of the processing process
   - in general, with complex sensors and without automation, I_9= 1,
10) the type of primary energy used to carry out the processing
   - in general, medium carbon footprint, I_{10}= 3,

The sum of the values of the indicators is calculated: I_s= 7.

Impact index is thus $\boldsymbol{I_{imp} = 0.7}$.

3.3. Comparative analysis with other processing processes
The hydroforming process was compared with the other processing process, based on the value calculated for the impact index for each of these processes. Synthetically, the results are presented in table 2.

It is observed that of all processes, the least negative impact on the environment has the hydroforming process, implicitly also the hydroforming process, which occurs under the same conditions regarding the impact determinants. This means that hydroforming and hydroperforation are processes that can be used and even improved in the future, in manufacturing processes, being sustainable processing processes.

Table 2. Impact index values for the considered processing processes.

Road sign	Drilling	Panching with associated edges	Single edge punching	Hydroforming	Ultrasonic perforation	Laser perforation	Perforation by electrical erosion	Perforation by thermal erosion
I_1	2	1	1	0	1	1	0	3
I_2	5	3	3	0	1	3	1	1
I_3	3	1	1	1	1	1	5	0
I_4	2	0	0	0	0	0	1	1
I_5	2	1	1	1	1	1	2	1
I_6	2	1	1	0	1	1	2	0
I_7	3	1	1	0	1	1	3	0
I_8	3	2	2	0	1	1	3	2
I_9	5	3	5	2	3	5	2	2
I_{10}	3	3	3	3	3	3	3	3
I_s	30	16	18	7	13	17	22	13
I_{imp}	**3.0**	**1.6**	**1.8**	**0.7**	**1.3**	**1.8**	**2.2**	**1.3**

4. Conclusions

Regarding the advantages, which the hydroforming and hydroperforation processes used in the manufacture of the car present, one can be concluded:

1) There are sustainable processing processes;
2) They contribute to reducing the weight of half-finished products for certain components of the car body and chassis, leading to material savings;
3) They determine the reduction of the number of components, obtaining parts in one piece, replacing certain complex parts obtained by other processes by the assembly;
4) They might reduce the costs of some components and that of the vehicle, as a whole;
5) They allow obtaining high resistance and rigidity of the finished parts obtained by hydroforming;
6) They contribute to reducing equipment costs;
7) Through hydroperforation, holes with good dimensional accuracy are obtained;
8) A good geometric and positioning tolerance of the hydro-perforated holes is obtained;
9) High dimensional stability of the finished parts obtained by hydroforming is obtained;
10) The possibility to use light materials, aluminum and magnesium alloys, reducing the weight of the car, is achieved.

Acknowledgement
The paper is accomplished in the frame of the PhD research activity of the first author, at the Politehnica University of Timisoara.

References
[1] Koç M 2008 *Hydroforming advanced manufacturing*, Woodhe Publishin and Maney Publishing on behalf of The Institute of Materials, Minerals & Mining, (CRC PressBoca Raton Boston New York Washington, DC, 2008).
[2] Singh H 2003 *Fundamentals of Hydroforming*, Copyright © 2003 by the Society of Manufacturing Engineers 987654321, 2003
[3] Neugebauer R 2007 *Hydro-Umformung*, ISBN-13978-3-540-21171-6, Springer Berlin, Heidelberg, New York, http: / dnb.ddb.de, Springer-Verlag Berlin Heidelberg, 2007
[4] Kumar E and Hynes RJ 2019 Thermal drilling processing on sheet metals: A review, *International Journal of Lightweight Materials and Manufacture*, journal homepage: https: //www.sciencedirect.com/journal/international-journal-of-lightweight-materials-and - manufacture, Journal homepage: www.elsevier.com/ locate / jestch, pp193-205 (accessed 01.06.2021)
[5] Kumar R and Hynes RJ 2019 Prediction and optimization of surface roughness in thermal drilling using integrated ANFIS and GA approach, *Engineering Science and Technology an International Journal*, Journal homepage: www.elsevier.com/locate/jestch,(accessed 01.06.2021)
[6] Kumar R et al 2019 Multi-objective optimization of green technology thermal drilling process using gray-fuzzy logic method, *Journal of Cleaner Production*, Journal homepage: www.elsevier.com/ locate / jclepro (accessed 01.06.2021)
[7] Pellegrini G and Ravasio C 2019 A sustainability index for the micro-EDM drilling process, *Journal of Cleaner Production*, https://doi.org/10.1016/j.jclepro.2019.119136, (0959-6526/© 2019 Published by Elsevier Ltd.)
[8] Byrne G and Scholta E 1993 Environmentally Clean Machining Processes-A Strategic Approach, *Annals of the CIRP* **42**/1/1993, 14 January 1993

Analysis of the cylinder head for a spark ignition engine using zonal cooling principle

T M Ulian, E Rakosi and A I Donțu

Technical University "Gheorghe Asachi" of Iași, Department of Mechanical and Automotive Engineering, Blvd. Dimitrie Mangeron, No. 43, Iași, România

uliantudor@yahoo.fr

Abstract. In the paper, it is performed a detailed thermo-structural analysis of the cylinder head for a spark ignition engine using the zonal cooling principle. The zonal cooling principle assumes two cooling circuits, one for the cylinder head and one for the cylinder block and a continuous adjustment of cooling temperature in accordance with the working principle. By using this principle, it is considered the possibility of using higher values of the volumetric compression ratio, leading to the increase of the engine performance. Thus, the reduction of the temperature of the fresh air fuel mixture in the intake channel and the lower temperature of the fresh air fuel mixture, when getting in contact with combustion chamber walls, leads to a delay of self-ignition and finally the detonation can be suppressed. Due to the possibility of using a higher compression ratio, the thermo-structural analysis was performed using ANSYS Workbench.

1. Introduction

The thermo-structural analysis of the cylinder head was performed due to the possibility of using a higher compression ratio by using ANSYS Workbench. The 3d model of the cylinder head, which will be used for FEA analysis, is presented in figure 1 [1].

Figure 1. Cylinder head block [2].

As everyone knows, cylinder heads are usually made of the aluminum alloy due to the lower weight and higher heat transfer coefficient [3].

After importing the geometry into ANSYS, the first step is to define the material properties of the cylinder head [4]. Therefore, the cylinder head uses an Al-Si-Cu alloy with the following chemical composition presented in table 1.

Table 1. Chemical composition % of grade ENAC-AlSi8Cu3.

Fe	Si	Mn	Ni	Ti	Cu	Pb	Mg	Zn	Sn	Others
max. 0,8	7,5 - 9,5	0,15 - 0,65	max 0.35	max 0.25	2 - 3,5	max 0,25	0.05 - 0.55	max 1.2	max 0.15	total 0.25

The aluminum alloy used for the cylinder head has high strength at the intermediate temperature range and has also good castability properties.

The mechanical properties of this material are presented in table 2.

Table 2. Material property parameters.

Material	Tensile strength R_m [MPa]	Yield Strength $R_{p0,2}$ [MPa]	Thermal Conductivity λ [W/m^2·K]	Thermal Expansion α [1/°C]
ENAC-AlSi8Cu3	140	120	103	$21 \cdot 10^{-6}$

2. Thermo-structural analysis

The thermo-structural analysis was done in two steps. The first step implies a simulation to determine the influence of the coolant liquid on the temperature of the cylinder head combustion chamber wall. The second step implies a simulation to determine the mechanical stress distribution, total deformation and safety factor. We mention the fact that in this paper, there are presented the simulation results obtained for the extreme values of the coolant temperature, namely T_{lr} = 50 [°C] and T_{lr} = 90 [°C].

The flow chart of the thermo-structural analysis will be presented in figure 2 [1]:

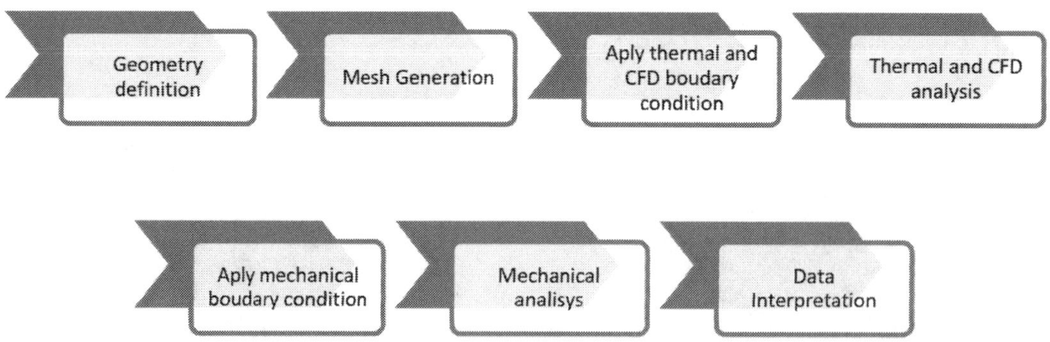

Figure 2. Thermo-structural analysis flow chart.

2.1. Thermal simulation

During the simulation, for verifying the temperature dependence hypothesis of the cylinder head inner surface which was demonstrated in the previous paper [5], the zonal distribution of the temperature values on the inner surface of the combustion chamber, T_{chi}, was obtained, for various values of the coolant temperature, T_{lr}. These values are represented in figure 3 (a), (b).

Thus, the reduction of the temperature values on the inner surface of the combustion chamber of the fresh air fuel mixture, when getting in contact with combustion chamber walls, leads to a delay of self-ignition and finally the detonation can be suppressed.

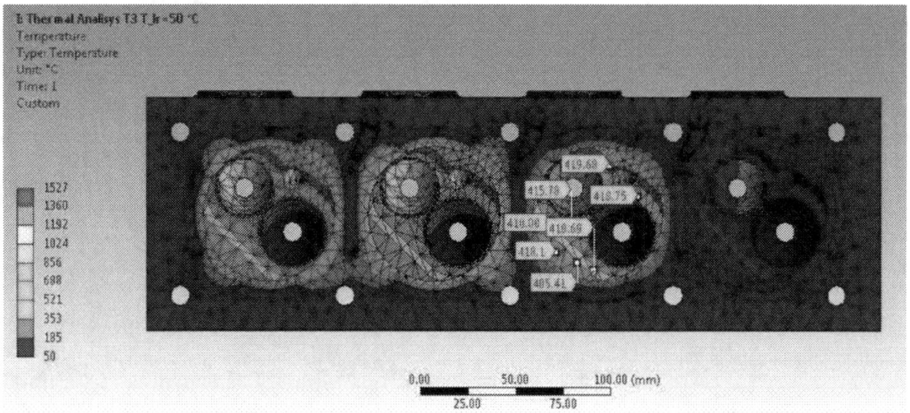

(a)

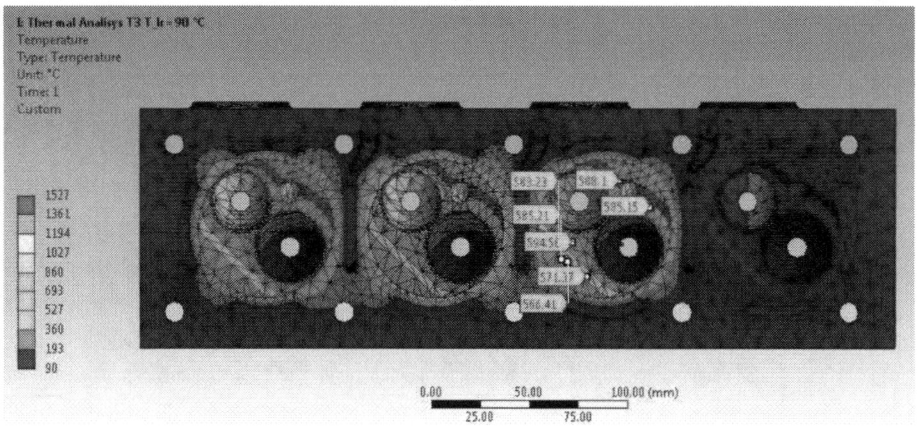

(b)

Figure 3. (a), (b) Zonal distribution of the temperature values on the inner surface of the combustion chamber, T_{chi} for various values of the coolant temperature, T_{lr} [1].

Based on these temperature values obtained, was determined an average value of the inner surface of combustion chamber temperature, which is represented in figure 4 (a), (b). Thus, for the coolant temperature, T_{lr} = 50 [°C], temperature values on the inner surface of the combustion chamber drop by approximately 27% from the value, T_{chi} = 423,8 [°C], corresponding to the temperature T_{lr} = 90 [°C] at T_{chi} =583,4 [°C] [1].

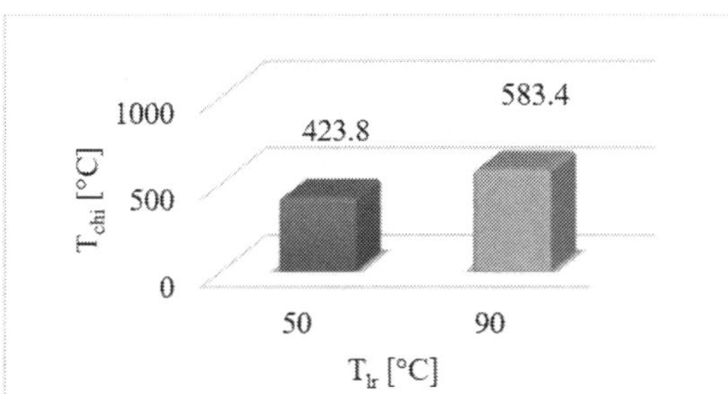

Figure 4. The average temperature values of the combustion chamber inner wall, T_{chi}, depending on the coolant temperature, T_{lr}, determined by simulation [1].

2.2. Structural simulation

As was mentioned, a low coolant temperature ensures the possibility of increasing the compression ratio, ε, from ε = 8,5 to ε = 10,5. This implies the increase of the maximum gas pressure during the combustion process. The variation of the maximum pressure in the cycle, p_z, depending on the temperature of the coolant obtained for different compression ratios is represented, in figure 5.

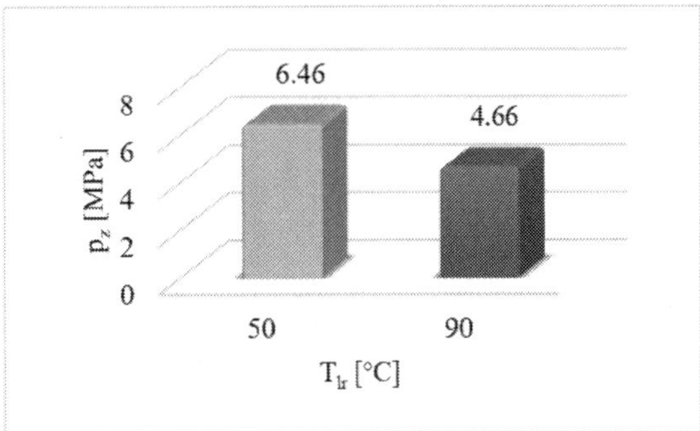

Figure 5. The maximum pressure of the combustion gases, p_z, based on the coolant temperature, T_{lr} determined by simulation [1].

Increasing the maximum combustion gas pressure during the combustion process involves performing a structural analysis of the cylinder head. The results of this simulation regarding the deformations are presented in figure 6.

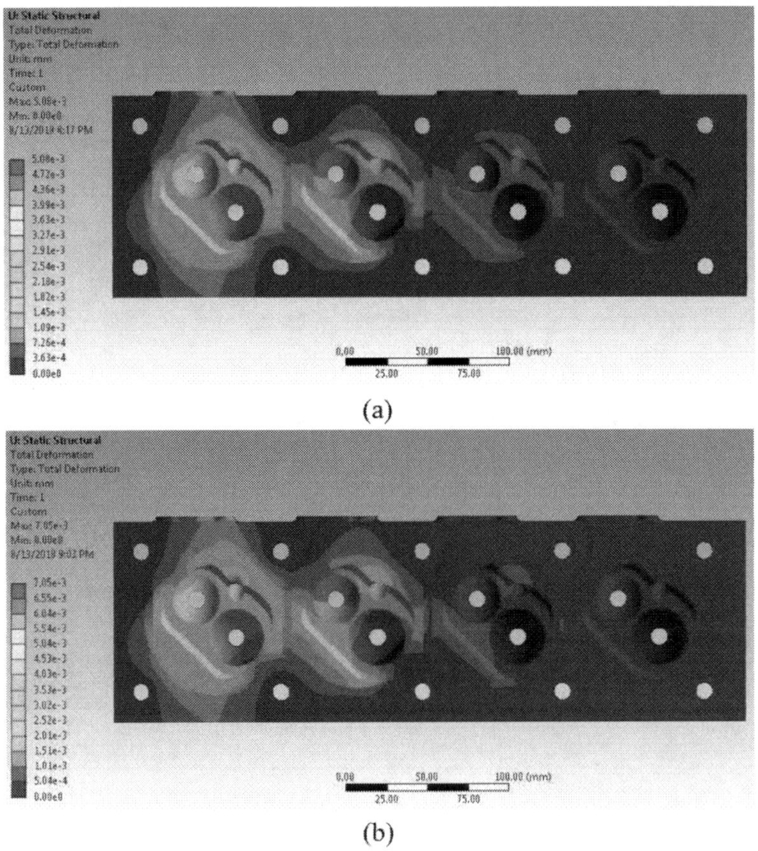

Figure 6. (a), (b) Total deformation of the cylinder head for various combustion gases pressure values [1].

For a general overview of the results, in figure 7 the maximum values of the total deformations of the cylinder head, θ, were centralized for various values of the combustion gas pressure.

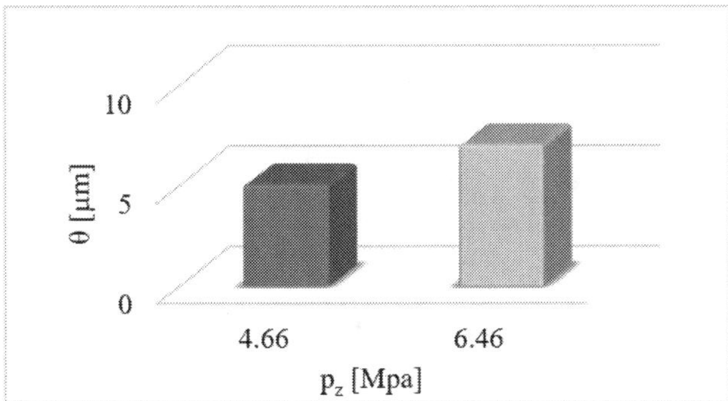

Figure 7. Total cylinder head deformation, θ, for various values of combustion gas pressure, p_z [1].

As can be seen, the values of these deformations are low, between 5 [µm] and 7 [µm] that is a positive aspect of the use of this solution.

Next, in figure 8, it is showed the equivalent stress distribution in the cylinder head structure, for various values of the combustion gas pressure.

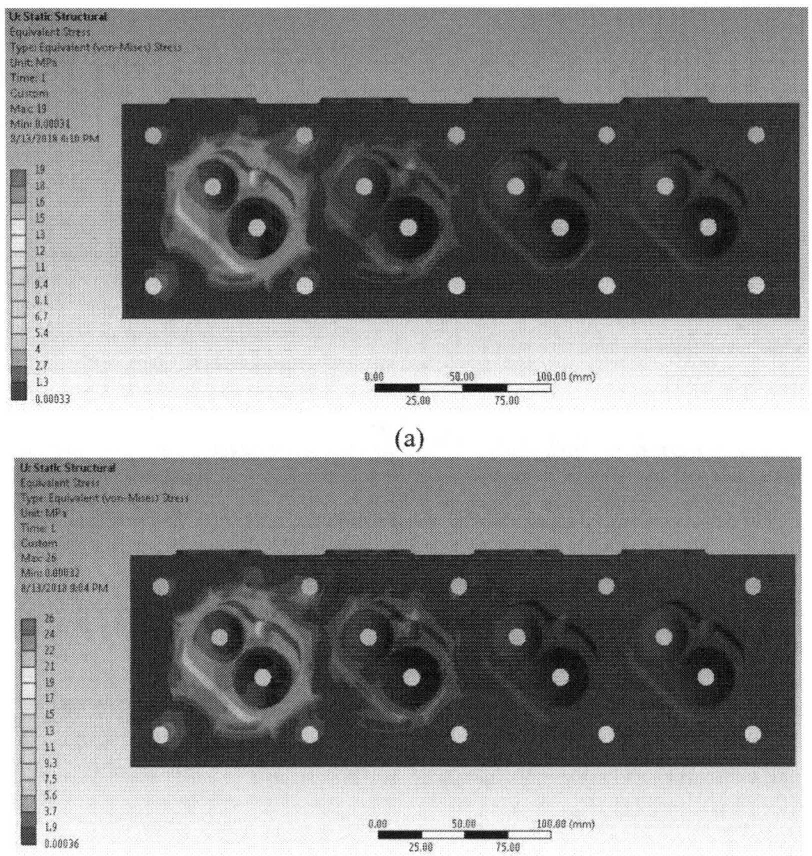

(a)

(b)

Figure 8. (a), (b) Equivalent stress distribution in the cylinder head structure for various combustion gases pressure values [1].

For the situations presented above, in the graph from figure 9, it is shown maximum values of the equivalent stress, σ_{ech}, at various values of the maximum pressure in the combustion cycle.

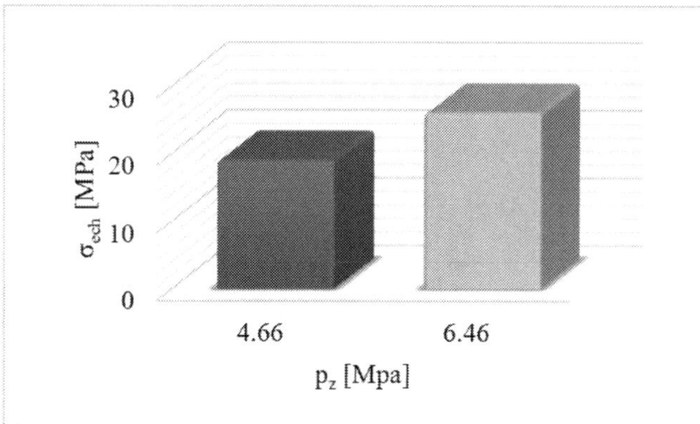

Figure 9. The maximum values of the equivalent stresses in the cylinder head structure, σ_{ech}, for various values of the combustion gas pressure, p_z [1].

In the graph presented above, at the maximum gas pressure in the cylinder of 6.46 [MPa], the maximum equivalent stress, σ_{ech}, appeared at the cylinder head level is about 26 [MPa], a stress that cannot be considered dangerous.

3. Conclusion

The simulations, led to a detailed analysis of the temperature and stress distribution in the cylinder head mass, generated by the operation with higher pressures in the engine cycle due to the possibility of using the zonal thermal regime.

In conclusion, based on the structural finite element analysis performed at the cylinder head level, we can say that the main stress that acting on the tested parts are compression.

The maximum equivalent stresses are located on the inner surface of the combustion chamber, which is not prone to deformations that would cause the engine to malfunction.

References

[1] Ulian T., Contribuții la ameliorarea performanțelor motorului cu aprindere prin scânteie prin utilizarea unui regim termic zonal, teză de doctorat, Universitatea Tehnică "Gheorghe Asachi", Iași, 2020.

[2] Liu LEI, Chen BING-WEI, *Finite Element Analysis Of Thermal Load On Cylinder Head Of Small-Sized Piston Engine*, U.P.B. Sci. Bull., Series D, Vol. 81, Iss. 2, ISSN 1454-2358, 2019

[3] M. Fadaei, H. Vafadar, A. Noorpoor, *New thermo-mechanical analysis of cylinder heads using a multi-field approach*, Scientia Iranica, Transactions B: Mechanical Engineering 18 (2011) 66–74, 2011.

[4] https://www.rupbox.com/material-properties/information/cast-aluminium-alloy-g-alsi8cu3

[5] Ulian T., Vornicu V., Rakosi E., Agape I., The physically-mathematically model for studying the thermal regime of the spark ignition engine concerning the filling coefficient, 11th International Congress of Automotive and Transport Engineering - Mobility Engineering and Environment (CAR), November 08-10, 2017, Pitești, Romania, Book Series: IOP Conference Series-Materials Science and Engineering Volume: 252, UNSP 012076, doi: 10.1088/1757-899X/252/1/012076, 2017.

Study on a test bench of a vehicle rear axle fatigue behavior

V Ionica, I Dumitru, M Bogdan and I Geonea

University of Craiova, Faculty of Mechanics, 107, Calea Bucuresti, no. 107, Craiova, Romania

igeonea@yahoo.com

Abstract. This paper presents a study with finite elements of the rear axle of a car. The virtual model of the rear axle assembly of a vehicle is made. Based on this model we made a study of the state of stresses and strains that occur in the components of the rear suspension. When constructing the analysis model, we considered the bearing conditions, as well as the loads from the real operation of the assembly. The purpose of this finite element analysis is to determine the most requested areas of the rear axle assembly, as well as to compare the results obtained for specific deformations and stresses experimentally with those obtained by finite element analysis. Also, through the analysis with finite elements, we obtained and presented in the paper results regarding the fatigue behavior of the elements of the ensemble.

1. Introduction

Stands for testing vehicle axles are made and presented in the literature. For analysis of the suspension characteristics, the axles of the vehicle are attached and measured on the new extremely dynamic axle test installation from TU Dresden. This test device allows the analysis of static and dynamic suspension properties [1].

The paper [2] deals with the lifetime for new construction of the rear axle of a light commercial vehicle (LCV) that was on the market. Simultaneously, the ADAMS / View model of the rear axle was created. The force-displacement graphs obtained from the bench tests and the ADAMS / View simulation were compared with each other. Fatigue tests were performed under 2.5 g swelling conditions and it was possible to simulate small and large loads. The purpose of this study was to validate the test device and test conditions. Numerous patents have also been issued for the fatigue testing of car axles [3].

The suspension system is designed to have the comfort of driving on the road. Vibrations are not transferred from the tire to the passenger if the suspensions are good. In this case, the most important factor is the vertical acceleration produced in the suspended mass. So, the reduction of vertical acceleration provides better walking comfort. Thus, the analysis of the suspension must be made to determine the vertical acceleration. The paper [4] emphasizes the effect of suspension parameters, i.e. sprung mass, unsprung mass, damping value, suspension spring stiffness and tire stiffness on vertical acceleration. For the purpose of the analysis, the simple model of a quarter suspension is considered. The equation of motion is obtained in differential form. These equations of motion underlie the Simulink model in MATLAB. This analysis will help to simulate the suspension parameters.

For many safety issues, laboratory tests remain indispensable. In the early stages of development, when physical prototypes are not yet available, numerical simulation is used to analyse and optimize the design. The paper [5] shows how to construct numerical simulation models of complex servo-hydraulic test systems and their test specimen using multi-body simulation for mechanics in combination with simulation models for hydraulics and controls. New test platforms for axle dynamometer vehicles are presented in [6, 7, 8].

In this paper, we will present the construction of a stand for fatigue testing of the rear axle. We will perform a simulation of the stand operation in Adams and a finite element analysis of the rear axle in ANSYS.

2. Dynamic analysis in ADAMS of the stand

In this paragraph, we modelled in SolidWorks the stand for fatigue testing of the rear axle of a car. In the first stage, we modelled all components of the stand structure for the fatigue testing of the rear axle of a car. The complete assembly of the test stand is shown in figure 1.

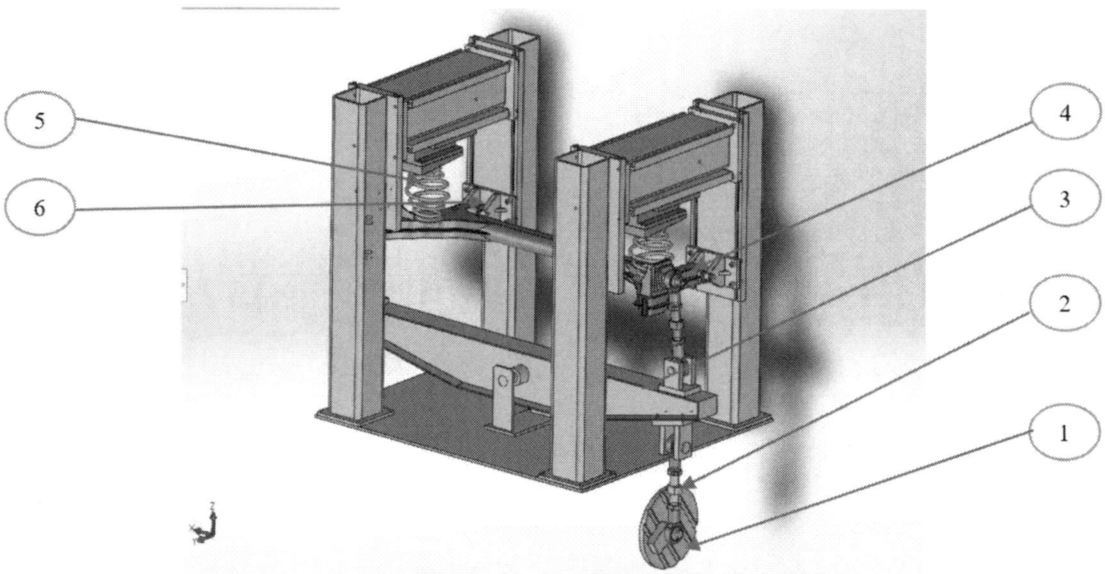

Figure 1. Virtual model of the rear axle fatigue test stand:
1 – flywheel, 2 – connecting rod, 3 – rocker, 4 – rear axle, 5 – helical springs, 6 – support.

This modelling allows us to simulate the model made in a dynamic regime. The results of the simulation in ADAMS are mainly: the kinematic and dynamic parameters of the system, the deformed representation of the elastic elements for different vibration modes, the frequencies at which the vibration modes appear, the translation deformation, the deformation speed as well as the deformation acceleration of the centers of mass of deformable elements.

The steps taken for the elastic-dynamic modelling in ADAMS of the rear axle fatigue test stand of cars are described below.

The CAD model of the stand made in Solid Works, is transferred to the ADAMS program, as shown in figure 1.

The next step is to define the material properties of the elements. The window for defining these properties is shown in figure 2.

The next step in making the ADAMS model is to define the kinematic joints. In the structure of the stand, there are, according to the kinematic scheme, rotational couplings, spherical couplings. The spherical coupling definition window contains the selection of the two bodies and the definition of the coupling location.

A very important aspect of modelling this system is that it has elements in the structure that are deformable. These are the springs, the stabilizer bar and the rear axle.

To turn them into flexible bodies, the well-known procedure in ADAMS is followed. Thus, in the dialog box for defining a flexible body, by discretizing into tetrahedral finite elements, the size of the finite element is specified, the connections with the adjacent elements.

By constructing them as deformable elements, their own modes of vibration are also calculated, as well as their own frequencies. The deformed representation of the left spring, for different eigen vibration modes, from which we selected the vibration mode 7 is shown in figure 2.

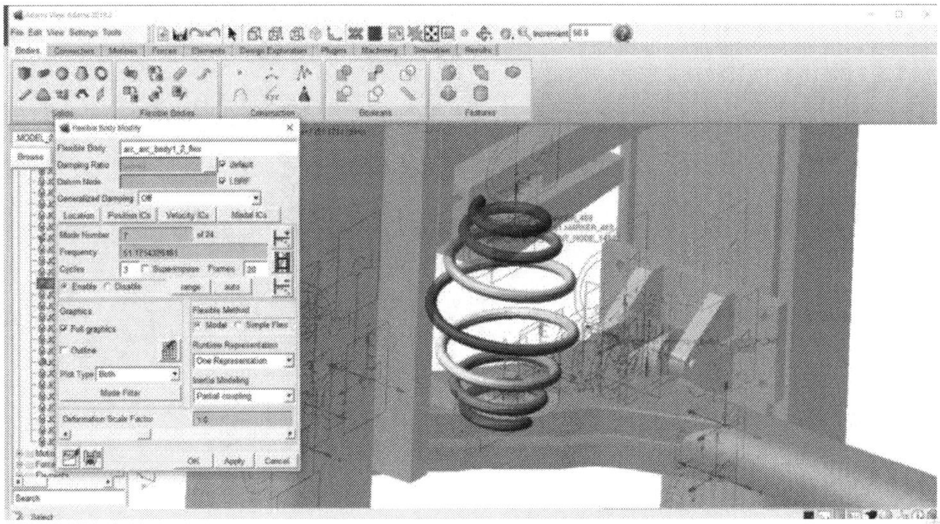

Figure 2. Deformed representation of the spring, for the vibration mode 7.

We propose to study, in the first part, the motion as a deformable solid of the stabilizer bar, from figure 1.

Figure 3 shows the graphs obtained for the displacements of the centre of mass of the stabilizer bar, along the X and Y axes. The graphs represented are obtained in relation to the global reference system.

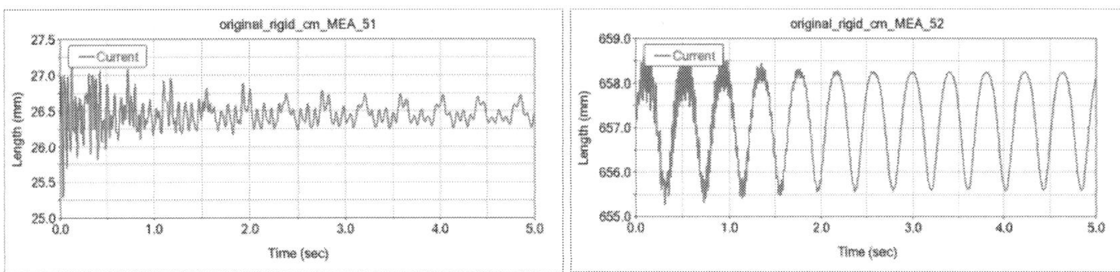

a) b)

Figure 3. Displacement of the centre of mass of the stabilizer bar along the axis: a) X; b) Y.

The values obtained for the translation deformations of the centre of mass of the stabilizer bar have values within the limits of normal operation.

The deformed representation of the stabilizer bar corresponding to the vibration modes 9 is represented in figure 4.

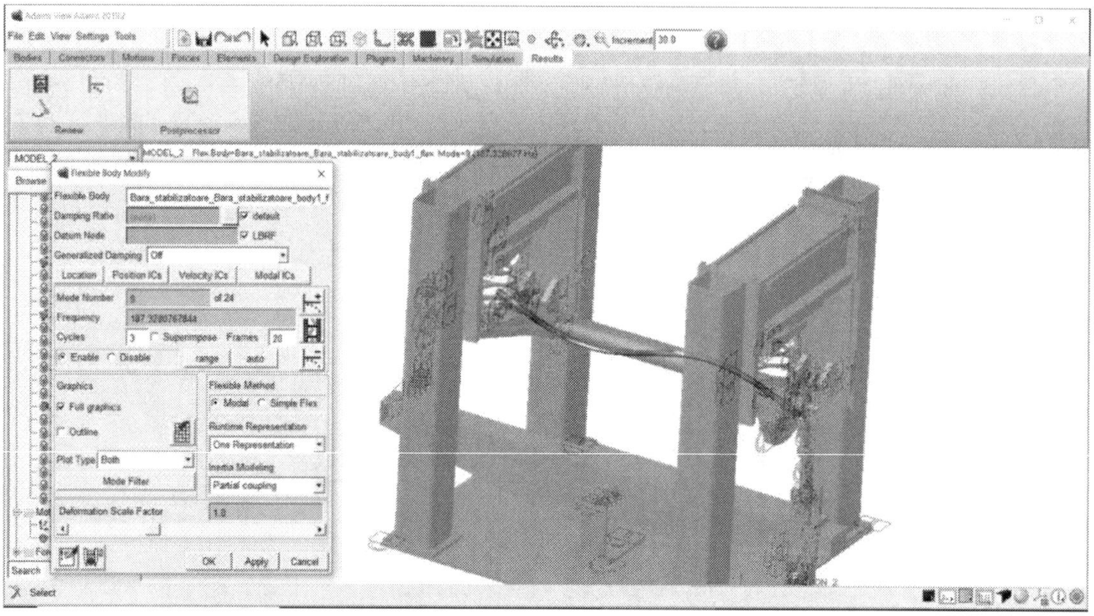

Figure 4. Deformed stabilizer bar representation for vibration mode 9.

3. Rear axle finite element analysis

In this paragraph, we aim to determine the state of stresses and strains in the whole rear axle tested experimentally on the test stand.

We need to do this for the following reasons:

- ✓ We will determine which are the critical areas, with stress concentrators in which the maximum stresses appear;
- ✓ Mechanical stresses and specific deformations obtained by simulation; we will compare the values obtained by the experimental method.

For finite element analysis, we will use a licensed program, namely Ansys Workbench. This program has the possibility to perform several types of analysis: in static structural, transient or dynamic regimes. Because the rear axle model is particularly complex, it consists of elements with nonlinear behaviour (such as rubber flexiblocks) that are suitable for use in this case, the Ansys Explicit Dynamics module. This module allows the realization of explicit dynamic analyses, allows us to use materials with nonlinear characteristics (steel with nonlinear behaviour, rubber).

The geometric model is a complex one, which consists of several pieces assembled by continuous cord welding or by points in some parts.

The components of the rear axle assembly are modelled separately, to be subsequently assembled in SolidWorks. Thus, the axle consists of a central beam, U-shaped, as shown in figure 5, the side elements of figure 5, which are made of stamped sheet, 2.5 mm thick and are welded to the side surfaces. The stabilizer bar in figure 5 is mounted with screws in the slot in the side members. It passes through the central U-shaped cross member. By welding, the end elements of figure 5 are assembled, which are made of sheet metal by bending. They are provided with 4 equidistant holes, with a diameter of 80 mm, in which the conical spindle on which the wheel bearings are mounted by nut screw assemblies. The wheel axle, in figure 5, has two sections for mounting the bearings.

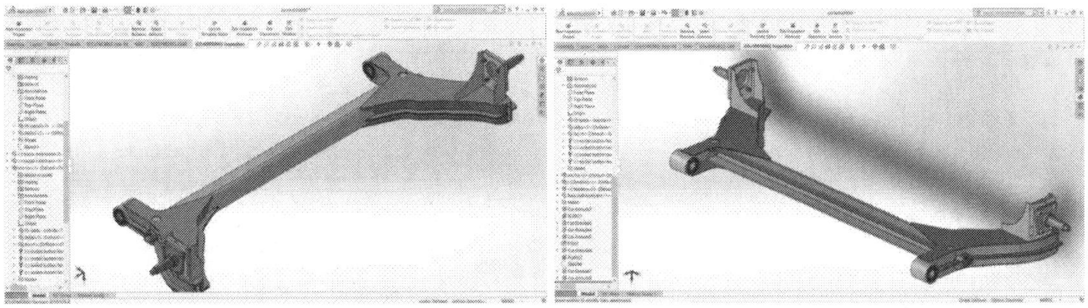

Figure 5. Rear axle assembly (front view and rear view).

The rear axle assembly is shown in figure 5. For better visibility, I assigned the stabilizer bar red.

For correct modelling, it is necessary to check for interference. If there is no interference, the success of finite element discretization is ensured, as well as the correctness of the results obtained in the finite element analysis. We checked the interferences on the whole assembly, as shown in figure 5, where it is observed that there are no interferences.

The steps to perform the finite element analysis of the rear axle assembly are described below. The model developed in Solid Works was transferred to the ANSYS Workbench software database. To check the correct import of the file, it is opened with the Ansys Design Modeler pre-processor. Geometry corrections can be made here, which lead to erroneous analysis results. For example, connecting beams can be made, several bodies can be joined in a kinematic element, the surfaces of the parts can be divided into several surfaces. The last utility described is useful for defining couplings between bodies, for placing loads. The built model of the rear axle contains 26 bodies (parts). We joined some of the bodies to form a single element, namely: the U-shaped cross member, the arms of the stamped sheet, the end elements of the bent sheet were joined to form the axle element, as shown in figure 5.

Editing geometry in Ansys Design Modeler is useful, for the reasons explained above, namely:

- ✓ It gives us the possibility to unite several bodies that form an ensemble;
- ✓ It offers us the possibility to make connecting radius, chamfers, to divide an area into several surfaces, which is useful for defining fixed or other joints, for defining loads.

We can now edit the model in Ansys Explicit Dynamic Mechanical. In the next step, we will define the material corresponding to each element. For the component elements, we used the steel material with nonlinear characteristics, as presented in the ANSYS database. The material has the following mechanical characteristics:

- Longitudinal elasticity module, Young, 2E+2 Pa;
- Coefficient of transversal contraction, Poisson, 0,3;
- Compression strength module, 1,6667E+2 Pa;
- Transverse elasticity module, G=7,6923+1 Pa;
- Characteristic curve, stress-specific deformations, is presented in the database.

For the elastic flexible blocks, which have rubber bushes in the structure, we used a material, which has the characteristics selected according to the database.

The next step in building the analysis model is to define the kinematic joints. It is possible to define the basic kinematic joints in Ansys, but also the complex joints (point on a curve). In our case, the model includes in the structure fixed couplings and rotational couplings. Thus, we considered the flexiblocks connected to the base by rotating couplings, as they are mounted on the test stand but also on the vehicle, by screw and nut. We defined a rotating coupling (in this case between the inner metal sleeve of the flexiblock and the base). It is observed that the program shows us that the only movement allowed in the joint is the rotation along the Z axis of the joint.

From the model, it is observed that we have defined two rotation couplings, which correspond to the two flexiblocks connected to the base. The rest of the couplings fix the elements between them (the elements of the flexiblock are fixed, the side plates bent with the spindle).

The discretization in finite elements of the assembly is further performed. Because it is an assembly of parts with complex shapes, it is important to specify an optimal size of the finite element used for discretization. In the first stage, we specified various sizes, for discretization, namely 2 mm, 4, 6 and 10 mm. The type of the finite element is tetrahedral, because the model is easier to solve. At the size of 10 mm of the element, the discretization is coarse and the shapes of the parts are distorted. So, a reduction in the size of the element is needed. I finally used the 4 mm size for the finite element.

With these considerations, the finite element analysis model is finalized, and we can run the analysis. Because the model is particularly complex, containing a number of 5 million elements, running takes quite a long time, namely 1250 hours. I used a graphics station with 2 Intel OCTA Core Xeon E5-2650 v2 2.60 GHz processors, 32GB DDR3 ECC, 250GB SSD, nVidiaQuadro K2000.

The processing and interpretation of the results are performed in the next paragraph.

4. Presentation of the results of the analysis with finite elements

The total displacements of the ensemble are presented in figure 6. It is observed that the maximum displacements occur in the area where the loading force is placed.

The maximum value recorded for total deformation at end of the axle is 34.189 mm.

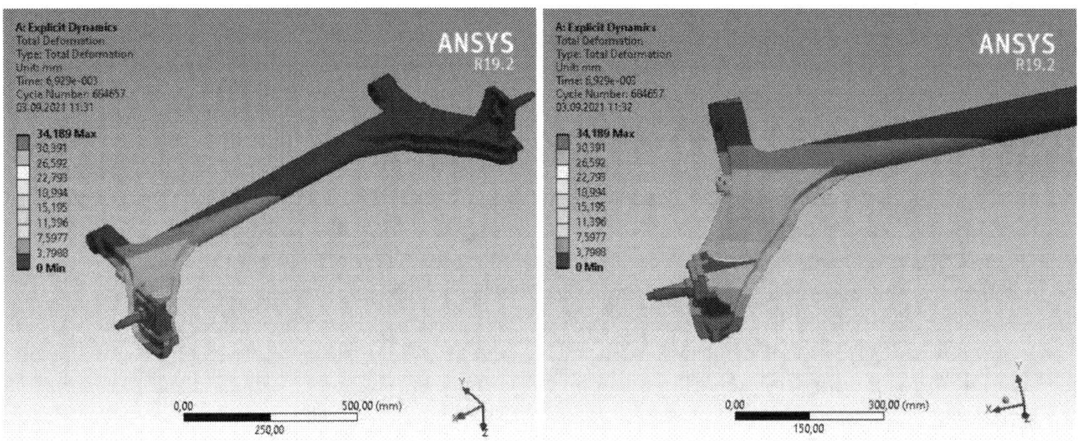

Figure 6. Map of the distribution of the resulting displacements.

A special interest is represented by the equivalent stresses according to the von Mises method obtained, which are presented in figure 8. The critical points were numbered on the axle, as shown in figure 7.

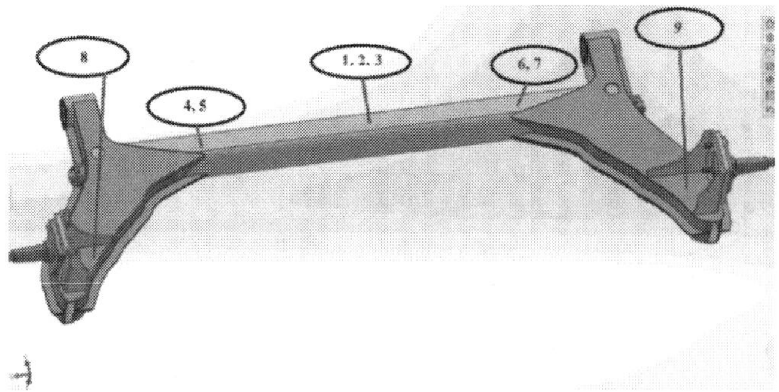

Figure 7. Mechanical stress measurement points.

If we analyse figures 7 and 8, we can draw the following conclusions:

- The maximum stress that occurs is 122 MPa, as seen in figure 8; It appears in the area of the joint corner of the side sheet bent with the stamped side arm. This area is the most requested in this case;
- Equivalent stresses with values of 54 MPa appear on the central crossbeam, which decrease to 40 MPa in the area adjacent to this portion;
- The stresses obtained by simulation in the central part, measured in points 1, 2 and 3 have values of 50.74; These values are close to those obtained by experimental analysis;
- Points 4, 5 and 6, 7, are observed to be placed symmetrically; The values of the equivalent stresses in these areas have different values, in the analysis with finite elements; This result is correct because the force is applied only on one spindle, i.e. on the left spindle of points 4, 5; Thus in the area of points 4, 5 by simulation we obtain stresses of 27 MPa, and in the area of points 6, 7 the equivalent stresses have values of 40 MPa; A portion with stresses of 54 MPa indicated in figure 8 is also observed.

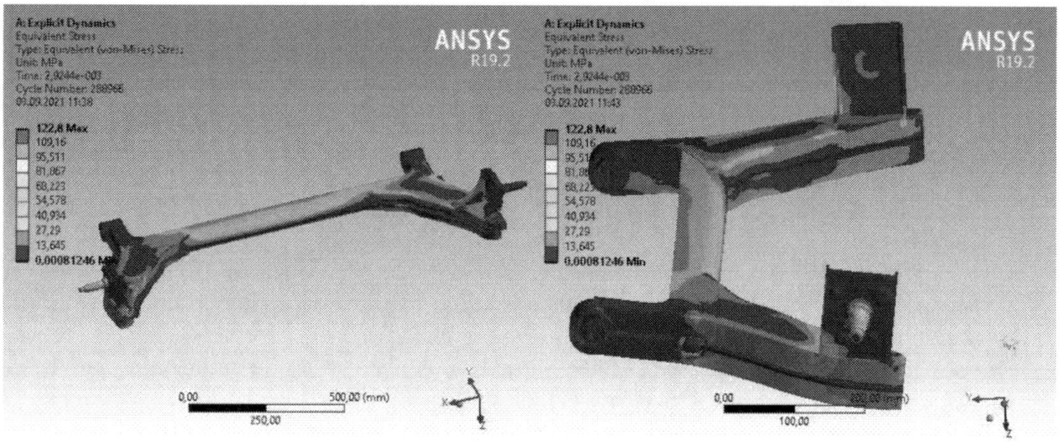

Figure 8. Von Mises equivalent stress distribution map.

Very important is also the safety coefficient at the fatigue of obtained bridge, which has values in the normal limit, as seen in figure 9.

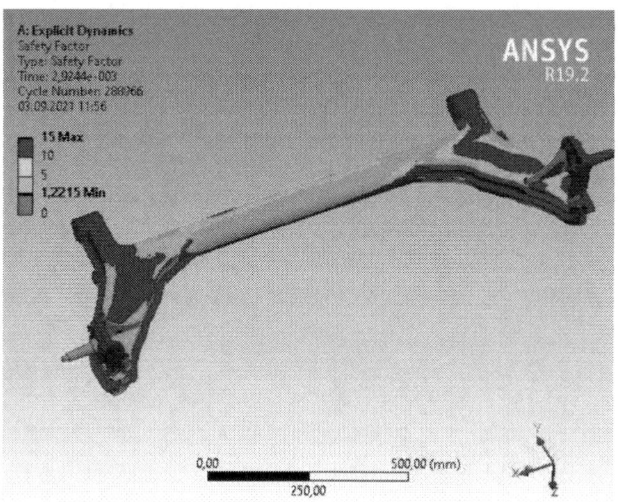

Figure 9. Map of the distribution of the safety coefficient to fatigue.

IOP Conf. Series: Materials Science and Engineering 1220 (2022) 012040 doi:10.1088/1757-899X/1220/1/012040

5. Observations and conclusion

The model built with the ADAMS program allows dynamic analysis for each kinematic element of the mechanism and also for the whole assembly when the kinematic elements are considered to be deformable. We performed the dynamic modelling in ADAMS of the test stand, and we presented the obtained results, namely displacements, of the centre of mass of the stabilizer bar of the rear axle. We performed the analysis with finite elements in a dynamic regime for the whole tested rear axle. This is important because we can compare the stresses obtained by finite element analysis with those determined experimentally. We must specify that experimentally we can measure the stresses in a limited number of points, in our case nine points. Instead, by finite element analysis, we obtain a map of the distribution of specific stresses, displacements and deformations in all points of the assembly.

References

[1] Bindauf A, Angrick C and Prokop G 2014. Suspension Characterisation on a Highly Dynamic Axle Test Rig. *ATZ worldwide*, 116(12), 48-53.

[2] Kuris S, Gungor E and Aykent B 2018. The Durability Test Setup of Test Rig for the Rear Axle of a LCV and Its Validation in Adams/View (No. 2018-01-1229). *SAE Technical Paper*.

[3] Muller A, Grubisic V and Fischer G 1996. U.S. Patent No. 5,487, 301. Washington, DC: U.S. Patent and Trademark Office.

[4] Ashtekar J B and Thakur A G 2014. Simulink model of suspension system and it's validation on suspension test rig. *International journal of mechanical engineering and robotics research*, 2278-0149.

[5] Chindamo D, Gadola M and Marchesin F P 2017. Reproduction of real-world road profiles on a four-poster rig for indoor vehicle chassis and suspension durability testing. *Advances in Mechanical Engineering*, 9(8), 1687814017726004.

[6] Dressler K, Speckert M and Bitsch G 2009. Virtual durability test rigs for automotive engineering. *Vehicle system dynamics*, 47(4), 387-401.

[7] Von Thun, H J Pfeiffer, M and Etschmaier L 1988. A new dynamometer test rig to develop drive lines for all-wheel driven vehicles. *SAE Transactions*, 1322-1334.

[8] Zaisheng L, Fuxiang H, Changming W, Xiang L and Jin Z 2012. Development of Multi-axis Fatigue Test Rig Load for Car Suspension Components [J]. *Automobile Technology*, 6.

The Study of a Cost-Effective Developed System for the Comfort Testing of the Motor-Vehicle Driver and the Passengers

Ş Voloacă and **M F Toma**

Automotive Department, Faculty of Transports, University POLITEHNICA of Bucharest, 313 Splaiul Independenţei St., Sector 6, Bucharest, Romania

stefan.voloaca@upb.ro

Abstract. The main goal of the actual research is to develop a cost-effective Data Acquisition System. The system can be used to measure accelerations, inside the vehicles, that can affect the comfort of the passengers or the driving pleasure, known as drivability. The study of comfort, influenced by the accelerations, is currently conducted on different types of vehicles equipped with thermal, electric engines or hybrid systems. This study can have two approaches, a subjective one and an objective one. The subjective study of the drivability must be made by a specialized person, with good skill and vast experience in this domain, but the results, sometimes, can be influenced by that person preferences or its moment feelings. The objective of the study is made by analysing the recoded accelerations by a data acquisition system. The paper starts from the main tendency on every industry, represented by the indirect ratio of growing performances versus the size reduction of the measuring devices. In the beginning, this tendency can have high costs but with good optimization will have a significant cost reduction. Thus, the paper combines three main parts of engineering like electronics, automatics and not least the automotive one. The research starts from a system developed inside the automotive engineering laboratories that was used to record data in real time, on a motor-vehicle, for different running regimes, met in the real life.

1. Introduction

Starting from the physical combination of inexpensive components up to the software development, the authors intended to create and develop a low-cost data acquisition system for measuring accelerations. With the created system it is possible to measure, record and follow in real time the accelerations directly on the vehicles, in any dynamic regimes.

The articles started from the current tendency in any research domain, meaning the development of an economical solution with low dimensions for ergonomic use, without affecting the quality of the product.

The utility of the system is for the university research and for the students use. They could use it for their training for competitions like Formula Student, Shell Eco-marathon, or Kart Low Cost.

Its utility results from the multidisciplinary character, that combines electronic part, automatic and automotive engineering. In the automotive domain, it can be used, especially, for the study of the influence of accelerations over the passengers' and the driver's comfort.

2. Related research

From the study of drivability, meaning - the degree of smoothness and steadiness of the acceleration of a motor-vehicle, the authors developed a measuring system necessary to study it. The drivability of a motor-vehicle influences the passenger's comfort and the pleasure of driving that depends on the driving style.

The drivability characteristics are influenced by the longitudinal acceleration felt by the driver and the passengers. The acceleration analysis can be based on some attribute parameters [1] like duration of acceleration, duration of load reversal, stumble, jolt, backlash, and surging.

Measuring and recording the acceleration variation can be done by a professional Data Acquisition System or following the trends in the last decay, where the price is the main acquisition factor. In recent years, there were built and developed lots of cost-effective systems. With those low-cost systems, it is possible to study the passengers' comfort according to driving style [2] and makes it possible to evaluate the dynamic movement factors that can influence a comfort index [3].

The driving style influences the gear shifting process or the brake manners. Again, the accelerations can be studied by low-cost tri-axial accelerometers connected to a data acquisition platform [4, 5].

Regarding the passengers' comfort that has an international standardization ISO 2631, the main interest is focused on vertical acceleration. Studies using those kinds of accelerometers were done [6], where the road pavement state is the main influence factor.

The main advantage, of those platforms, are their dimensions, which make them easy to install or detach, without an aggressive modification of the tested vehicles [7].

The studies were made with professional equipment (produced by National Instruments) in parallel with low-cost micro-controllers and accelerometers available on the market (Arduino and ADXL335). Comparing the measuring results from the accelerometers, a maximum 2% end of scale error was found. This percentage confirms that the two criteria like precision and sensibility can be achieved with those systems [8].

Another low-cost accelerometer available on the market is MPU6050 model, which has excellent performance up to 80 Hz, achieving extremely accurate results with a maximum normalized root-mean-square deviation error (NRMSE) of 1.85% in laboratory tests and 2.19% in road tests compared to professional piezoelectric accelerometers [7].

3. Technical part

3.1. The equipment specifications

The purpose of choosing technical components was the price and their small dimensions. Thus, the created system has in its componence a Single Board Computer (SBC), a power module and one accelerometer. Their main characteristics are presented below.

For an easy way of recording data and use, it was used a Raspberry Pi 4 Single Board Computer (SBC) [9] that has small dimensions and is equipped with a quad-core 64-bit SoC @ 1.5GHz processor. It has a SDRAM memory of 4GB. Regarding the connectivity, it has a wireless LAN, Bluetooth 5.0, Ethernet, 2 ports of USB 3.0 and two of USB 2.0. It has 40 pins of General-Purpose Input Output used for hardware configurations.

To measure the accelerations, it was used an InvenSense's 3-axis accelerometer type MPU-6050 with a 16-bit Analog to Digital Converter (ADC) that does not require an external multiplexer. The main reason for choosing this type of accelerometer was the fact that it was tested according to the JESD47I Standards of "Stress-Test-Driven Qualification of Integrated Circuits". Thus, it can be used for different measurement ranges with different Sensitive Scale Factor (SSF) of 16384 LSB/g @±2g, 8192 LSB/g@±4g, 4096 LSB/g@±8g, 2048 LSB/g@±16g [10]. The change of sensitivity with temperature from -40°C to +85°C is ±0.02 %/°C. Its nonlinearity that influences the best fit of the straight line is 0.5% and the Cross-Axis Sensitivity is ± 2%.

For the power supply, it was used a Voltage Adjustable Step-Down Power Module LM2596 BUCK 3A DC-DC that transforms the direct current of 12V vehicle battery voltage into a direct current of 5V@3Ah. For easy use, the values of output/input voltage are displayed on its small display.

3.2. Physical design of the measuring system

The modules of the accelerometer (ACC) and the remote controller (RC) are coupled to the Single Board Computer (SBC) through the General-Purpose Input Output (GPIO) (figure 1).

The accelerometer is coupled to the computer by 4 wires making a connection between its Source Data Acquisition line (SDA), Serial Clock Line (SCL), The Ground (GND) and the Power supply voltage (VCC).

To make easy control of the system, it was designed a Remote Controller (RC) that has two Push Buttons (PB) and a Green Light LED for the state of the activity.

The Power Module (PM) is coupled to the vehicle's 12V battery and supplies a voltage of 5V at 3Amps to the SBC.

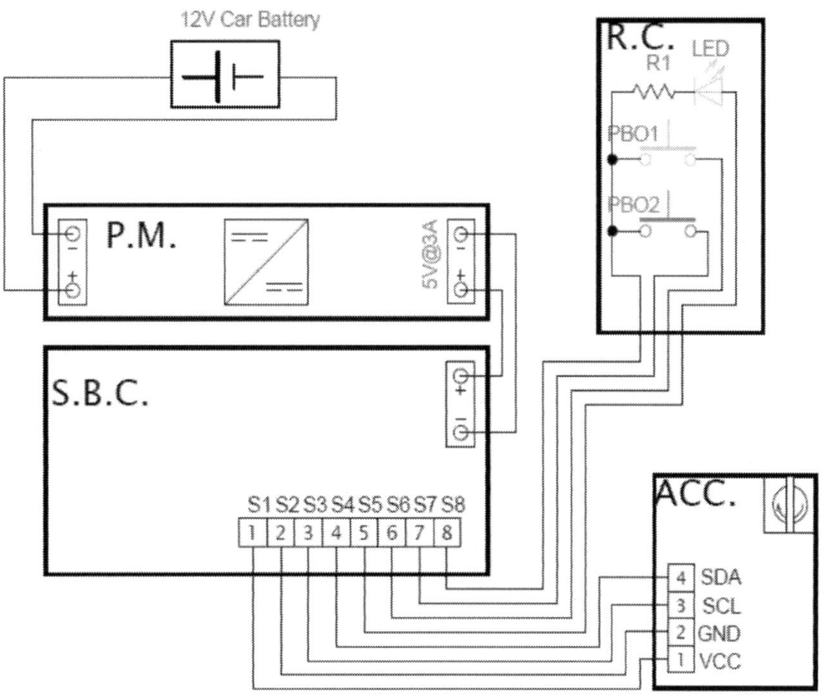

Figure 1. The Electronic Scheme of the Designed Data Acquisition System.

For a high endurance, for every component, it was built a protective housing.

3.3. The Motor-Vehicle Equipping

The low dimensions of the system components make possible the positioning of the accelerometer in a safe place, inside the vehicle, on the right side of the driver, or even inside the glove compartment. For a facile observation and use, the remote controller can be placed near the driver, being possible to be stuck even on the vehicle's board.

The accelerometer can be mounted on a chassis element. In this case, it was glued on the passenger chair's rail (figure 2).

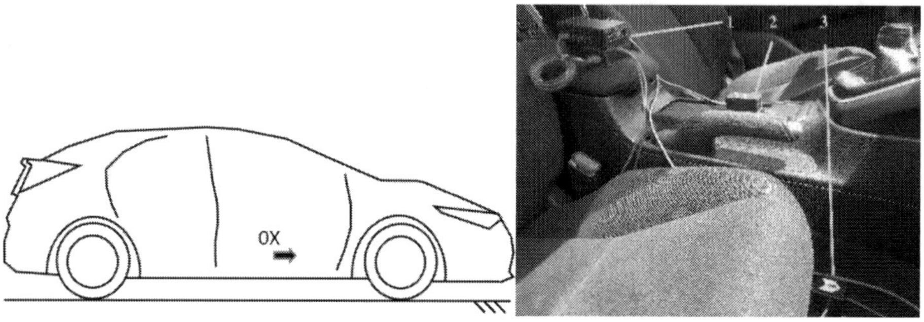

Figure 2. The position of the accelerometer and the data acquisition system:
1 – Data Acquisition System; 2 – The remote controller; 3 – The accelerometer.

Before gluing the accelerometer over the chair's rail, it is necessary to calibrate it. To avoid any influence of the vibrations produced by the engine, the engine must be turned off.

3.4. The Accelerometer Calibration
For a correct acceleration measuring it is mandatory to calibrate the measuring system.

At the beginning, the indicated values are like a constant virtual acceleration [8]. They are influenced by the Electromagnetic Noise Insulation (EMI) due to the cables' length and quality.

The system measures the current voltage, and the software transforms them into accelerations. The calibration procedure is split into two parts (figure 3). The first one is for the determination of the gravitational acceleration's voltages for 0G and the second one – for 1G.

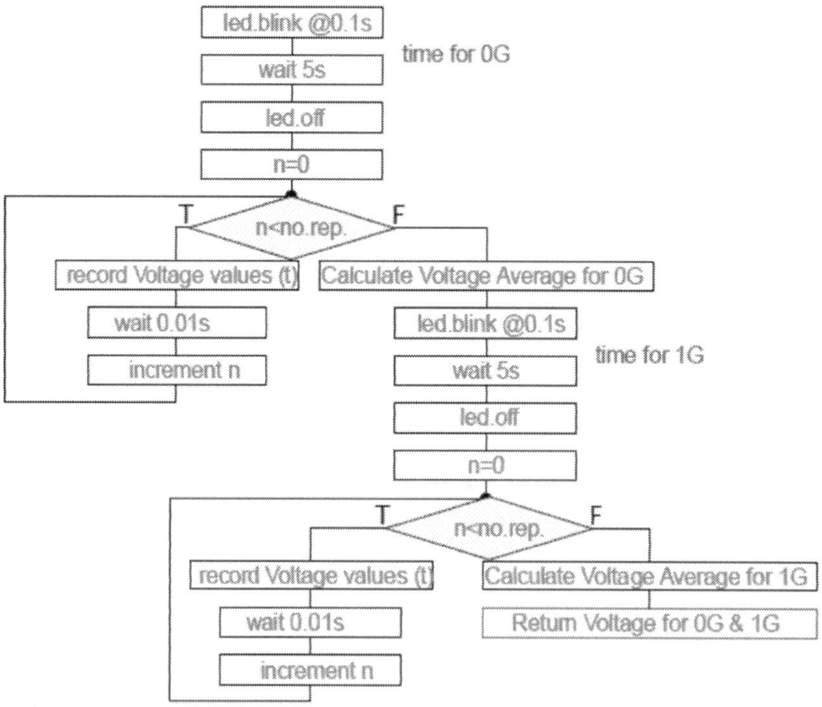

Figure 3. The logical scheme of the calibration process.

At the procedure activation, the remote-controller's LED will blink for 5 seconds at every 0.1 seconds. The user will place the accelerometer in its normal position. When the LED is off, the software, at every 0.1 seconds, will record voltage values, to extract the average. This average will represent the value of 0 m/s^2. The LED will start to blink again for 5 seconds, the time when the user should turn the accelerometer with 90°. When the LED turns off, the several voltage values will give the acceleration value for 1G.

In our case, the average was made for a number of 1000 values. Depending on the complexity of the measuring situation, the number of repetitions and the frequency can easily be modified in the code.

3.5. How to use the system

After the system is mounted and calibrated, it can be easily used as presented in figure 4.

When the electric current is supplied to the system and the software is loaded, on the remote controller, the green LED stays continuously on.

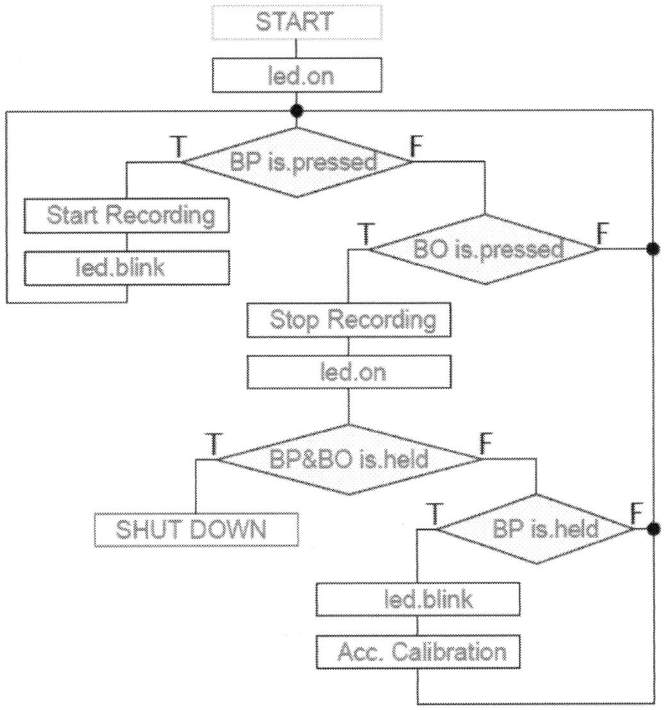

Figure 4. The logical scheme of using the system.

If the green push button is pressed, all the acceleration values are recorded. The developed code makes it possible to record values at every 10ms. During the recording process, the LED will blink at a 1 Hz frequency.

To stop the recording, it is necessary to press the red push button and all the data will be saved in a file with the name containing the time when the tests were done. After that, the LED will stay on, again.

For a system calibration, at the beginning or during the tests, it is necessary a short push of the red button followed by keeping pressed the green button. To avoid confusing recording and calibration, the LED blinks faster, at a frequency of 10 Hz.

At the end of the tests, if the system must be shut down, the user must keep pressed, at the same time, both push buttons. The LED on the remote controller will turn off too.

The data downloading process can be easily done directly from the memory card, where all the data were recorded in comma separated values (.CSV) files.

The Single Board Computer has a wireless function making it possible to connect to any mobile device like phone, tablet, or laptop. For this reason, the code written by the authors has the possibility of viewing the acceleration variation as values and graphs and the state of the process.

4. Performing the Measurements in a Dynamic Regime

When performing the tests, the condition of the vehicle, the condition of the road, the weather conditions and the geographical position were considered.

The tests were done at an altitude of 190 m and a northern altitude of 44°50' where the gravity is 9.806 m/s^2. The road state was dry asphalt with a slope of 0%.

The tested vehicle was a compact one of the mid-size class with a hatchback body. Its energetic characteristics are the torque net of 300Nm@2000rpm and the horsepower net of 88kW@4000rpm. The curb weight was 1346 kg with a load of 75 kg from the driver and 60 kg from the passenger.

4.1. Test procedures

For better coverage, two main functioning regimes were pointed, the vehicle Tip-In acceleration and intensive braking, both influencing the comfort and the drivability. With them is possible to obtain conclusive results regarding longitudinal acceleration variation. For every regime, there were made at least 4 measurements.

For the Tip-In regime, initially, the motor-vehicle is running in the second gear at idle speed, to obtain a constant speed. When the speed is stabilized, the recording will be started and after a couple of seconds, the acceleration pedal is firmly pressed down up to 100%. The vehicle will run for at least 5 seconds and after that, the recording will be stopped. This procedure can be repeated.

For intensive braking, the initial speed is 50 km/h. To obtain this constant speed, the cruise control was activated and thus at every recording, the initial regimes are almost the same. When the speed is stabilized, the recording will be started and after that, the brake pedal will be firmly pressed down. After the vehicle is stopped the recording can be stopped.

5. Results and Interpretation

After the data extraction and processing, a series of graphs are obtained. By overlapping 4 test results they present the longitudinal acceleration variation in time for both launch and brake regimes.

For the Tip-In test (figure 5), the maximum acceleration was 4.4 m/s^2, and for the intensive braking (figure 6), a deceleration of -16.81 m/s^2 was reached.

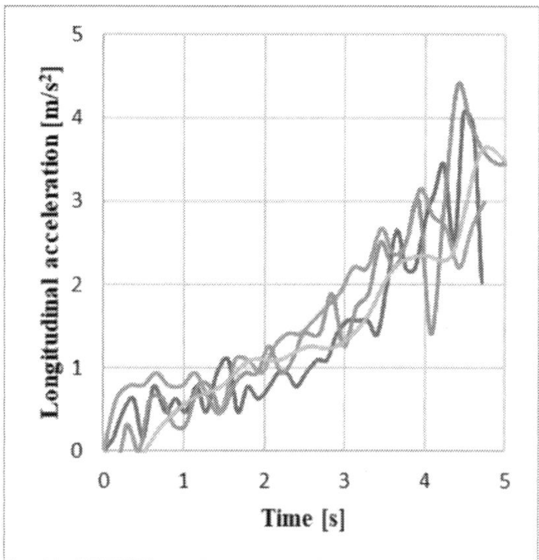

Figure 5. The variation of longitudinal acceleration for Tip-In process.

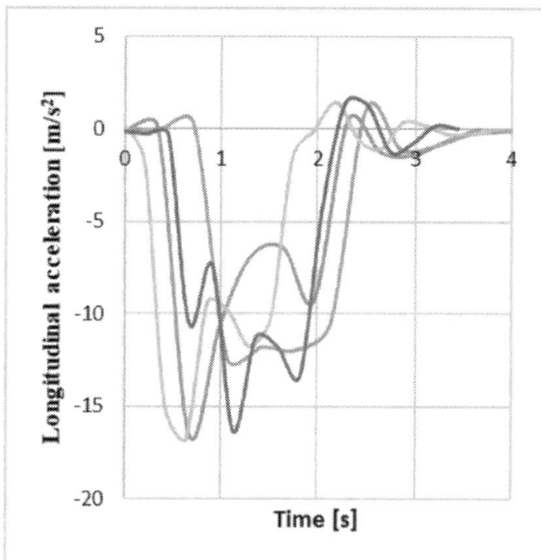

Figure 6. The variation of longitudinal acceleration for the intensive braking.

Further analysis will be made for the acceleration, in the case of a Tip-In (figure 7) and intensive braking (figure 8).

To facilitate the result interpretation, necessary at the study of the comfort and the drivability, it can be inserted a moving average curve. It will reduce the influence of external factors like the road state, engine vibrations, atmospheric conditions, or the Electromagnetic Noise Insulation (EMI). This moving average is useful when we want to highlight the functioning regime of auxiliary systems of the vehicle. Thus, in figure 7 is possible to analyse three areas of longitudinal acceleration (continuous line) and moving average (dash line). In the first area (I) the acceleration rises slow. Starting with the second area (II) the acceleration starts to rise fast due to the turbocharger. The third area (III) presents a fluctuation possibly caused by the elasticity characteristics of the powertrain. This fluctuation can create discomfort producing the "Salut" effect felt by the driver and the passengers.

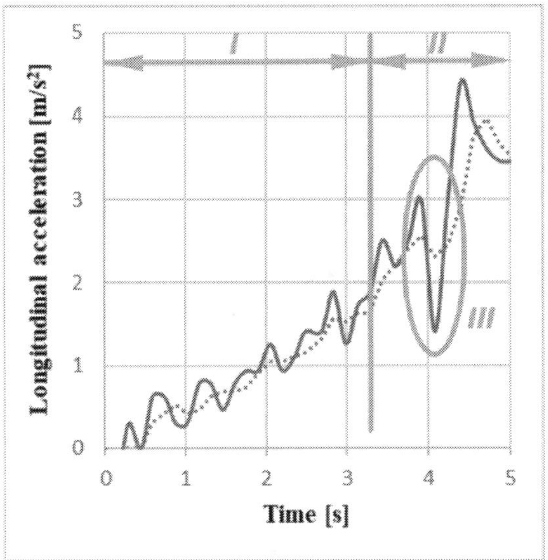

Figure 7. The variation of longitudinal acceleration for a Tip-In.

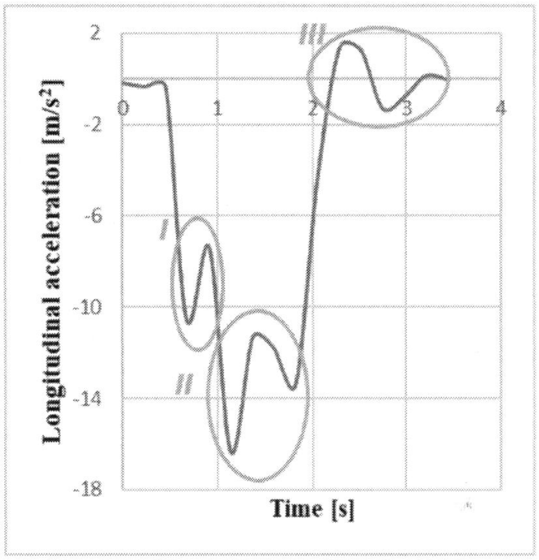

Figure 8. The variation of longitudinal acceleration for intensive braking from 50-0 km/h

For the intensive braking, from 50 to 0km/h, presented in figure 8, is easy to highlight the Anti-lock Braking System operation and the suspension elasticity. The first two areas (I and II) represent the ABS operation and the last one (III) shows the effect of suspension on the passengers' comfort. Is easy to see that the wave duration is less than 1 second.

The maximum values of deceleration can be influenced by the position of the accelerometer that was fixed in front of the passenger's chair.

6. Conclusions

Using some low-cost electronic components, found on the market, was possible to obtain a Data Acquisition System, necessary to measure the accelerations felt inside the vehicle to study the comfort and the drivability. This system has small dimensions, being easy to be mounted, used, or stored. Thus, by positioning it under the passenger's chair or inside the glove compartment, it can be used as a black box to study the driving style and the driver's behaviour, which can differ from one world region to another.

For the beginning, it was obtained a calibration function, a physical remote controller or even distance control through mobile devices (phone, laptop, or tablet), where is possible to follow in real time acceleration variation and all the data acquisition state.

Depending on the measuring needs, the source code, developed by the authors, can be modified by adding new functions or modifying data acquisition parameters.

The recorded data could be influenced by the mounting position of the accelerometer, the electro-magnetic noise and the endurance of the system, things that could be studied in the next research.

As a conclusion, the developed system has a low cost and can be used and adapted to a wide range of applications, from the passengers' comfort study to the students training for different competitions.

References

[1] Scharf J, 2019, *FEV - Corporate Magazine,* "Objectified Evaluation at Engine-in-the-Loop Test Bench - Drivability Characteristics", **2019**

[2] Vaitkus V, Lengvenis P, Žylius G, 2014, *Proceedings of the 2014 19th International Conference on Methods and Models in Automation and Robotics (MMAR 2014)*, "Driving style classification using long-term accelerometer information", pp. 641-644, DOI: 10.1109/MMAR.2014.6957429, 2014

[3] Castellanos J, Fruett F, 2014, *Measurement,* "Embedded system to evaluate the passenger comfort in public transportation," vol. **47**, pp. 442-451, DOI: 10.1016/j.measurement.2013.08.068

[4] Andria G, Attivissimo F, Di Nisio A, Lanzolla A M L, Pellegrino A, 2016, *Measurement,* "Development of an automotive data acquisition platform for analysis of driving behaviour", vol. **93**, pp. 278-287, DOI: 10.1016/j.measurement.2016.07.035

[5] Catur Edi Widodo K A, 2017, *International Journal of Innovative Research in Advanced Engineering (IJIRAE),* "Monitoring for Vehicle Velocity and Acceleration Using An Accelerometer", vol. **4**, no. 11, DOI: 10.26562/IJIRAE.2017.NVAE10086

[6] Loprencipe G, de Almeida Filho F G V, de Oliveira R H, Bruno S, 2021, *Sensors 2021,* "Validation of a Low-Cost Pavement Monitoring Inertial-Based System for Urban Road Networks", vol. Sound Monitoring Acoustic Sensor Network Design for Urban and Suburban Environments, no. **3127**, DOI: 10.3390/s21093127

[7] González A, Olazagoitia J L, Vinolas J, 2018, *Sensors 18(2)* "A Low-Cost Data Acquisition System for Automobile Dynamics Applications", No.**366**, DOI: 10.3390/s18020366

[8] Afonso J D R, 2010, *Semantic Scholar,* "Development of an Instrumentation and Data Acquisition system based on a low cost microcontroller Arduino,", no. **128337011**

[9] Raspberry Pi Trading Ltd., 2019, *Raspberry Pi 4 Computer*

[10] InvenSense Inc., MPU-6000 and MPU-6050 Product Specification Revision 3.4.

Diagnosis of mechanical parts by analyzing vibrations measured during operation

M M Udrea[1], D Voicu[2], R M Stoica[1], L Barothi[1]

[1]Department of Military Vehicles and Transportation, Military Technical Academy "Ferdinand I", 39-49 George Cosbuc bld., Bucharest, Romania
[2]"Dunărea de Jos" University of Galati, 47 Domneasca Street, Galați Romania

madalina.udrea@mta.ro

Abstract. The present study aims to measure vibrations in machine parts, perform the corresponding spectral analysis and, finally, obtain a diagnosis. The test was performed on a cardan shaft, measuring the vibrations with a uni-axial accelerometer, and follow-up, with the use of the Adash acquisition board, data were recorded and analyzed. It is known that each mechanical part has its own vibration frequency and that the operation at critical speed determines the occurrence of the resonance phenomenon. When this phenomenon appears, damage occurs, which is difficult to remedy and in some cases could endanger the safety of operators. Therefore, it was necessary to perform modal analysis in the Ansys program in order to be able to analyze the own vibration modes of the cardan shaft.

1. Introduction

Due to the technical progress obtained during recent years, there have been developed new specializations, such as: mechanization, automation, IT, thus replacing the human operator with a robotic substitute. However, among vehicles, even if there is a computer that monitors the operation of various components and controls the multiple actuators to achieve both maximum performance and energy savings, maintenance still requires the human operator.

The diagnostic part has developed a lot, reaching even to self-diagnosis, but some defects still require extensive engineering knowledge and the involvement of qualified personnel.

The present study aims to measure vibrations in machine parts, perform spectral analysis and, finally, obtain a diagnosis. It is known that each body has its own vibration frequency and that the operation at critical speeds involves the occurrence of the resonance phenomenon. When this phenomenon appears, damage occurs, which is difficult to remedy, and in some cases, it could endanger the safety of people.

There are finite element analysis programs, such as the Ansys program, which allow obtaining the maximum deformations, the maximum stresses, and their distribution on the whole ensemble, based on already known loads, but also to perform a modal analysis for determining its own vibration modes.

In this paper, the proper modes of vibration of a cardan shaft are analyzed and followed-up, the measured vibrations are verified so that to avoid the occurrence of the resonance phenomenon.

IOP Conf. Series: Materials Science and Engineering 1220 (2022) 012042 doi:10.1088/1757-899X/1220/1/012042

2. Experimental research

The imbalance occurs when the center of gravity of the rotating component does not coincide with the center of rotation. The manufacture of a perfectly balanced component is practically impossible and even after balancing there is a residual imbalance in the rotating components. The causes of the imbalance can be the mass plus or minus on one side of the component, resulting from dimensional errors during the manufacturing process (casting, cutting, assembly), inhomogeneities of materials (gaps, porosity, inclusions), asymmetrical construction shapes and temperature variations.

The imbalance produces a periodic vibration signal, with the same amplitude at each rotation of the shaft [1, 3].

Misalignment of an in-line system occurs when the axis of the drive shaft does not coincide with that of the driven shaft. This misalignment can be angular or parallel. Misalignment produces a constant radial preload that pushes the shaft sideways. As an example, there can be mentioned the preloads produced by bearings which are forced obliquely into the housing or on the shaft, by the contact forces of the gear teeth.

Angular misalignment has a predominant peak at 1X in the axial direction; and in the radial direction, peaks at 1X and 2X will appear in the spectrum if parallel misalignment is also present [2].

For many components, the measurements made on the fixed parts are sufficient for the correct characterization of the operating state. Measurements must be made on the bearings, on the bearing housing or on other parts of the machine which have a measurable response to the action of dynamic forces and characterize the overall level of component vibrations. Typical measuring points are established according to standard ISO 10816 and are shown in figure 1 [1]:

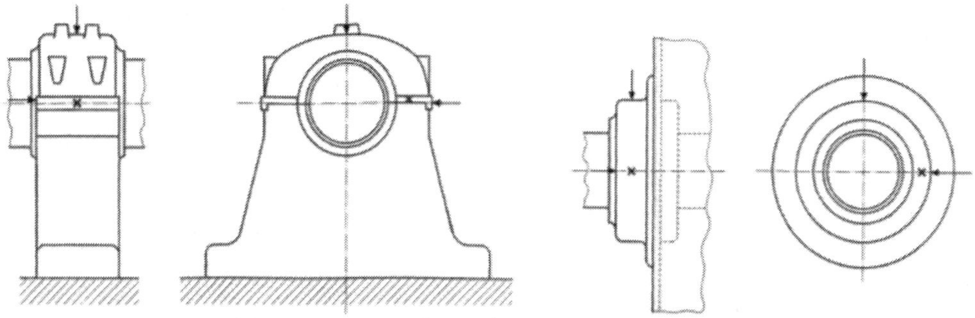

Figure 1. Measuring points on the bearing [4].

For the experimental research, measurements were performed on the VAMTAC S3 vehicle. This is a 4x4 tactical vehicle, produced by the Spanish company UROVESA, equipped with a Diesel engine, with 6 in-line cylinders, which provides a power of 188 hp and a maximum torque of 410 Nm. The transmission is automatic, with 5 forward gears and one for reversing the movement. It can reach a maximum speed of 135 km/h [5].

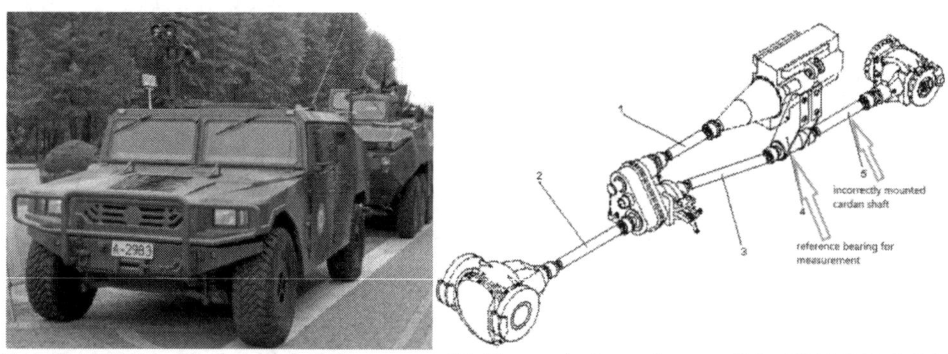

a) VAMTAC S3 tactical vehicle [6]. **b)** Transmission scheme of the VAMTAC S3 [7].

Figure 2. The vehicle and the assembly used during the experimental research.

Figure 2.b) shows the transmission scheme, being mentioned the bearing on which the accelerometer is mounted and the incorrectly mounted cardan shaft. The components of the cardan shaft are as follows [7]:

1 - the shaft that makes the connection between the automatic transmission and the transfer case;
2 - the shaft that makes the connection between the transfer case and the conical group of the rear axle;
3 - the shaft that makes the connection between the transfer case and the intermediate bearing;
4 - intermediate bearing;
5 - the shaft that makes the connection between the intermediate bearing and the differential group of the front axle.

Working procedure:

- the cardan shaft was disassembled and a grooved part was rotated by 47 degrees (3 grooves), so that there was an angular difference of 47 degrees between the planes of the forks;
- engine was started, the Drive mode was selected for the gearbox and the front axle was engaged so that the engine torque reaches the reference shaft;
- in the "Spectral Analysis" module of the Adash 4400 VA4 Pro II acquisition board, was set the measurement direction (axial / horizontal / vertical);
- the accelerometer was placed in the set direction of the acquisition board (connected to IN1);
- the measurement was started from the acquisition board, measuring the speed of the cardan shaft with the laser tachometer (figure 4).

Figure 3. Vibration measurement on the intermediate bearing.

It is necessary to check if the speed measured with the digital tachometer is the same as the calculated one. Knowing the following technical details:

- idle speed: $n_{idle} = 500\ rpm$;
- gearbox transmission ratio in first range: $i_{CV,I} = 3{,}51$;
- transmission ratio in the transfer case for high gears: $i_{CD,N} = 1$;

the speed of the reference cardan shaft can be calculated as follows:

$$n_{cardan.s} = \frac{n_{idle}}{i_{CV,I} \cdot i_{CD,N}} = 142 \quad [\text{rpm}] \tag{1}$$

Figure 4. Cardan shaft speed measurement with digital tachometer.

It can be seen that the calculated speed of the cardan shaft is approximately equal to the speed measured with the digital tachometer (figure 4). The difference between the two values is due to the slip in the hydrotransformer.

Figure 5 shows how the accelerometer was placed in 3 directions of measurement.

a)horizontal direction. **b)** vertical direction. **c)** axial direction.

Figure 5. Cardan shaft speed measurement with digital tachometer.

Thus, we used the Catia program to model the cardan shaft. Figure 6 shows the modeled cardan shaft, consisting of 4 components: 2 cardan joints, the part of the shaft that has outer grooves and the one that has inner grooves. Using constraints, it was assembled, assuming that the contact between the grooves is made on the entire side surface, and between the axes of the forks, the condition of parallelism was respected.

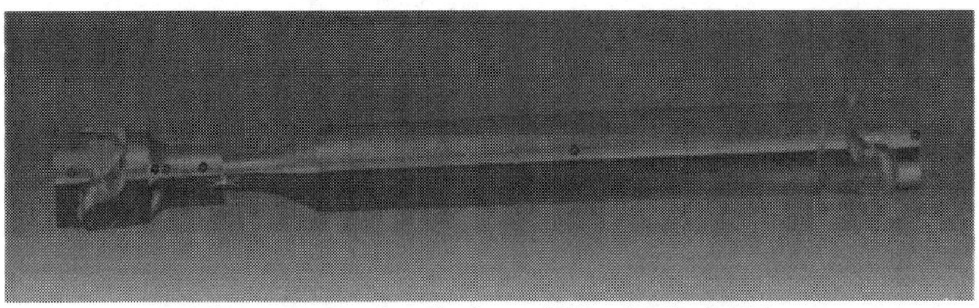

Figure 6. Cardan shaft modeling in Catia V5.

For a more detailed analysis and to observe the natural frequencies of the cardan shaft, a modal analysis is required.

The use of the finite element method requires engineering knowledge of the materials strength and the mechanical vibrations. It is necessary to identify the type of load and to establish other environmental influences [8].

In order to perform an analysis in the Ansys program, it is necessary that the assembly is discretized. This consists in dividing the whole into smaller elements. Automatic discretization cannot predict phenomena, but only the possible locations of voltage concentrations. Therefore, it is necessary for the user to specify certain amounts of discretization based on the accumulated knowledge (in the areas where the highest loads are known to occur/ in areas with stress concentrators, a finer discretization is required, for capturing all phenomena) [8]. Following the discretization, there were resulted 13.316 nodes and 7.168 elements.

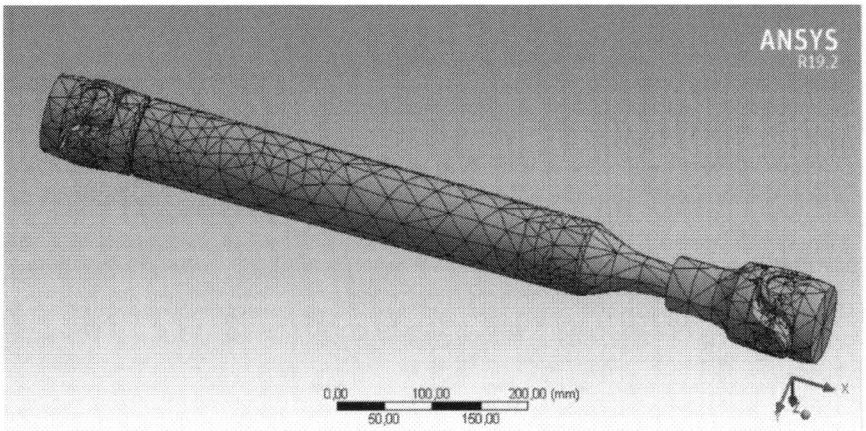

Figure 7. Cardan shaft discretization.

The modal analysis was performed using the MODAL mode from the WORKBENCH platform. In figure 11, the first 10 own modes of vibration of the cardan shaft are shown. Knowledge of the modes helps to analyze the overlap of the own vibrations over those that occurred during operation (excitation frequencies), in order to avoid the resonance phenomenon.

3. Results

The accelerometer was placed in a horizontal direction. After performing the measurement for 30 seconds, there were recorded the data which is depicted in figure 8.

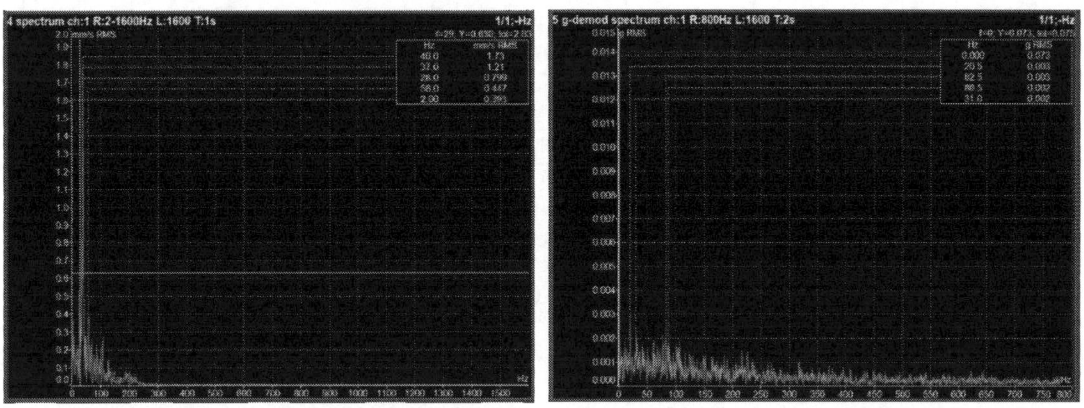

a) Spectrum 1 (Speed of vibration measured in the horizontal direction on the intermediate bearing).

b) Spectrum 2 (Acceleration of vibrations measured in the horizontal direction on the intermediate bearing)

Figure 8. The results obtained on the horizontal direction

Table 1. Operating rating according to ISO 10816-1:1995

Standard		ISO 10816-1
Power		Medium-sized machines 15<P<300 [kW]
Foundation		rigid
Measurement parameter		RMS vibration velocity [mm/s]
Operating rating	Good	<1.8
	Usable	1.8 – 4.5
	Admitted under supervision	4.5 – 11.2
	Forbidden	>11.2

The highest effective RMS value is 1.73 mm/s at a frequency of 40 Hz. According to ISO 10816-1, the operating rating will be "good", but still, the value is to the upper most limit of this rating.

According to the bibliographic source [9], the diagnosis for dynamic imbalance is based on the detection of amplitude peaks at 1X in the radial direction.

The reference speed is 137 rpm (2.28 Hz), and as seen in figure 8, there is no signal peak at this frequency. It is concluded that the system is properly balanced.

The accelerometer was placed in the vertical direction. After performing the measurement for 30 seconds, there were recorded the data depicted in figure 9.

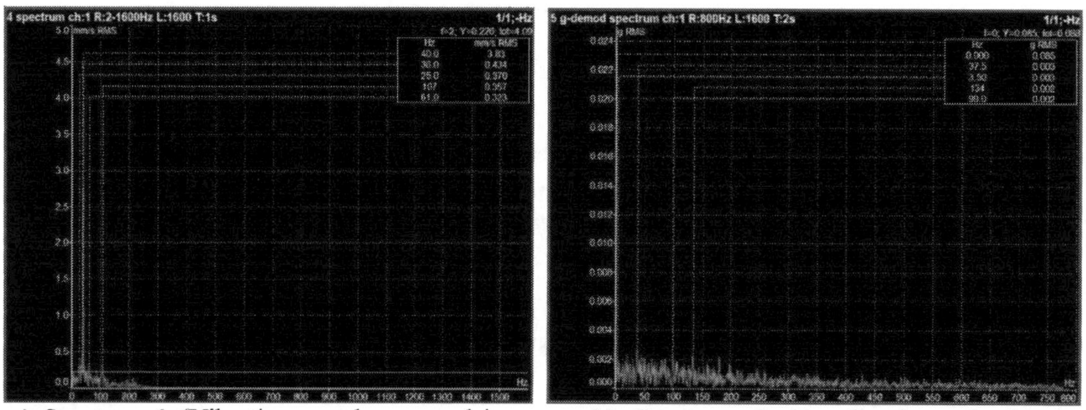

a) Spectrum 1 (Vibration speed measured in the vertical direction on the intermediate bearing).

b) Spectrum 2 (Acceleration of vibrations measured in the vertical direction on the intermediate bearing).

Figure 9. The results obtained for the vertical direction.

The highest RMS value is 3.83 mm/s at 40 Hz. According to ISO 10816-1, the operating rating will be "usable". This high value of the vibration speed could be due to the wrong mounting of the cardan shaft. It is known that the speed of the cardan shaft is not constant and it is transmitted in the same way, if the homokinetic conditions are not respected (it rotates "jerky").

According to the bibliographic source [9], the diagnosis for dynamic imbalance is based on the detection of peaks of higher amplitude at 1X in the radial direction. As with the help of the acquisition board there can be measured and analyzed the vibrations in the horizontal and vertical direction, by analysing both spectra.

The reference speed is 137 rpm (2.28 Hz), and as seen in figure 9, there is no signal peak at this frequency. It is concluded that the system is properly balanced.

For the third measurement, the accelerometer was placed in the axial direction. The results are depicted in figure 10.

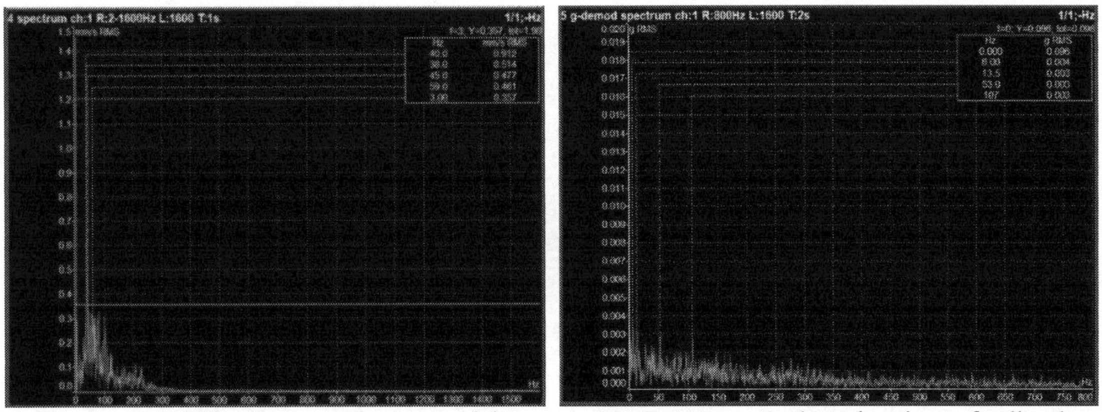

a) Spectrum 1 (Vibration speed measured in the axial direction on the intermediate bearing).

b) Spectrum 2 (Acceleration of vibrations measured in the axial direction on the intermediate bearing).

Figure 10. The results obtained for the axial direction.

The highest RMS value is 0.912 mm/s at 40 Hz. According to ISO 10816-1, the operating rating will be "good".

The reference frequency is 2.28 Hz (137 rpm). Figure 1 shows no frequency of 1X (2.28 Hz) or 2X (4.56 Hz), which indicates a good alignment of the assembly.

As mentioned above, it is necessary to compare the values of the cardan shaft's own frequencies with those measured using the accelerometer. The lowest frequency value is for vibration mode 1 and has a value of 377.39 Hz (figure 11). The highest value of the frequency is in the vertical direction and has the value of 134 Hz at the acceleration of 0.002 g. Thus, the values of the natural frequencies of the cardan shaft are not reached in any of the situations.

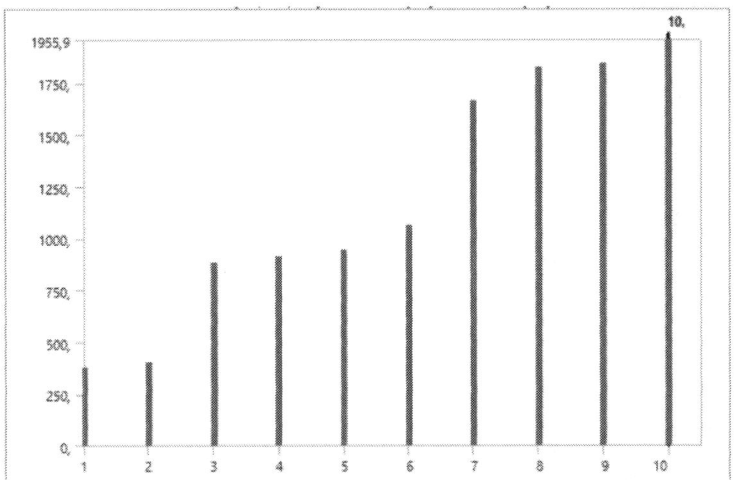

Mode	Frequency [Hz]
1,	377,39
2,	404,28
3,	882,83
4,	911,11
5,	940,67
6,	1063,5
7,	1660,3
8,	1822,8
9,	1842,3
10,	1955,9

Figure 11. The cardan shaft's own vibration modes.

Figure 12 shows the deformation distribution at the frequency of 1822.8 Hz (own vibration mode 8). It is observed that the resonance deformations have high values of up to 14.254 mm, which leads to the conclusion that in the case of resonance phenomenon, the shaft would break.

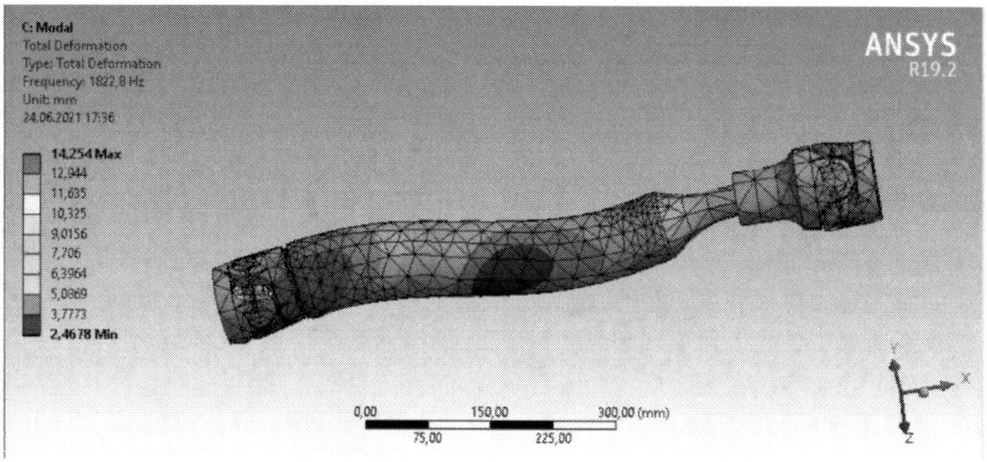

Figure 12. Deformation distribution at the frequency of 1822.8 Hz.

4. Conclusions

Non-destructive diagnosis methods allow the analysis of the technical condition of an assembly without disassembling or destroying the components.

For the correct operation of an ensemble, it is necessary to avoid reaching its own frequencies, in order to avoid the resonance phenomenon and to avoid major failures.

In the performed measurements, the vibration analysis allowed verification of the whole and the comparison with the reference values from the standards and with the information provided by the specialized literature.

Determining misalignment and dynamic imbalance with vibration analysis can provide an opportunity to apply preventive maintenance.

Based on the standards and by measuring the vibrations, the operating status of a system can be assessed, using ratings.

The modal analysis part in the Ansys program provides the user with the necessary information about the system's own vibration modes.

Acknowledgement

This work is supported by the project ANTREPRENORDOC, in the framework of Human Resources Development Operational Programme 2014-2020, financed from the European Social Fund under the contract number 36355/23.05.2019 HRD OP /380/6/13 – SMIS Code: 123847.

References

[1] Radeş M *Dinamica maşinilor III* Editura Printech Bucureşti 2008
[2] https://ro.scribd.com/doc/59581509/Ghid-analiza-vibratii
[3] Vilau R, Marinescu M, Alexa O, Truta M and Vinturis V *Diagnose method based on spectral analysis of measured parameters* Advanced Materials Research octombrie 2014
[4] ISO 10816-1:1995 (E) *Mechanical Vibration - Evaluation of machine Vibration by measurements on non-rotating Parts - Part 1: General Guidelines*
[5] *Instructions and maintenance manual VAMTAC S3* URO VEHICULOS ESPECIALES S.A.
[6] https://www.rumaniamilitary.ro/vanator-de-tancuri/rafael-spike-lr-uro-vamtac-s3-2
[7] *Workshop Manual URO VAMTAC S3,* URO VEHICULOS ESPECIALES S.A.
[8] *** https://issuu.com/nicolaeiacob6/docs/analiza_dinamic___a_sistemelor_de_t
[9] *** http://www.mobilindustrial.ro/current_version/online_docs/COMPENDIU/ rezonanta_1.htm? mw=NDY4&st=MA==&sct=MA==&ms=AAAAAAA=

IOP Conf. Series: Materials Science and Engineering 1220 (2022) 012043 doi:10.1088/1757-899X/1220/1/012043

The influence of the air filter location on the intake system temperature in the case of engines for drifting

C Birtok Baneasa, A Budiul Berghian, S A Ratiu

Politehnica University Timişoara, Faculty of Engineering Hunedoara, 5 Revolutiei Street, Hunedoara, Romania

corneliugroup@gmail.com

Abstract. This paper presents a comparative study regarding the influence of the air filter location on the intake system temperature in the case of engines for drifting. The heat flow dispersion map at the engine compartment was determined for four different cases. The measurements were performed with a thermographic camera in the area of the air filter and intake manifold. The results obtained after the test contribute to the efficiency of the thermal management of the engine by reducing the temperature of the intake air.

1. Introduction

Universal concerns about the degradation of ambient environmental conditions, stringent emission legislations, depletion of petroleum reserves, security of fuel supply, and global warming have motivated R&D of engines operating on alternative combustion concepts, which have the capability of using renewable fuels [1]. Homogenous Charged Compression Ignition (HCCI) is one of the new combustion strategies developed recently for Internal Combustion (IC) engines to fulfill the stringent emissions regulations while it can keep the thermal efficiency of the engine [2].

The amount of NOx in the exhaust increases evenly with the increase of the intake temperature 437 K to 477 K of the HCCI combustion engine powered by natural gas [2, 3].

A common fuel injection strategy in RCCI context is to inject the low-reactivity fuel (LRF) to the intake manifold while injecting the high-reactivity fuel (HRF) directly to the cylinder as a liquid fuel spray with a relatively long premixing time, i.e. early injection [4, 5].

The study shows the availability and potential of residual heat from internal combustion. It is known that over 50% of the heat from the combustion of internal combustion engines is residual heat in the main form of exhaust gases. Therefore, waste heat should be used for useful work [6].

2. Conditions and measurements

The thermal map was established through a thermal imaging camera, after the engine reaching the operating temperature at an ambient temperature being 16^0C within 20 minutes, after the drift cars left the track. Comparative measurements of the maximum temperature were performed at the engine compartment level, the air filter region, the intercooler (supercharged engines) and the intake manifold surfaces.

2.1. The cases studied

Four Drift engines with an engine displacement between 3.2l and 4.4l were considered for this study.

Case I – BMW E36 supercharged engine L 6 with an engine displacement of 3,2l, equipped with air filter of sport conical type;

Case II – Nissan Skyline r32 supercharged engine L 8 with an engine displacement of 4,0l, equipped with air filter of sport conical type;

Case III – BMW E30 supercharged engine V 8 with an engine displacement of 4,4l, equipped with air filter of sport conical type;

Case IV – BMW E36 aspirated engine V 8 with an engine displacement of 4,0l, equipped with air filter of sport conical type.

In all cases, the air filters are sport conical type with the filter element from textile or sponge, placed longitudinal or transversally to the car axis. The intake manifolds are positioned in the longitudinal side (left or right side) in the case of engines with the cylinders placed in a line (cases I and II) and in a central longitudinal plan in the case of engines with cylinders placed in V shape (cases III and IV). The materials used in the making of intake manifolds are Aluminium alloys AL319F (cases I and II) and plastics PA66 GF35 (cases III and IV).

2.2. Heat sources

The heat sources are:

- exhaust manifold;
- supercharging group;
- engine (cylinder head, engine block);
- cooling radiator;
- intercooler.

3. The heat flow dispersion

Case I BMW E36

The unprotected air filter is placed longitudinally on the right side of the car directly on the supercharger group. The heating sources are in this case the cooling radiator, intercooler and engine.

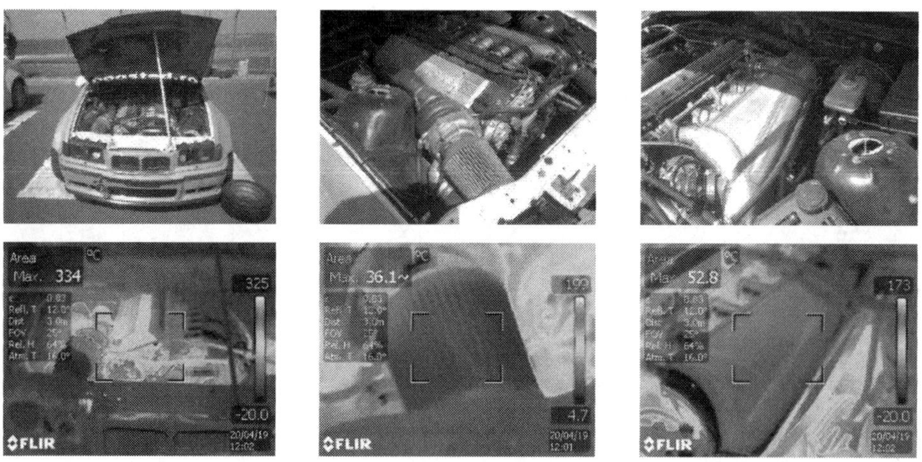

a b c

Figure 1. BMW E 36 heat flow dispersion: a-level of the engine compartment; b-surface of the air filter; c-surface of the intake manifold.

Case II Nissan Skyline r32

Compared to case one, in case two, the only difference is that the air filter is positioned on the left side of the car.

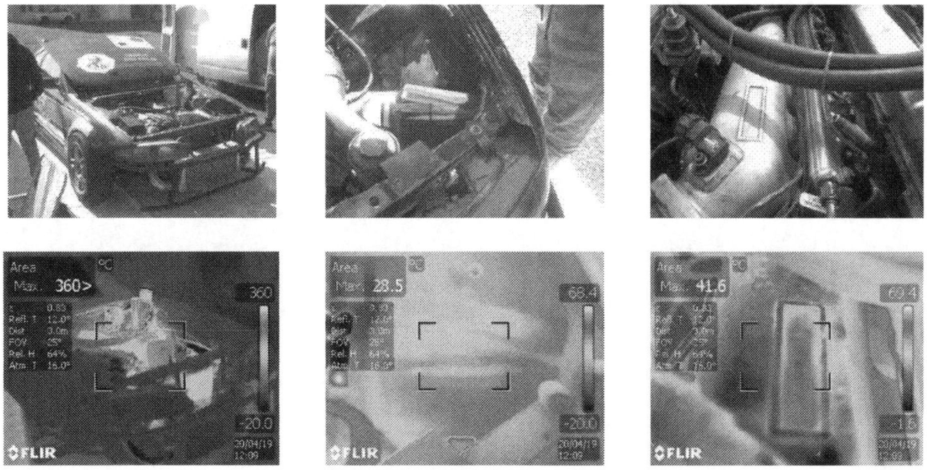

a b c

Figure 2. Nissan Skyline r32 heat flow dispersion: a – level of the engine compartment; b – surface of the air filter; c – surface of the intake manifold.

Case III BMW E30

The unprotected air filter is located centrally, directly on the supercharger group. The supercharging unit and the air filter are mounted in front of the engine due to the relatively small space of the engine compartment. The heating sources, in this case, are the cooling radiator, the intercooler, the engine and the exhaust manifold.

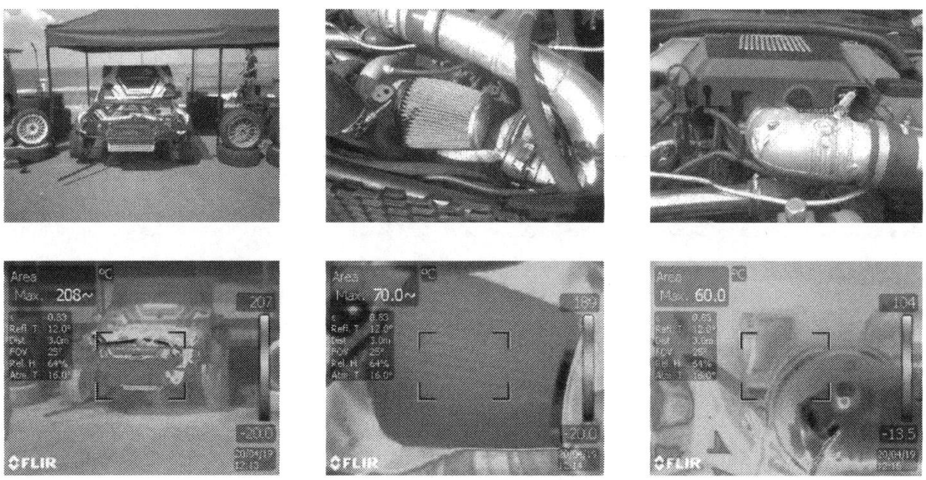

a b c

Figure 3. BMW E 30 heat flow dispersion: a – level of the engine compartment; b – surface of the air filter; c – surface of the intake manifold.

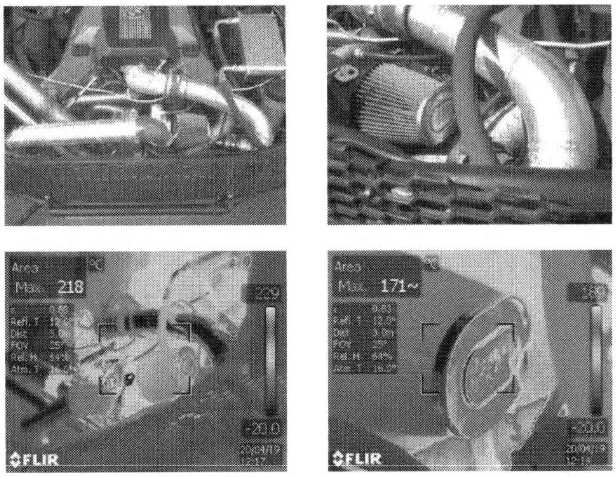

a b c

Figure 4. Air filter heat flow dispersion: a – air filter aria; b – frontal surface of the air filter; c – lateral surface of the air filter.

For constructive reasons, the route of the exhaust system is located under the air filter, which determines an additional heat input.

Figure 4 shows the following temperature values:

- 218^0C at the air filter aria;
- 171^0C on the frontal surface of the air filter;
- 70^0C on the lateral surface of the air filter.

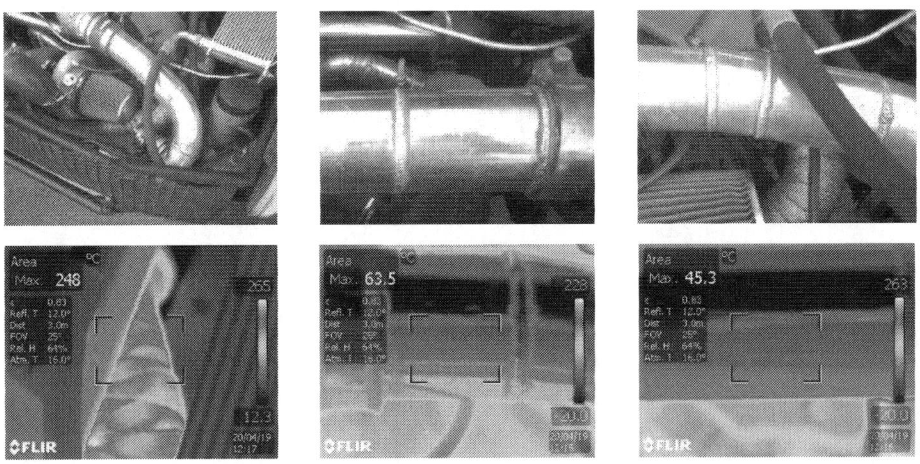

a b c

Figure 5. Pipes heat flow dispersion: a – exhaust pipe; b – intercooler inlet pipe; c – intercooler outlet pipe.

Figures 5 and 6 show the temperature values recorded on the surface of the exhaust pipes (figure 5a), on the supercharging route (intercooler inlet/outlet pipes figure 5b, c), aria of intercooler pipes (figure 6a), the surface of the intercooler (figure 6b) and the surface of the intake pipe (figure 6c).

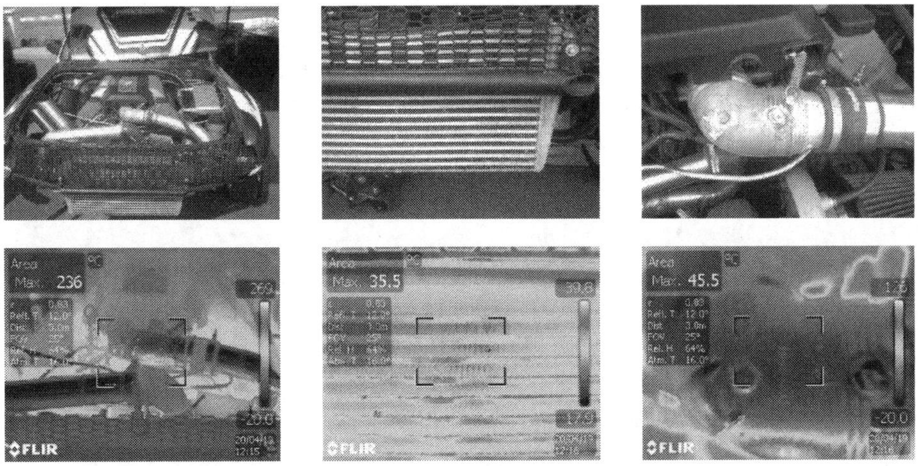

a b c

Figure 6. Intecooler heat flow dispersion: a – aria of intercooler pipes; b – surface of the intercooler; c – surface of the intake pipe.

Case IV BMW E36

The unprotected air filter is located transversely on the right side of the car. The heating sources are in this case the cooling radiator and the engine.

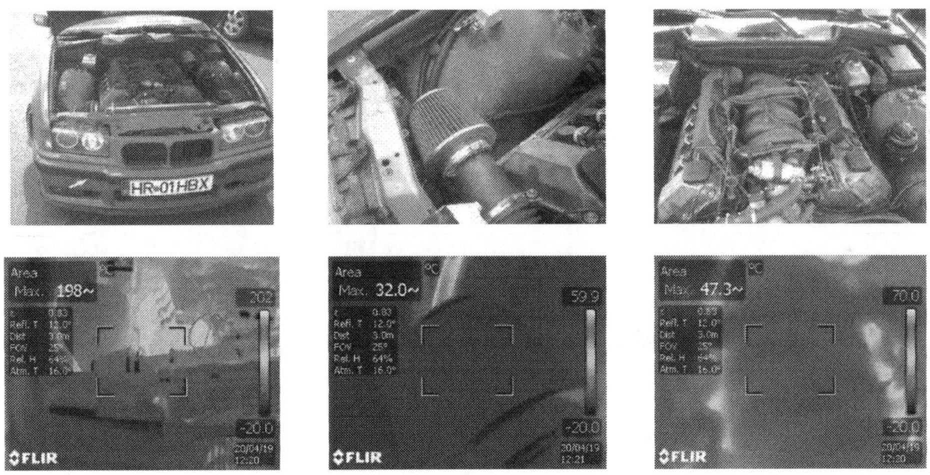

a b c

Figure 7. BMW E 36 heat flow dispersion: a – level of the engine compartment; b – surface of the air filter; c – surface of the intake manifold.

The temperature values measured in the case of the BMW E36 with aspirated engine are shown in figure 7.

4. Result and Conclusions

Following the execution of the measurements, for each case, three thermally stressed areas were taken into consideration engine compartment, air filter and intake manifold whose maximum values of temperature are presented in table 1.

Table 1. Cases maximum temperatures values

Cases	Temperatures (^{0}C)		
	engine compartment	air filter	intake manifold
1	334	36.1	52.8
2	360	28.5	41.6
3	208	70	60
4	198	32	47.3

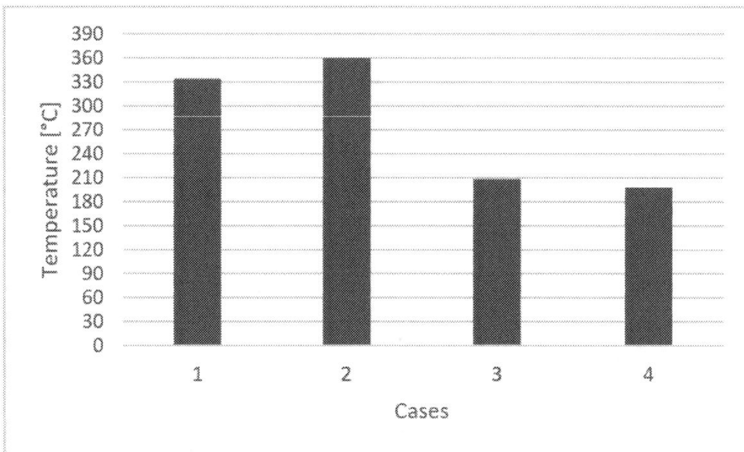

Figure 8. Cases variation of temperature fields in engine compartment.

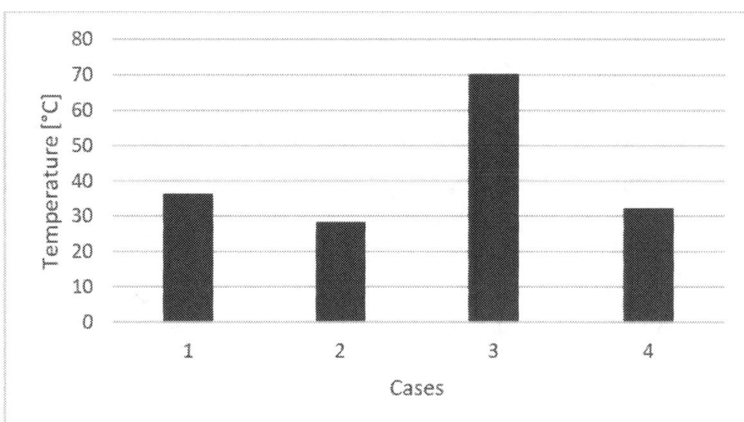

Figure 9. Cases variation of temperature fields on air filter.

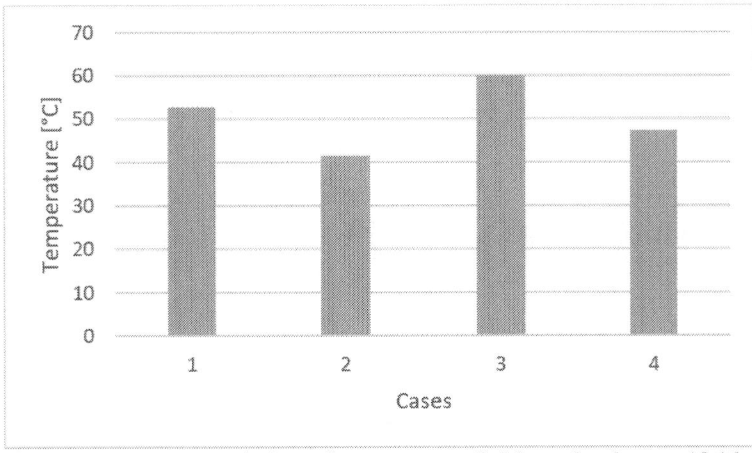

Figure 10. Cases variation of temperature fields on intake manifold.

Figures 8, 9 and 10 reveal that the maximum temperature measured at the engine compartment for case 2 is value 360 ^{0}C, regarding the air filter and the intake manifold, case 3 registers the highest values compared to the other cases.

In conclusion, the relatively high values of the temperature recorded on the intake path in the case of the 4 engines studied, mainly case 3 is due to the organization of the supercharging group, the air filter location and the lack of protection in the filter area.

A solution in order to reduce the temperature on the intake system consists in the implementation of an Air by Corneliu system composed of the super-aspiring air filter YXV, dynamic system of air transfer (STDA) and integrated thermal deflector. The researches have shown that temperatures have been reduced by up to 50% [7, 8, 9].

References

[1] Singh A P and Agarwal A K 2018 Low-Temperature Combustion: An Advanced Technology for Internal Combustion Engines *Singapore Springer*

[2] Chang D 2018 Parametric study of intake manifold temperature on a HCCI combustion engine. Combustion Theory and Modelling *Taylor & Francis* hal-01763206

[3] Chen R-H Chiang L-B Wu M-H and Lin T-H 2010 Gasoline displacement and NOx reduction in an SI engine by aqueous alcohol injection *Fuel 89* **3** 604-610.

[4] Li Y Ming J, Liu Y and Xie M 2013 Numerical study on the combustion and emission characteristics of a methanol/diesel reactivity controlled compression ignition (RCCI) engine *Applied energy* **106** 184-197

[5] Maurya R K, and Avinash K A 2012 Statistical analysis of the cyclic variations of heat release parameters in HCCI combustion of methanol and gasoline *Applied Energy 89* **1** 228-236

[6] Shahid Z Rehan H Ahmed K 2020 Thermodynamic analysis of increasing efficiency of a 4-Stroke Engine using Otto Cycle *researchgate.net DOI: 10.13140/RG.2.2.18217.21608*

[7] Birtok Baneasa C Ratiu S 2011 Admisia aerului în motoarele cu ardere internă – Filtre supraaspirante, sisteme dinamice de transfer *Timisoara Editura Politehnica*

[8] Birtok Baneasa C Heput T 2017 The flow heat dispersion in the engine compartiment. Case study on BMW 4.4L V8, *Annals of the Faculty of Engineering Hunedoara - International Journal of Engineering* **15(3)** pp.115-119

[9] *** www.corneliugroup.ro

Dynamics of frontal crash in/without the presence of passive safety systems

Garnita Ionut Alexandru, Corbu Gheorghe Marius, Toma Florin Daniel and Trusca Daniel Dragos

Automotive and Transport Department, Faculty of Mechanical Engineering, University "TRANSILVANIA" of Brasov, Brasov, Romania.

ionut.garnita@unitbv.ro

Abstract. The paper aims to debate on experimental retrieved accelerations from real crash tests, based on different scenarios, to enable assessment on driver potential injury. Scenarios included the presence of passive safety systems (e.g., airbag, seat belt) and test data are being processed to estimate the head injury criteria (HIC) followed by its correlation with abbreviated injury scale values (AIS). A dummy, average male resemblance, was developed in-situ and used in the experimental configuration. Signals were being acquired by few accelerometers disposed on the dummy's head and thorax using a Pic DAQ system (DSD, Linz, Austria). Data processing and values of injury criteria underline the importance of passive safety systems proven HIC experienced a consistent decrease in the presence of both air-bag and seatbelt.

1. Introduction

The increasing number of vehicles during the last decades in the urban area, as well as the intensification of congestions in intersections, has led to considerable road-accident recordings. Because of the increasing number of motor vehicles on the roads, the number of accidents has risen as well, mainly in the urban areas where congestions frequently occur. To counteract this massive increase, automotive manufacturers implemented active and passive safety systems to reduce the high level of traffic accidents and to prevent injuries not only to the occupants of the vehicles, but also to the pedestrians, that are usually involved in urban traffic accidents [1-2]. The research in road traffic motor vehicle accidents shows the study of the kinematic occupant behavior of the vehicle occupants during the collision but also the risk evaluation of the injury level [3]. Frontal collisions are considered deadly since both vehicles are traveling in opposite directions cumulating the impact velocity [4]. This type of collisions occurs very often on two-lane roads when a vehicle is engaged in a passing maneuver, the driver is not estimating the correct distance and is unable to retreat in time and colliding with the vehicle, which comes in the opposite direction [5].

Since its introduction on passenger vehicles, seat belts are very effective systems to protect the occupants in case of a collision. In Europe, the standard setup of the seat belt is the three-point configuration, covering the lap and shoulder of the passengers [6]. The most important motor vehicle crash safety innovation, which contributed to the reduction in mortality has been the installation and proper use of seatbelts [7]. Seatbelts reduce morbidity and mortality. About 50% to 80% of all deaths of RTC (Road Traffic Collision) could have been prevented by properly used seatbelt [8-9]. It has been proven that the seatbelt will save the passenger's life if it used, and not using it in the same circumstances will lead to death [10].

Along with the seatbelts, airbags have shown a good protection of the occupant head [11]. Airbags have been demonstrated to reduce car accident mortality by 19% and reduce morbidity from facial fractures and chest injuries. In purely frontal crashes, airbags reduce the fatality risk by 34% [12]. The airbag provides a good protection for all kinds of crashes (e.g., frontal, rear or lateral) that could seriously injure or kill the occupants of the vehicle [13]. Huère et al. has shown in a study that airbags can reduce head injuries by up to 82% for the 56-65 km/h range. Using the airbag, it was demonstrated that in 85% of cases, head injuries were classified as minor injuries [14].

The paper aims to present the experimental results gathered from controlled front collision tests, in the presence/absence of passive safety systems (e.g., air bag, seat belt, etc.). An in-situ manufactured dummy was used along with accelerometers to build in the recording system enabling assessment of one of the most important parameters in resulted injuries, namely the head injury criterion (HIC).

2. Experimental procedure and data processing

Two different scenarios were accounted on the controlled collision tests undertaken. The car body was being mounted on an in-situ developed testing stand (i.e., sleigh) with the Transilvania University of Brasov. *Scenario 1* envisaged the body section of the vehicle, the presence of an airbag incorporated in the steering wheel and the occupant wearing a seat belt, while *scenario 2* envisaged the body section of the vehicle, without airbags, without the occupant wearing a seat belt. Both scenarios were performed at a velocity of 17 km/h.

2.1. Equipment

An anthropomorphic dummy was developed *in-situ*, resembling the anatomy of a senior male and positioned on the driver seat. The acquisition system includes PIC DAQ accelerometers displayed on the vehicle body and Pocked DAQ accelerometers (DSD, Linz, Austria) mounted inside of the dummy's head and thorax. A high-speed camera, HiSpec 5 (Fastec Imaging, USA), enabling image acquisition at 1400 frames/second, along with a GPS velocity sensor to measure vehicle's velocity was completing the system.

All tests have been conducted in the absence of other vehicles or pedestrians nearby to ensure safe environments. Figure 1 captures the car body structure (left) and the sleigh for front collisions (right). The movement of vehicle body is free as it is being launched from the top of the test stand whereas the impact zone is with a rigid barrier.

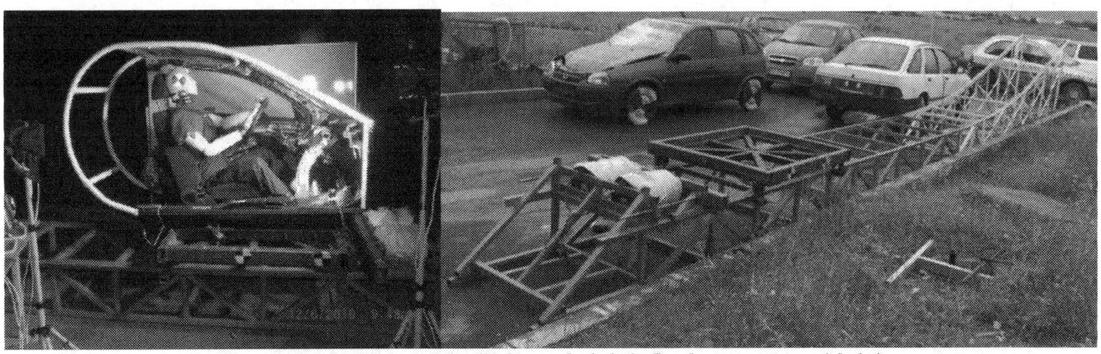

Figure 1. Vehicle body (left) and sleigh for impact tests (right).

The collision phases are presented in figure 2 for both scenarios. The *pre-impact phase* where the male dummy is positioned before the contact with airbag or body section of vehicle, the *impact phase* where the energy is transferred and the occupant moves inside of body section of the vehicle and the *post-impact phase* when the test dummy remains motionless. The experimental configuration of *scenario 1* (left) is more suggestive accounting on the presence of the airbag shutter. Additionally, the seatbelt has a pre-tension system that activates during the impact.

2.2 Data processing

Experimental data were imported in Microsoft Excel and subjected to additional processing to retrieve one of the impact injuries parameters, namely HIC. The HIC criterion is the main route of assessing the risk of a cranial trauma injury as a result of an accident or for testing individual protection equipment [15]. HIC is being estimated by performing the integral on the recorded accelerations [16]:

$$\text{HIC} = \left\{\left(\left[\frac{1}{t_2-t_1}\cdot\int_{t_1}^{t_2} a\cdot dt\right]^{2.5}\cdot(t_2-t_1)\right)\right\}_{MAX} \quad (1)$$

HIC is time range dependent, such as:

- HIC 15 – maximum 15 ms;
- HIC 36 – maximum 36 ms;
- HIC – unlimited.

Alternatively, the abbreviated injury scale (AIS) is being used to assess the degree of injury of occupants involved in traffic accidents by ranking them from 1 to 6, the function of their severity, minor injuries are ranked as 1 while profound lesions leading to death are ranked as 6 [17, 18].

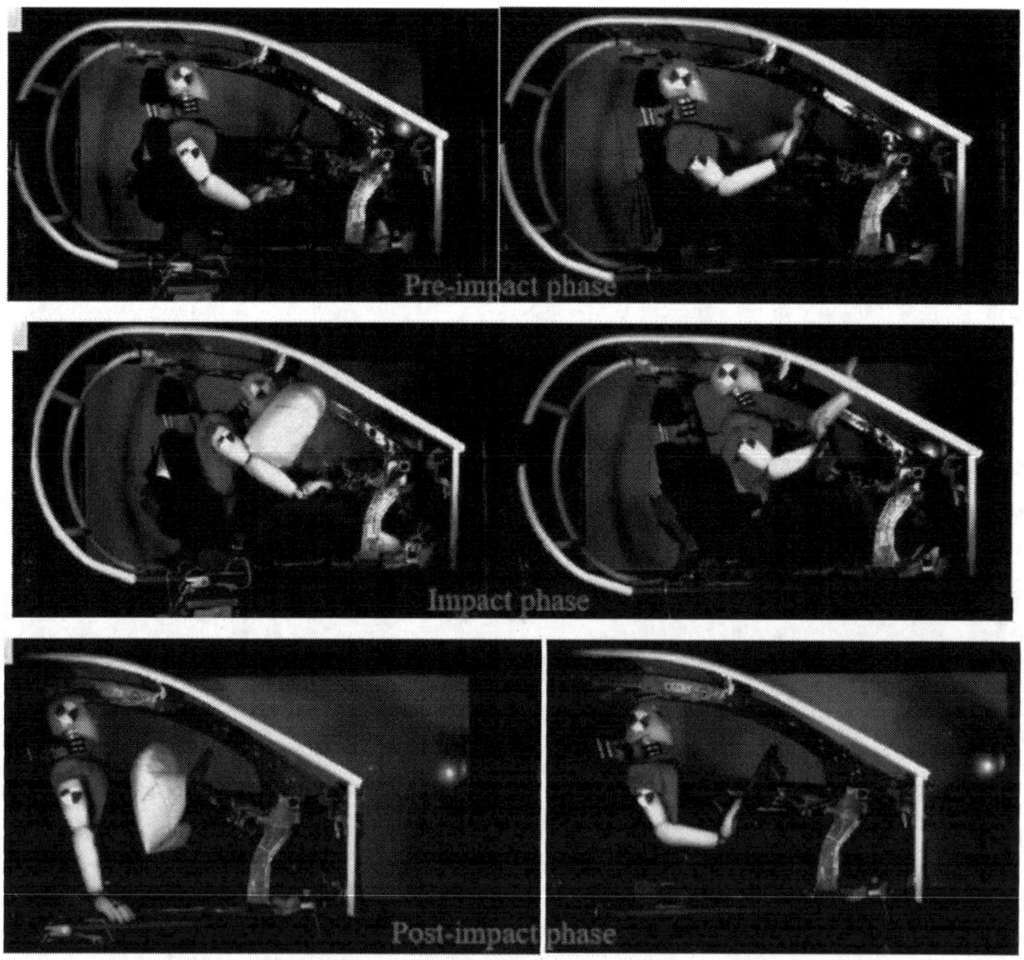

Figure 2. Frontal crash phases – *scenario 1* (left) and *scenario 2* (right).

3. Results and discussions

In figures 3 and 4 are being represented the experimental data recorded by the accelerometers disposed inside dummy's head while figures 5 and 6 from those disposed inside dummy's thorax, in all directions, accounting on both scenarios. Directions Ox and Oy correspond to the accelerometers main section while Oz axis is normal to the dummy's body.

As it can be seen from figure 3 and figure 4, recorded deceleration values are higher in the conditions of scenario 2 corresponding to the no seat belt, no air bag safety systems, 140 ms^{-2} on the Z axis during the first 200 ms of the collision compared with 73 ms^{-2} around 650 ms from the conditions of the *scenario 1*. The reported values correspond to the peak values. With respect to the values on the Ox direction, these are comparable for both scenarios and about 65 ms^{-2} for the same recorded time.

In the context of the first scenario, the values of accelerations recorded at the level of the thorax were much lower than their counterparts, because of the presence of the seat belt (see figure 5). Thus, the acceleration along the Ox axis is around 125 ms^{-2} at 120 ms after impact. Conversely, the maximum value of acceleration along the Ox axis, in the scenario 2 condition, is about 190 ms^{-2} at 200 ms after impact. The latter value is higher compared with those in the first scenario conditions due to the free movement condition.

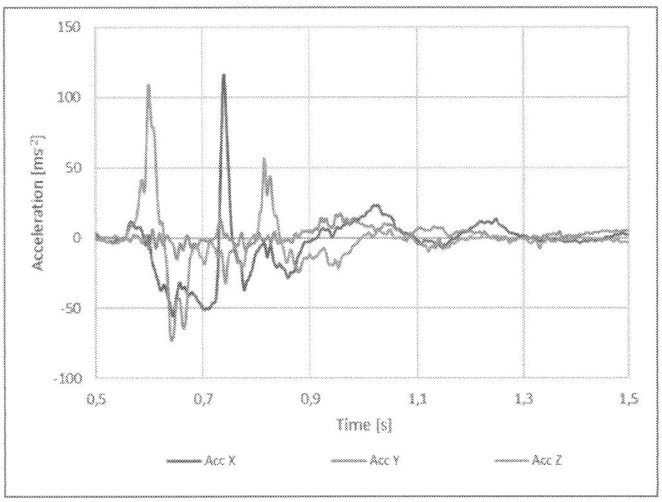

Figure 3. Acceleration values recorded at occupant's head - *scenario 1.*

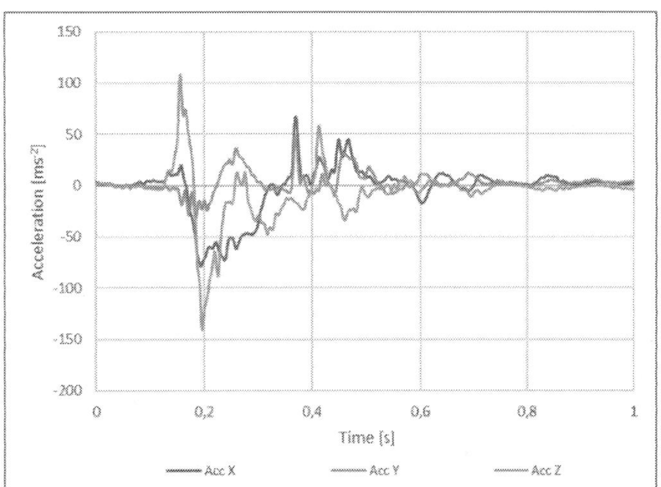

Figure 4. Acceleration values recorded at occupant's head - *scenario 2.*

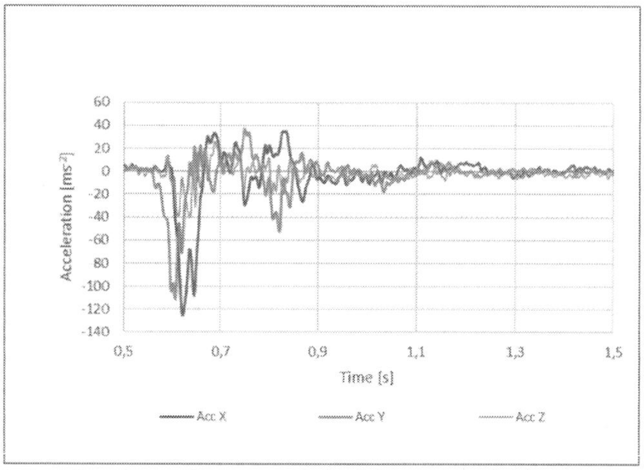

Figure 5. Acceleration values recorded with occupant's thorax – *scenario 1.*

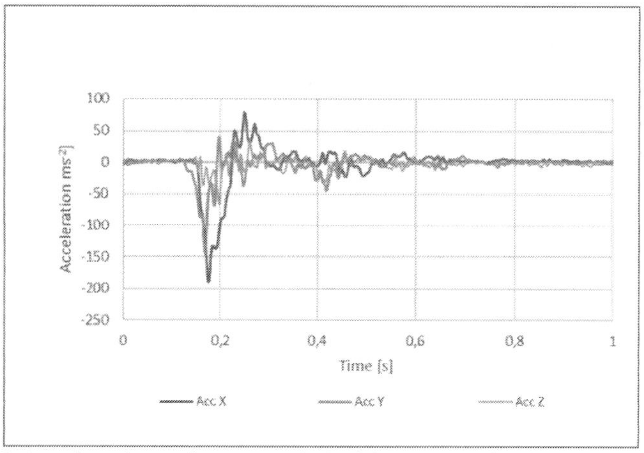

Figure 6. Acceleration values recorded with occupant's thorax – *scenario 2.*

Next, for HIC estimation the resultant acceleration is needed. In the present study, only the HIC 36 was estimated in all scenarios, corresponding to a time interval of 36 ms. Thus, figure 7 and figure 8 are capturing the acceleration resultant of values recorded at head level in the conditions of scenarios followed.

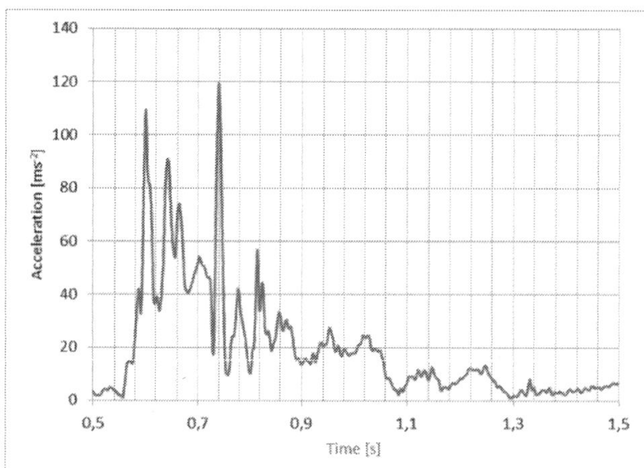

Figure 7. Acceleration resultant – *scenario 1.*

As it can be seen from figure 7, the maximum peak value is about 119 ms^{-2} caused by the impact of the head with the airbag. In the conditions of *scenario 2*, with no set belt and no air bag, the first peak of the resultant of accelerations in all directions of the cartesian system is about 162 ms^{-2} as recorded at 200 ms after impact (see figure 8).

HIC parameter estimation follows in data processing steps. Acceleration resultants were subjected to a smoothing process using an FFT filter to remove the noisy data from the entire signal. Next, by numerical integration accounting for a time range of 36 s the values of HIC36 for both scenarios were gathered to enable head injury's assessment.

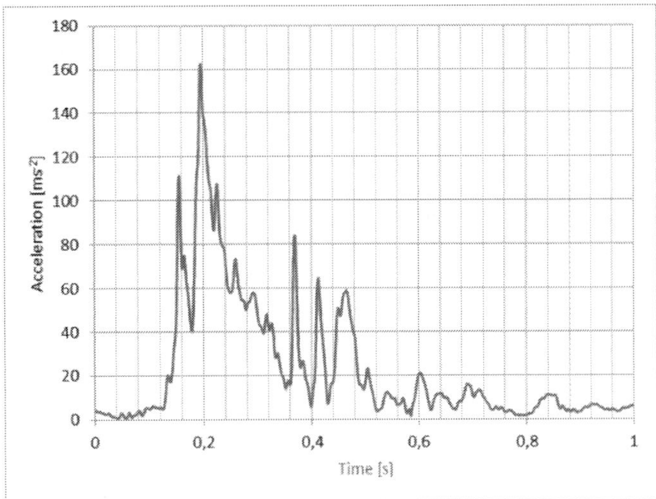

Figure 8. Acceleration resultant – *scenario 2.*

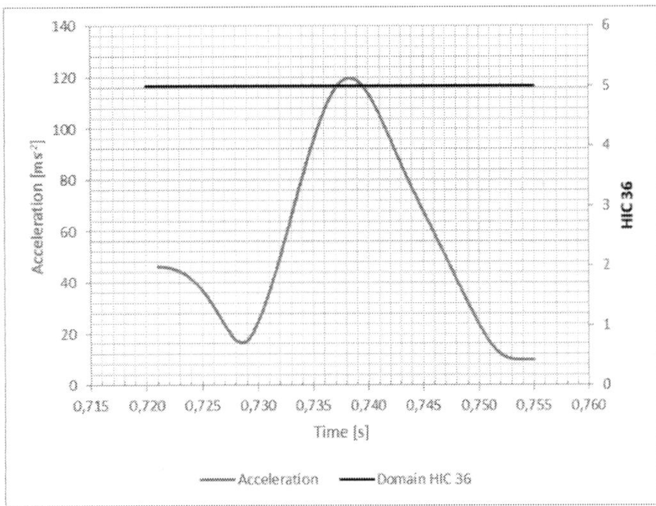

Figure 9. HIC 36 estimations in the *scenario 1* conditions.

In the conditions of *scenario 1*, in the presence of passive safety systems (e.g., seat belt, air bag) data recorded lead to a HIC36 value of 5.00 (see figure 9). A correlation with the AIS scale enables head injury ranking to 1 that corresponds to a minor injury. Even in the controlled conditions of low velocity tests such as the herein approach, the presence of passive safety systems reveal their importance in the injury assessments.

On the other hand, as it might been foreseen, the HIC36 value in the conditions of *scenario 2*, without deployment of safety systems is higher than above. Numerical integration was leading to the value of 18.85 for the HIC36 parameter. Translating to the AIS scale, the injuries can be ranked as minor injuries since the corresponding value is 1 (see figure 10).

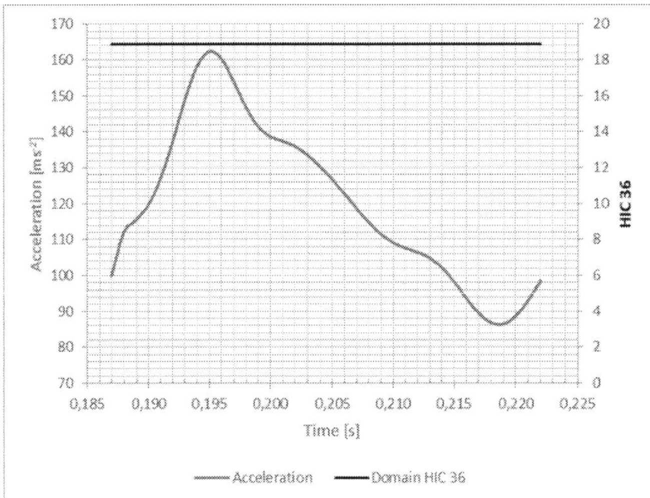

Figure 10. HIC 36 estimations in *scenario 2* conditions.

4. Conclusions

Following the processing of the experimental data and the calculation of the injury criteria, the degree injury of the occupant was determined by correlating the results with the values corresponding to the AIS scale diagram. Based on the obtained results it can be concluded that in low-speed impacts such as the ones presented, passive safety systems are very important and can make the difference in reducing the injury degree of the occupants. Even if the probability of injury was minor, the tests performed demonstrated that the passive safety systems, which today's vehicles are equipped will reduce the severity level for the occupants of the vehicle even at low speeds such as 17 km /h.

Head Injury Criteria (HIC) values for frontal collision tests recorded low values in both scenarios enabling to rank the risk of injury of the occupant as minor (1 on the AIS scale).

Regarding the accelerations recorded by thorax level, we can observe that by wearing the seat belt we can reduce the potential injuries of the occupant.

References

[1] Otat, O.V., 2015, Modeling The Frontal Collison In Vehicles And Determining The Degree Of Injury On The Driver. ACTA Universitatis Cibiniensis, 67(1):115-20.

[2] Tolea, B., Trusca, D., Antonya, C., and Beles, H., 2015, The influence of the frontal profile design of a vehicle upon the pedestrian safety at low velocity, Annals of DAAAM & Proceedings, 26(1).

[3] Otat, O.V. and Dumitru, N., 2016, Vehicles Passive Safety Systems Influence on Driver's Thorax Injuries, In Applied Mechanics and Materials (Vol. 823, 187-192). Trans Tech Publications.

[4] Jarasuniene, A. and Jakubauskas, G., 2007, Improvement of road safety using passive and active intelligent vehicle safety systems, Transport, 22(4), 284-289.

[5] Elmarakbi, A. M. and Zu, J. W., 2006 Crash analysis and modeling of two vehicles in frontal collisions using two types of smart front-end structures: an analytical approach using IHBM, International Journal of Crashworthiness, 11(5), 467-483.

[6] Van Houten, R., Malenfant, J. E., Austin, J. and Lebbon, A., 2005, The Effects of a Seatbelt-Gearshift Delay Prompt on the Seatbelt Use of Motorists Who Do Not Regularly Wear Seatbelts, Journal of applied behavior analysis, 38(2), 195.

[7] Abbas, A. K., Hefny, A. F. and Abu-Zidan, F. M.,2011, Seatbelts and road traffic collision injuries, World Journal of Emergency Surgery, 6(1), 18.

[8] Hodson-Walker N.J., 1970, The value of safety belts: a review. Can Med Assoc J., 102: 391-393.

[9] Cummings P., 2002, Association of seat belt use with death: a comparison of estimates based on data from police and estimates based on data from trained crash investigators. Inj Prev.

[10] Høye, A., 2016 "How would increasing seat belt use affect the number of killed or seriously injured light vehicle occupants?." Accident Analysis & Prevention, 88, 175-186.

[11] Gabauer, D. J. and Gabler, H. C., 2010, The effects of airbags and seatbelts on occupant injury in longitudinal barrier crashes, Journal of safety research, 41(1), 9-15.

[12] Toganel, G.R. and Soica, A.O., 2017. A late and failure of airbag deployment case study for drivers of passenger cars in rear-end collisions. In IOP Conference Series: Materials Science and Engineering (Vol. 252, No. 1, p. 012020). IOP Publishing.

[13] McFeely Jr, W.J., Bojrab, D.I., Davis, K.G. and Hegyi, D.F., Airbag Deployment Study-Otologic Injuries Secondary to Airbag Deployment Otologic Injuries Secondary to Airbag Deployment.

[14] Huère, J.F., Foret-Bruno, J.Y. and Faverjon, G., 2001 Airbag efficiency in frontal real world accidents. In Proceedings: International Technical Conference on the Enhanced Safety of Vehicles (Vol. 2001, pp. 6-p). National Highway Traffic Safety Administration.

[15] Gao, D. and Charles W. W., 2009, "Head injury criterion." IEEE robotics & automation magazine 16.4: 71-74.

[16] McHenry, B. G., 2004, Head injury criterion and the ATB, ATB Users' Group, 5-8.

[17] Greenspan, L., McLELLAN, B. A. and Greig, H., 1985, Abbreviated Injury Scale and Injury Sever-ity Score: a scoring chart, The Journal of trauma, 25(1), 60-64.

[18] Shojaati, M., 2003, Correlation between injury risk and impact severity index ASI, Swiss Transport Research Conference.

The influence of the Advanced Emergency Braking System in critical scenarios for autonomous vehicles

Laurențiu Carabulea[1, 2], Claudiu Pozna[2], Csaba Antonya[3], Călin Husar[1] and Alexandra Băicoianu[1, 4]

[1]Siemens Industry Software SRL, 13A Gării Blvd., Brașov, Romania
[2]Department of Automation and Information Technology, Transilvania University of Brasov, 29 Eroilor Blvd., Romania
[3]Department of Automotive and Transport Engineering, Transilvania University of Brasov, 29 Eroilor Blvd., Romania
[4]Department of Mathematics and Computer Science, Transilvania University of Brasov, 29 Eroilor Blvd., Romania

laurentiu.carabulea@siemens.com, ionut.carabulea@unitbv.ro

Abstract. This paper proposes use indications for an Advanced Emergency Braking System (AEBS) module developed in Matlab and included in a Simcenter Prescan simulation. The analysed indications depend on a Long-Range Radar (LRR) and Short-Range Radar (SRR) beam range, on the speed of the autonomous vehicle and on the time between the detection of the obstacle in front of the autonomous vehicle and the time the obstacle starts to move. The simulated scenario consists of an autonomous vehicle that travels at a certain speed, where another vehicle starts to move from a bus station after a sudden stop, challenging the autonomous vehicle to stop in a safe manner and in a short time.

1. Introduction

Safety on public roads is a global concern of car manufacturers and national/European authorities. Future trends in urban mobility, given by the appearance of new technologies, propose autonomous vehicles as a solution for classic road transport. Although the Advanced Driver-Assistance Systems (ADAS) technology improve significantly the safety of a vehicle compared to a human-only driven vehicle, there are some challenges for engineers and software developers for which they must find the proper solution in order to progress on the car autonomy levels. The Autonomous Emergency Braking System (AEBS) is one of the ADAS that gathered more attention in the last 10 years. The AEBS mounted on automobiles use radar (radio detection and ranging), lidar (light detection and ranging) and camera altogether, a combination of two of these or single detection system (e.g. 2 radars) [1, 2] as studied in this paper, to detect and avoid an imminent potential collision. Figure 1 and figure 2 describe the role and the structure of this AEBS.

The hardware and software of the ADAS have been consistently tested over the years, and their stability has been proven not only for them but for many other applications. The interest is therefore more acute on the verification and validation of the systems. To show that an autonomous (ego-driven) vehicle performs better than a human-driven vehicle, they have to be driven for 440 million kilometres [3]. Considering the number of kilometres to be driven to prove the autonomous vehicles are safe, the real-world testing alone is impossible in a short period of time, therefore the virtual development of critical scenarios became a common practice to test many aspects of a critical scenario even faster than in real-time, at much lower costs.

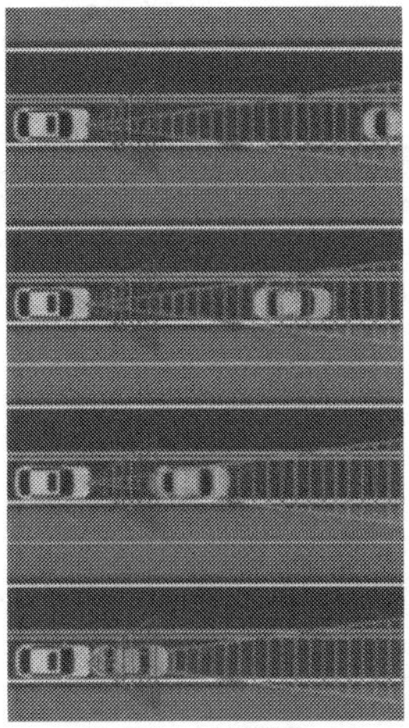

Figure 1. The AEBS operation steps.

Step 1 – Object detection (TTC warning)
Step 2 – Partial braking phase (TTC 40% brake)
Step 3 – Full braking phase (TTC 100% brake)
Step 4 – Collision avoidance or mitigation
TTC – Time To Collision (explained in Chapter 4)

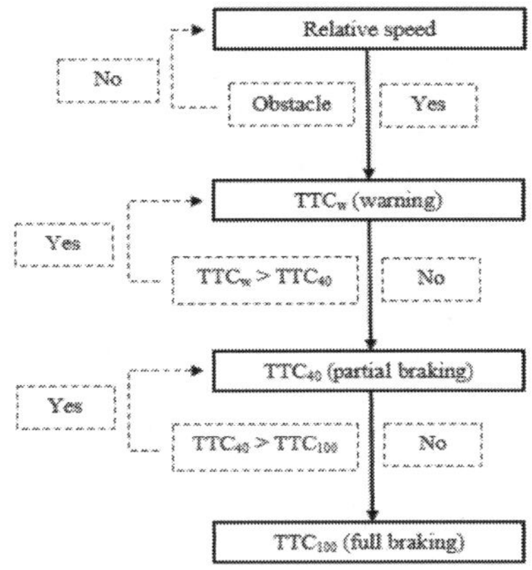

Figure 2. The AEBS structure.

Its performance, in terms of prevention or reduction the severity of a collision, can be seen in a study conducted by the Ministry of Land, Infrastructure, Transport and Tourism in Japan, summarized in figure 3, showing that more than 40% of the vehicles sold in 2016 were equipped with AEBS [4]. Another study provided by Technavio [5] confirms the growth of AEBS in the 2021 – 2025 interval at a Compound Annual Growth Rate (CAGR) of 8%, where a 2.8 million units growth in global commercial vehicles is expected.

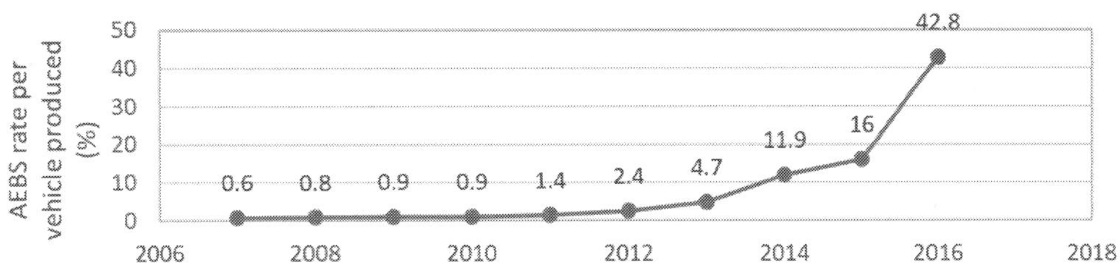

Figure 3. AEBS trends in Japan [3].

2. Software platforms used

The current study involves the capabilities of Simcenter Prescan and Matlab Simulink software to model and simulate autonomous vehicles' safety system architectures, which are then placed in various critical traffic situations.

2.1. Simcenter Prescan

Simcenter Prescan software is an integrated physics-based simulation platform for ADAS and automated driving, helping the simulation engineers to develop, simulate and validate advanced driver assistance systems. A 2D image (Simcenter Prescan view) and a 3D image (Simcenter Prescan Viewer) are shown in figure 4 and figure 5, respectively.

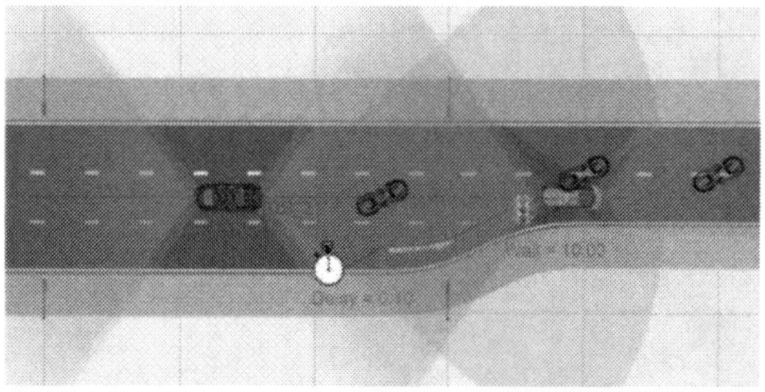

Figure 4. Simcenter Prescan.

Figure 5. Simcenter Prescan Viewer.

2.2. Matlab Simulink

Simulink, the graphical and more engineer-oriented part of *Matlab*, is a block diagram environment for Model-Based Design. It supports system-level design and simulation for modelling and simulating dynamic systems. The Compilation Sheet attainment contains all the information written by Prescan in the pre-processing phase, that can be processed by Matlab. An existent AEBS model block has been added to the Simulink model. The block scheme shown in figure 6 exposes the inputs and outputs of the two (Prescan for simulation, Matlab Simulink for calculus) platforms used.

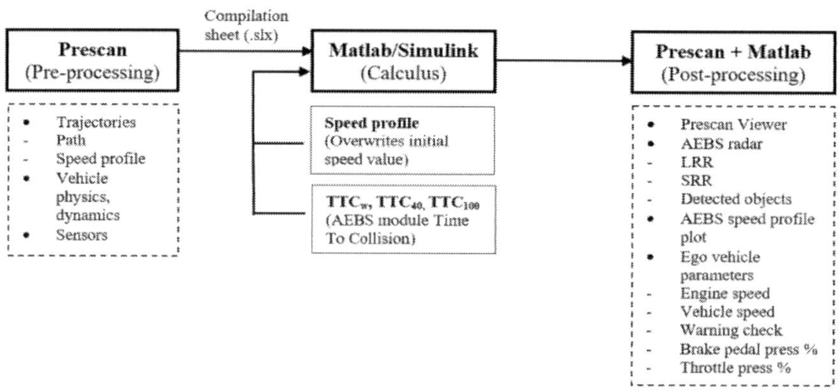

Figure 6. Input-output Prescan-Simulink simulation.

The simulation starts with Simcenter Prescan. In this pre-processing phase, the user sets as input the following parameters:

- **UI** (User Interface) **parameters** – not considered during the calculus phase. These parameters are used only for visualization purposes. As example: environment, which is the scene base layer (e.g. grass, concrete), infrastructure (e.g. roads, traffic signs), actors (e.g. vehicles, including sensors, trajectories), paths [6]. The actors and the paths are taken into consideration also in the calculus phase.
- **Calculus parameters**. In this category of parameters are included: actors (vehicles) with their specified trajectories (i.e. trajectories subdivided to path and speed profile), physics and dynamics. Another parameter considered during the calculus phase are the sensors and their setup.

The Prescan setup of the simulation is then exported as a file called Compilation Sheet (pre-processing output), having the .slx Simulink extension. By opening the Simulink .slx file, other parameters can be set: speed profile (can be overwritten the initial speed of the ADAS vehicle, and the time-to-collision TTC for the warning phase, partial brake and full brake. Thus, the input for the calculus phase consists of the output from the pre-processing phase (i.e. the Compilation Sheet) and of the parameters set in Simulink. Once all parameters are set, the simulation can start. There are multiple outputs sent to the post-processing phase, as braking system pressure[u2], throttle pedal %, vehicle deceleration, depending on the ADAS vehicle speed can be triggered the warning flag, the partial brake and, if necessary, the full brake. Other outputs are the indicator lights, brake lights, but these outputs have only visualization purposes.

The post-processing phase has as starting point the output resulted from the simulation. The post-processing is divided into a Prescan post-processing, the Prescan Viewer being used to visualize the scene in a three-dimensional perspective, and the Simulink plots. The AEBS radar plot (figure 8) points out the SRR and LRR influence over the obstacles detected (on both road and sidewalk), the AEBS speed profile plot shows the deceleration and velocity of the vehicle at TTC_w, TTC_{40} and TTC_{100}, and the third plot shows the ego vehicle parameters as it approaches the obstacle: engine speed, vehicle speed, the warning flag and the brake and throttle pedals pressing percentages.

3. Critical scenario under study

The studied critical scenario consists mainly of an automated vehicle and a vehicle considered obstacle. The autonomous vehicle has mounted a short-range radar (SRR) [7] and a long-range radar (LRR) [8]. Data read by the two radars is sent to an integrated solution for ADAS data collection, used further by the autonomous emergency braking system to detect potential dangers (figure 7). The second vehicle starts to move from a bus station, then suddenly stops in front of the automated vehicle which has different speed and AEBS settings (i.e. a Prescan and a Prescan Viewer example are shown in figure 4 and figure 5). The speeds of the vehicles are set to their speed profiles in the pre-processing phase, as well as their paths.

To be noted the two detected objects sketched by the two blue stars, which are in the SRR-only range. They are not detected as warnings (red stars) because they are not in front of the autonomous vehicles, therefore are not considered dangerous obstacles. The vehicle in front, later seen as an obstacle, is first seen as a detected object. As the autonomous vehicle approaches the vehicle in front, the ADAS goes into the warning phase (the blue start turns to a red star). Taking into account the TTC_{40}, the autonomous vehicle starts to brake. This is the partial braking phase. As it approaches more to the obstacle, the full braking phase is initiated, but only when the obstacle is in the SRR range. For this reason, the speed of the vehicle, the SRR range and the TTC_{100} have to be correlated, to avoid the case when the TTC_{100} is set to a value, but the SRR does not observe the obstacle in its range.

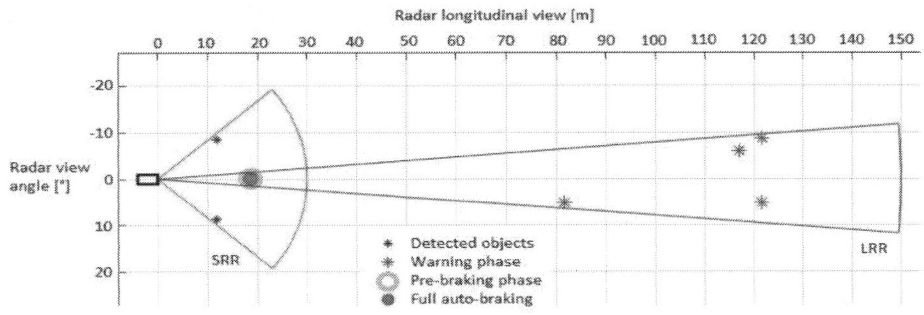

Figure 7. Vehicle sensors and objects detections.

4. Autonomous vehicle model

The autonomous vehicle developed and used in this paper incorporate two radar-based sensors. As mentioned in the critical scenario description chapter, the LRR [7] has a 150 metres view range, while the SRR [8] has two different ranges, depending on the case of study: a 30 metres range and a 50 metres range. Even if the 50 metres range SRR detects the obstacle earlier, a 30 metres version has been used to highlight the importance of a longer-range radar when the vehicle is traveling at higher speeds. For lower speeds, there is no difference between the 30 and the 50 metres radars, as example the vehicle has a speed of 20 metres per second, and the TTC_{100} is 1.5 seconds. In this example, the 50 metres SRR is not necessary, as the full brake is applied at the moment the obstacles are detected in the 30 metres range.

The configuration of these sensors is:

- Long-range radar: beam range: 150 m, beam type: pyramid, beam $\Delta\theta$: 9°, beam $\Delta\varphi$: 9°, scan direction: left-right/top-bottom, number of beams: 1, capture frequency: 25 Hz
- Short-range radar: beam range: 30/50 m, beam type: pyramid, beam $\Delta\theta$: 80°, beam $\Delta\varphi$: 9°, scan direction: left-right/top-bottom, number of beams: 1, capture frequency: 25 Hz

Vehicles have pre-defined paths of movement and speed profiles, configured to generate the critical scenarios (figure 6 – Pre-processing phase). Moreover, the automated vehicle has set a 2D simple dynamics configuration. The dynamic configurations of the vehicles have existent (default) Prescan values, specific for each vehicle used in simulation.

The pre-defined AEBS module has been set to fulfil the studied critical scenarios, where the automated vehicle is driven at different speeds (40 m/s to 10 m/s, with a 5 m/s decrement), considering different warning, partial brake and full brake times. As seen in figure 7, the TTC for the warning and the two braking phases is set as input directly to Simulink, as the AEBS block (Matlab .mdl file) is inserted separately into the Simulink file exported from Prescan. The calculated time from detection to collision is the pre-mentioned time-to-collision (TTC). Warning TTC occurs when a possible dangerous situation in front of the vehicle is detected, followed by a pre-set partial braking time (40% of maximum braking pressure measured in bar), and if the vehicle cannot avoid the collision, then the full brake is applied (100% braking pressure).

5. Results

During the simulations, several parameters of the critical scenario have been modified, in order to find the correct configuration of the radars, of the warning TTC, partial braking TTC and full braking TTC, considering different speeds of the automated vehicle. The four iterations that have been studied are presented in table 1.

Observing the cases from each of the four iterations, the only difference consists in the initial conditions, more specifically in the vehicle speed. All the other initial parameters of the cases contained by an iteration are identical.

At the iterations level, the scenarios of the first iteration differ from the second and third iteration by the TTC set up. To be observed the TTC for the warning, partial- and full-braking phases are set to 2.5 seconds for TTC_w, 1.5 seconds for TTC_{40} and 0.5 seconds for TTC_{100}. The highest speed of the vehicle (and considering the first iteration initial conditions) where there is no impact to the obstacle is 15 metres per second (see table 1 – Case 6).

During the second iteration, the TTC has been increased for the 3 indicators by 1 step for each of them: TTC_w: 2.5 s --> 3.0 s, TTC_{40}: 1.5 s --> 2.0 s, TTC_{100}: 0.5 s --> 1.25 s.

Thus, by increasing the TTC, the initial speed of the vehicle can be increased to 25 metres per second without impact to the obstacle. This means the earlier the vehicle observes the obstacle and brakes, factors correlated to the speed of the vehicle, the smaller the impact, or no impact at the same speed in contrast with the first iteration

For the third iteration, only the TTC_{40} has been increased from 2.0 seconds to 2.5 seconds. This iteration compared to the second iteration sets more highlight on the importance of a sooner detection of the obstacle. figure 9 represents Iteration 3 – Case 15, where the acceleration and the velocity of the autonomous vehicle are plotted, pointing out the warning, partial braking, full braking and the impact of 67 kilometres per hour to the obstacle.

The approach for the fourth iteration is different than for the first three iterations. Instead of increasing the TTC, the beam range of the SRR has been increased, from 30 metres to 50 metres. Even so, there is an impact of 42 kilometres per hour to the obstacle for a vehicle speed of 40 metres per second. Compared to the first three iterations, the impact when the autonomous vehicle has the highest simulated speed decreases significantly. Thus, we can consider the impact can be accepted in this high vehicle speed case.

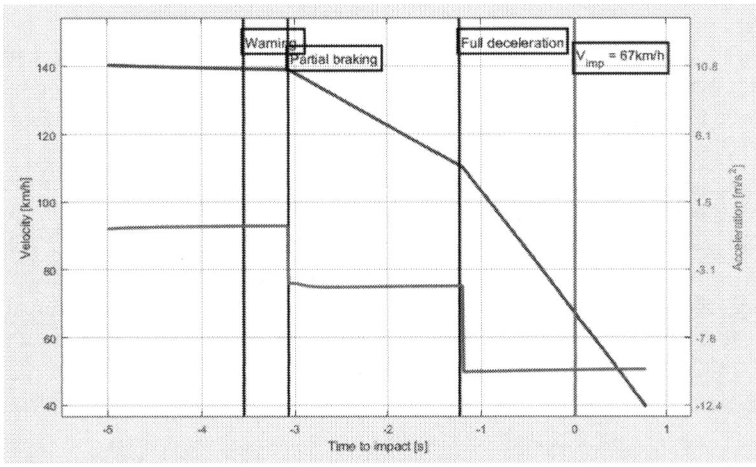

Figure 8. Velocity and acceleration plots from TTC warning to collision (Case 15).

To be noticed in figure 8, the deceleration slope of the vehicle is constant after the impact. The reason is that the collision is detected, but the vehicles are not modelled to absorb the energy of the impact.

Table 1. Simulation results.

Parameters values / Cases	Vehicle speed							Beam range		TTC warning		TTC 40% brake		TTC 100% brake		Impact
	144 km/h	126 km/h	108 km/h	90 km/h	72 km/h	54 km/h	36 km/h	SRR= 30m LRR= 150m	SRR= 50m LRR= 150m	3.0 s	2.5s	2.5 s	2.0 s	1.5 s	1.25 s	0.5 s	v_imp (km/h)
Case 1	x	-	-	-	-	-	-	x	-	-	x	-	-	x	-	x	103 km/h
Case 2	-	x	-	-	-	-	-	x	-	-	x	-	-	x	-	x	86 km/h
Case 3	-	-	x	-	-	-	-	x	-	-	x	-	-	x	-	x	68 km/h
Case 4	-	-	-	x	-	-	-	x	-	-	x	-	-	x	-	x	49 km/h
Case 5	-	-	-	-	x	-	-	x	-	-	x	-	-	x	-	x	29 km/h
Case 6	-	-	-	-	-	x	-	x	-	-	x	-	-	x	-	x	NO IMPACT
Case 7	-	-	-	-	-	-	x	x	-	-	x	-	-	x	-	x	NO IMPACT
Case 8	x	-	-	-	-	-	-	x	-	x	-	-	x	-	x	-	77 km/h
Case 9	-	x	-	-	-	-	-	x	-	x	-	-	x	-	x	-	59 km/h
Case 10	-	-	x	-	-	-	-	x	-	x	-	-	x	-	x	-	36 km/h
Case 11	-	-	-	x	-	-	-	x	-	x	-	-	x	-	x	-	NO IMPACT
Case 12	-	-	-	-	x	-	-	x	-	x	-	-	x	-	x	-	NO IMPACT
Case 13	-	-	-	-	-	x	-	x	-	x	-	-	x	-	x	-	NO IMPACT
Case 14	-	-	-	-	-	-	x	x	-	x	-	-	x	-	x	-	NO IMPACT
Case 15	x	-	-	-	-	-	-	x	-	x	-	x	-	-	x	-	67 km/h
Case 16	-	x	-	-	-	-	-	x	-	x	-	x	-	-	x	-	41 km/h
Case 17	-	-	x	-	-	-	-	x	-	x	-	x	-	-	x	-	NO IMPACT
Case 18	-	-	-	x	-	-	-	x	-	x	-	x	-	-	x	-	NO IMPACT
Case 19	-	-	-	-	x	-	-	x	-	x	-	x	-	-	x	-	NO IMPACT
Case 20	-	-	-	-	-	x	-	x	-	x	-	x	-	-	x	-	NO IMPACT
Case 21	-	-	-	-	-	-	x	x	-	x	-	x	-	-	x	-	NO IMPACT
Case 22	x	-	-	-	-	-	-	-	x	x	-	x	-	-	x	-	42 km/h
Case 23	-	x	-	-	-	-	-	-	x	x	-	x	-	-	x	-	NO IMPACT

6. Conclusions

Aimed on the critical scenarios constructed for testing purposes of the ADAS, Simcenter Prescan together with Matlab/Simulink, offer the solution for verification and validation of the scenarios. Prescan provided the environment of the simulation, and in close correlation with Simulink, an ADAS Simulink model (figure 9) has been introduced, obtaining different results for each specific configuration. figure 10 illustrates the input and outputs of the Simulink AEBS module.

An analysis of the obtained results conducted to guidelines for choosing the most appropriate configuration of the radars and AEBS, considering different speeds of the studied automated vehicle. Therefore, two main guidelines may be concluded by the interpretation of the results. As a general guideline, the sooner the obstacle is detected (high TTC values), the chances for an impact to the obstacle decreases exponentially. The second important guideline is represented by the importance of the early detectability of the obstacle by the radars. This aspect is important because the high TTC values do not have any effect on the autonomous vehicle behaviour to detect an obstacle if the obstacle is not detected first by the radar.

Intuitively, as a vehicle starts to break as soon as the obstacle is detected, the impact (i.e. if exists), will have a significantly smaller magnitude. Thus, the most significant parameters that determine the automated vehicle to stop in a shorter time up to the possible collision are the TTC parameters of the AEBS module.

As a general aspect of the studied results, it has been remarked, as stated before, the TTC parameters play a vital role in the vehicle's collision. The sooner the obstacle is detected, and the brake is applied, the chances of collision decrease significantly. This approach leads to some advantages, but also to some limitations.

One noticeable limitation of the warning, partial- and full-braking TTC in terms of radar beam range is in the case of large values of the braking time, speed, and short beam range of the radars. This means the obstacle will be observed later than it can be, due to the radar limitation. Therefore, a mathematical calculation that involves times to collision, vehicle speed and radar beam range must be considered, otherwise, the radar range will limit the TTC values.

As an advantage, it has to be noted the guideline of detecting the obstacle sooner and applying the brakes right after the detection involves no additional costs for the manufacturer, the adjustment being mainly at the AEBS parameters level.

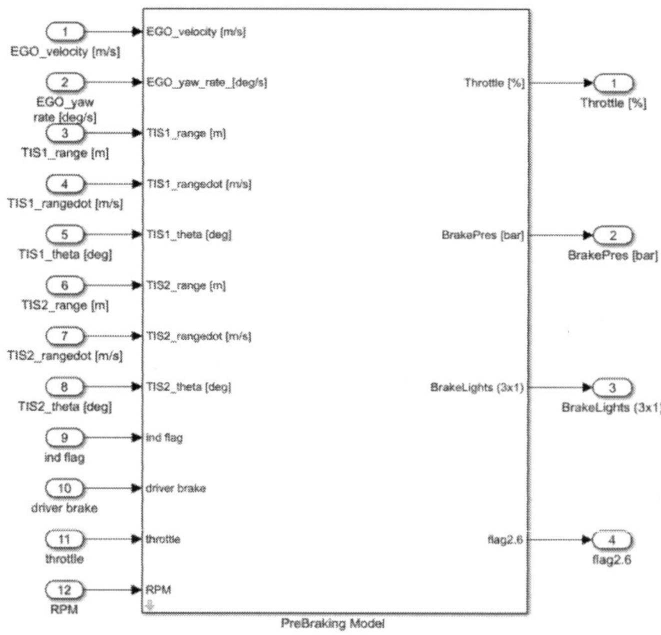

Figure 9. Inputs and outputs of the AEBS module

Acknowledgments
"This work was supported by a grant of the Ministry of Research, Innovation and Digitization, CNCS/CCCDI – UEFISCDI, project number PN-III-P2-2.1-PED-2019-4366 within PNCDI III".

References

[1] Mimura Y, Ando R, Higuchi K and Yang J 2020 Recognition on trigger condition of autonomous emergency braking system *Journal of Safety Research* **72,** 239-247 [https://www.sciencedirect.com/science/article/pii/S0022437519306760]
[2] Ivanov A, Kristalniy S, Popov N, Toporkov M and Isakova M 2018 New testing methods of automatic emergency braking systems and the experience of their application *IOP Conference Series: Materials Science and Engineering* **386**
[3] Kalra N and Paddock S M 2016 How many miles of driving would it take to demonstrate autonomous vehicle reliability? *RAND Corporation Tech. Rep.* [http://www.rand.org/pubs/research reports/RR1478.html]
[4] Ministry of Land, Infrastructure, Transport and Tourism in Japan 2016
[5] [https://www.technavio.com/]
[6] Wang C S, Liu D Y, Hsu K S 2021 Simulation and application of cooperative driving sense systems using prescan software *Microsyst Technol* **27** 1201–1210
[7] ***https://www.continental-automotive.com/getattachment/e98ad36c-49d6-400f-8b99-32ff1ef4e45b/SRR20X-Datasheet-EN.pdf.pdf
[8] ***https://www.xilinx.com/publications/presentations/continental-ARS540-powered-by-xilinx.pdf

Comparative analysis of driver's motion using different headrest positions during the rear-end collision

Corbu Gheorghe Marius, Garnita Ionut Alexandru, Trusca Daniel Dragos and Toma Florin Daniel

Automotive and Transport department, Faculty of Mechanical Engineering, University "TRANSILVANIA" of Brasov, Brasov, RO

gheorghe.corbu@unitbv.ro

Abstract. The biggest problem regarding the safety of occupants during the rear-end collision is the occurrence of "Whiplash" phenomenon, which means that the human head moves backwards in relation to the thorax movement which is in the sitting position. In this case, only the headrest opposes the head movement. An important role in the attenuation of this phenomenon is played by the correct position of the headrest relative to the head position. If the headrest is adjusted correctly, to limit the angle and movement of the head and neck during a rear-end collision, it should reduce the injury risk caused by "Whiplash", a phenomenon that can cause easy to dangerous injuries. This paper presents a comparative analysis during the rear-end collision between a driver with the headrest adjusted in optimal and wrong position. The analysis is based on the front quarter driver side of body car, using an anthropomorphic dummy. The quarter of car was positioned on a sledge that was located on the test stand. The data acquired during the tests were processed in order to evaluate the most common injury criteria, called HIC (Head Injury Criteria).

1. Introduction

Road safety is considered a major public health problem. According to the World Health Organization (WHO) approximately 1.35 million people are killed and 50 million injured in traffic collisions on the roads around the world each year. Road traffic injuries are now the leading killer of people aged 5-29 years [1]. Among all the traffic collisions, rear-end collisions are most frequent, accounting for about 33% of the total collisions in the United States [2], 33% in China [3], 35% in Japan [4] and 27% in Europe [5].

Rear impact collisions are mostly low severity, but carry a very high societal cost due to reported symptoms of whiplash and related soft tissue injuries [6]. In rear end collisions between two vehicles, the most common type of injury for the occupants is the fracture of the cervical spine. This phenomenon goes by the name of "Whiplash" [7]. Even lower speed collisions can lead to prolonged neck injuries [8]. The best way to reduce the whiplash effect is the headrest. The headrest is an automotive safety feature, attached or integrated into the top of each seat to limit the rearward movement of the occupant's head, relative to the torso, in a collision – to prevent or mitigate whiplash or injury to the cervical vertebrae [9, 10]. Another advantage is that they provide comfort when the people support their head by the headrest [11]. Adjustable headrests can reduce the injury risks by 10%. In theory, if the headrest is adjusted correctly, to limit the angle and movement of the head and neck during a rear end collision, it should reduce the injury risk caused by whiplash phenomenon [12].

The main purpose of this paper is to perform 2 controlled rear-end collision tests with the headrest in different positions (wrong and optimal position) to see the influence of the headrest position, one of the most important parameters that can occur during a rear-end collision, namely the head injury criterion (HIC).

2. Methodology

To realize this study we made two similar tests. Both tests were realized with the dummy in the same position at the same speed but the only change was the headrest position to see what are its benefits in case of a rear-end collision.

First test (T_1) – wrong position of headrest;

Seceond test (T_2) – optimal position of headrest.

2.1. Equipment used

The experiments were conducted using a quarter of car was positioned on a sledge that was located on the test stand, as shown in figure 1. A test dummy (male) positioned on driver seat and equipped with an acquisition system Pocked DAQ (DSD, Linz, Austria) inside the head and thorax to acquire the triaxial accelerations values during the rear-end collision. To be able to analyze the tests we used a high-speed camera HiSpec 5 (Fastec Imaging, USA) capable to recording 1400 FPS. To measure the velocity of the sledge we used a GPS (DL-10) velocity sensor.

Figure 1. Test stand and sledge with the left body section of the vehicle.

2.2. Experimental procedure

First, the stand was positioned in the testing area (inside the Research & Development Institute of Transilvania University from Brasov) where only the test team members have access. The sledge with vehicle body section was positioned on the stand. We performed some tests to adjust the stand position and calibrate the equipment needed for the experimental tests. We set the tasks for each team member during the tests. We established the velocity, which the tests will be performed, and the testing scenario.

2.3. Testing scenario

T_1 – The test performed at the velocity around 3.90 ms^{-1} with the headrest in wrong position and the dummy was wearing the seat belt;

T_2 – The test performed at the velocity around 3.90 ms^{-1} with the headrest in optimal position and the dummy was wearing the seat belt;

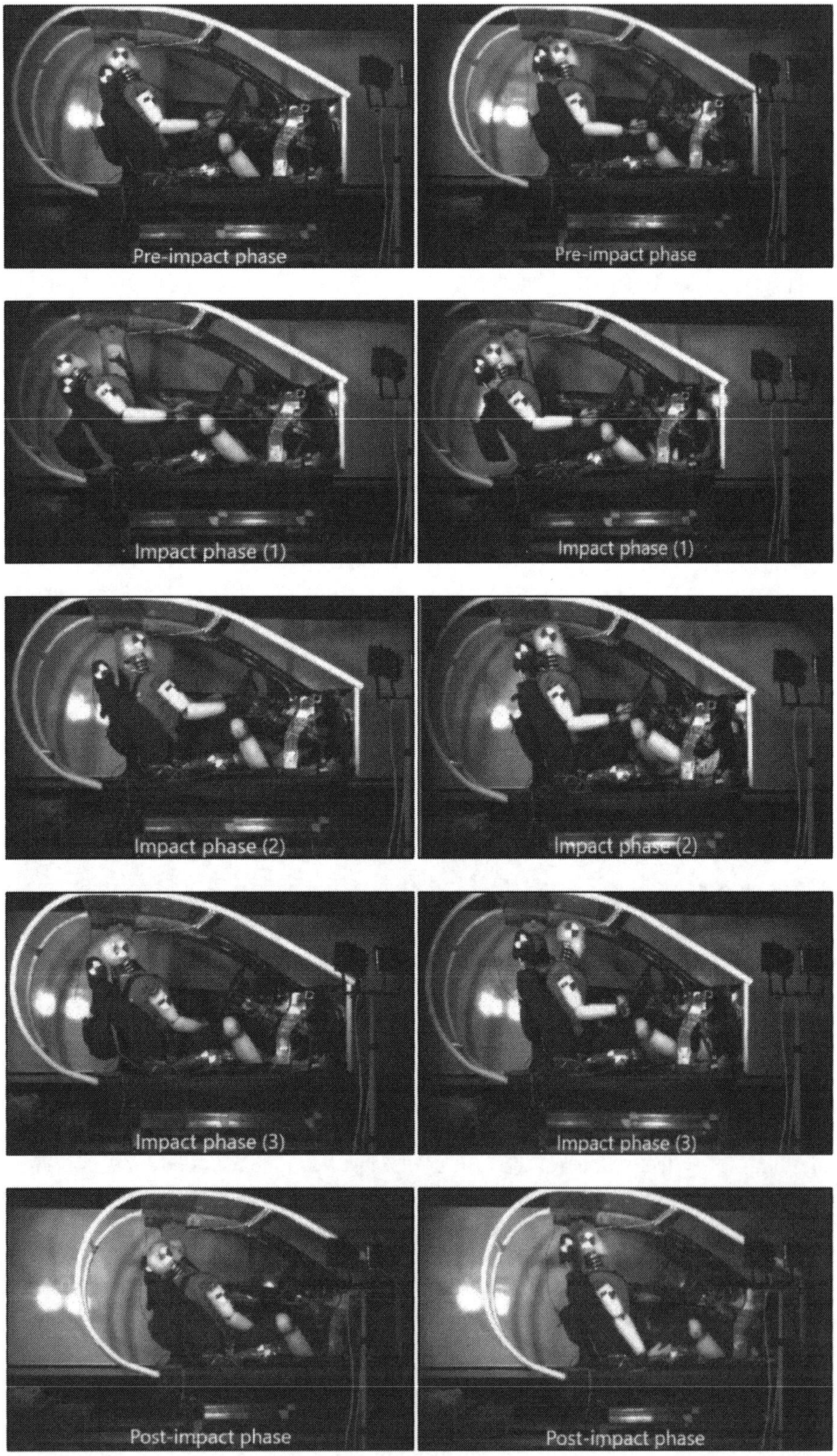

Figure 2. Collision phases
(Left side – wrong position of headrest / Right side – optimal position of headrest).

The dummy (driver) motion during the rear-end collision for both scenarios can be observed in figure 2 where the impact's phases are presented.

The *pre-impact phase* is the period before the collision (it occurs until the two elements come in contact). The dummy is in the initial position before contact.

The *impact phase* corresponds to the period in which the two elements are in contact. During this time the energy is transferred and the dummy (driver) moves inside the body section of the vehicle.

The *post-impact phase* begins from the moment the two elements come off until they stop. The test dummy remains standstill.

2.4. Data processing

Experimental data were imported in Excel and subjected to additional processing to highlight one of the most important injuries parameters during the collisions, namely HIC. The HIC criterion is the principal way of assessing the risk of a cranial trauma injury as a result of an accident or for testing individual protection equipment [13]. The HIC criterion represents the maximum standard value of the integral of the head acceleration and depends on what interval it`s calculated: HIC 15 – maximum 15 ms; HIC 36 – maximum 36 ms; HIC – unlimited.

The mathematical formula by which the HIC criterion is determined: [14]

$$\text{HIC36} = \left\{\left(\left[\frac{1}{t_2-t_1}\cdot\int_{t_1}^{t_2} a\cdot dt\right]^{2.5}\cdot(t_2-t_1)\right)\right\}_{MAX} \tag{1}$$

where: "a" is a resultant head acceleration.

3. Results and discussions

The experimental data recorded with the Pocked DAQ system (DSD, Linz, Austria) disposed inside the dummy's head is presented in figure 3 for T_1 respectively in figure 4 for T_2. These data represent the triaxial accelerations values recorded by the dummy's head during the collisions.

As we can see in figure 3 and figure 4 the maximal values of accelerations/decelerations were recorded on the X axis for both tests. For testing scenario T_1, the maximal value of acceleration is 85,84 ms^{-2} and the maximal value of deceleration is 35,15 ms^{-2}. In the same conditions of velocity but for testing scenario T_2 the maximal value of acceleration is 77,43 ms^{-2} and the maximal value of deceleration is 28,97 ms^{-2}. According to the data above mentioned the values of T_1 are higher than values of T_2.

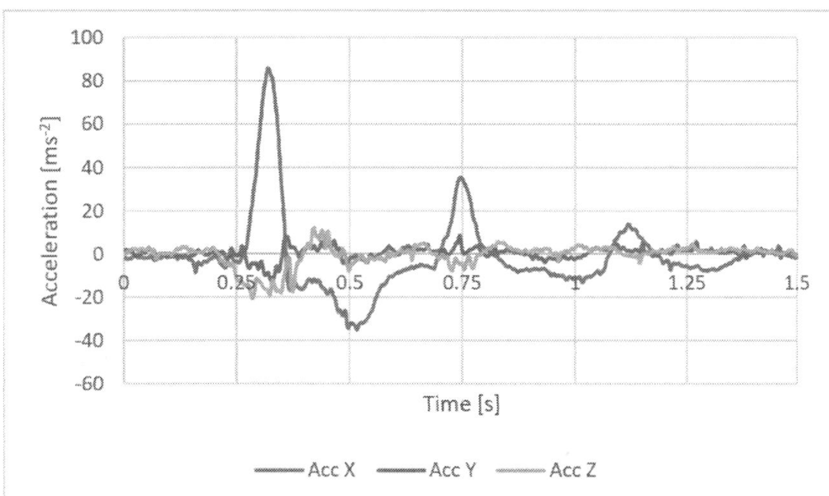

Figure 3. Acceleration values recorded at occupant's head - T_1.

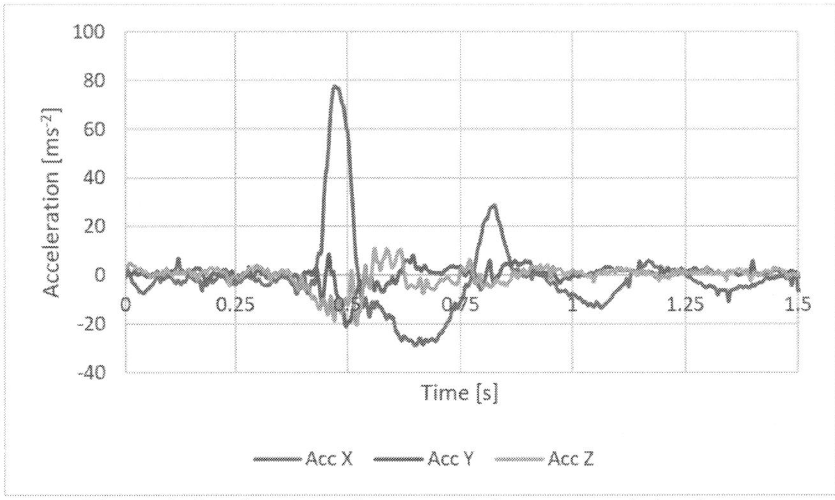

Figure 4. Acceleration values recorded at occupant's head – T_2.

In this paper is estimated for both tests only the HIC 36, corresponding to a time interval of 36 ms. For HIC estimation the resultant acceleration is needed. Thus, figure 5 and figure 7 are capturing the acceleration resultant of values recorded at head level in the conditions of scenarios followed. HIC parameter estimation follows data processing steps. Acceleration resultants were subjected to a smoothing process using an FFT filter to remove the noisy data from the entire signal. Next, by numerical integration accounting for a time range of 36 s the values of HIC36 for both testing scenarios were gathered to enable head injury's assessment.

As it can be seen from figure 5 the peak value during the collision in testing scenario T_1 is about 86 ms^{-2}. In the conditions of T_1 scenario, where the headrest is in the wrong position, the value of the HIC 36 parameter is 13,85 as shown in figure 6. The value of HIC 36 parameter in T_1 scenario is not high but considering the testing conditions, where the velocity was around 3.90 ms^{-1}, it is a considerable value.

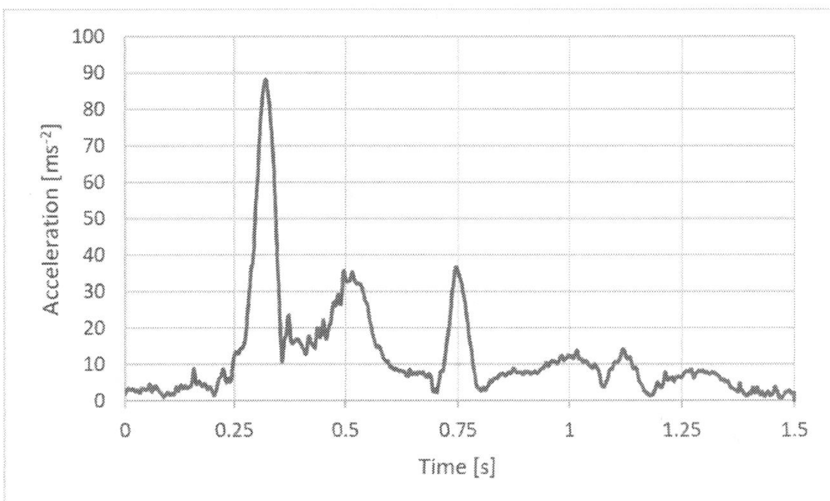

Figure 5. Head acceleration resultant - T_1.

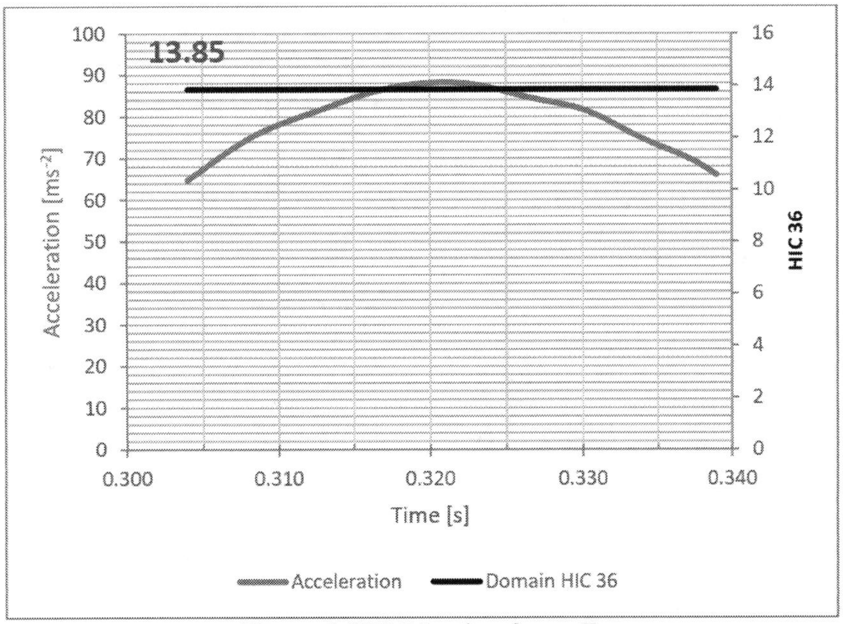

Figure 6. HIC 36 estimations - T_1.

In figure 7, it is represented the resultant of head accelerations for T_2 testing conditions with a maximum value about 79 ms^{-2}. It can be seen that the peak value of head acceleration resultant for T_2 is smaller than T_1. The difference between T_1 and T_2 is about 7 ms^{-2} in conditions of a relatively low velocity but the difference can increase considerably in case of high values of velocity.

For the same test T_2, where the headrest is in optimal position, the value of HIC 36 parameter can be observed in figure 8 having a value of 11,70. In T_1 testing conditions, the value of HIC 36 parameter is higher than T_2.

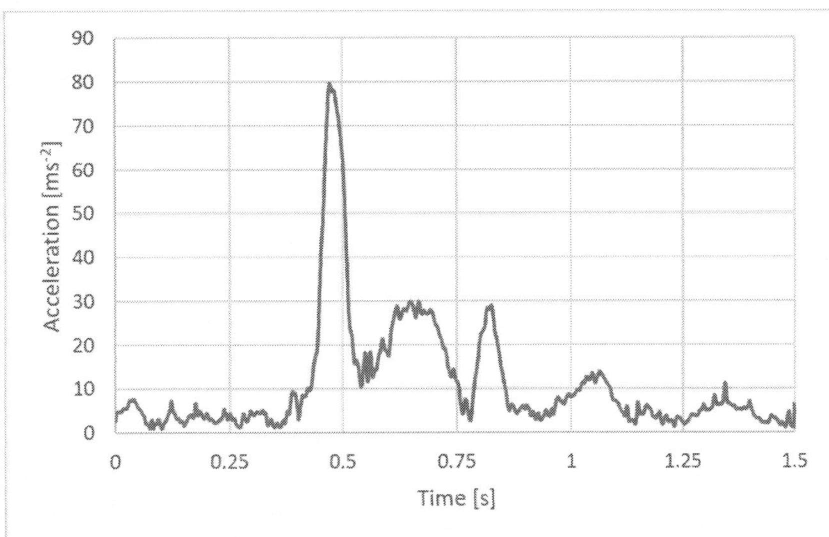

Figure 7. Head acceleration resultant – T_2.

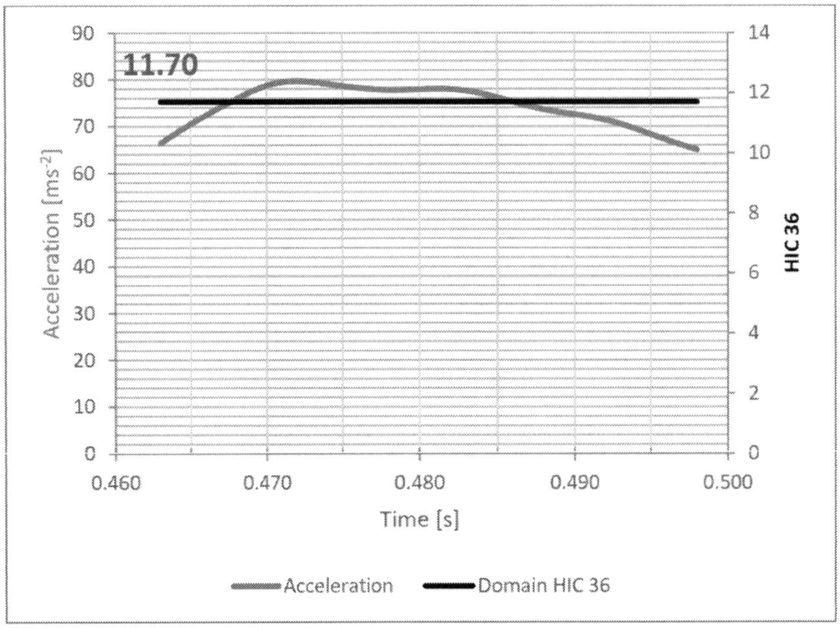

Figure 8. HIC 36 estimations – T_2.

4. Conclusions

Following the processing of the experimental data and the calculation of the head injury criteria (HIC), the obtained results were compared to highlight how the headrest position influence the injury degree of the occupants. Analysing the comparisons between head acceleration resultants and HIC 36 parameters we can see that the higher values are found in the testing conditions T_1 which means that the injury risk is higher when the headrest isn't in an optimal position. Even if the probability of injury was minor because the velocity was low, the tests performed demonstrated that the optimal position of headrest with which the vehicles are equipped nowadays will reduce the head injury degree for the occupants of the vehicle.

Based on the obtained results it can be concluded that in rear-end collisions with low-speed such as the ones presented, the headrest plays a very important role and can make the difference between a major head injury and a minor head injury of the occupants.

References

[1] World Health Organization, (2018), *Global status report on road safety 2018.* https://www.who.int/publications/i/item/9789241565684.

[2] National Highway Traffic Safety Administration, (2015), *Traffic safety facts.* https://crashstats.nhtsa.dot.gov/Api/Public/ViewPublication/812384, 70-71.

[3] Yan, X., Li, X., Liu, Y., & Zhao, J. (2014), *Effects of foggy conditions on drivers' speed control behaviors at different risk levels*. Safety Science, 68, 275–287.

[4] Watanbe, Y., & Ito, S. (2007), *Influence of Vehicle Properties and Human Attributes on Neck Injuries in Rear-Collisions.* In: 20th International Technical Conference on the Enhanced Safety of Vehicles (ESV), Lyon, Paper No. 07-0160.

[5] Economic Commission for Europe, (2017), *Statistics of Road Traffic Accidents in Europe and North America. Volume LIV.* https://unece.org/fileadmin/DAM/trans/main/wp6/publications/RAS-2017.pdf.

[6] Nishimura, N., Simms, C.K., Wood, D.P. (2015), *Impact characteristics of a vehicle population in low speed front to rear collisions*. Accid. Anal. Prev. 79, 1–12.

[7] Krafft, M., Kullgren A., Lie A., Tingval, C., (2005-04-01). "*Assessment of Whiplash Protection in Rear Impacts*" (PDF). Swedish National Road Administration & Folksam. Retrieved 2016-

01-10.

[8] Clerk Maxwell, J., (1892), *A Treatise on Electricity and Magnetism*, 3rd ed., vol. 2. Oxford: Clarendon, pp.68-73.

[9] UNECE Regulation 17, (2010-08-31), *Uniform Provisions Concerning the Approval of Vehicles With Regard to the Seats, Their Anchorages and Any Head Restraints.* https://op.europa.eu/en/publication-detail/-/publication/4d5ab93c-7d45-4b3a-b49f-b10b6476b5df.

[10] FMVSS No. 202, (1989-10-12), *Head Restraints for Passenger Vehicles: Preliminary Economic Assessment and Regulatory Flexibility Analysis*. https://www.nhtsa.gov/document/laboratory-test-procedure-fmvss-202-head-restraints

[11] Franz, M., Durt, A., Zenk, R., Desmet, P.M.A., (2012), *Comfort effects of a new car headrest with neck support*. Appl. Ergon. 43, 336–343.

[12] Radu, A.I., Cofaru, C., Tolea, B. (2017). *Influence of the headrestrain position in case of rear impact and its effects upon the whiplash phenomenon*.

[13] Gao, D. and Charles W. W., (2009), "*Head injury criterion.*" IEEE robotics & automation magazine 16.4: 71-74.

[14] McHenry, B. G., (2004), *Head injury criterion and the ATB*, ATB Users' Group, 5-8.

IOP Conf. Series: Materials Science and Engineering 1220 (2022) 012047 doi:10.1088/1757-899X/1220/1/012047

Improving road traffic by rerouting flows based on various regulation scenarios analysis

A A Boroiu, E Neagu, A Boroiu and I Vieru

Department of Automotive and Transports, University of Pitesti, 1 Targul din Vale Street, 110040 Pitesti, Romania

elena.neagu@upit.ro

Abstract. One of the major road traffic problems in Pitesti is manifested in the roundabout at Podul Viilor, which, although occupying a fairly generous space, has a very low level of service during peak hours, due to difficult access from one of the arms. The research, based on traffic measurements performed with DataFromSky software and micro-simulation traffic analyses performed with Vissim PTV software, indicated as the only solution the redirection of road flows from the intersection, which required traffic analysis for a larger area. The results obtained for the proposed scenarios - by estimating the degree of road congestion not only in the respective intersection, but also in the adjacent intersections, through which some road flows are reoriented - indicate that in this way a substantial improvement of the road traffic can be obtained for the intersection of Podul Viilor, without significantly affecting the traffic in the adjacent area. Thus, it is found that the service level of an intersection can be increased not necessarily through new road arrangements, but only through better regulation of road traffic in the area, based on the analysis of various scenarios using measurement software and traffic modelling.

1. Introduction

It is well-known that one of the biggest problems large cities in Romania have been facing is road congestion, which for several years has occurred not only during the hours traditionally considered "traffic peak hours", but even throughout the day, especially for the areas that constitute important nodes for the traffic flows in the city.

These road traffic problems are also manifested in the city of Pitesti, which is one of the medium-sized cities in Romania and has a special configuration of the road network [1, 2], given the longitudinal arrangement of the city on the terraces on the right side of the Arges river.

The regulation of road traffic in intersections is achieved through the 3 known ways (by indicators regulating the priority of passing, through traffic lights or roundabouts), but it should be noted that in recent years the number of road intersections with organized traffic has increased constantly, at this moment the street network of Pitesti municipality having 16 traffic light intersections and 13 roundabout intersections [3].

Regarding the spatial efficiency of these two types of intersections, it should be noted that the differences are not major for the same number of lanes on the access roads, instead, calculating the spatial efficiency by relating the capacity of the intersection to the area required for development, specialized literature [4, 5, 6], as well as the results of the research carried out by the group of authors [7], reveal that spatial efficiency decreases with the increase of the number of lanes on the circular path.

But a drastic reduction in capacity for these intersections is caused by the indiscipline of drivers on the circular path, because in general, they drive without observing the separation of lanes, which makes the use of the 2 or, especially, the 3 ring lanes not to be at full capacity [8].

All this drastically limits the capacity of roundabout intersections, despite the extremely generous space they have. The existence of a 3rd circular lane in an intersection is not capitalized at all, it even impedes the performance of the intersection – in terms of capacity, level of service, road safety. As a result, there are great resources for increasing the capacity of the intersection (and for the disappearance of the phenomenon of road congestion as well as for increasing road safety) by correctly organizing traffic at the intersection, but this requires theoretical and experimental research of traffic at such intersections.

2. Analysis of the possibilities to increase the service level by micro-simulating the traffic based on the data collected with the DataFromSky equipment

The DJI Mavic Air Fly More Combo drone [9] was used to perform traffic measurements by video recording traffic, and the associated data processing software "DataFromSky" was used for data analysis. The two types of equipment represent an integrated system for collecting traffic data, extremely important, especially in the situation when it is desired to analyse the road traffic for an area that includes several neighbouring road intersections [10]. Thus, the work stages of the traffic analysis at the Podul Viilor intersection, carried out after the purchase of these pieces of equipment, were the following:

- collection and processing of traffic data at peak hours at the Podul Viilor road intersection, by using the purchased *drone* with which the road traffic was filmed and the dedicated software for the processing of the traffic data, which carries out the identification, counting and classification of the vehicles, the drafting of the Origin-Destination matrix of the traffic flows;
- road traffic modelling (using the dedicated Vissim software from the PTV Vision package for traffic modelling, which is owned by the Automotive Engineering Research Centre and for which there is an educational license);
- evaluating the level of service for the various possible scenarios;
- choosing the solution for organizing traffic at the road junction, based on the values obtained for the service level in the various scenarios under analysis.

In order to collect the traffic data, a first evaluation of the Calea Bucuresti - Calea Bascovului roundabout intersection was performed using the program https://www.google.com /maps [11]. This identifies the periods of the week days when traffic is difficult. Video recordings were made on the targeted intersection by using the drone and the recordings were then processed using the DataFromSky software (figure 1).

Figure 1. Video processing with DataFromSky software.

In the first stage, there were analysed, only at the level of the intersection, the scenario that reflects the current situation and 3 more possible scenarios for traffic improvement.

For traffic scenarios simulation we used the Vissim PTV software, which allows the calculation of the performance of an intersection according to the scenario created, by analysing several factors:

- average waiting time of vehicles/control delays (seconds);
- length of waiting lines (number of vehicles);
- average vehicle travel time for one km (seconds);
- duration of stops (seconds);
- average vehicle speeds (km/h);
- level of service.

The level of service in size classes characterizes the traffic conditions in an intersection (with letters from A to F), and for non-traffic light intersections, therefore for the roundabout as well, the service level of a non-traffic light intersection is determined by the values of the aggregate control delays at the level of the entire intersection presented in the previous chapters [12] – table 1:

Thus, using the Vissim software, for the traffic values obtained after video recording and then processing with the DatafromSky software, the following scenarios were simulated and analysed:

1. Current stage (used for model calibration);
2. Building of an overpass or underpass of the intersection for vehicles coming from Calea Bascov;
3. Installation of speed bumps for limiting the speed of the flow of vehicles coming from the Arges Bridge;
4. Installation of traffic lights at the intersection.

Table 1. Level of service of non-traffic light intersections in relation to control delays.

Level of service	Control delays (sec/veh)
A	< 10
B	10 - 15
C	15 - 25
D	25 - 35
E	35 - 50
F	> 50

Scenario 1. Current stage

Simulating the situation corresponding to the current stage with the Vissim software, there resulted an average waiting time of the vehicles (control delay) of 32.62 sec/veh, resulting in a D - level of service of the intersection.

As in the case of physical observations, the simulation revealed that long waiting lines are formed on Calea Bascovului. This happens because of the high speed with which the vehicles from Podul Viilor enter the roundabout, the access of the vehicles from Calea Bascovului being extremely difficult, in conditions of insecurity.

In this regard, the following scenarios have been developed to address this issue.

Scenario 2. Building of an overpass or underpass of the intersection for vehicles coming from Calea Bascov.

Simulating this scenario with the Vissim software, there resulted an average waiting time of vehicles (control delay) of 24.21 sec/veh, resulting in a C-class service level of the intersection. However, this scenario has some major disadvantages:

- the need to perform feasibility studies on the geometry of space to achieve the overpassage;
- high costs for the construction of the overpassage.

Scenario 3. Speed limitation for the flow of vehicles coming from Arges Bridge

Within this scenario, it is proposed to mount speed bumps to limit the speed for the flow of vehicles coming from Argeş Bridge, this way the entry of the vehicles coming from Calea Bascovului into the roundabout being made more easily.

Simulating this scenario with the Vissim software (with the traffic volumes obtained with the Data from Sky software), there resulted an average waiting time of the vehicles (control delay) of 27.45 sec/veh, resulting in a class-D service level of the intersection. This scenario brings a very important advantage: increasing the safety of traffic participants, but also the disadvantage that these bumps are annoying for the vehicles crossing them.

Scenario 4. Installing traffic lights at the intersection

In this scenario, it is proposed to install a traffic light to limit the speed for the flow of vehicles coming from the Argeş Bridge, so that, in this way, the vehicles coming from Calea Bascovului could enter the roundabout more easily.

Simulating this scenario with the help of the Vissim software, there resulted an average waiting time of vehicles (control delay) of 34.85 sec/veh, resulting in a D-class service level of the intersection, but with values of the average vehicle waiting time higher than in the case of the current scenario, forming long waiting lines for the vehicles coming from the Arges Bridge.

Table 2 shows the values obtained for each scenario studied:

Table 2. Simulation-obtained service levels for the 4 scenarios.

Scenario	Resulting level of service/ Control delays, (sec/veh)
1. Current stage	**D** / 32,62
2. Building an overpass	**C** / 24,21
3. Installing speed bumps for speed limitation	**D** / 27,45
4. Installing traffic lights at the intersection	**D** / 34,85

The only variants that result in a traffic improvement are scenarios 2 and 3, for which the control delay values obtained were lower than in the current scenario, but the differences are not very large, and the disadvantages of these solutions weigh heavily in taking some decisions in this regard.

3. Traffic analysis in a larger area near the intersection

It is concluded that, in order to formulate a sustainable proposal to solve the problems at the intersection, one that would not involve very high costs and which would be sustainable, an analysis is needed to identify possibilities for reorienting the road flow for which a satisfactory solution for access to the intersection was not found - the entry flow on the North arm. Thus, a direction of interest is the analysis of the possibility to give up the access of vehicles at the intersection through the North arm by changing the current route of vehicles coming on the North arm towards the intersection (with red in the figure), orienting them on the route represented in blue (figure 2).

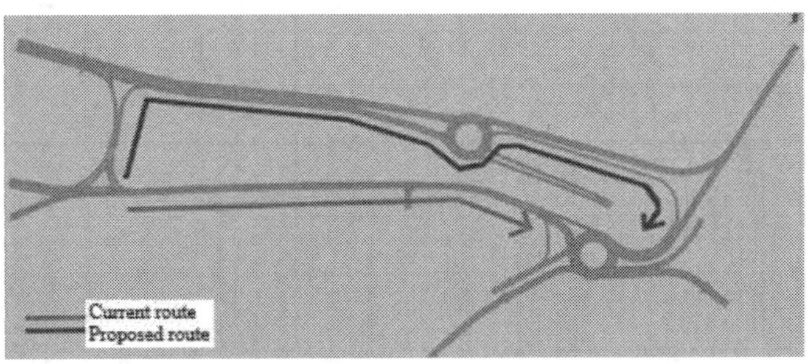

Figure 2. The proposed route change.

As can be seen, the length of the route increases, but the answer is decisive regarding the length of the waiting lines, which characterizes the degree of congestion. Examining the area from which the vehicle flows come on a broader level, one can notice that there are 7 origins/destinations of the road flows, written from A to F in figure 3, so the situation can be analysed at this level to observe the consequences of the reorientation of the flow on the North arm where the traffic problem could not be solved through the previous scenarios imagined.

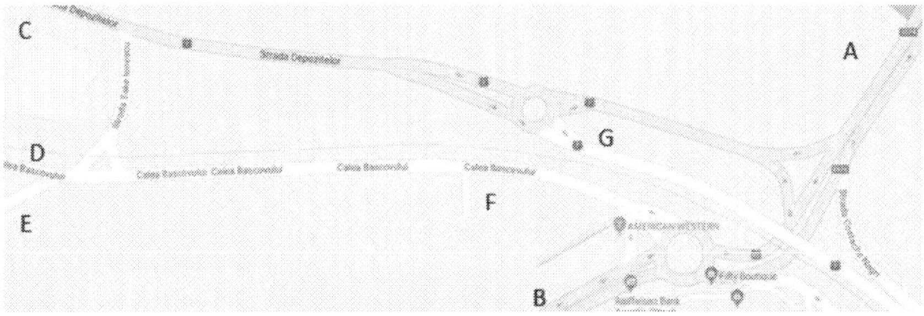

Figure 3. The new traffic analysis area.

The area under analysis was expanded, which required new measurements, in other parts of the area, in order to model traffic. The distribution of traffic at the level of the entire area, for the two categories of vehicles, is illustrated in figures 4 and 5.

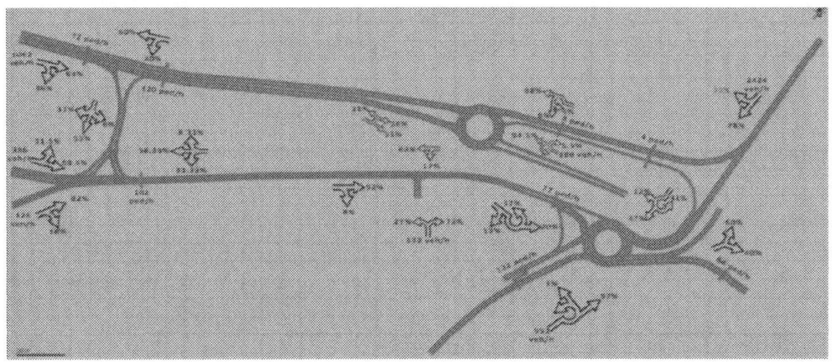

Figure 4. Road traffic distribution in the area under analysis.

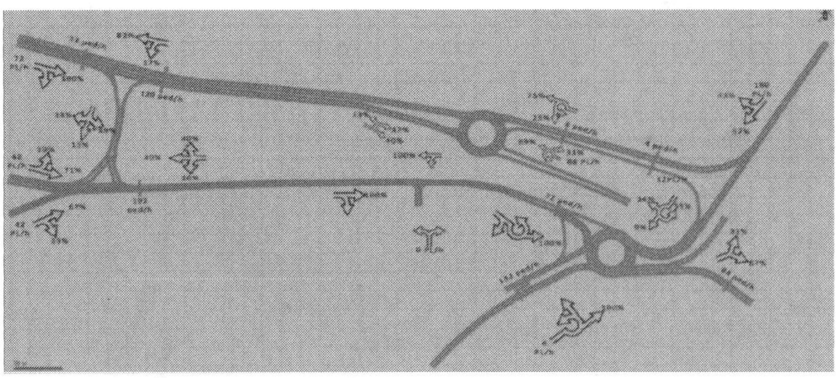

Figure 5. Traffic distribution of vehicles of over 3.5 tons in the area under analysis.

The results of the new measurements at the entry on the East arm also revealed traffic variation during peak periods, which is an argument for an analysis that takes into account this variation, possibly through the PHF indicator - Peak Hour Factor, as used in the calculation programs presented in the previous chapters. After entering all the data in the Vissim microsimulation software, the congestion sensors were placed in the points of interest (the 3 entries into the Podul Viilor roundabout - RP1, for which microsimulations were performed even earlier, and the 3 entries into the roundabout from Calea Depozitelor - RP2, where the traffic flow from the North arm was directed) – figure 6.

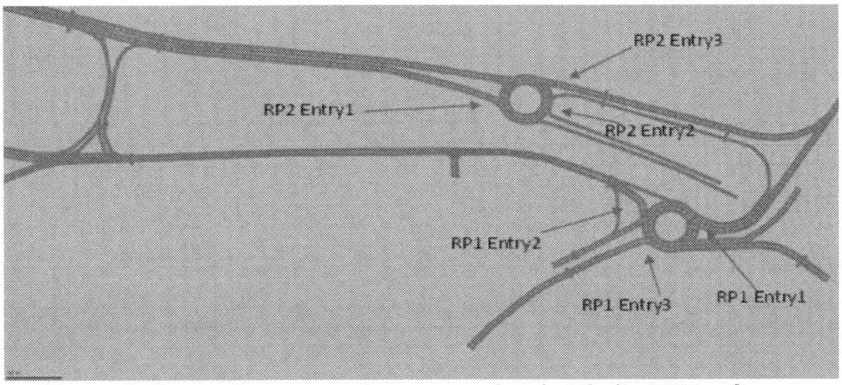

Figure 6. Congestion sensor placement points in Vissim PTV software.

Hence the values for the average line length (road congestion indicator) specified in table 3. Analysing the values obtained, it is obvious that, in the current situation of road traffic regulation in the area, the values for the average length of the waiting line (indicator of road congestion) during peak hours is unacceptable, while at all other points of congestion measurement, traffic was found to be free.

Table 3. The average length of the waiting line in the 6 points in the current situation.

Congestion sensor	Average length of the waiting line (m)
RP1 Entry 1	1,76
RP1 Entry 2	226,75
RP1 Entry 3	8,27
Entry D	0,54
Entry E	0
RP2 Entry 1	0,20
RP2 Entry 2	0,10
RP2 Entry 3	0

As a result, it is necessary to propose the diversion of the incoming traffic from the North arm on the new identified route, so that the access road from the North arm to the intersection will disappear (figure 7).

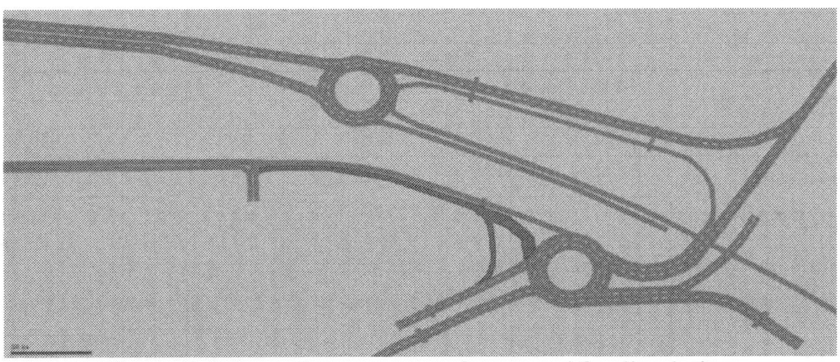

Figure 7. Intersection access road proposed for dissolution.

Performing once again a simulation for the current and the new organization of the road traffic using the Vissim software, the new values of the length of the waiting line are obtained in the 6 points PLUS 2 other points of interest (D and E from figure 3), and the results can be analysed comparatively by presenting them in the two scenarios: the current stage and the scenario with the prohibition of road access on the North arm to the Podul Viilor intersection – table 4.

Table 4. The average length of the waiting line in the 8 points in the 2 situations.

Congestion sensor	Average length of the waiting line (m)	
	Current stage	Redirecting traffic on the North Arm
RP1 Entry 1	1,76	5,93
RP1 Entry 2	226,75	0
RP1 Entry 3	8,27	35,49
Entry D	0,54	50,02
Entry E	0	0
RP2 Entry 1	0,20	0,57
RP2 Entry 2	0,10	111,60
RP2 Entry 3	0	0

The table reveals that the congestion on arm 2 (the North arm) disappears, which is normal because the road traffic is now forbidden on this direction of access in the intersection, but it does not appear on Entry 1 (the East arm), although the road flow was directed in that direction, which is a very good result.

However, low congestion occurs at the D Entry in the area, because the left turn made by vehicles, in conditions of giving the right of way, presupposes the intersection with higher traffic flows in the new conditions for regulating road traffic. Reduced congestion occurs at Entry 3 (South arm) because the return flow from arm 1 has remained continuous.

The biggest problem occurs at the RP2 road junction, where the road flow in question was redirected, not on its access arm, but rather on the opposite arm, where vehicles have to give way to the road flow (now increased by the new diverted flow) from the circular path. Especially the trucks (which come from the road artery that crosses Calea Bucuresti) have problems with access at the intersection.

However, the length of this waiting line is less than half as compared to the length of the waiting line that disappeared from the Podul Viilor intersection, and at the level of the entire area, the congestion is lower by 14.3%. So this can be a viable solution at a very low cost.

It is also interesting to check whether the travel times have increased substantially, so that the travel time between origin D and destination B was checked (see figure 10), in the two ways: by F (current situation), respectively G (scenario with traffic flow redirection on arm 2 of the RP1 intersection) – table 5.

Table 5. Comparative results for the two intersection access variants.

Route	Length of route (m)	Average travel time (s)	Average speed (km/h)
Before the deviation (Route D – F – B)	561	159.34	12.67
After the deviation (route D – G – B)	958.6	123.83	27.87

One can notice that although the length of the route (the distance between the two sensors placed in the Vissim micro-simulation program) has almost doubled, the travel time has been significantly reduced, given that the average speed has increased by more than 2 times.

IOP Conf. Series: Materials Science and Engineering 1220 (2022) 012047 doi:10.1088/1757-899X/1220/1/012047

4. Conclusions

It was found that the organization of road traffic by traffic lights at the roundabout did not lead to results of interest in any of the analysed options, because it cannot reduce congestion on access arm 2 without worsening traffic at other points in the intersection.

The proposed traffic redirection is certainly the most efficient solution, both in terms of reducing road congestion in the area and as an investment, basically requiring only the modification of road signs for the two routes.

In the future, if the commitments of the current local administration are implemented (two new bridges over the Arges River - one to the North and a second one to the South of the Podul Viilor intersection), the demand on the Podul Viilor area from the point of view of road traffic will be much reduced, because the road flow on the East arm will be substantially reduced under the circumstances in which there appear direct road connections with the industrial areas across the Arges river, so that the service level of the intersection will improve considerably.

The existing micro-simulation programs have evolved and have reached the performance of carrying out a simultaneous analysis of several intersections in an area even larger than those presented in the paper. This creates the possibility of a complete analysis of traffic in an area large enough to capture all road flows generated or attracted by the intersection.

References

[1] Rodrigue J P et al 2013 - *Site Web Géographie des Transports* (Hofstra University, Department of Economics and Geography)

[2] Boroiu A A and Boroiu A 2020 – *Geografia sistemelor de transport* (Pitesti: University of Pitesti Publishing House)

[3] Vilcan A et al 2017 *The Plan of Sustainable Urban Mobility of the Municipality of Pitesti* Beneficiary S.C. SYNERGESTICS CORPORATION SRL Bucharest

[4] Hall F L 1995 Fundamentals of transportation. Traffic stream characteristics (Ontario: McMaster University)

[5] Kimber R M 1980 *The traffic capacity of roundabouts* (Crowthorne: TRRL Laboratory Report LR 942. Transport and Road Research Laboratory)

[6] Boroiu A A, Neagu E, Boroiu A and Vieru I 2018 Proposals for the reorganization of road traffic in the central area of Pitesti municipality based on micro-simulation performed traffic analyses IOP Conf. Series: Materials Science and Engineering 444(2018)072021

[7] Boroiu A A, Neagu E, Boroiu A and Parlac S 2019 *Study of the Possibilities to Improve the Service Level of Traffic Light Intersections by Road Traffic Micro-simulation* (Proc. 30th SIAR International Congress of Automotive and Transport Engineering Science and Management of Automotive and Transportation Engineering SMAT 2019, Craiova, DOI 10.1007/978-3-030-32564-0_41 ISBN: 978-3-030-32563-3, pp. 349-358, WOS 000528526600041)

[8] *** 2000 *Roundabouts: an informational guide* (U.S. Department of Transportation: Federal Highway Administration, Publication No FHWA RD-00-067)

[9] *** *https://phantom.net/products*

[10] Blion A 2018 - *Scientific research to improve road traffic in the roundabout of Podul Viilor* University of Piteşti, Fourth year internship report traffic dynamic study

[11] *** *www.google.com/maps*

[12] CNADNR 2010 *Standard for the Development of Intersections at Public Roads* Road Technical Bulletin, no.4/2010, Designer Search Corporation

The connection between the quality of the periodic technical certification of lorries and the road traffic safety and other ecological and commercial aspects

T D Ioniță[1], N Ispas[2], A Chiru[2], D L Motoc[2]

[1]Romanian Automotive Register, Vrancea Office,
[2]Automotive and Transportation Department, Transilvania University of Brașov, Brașov Romania

businesstransport@yahoo.com

Abstract. The technical diversification of the current road traffic units in Romania is enhanced by the growth of the national fleet of road vehicles. According to APIA statistics, cars have the largest share in the volume of the car fleet in Romania. In 2019, the national car park increased by 7% compared to the previous year - 606,163 units, including 161,562 new and 444,601 used. The fleet of commercial vehicles registered an increase of 80,895 units (25,936 new and 54,959 used) or 6.1%, reaching a total nationwide at about 8 million units. The technical condition of these vehicles significantly determines road safety as well as the impact that road transport has on the environment. For each vehicle, the technical condition is validated by periodic technical inspection. In this paper, the findings of periodic technical inspections allowed the authors to develop significant scenarios related to road safety, environmental influences, but also related to commercial issues associated, for example, with the change of owner of the vehicle. The objectives we achieve in this paper are directly related to the direct link between the scope of European Union regulations certifying the compliant technical condition of a particular vehicle (result of periodic technical inspection) and road traffic safety in Romania today corroborated with projections of environmental influences and changes climate.

1. Introduction

9246710 vehicles have been registered nationally at the end of 2020 year [1]. The increase in the number of vehicles registered in Romania (new and second hand) also meant a rise in the total volume of technical problems present on vehicles on the national road network. The lower technical quality of vehicles with a certain age influences the efficiency of their use and substantially determines the safety of road traffic.

In respect with [2] "Vehicle checks are fundamental to road safety. Technical defects contribute heavily to accidents. Previous EU rules setting minimum standards for vehicle checks dated back to 1977, with only minor updates. Cars, driver behavior and technology have developed a lot since then. Moreover, many technical defects with serious implications for safety (such as anti-lock braking system and electronic stability control) were not even checked under previous rules. There are three parts to the roadworthiness package: Directive 2014/45/EU on periodic roadworthiness tests, Directive 2014/47/EU on technical roadside inspections for commercial vehicles and Directive 2014/46/EU on vehicle registration documents".

Due to intensive use, most often at maximum load or overloaded, light commercial vehicles (lorries) are subject to detailed periodic technical inspections. The minimum content of inspection areas comprises: identification, braking equipment, steering, visibility, lighting equipment and electric system, axles, wheels, tires, suspension, chassis and chassis attachments, other equipment, including speed limitation devices, nuisance including emissions and spillage of fuel and/or oil.

The use of light commercial vehicles for the transport of goods, often at maximum load or poorer dynamic performance, compared to other smaller vehicles on the road, make such vehicles often involved in serious road accidents. Figure 1 shows the evolution of the annual number of fatal accidents, in which light commercial vehicles (lorries) were involved in Romania, from 2010 to 2019 [3].

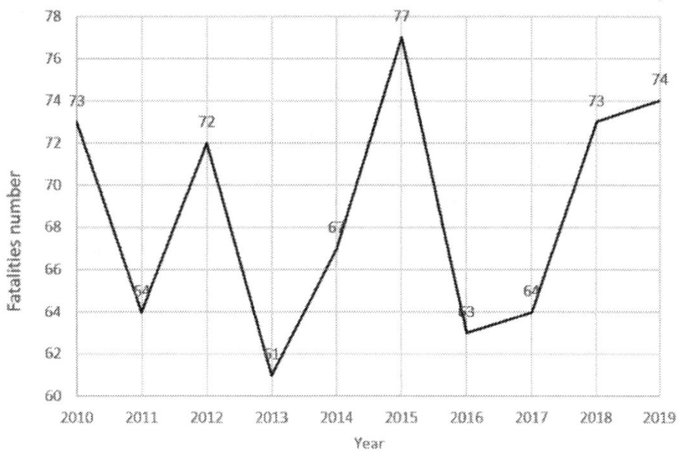

Figure 1. The annual number of lorry fatalities in Romania for the period 2010 – 2019.

The number of these fatal accidents in Romania is influenced, in many cases, by unauthorized and unsafe constructive changes that were not detected during the periodic technical inspections [4]. Due to regulations other than those with vehicles with a maximum authorized mass greater than 3.5 t, these changes cannot be the subject of road traffic detection by the State Inspectorate for Road Transport Control - ISCTR.

Often exceeding the maximum authorized load is related to the unsafe and unauthorized modification of the lorry rear suspension. Figure 2 shows such a non-compliant suspension modification for the overloaded vehicle in figure 3.

Figure 2. Non-compliant modified rear lorry suspension.

Figure 3. Lorry overloaded with timber.

Unauthorized and unsafe modification of the lory suspension to transport overloads has multiple consequences on the speed of road degradation, the degradation of the technical condition of the vehicle, the increase of fuel consumption and pollutant emissions and the reduction of the level of road safety. The effects of the presence of such vehicles in the road traffic will be estimated and evaluated in the following.

2. Modeling, simulation and discussions

The main purpose of this chapter is to model and simulate the operation of a light commercial vehicle equipped with a modified suspension (made after a positive vehicle periodic technical inspection) and progressive overload when accelerating through a transient test cycle. The simulation was done using a specialized software package called ADVISOR (we use version 003-00-r0116 from March 2019). ADVISOR is a software product, originally developed within the National Renewable Energy Laboratory (an American institution under the direct coordination of the United States Department of Energy), based on MATLAB and SIMULINK.

ADvanced **V**eh**I**cle Simulat**OR** also represents a set of models, a database and a collection of routines (scripts) used for the rapid analysis of the performance and economy of conventional, electric and hybrid vehicles. ADVISOR is a solution for detailed simulation and virtual analysis of user-defined powertrain models, a starting point for verifying experimental data from real vehicles, which highlights the flexibility of SIMULINK-defined models and the analytical power offered by MATLAB [5].

The following may be considered benefits of using the ADVISOR product: estimating the fuel consumption and the levels of noxious emissions of the vehicles that have not yet been built, analysis of energy use in the propulsion process of conventional, electric and hybrid vehicles, allows the assessment of the emissions produced by a motor vehicle following a certain test cycle, - evaluation of the energy management strategy in case of simulation of a hybrid vehicle, optimization of transmission ratios (in the case of conventional transmissions) in order to minimize fuel consumption and/or maximize dynamic performance [6].

In the following, we modeled a lorry dump truck similar to the 2007 Iveco Daily and simulated a transient drive cycle with it. Figure 4 shows the configuration screen of the application with the main options chosen for the structural subsystems of the vehicle. Figure 5 shows the block diagram of the same model vehicle and figure 5 shows the block diagram of the front and rear axe.

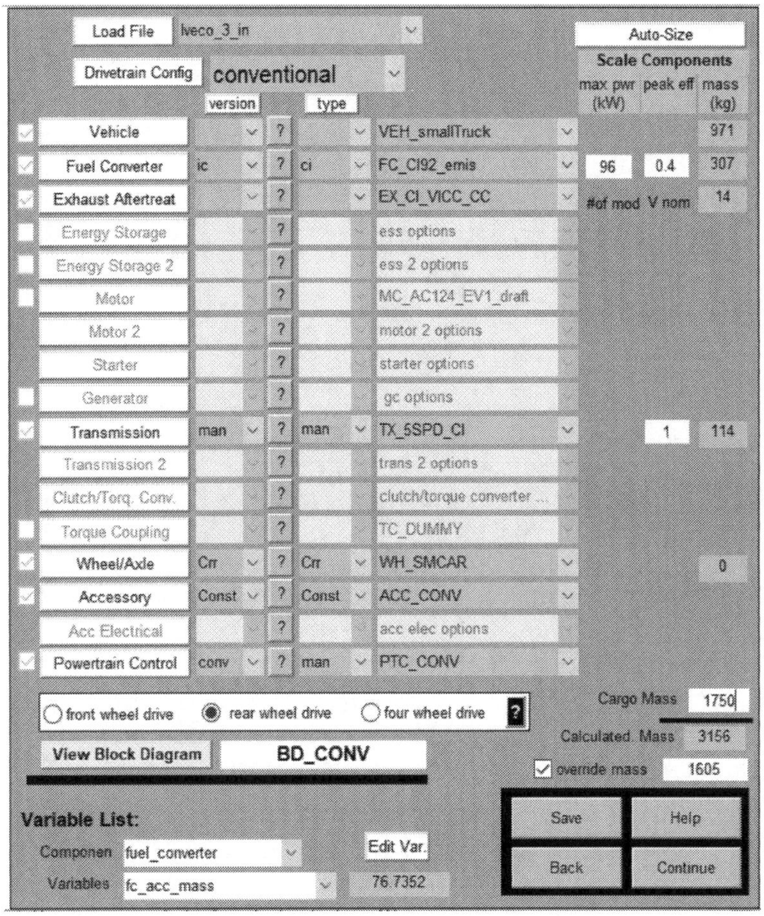

Figure 4. Configuration of the vehicle model similar to the 2007 Iveco Daily dump truck.

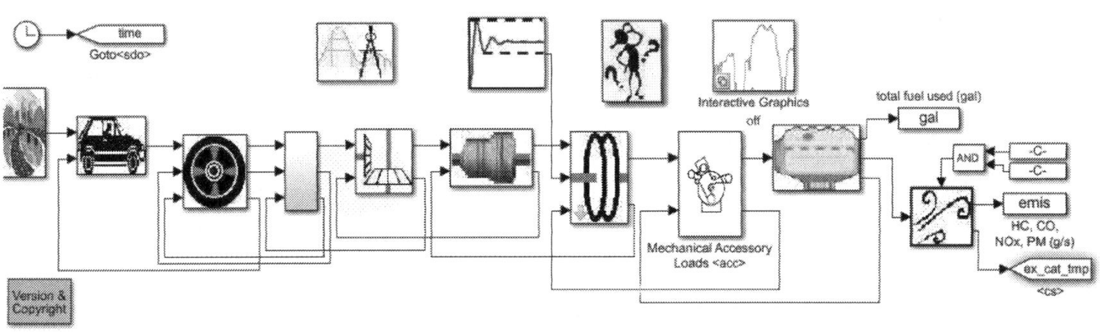

Figure 5. Dump truck block diagram.

Figure 6 shows the block diagram of the front and rear axles of the vehicle (including suspensions). The suspension with lamellar springs of the rear axle has been modified for a hypothetical better behavior of the axle when overloaded.

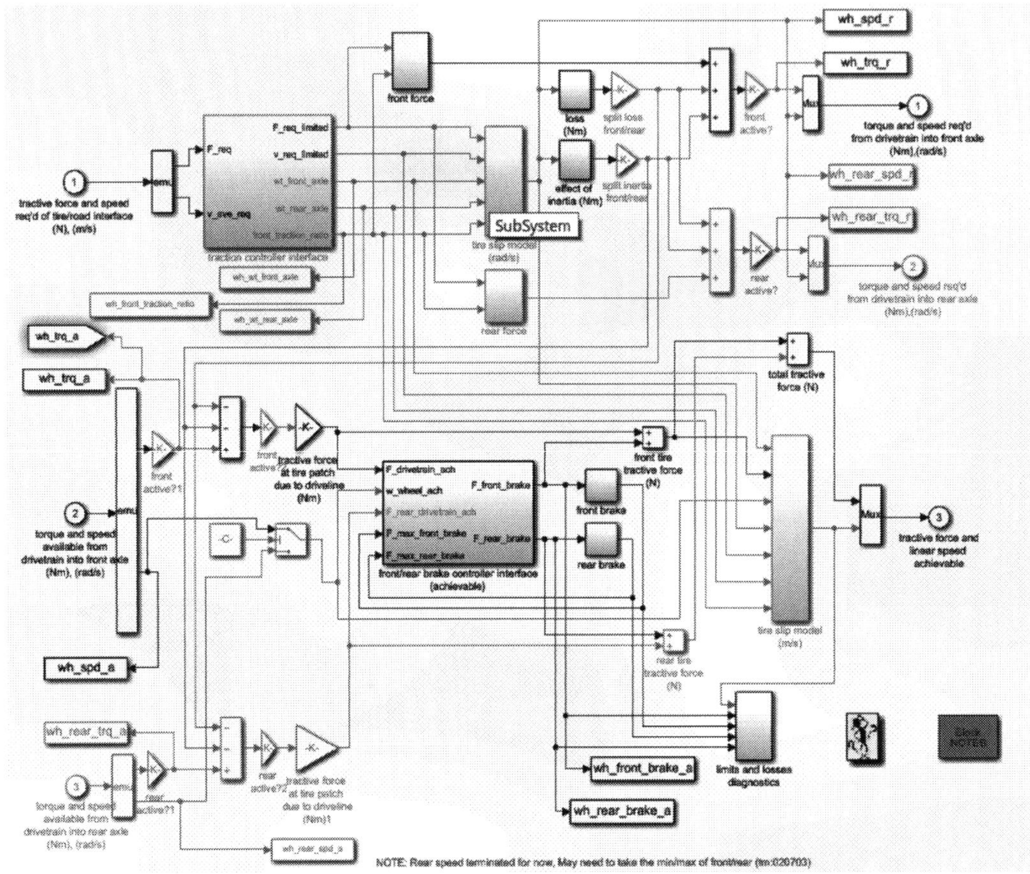

Figure 6. Vehicle axles block diagram.

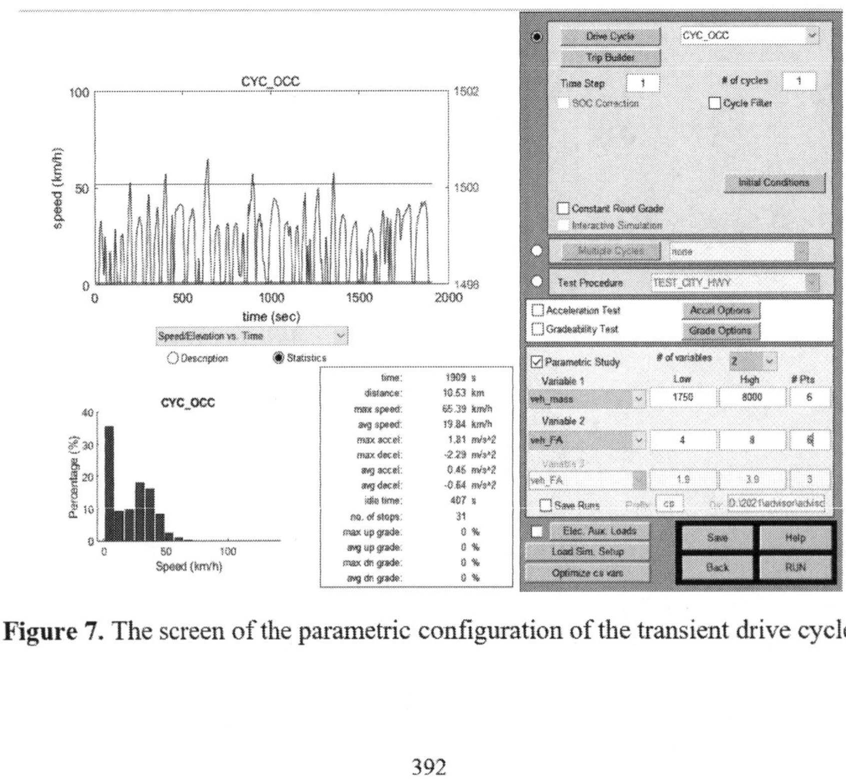

Figure 7. The screen of the parametric configuration of the transient drive cycle.

The simulation of the passage of a drive cycle by the modified commercial vehicle after the successful passing of the periodic technical inspection was performed parametrically by the variation of the total mass of the vehicle (by overloading) and the simultaneous variation of the surface of the front section of the same vehicle. Even if the performance of the periodic technical inspection did not exceed the upper limits for emissions corresponding to Euro 4, respectively: CO = 0.50 g/km, NOx = 0.25 g/km, HC + NOX = 0.30 g / km, and PM = 0.025 g/km, however, overloading can lead to configurations that cause higher emissions. Figures 8 – 12 show engine fuel consumption, CO, NOx, HC and PM emissions in the same parametric lorry simulation.

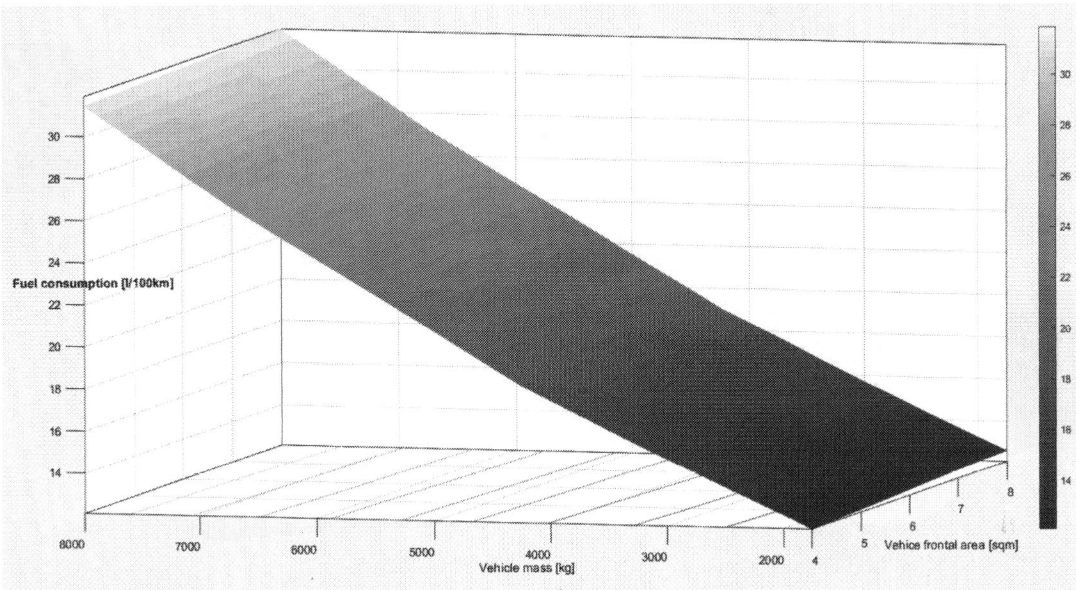

Figure 8. Engine fuel consumption as a function of vehicle mass and front section area.

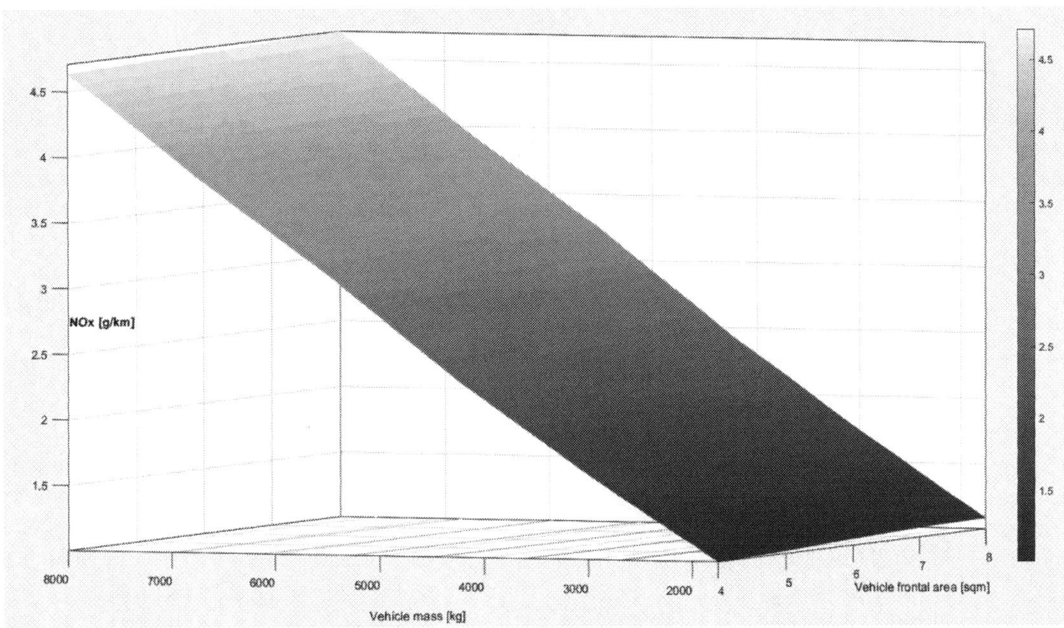

Figure 9. NOx emissions as a function of vehicle mass and front section area.

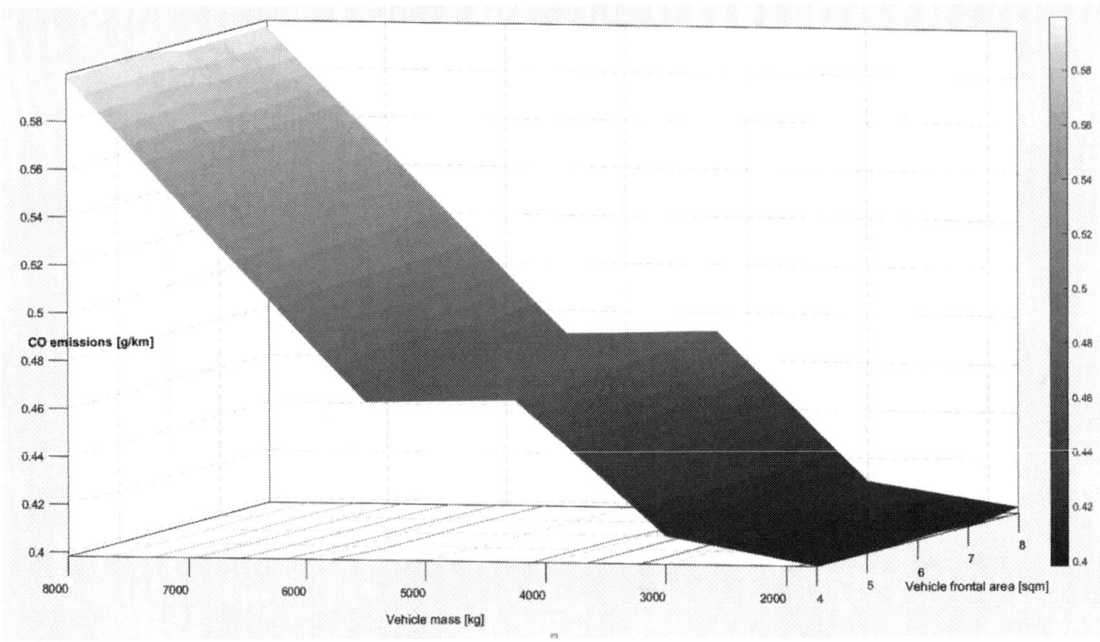

Figure 10. CO emissions as a function of vehicle mass and transversal section area.

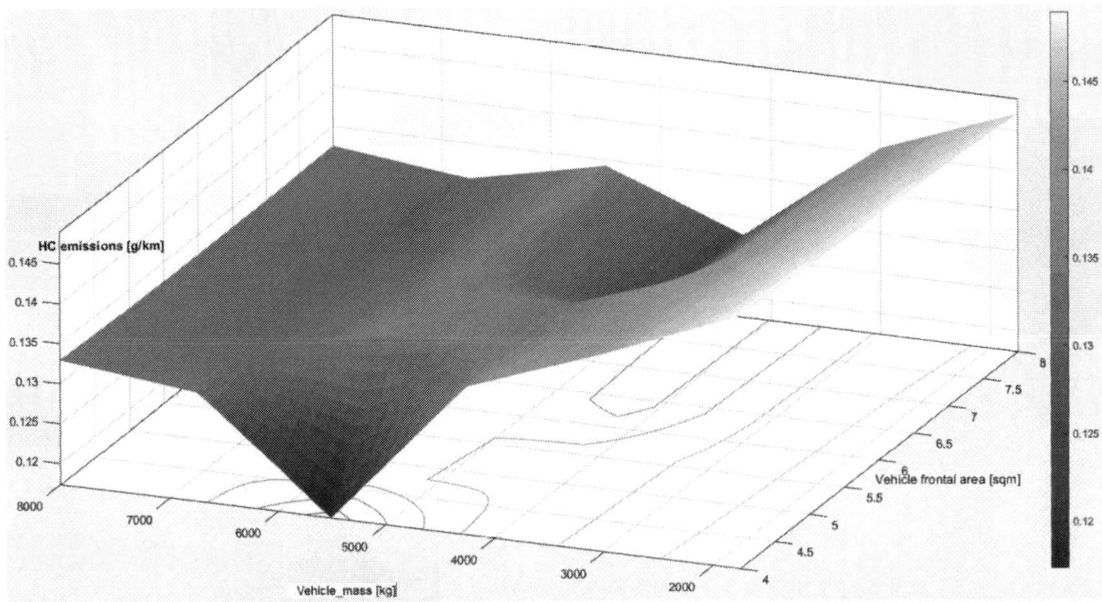

Figure 11. HC emissions as a function of vehicle mass and transversal section area.

The parametric simulation of the passage of a transient drive cycle by lorry modified after passing the periodic technical inspection revealed significant increases in fuel consumption and significant exceedances of the upper emission limits imposed by European legislation (Euro4). These exceedances depend on the configuration of the overload variant – the cross-sectional area of lorry, as shown in figures 9 - 12. These unsafe and illegal changes are frequently encountered in Romania in the transport of wood, according to figures 2 and 3.

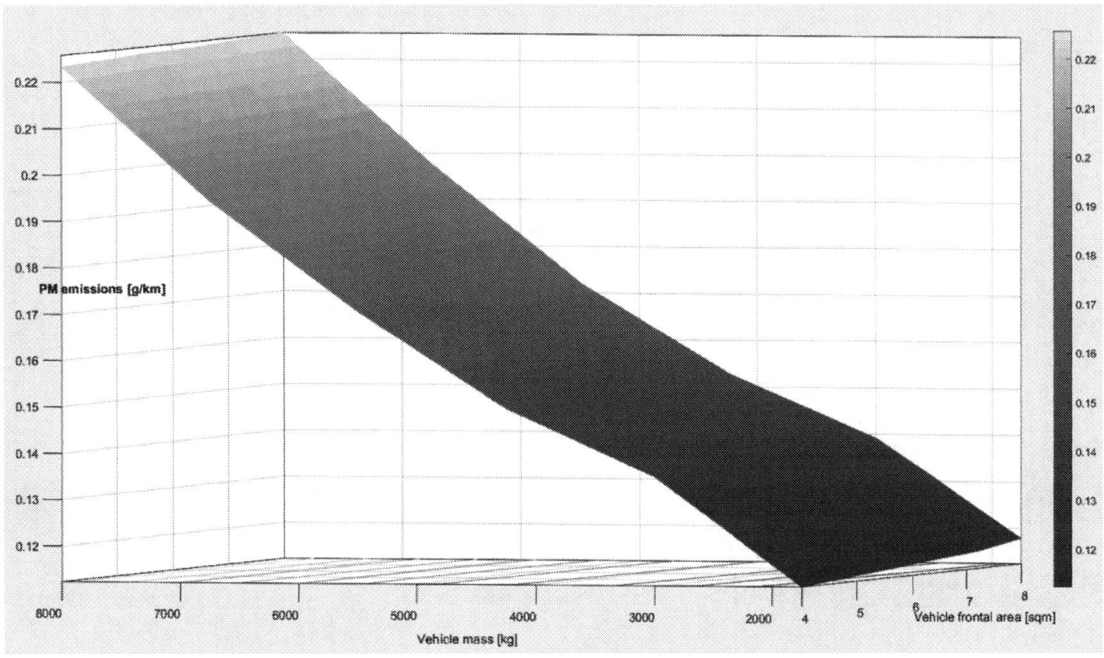

Figure 12. PM emissions as a function of vehicle mass and transversal section area.

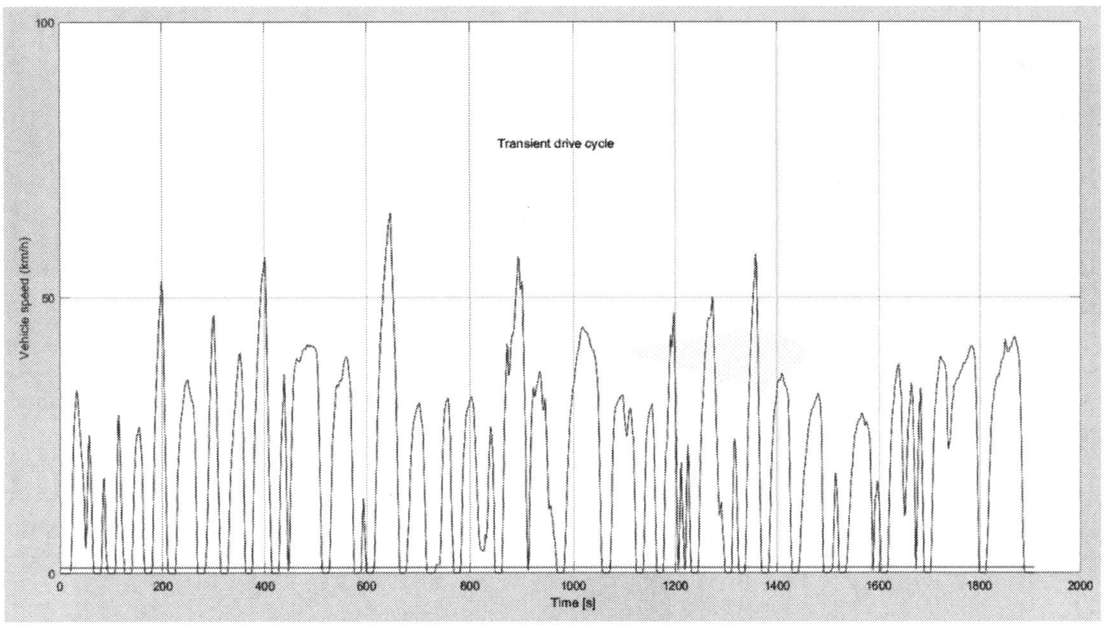

Figure 13. Transient drive cycle used in simulations.

The structure of the test cycle used in the simulations is presented in figure 13, and in figure 14, we presented the synthesis regarding the emissions of the engine that equipped lorry in the extreme situation in which its total mass was 8000 kg (exceeding the legal mass of 4500 kg).

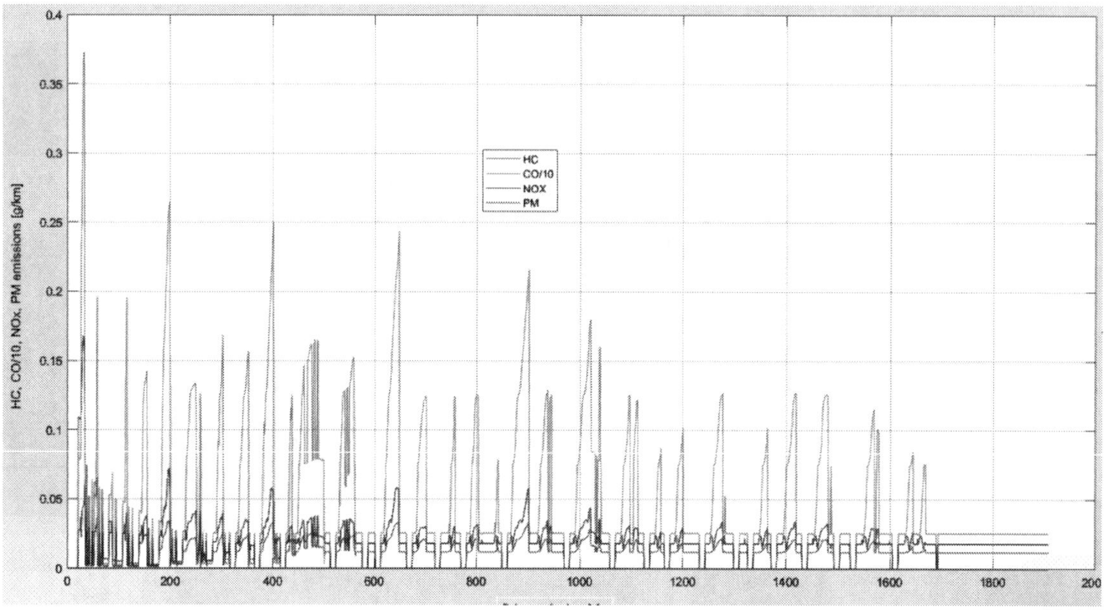

Figure 14. Synthesis of lorry engine emissions at maximum overload during transient drive cycle.

3. Conclusions

The increase in the number of vehicles registered in Romania (new and used) also meant an increase in the total volume of technical problems present in vehicles on the national road network. The lower technical quality of vehicles of a certain age influences the efficiency of their use and substantially determines the safety of road traffic. The technical diversification of the current road traffic units in Romania is enlarged by the increase of the national road vehicle fleet.

The technical condition of these vehicles significantly determines road safety as well as the environmental impact of road transport. For each vehicle, the technical condition is validated by periodic technical inspection.

In this scientific paper, the findings of the periodic technical inspections allowed the authors to develop significant scenarios related to road safety, environmental influences, but also related to commercial problems associated, for example, with the change of vehicle ownership.

In the paper, we emphasize the direct link between the scope of European Union regulations certifying the compliant technical condition of a particular vehicle (the result of a periodic technical inspection) and road safety in Romania today in conjunction with projections of environmental influences.

Due to intensive use, most often at maximum load or overloaded, light commercial vehicles (lorries) are subject to detailed periodic technical inspections. The minimum content of inspection areas comprises: identification, braking equipment, steering, visibility, lighting equipment and electric system, axles, wheels, tires, suspension, chassis and chassis attachments, other equipment, including speed limitation devices, nuisance including emissions and spillage of fuel and/or oil.

To model parametrically modified lorry unsafety after successfully passing the periodic technical inspection we used the specialized software ADVISOR. (version 003-00-r0116 from March 2019). ADVISOR is a software product, originally developed within the National Renewable Energy Laboratory (an American institution under the direct coordination of the United States Department of Energy), based on MATLAB and SIMULINK.

The following can be considered the main conclusions of this paper:

1. The presence in traffic of light commercial vehicles, apparently technically safe, presents an appreciable risk on road safety;
2. By overloading these vehicles exceed the N1 category of commercial vehicles with a total mass

of not more than 3500 kg;

3. Due to current regulations other than those with commercial vehicles with a maximum authorized mass greater than 3.5 t, the unsafe lorries changes made after a successful periodical technical inspection, cannot be the subject of road traffic detection by the State Inspectorate for Road Transport Control – ISCTR;
4. Unauthorized and unsafe modification of the lorry suspension, to transport overloads, has multiple consequences on the rate of road degradation, the degradation of the technical condition of the vehicle, the increase of fuel consumption and pollutant emissions and the reduction of the level of road safety [7];
5. The simulation of the passage of a drive cycle by the modified commercial vehicle after the successful passing of the periodic technical inspection was performed parametrically by the variation of the total mass of the vehicle (by overloading) and the simultaneous variation of the surface of the front section of the same vehicle;
5. Even if through accost/benefit analysis it can be found that in certain situations it is efficient to use overload [8], for reasons related to the legality and safety of road traffic, these uses should be avoided;
6. Even if the performance of the periodic technical inspection did not exceed the upper limits for emissions corresponding to Euro 4, respectively: CO = 0.50 g/km, NOx = 0.25 g/km, HC + NOX = 0.30 g / km, and PM = 0.025 g/km, however, overloading can lead to configurations that cause higher emissions.

References

[1] *** https://www.drpciv.ro, last accessed 2021.08.05

[2] *** https://ec.europa.eu/transport/road_safety/topics/vehicles/vehicle-inspection_en, last accessed 2021.08.06

[3] ***https://ec.europa.eu/transport/road_safety/sites/default/files/pdf/statistics/dacota/asr2020.pdf European Road Safety Observatory Annual statistical report on road safety in the EU 2020, last accessed 2021.08.05

[4] *** https://op.europa.eu/en/publication-detail/-/publication/037bf4a1-242f-11eb-9d7e-01aa75ed71a1/language-en, Report from the commission to the European parliament on periodic roadworthiness, Brussel 2020, last accessed 2021.08.05

[5] Same A, Stipe A, Grossman D, Park J W *A study on optimization of hybrid drive train using Advanced Vehicle Simulator (ADVISOR),* Journal of Power Sources 195 (2010) pag. 6954–6963

[6] Varnhagen S, Same A, Remillard J, ParkJ W A *numerical investigation on the efficiency of range extending systems using Advanced Vehicle Simulator,* Journal of Power Sources 196 (2011) pag. 3360–3370

[7] Schmidt F, Jacob B, Domprobst F *Investigation of truck weights and dimensions using WIM data,* Transportation Research Procedia 14 (2016) pag. 811 - 819

[8] Zhang H, Lu Y, Shi Y, Zhu D *Overloaded Vehicle Choice Behavior Analysis Based on Nested Logit Model,* Journal of Transportation Systems and Information Technology, Volume 12, Issue 6, December 2012, pag. 113 – 118

Crash testing and evaluation of W-beam guardrail using finite elements method

M I Gheres and I L Scurtu

Department of Automotive Engineering and Transports, Technical University of Cluj-Napoca, Romania

marius.gheres@auto.utcluj.ro; liviu.scurtu@auto.utcluj.ro

Abstract. On public roads, the guardrails represent the most used passive protection devices in the case of road accidents. Their role is to absorb the car impact energy. This paper presents and analyses tests of a W-beam guardrail type placed on the roadside using the finite element method. The introduction of the paper presents the state of the art, the requirements and the standards used for guardrails testing and crash test methods. In the second part of the paper is achieved the CAD model of the parapet and the impactor used to create the crash test. In the third part of this study, the boundary conditions of the guardrail structure and impactor are created for two cases of speed (80 and 110 km/h) at 20 degrees angles of impact, according to the SR EN 1317 standard. The fourth part proposes a new guardrail model changed by adding a new shock-absorber element and the distance between the poles is increased after visualization and interpretation of the obtained results of the guardrail structure. The new guardrail structure is tested at the same boundary condition as the base structure. The conclusions are highlighted in the last part of the study.

1. Introduction

Due to the large increase of the vehicles number and the required safety standards, the road restraint systems are intensively studied. In the unfortunate vehicle crash, the guardrail structure must be capable to reduce the kinetic energy of the vehicle without many injuries, also reducing the shape deformation of the car bodywork. Nowadays, the safety evaluation of the W-beam guardrail is the subject of many studies regarding the public road's safety of the vehicles [1]. A detailed crash study between a small vehicle and W-guardrail is performed to evaluate the different configurations of the guardrail supporting post [2]. Four types of the supporting post sections are tested: C-shaped, I-shaped, U-shaped and Sigma-shaped. Following this crash study, the best performance is demonstrated by the barrier with sigma-shaped and I-shaped, which have a lower ASI value and higher energy crash [2].

Distance optimization between the supporting post is an important aspect in the design of the road passive safety systems. A crash study performed in accordance with TB11 impact test descriptions on highway W-guardrail establishes that a 700 mm height of the guardrail and a distance between the supporting post of 2000 mm fulfil the road safety requirements [3].

Traffic safety on bridges involves the design of safety guardrails that are assembled in the bridge structure through the bolted assembly. The design of these types of guardrails involves minimum lateral deformations and bringing the vehicle back on the road with minimal damage. A comparative crash impact study is performed in the real environment and simulated in Ls-Dyna, respecting the impact conditions described according to the TB11 for impact with a vehicle with a 900 kg mass and TB31 on impact with an articulated HGV with a mass of 37390 kg [4].

2. Guardrail system

The guardrail system used for vehicles road safety is defined by the impact performance classes in accordance with the European standard SR EN 1317. The established of the impact performance classes is determined by the level of protection, the severity of the passenger impact and the deformation of the protective system.

Table 1. Vehicle impact crash test descriptions.

Test	TB 11	TB 21	TB 22	TB 31	TB 32	TB 41	TB 42	TB 51	TB 61	TB 71	TB 81
Impact speed (km/h)	100	80	80	80	110	70	70	70	80	65	65
Impact angle (°)	20	8	15	20	20	8	15	20	20	20	20
Total mass (kg)	900	1300	1300	1500	1500	10000	10000	13000	16000	30000	38000
Type of vehicle			Car			Rigid HGV		Bus	Rigid HGV	Articulated HGV	

The efficiency of a guardrail system is given by the ability to attenuate the crash impact and retain or redirect the vehicles under safe conditions. For each vehicle's classes performances of the guardrail system, depending on the vehicle mass, impact angles and the various speed are presented in table 1 the impact test scenario established in EN 1317 standards [5]. Impact severity is a theoretical evaluation parameter of the mechanical stress, severity of injuries and other risks for the inside passenger from the vehicle. The index reference values of the impact severity level are presented in table 2 [5].

Table 2. Impact severity level.

Impact severity level	A	B	C	
Index values	ASI ≤ 1.0	ASI ≤ 1.4	ASI ≤ 1.9	THIV ≤ 33km/h

Guardrail assembly deformation according to the working width classes is presented in table 3. The working width is represented by the maximum lateral distance between the guardrail parts on the undeformed traffic side and the maximum deformed positions after impact.

Table 3. Guardrail deformation.

Working width classes	W1	W2	W3	W4	W5	W6	W7	W8
Working width W_n (m)	≤ 0.6	≤ 0.8	≤ 1.0	≤ 1.3	≤ 1.7	≤ 2.1	≤ 2.5	≤ 3.5

It is desired that in case of a vehicle crash the guardrail deformation not be larger than the available space behind the guardrail.

2.1. W-beam guardrail model

The CAD model of guardrail was performed in SolidWorks modelling environment. As can be observed in figure 1, the guardrail parts are assembled by bolted assembly. The parts are made of sheet metal of different thicknesses to result a flexible assembly for all classes of the impact vehicles. The supporting post is fixed in soil at a depth of 960 mm.

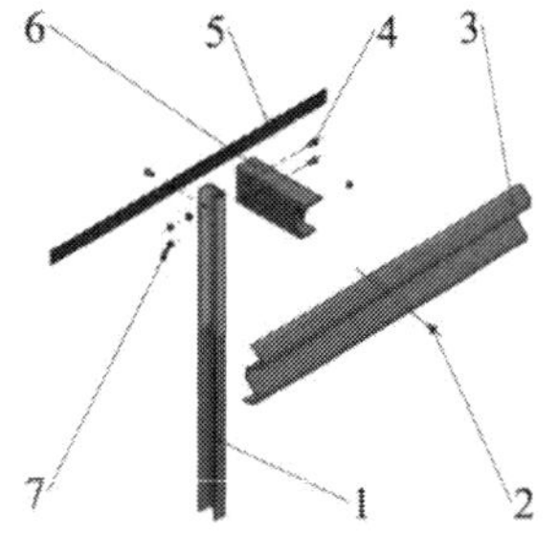

1. Supporting post
2. Bolt for W-beam connection
3. W-beam
4. Bolt between spacer element and supporting post
5. Rear plate
6. Spacer element
7. Nuts for bolt assembly

a) Parts descriptions

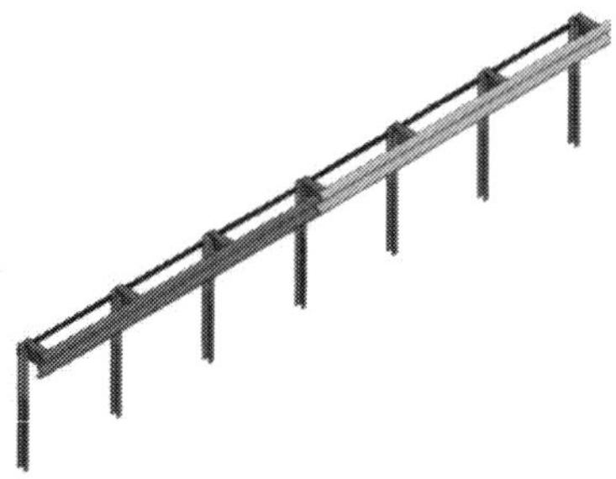

b) W-beam assembly

Figure 1. W-beam guardrail parts.

Figure 2 shows the overall dimensions of the W-beam guardrail assembly and the distance between the guardrail supporting post.

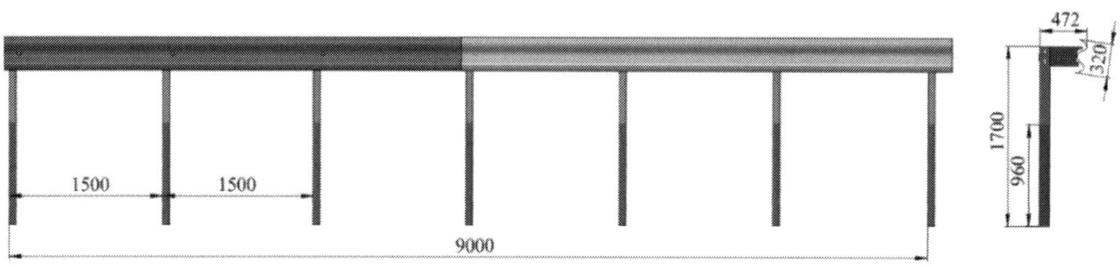

Figure 2. W-beam guardrail overall dimensions.

2.2. Impactor model

In case of a crash, the guardrail behavior also depends on the vehicles body structure. To imitate the real vehicle in this study is proposed a generic vehicle model presented in figure 3, where can be observed the shape, the parts of the impactor and the overall dimension of the assembly that respect the real vehicle dimensions. The middle shock absorber components have the role to attenuate the crash impact, allowing behavior specific to a real vehicle.

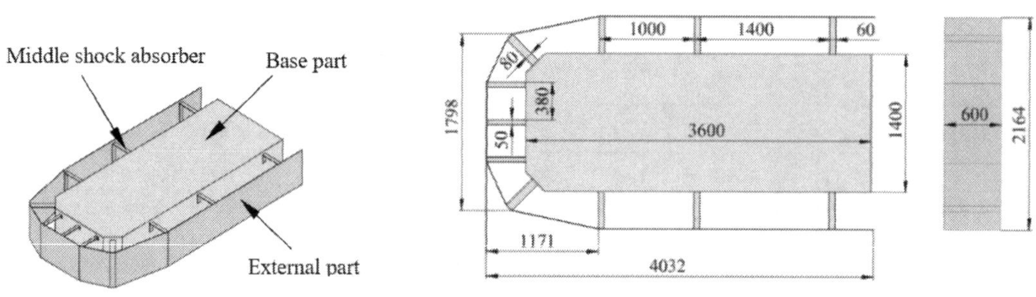

Figure 3. Impactor model shape and overall dimensions.

3. Finite element model and crash test

To create the crash scenario, it is necessary to follow several stages such as: prepare the guardrail and impactor geometry to create the mesh, material assign and load case creation.

3.1. W-beam guardrail finite element model

The current guardrail system comprises four parts meshed into mixed quads and trials shell elements. The bolt connection between W-beam, spacer elements and the supporting post are modelled using rigid elements. The continuations of the W-beam guardrail are represented by the rigid connections at both ends of the protection device.

Table 4. Thickness and mass of the guardrail.

W-beam guardrail	Thickness (mm)	Target length (mm)	Number of shell elements	Mass (kg)
Supporting post	5	10	4088	14.4
W-beam	3	10	25355	56.6
Rear plate	5	10	6476	12.6
Spacer element	3	10	1611	7.14

The properties of the guardrail components model are presented in table 4, where it is presented the shell thickness, the target length of the elements, the resulted elements number and mass for each part of the protection device.

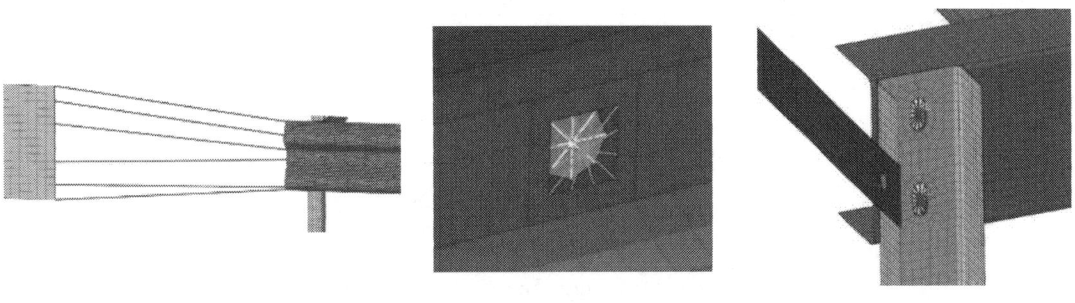

a) Guardrail continuation connections b) W-beam bolt connections c) Supporting posts assembly

Figure 4. W-beam guardrail parts.

The connection method between the parts of the guardrail device is presented in figure 4.

3.2. W-beam guardrail finite element model

The mass and the thickness of the impactor parts are presented in table 5. The mass of the finite elements is 1150 kg. To fulfil the mass requirement by the EN 1317 standard about the vehicle used for testing, an additional mass of 350 kg was attached to impactor body. Considering the simplified construction of this vehicle, the friction between the tires and the road and the behaviors of the real vehicle fluids is neglected.

The connections between all impactor components are done by using the rigid elements that are linked to finite elements nodes. Steel is the assigned material properties for all the parts. In figure 5 shows the finite model of the impactor model prepared for simulations.

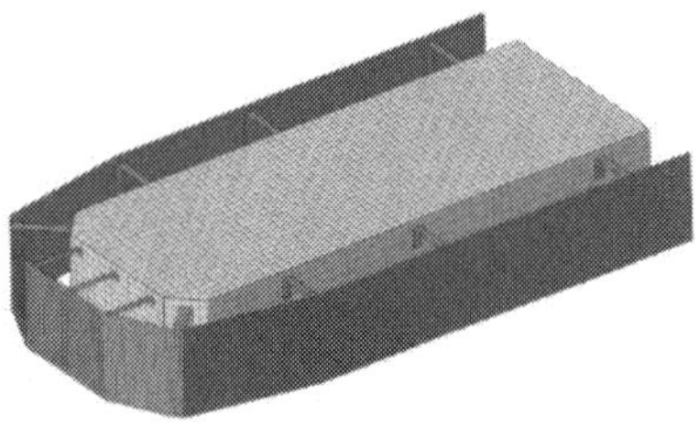

Figure 5. The impactor model.

Table 5. Thickness and mass of the impactor parts.

Impactor element	Thickness (mm)	Target length (mm)	Number of shell elements	Mass (kg)
External part	4	30	6340	178
Middle shock absorber parts	5	20	2623	40
Base part	8	60	4256	932

3.3. Material model

The used material in this study for all parts is an isotropic elasto-plastic, material that uses the Zerilli-Amstrong plasticity model. This material law is applicable to the components meshed into shell and solid elements. In the next equation is described the stress during plastic deformation:

$$\sigma = C_0 + \left(C_1 \, exp\left(\left(-C_3 T + C_4 \, T ln\left(\frac{\dot{\varepsilon}}{\dot{\varepsilon}_0}\right)\right)\right)\right) + C_5 \varepsilon_p^n \quad (1)$$

where: C_0 is the plasticity yield stress, C_1 is strain rate formulation coefficient, C_3 is temperature effect coefficient, C_4 is temperature effect coefficient without strain rate effect, C_5 is plasticity hardening parameter, ε_p is the plastic strain, $\dot{\varepsilon}$ represent the strain rate, T is a temperature, n is plasticity hardening exponent and it must be less than 1, $\dot{\varepsilon}_0$ represent reference strain rate.

The mechanical properties of the steel assigned for the guardrail parts and the impactor are presented in table 6 [6].

Table 6. Mechanical properties of the assigned steel.

Initial density	$7.85e^{-6}$ kg/mm3
Poisson ratio	0.3
Hardening exponent	0.5
Hardening parameter	0.5 GPa
Failure plastic strain	0.3
Young modulus	210 GPa

3.4. Crash test simulation

In this crash study, the guardrail device is presented in two impact cases, the first case is applied for an actual road safety device and in the second case, the geometry of the device is improved. All the simulations are done using the explicit Radioss solver [7]. The impact distance between the impactor model and W-beam guardrail is approx. 2300 mm from the start guardrail extremity. A general interface TYPE 7 has been chosen for the contact simulation during both simulations' cases and a Columb friction coefficient of 0.2 has been applied [8].

3.4.1. Base model. In the first simulation case, the guardrail device is hit by the impactor model at an initial velocity of 80 and 110 km/h and 20° impact angle corresponding to the TB31 and TB32 vehicle crash test descriptions from SR EN 1317 standard. In figure 6 is presented the base model simulation scenario.

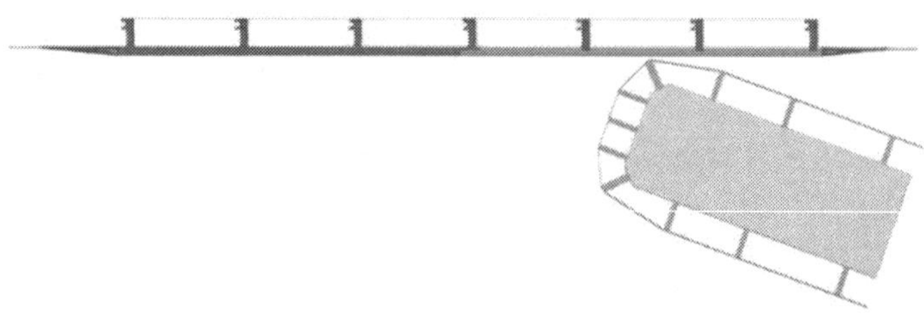

Figure 6. The base model simulation.

3.4.2. Improved model. In the second simulation case, all the boundary conditions are the same as the first case. The distances between the supporting posts are changed from 1500 to 2250 mm and a curved shape element whit a thickness of 5 mm is added in the new guardrail configuration, as can be seen in figure 7.

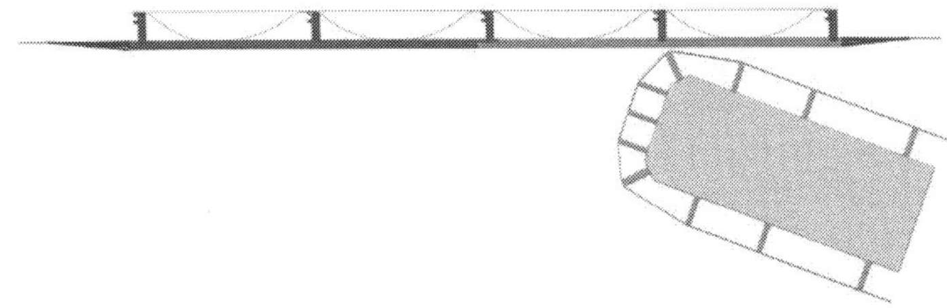

Figure 7. The base model simulation.

4. Visualizations and analysis of computational results

In this section are presented the results of the simulations for both cases, for each velocity. The resulted graphs of the kinetic energy curve are interpreted, and the parts deformations are compared.

4.1. Results-base model

First case simulations results are presented in figure 8 where can be observed the behavior of the guardrail and impactor at 80 km/h initial velocity of the impactor, at three-time phases of 100 milliseconds, 250 milliseconds and at 400 milliseconds. The maximum value of the von Misses stress is 6.123E-1 GPa.

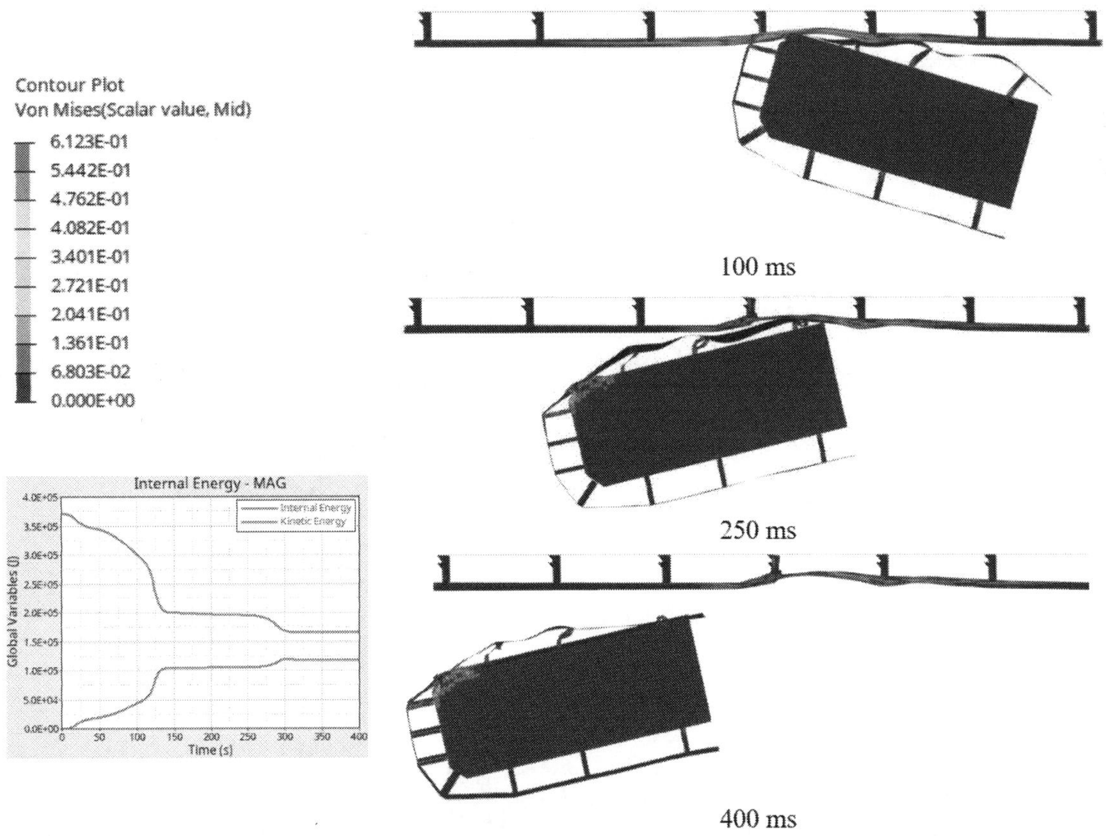

Figure 8. Simulation results at 80 km/h (22.23 m/s) of the base guardrail model.

After the crash simulations are performed the energy curves are plotted as shown in figure 8. The kinetic energy curve starts from the initial energy value of 3.70726E+05 joules and remains constant to the first 5.5 milliseconds, time when the impactor gets in contact to the guardrail device. The kinetic energy drops continuously until 145 milliseconds, time when the impactor elements and W-beam are deformed by the crash impact. After 145 milliseconds the kinetic energy curves has a smoothly decrease until 307 milliseconds at the values of 1.19054E+05 joules. The crash simulations have stopped at 400 milliseconds, during which time it can be observed that the safety barrier deflects the impactor back into the roadside.

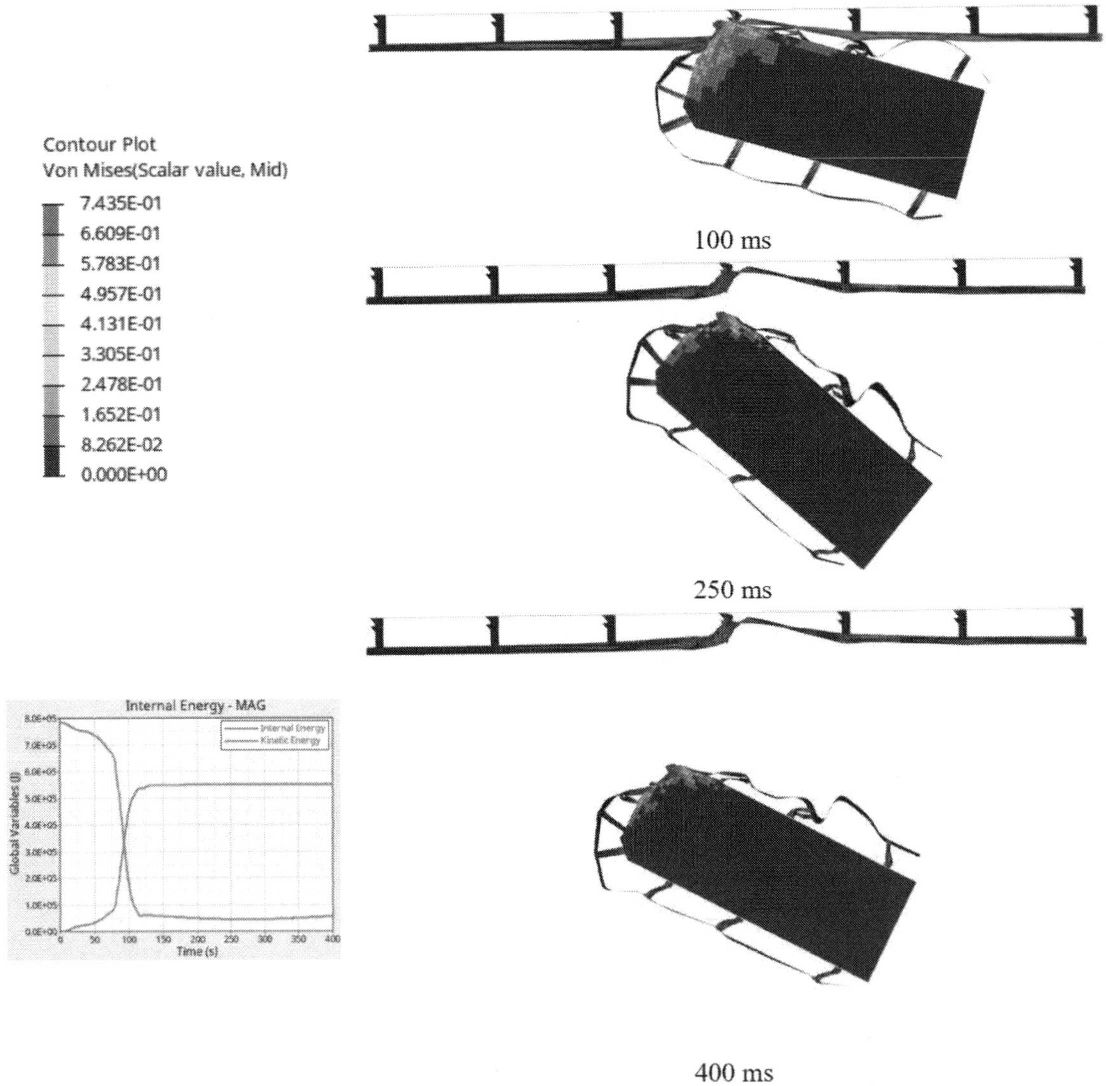

Figure 9. Simulation results at 110 km/h (30.06 m/s) of the base guardrail model.

In figure 9 are presented the results at the 110 km/h initial velocity of the impactor. The maximum value of the von Misses stress is 7.435E-1 GPa. It can be observed that the impactor is crashed faster into guardrail compared to the 80 km/h case. At the 100 milliseconds, the impactor is more deformed during the crash impact compared to the first simulation, this being suddenly deviated back on the roadside and a more distance as the first simulation that can be dangerous because the vehicle can be easily reached at the opposite side of the road. The kinetic energy curve starts from the initial energy value of 7.85201E+05 joules and remains constant to the first 4.2 milliseconds. The kinetic energy drops continuously until 117 milliseconds. After 117 milliseconds the kinetic energy curve remains constant to the end of simulations.

The guardrail and impactor deformations stop when the kinetic energy of the impactor is completely absorbed in plastic deformations and gets converted into internal energy. Due to the increase the impactor velocity the impactor and guardrail deformations are higher compared whit the first simulation case.

4.2. Results- improved model

The results of the improved model at 80 km/h are presented in figure 10. The maximum value of the von Misses stress is 5.766E-1 GPa. The kinetic energy curve starts from the initial energy value of 3.70800E+05 joules and remains constant to the first 5.8 milliseconds. The energy curves drop continuously smooth until 330 milliseconds, then the curves remain constant.

After simulation, it can be observed that the guardrail deformation is higher over a longer distance but is more smoothly compared to the base model at this simulation case. The impactor deformation and roadside deviation is lower compared to the simulation of the base guardrail model.

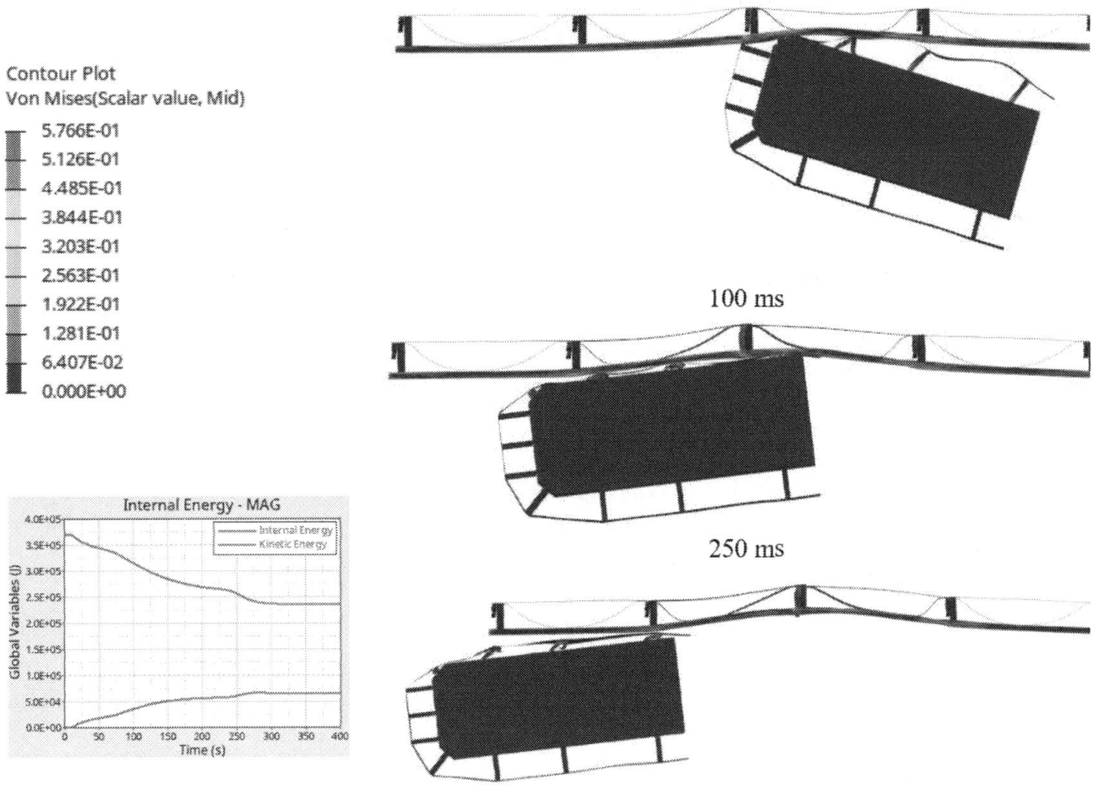

Figure 10. Simulation results at 80 km/h (22.23 m/s) of the improved guardrail model.

The simulation results at the impactor velocity of 110 km/h are presented in figure 11. It can be observed that the guardrail deformation is higher compared to the guardrail base model, but the impactor deviation on the roadside is smooth without many damages.

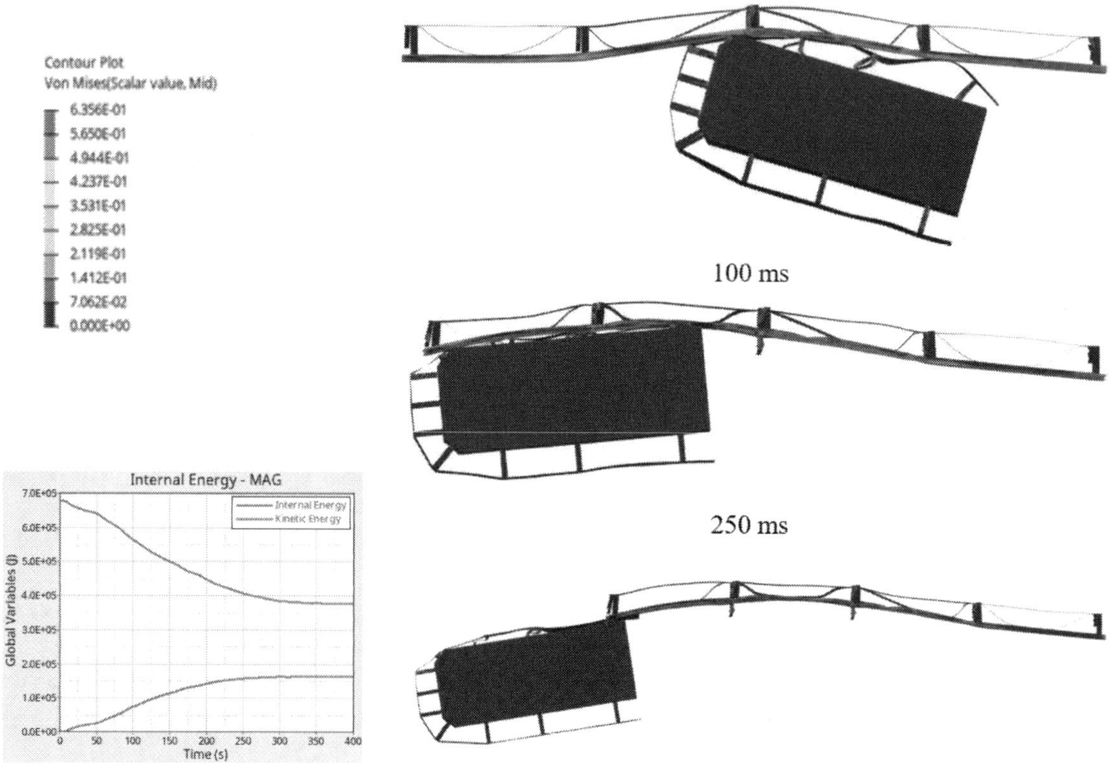

Figure 11. Simulation results at 110 km/h (30.06 m/s) of the improved guardrail model.

The energy balance during the guardrail impact is presented in figure 11. The kinetic energy curve starts from the initial energy value of 6.779954E+05 joules and remains constant to the first 4.5 milliseconds. The energy curves drop continuously smooth until 340 milliseconds, then the curves remain constant.

a) Deformations at 80 km/h b) Deformations at 110 km/h

Figure 12. Improved guardrail deformation.

The improved guardrail parts deformation at the 80 and 110 km/h is presented in figure 11, where it can be observed behavior of this after impact. At the 80 km/h impactor velocity just middle supporting post have a major deformation and the W-beam is deformed in the impact area. At the 110 km/h velocity of the impactor, the guardrail device has a large deformation compared to the 80 km/h case. It can be observed that the two supporting posts are much bent and two adjacent supporting posts have a minor bent. The spacer elements placed on the most bent supporting post and the W-beam are very damaged at the impact velocity.

5. Conclusions

The present study presents a methodology to analyze the crash evaluation and improvement of a W-beam guardrail used to increase safety on the roadside. To evaluate the guardrail behavior two simulations cases are done in accordance with vehicle impact test TB 31 and TB 32 from SR EN 1317 standard. After visualization of the simulation results, it can be observed that the improved model of the guardrail has a more working width compared to the base model, but the kinetic energy of the impactor decreases smoothly and is deviated on the roadside without many damages. The most affected parts of the guardrail are the W-beam, supporting post and the spacer elements that absorb the impactor kinetic energy. The numerical evaluation of the guardrail is validated by the energy balance during the simulation process. A weakness point of this impact study is that the supporting posts are rigidly fixed, as in concrete structure. Because the working width of the improved guardrail is longer, it is necessary to eliminate the object placed behind that can be present the working deformation of the guardrail. The main advantages of this evaluation method are the possibility to create a detailed study of the crash damages, to perform the simulation at the different vehicle velocity and the different impact angles, the possibility to optimization of the guardrail shape from the project stage, without necessary to manufacturing a prototype.

References

[1] Sassi, S., Sassi, A., & Ghrib, F. 2017 *Effect of crushable blockouts on a full-scale guardrail System* International Journal of Crashworthiness 22(1) 63-82 DOI:10.1080/13588265.2016.1223525

[2] Tso-Liang Teng, Cho-Chung Liang, Ching-Yu Hsu, Chien-Jong Shih, Thanh-Tung Tran. 2016 *Impact Performance of W-beam Guardrail Supported by Different Shaped Posts* International Journal of Mechanical Engineering and Applications. Vol. 4, No. 2, 2016, pp. 59-64. doi: 10.11648/j.ijmea.20160402.1

[3] Teng, T. L., Liang, C. C., & Tran, T. T. 2015 *Effect of various W-beam guardrail post spacings and rail heights on safety performance* Advances in mechanical engineering 7(11) DOI: 10.1177/1687814015615544

[4] Atahan, A. O. 2018. Development of a heavy containment level bridge rail for Istanbul. Latin American Journal of Solids and Structures 15 https://doi.org/10.1590/1679-78254684

[5] European Committee for Standardization, European *Standard EN 1317-1, EN 1317-2, Road Restraint Systems*, 2010

[6] Scurtu I.L. and Lupea I. 2014 *Frontal Crash Simulation of a Chassis Frame,* vol 57-III, Acta Technica Napocensis-Series: Spplied Mathematics, Mechanics, and Engineering, pp 411-414

[7] Radioss Theory Manual, *Large Displacement Finite Element Analysis,* 2021

[8] Scurtu, I.L., Szabo, I., Mariasiu, F., Moldovanu, D., Mihali, L., and Jurco, A. 2019 *Numerical analysis of the influence of mechanical stress on the battery pack's housing of an electric vehicle* IOP Conference Series: Materials Science and Engineering (Vol. 568, No. 1, p. 012054). IOP Publishing

IOP Conf. Series: Materials Science and Engineering 1220 (2022) 012050 doi:10.1088/1757-899X/1220/1/012050

Experimental test to determine the strength of the seat belt anchorages

A M Manea, D Iozsa, C Stan and A Ioniță

Automotive Engineering Dept., Faculty of Transports, POLITEHNICA University of Bucharest, Splaiul Independentei nr. 313, Sector 6, Bucuresti, ROMANIA, RO-060042

mannea_annamaria@yahoo.com, daniel_iozsa@yahoo.com, corraista@yahoo.com, alexandra.ionita930111@gmail.com.

Abstract. Experimental tests on the type-approval of motor vehicles with regard to safety belt anchorages, ISOFIX anchorages and ISOFIX top anchorages have been carried out in accordance with Regulation No. 14 UNECE [1]. From the design stage, the safety belt anchorages must comply with the requirements for the installation of a safety belt and reduce slipping and damage to the belt while wearing it. These can be components of the seat structure or of the body's resistance structure, which have the role of fastening the seat belts. The position of the anchor points is determined by the R point. The R point is defined as the seat reference point on the seat. This point is set by the vehicle manufacturer for each seat in the vehicle. In this paper, two models of chairs were tested. Both seat models are equipped with seat belts anchored at three anchor points. Seat belt anchorages were tested simultaneously using parallel force devices. These traction devices are placed on the seat cushion and are pressed into the seat back. The pulling device has a vertical pulling arm, which has the role of anchoring the steel wire cables. Two test forces are applied simultaneously to the anchor points during the test.

1. Introduction

From the project stage, according to Regulation no. 14 UNECE [1], safety belt anchorages must comply with the requirements for the conditions of installation of a safety belt and reduce slipping and damage to the belt while wearing it. These can be components of the seat structure or of the body's resistance structure, which have the role of fixing the seat belts.

Seat belts are designed to increase safety. They are equipped with a pretensioning mechanism and a force transducer, which contributes to improving the protection of the occupants by maintaining the movement of the passenger's chest during the impact. The pretensioning mechanism restrains the occupant, applying a restraining force to the chest area and stops the forward movement of the occupant. When the holding force reaches a maximum value, the mechanism can control the movement of the chest, with the help of a force transducer, further applying constant pressure on the chest. These restraint systems of the occupant slow down the movement of the body and increase its stopping distance to the limit of contact with the elements of the interior components of the passenger compartment [4, 5, 6].

2. Experimental test to determine the strength of the safety belt anchorage points

The actual test was performed on a test stand consisting of universal devices and specific to the type of test in accordance with Regulation no. 14 UNECE [1].

During the work, two models of medicine seats were tested for the strength of the anchorage points of the seat belts. Both seat models are equipped with seat belts anchored at three anchor points.

The experimental test was performed only on one area of the vehicle structure. The seat structures are mounted at the bottom on a metal structure. The metal structure is constructed in accordance with the specific requirements imposed on the construction of a vehicle floor. It features ribs, embossed reliefs, and other reinforcing elements to increase stiffness and mechanical strength in the seat grip areas [2].

Seat belt anchorages were tested simultaneously using parallel force composition devices (figure 1 and figure 2). These traction devices are placed on the seat cushion and are pressed into the seat back. The seat belt in the pelvic area is positioned above the device pressed into the seat back, tightening tightly around it. The role of the upper device is to support the seat belt in the chest area. Also, after the belt has been placed over the device, it is tight. Throughout this process of positioning the towing device, no force is applied to the lashing points for the safety belts. The pulling device has a vertical pulling arm, which has the role of anchoring the steel wire cables [1].

Figure 1. Mounting the traction device on the seat model 1 [3].

Figure 2. Mounting the traction device on the seat model 2 [3].

The test stand consists of two hydraulic cylinders (figure 3), which can determine a force of 60 kN, two tensometric transducers of 60 kN and 100kN, respectively, and an electrical control system consisting of electronic modules for amplifying and conditioning the signals taken from the transducers for measuring and recording force and displacement.

Figure 3. Hydraulic cylinders of the test installation [3].

The conditioning of the signals and the conversion of the analog-numerical data of the signal taken from the test installation is performed with the help of a data acquisition system with 8 data acquisition channels.

2.1. Initial conditions

In the experimental tests [1], actual measurements were made to determine the position of the R point and the gripping positions of the anchor points. It was also established point C which is the point located vertically 450 mm above the point R.

The traction force is applied in a direction corresponding to the passenger's seating position at an angle of 10 ° ± 5 ° to the horizontal axis.

The seat backrests were locked at a 15° angle to the vertical plane.

The anchor points must withstand the specified load for at least 0.2 s.

Two test forces are applied simultaneously to the anchor points during the test. The forces applicable to traction devices are for vehicles of categories M3 and N3. For these vehicles, the value of each traction force shall be 450 ± 20 daN. In addition, these traction forces will be supplemented by an additional force equal to the full mass of the seat, multiplied by 6.6 times.

Table 1 will present the main parameters used in the tests on the anchorage points of the seat belts located in the structure of the two types of seats.

Table 1. The main parameters of the seat models.

Model Type	Mass (kg)	*Weight (daN)*	F_s *(daN)*	F_t *(daN)*	F_{total} *(daN)*
Seat 1	28	27.4	181	2*450 ±20	1081±40
Seat 1	36.2	35.4	560	2*450 ±20	1134±40

Where: F_s = 6.6xG [daN] - additional force; F_t [daN] - traction force; Ftotal = F_s + F_t [daN] - total force.

Before the start of the test, the measuring system will be calibrated. Following the calibration of the installation, the diagrams showing the scope of the force to be used for testing the anchorage points of the seat belts will be displayed.

The preload of the hydraulic cylinders is done with the values of the corresponding forces for each seat model. An impact loading is applied during the test on the anchor points. By applying the impact loading, the rear impact of the vehicle is simulated.

Figure 4 shows the graph of the first force as a function of time, obtained from the calibration of the installation, for testing the anchorage points of the seat belt 1. Figure 5 shows a detail regarding on the application value of the traction force.

Figure 6 shows the force graph of the seat model 2 as a function of time, obtained from the calibration of the installation, for testing the seat belt anchorages. Figure 7 shows a detail regarding on the application value of the traction force.

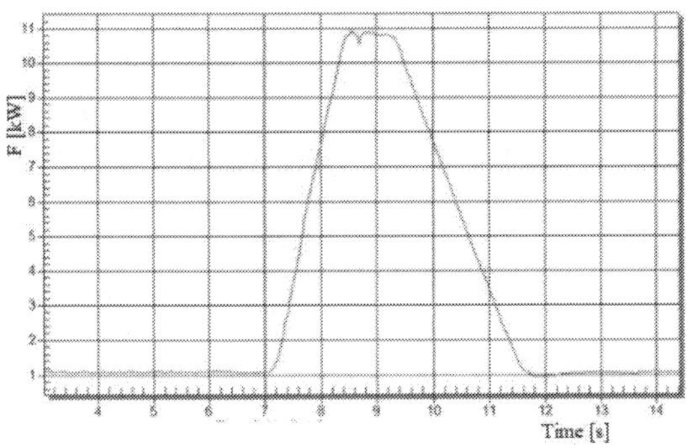

Figure 4. Traction force for testing seat belt anchorages of model 1 of the seat [3].

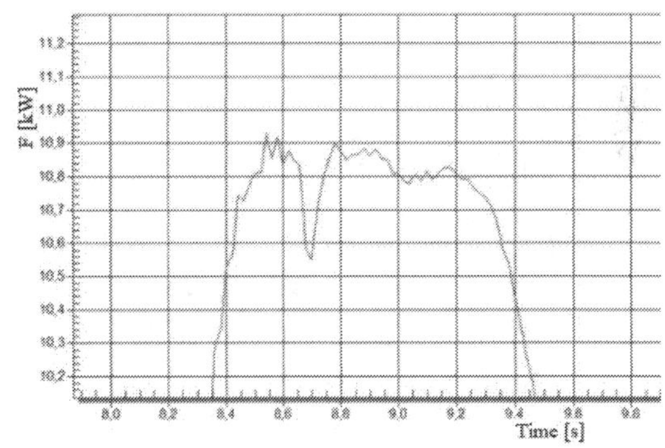

Figure 5. Detail on the application value of the traction force.

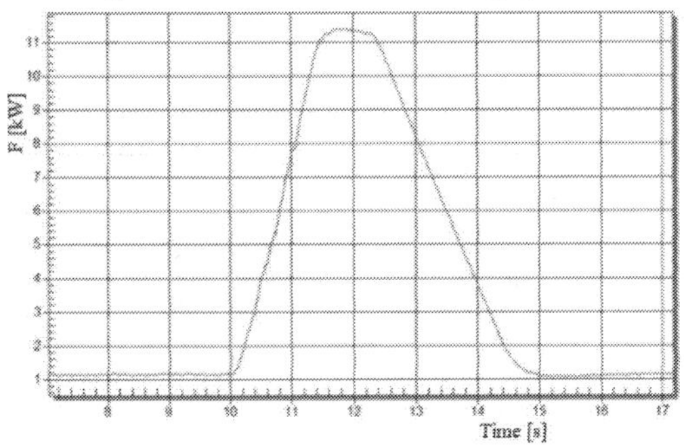

Figure 6. Traction force for testing seat belt anchorages of model 2 of the medical seat [3].

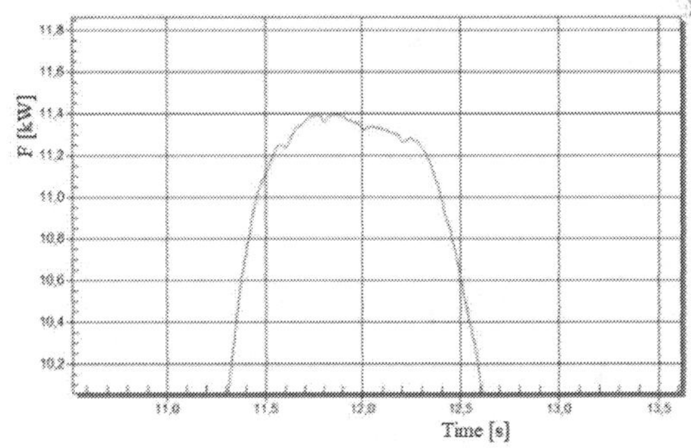

Figure 7. Detail on the application value of the traction force.

3. Experimental results

Following the experimental tests on the anchorage points of the seat belt of the two seat models, all the anchorage points withstood the application of traction forces on them throughout the period.

According to [1], the upper anchorage point of the seat belt must not exceed during the test a transverse plane inclined 10 ° forwards and passing through the R point of the seat.

Thus, following the experimental tests, the upper anchorage points of the seat belts of the two seat models did not exceed during the test the transverse plane inclined 10 ° forwards and passing through the R point of the seat.

IOP Conf. Series: Materials Science and Engineering 1220 (2022) 012050 doi:10.1088/1757-899X/1220/1/012050

4. Comparison between the results obtained by modeling and simulation with those determined experimentally

To highlight the degree of confidence of the realized models, the results obtained by modeling, and simulation were compared with those obtained experimentally. The modeling and simulation of the seat structures was performed in the Ansys software.

The following figures show the angular deformations of the seat model 1. Figure 8 shows the experimentally determined angular deformation of the seat model 1. In this case, during the experimental test, the value of the angular deformation is 8.369°.

Figure 9 shows the angular deformation of the seat model 1 determined by modeling and simulation. In this case, the value of the angular deformation is 9.608°.

The following figures also show the angular deformations of the seat model 2. Figure 10 shows the experimentally determined angular deformation of the seat model 2. In this case, the value of the angular deformation is 5.446°.

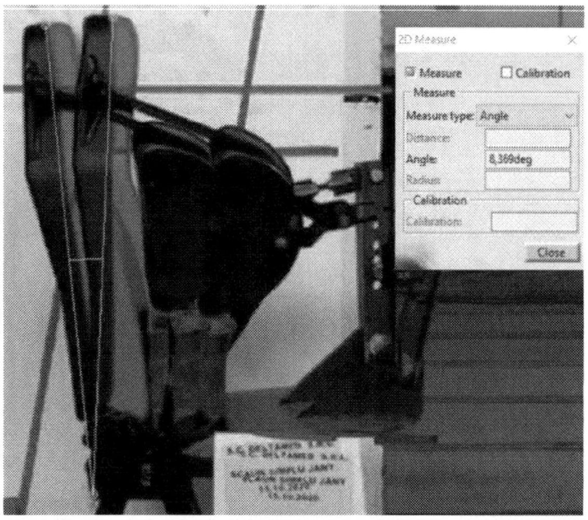

Figure 8. Angular deformation of the seat model 1 following the experimental test anchorages.

Figure 9. Angular deformation of the seat model 1 obtained by modeling and simulation [3].

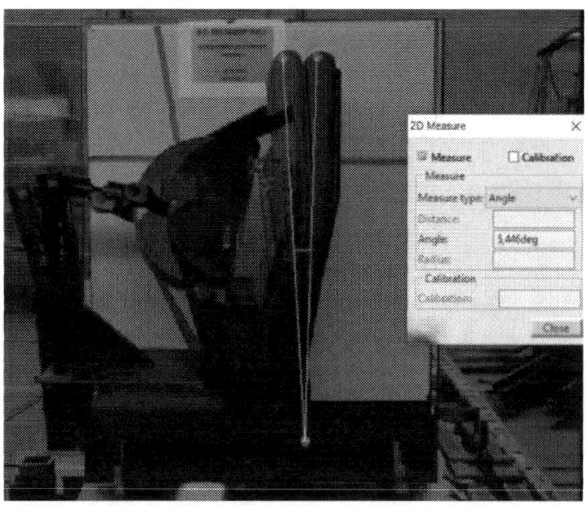

Figure 10. Angular deformation of the seat model 2 following the experimental test.

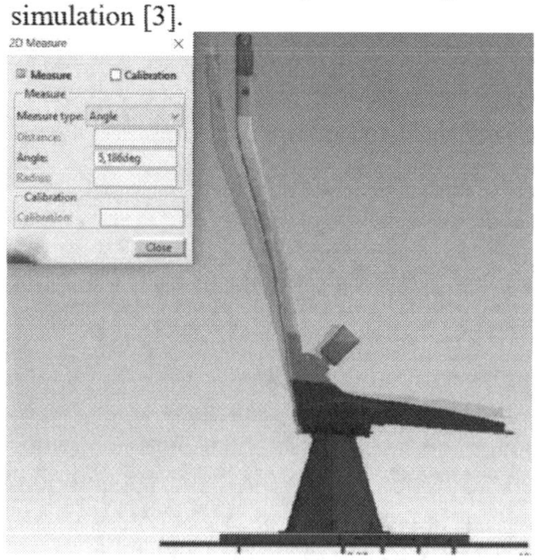

Figure 11. Angular deformation of the seat model 2 obtained by modeling and simulation [3].

Figure 11 shows the angular deformation of the seat model 2 determined by modeling and simulation. In this case, the value of the angular deformation is 5.186°.

5. Conclusions

Following the experimental tests on the seat belt anchorages, the upper seat belt anchorages did not exceed during the test the transverse plane inclined 10° forwards and passing through the R point of the seat.

During the experimental tests, the applied force was supported by the anchor points of the two seats throughout the specified period.

Comparing the results obtained regarding the angular deformations of the seat model 1, it is observed that the value of the angular deformation determined experimentally is 8.369°, and the angular deformation determined by modeling, and simulation is 9.608°.

Also, comparing the results obtained regarding the angular deformations of the seat model 2, it is observed that the value of the angular deformation determined experimentally is 5.446°, and the angular deformation determined by modeling, and simulation is 5.186°.

In conclusion, the differences obtained between the values of the angular deformations of the two models of seats determined experimentally and those determined by modeling, and simulation are reduced, the models being thus validated.

References

[1] *** Regulation No 14 of the Economic Commission for Europe of the United Nations (UNECE) - Uniform provisions concerning the approval of vehicles with regard to safety-belt anchorages, ISOFIX anchorages systems, ISOFIX top tether anchorages and i-Size seating positions.
[2] Huang M, *Vehicle Crash Mechanics*, 2nd ed., SAE International, CRC Press, Boca Raton, FL, 2002;
[3] Manea, A.M., Doctoral Thesis - Automotives passive safety research, POLITEHNICA University of Bucharest, 2021.
[4] *** *Bosch Automotive Handbook*, SAE, USA, (Automotive (BOSCH) Handbook - PDF Drive);
[5] *** https://www.toyota-europe.com/world-of-toyota/safety-technology/passive-safety;
[6] Du Bois, P., Chou, C.C., Fileta, B.B., Khalil, T.B., King, A.I., Mahmood, H.F., Mertz, H.J., Wismans, J., VEHICLE CRASHWORTHINESS AND OCCUPANT PROTECTION, American Iron and Steel Institute, 2000 Town Center Southfield, Michigan, 2004.

IOP Conf. Series: Materials Science and Engineering 1220 (2022) 012051 doi:10.1088/1757-899X/1220/1/012051

Frontal impact on truck cabin structure using FE modeling

A Ioniță , M D Iozsa, C Stan and A M Manea

Automotive Engineering Dept., Faculty of Transports, POLITEHNICA University of Bucharest, Splaiul Independentei nr. 313, sector 6, Bucuresti, ROMANIA, RO-060042

alexandra.ionita930111@gmail.com, daniel_iozsa@yahoo.com, corraista@yahoo.com, mannea_annamaria@yahoo.com.

Abstract. This study aimed to investigate the component structural behaviour and response under overturned truck crash that subsequently collides with road-side pole or tree. The safety of the truck occupants in an accident is very important and should be considered in the process of cabin design. Specific focus will be on the structural integrity of the truck cabin in response to a collision with a cylindrical pendulum. In the past, dynamic analysis was done by real world testing. Nowadays crash simulation is proven to be much more efficient by being a virtual representation of destructive crash impact and providing opportunities to reduce the number of physical prototypes consumed for design verification, thus reducing development time. According to cab strength tests (ECE R29), the cabin body of the truck shall be attached to the chassis in such a way that in the event of impact it ensures a sufficient survival space and eliminates the risk of injury. This paper investigates the FE (finite element) simulation of a frontal impact on the cabin structure to validate the cabin according to the ECE regulation 29 tests. The study also includes the assessment of the energy absorption capabilities of the cabin during the pendulum impact. The pendulum impact was carried out using Ansys LS-Dyna explicit solver to evaluate the cabin strength under the loads.

1. Introduction

Lately, road transport has acquired great importance, being a cheaper way of transporting goods from one country to another and having a strong impact on the country's economy. This has led to an increase in the number of trucks and implicitly to the increased attention given to accidents involving trucks. Although most fatal accidents are between a truck and another traffic participant, such as pedestrians, cars, bicycles, motorcycles, etc. accidents in which truck drivers suffer personal injury must not be neglected, so in this case, a survival space must be provided even when accidents happen. This is taken into account during the design of the cabin resistance structure. This represents the passive safety of the truck. The manufacturers put great emphasis on active safety which has the role of preventing the accident, but when the accident could not have been avoided, passive safety must ensure the survival space [1].

In this study, we carried out the measurement of a truck cab already existing on the market, the realization of the geometry in virtual space with the help of the CATIA V5 program and the analysis of its structural behavior following the type B impact from the ECE R29 regulation made in Ansys program. This regulation was created in order to homologate safe and robust trucks. The ECE R29-03 Regulation revised in 2011 includes three types of structural impact, namely: type A which represents the frontal impact, type B impact on the A-pillars and type C overturned on the roof.

The Type B impact is the subsequent impact of the structure with an object from the side of the road such as a pole or a tree. This type of impact is represented in figure 1 and in order to achieve this, a rigid impact element is needed, with a mass of at least 1000 kg evenly distributed. The impact element is a cylinder with a diameter d of minimum 550 mm and maximum 650 mm and with a length b of minimum 2500 mm. The impactor is freely suspended by means of two rods rigidly fixed to it at a distance of at least f = 1000 mm. The rods have a minimum length L = 3500 mm, measured from the suspension point to the geometry center of the bob impactor [2].

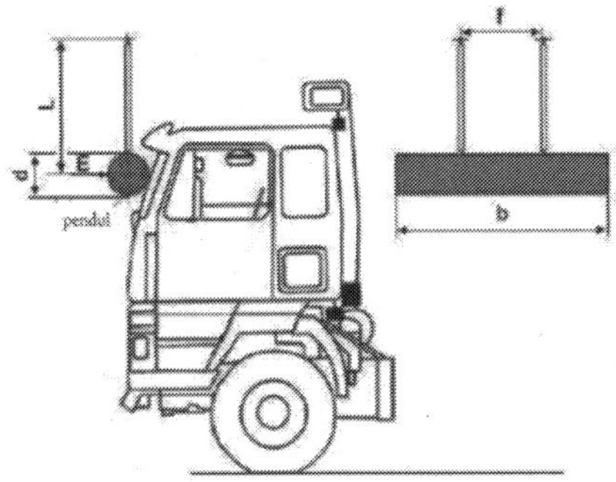

Figure 1. Roof strength test (Test B) ECE Regulation No. 29-03 [2].

The impact element hits the cabin from the front to the rear. The direction is horizontal and parallel to the median longitudinal plane of the vehicle. It will have an impact energy of 29.4 kJ [2].

2. Numerical simulation

2.1 Truck cabin CAD model

The CAD model was made in CATIA V15 software using sheet metal elements later imported in the Ansys numerical program where these have received differentiated thicknesses. Due to the low computing capability of the computer on which the modeling was performed and due to specialized studies[3-5], that showed that in such an impact the floor did not suffer major deformations, we opted for a truck resistance structure without floor, this one being composed only of the front wall, the side walls, the roof and the back wall.

The study was carried out on two models, the first model being composed only of the safety structure of the cabin, without doors, and the second model includes the side doors also. After designing the two models in the CATIA V15 software, these were imported into the Ansys calculation program where the finite element analysis FE was performed (figure 2 and figure 3).

2.2 Finite element methods

Finite element analysis is a process of simulating the behavior of the material following the pressure of forces according to the imposed conditions. FE is widely used in the field of engineering to help simulate physical phenomena on the computer, thus reducing the need for physical prototypes and at the same time allowing the optimization of components as part of the design process.

FEA uses mathematical models to quantify and understand the effects that certain real-world conditions have on assemblies. These simulations allow the localization of potential problems from the design stage, including weaknesses and stressed areas.

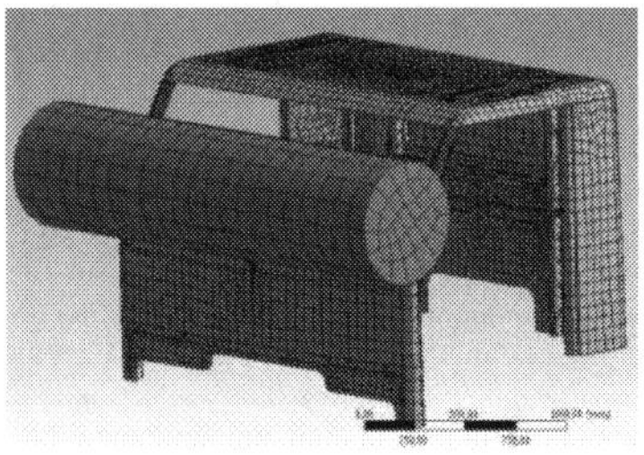

Figure 2. Mesh in the case of cabin structure without doors.

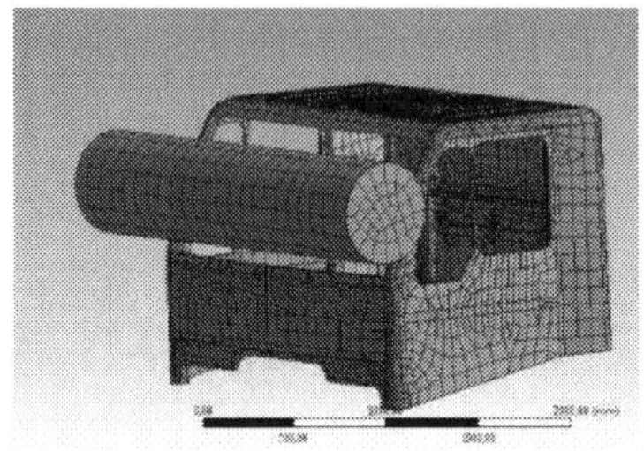

Figure 3. Mesh in the case of cabin structure with doors.

In order to perform an analysis with finite element, the structure meshing is needed. The meshing elements are shell type, because the structure consists of sheets with variable thicknesses. A fine meshing leads to high accuracy results, but also to a very long calculation time. The studied model was meshed in the impact area of the A-pillars, using elements with a size of 20mm, and on the rest of the structure, there were used elements with a dimension of 80 mm, thus resulting in a calculation time of approximately 50 hours in the Ansys program for the first case (figure 2). For the case of cabin structure with doors, there were used elements with the size of 30 mm in the impact area of the A-pillars and on the rest of the structure, there were used elements with a dimension of 90 mm and the time of calculation was approximately 60 hours because of the increased complexity of the model (figure 3).

2.3. Material modelling

For these studies, it was used the same material for all profiles of the strength structure. The material was selected from the Ansys program library, named "Structural Steel Non-linear" with the following properties: Young's modulus $E = 2 \cdot 10^7$ N / m^2; Tensile Yield Strength: $\sigma_c = 370$ MPa. The impactor is rigid, with a mass of 1039 kg.

3. Initial conditions

In order to perform the type B test from the ECE R29 Regulation, the strength structure was imported in the Ansys program where the impactor was drawn. The impact element has been positioned so that the impact surface is in contact with the A pillars of the cabin, its longitudinal median line is horizontal and perpendicular to the median longitudinal plane of the cabin and the center of gravity is halfway between the lower frame and the upper frame of the windshield. The impact element hits the cabin from the front to the rear and will have an impact energy of 29.4 kJ.

$$E_i = \frac{m \cdot v^2}{2} = 29{,}4\ kJ \tag{1}$$

Where m is the mass of the impact element, m=1039kg, therefore the speed will be:

$$v = \sqrt{\frac{2 \cdot E_i}{m}} = 7{,}522\ m/s \tag{2}$$

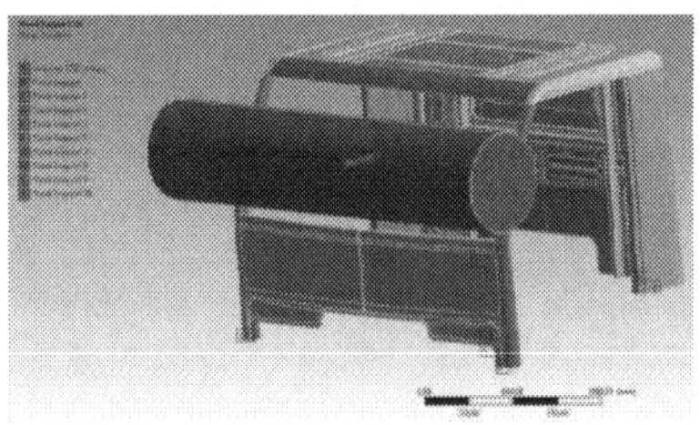

Figure 4. Initial conditions of UNECE R29 impact type B for the first case.

Figure 5. Initial conditions of UNECE R29 impact type B for the second case.

4. Results and discussion

The impact represents hitting a tree or a pole on the side of the road. In the initial phase, the cabin turns down to 90 degrees which could be due to unbalanced load distribution and/or a sudden movement of the wheel. Next is the B impact with a tree or a pole shown in figure 4. The objective of the study is to verify if the strength structure of the cabin can pass the impact while still ensuring survival space for the occupants.

The first study was tested on a structure with no doors. In the image below it is visible that the impact resulted in a deformation of 231 mm which allows for enough survival space. The dashboard has been deformed which can lead to legs injury for the driver.

The maximum deformation happened by pillar A exactly in the impact point with the truck cabin, and the maximum displacement was in the same area, while not invading the occupants' survival space.

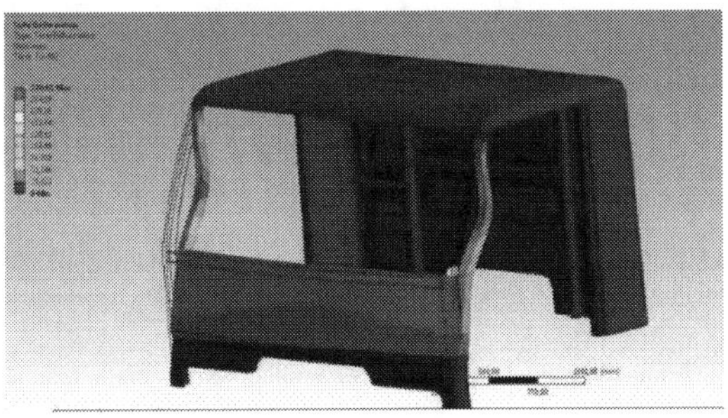

Figure 6. Total deformation of the cabin structure in the first case.

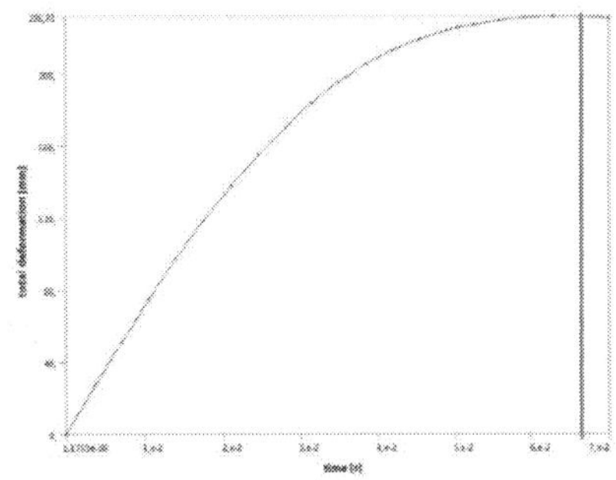

Figure 7. Graph of total deformation vs time.

By analysing the total deformation graph, it is visible that the deformation at the moment of impact goes up to the maximum 231 mm and after that point the deformation decreases, thus meaning the material is self-reversing and the impact is finished.

The equivalent stress analysis can be useful to determine the areas where the surface fails. The equivalent stress in the cabin structure at the moment of type B impact can be seen in the figure below. It reached a value of 450 MPa, getting to and passing the tensile yield strength limit (which is 370 MPa). From this point on, plastic deformations will happen – see in red in figure 8.

In figure 9, the maximum value of the equivalent plastic strain is noticed at the base of A-pillar in the zone of the bottom area of the dashboard and the value is ε=0.3303, at that point a permanent deformation occurs.

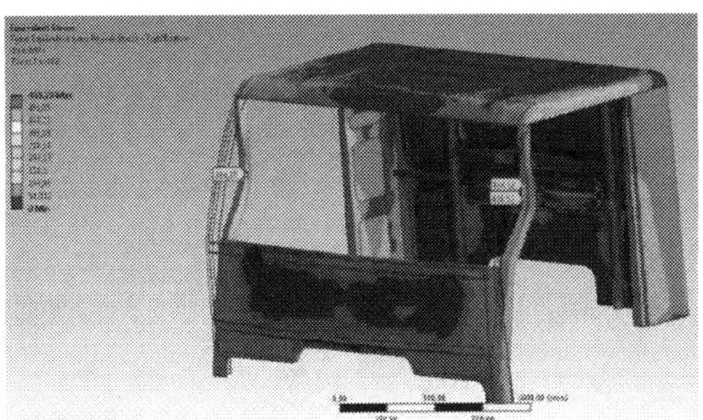

Figure 8. Equivalent stress in the cabin structure.

Figure 9. Equivalent plastic strain in the cabin structure.

In the second case, the total deformation is 131 mm which is a lower value than in the first case, as shown in figure 10. The maximum deformation is in the same A pillar zone, like in the first case, but the displacement of the front wall in the zone of the dashboard is lower, which is very important for the survival space of the occupants' legs. Also in figure 11, it is shown the maximum value of the equivalent plastic strain that is noticed this time on the A-pillar in the zone of the top area of the dashboard and the value is ε=0.3114. In both cases, the value is zero, with means there isn't specific plastic deformation.

Figure 10. Total deformation of the cabin structure in the second case.

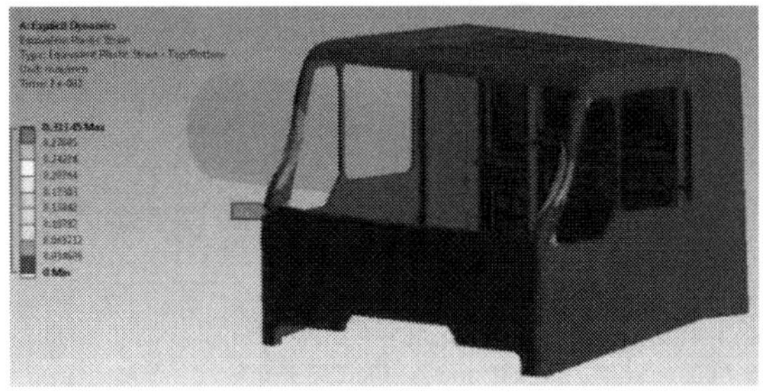

Figure 11. Equivalent plastic strain in the cabin structure in the second case.

5. Conclusions

The main objective of this study is to determine the cabin's degree of deformation and if it would penetrate the truck driver's survival space. The truck structure on which the test was made is an existing structure that was measured and then introduced in the CATIA software (reverse engineering). Following the initial case of the cabin structure with no doors, the cabin suffered larger deformations than in the case of the cabin structure with doors. The doors add strength to the cabin structure. In the second case (with doors) we can notice a much smaller movement at front wall in the zone of the dashboard, which leads to a larger survival space for the driver by the legs.

These two studies were done to see if the strength structure passes the tests imposed by ECE. Based on the results of this study, it can be validated that the truck cabin structure passes and follows the R29 requirements.

References

[1] Iozsa D, *FE Analysis of a Truck Structure in Order to Predict its Crash Behavior,* 12th Automotive 'European Congress, EAEC 2009, Bratislava, Slovakia, 2009;

[2] *** Regulation No 29 of the Economic Commission for Europe of the United Nations (UN/ECE) Uniform provisions concerning the approval of vehicles with regard to the protection of the occupants of the cab of a commercial vehicle

[3] Woodrooffe, J., and Blower, D. *Heavy truck crashworthiness: injury mechanisms and countermeasures to improve occupant safety.* (Report No. DOT HS 812 061). Washington, DC: National Highway Traffic Safety Administration, 2015.

[4] Herbst, Brian R; Forrest, Stephen M; Meyer, Steven E; Clarke, Christopher C; Bell, Lauren D; Nelson, Arin Oliver *Test Methods for Occupant Safety in Heavy Truck Rollovers* Conference: 2015 IRCOBI;

[5] Lilin Zhou, Yanhui Wang, Shifei Wang, Wei Li, Shuli Zhao, *The New ECE Regulation Brings Challenge on Commercial Vehicle,* Proceedings of the FISITA 2012 World Automotive Congress.

The analysis of kinematic parameters of the vehicle occupants on impact with a rigid barrier

C Nedelescu[1], A Chiru[1], P Vrabie[2] and D Trusca[1]

[1]Transilvania University of Brasov, Department of Automotive and Transport Engineering, Str. Politehnicii, Nr.1, 500024 Brasov, Romania
[2]Transilvania University of Brasov, Department of Mechanical Engineering, Str. Politehnicii, Nr.1, 500024 Brasov, Romania

claudiu.nedelescu@unitbv.ro

Abstract. This paper studies the kinematic parameters of the vehicle occupants in case of a frontal impact between a vehicle and a concrete barrier. The main objective is to determine the head and thorax acceleration of vehicle occupants during the impact with a rigid barrier. To achieve the scope of the study a frontal damping system was used.

1. Introduction
Road accidents are one of the most important causes affecting health in several countries through a very high number of injuries to occupants or pedestrians, some of which are very serious [1]. In the event of traffic accidents, it is estimated that the use of a seat belt can reduce the risk of irreversible injury or even death of the occupants by about 50%, which is the most important element of passive safety for passengers [2].

The main objective of this study is to determine the acceleration of the head and thorax of the occupants of a car on impact with a rigid barrier. A frontal damping system was used to achieve the purpose of this study.

2. Methodology

2.1. General aspects regarding the injuries of vehicle occupants in frontal impact with an obstacle.
In the event of a frontal impact when the driver is not wearing a seat belt, serious injury can occur as the driver hits the steering wheel with his chest and afterwards the windscreen with the forehead. The main injury of the head is to the brain. Due to backward neck bending there is a risk of fracture of the C2 vertebra (axis) and the injury is similar to that produced by hanging. If the occupant is equipped with a standard seatbelt the severity of the injuries decreases substantially, but the body movements are still large.

It is noted that in the case of correct positioning of a safety belt with pre-tension the risk of injury is the lowest.

The severity of injuries is influenced by the impact velocity and the correct positioning of the safety belt. The acceleration magnitude is not directly proportional to the contusive effect. In this case, the duration of the impact has the greatest influence on the resulting injury [3].

The determination of the kinematic parameters of the head, respectively of the thorax, is necessary in order to establish the human tolerance limits in case of impact. The severity of injuries caused by traffic accidents is assessed using the AIS scale (Abbreviated Injury Scale).

2.2. Procedures and techniques for the organization of the experimental research

2.2.1. Testing procedures.
The test procedure consists of simulating the frontal impact of a Peugeot 306 vehicle with a concrete barrier shown in figure 1 and figure 2 at a velocity of 40 km/h. Two dummies are positioned inside the vehicle.

Figure 1. Top view of the rigid barrier.

Figure 2. Side view of the rigid barrier.

The frontal impact with the rigid barrier was performed on a test track by using a quick-load release hook, a metal plate that contains a cable return roller for vehicle towing, a cable with a length of 85m, a Peugeot 306 vehicle, a front damping system mounted on the front bar and a Toyota Land Cruiser towing vehicle.

The collision scenario is shown in figure 3:

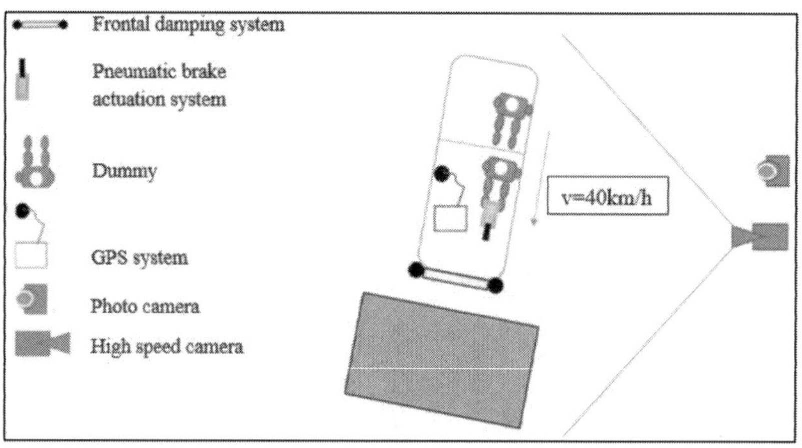

Figure 3. Collision scenario.

2.2.2. Equipment features and performance.

To accurately determine the velocity of the motor vehicles during the experiment taking into account the global coordinates, latitude and longitude, the Garmin GPS system was used on both vehicles. Based on the Garmin GPS 18x-5Hz receiver, a dedicated tool is developed to analyze the dynamic behavior of the vehicle. This system has two main components: equipment (hardware) and software. The equipment consists of the GPS receiver, a computer (mini-notebook or tablet-PC) and the DS-X real time GPS data Processor 1.1.0.19 connection interface.

In order to observe the shocks and forces suffered from the impact, we need to acquire the accelerations and decelerations which could be obtained using PIC DAQ devices and their related sensors. The PIC DAQ DSD is a data acquisition platform for recording kinematic data. Using this platform the accelerations and velocities can be measured. This platform can be used for braking, steering or impact tests.

A pneumatic brake actuation system is used to stop the vehicle safely, working at a pressure of 8 bar. It consists of a pressurized air chamber, Festo piston mounting bracket, connecting hose and an electric valve operated by a remote electronic module via wireless technology. A remote steering system has also been used.

Video recording was done using a CASIO Exilim EX-F1 camera which allows digital video recording at speeds up to 1200 frames per second (1200 fps) and images capturing at a resolution up to 2816 x 2112 pixels. The CASIO Exilim EX-F1 was also used as the main camera for high-speed video recording and for taking photos during the experiment. The Nikon Coolpix L22 digital camera and DJI Phantom drone were also used.

2.2.3. Data acquisition and processing.

The instantaneous velocity v can be expressed as the derivative of the position dS with respect to time dt as follows:

$$v = dS/dt \quad [m/s] \tag{1}$$

The velocity derived in the considered time interval is called acceleration and is expressed by the following formula:

$$a = dv/dt \quad [m/s^2] \tag{2}$$

The video impact analysis is processed using the Tracker software and the vehicle velocity and acceleration will be graphically interpreted using relationships (1) and (2). The acquired data will be processed in the Origin and Excel softwares. In the case of the occupant, the values obtained from the accelerometers mounted inside the dummies were introduced into the Pocket Daq program and the graphs were performed in Excel using the calculation relationships (1), (2).

3. Results analysis

The frontal impact is analysed in three phases: The pre-collision phase (figure 4), the collision phase (figure 5) and the post-collision phase (figure 6).

Figure 4. Pre-collision phase.

Figure 5. Collision phase.

Figure 6. Post-collision phase.

The material traces on the road and inside the vehicle are presented in figure 7 and figure 8.

Figure 7. Material traces on the road.

Figure 8. Material traces inside the passenger compartment of the vehicle.

After the collision was resulted a 100% coverage on the front of the impacted vehicle. The damage is distributed on a length of approximately 1/3 of the front structure of the vehicle as shown in figure 9 and figure 10.

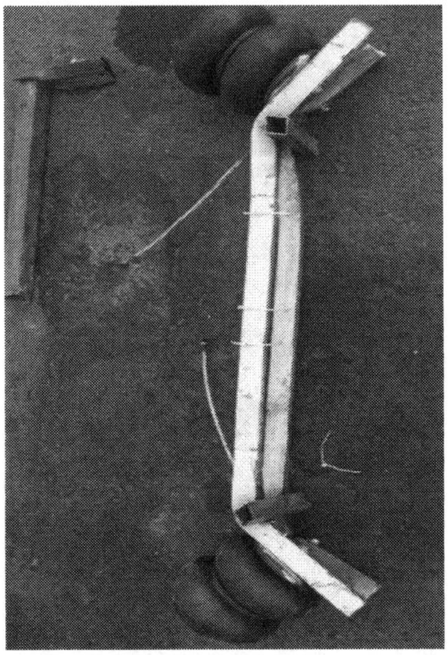

Figure 9. Deformations resulting on the damping system.

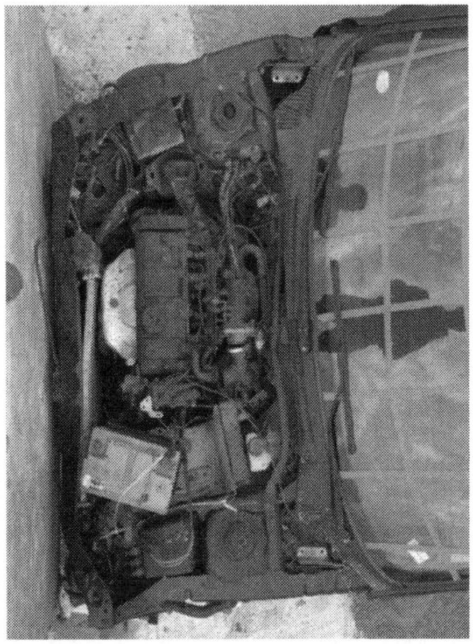

Figure 10. Deformations resulting from impact.

By establishing the final position of the vehicle after impact, the stopping space of the vehicle is determined. Afterwards, the initial velocity can be determined by simulating the virtual accident or using calculation methods.

The velocity of the Peugeot 306 vehicle measured by the GPS device was 11 m/s (39.6 km/h) and the deceleration of the moment of impact is 192 m/s^2. The results obtained from the data processing are presented in figure 11 and figure 12.

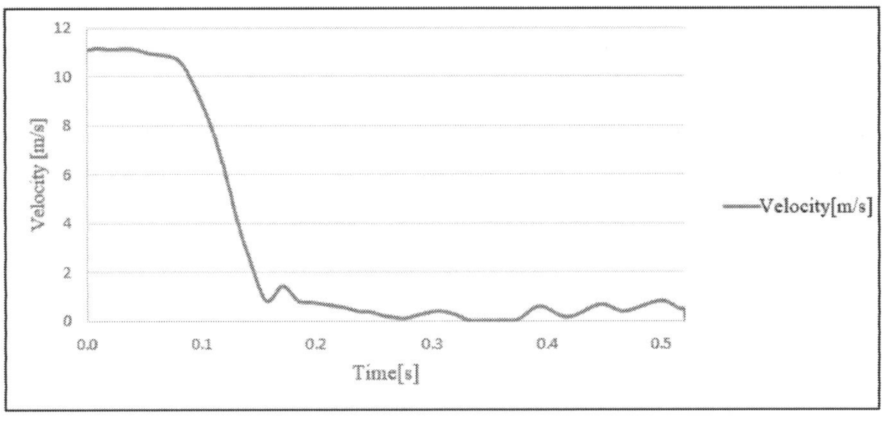

Figure 11. The velocity of vehicle measured by the GPS.

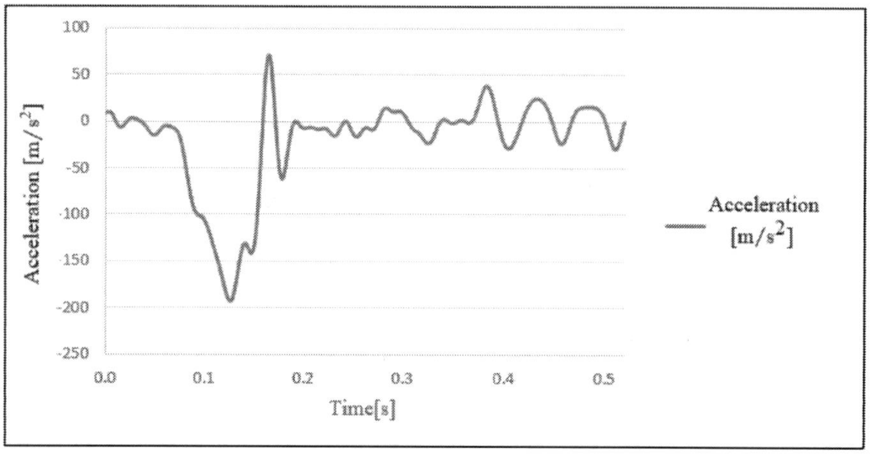

Figure 12. The deceleration of the impact moment.

For the rear passenger, the maximum value of deceleration was -230 m/s^2 on the X axis in the thorax area of the dummy. The variation of accelerations is presented in figure 13.

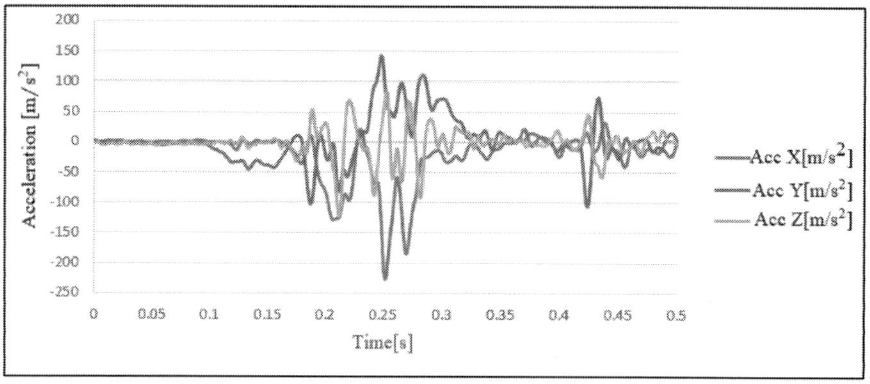

Figure 13. Thorax accelerations for the rear dummy.

For the head of the rear dummy, the maximum variation of the acceleration on the X-axis is from 200 m/s^2 (20g) to a deceleration of -400 m/s^2 (-40g). The results are shown in figure 14.

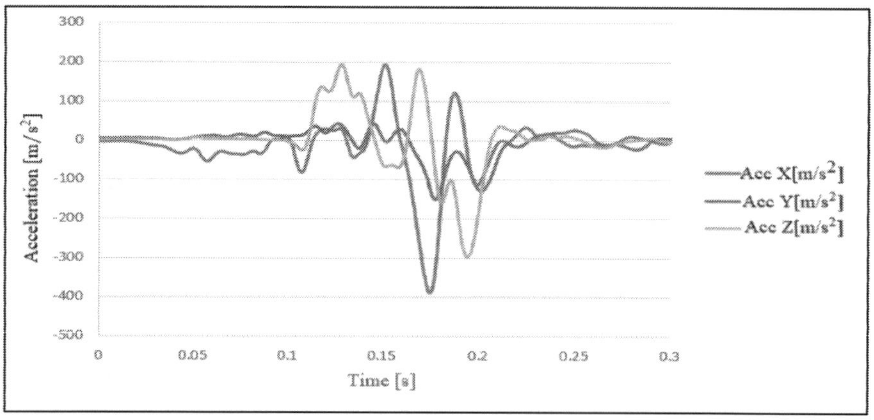

Figure 14. Head acceleration for the rear dummy.

In the case of the front occupant in the driver's seat, the maximum deceleration on the X-axis was 510 m/s^2 (51g) in the thorax area of the dummy presented in figure 15.

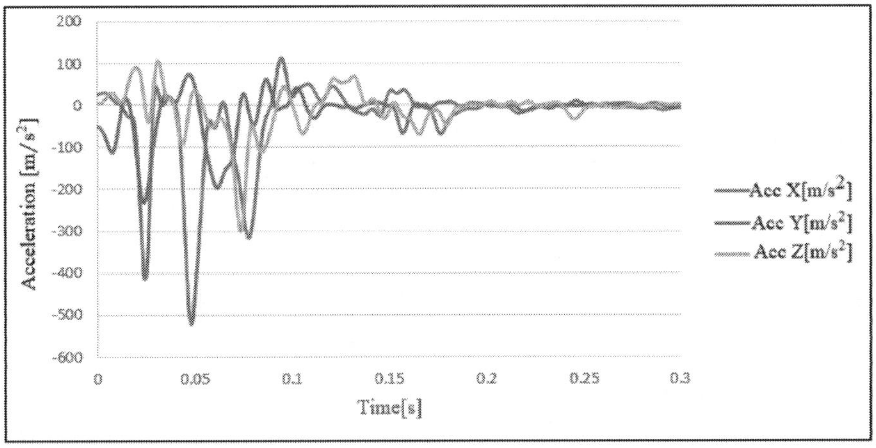

Figure 15. Thorax accelerations for the front dummy.

The maximum head acceleration of 210 m/s^2 (21g) was achieved for the front dummy on the X-axis as shown in figure 16.

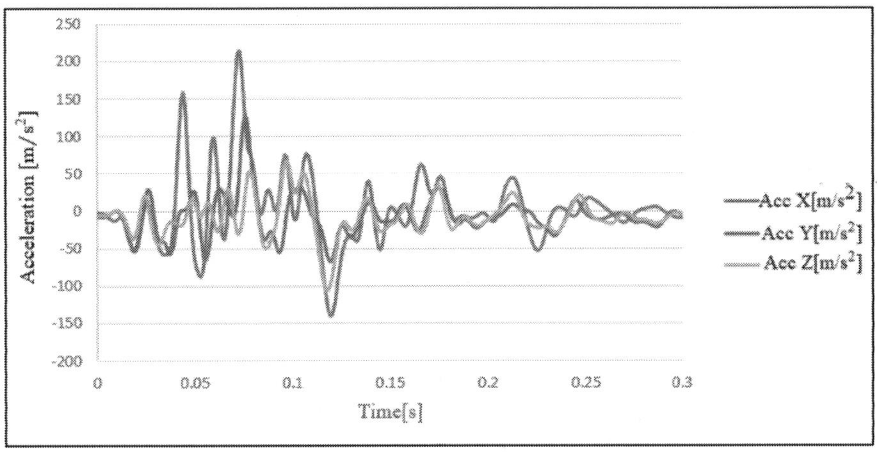

Figure 16. Head acceleration for the front dummy.

4. Conclusions

After performing the frontal impact with the rigid barrier is observed a decrease of the deformation of the vehicle in the frontal area compared to other similar collision situations. The 100% damage coverage in the front of the vehicle area is distributed on a relatively short length of approximately 1/3 from the front structure of the vehicle. The reduced deformation is mainly due to the damping system used.

If the occupants do not wear a three-point seatbelt or it is incorrectly positioned they may be injured as shown in the experimental study.

For the rear passenger, a maximum head deceleration of -400 m/s^2 (-40g) is obtained. In this case, the head deceleration limit of 80 g in the anterior-posterior direction according to the international regulations ECE 21 is not exceeded [4]. By variation of the acceleration of the occupant's head from

20g to -40g the passenger may suffer damage to the soft tissue of the neck due to flexion and extension motion.

In the case of the driver due to a variation of acceleration, the occupant hits his thorax in the steering wheel and his head in the windscreen, thus the driver is injured. A maximum deceleration of - 51 g was achieved in the thorax area of the driver and it does not exceed the tolerance limit of 60 g according to FMVSS 208. Due to the high deceleration, there is a risk of injury to the ribs or even to the internal organ [5].

The experimental research has shown how important it is to wear a seatbelt even in the case of an impact of 40 km/h.

References

[1] Todoruț A 2008 *Dinamica accidentelor de circulatie* U.T. Press Cluj-Napoca chapter 1 p 12
[2] Zuchowski A and Jackowski J 2011 Analysis of properties of operation of the supporting equipment for the seat belts *Journal of Kones* **18** pp 697-704
[3] Gaiginschi R 2009 *Reconstrucția și expertiza accidentelor rutiere* Editura tehnică București chapter 2 pp 75-86
[4] Ilie S 2012 *Modelarea fenomenelor de impact ale autovehiculelor* Editura Agir București chapter 2 pp 66-74
[5] Cavanaugh J M and Yoganandan N 2015 Thorax injury biomechanics *Accidental Injury: Biomechanics and Prevention* Springer New York pp 331-372

IOP Conf. Series: Materials Science and Engineering 1220 (2022) 012053 doi:10.1088/1757-899X/1220/1/012053

Road safety-related determining of time and space braking under different travel conditions

Oţăt Victor[1], Oţăt Oana Victoria[1] and Oţăt Diana Alexandra[2]

[1]University of Craiova, Faculty of Mechanics, Craiova, Romania
[2]University of Craiova, Faculty of Letters, Craiova, Romania

otatoana@yahoo.com

Abstract. The present paper aims to determine the vehicle kinematic parameters during braking, namely the braking time and space needed. Considering different travel conditions, dry and wet roads, we put forward two methods to determine the braking capability. First, we resort to the analytical method; hence the total braking time needs to be considered as obtained from the necessary braking time until the vehicle stops, or the braking process is completed, i.e. the sum of the time due to physiological delays, the time due to involuntary delays and the actual braking time until the vehicle stops. Next, through the Virtual Crash software platform, we carried out a case study to determine the kinematic parameters of a braking vehicle under different travel conditions. The virtual modelling developed for our case study features braking analyses on a track in alignment. The initial parameters of the vehicle and the travel conditions, i.e. the adhesion coefficient value, are similar as in the analytical method. By processing the numerical results, we developed a comparative analysis of the vehicle kinematic parameters for the two methods used, focusing on validating the performance and effectiveness of the modelling method in terms of the time needed to set new initial parameters.

1. Introduction

The steady development of the automotive field has also brought a series of issues for which solutions are still being sought, particularly the innovative attempts to improve road safety. In conjunction with the cutting-edge dynamic and economic performances of vehicles, considerable effort has been put into increasing vehicles' braking performance to reduce the number of traffic accidents significantly. Under the circumstances, the vehicle's braking system remains a key factor among the active safety systems.

The determining factor in escalating the attempts to improve road safety is generated by the dramatic increase in the number of accidents and implicitly the number of road casualties. The increasing number of road accidents grows directly proportional to the increase in the engine index, the constant upgrades of vehicles' performances, and, predominantly, the travelling speed - as one of the most common causes of road accidents.

Road safety requires optimal collaboration and coordination between several factors directly or indirectly involved to ensure safety conditions for all road users within a utilitarian national transport system. Such a system must efficiently frame the tasks of each component, clearly defined and regulated for the entire system to operate properly.

Road safety aims to detect, analyse, and model the factors that contribute to avoiding the occurrence of road traffic accidents or, in extreme cases, reduce the consequences of these accidents.

Vehicle traffic safety is achieved via two major groups of measures:

- active safety measures, which aim to improve vehicle features related to avoiding accidents;
- passive safety measures, which aim at reducing the consequences of traffic accidents.

In terms of active safety systems research, the experimental studies put forward in [1, 2, 3] use the "quarter" vehicle model to validate the ABS/ASR system working algorithm over the braking or starting processes.

Studies on vehicle dynamics in terms of stability during the braking or starting process are approached in [4]. Experimental determinations have been applied in [5], as well, in order to develop an analysis of vehicle behaviour using different complex data acquisition systems.

The safety system that can increase the braking performances was studied in [6, 7] by implementing an innovative solution that improves the dynamic of ABS regulator of the vehicle, allowing an increase of frequency of work and implicit reducing the total braking distance.

2. The analytical method

In establishing the analytical method for determining the braking performance of a vehicle, aspects not directly related to the vehicle's performance must be taken into account, such as the time of perception, i.e. the reaction of the driver, and road conditions.

The state of danger, triggered before an accident, is a new situation; each time, the driver has to adopt various manoeuvres in an attempt to avoid that accident. Reaction time and the adopting of an accident avoidance maneuver are largely influenced by the individual skills of each driver, namely: perception, reasoning, decision, physical capacity, state of emotions, physiological delay and delay in braking. Given the very short time between the occurrence of the state of danger and the actual occurrence of the accident, the driver has an even shorter time to perceive the situation, analyse the measures available, and operate the systems accordingly.

Vehicle braking is achieved by the interaction between the braking system, the vehicle's tires the road surface. The braking forces that reduce the vehicle's speed are applied to the contact points between the tires and the road. All forces that control the movement of the vehicle are developed in the contact patch. The braking system capacity is directly influenced by the condition of the road, e.g. dry, wet, snow-covered road.

At a higher speed (critical speed), the water can no longer be evacuated entirely, facilitating the formation of a water feather at the front of the tire that decreases grip and which, as the speed increases, enters under the tire and then passes entirely at the back of the tire [4].

This phenomenon of apparent floating of the tire on a film of water, referred to as hydroplaning or aquaplaning, leads to a total loss of traction, braking and guidance of the tire.

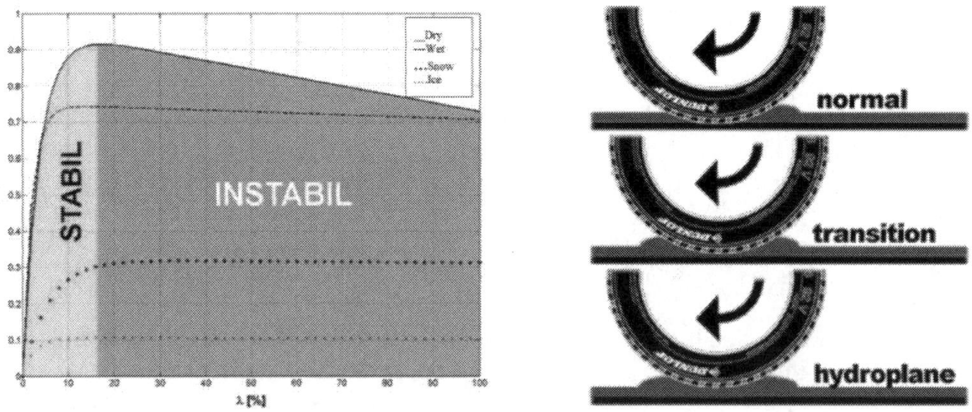

a) b)

Figure 1. a. The dependence of the coefficient of friction subject to vehicle tire slippage φ - the coefficient of friction, λ - tire slippage; **b.** aquaplaning [8]

2.1. *Analytical determination of the braking time*

The braking process commences when a specific event to stop appears, initiating the vehicle to brake, and ends when the vehicle stops. The process involves a series of phases [9]: T_{if} – the duration of physiological delays, T_{im} – the duration of mechanical delays.

In accordance with these constituent elements, we can establish the time elapsed throughout the involuntary delays – T_{ii} [9]:

$$T_{ii} = T_{if} + T_{im} + 0{,}5 \cdot t_3 \tag{1}$$

where: t_3 is the time elapsed between the commencement of the braking process and the tire lock.

We can establish the time elapsed between the commencement of the braking process and the moment the vehicle stops, if the vehicle velocity is rendered in km/h, by applying the relation [9]:

$$T_{fi} = \frac{K_e v_a}{3{,}6\, g\, \varphi} \tag{2}$$

To determine the total stop time, we can rely on the relation below if the initial speeds are given [9]:

$$T_t = T_{if} + T_{im} + 0{,}5 \cdot t_3 + \frac{K_e v_a}{3{,}6\, g\, \varphi} \tag{3}$$

where: φ – the adhesion coefficient when braking;
v_a[km/h] – the vehicle speed at the start of braking.

We carried out the analytical determination of the braking time for nine specific situations, taking into account different speed parameters, i.e. 50 km/h, 80 km/h and 100 km/h. Also, different values for the adhesion coefficient were considered, namely: 0,8 (dry road), 0,5 (wet road) and 0,3 (snow-covered road). Via Microsoft Excel, we developed a mathematical model with a flexible structure, a sequence that allows users to rapidly adjust the input parameters and the new braking analysis. A section of the sequence developed is summarized in table 1 below.

Table 1. Microsoft Excel sequence to determine the braking time.

Determining the braking time

Tif=	0.7	ke=	1		
Tim=	0.2	f=	0.8	0.5	0.3
t3=	0.2	g=	9.81		
		va=	50		
Tii=	1		80		
			100		
		Tfi	0.8	0.5	0.3
		50 km/h	1.77	2.83	4.72
		80 km/h	2.83	4.53	7.55
		100 km/h	3.54	5.66	9.44
		Tt	0.8	0.5	0.3
		50 km/h	2.77	3.83	5.72
		80 km/h	3.83	5.53	8.55
		100 km/h	4.54	6.66	10.44

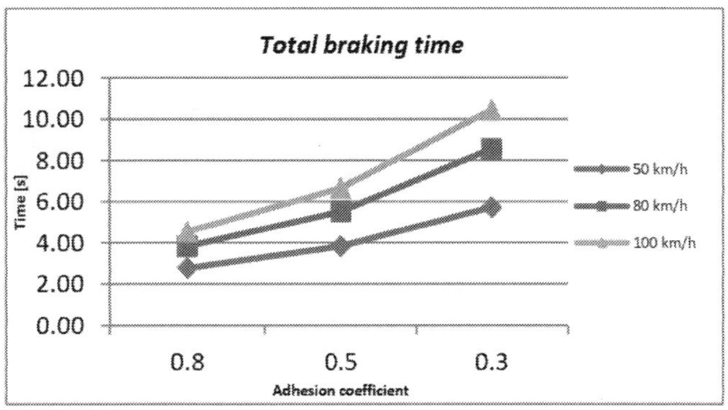

Figure 2. Analytical determination of the total braking time.

2.2. *Analytical determination of the braking space*

After determining the initial travel speeds of the vehicle [km/h], in compliance with the elements previously set, we can further establish, via calculus:

- the space due to physiological delays, S_{if};
- the space due to involuntary delays, S_{ii};
- the actual braking space until the vehicle stops, S_f;

○ the total braking space results from the sum of the space run during the duration of the involuntary delays and the actual braking space [9], S_T:

$$S_T = \left(T_{if} + T_{im} + 0{,}5t_3\right)\frac{v_a}{3{,}6} + \frac{K_e v_a^2}{26 g f_{red}} \tag{4}$$

Similarly to the braking time, we established the total braking space employing the analytical method. The same input parameters were considered i.e. the speed and the adhesion coefficient values. Figure 3 below shows the Microsoft Excel sequence designed to determine the braking space.

Determining the braking space

Sif	0.8	0.5	0.3
50 km/h	9.72	9.72	9.72
80 km/h	15.56	15.56	15.56
100 km/h	19.44	19.44	19.44

Sii	0.8	0.5	0.3
50 km/h	13.89	13.89	13.89
80 km/h	22.22	22.22	22.22
100 km/h	27.78	27.78	27.78

Sf	0.8	0.5	0.3
50 km/h	12.25	19.60	32.67
80 km/h	31.37	50.18	83.64
100 km/h	49.01	78.41	130.69

St	0.8	0.5	0.3
50 km/h	26.14	33.49	46.56
80 km/h	53.59	72.41	105.86
100 km/h	76.79	106.19	158.47

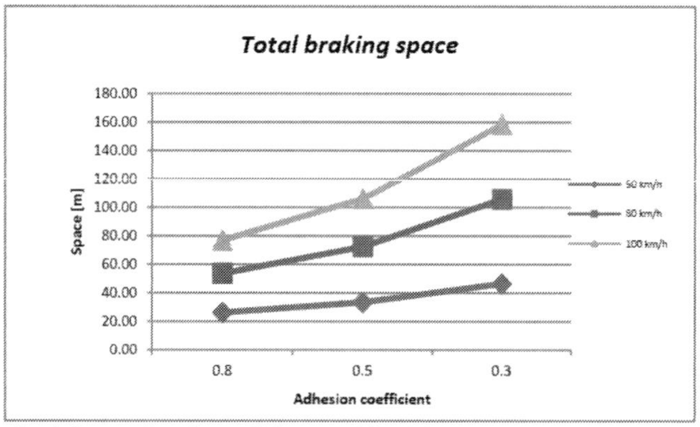

Figure 3. Analytical determination of the total braking space.

3. Determining vehicle braking performances through computer simulations

In order to determine the braking performance of a middle-class vehicle, we resorted to the Virtual Crash software computer simulation.

We set out to determine the parameters that quantify the monitored performances of the vehicle, with the help of the Virtual Crash software; thus, a series of tests were performed, as follows:

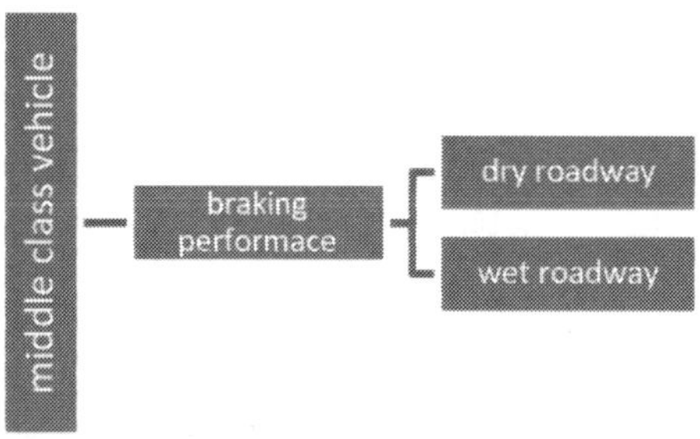

Figure 4. Determination of braking performances via computer simulations.

In order to input the vehicles and their parameters into the software, we need first to define and design the road sections on which the studied vehicles will run.

Then, we have carried out the analysis of a vehicle's performance both in dry road conditions and for wet road. Under the circumstances, the two road sections were designed with different adhesion coefficients, i.e. with dimensions of 4 m wide and 100 m long.

3.1. Selection of the studied vehicle

The Virtual Crash software provides the users with a series of vehicles that display a series of preset dimensions and parameters; the vehicles are classified according to the manufacturer, model, year of manufacture, and engine type.

In order to carry out the simulation, the vehicles are selected and positioned on the roadside corresponding to an adhesion coefficient for dry or wet roads, respectively.

From the database, we selected a middle-class vehicle. Aiming to develop a detailed comparative analysis and better reflect the vehicle's condition and performance, we set for the two road sections (dry and wet) the same vehicle model, according to table 2.

Table 2. Parameters for the studied vehicle.

Vehicle / ***Parameter***	***Middle-class vehicle***
Cylinders [cm^3]	1390
Power [kW]	55
Own vehicle mass [kg]	1020
Length [m]	4.25
Width [m]	1.74
Height [m]	1.53
No. of axles	2
Front / rear wheelbase [m]	1.48 / 1.47
Gauge [m]	2.63
Front console [m]	0.78
Rear console [m]	0.86
Braking system	with ABS
Coefficient of adhesion	0.8 – dry road 0.5 – wet road

3.2. Determination of the braking performance

The comparative analysis of the braking performances for the selected vehicle model was carried out, taking into account the value of the adhesion coefficient of the road on which the two vehicles run. In order to easily identify the type of road on which the vehicles travel, we resorted to colour-based labelling: green for the vehicle running on the dry roadside, respectively blue for the vehicle moving on the wet roadside.

The initial speed at t0=0 for both vehicles is 100 km/h. The dynamic braking state is selected, with a deceleration value of 5 m/s^2. The adhesion coefficients are 0.8 - corresponding to a dry road surface, respectively 0.5 - corresponding to a wet road surface (see figures 5 below).

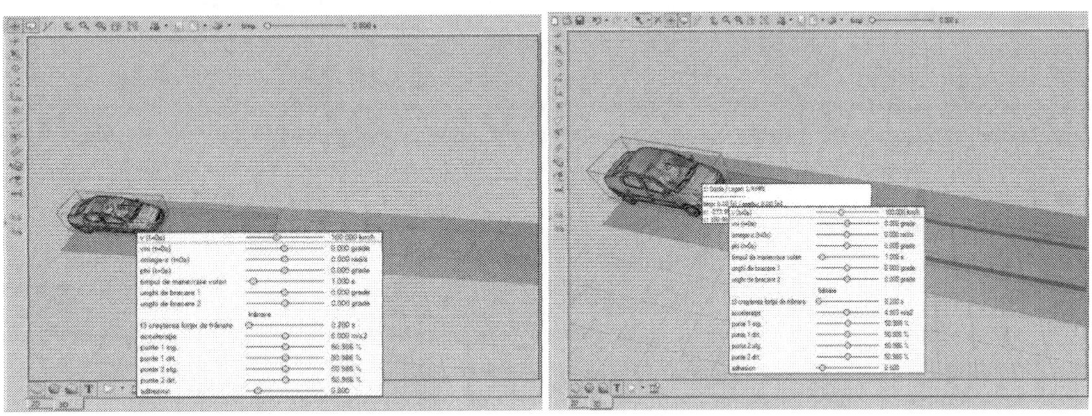

Figure 5. Initial parameters for vehicle 1– dry roadway and vehicle 2– wet roadway.

During the computer simulation, different intermediate positions of the vehicles were captured during the braking process. The time moment *t=2.345 s* and the values of the parameters at that moment are illustrated in figure 6 below.

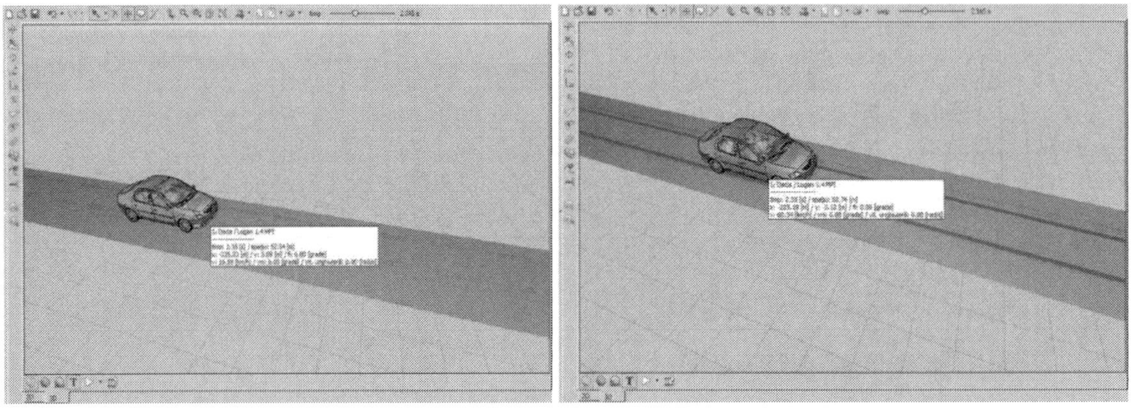

Figure 6. Intermediate positions for vehicle 1– dry roadway and for vehicle 2– wet roadway.

The final position of the two vehicles, corresponding to the time moment $t_f = 6s$ is illustrated in figure 7 below.

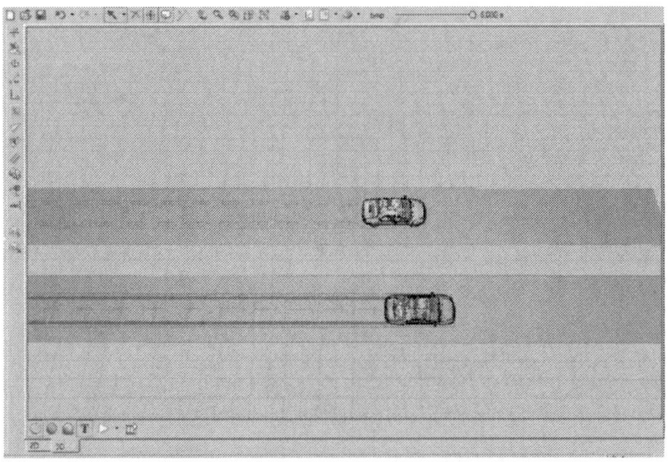

Figure 7. Final braking position for the two vehicles.

3.3. Results obtained from the computer simulations

The Virtual Crash software [10] is mainly designed to simulate road accidents; nevertheless, the options provided by the software enable the users to carry out various simulations of the vehicle dynamics. The simulation results can be viewed in the form of diagrams or can be entered in tabular form in a report generated automatically by the software. The following results were extracted for our case study:

- ✔ Speed variation diagram as related to time / to space
- ✔ Diagram of deceleration variation as related to time / to space
- ✔ Diagram of space variation as related to time, which represents the braking time of the vehicles

To better illustrate the results in the form of a diagram, we resorted to the colour-based labelling assigned to each vehicle in the simulation.

The extracted diagrams show the green curves corresponding to the vehicle travelling on a dry road section, respectively, the blue curves - corresponding to the vehicle travelling on a wet road section.

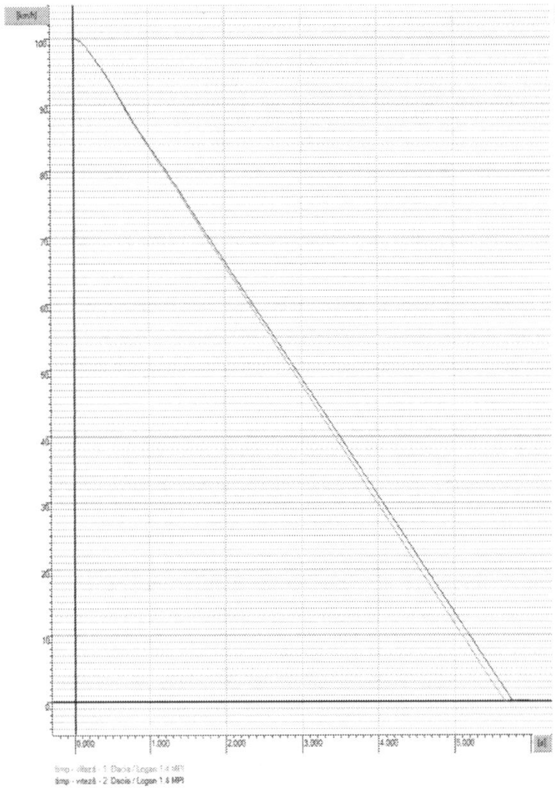

Figure 8. Braking time comparison

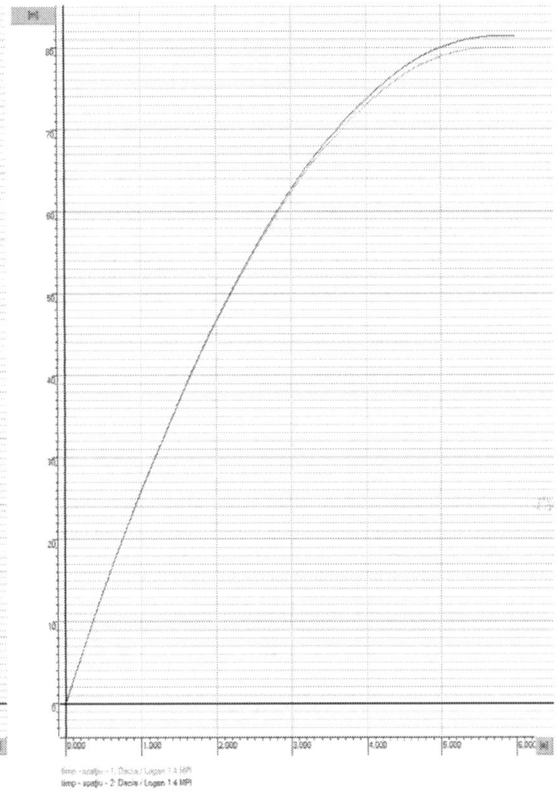

Figure 9. Braking space comparison

According to the variation of speeds related to time (figure 8), at time t_0 both vehicles were travelling at a speed of 100 km/h. The vehicle travelling on the dry road section needs 5.5 seconds to brake until it stops, while the speed of the vehicle travelling on the wet road section reaches 0 km/h after 5.7 seconds.

According to diagram 9, we can determine the total braking space to stop the vehicles: the vehicle travelling on the dry road section requires a space of approx. 77 meters to break from 100 km/h to 0 km/h, while for the vehicle travelling on the wet road, the braking space is more significant, about 85 meters.

4. Conclusions

Our research study brings forward another method to determine the braking performance, namely via computer simulation. In order to determine the starting and braking performance of a middle-class vehicle, considering the technical characteristics similar to the vehicle for which the analytical method was performed, we resorted to a Virtual Crash software computer simulation. Set out to determine the parameters that quantify the monitored performances of the vehicle, with the help of the Virtual Crash software, we had to carry out a series of tests with different values of the adhesion coefficient. Based on the virtual modelling, we could evaluate the performances of the considered vehicle for several travelling situations.

By developing a comparative analysis between the results obtained following the two methods applied in the present paper, i.e. the analytical method vs. computer simulation, we could establish that the values obtained for the braking performances (braking time and space) are similar, especially in the situation the vehicle travels with a speed of 100 km/h, on a dry road section. For this situation, we obtained 4.54 seconds and 76.79 m via the analytical method, respectively 5.5 seconds and 77 m through simulation.

The relative error obtained when travelling on wet roads via the two methods is 0.19. Of course, the time of physiological delay must also be taken into account, though more challenging to simulate.

In conclusion, with a view to the values obtained for the braking performance, highly similar values obtained via the two methods applied, we can validate the computer simulation method based on the analytical method, thus fulfilling our research goal.

References

[1] Beles H., Rus A., Dragomir G., Mitran T., Trusca D., Tolea B., Researches regarding the continuous improvement of the abs (anti-lock braking system) operation for the passenger cars, Annals of DAAAM and Proceedings of the International DAAAM Symposium, University of Zadar, Croatia, Publisher: Danube Adria Association for Automation and Manufacturing, DAAAM, DOI: 10.2507/26th.daaam.proceedings.027, ISSN: 17269679, ISBN: 978-390273407-5, 2015, pp. 196-205.

[2] Beles H., Chioreanu N., Mitran T., Researches Regarding the Development of a Mathematical Model to Optimise the Operation of the Anti-Lock Braking System, Annals of the Oradea University, Fascicle of Management and Technological Engineering, ISSUE #2, Publisher: University of Oradea Publishing House, ISSN 2285-3278, ISSN-L 2285-3278, ISBN 978-606-10-1537-5, 2015, pp. 11-16.

[3] Beles H., Preda I., Dragomir G., Blaga V., Ardelean F., 2008 Theoretical Research Regarding the Active Safety Systems. Annals of DAAAM for 2008 & Proceedings of the 19th International DAAAM Symposium, Pages: 87-88, Published: 2008. Publisher: DAAAM Int Vienna, Austria, ISBN:978-3-901509-68-1, ISSN: 1726-9679

[4] Oţăt V., Bolcu D., Thierheimer W., Loreta Simniceanu, Dinamica autovehiculelor, Editura Universitaria, Craiova, 2005

[5] Dumitru I., Racila L., Tutunea D., Matei L., Dima A. and Oprica A., 2017 Experimental Investigation of a Vehicle Behavior Using Different Complex Data Acquisition Systems CONAT 2016: Int. Congress of Automotive and Transport Engineering pp 136-143

[6] Thierheimer W., Alexandru C., Thierheimer A. and Crauciuc D., Performance of Diaphragm Elastic Elements in ABS, Applied Mechanics and Materials, Volume 896, pp. 241-248, 2020.

[7] Thierheimer W., Tane N., Thierheimer Diana and Danila D., Study of using diaphragm like elastic elements in development of ABS controller, Annals of DAAAM & Proceedings, DAAAM International Vienna, pp. 245+ , 2009.

[8] ***https://www.e-automobile.ro/categorie-dinamica/41-sistem-abs-frane-auto.html

[9] Nistor N., Stoleru M., 1987 Expertiza tehnică a accidentelor de circulaţie (Editura Militară, Bucureşti)

[10] Virtual Crash – User's Manual

Corroboration of the data obtained by virtual methods with the real ones in the study of road accidents

Clinciu Mihai[1], **Ciunel Stefanita**[2]

[1] Transilvania University of Brasov, Faculty of Mechanical Engineering, Politehnicii Street, no.1, Brasov, Romania
[2] University of Craiova, Faculty of Mechanical Engineering, Calea Bucuresti Street, no. 107, Craiova, Romania

ciunel_stefanita@yahoo.com

Abstract. Even though the frontal collisions occur in a much larger percentage than the lateral collisions, the lateral collisions consequences can be much more serious than the frontal collisions. This paper is meant to be a study on the importance of crush energy, which determines the mechanism that produces the severity of injuries. Using Newton laws of mechanics, we can discuss about the conservation of energy and linear momentum law which determine the impact force, speed and deformation. After, we compared the obtained data from the laboratory virtual crash tests with the determined values from the polygon experimental tests. Analytical techniques that determine the occupant and vehicle behaviour from this stage allow determining the mechanism that produces the damages and biomedical parameters, which are representative and correspond to the real applications from the practice investigation of the traffic events. The information is extremely useful to expert investigators who activates in the field of traffic events investigation and reconstruction.

1. Introduction

The most common road events consist in frontal or lateral followed by rear end collision, respectively by rollover. Annually road events with side-impact car crashes generate victims whose mortality accounts for 25% of their total. The load force that is acting in these types of collisions on the occupants is violent, because the space between the victim and the vehicle parts (known as non-lethal space) is limited. After the collision, serious injuries result because the lateral structure of the vehicle is less rigid (stiffness) than the front [2]. Passenger safety is one of the biggest concerns of those involved in the automotive industry. Vehicle manufactures are required to carry out a variety of tests for each model that is on the market. Even though these tests can be easily evaluated, they are complex and complicated requiring appropriate facilities, data acquisition systems, qualified personnel, not to mention the complete destruction of motor vehicles.

Because there is a long way to go from the design phase to the final prototype, for this reason, to prove if a vehicle meets the initial safety requirements, tests were carried out at different design phases, which were quite expensive. It is therefore advisable to establish a type of standard experimental test and to use large scale measurements to determine the behaviour of the vehicle and passengers during a collision. Methods for reconstructions of road events have evolved in line with the progress in related fields, such as modern methods of investigation and analyses of the scene accident, which have taken topometry techniques, car maintenance, satellite surveillance, biomechanics, etc. [1, 5, 7].

All these models aim to simulate the behaviour of the victim and the vehicle during the road event and the degree of accuracy is validated by comparing the data of interest offered by the simulation with the data obtained from the experimental laboratory tests or under real conditions.

2. Evaluation of deformation energy

An important objective of investigating road events is to determine the principal direction of force, the energy consumed in the collision at the deformation of the vehicle, the injury severity, respectively initial and final speeds. So, the kinetic energy during the road event will be consumed by deforming the vehicle and by solicitation and injuring the car occupants. The size of a specific load force, which acts on different parts of the human body, is determining the factor that causes injuries and if the tolerance limits are exceeded [4, 6].

However, the only disadvantage would be that the collision data shows that this linear relationship fails to correlate with the results for extremely high speeds and deformations. However, despite this argument, the solution has been widely accepted throughout the scientific community. The energy determination in CRASH 3 program is based on the residual deformation of the vehicles involved in the collision. This is based on the hypothesis that the impact force is proportional to the residual deformation and the residual deformation is proportional to collision speed. The first hypothesis is presented in figure 1.

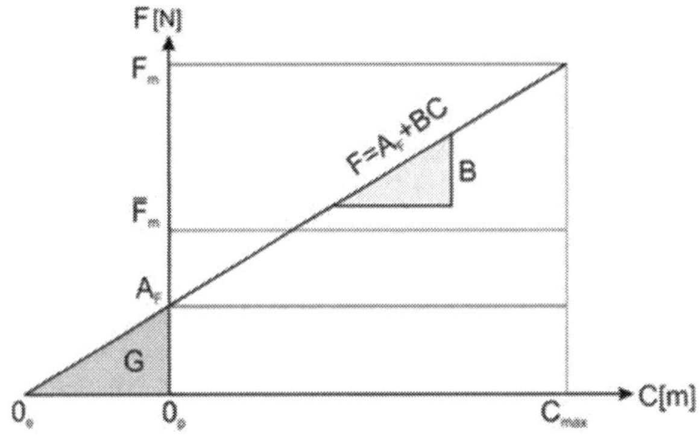

Figure 1. Linear characteristic force – deformation.

$$F = A_F + BC \quad (1)$$

where: AF is a stiffness coefficient $[N/m]$, which ensures that a force is taken without the contact area being deformed residually; B – stiffness coefficient $[N/m^2]$, which represents the ratio of the maximum force over the unit width of the deformed area and the amplitude of the deformation and C – the amplitude of the residual deformation $[m]$.

The energy needed for the deformation of the vehicle is:

$$E = A_F c + \frac{BC^2}{2} + \frac{A_F^2}{2B} \quad (2)$$

The energy lost in the collision can also be expressed by the double integration of the relation (1), as follows:

$$E = \int_0^{L_z} \int_0^C (A_F + BC) dc dl \quad (3)$$

$$E = \int_0^{L_z} \int_0^C \left(A_F C + \frac{BC^2}{2} + \frac{A_F^2}{2B}\right) dl \quad (4)$$

where: L_Z is the width of the deformed profile.

This model allows simulation of the vehicle's behaviour respectively cars occupant behaviour during the collision process, and the degree of accuracy is validated by comparing the data of interest obtained from the simulation process with those obtained experimentally under laboratory test or real conditions, showing the link between the amplitude of the deformation, injury severity and initial speed.

The kinematic, dynamics and modeling analysis processes of deformation are interdependent because the parameters of pre-crash phase condition the body deformation, which also influences the kinematic parameters of the post-crash phase. In the side impact collision, the driver and occupants are subjected to high solicitation loads in comparison with other types of collision with the same level of crush energy absorbed [2, 3].

During lateral collision with a vertical cylindrical obstacle of unique radius, the central forces induce to the vehicle a particular rotating angular velocity, figure 2.

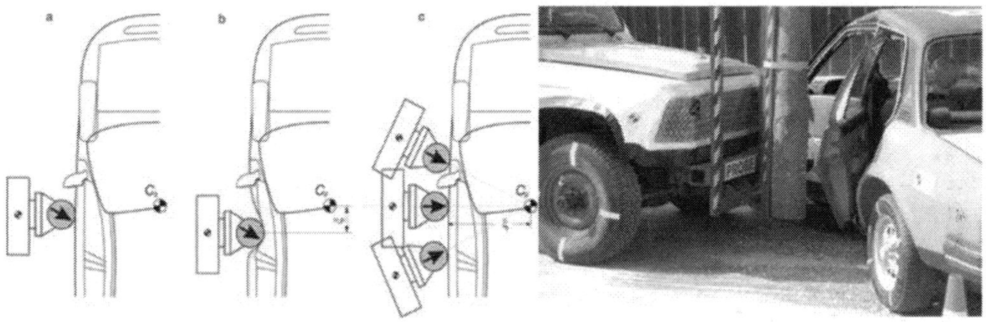

a b

Figure 2. (a) trajectory of collision, through center of mass; (b) real test

As a result, the vehicle will move in the direction of the resulting force, which acts on it, describing a rotation motion around its center of mass, figure 3:

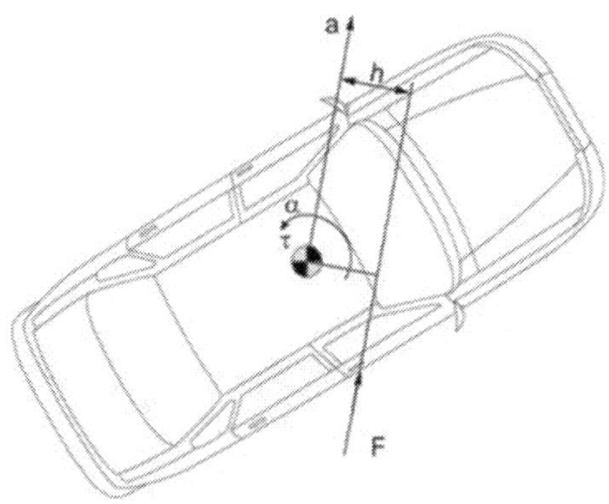

Figure 3. Linear characteristic force – deformation, where:

F is the impact force $[N]$; a – mass center acceleration $[m/s^2]$, which overlaps with the direction of the resulting force of collision; h – normal distance to the trajectory of the collision force $[m]$; τ – the moment of gyration $[Nm]$, being the product of F and h; α – angular acceleration of gyration $[rad/s^2]$.

This induced angular acceleration will continue to grow, with the deformation of the vehicle body elements, until it reaches the maximum level. For calculating the energy absorbed by the deformation of the vehicle structure resulting rotational energy must be subtracted from the total energy of the impact.

The rotation induced as a result of vehicle impact is composed of:

- rotating phase of the vehicle around the obstacle collided;
- rotation phase of the vehicle after separation from the obstacle collided [6].

At this stage, the rotational speed decreases thanks to the interaction between tire and road surface area.

In general, the angular velocity reaches the maximum value at the end of the collision phase.

For the moment of gyration of Newton´s second law, we have:

$$\tau = I_{\Psi}\alpha \tag{5}$$

$$I_{\psi} = \text{moment of inertia} = m\, r_g^2;\; r_g^2 = \frac{I_{\Psi}}{m} \tag{6}$$

For motor vehicles, the moment of inertia is: $I_\psi = 3{,}079 \cdot m - 1635$
with r_g - radius of gyration; from which it results:

$$\tau = m r_g^2 \alpha \tag{7}$$

or the moment can be written as follows:

$$\tau = F \cdot h \tag{8}$$

if:

$$a_c = a + h\alpha \tag{9}$$

$$\alpha = \frac{a_c - a}{h} \tag{10}$$

then:

$$F \cdot h = m \cdot r_g^2 \cdot \alpha \tag{11}$$

$$F \cdot h = m \cdot r_g^2 \cdot \left(\frac{a_c - a}{h}\right)$$

$$\frac{F}{m} \cdot h^2 = r_g^2 \cdot (a_C - a) \tag{12}$$

knowing that: F=m·a or $a = \frac{F}{m}$ it results:

$$ah^2 = r_g^2(a_C - a)$$

$$a = \left(\frac{r_g^2}{r_g^2 + h^2}\right) \cdot a_C \tag{13}$$

We will define the dynamic ratio as being $\gamma = \frac{r_g^2}{r_g^2+h^2}$. Substituting $\frac{r_g^2}{r_g^2+h^2}$ with γ in (13) is obtained the mass center acceleration of the vehicle, which is equal to the proportion γ from the acceleration of the components located in the deformed zone:

$$a = \gamma \cdot a_c \tag{14}$$

The dynamic ratio will always be subunit or at most equal to 1 in case of a centric collision without gyration. To correlate the dynamic ratio γ with change in velocity Δv, we can say that: $\Delta v = a \cdot \Delta t$, and implicit: $a_C = \frac{\Delta v_C}{\Delta t}$, so, from (14) result:

$$\frac{\Delta v}{\Delta t} = \gamma \frac{\Delta v_c}{\Delta t} \tag{15}$$

$$\Delta v = \gamma \, \Delta v_c \tag{16}$$

In noncencentric collisions, the common acceleration a_C and change in velocity Δv of the deformed volume elements will show values higher than those of the mass center. It is known that the at centric collisions, the direction of the resulting collision force passes through the mass center of the vehicle and through the mass cener of the deformed volume, which allows its determination of Δv, therefore the mass of the deformed volume and the mass center of the vehicle will have the same change in velocity [6].

$$\Delta v = \sqrt{\frac{2E}{m_1\left(1+\frac{m_1}{m_2}\right)}} \tag{17}$$

$$\Delta v = \sqrt{\frac{2E}{\gamma m_1 (1+\frac{\gamma m_1}{\gamma m_2})}} \tag{18}$$

$$\Delta v = \gamma \sqrt{\frac{2E}{\gamma\, m_1(1+\frac{\gamma_1 m_1}{\gamma_2 m_2})}} = \sqrt{\gamma^2 \frac{2E}{\gamma m_1(1+\frac{\gamma_1 m_1}{\gamma_2 m_2})}} = \sqrt{\frac{2\gamma E}{m_1 (1+\frac{\gamma\, m_1}{\gamma\, m_2})}} \tag{19}$$

3. Modeling-simulation of lateral impact

In the paper, the kinematic and dynamic behavior of the driver in all phases of the road event was realized and studied. Several collision scenarios were considered, starting from the front, side and rear collision with another vehicle, followed by the collision with a fixed rigid vertical cylindrical obstacle (tree trunk), following the performance of the occupant's motion limitation systems.

Thus, the behavior of the 50% male mannequin contained in the LS-Dyna subroutine models was modeled kinematically and dynamically, for the side impact.

The simulations considered a side impact between two cars, figure 4a, or between a car and a tree trunk, figure 4b, with the direction of the resulting collision force (PDOF) perpendicular to the side surface and with partial coverage.

a) b)

Figure 4. a) car - car collision; b) collision with an obstacle cylindrical vertical rigid fixed.

The initial impact velocities of the pusher vehicle were different, and the distance between the vehicles or between the vehicle and the fixed rigid vertical cylindrical obstacle was approximately 0.350 m. By means of the option of creating the speed-time diagram, it is possible to follow the speed variation before and after the lateral impact, figure 4.b, being considered as representing the speed values in the time interval 0.0 - 0.1 s.

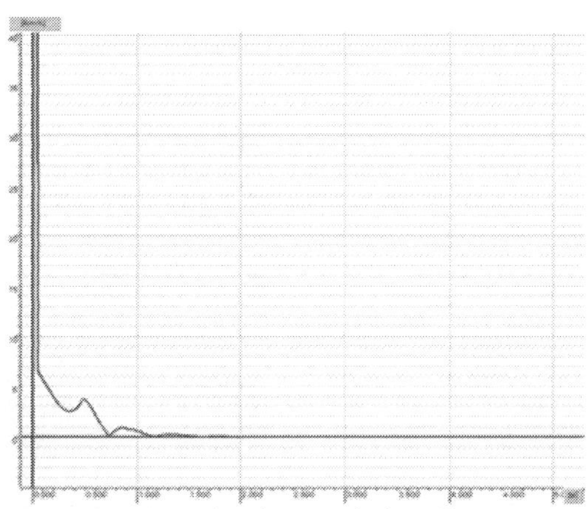

Figure 5. Speed – time variation diagram.

In order to determine the speed variations, in the desired time interval, the speed values from the whole collision simulation period (2s) were used. After that, the values were entered in a calculation sequence of the Mapple program, realizing the transformations of the measurement units from s to ms, respectively from km/h to mm/ms.

In the dynamic study of the side impact, an analytical model was used for the virtual simulation of the vehicle-dummy-safety device system. The virtual modeling-simulation of this dynamic system aims at the dynamic behavior of the dummy, but also of some components of the vehicle, for the side collision. Thus, the CAD models were elaborated, the kinematic and dynamic models having as input data the sizes characteristic of the car movement, the geometric, mass, inertia characteristics, the active loads and the properties of the materials.

With the help of the option available in LS-DYNA, the speed variation diagram is made as a function of time, through which the speed variation is followed before and after the side impact, figure 6, where curve 1 represents the centric collision, curve 2 - eccentric collision, curve 3 - acceleration, curve 4 - displacement and curve 5 – speed.

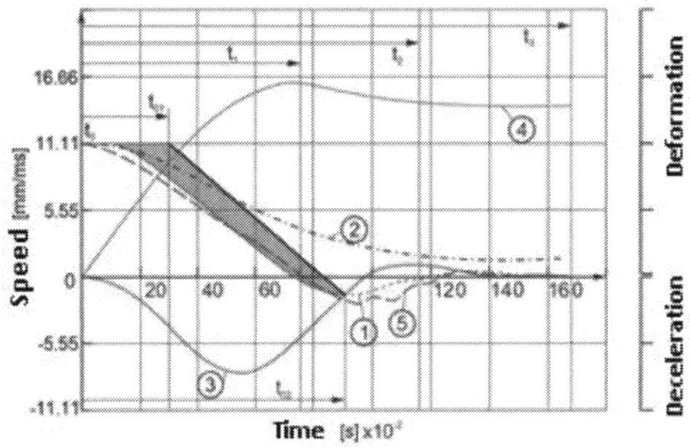

Figure 6. Speed - time variation diagram.

4. Models used for virtual simulation

Modeling-simulation determines the demands and movements performed by an occupant of the front seat, during a side collision, being built a kinematic and dynamic model for analytical study and a stand for experimental laboratory tests that includes the following elements: mannequin, chair, board dashboard, steering wheel and steering column and pedals, figure 7. The elements were introduced in the construction of the experimental model to position the dummy in accordance with the actual position of the driver [5, 7].

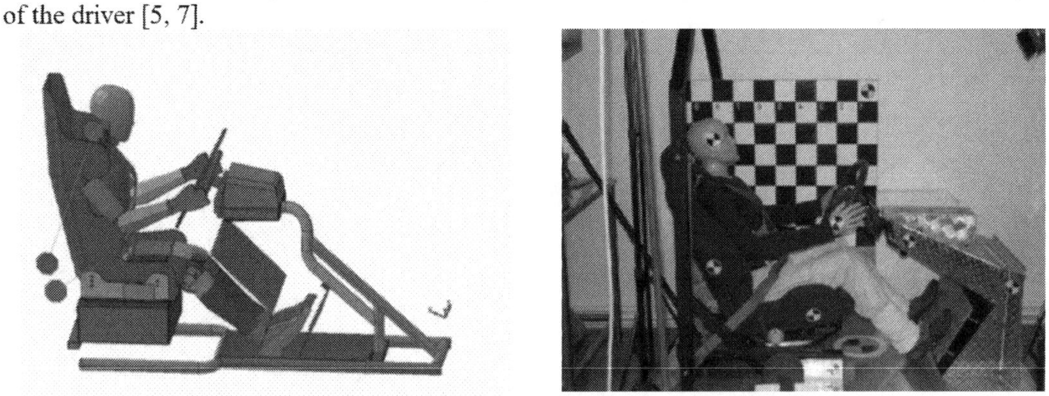

Figure 7. The dummy - seat - steering wheel-board-pedal assembly.

These elements also allow the study of their interaction with the victim, during which time the occupant may be injured. The passive safety features introduced in the model are compatible with those on the standard vehicles involved, the three-point seat belt and the front airbags.

5. The experimental part

From specialized studies, the severity or quantification of brain damage or a skull fracture is given by both the size of the acceleration and its duration. Accelerations of 200 [g] exercised for 2 [ms], or accelerations of 80 [g] exercised for up to 200 [ms] may be tolerated. The results obtained theoretically and experimentally allowed the establishment of tolerance curves, in which the severity of the lesions is defined by the variation of acceleration depending on the load.

In this article, there are inserted the data obtained experimentally at the level of the chest of the road accident test dummy.

For the analysis of the kinematic behavior of the dummy in the side collision time variations of the components of displacement, velocity and linear acceleration were followed.

The variations of the accelerations at the driver's chest level, determined by the collision simulation, are included in the injury standards. figure 8 shows the variation of the linear displacement of the dummy thorax along the x, y and z axes. It is observed that the maximum value of 300 [mm] is recorded at time t = 100 [ms], in the direction of the longitudinal axis y.

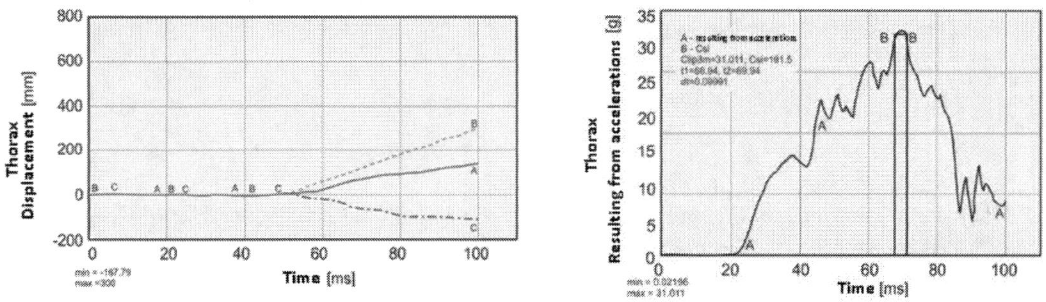

a) b)

Figure 8. a) Variation of displacement components b)Variation of speed components

The variations of the displacement along the z axis are smaller, which is explained by the type of collision and the action of the motion limitation system, the movement of the dummy at the level of the thorax having a large amplitude in the direction of the transverse axis.

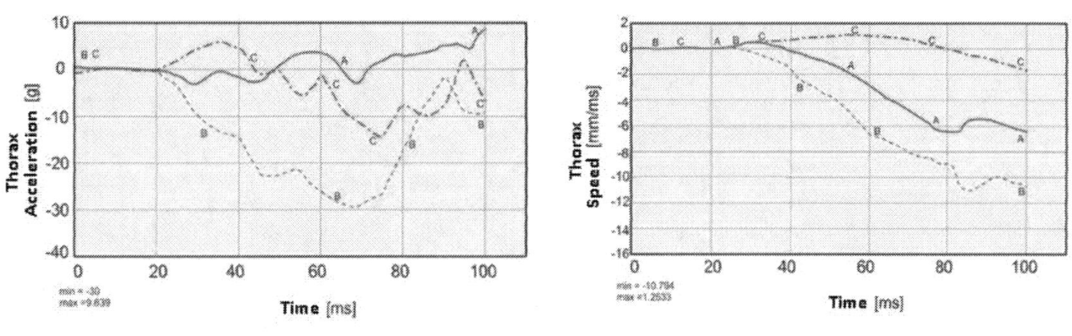

a) b)

Figure 9. a) Thorax time-acceleration; b) CSI value

The maximum values of the linear velocity are described by the velocity component along the y axis, reaching at the final moment of the simulation t = 100 [ms] a maximum velocity value of 10.794

[mm/ms], in the transverse direction, figure 9. The explanation of the low values of the speed along the x and z axes, as in the previous case, consists in the type of collision studied, but also the fact that the movement of the dummy is limited by the seat belt. Following the impact simulation, the variations of the linear acceleration components were represented, along the x, y and z axes, registered at the level of the driver's chest.

The transverse acceleration component, along the y-axis, has the largest variations. The acceleration at the level of the thorax reaches the maximum value of 30 [g] at the time t = 70 [ms].

FMVSS Directive No. 208 sets the maximum value of the chest acceleration to 60 [g]. It is observed that the maximum acceleration limit is not reached, which means a low degree of injury to the chest. According to the FMVSS directives Nr. 208 and CMVSS 208, in case of side impact, the maximum CIS value, compatible with survival, must be below 1000 (CSI <1000).

6. Conclusions

The virtual study for determining the limit values for different parts of the body of the test dummy was confirmed in the polygon. In the paper are inserted the data obtained only for the thoracic area. Similar were obtained for the cranio-cervical area, pelvis, lower limbs.

The data obtained from the virtual simulations in LS Dyna were superimposed with the data obtained after performing the laboratory experimental tests (the experimental model presented above) and those obtained in the test site, respecting the conditions of a frontal-lateral collision, for the cranio-cervical area, thorax, respectively to the pelvis of a crash test dummy, the error being small between the three test methods.

References

[1] Brach R. M., Brach R. M. Vehicle Accident Analysis and Reconstruction Methods, Pennsylnvania, USA: SAE International, 1998.

[2] Brach R. M. Comments on Energy loss in vehicle to vehicle impact by Dario Vangi. (2009) International Journal of Impact Engineering. May 2009.

[3] Bourdet N., Willinger R., Coupled head–neck–torso and seat model for car seat optimization under rear-end impact, Journal of Sound and Vibration 313, Elsevier, pp.891–907, 2008.

[4] Feng, Ch., Suren, Ch.: Injury severities of truck drivers in single and multi vehicle accidents on rural highways, In Accident Analysis & Prevention, September 2011,Volume 43 issue 5, pp. 1677-1688, Published by Elsevier,, 2011, USA,doi:10.1016/j.aap.2011.03.026.

[5] Burg H., Moser A. Handbook of Accident Reconstruction, Kindle Edition, 2013.Franch, H., Franch, D. Mathematical Methods for Accident Reconstruction. CRC Press, 2010.

[6] Maltese M.R, et al, Injury Mechanisms in Seat Belt-restrained Occupants in Side Impact Crashes, Crash Injury Research and Engineering, Network (CIREN) Public Meeting Washington, DC, USA, March 28, 2007.

[7] McHenry Software, CRASH3 Technical Manual.

A review of road traffic accidents reconstruction methods and their limitations with respect to the national legal frameworks

Irina Duma, Nicolae Burnete and Adrian Todoruț

Technical University of Cluj-Napoca

dumairina.d@gmail.com

Abstract. Road traffic accidents are the main issue in terms of public safety among Romania and Europe. Reconstructionists aim to ease their work and to offer results as exact as possible, by using a variety of reconstruction methods, from numerical modelling to software simulations of the collisions in which vehicles are involved. Depending on the conditions of a road traffic event, several methodologies can be used, with proper validation. However, the national legal frameworks may be considered as barriers in using a large variety of reconstruction methods. Therefore, this area of research should be further developed, in order to enlarge experts' choice, especially in the context of rapid development of the automobile industry.

1. Introduction

The present paper aims to offer an overview of road accidents reconstruction techniques, based on the methodologies that are considered valid at the moment according to the specialty literature. Furthermore, these methods of reconstruction are then analyzed from a legal perspective, in terms of general use among Romania, compared to other countries. In this regard, it is also important to have an overview on the general numbers when it comes to road traffic accidents fatalities, thus highlighting the need for alternative accidents reconstruction methods.

Road traffic accidents still cause a major number of fatalities among the European Union and Romania, as the country with most casualties resulted from this kind of events. The most recent data show that, for example, 2020 is the first year with a dropping rate of fatalities per million inhabitants, by 12%, since 2014 [19].

However, this might also be a direct result of national lockdowns due to the outbreak of COVID-19 pandemic, when travelling was limited to the immediate needs, and not as a result of improving infrastructure or awareness campaigns as part of public policies promoted at the national level. The same trend regarding 2020 is also noticed at European level, with a 17% drop on average while, on the contrary, preliminary data in the United States show that the number of deaths from road traffic accidents increased during the pandemic [20].

The average evolution of road traffic fatalities in the EU is depicted in figure 1, which shows that even if there is a descendent trend, the Union's targets for 2020 have not been reached [20]. Also due to the pandemic situation, many countries have opted for increasing the possibilities of green mobility and avoiding crowded transport means, thus the number of bicycle users on a daily basis has increased. However, it is also known that about 70% of road traffic accidents fatalities involve pedestrians, motorcyclists and cyclists, as vulnerable road users [21]. In this regard, there is a clear urge for promoting road safety policies and improving the infrastructure and road traffic monitoring means in order to offer vulnerable road users a safe environment for their travels, especially in the urban areas.

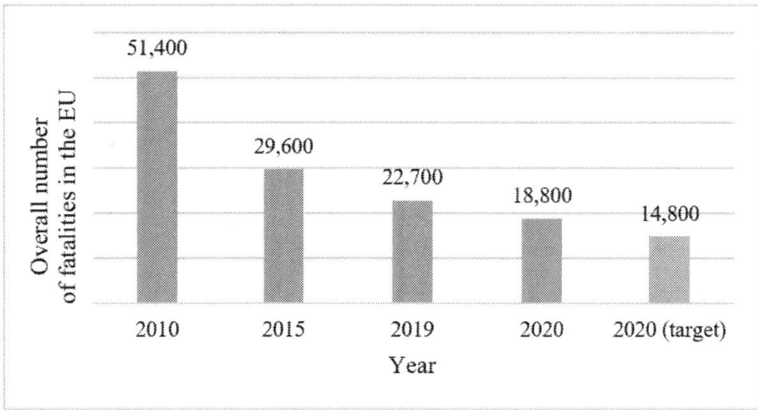

Figure 1. Evolution of road traffic fatalities in the last decade [20].

A general situation in terms of fatalities per million of inhabitants in the EU countries, according to European Commission's annual report for 2019 and preliminary data for 2020, is depicted in figure 2. Even though in overall terms reported numbers are higher than the proposed targets, Europe remains to have the safest infrastructure in terms of road mobility [20].

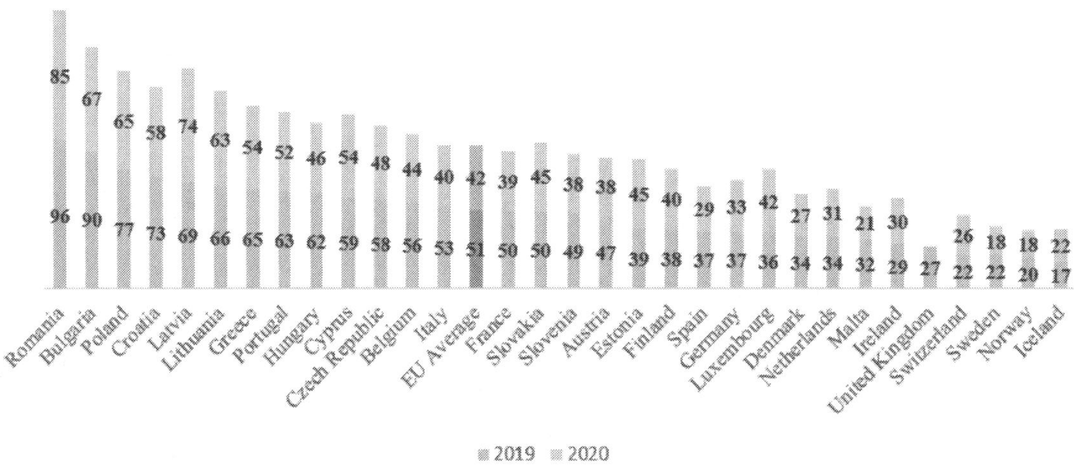

Figure 2. Number of fatalities per million inhabitants among EU countries, reported in 2020 [18, 20].

2. Methods and means for road traffic accidents reconstruction

2.1. Numerical modelling

According to the data provided from the crash site (such as the final positions of the involved vehicles, tire tracks from the braking process, broken pieces coming from the vehicles or other crashed objects, plastic deformation of vehicle's parts), the algorithm for analytical determination of vehicle's kinematic and dynamic parameters is chosen and adapted. Thus, the following methods are usually used [5-8, 12-16]: use of mechanics fundamental laws for road traffic accidents reconstruction; determination of pre- and post-collision vehicle's travel velocities using the conservation of quantity of motion law; determination of vehicle's velocity using the energy conservation principle; evaluation of vehicle's pre-collision and traveled distance using braking tracks; evaluation of traveled distance and time needed in different stages of the event; numerical modelling of vehicle-pedestrian collision; numerical modelling of vehicle-motorcycle collision; determination of vehicle's velocities based on projection distances of the detached parts from the vehicles involved in the collision; determination of vehicle's velocities based on the injuries suffered by vulnerable road users (pedestrians, bikers and motorcycle drivers).

2.2. Computational simulations

The very first simulation programs of a vehicle's dynamic behavior date back ever since 1960, when the proposed models gave a more or less simplified simulation of the interaction between a vehicle and other elements from the environment, in some cases even other vehicles [13]. Then, other such programs were developed in order to simulate the vehicle's behavior at first without interferences and then, with gradually introducing a series of parameters in order to give a more accurate picture of a real collision [13]. The urge for developing a software program for vehicle crash simulations came from the fact that both automobile manufacturers and Euro NCAP keep the confidentiality of data regarding vehicle's structural behavior, as opposite to the United States where approximate results from vehicles testing are made public [13]. In this regard, the most commonly used simulations program in Europe is PC-Crash, which uses the impulse conservation methods, together with simulating the pre- and post-collision vehicle's trajectory. The working mode consists of several simulation cycles until the obtained results match the traces found at the crash site.

The accuracy of such simulations is usually given by the input parameters chosen by the user and the results validation should always be made based on the crash site scene and, eventually, registered values of vehicle's dynamic parameters.

2.3. Video recordings

Another method for road traffic accidents reconstruction is represented by the analysis of video recordings during the impact. In this way, it is possible to determine the dynamic parameters of the collision, preceded by the calibration of the acquired images [11]. References in the field of automotive crash reconstructions approach this direction from the perspective of the analysis of the data acquired by the road traffic surveillance cameras [3, 9]. The video recordings analysis has also been validated in experimental conditions by previously conducted researches [4, 9].

2.4. Crash data retrieval

According to the National Highway Traffic Safety Administration (NHTSA), the Event Data Retrieval (EDR) is defined as a device mounted on road vehicles which records several parameters for a time given in seconds prior to a road traffic collision, such as pre-collision vehicle dynamics and system status, driver inputs, vehicle crash signature, restraint usage/ deployment status, or activation of an automatic crash notification [24]. The importance of EDRs is recognized for more than 20 years, both in the United States and the European Union, together with highlighting the potential of preventing road traffic accidents [25].

The Crash Data Retrieval (CDR) is a tool used for extracting data from EDR, being now manufactured on large scale by three producers, with mostly the same working principles. The CDRs are, in fact, software tools able to extract reports from EDRs in order to have a clear view on the vehicle's dynamic parameters in the seconds prior to a crash [19]. In this regard, several researches have been conducted in order to validate the accuracy of pre-collision data recorded by EDRs and the efficiency of extracting them by CDRs [2, 10].

3. Legal limitations for the use of road traffic accidents reconstruction

The national legal framework in Romania covers the responsibilities, attributes and conditions for conducting the technical ascertainment and judicial expertise for the domain of road traffic accidents [1, 6, 22]. Thus, the road vehicle accidents reconstruction represents the main probation mean, even though it is not mandatory and can be requested by either of the parties involved - attorney's office, judge or even natural persons [6].

According to the National Institute of Forensic Expertise (*NIFE*), the following data should be evaluated, calculated or described [23]: determination of accident's dynamics based on probation means found and analyzed at the crash site, evaluation of involved vehicle's travel velocity prior to the collision, direction(s) of involved vehicles' displacement, establishment of collision's point and the crash site; establishment of timeframes in which several sequences of the road traffic event have taken place (acceleration, braking, cornering in order to avoid the collision, etc.). According to the same procedure, the means for avoiding the road traffic accidents are also analyzed in order for the guilt and litigations to be established by the judicial bodies.

From an international perspective, a study conducted by the European Association for Accident Research and Analysis [17] gives a broad image on the general aims of reconstructionists and their profile in different European countries, which vary from the determination of the accident's details (such as velocities, time and distances) and may go into in-depth analysis of the collision (technical condition of the involved vehicles, biomechanical loading during the impact, technical failures of the involved vehicles' systems or even malfunction of road monitoring systems in order to evaluate the liability of the road authority in the case of an accident) or, in some cases, the coverage of costs evaluation for vehicles' repair or court litigations caused by a road traffic collision. However, solely for the purpose of road accident reconstruction, it is noticeable that in countries like Estonia, France, Hungary and The Netherlands, reconstructionists usually make use of software programs such as PC-Crash, Carat, Analyzer, Rollei-Metric or PC-Rect [17]. Furthermore, the same study emphasizes that there are countries such as Finland, where the recorded accident data is actively used as a mean for promoting road safety policies.

4. Conclusions

Given the variety of means and methodologies meant to offer an accurate picture in the case of a road traffic event, reconstructionists may use them in order to provide precise results in judicial expertise. Even though the Romanian legal framework is not yet open to accept other means than numerical modelling, we believe that further research and argumentation could be done in order to ease the work of reconstructionists and to initiate public policies for the mass use of tools that aim both to analyze the cause of road traffic accidents and to prevent them from happening.

In this regard, the usage of vehicle's technical data and driver's behavior prior to a road traffic collision would be a step further for improving road traffic safety, together with an in-depth and evidence-based analysis of all factors that contribute to a collision, especially when severe injuries or even fatalities occur.

A proposed algorithm for enhancing the activity of road traffic accidents reconstructionists is depicted in figure 3. In this case, for the proper validation of reconstruction techniques and their feasibility in real case usage, it is considered that an algorithm, which would take into consideration both the experimental testing under controlled conditions and an analysis of a real traffic event, would be suitable. In this regard, the tested vehicle(s) should be equipped with data acquisition means, for the error calculation between registered parameters and calculated ones through the proposed reconstruction methodologies. Similarly, in the real case traffic event, vehicles should be equipped with EDRs for the same validation purpose. Ultimately, the calculated errors should be compared between the two scenarios (laboratory and real conditions), for proper validation of the methods' feasibility.

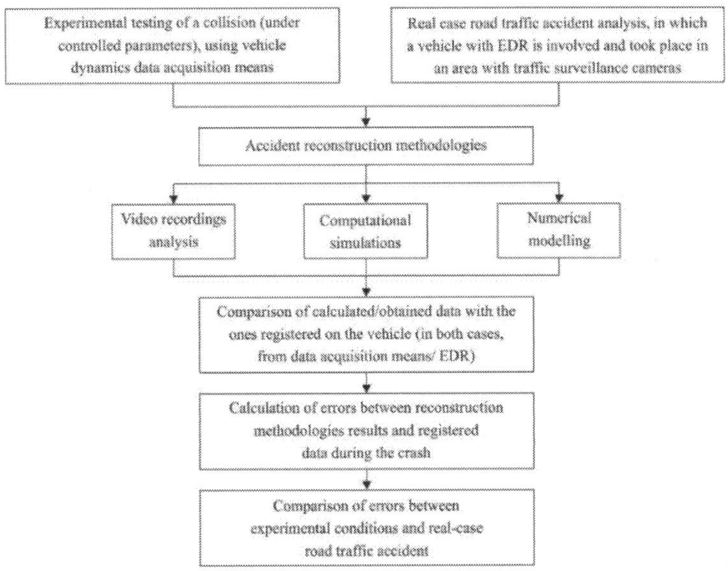

Figure 3. Proposed algorithm for further research in terms of road traffic accidents reconstruction methodologies validation and implementation for real cases.

In more details, it is proposed that a crash (vehicle-to-vehicle, vehicle to fixed object or vehicle to vulnerable road user) is staged in experimental conditions. The tested vehicle(s) should be equipped with GPS and accelerometers, for the extraction of dynamic parameters (travel velocity and distance, accelerations), and with anthropometric testing devices (ATDs) also equipped with dynamic parameters acquisition means (e.g. accelerometers). At the same time, the collision should be registered using video equipment and, then, based on the final position of the vehicle, the reconstructionist should proceed with the numerical modeling and computational simulation based on the data collected from the crash site (size of deformations, type and materials of the involved vehicle, etc.), according to the type of the collision. Then, the results obtained using each one of these methods in terms of accelerations and decelerations at vehicle and ATDs level, traveled distance and collision velocities, should be compared with the parameters measured during the test. Then, an error calculation is recommended in order to give a more exact perspective on these methods' feasibility. In a similar way, the usage of such methods for a real case scenario can be conducted in order to facilitate the comparison of reconstruction means between the real case and the experimental scenario.

References

[1] Boboş, B., *Cercetări privind reconstituirea coliziunii autovehiculelor,* ISBN 978-973-751-796-8, Risoprint Publishing House, Cluj-Napoca, Romania, 2008;

[2] Brown, R., Lewis, L., Hare, B., Jakstis, M. et al., *Confirmation of Toyota EDR Pre-crash Data,* SAE Technical Paper 2012-01-0998, 2012, https://doi.org/10.4271/2012-01-0998;

[3] Crouch, M., Cash, S., *Video analysis in collision reconstruction,* ISBN-13: 978-1788089302, Independent Publishing Network, 2017;

[4] Duma, I., Burnete, N., Todoruţ, A., Truşcă, D., *Determination of pre-collision travel speed in the event of a frontal collision between a vehicle and a fixed obstacle, using video recordings,* în: XXIX International Scrientific Conference on Transport, Road-Building, Agricultural, Hoisting & Hauling and Military Techniques and Technologies 'trans&MOTAUTO2021', print ISSN: 1313-5031, online ISSN: 2535-0307, Year IV, Issue 1 (8), Sofia, Bulgaria, 2021;

[5] Franck, H., Franck, D., *Mathematical methods for accident reconstruction. A forensic engineering perspective,* CRC Press, 2010;

[6] Gaiginschi, R., *Reconstrucţia şi expertiza accidentelor rutiere,* ISBN 978-973-31-2345-3,

Technical Publishing House, Bucharest, Romania, 2009;
[7] Huang, M., *Vehicle crash dynamics,* CRC Press, 2002;
[8] Huang, M., *Vehicle crash mechanics,* CRC Press, 2002;
[9] Kolla, E., Ondruš., J, Vertaľ., P., *Reconstruction of traffic situations from digital video-recording using method of volumetric kinetic mapping,* The Archives of Automotive Engineering – Archiwum Motoryzacji, DOI: 10.14669/AM.VOL84.ART11, Slovakia, 2019;
[10] Lawrence, J. and Wilkinson, C., *The Accuracy of Crash Data from Ford Restraint Control Modules Interpreted with Revised Vetronix Software*, SAE Technical Paper 2005-01-1206, 2005, https://doi.org/10.4271/2005-01-1206;
[11] Neale, W., Fenton, S., McFadden, S., Rose, N., *A video tracking photogrammetry technique to survey roadways for accident reconstruction,* SAE Technical Paper, ISSN: 0148-7191, United States, 2004;
[12] Randall, K.N., *Forensic engineering investigation,* ISBN 0-8493-0911-5, CRC Press, 2001;
[13] Struble, D.E., *Automotive accident reconstruction. Practices and principles,* CRC Press, 2014;
[14] Todoruț, A., *Dinamica accidentelor de circulație,* ISBN 978-973-662-383-7, UT Press Publishing House, Cluj-Napoca, Romania, 2008;
[15] Todoruț, A., Cordoș, N., *Modele fizico-matematice în dinamica accidentelor de circulație rutieră,* ISBN 978-606-737-267-0, UT Press Publishing House, Cluj-Napoca, Romania, 2017;
[16] Van Kirk, D.J., *Vehicular accident reconstruction,* ISBN 0-8493-2020-0, CRC Press, 2001;
[17] Weber, M., *The Query Project. Developing guidelines for a best practice qualification of accident analysts,* European Association for Accident Research and Analysis, ISBN 978-3-00-019328-6, Germany, 2006;
[18] *** European Commission (2020), *Road safety targets - Monitoring report June 2020,* European Road Safety Observatory, Brussels, European Commission, Directorate General for Transport,https://ec.europa.eu/transport/road_safety/sites/default/files/pdf/monitoring_report_novembe r_2020.pdf (accessed on 12/08/2021);
[19] *** Crash Data Group, https://crashdatagroup.com/ (accessed on 12/08/2021); [20] *** European Commission, *2020 Road Safety Statistics,* https://ec.europa.eu/transport/modes/road/news/2021-04-20-road-safety-statistics-2020_en (accessed on 03/08/2021);
[21] *** European Commission, *Road Safety,* https://ec.europa.eu/transport/modes/road/news/2021-04-20-road-safety_en (accessed on 12/08/2021);
[22] *** Government Ordinance no. 75/2000 on the organization of the activity of forensic expertise, published in the Official Gazzette Part I, no. 407/2000, amended and completed with Law no. 488/2002 and Law no. 156/2011, https://ec.europa.eu/growth/tools-databases/regprof/index. cfm?action=regprof&id_regprof=40287; http://legislatie.just.ro/Public/DetaliiDocument/ 24022 (accessed on 11/10/2021);
[23] *** National Institute of Forensic Expertise - NIFE (Institutul Național de Expertiză Criminalistică - INEC), https://www.inec.ro/index.php/activitate/expertize, Romania (accessed on 12/08/2021);
[24] *** National Highway Traffic Safety Administration (NHTSA), *Event Data Recorder,* https://www.nhtsa.gov/research-data/event-data-recorder#overview-10516 (accessed on 12/08/2021);
[25] *** National Highway Traffic Safety Administration (NHTSA), *Summary of findings by the NHTSA Event Data Retrieval Working Group,* August, 2001.

IOP Conf. Series: Materials Science and Engineering 1220 (2022) 012056 doi:10.1088/1757-899X/1220/1/012056

Considerations on reducing fuel consumption and emissions by optimizing traffic control in congested urban areas

Adrian Vilcan[1], Ivana Martin[1], Viorel Nicolae[2], Mihaela Istrate[2]

[1]AV TRANSPORT PLANNING SRL, Ionel Perlea 11, Apt. 15, S1, Bucharest, Romania
[2]University of Pitești, Faculty of Mechanics and Technology, Târgu din Vale Street, no.1, Pitești, Romania

adrian@vilcan.com, viorel.nicolae@upit.ro, ivanamartin.mi@gmail.com, mihaela.istrate@upit.ro

Abstract. The balance between city development, as, for example, an increasing number of working places and emissions control, is a challenge in our days in the urban environment. All new developments in the urban area, either commercial, office or residential, will generate a specific demand for mobility. This new demand will put supplementary pressure on the urban transport system in the influence area. Therefore, approaching it from a macro to micro dimension using transport models would be a benefit for both, as optimizing the traffic control in main junctions in the area of the development will reduce the traffic congestion and delays. The paper deals with effects on reducing fuel consumption and emissions as an effect of optimization of traffic control in main junctions in the areas of new developments.

1. General considerations

This paper is based on the application and implementation of Pitesti Multi-modal Urban Transport Model [1] regarding the optimization of traffic control. The City of Pitesti is looking for the development and implementation of the TMS – Traffic Management System [2], in order to optimize the traffic performance and to increase the quality of life in the urban area. What a TMS does is very well known, thus it optimize in real time the junction traffic control, in order to minimize delays of vehicle traffic flows all over the control area, which usually is an urban area or an important urban transport corridor.

In this paper, it is shown how this process can be simulated with a transport model in VISUM [3], combining macro-simulation with detailed junction modelling (ICA – Intersection Capacity Analysis).

Nevertheless, as also well known a transport model for an urban area, as for other areas, includes Origin-Destination matrices that are assigned on the network. In our case, for scientific purposes, the OD's by type of vehicle have been all converted into PCU – Passenger Car Units.

Results are considered for both AM and PM peak hours, for time horizon 2030.

2. Basic forecasts 2030

The basic 2030 forecasts are based on the foreseen socio-economic pattern in the urban area, thus population, workplaces, employment, car ownership and activities.

Thus, the transport forecasts are based on the Urban Transport Model developed for the city of Pitesti and the influence area. The Urban Transport Model is a 4-step model. The zoning system is shown below:

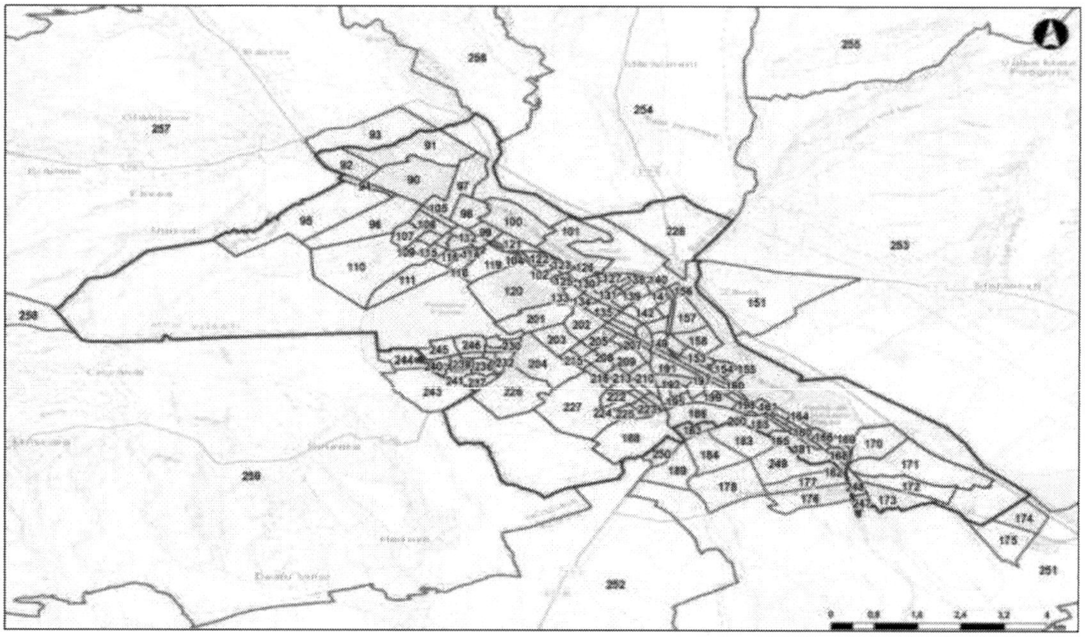

Figure 1. Zoning system Urban Transport Model Pitesti.

As an example, the number of working places by zone is shown in the figure below.

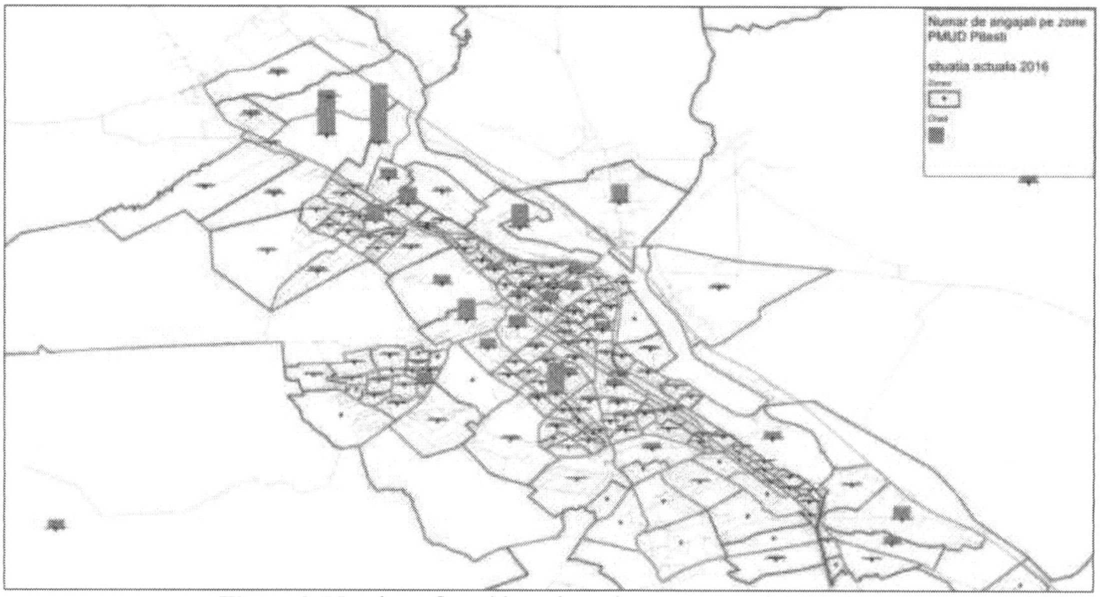

Figure 2. Number of working places base year model, Pitesti.

The basic structure of the demand is shown in the table below, namely the trip purposes considered in the transport model.

Table 1. The basic structure of the demand.

Trip Purpose	Home – Work	HBW
	Home – Secondary school and University	HBCol
	Home – Primary school	HBSch
	Home - Shopping	HBSh
	Home - Others	HBOth
	Origin other than home	NHB

It is specified that the demand for mobility was disaggregated considering 18 different segments of the population, based on car ownership, age and occupation.

Regarding the 2030 forecasts and model application for this case, the figures below show the Volume/Capacity ratio and LOS – Level of Service in junctions [4] for AM and PM peak hours, in a normal working day, without the implementation of the TMS.

Figure 3. Volume/Capacity ratio and LOS – Level of Service in junctions for AM peak, without TMS, 2030.

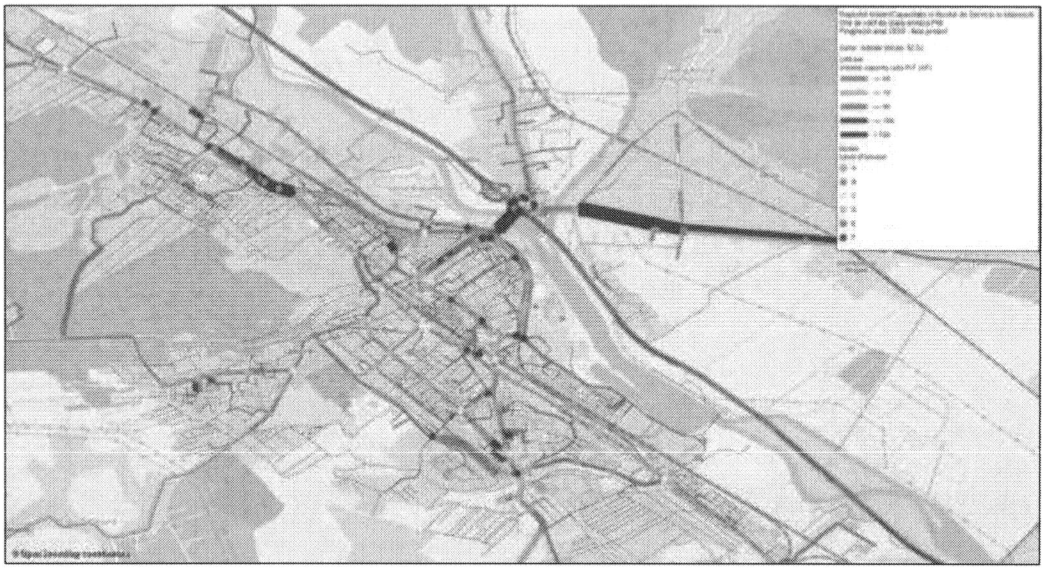

Figure 4. Volume/Capacity ratio and LOS – Level of Service in junctions for PM peak, without TMS, 2030

A high level of congestion is observed in the specific area of the city and on the busiest corridors.

3. Forecasts 2030 with TMS

The 2030 forecasts with TMS are based on the previous assumptions, plus the implementation of the TMS. The figures below show the Volume/Capacity ratio and LOS – Level of Service in junctions for AM and PM peak hours, in a normal working day, with the implementation of the TMS.

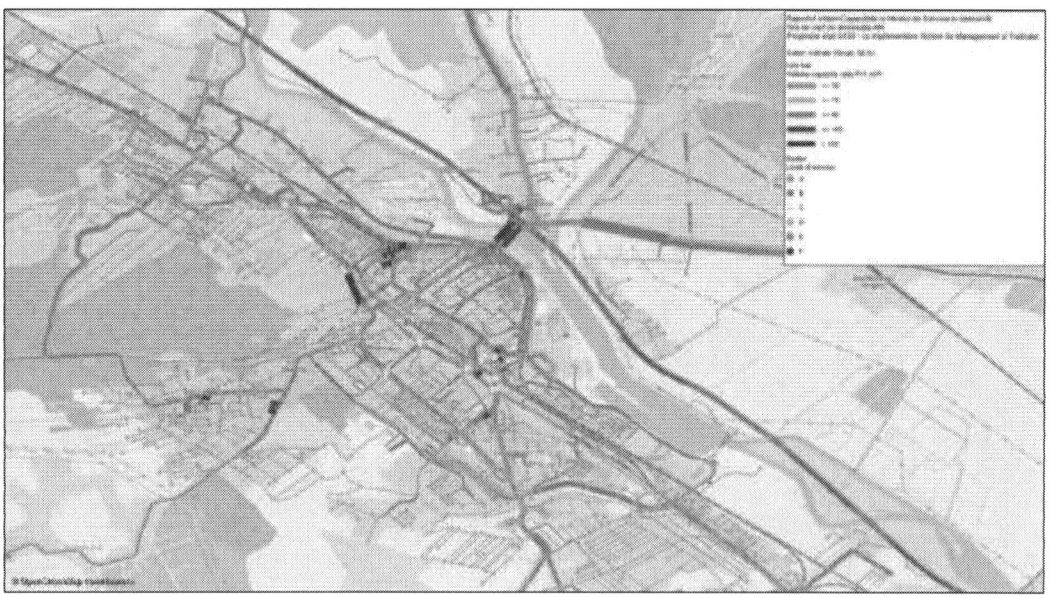

Figure 5. Volume/Capacity ratio and LOS – Level of Service in junctions for AM peak, with TMS, 2030

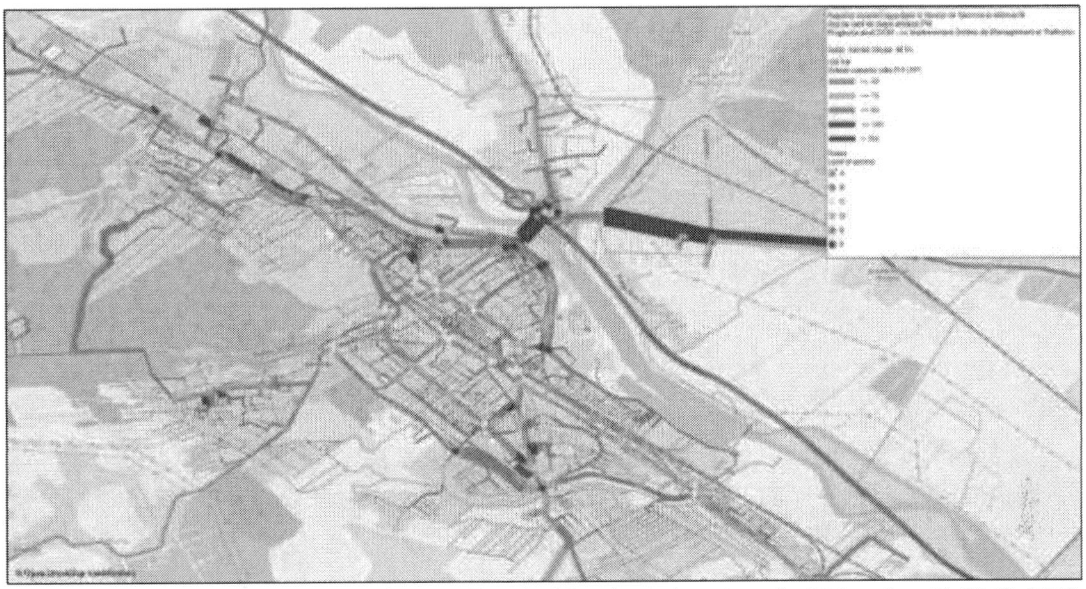

Figure 6. Volume/Capacity ratio and LOS – Level of Service in junctions for PM peak, with TMS, 2030

4. Main conclusions

It is observed that the main benefit, which consists in saved time in the network [5], is for both peak hours estimated at 1,225 veh-hours per day in 2030. The total traffic performance expressed in veh-kms is almost constant for the AM peak, and shows a decrease for the PM peak in the scenario "with project". This shows that the positive impact of the implementation of the integrated traffic management system is relevant.

Quantified for the whole year, it can be estimated a total effect of 306.250 veh-hours time saving for both peak hours, Considering a car occupancy of 1.25, this means 382.812 person-hours per year, thus in monetary terms, this would be 4.6 million Euros per year for the peak hours only, as an overall benefit for the users per year.

In the following, it is shown the overall reduction of fuel consumption and emissions in the study area.

Table 2. Time benefit

Forecast 2030		
SCENARIO	Total time in the network, veh hours	Total veh-kms
AM peak		
Basic scenario without TMS, 2030, AM	16062	240799
Scenario with TMS, 2030, AM	15619	240878
Time benefit AM	443	
PM peak		
Basic scenario without TMS, 2030, PM	17244	271523
Scenario with TMS, 2030, PM	16463	270916
Time benefit M	781	
AM & PM time reduction, per working day	1.225	

Table 3. Emissions time horizon 2030.

Fara proiect 2030											
	AM si PM	MZA		Reducere emisii GES							
Emisiile totale GES (tCO2e)	126,305	421015		4572							
Emisii totale de GES pentru intregul model de trafic pentru anul											
			COMBUSTIBILI CONVENŢIONALI					ELECTRIC			
Clasa	LDV (integrate la Autoturisme)	HDV	Autoturisme	LGV (integrate la OGV2)	OGV1 (integrate la OGV2)	OGV2 (integrate la OGV2)	PSV	Autoturisme electrice	Troleibuz	Autobuz electric	Tramvai
Emisii GES (tCO2e)	41,487	77,040	41,487			77,040	7,777	0	0	0	0
Cu proiect 2030											
	AM si PM	MZA									
Emisiile totale GES (tCO2e)	124,933	416443									
Emisii totale de GES pentru intregul model de trafic pentru anul											
			COMBUSTIBILI CONVENŢIONALI					ELECTRIC			
Clasa	LDV (integrate la Autoturisme)	HDV	Autoturisme	LGV (integrate la OGV2)	OGV1 (integrate la OGV2)	OGV2 (integrate la OGV2)	PSV	Autoturisme electrice	Troleibuz	Autobuz electric	Tramvai
Emisii GES (tCO2e)	41,945	75,411	41,945			75,411	7,577	0	0	0	0

It is observed that the final results show a reduction of GHG with 4.572 tons CO_2 per year. Most of the reduction is shown for the public transport and for the freight transport vehicles. In monetary terms, considering the current market value of 50 Euro for one ton of CO_2-eq., this would mean 228.6 thousand Euros per year. However, it is expected that the CO_2-eq. transaction value will reach 100 Euros/ton in 2030.

Considering all above estimated effects it can be concluded that the effects of the implementation of the traffic management systems, estimated with transport urban models combining the macroscopic approach with detailed network representation, including junctions and junction control, and considering detailed junction modelling, can be estimated with high accuracy, and helps for preparing and assessing interventions in this area.

However, combining these measures with promoting the use of Hybrid and electric cars and public transport vehicles, as well as the electric freight vehicles for delivery, would contribute to improving the urban environment overall.

References

[1] * * * 2017, *SUMP – Sustainable Mobility Plan Pitesti 2017*, AV Transport Planning SRL / PANTEIA bv / Synergetics Corporation SRL

[2] * * * 2017 VISUM Manual, PTV Germany

[3] Adrian Vilcan, 2018, *Traffic study for the implementation of the Traffic Management System in Pitesti city 2018*, AV Transport Planning SRL

[4] * * *2000, Highway Capacity Manual, Institute of Transportation Engineers, ISBN 0-309-06681-6

[5] * * * 1999, Transportation Planning Handbook, Institute of Transportation Engineers, ISBN: 0-935403-33-7, Publication No. TB-011A, 1000/TA/0402.

Blockchain-based solutions for smart mobility sustainability assurance

Luisa Dungan[1] and Mădălin-Dorin Pop[2]

[1] Mechanical Machines, Equipment and Transports Department, Politehnica University of Timisoara, 300222 Timisoara, Romania
[2] Computer and Information Technology Department, Politehnica University of Timisoara, 300223 Timisoara, Romania

madalin.pop@upt.ro

Abstract. Smart mobility represents the future for many of today's cities and is part of the smart city concept. Besides all its characteristics, the smart city concept aims to use the IoT (Internet of Things) technology to improve the citizens quality of life through a secure and efficient administration of city resources. The exponential growth of population and urbanization creates many challenges, especially from a mobility perspective. The main challenge in this direction is to ensure a sustainable transportation system. The objective of this paper is to present a survey on the sustainability assurance for smart mobility systems by using the blockchain technology. Moreover, this study proposes a conceptual model for a sustainable smart mobility system that implies the usage of blockchain technology. Widely known through its connection with bitcoin and other cryptocurrencies, blockchain can serve the safety and security requirements that are specific to intelligent transportation systems (ITS). The main advantage of this technology is the capability to face against several security threats by storing securely the information. This approach will streamline and secure the sensors networks data handling by the TMCs (Traffic Monitoring Centers). The process of these data handling is challenging because of possible exposure to cyber-attacks that can lead for example to erroneous traffic lights phases and green-interval settings in crossroads. The conclusions will be issued based on a comparison with the current IoT approaches for ensuring the secure storage of traffic data retrieved from sensors networks.

1. Introduction

The new concept of cities development starts from the adaptation of community services to ensure a better quality of life for the citizens. For this reason, many local administrations apply IoT technology for the secure and efficient utilization of city resources. The offered services shall be just in time and at high quality. In this context appeared the smart city concept that aims to employ the IoT technology in various domains such as mobility, health, local governance, etc. to fulfill citizens needs and expectations.

This paper concentrates on the challenges of the urban transportation systems that are continuously affected by the exponential growth of population and the expansion of urban areas. Even if smart solutions are used in several cities, there are still vulnerabilities on the security assurance in the communication between smart devices and this shall be addressed to ensure the sustainability of a smart mobility system.

This research surveys many studies related to blockchain and highlights its benefits on secure communication. The use of this technology in the context of smart mobility can reduce significantly the cyber-attacks over the sensor networks used by ITS (Intelligent Transportation Systems). In this manner, the TMCs can securely handle the data received from the sensor networks and decide based on these new approaches to ensure a good road traffic flow and also to ensure secure V2X communication.

2. Smart mobility as part of the smart city concept

Smart city is a complex concept that combines the IoT, local administration decisions and the continuously participatory citizenship in the development of offered community services having as objective the improvement of the citizens quality of life.

Figure 1 shows the widely known Giffingers' smart city model. This model contains six characteristics: Smart People, Smart Governance, Smart Economy, Smart Environment, Smart Living, and Smart Mobility. A city is smart if the scoring for its components exceeds a specific threshold.

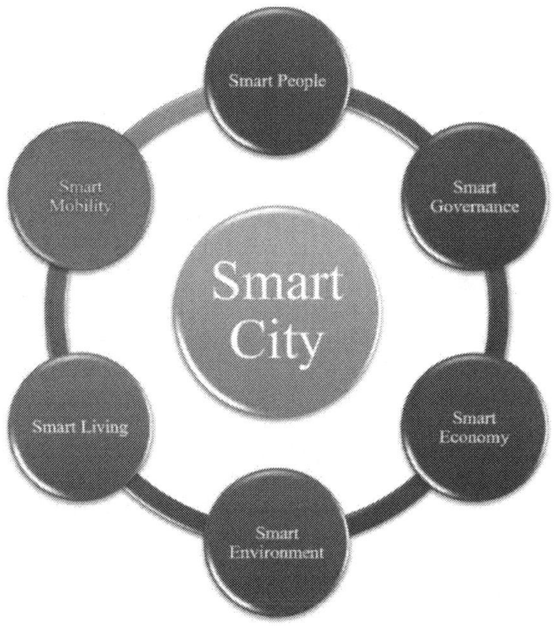

Figure 1. Giffingers' Smart City Model, adapted after [1-3].

Smart mobility is the characteristic that this research aims to address. This characteristic influence also the scoring for smart people, smart economy, smart environment, and smart leaving. A smart mobility system combines the usage of IoT with the road traffic policies and regulations to provide an intelligent road infrastructure capable of the efficient management of urban means of transportation schedules and routes, also ensuring a good traffic flow in the city for all vehicles, with reduced travel times and CO_2 emissions.

A smart mobility system consists of three levels of abstraction: areas of interest, indicators and metrics [3]. This survey addresses the highest level of abstraction represented by the areas of interest which incorporate the following directions: accessibility, ICT (Information and Communication Technologies), sustainability and focuses on the sustainability assurance based on the proper and secure use of IoT.

2.1. Vulnerabilities in smart mobility systems

Many of the vulnerabilities of a smart mobility system arise from secure data handling by different sensors networks. Security is a really important topic in V2X (Vehicle to Everything) communication because the intervention of unauthorized entities may endanger the lives of drivers and other participants in traffic by providing erroneous information. Biswas and Muthukkumarasamy identified the following five categories of security vulnerabilities directions in smart cities [4]:

- availability: this category refers to the unauthorized upholding of resources;
- integrity: related to the unauthorized changes of data, their usage and corruption of information;
- confidentiality: concerned with the break of confidentiality agreements by sharing sensitive data with unauthorized parties;
- authenticity: related to the permissions of authentication of unauthorized parties to the smart system resources and confidential data;
- accountability: refer to denial of communication by an entity from the smart system.

In addition to the previously listed vulnerabilities, an interesting case is the VANET (Vehicular Ad Hoc Network). In this case, the vulnerabilities arise from the needs of identity and location privacy protection considering the VANET decentralized architecture deployment [5, 6].

2.2. Sustainability in smart mobility

The evolution of smart cities shall be through an eco-sustainable direction that addresses mechanisms for efficient use of renewable resources and reducing wastes by safeguarding the environment [7]. Usually, the sustainability analysis follows three main directions: environmental, social and economic. Many of the environmentally related solutions refer to the impact of digital technologies on intelligent mobility infrastructure, water supply, smart buildings, waste disposal systems, and city administration services [8].

Regarding the mobility process the most frequently wastes are the usage of non-ecological fuels and the long travel times. A good measure to obtain energy savings is to use IES (Intelligent Energy Systems). In this case, an intelligent application advices the drivers to use not only the closest charging station for electric vehicles but routes them to stations based on the usage of renewable energy or equipped with the functionality of bi-directional exchange of electricity with the grid [7].

The car-sharing systems represent the most used forms of integrated sustainable mobility. The IoT grants access to the users for managing vehicles information such as localization, vehicle performance, fuel/battery (in case of electric vehicles) autonomy according to the travel plan. During the travels, V2X communication plays an important role in VANETs due to real-time data sharing that controls the driving strategies of vehicles in urban areas [6].

3. Blockchain technology

3.1. General overview

Haber and Stornetta created the basis of blockchain technology in 1991 from the idea of implementing a system where document timestamps could not be tampered with [9-11]. Although, the first practical implementation came in January 2009 with the launch of Bitcoin.

Blockchain is a type of database that stores data in blocks and chains them together. Each newly added block in the chain will receive a specific timestamp. In this manner, all blocks are stored chronologically and this approach, combined with a hash assigned to each block enhance the resilience at data altering from the previous blocks [10]. Besides its connections with cryptocurrencies, blockchain proved its utility in the banking domain to provide faster transactions, healthcare applications for secure storage of patients' data, supply chains, secure voting systems or in case of smart contracts that are used to facilitate, verify, or negotiate a contract agreement [10, 11].

The practical implementation of this technology faces many challenges. Jabbar et al. presented a detailed overview of these challenges in [12]. The main challenges are related to scalability and interoperability. Blockchain needs more time and computing resources compared to a normal database to retrieve and commit records. In the case of different types of systems integrated with blockchain is needed flexibility and good interaction between them [12].

3.2. Blockchain solutions for smart mobility

Blockchain technology represents the most appropriate solution for transportation decentralization. One direction in this regard is represented by blockchain-based applications for ride-sharing and carpooling. In this sense, the drivers can offer their vehicles as a disjointed group of service providers [11] that brings many benefits through the reduction in the traffic volumes, pollution level, traffic congestion or parking places needs in key places of a city [13]. Blockchain can provide a secure environment that addresses the privacy concerns, transparency and digital payments for these types of applications that enable the MaaS (Mobility-as-a-Service) [13, 14].

Many studies tried to evaluate the technical aspects of the integration of V2X with blockchain [13, 15, 16]. A possible solution is the integration of V2X with DSRC (Dedicated Short Range Communications) to enable blockchain technology. DSRC is a cost-effective technology that simplifies the integration between blockchain and V2X through a mesh of vehicular networks that allows the interactions between vehicles, smartphones, drones and other intelligent devices [13]. The main issues of V2X represented by centralization, lack of privacy and communication security can be solved also by applying an approach that combines smart contracts with Ethereum [16].

Autonomous vehicles are also a future direction where the blockchain can ensure secure communication between several sensors [11, 17]. The safety of the passengers and the other participants in traffic is directly dependent on the data referring to the surrounding environment, route information, speed control, power/fuels or parking services, traffic lights and signs recognition, etc.

Another interesting topic relates to the self-owning car concept. In this case, the cars become autonomous entities capable to manage transactions on their own with the help of the mixture between artificial intelligence and blockchain technology embedded in their system[18]. Combined with ridesharing and autonomous driving concepts can lead in future to a fully autonomous entity capable to manage the revenues achieved from carsharing, instantly process the payments for charging, needed repairs, specific roads taxes or vehicle insurance bills [18].

3.3. Conceptual blockchain-based model for a sustainable smart mobility system

Figure 2 illustrates a conceptual model for a sustainable smart mobility system that implies the usage of blockchain technology. This model considers the Smart sustainable city-blockchain (SSCB) integration framework proposed in [19] and highlights its benefits in the transportation field. Additionally, energy management for the transportation field has been added as an indicator. Recent works show a special interest in green transportation systems involving intelligent energy consumption management and the reduction of CO_2 emissions [20-21]. Blockchain aims to provide proper solutions for car-sharing services from privacy and security perspectives [22]. Moreover, the blockchain-based transactions process faster the payments for parking and transit, depending only on network congestion [10].

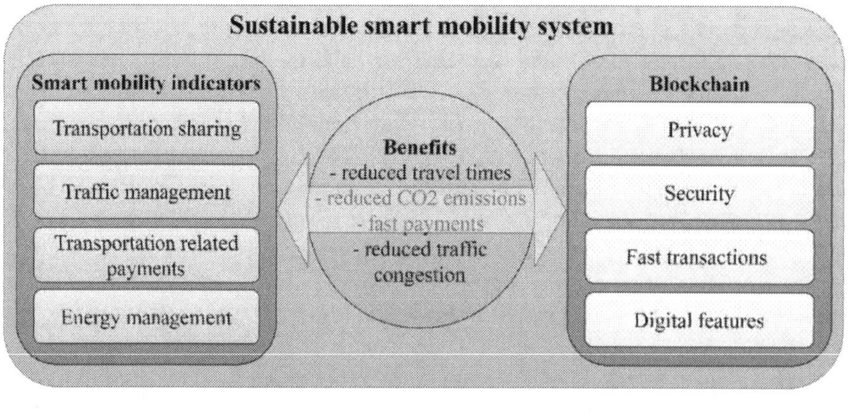

Figure 2. Sustainable smart mobility system - conceptual model.

4. Conclusions and future works

This research presented a survey on the usage of blockchain technology in improving the security of data shared between several sensor networks in a smart mobility system. The paper started with a presentation of the smart mobility concept that represents one of the six characteristics of the smart city model. Their aim was to increase the knowledge of the smart mobility concept and to identify their vulnerabilities. Once these threats were identified, it is easier to design a roadmap to a sustainable smart mobility system, especially from the proper and secure use of IoT perspectives. Further, this research proposed a conceptual model that aims to ensure the sustainability of a smart mobility system by employing the features of blockchain technology.

Blockchain technology proved its value regarding secure data handling in the field of cryptocurrencies. Although, this technology can be applied in the domain of ITS. The V2X communication is prone to cyber-attacks and in the worst-case scenario, unauthorized access to sensitive data can lead to traffic accidents endangering the lives of citizens.

Based on the reviewed literature, this study highlighted the advantages and disadvantages of the employment of blockchain technology by smart mobility systems. Further works can extend this research in the field of V2X communication to develop solutions for a cost-efficient gathering between these two concepts. Moreover, can be proposed solutions for the integration of blockchain technology with various traffic specific sensors such as inductive loops detectors, camera sensors, infrared sensors etc.

References

[1] Giffinger R, Kramar H, Haindlmaier G, and Strohmayer F. European smart cities. Retrieved August 4, 2021 from http://smart-cities.eu/

[2] Pop M-D and Proştean O 2018 A Comparison Between Smart City Approaches in Road Traffic Management *Procedia - Social and Behavioral Sciences* **238** 29–36.

[3] Pop M D and Proştean O 2019 Identification of significant metrics and indicators for smart mobility *IOP Conf. Ser.: Mater. Sci. Eng.* **477** 012017.

[4] Biswas K and Muthukkumarasamy V 2016 Securing Smart Cities Using Blockchain Technology *2016 IEEE 18th International Conference on High Performance Computing and Communications; IEEE 14th International Conference on Smart City; IEEE 2nd International Conference on Data Science and Systems (HPCC/SmartCity/DSS)* (Sydney, Australia: IEEE) pp 1392–3

[5] Li H, Pei L, Liao D, Sun G and Xu D 2019 Blockchain Meets VANET: An Architecture for Identity and Location Privacy Protection in VANET *Peer-to-Peer Netw. Appl.* **12** 1178–93

[6] Silva B N, Khan M and Han K 2018 Towards sustainable smart cities: A review of trends, architectures, components, and open challenges in smart cities *Sustainable Cities and Society* **38** 697–713

[7] Orecchini F, Santiangeli A, Zuccari F, Pieroni A and Suppa T 2019 Blockchain Technology in Smart City: A New Opportunity for Smart Environment and Smart Mobility *Intelligent Computing & Optimization* vol 866, ed P Vasant, I Zelinka and G-W Weber (Cham: Springer International Publishing) pp 346–54

[8] Toli A M and Murtagh N 2020 The Concept of Sustainability in Smart City Definitions *Front. Built Environ.* **6** 77

[9] Haber S and Stornetta W S 1991 How to time-stamp a digital document *J. Cryptology* **3** 99–111

[10] Conway L. Blockchain Explained. *Investopedia.* Retrieved August 4, 2021 from https://www.investopedia.com/terms/b/blockchain.asp

[11] Moorman J and Stricklen M 2020 Smart Cities Applications of Blockchain *Smart Cities in Application* ed S McClellan (Cham: Springer International Publishing) pp 101–17

[12] Jabbar S, Lloyd H, Hammoudeh M, Adebisi B and Raza U 2021 Blockchain-enabled supply chain: analysis, challenges, and future directions *Multimedia Systems* **27** 787–806

[13] Oliveira T A, Gabrich Y B, Ramalhinho H, Oliver M, Cohen M W, Ochi L S, Gueye S, Protti F,

Pinto A A, Ferreira D V M, Coelho I M and Coelho V N 2020 Mobility, Citizens, Innovation and Technology in Digital and Smart Cities *Future Internet* **12** 22

[14] Treiblmaier H, Rejeb A and Strebinger A 2020 Blockchain as a Driver for Smart City Development: Application Fields and a Comprehensive Research Agenda *Smart Cities* **3** 853–72

[15] Elagin V, Spirkina A, Buinevich M and Vladyko A 2020 Technological Aspects of Blockchain Application for Vehicle-to-Network *Information* **11** 465

[16] Jabbar R, Kharbeche M, Al-Khalifa K, Krichen M and Barkaoui K 2020 Blockchain for the Internet of Vehicles: A Decentralized IoT Solution for Vehicles Communication Using Ethereum *Sensors* **20** 3928

[17] Rotuna C, Gheorghita A, Zamfiroiu A and Smada D-M 2019 Smart City Ecosystem Using Blockchain Technology *IE* **23** 41–50

[18] Gösele M and Sandner P 2019 Analysis of blockchain technology in the mobility sector *Forsch Ingenieurwes* **83** 809–16

[19] Wong P F, Chia F C, Kiu M S and Lou E C W 2020 The potential of integrating blockchain technology into smart sustainable city development *IOP Conf. Ser.: Earth Environ. Sci.* **463** 012020

[20] Tarulescu S, Tarulescu R, Soica A and Leahu C I 2017 Smart Transportation CO_2 Emission Reduction Strategies *IOP Conf. Ser.: Mater. Sci. Eng.* **252** 012051

[21] Zhang X, Xu Q, Lu J and Xu J 2021 Conceptual characteristics and analysis of typical application scenarios of energy blockchain *J. Phys.: Conf. Ser.* **1738** 012113

[22] Creutzig F 2021 From smart city to digital urban commons: Institutional considerations for governing shared mobility data *Environ. Res.: Infrastruct. Sustain.* **1** 025004

Air quality analysis on a frequently congested road section

D M Mustață[1], R M Popa[1], I Ionel[1], R M Balogh[1], D Bisorca[1], D R Streinu[2]

[1]Faculty of Mechanical Engineering, Politehnica University of Timişoara, B-dul Mihai Viteazu, nr. 1, 300222, Timişoara, Romania
[2]General Surgery Department, Timiş County Emergency Clinical Hospital "Pius Brînzeu", Bulevardul Liviu Rebreanu 156, Timişoara 300723, Romania

dan.mustata@student.upt.ro, ioana.ionel@upt.ro

Abstract. This paper will focus on analysing the air quality on a frequently congested road section in the vicinity of Timisoara, Romania. The placement of the measurement equipment will be made next to a public transport station measuring the extent of different pollutants which affect the well-being of passengers waiting in these stations. The traffic congestion is present due to the unloading from the highway, reduced capacity of only a one lane road and because of the high amount of industrial area workplaces. The results presented will be part of the PhD thesis of the main author.

1. Introduction

Within the paper, the major focus represents measurements done in the vicinity of a public transport stop, in an urban area, on a highly congested road section, to analyze the air quality and the levels of different pollutants to which passengers are subjected, due to waiting in these stops.

One of the major environmental problems observed mostly in urban areas, the prevalence of toxic substances and greenhouse gases, is due to combustion processes of either fixed or mobile plants generating heat and/or energy, releasing several types of combustion products in the air.

From a qualitative perspective, these products of combustion are mostly the same, but when comparing from a quantitative aspect, results are very different when analyzing a stationary burner versus an internal combustion engine [1, 2].

Measured and analyzed pollutant levels within this paper were made for PM (particulate matter), CO (carbon monoxide), SO_2 (sulfur dioxide) and NO_2 (nitrogen dioxide).

Air pollution has a high impact on health, ecosystems, environment, and climate. According to a report provided by WHO (World Health Organization) in 2014, air pollution represents the largest environmental health risk in Europe along with the resulting diseases associated with it. The most common reasons for premature death due to air pollution are heart diseases and strokes [3].

Different pollutants are causing a wide range of health effects like respiratory diseases and inflammation of the airways(PM_{10}, $PM_{2.5}$, O_3, NO_2, SO_2), low hemoglobin production (Pb), low oxygen carrying capacity (CO) and spleen and liver associated diseases (NO_2) [4].

Several studies have been conducted worldwide to analyze the short-term air pollution as a cause for respiratory diseases. One of these studies performed in Lanzhou, China has determined that females and younger groups were more affected than male and elderly groups. $PM_{2.5}$, PM_{10}, SO_2 and CO have been associated with respiratory diseases, out of which SO_2 and CO had the greatest adverse effect [5].

Pulmonary exacerbations and a decline in lung function represent a major concern when it comes to air pollution especially in patients diagnosed with cystic fibrosis.

Additionally, children with asthma are at risk, several pollutants leading to reduced asthma control which lead to an increase in using rescue medication, medical services, and school absences [6].

When analyzing CVD (cardiovascular disease), CO levels measured in a study in China between 2014 and 2017 were associated with 11% increase out of the total CVD hospitalizations [7].

2. Initial data

The road section where the measurements have been executed is located under coordinates 45°47'54" N, 21°14'39" E, Petofi Sandor Street (DJ691), Dumbravita, Timis County, Romania.

The mobile air quality measuring station, accredited by RENAR (Accreditation Association of Romania) was placed near a public transport stop specifically to be able to measure the levels of different pollutants, which passengers waiting in these stops are exposed to (figures 1 and 2). The station was placed as close as possible in correlation with the wind direction (see table 1)to reduce the risk of obtaining corrupted data and get reading of the true values of emissions during measurements.

Figure 1. Mobile air quality measuring station, according RENAR standards (www.mediu.ro), placed next to public transport stop.

Figure 2. Inside view.

This road section was particularly chosen due to several impact factors which build to high traffic volumes and congestions in different time periods of the day.

One of these factors is represented by the location within the greater urban area of Timisoara, Romania. Dumbravita is a small town connected with Timisoara and the major highway (A1) connecting Hungary with the western part of Romania passes close by. The shortest connection between the highway and Timisoara is through Dumbravita, with a one lane per travel direction road, leading to one of the major factors that lead to high traffic volumes and very often congestions. See figure 3.

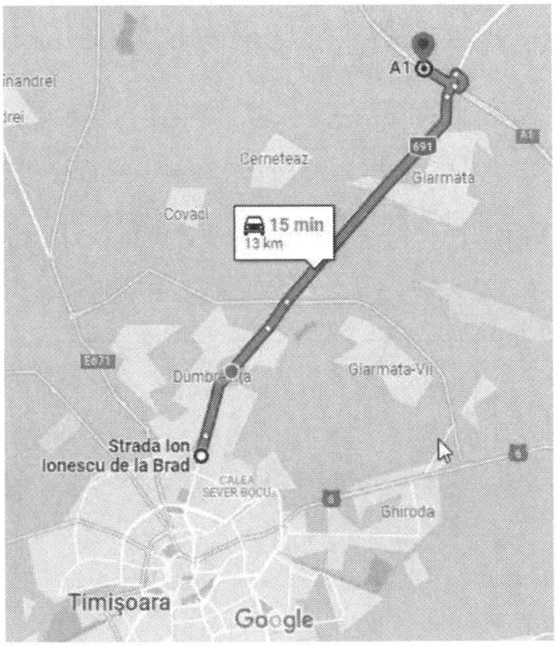

Figure 3. Connection to highway A1.

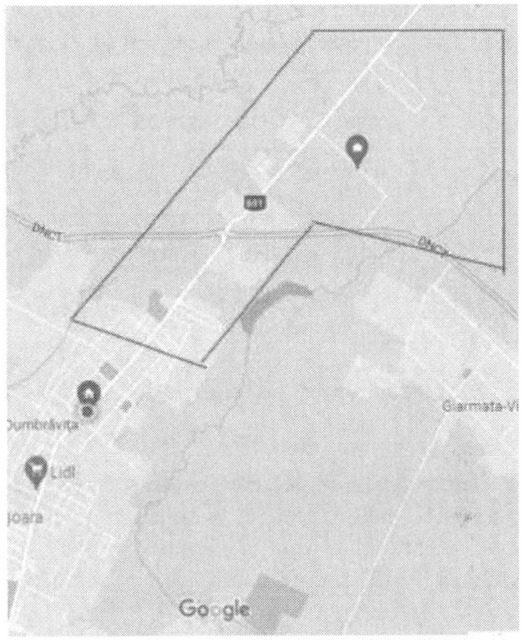

Figure 4. Blue grids represent the industrial area near Dumbravita.

Another factor is due to Dumbravita being a residential area without a high number of education facilities, which lead to children and students having to commute to Timisoara for schools. From a social perspective, in Romania it is more common to drive the students to school with personal vehicles than with public transport, leading to a high number of vehicles during rush hours.

Finally, it can be mentioned that the area next to Dumbravita is industrial, meaning that a high number of employees commute from the different residential areas of Timisoara using the same road section. See figure 4.

A detailed initial data analysis is found in table 1.

Table 1. Initial data.

Input factor	Value
Date	21.09.2021
Time interval for measurements [UTC+2h]	07:45-08:45, 12:00-13:00, 17:00-18:00, 21:00-22:00
Road travel direction	N→S (From highway A2 to Timisoara) S→N (From Timisoara to highway A2)
Number of lanes per travel direction	1
Maximum allowed speed on road section [km/h]	50
Average speed on road section [km/h]	45
Wind direction interval	0°NNE→17°NNE
Wind speed average	6 km/h

The chosen time interval represents the highest peaks of traffic usually recognized during the day on this road section. During the morning (07:45-08:45), people are traveling to their workplaces (starting point is either the residential area of Dumbravita or they are just in transit) or to drop off their children at school. Noon time (12:00-13:00), employees go to lunch, picking-up children from school or doing business in Timisoara. Afternoon period (17:00-18:00), people are leaving their workplaces (end point is either Dumbravita or they are transiting).

All measured number of vehicles were associated to their standard vehicle number based on size, volume, and weight. Weight is directly proportionate to the consumption of fuel; therefore, it was an important factor for the conversion. This equivalency is shown in table 2.

Table 2. Conversion table for realising the equivalence to standard vehicle.

Vehicle type	Vehicle weight maximum limit	Standard vehicle number
Car	<3.5 [t]	1
Van/lorry (2 axles)	3.5 –12 [t]	2
Public transport vehicles		3
Trucks (3 axles)	>12 [t]	3.5

3. Measurement method and results

3.1. Number of vehicles

Measured number of vehicles, expressed as standard vehicle, based on travel direction and period of the day is presented in table 3. A traffic light intersection is situated close by, therefore the number of vehicles passing in front of the measurement equipment was determined by measuring 5 traffic light cycles every 15 min of each time interval. The duration of the cycle is 1'25".

Table 3. Number of vehicles equivalent to standard vehicles.

Travel direction	Time interval	Cars	Vans/lorries	Public transport	Trucks
N→S	07:45-08:45	360	48	72	42
N→S	12:00-13:00	708	120	72	42
N→S	17:00-18:00	660	168	72	42
N→S	21:00-22:00	228	48	36	126
S→N	07:45-08:45	408	144	72	63
S→N	12:00-13:00	648	144	72	42
S→N	17:00-18:00	648	48	72	42
S→N	21:00-22:00	276	48	36	168

The highest number of cars (N→S = 708, S→N = 648)is around noon time (12:00-13:00) due to the several travel reasons explained previously, respectively the lowest number of trucks is present during this period. During evening (21:00-22:00) results are changed, the number of cars is lowest due to personal travels only for leisure and shopping activities, while the number of trucks is the highest due to transportation companies preferring to transit during the evening or even nighttime when the traffic is not congested.

Vans and lorries are mostly transiting during noon time, whereas public transport vehicles have a fixed number of vehicles per time interval, due to fixed schedules and passes per hour.

3.2. Concentration of particulate matter (PM2.5, PM10)

Particulate matter is represented by a mixture of solid particles and liquid droplets of dust, dirt, soot, or smoke which are generated by combustion engines, power plants, construction sites, unpaved roads, and fires. From size perspective, PM_{10} particles have a particle diameter of 10 micrometers and smaller, whereas $PM_{2.5}$ having 2.5 micrometers or smaller sizes resulting in a higher risk of pulmonary diseases due to the lower ability of the respiratory pathways to retain this particle size. As a comparison of size, the human hair is 50-70 micrometers in diameter [8].

Gasoline powered vehicles with a GDI (gasoline direct injection spark ignition) engine are more efficient in reducing the amount of CO_2 levels but show a higher level of PM emissions compared to GPI (conventional port-fuel injection) engine powered vehicles. This is due to high compression ratios in GDI engines by spraying fuel directly into the cylinder compared to GPI engine, where fuel is injected into the intake port [9].

Gravimetric measurement method was done according to SR EN 12341:2014 standard, in continuous measurement of 24h, using a $PM_{2.5}$ and PM_{10}analyzer, model LSV3, produced by Sven Leckel. During a 24h period, the equipment is aspirating a fixed volume of air V_{aspir}=2.3 Nm^3/h, passing it through a filter capturing these fine particles, resulting in a total value per 24h of V_N [Nm^3/h]. After measurement is finished, to compare between clean versus charged filters, they have been measured using a scale, model Ohaus DC215 CDM.

Environmental conditions are presented in table 4. Differences in the value of each factor are due to night versus day conditions of the 24h measurement interval.

Table 4. Environmental conditions present during the 24h measurements.

Factor	Value
Ambient temperature	8-19 [°C]
Atmospheric pressure	997.48-1012.69 [hPa]
Humidity	43-100 [%]
Wind	0-5.2 [m/s]
Sky	Clouded, scattered rain, sunny

Filters for both PM_{10} and $PM_{2.5}$ were each measured 10 times resulting in an average weight value presented in table 5.

Table 5. Measurement of filters before and after 24h.

Filter impactor	Time/date interval	m_1 [g]	m_2 [g]	m_{PM} [g]	V_{aspir} [m^3N/h]	V_N [m^3N/h]
PM_{10}	11:20/21.09-22.09.2021	28.71869	28.71893	0.00045	2.3	55.19
$PM_{2.5}$	12:10/21.09-22.09.2021	28.72381	28.72585	0.00020	2.3	55.17

Based on the weight measurement of filters before and after the 24h period, concentration levels of PM_{10} and $PM_{2.5}$ were determined and shown in table 6.

Table 6. Determined concentration levels of PM_{10} and $PM_{2.5}$.

Filter	Measured value [μg/m³]	Daily limit value [μg/m³]
PM_{10}	8.2	50
$PM_{2.5}$	3.6	35

The limit value for PM_{10} is given by European Union air quality standards. The limit value for $PM_{2.5}$ is given by EPA (United States Environmental Protection Agency). For $PM_{2.5}$ European Union standards do not offer a daily limit, only a yearly mean value [10, 11].

As for the concentration results, both PMs are present in the measured area mainly due to the result of the combustion process of vehicles passing close to the measurement equipment and the dust present next to the road which is mainly raised due to heavy and high-volume vehicles, like trucks, creating air currents when they pass.

Figure 5 represents how one of the filters looks like after the 24h measurement, charged with particles.

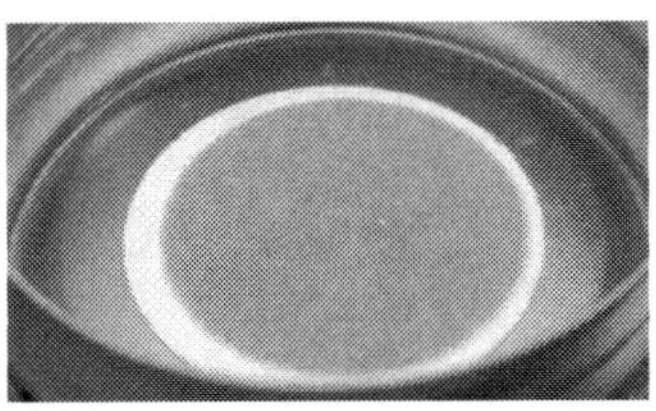

Figure 5. Filter charged with particles after 24h measurement.

3.3. Concentration of CO, SO_2, NO_2

The second part of the analysis was dedicated to measure the concentrations of chemical pollutants CO, SO_2 and NO_2.

Measurement of CO was done using a non-dispersion cross modulation infrared analysis method with a device model APMA-370. For SO_2 the device, model APSA-370 manufactured by the same company, measures using UV fluorescence. Finally, for determining NO_2 concentrations, the device with model APNA-370 is using a cross flow modulated semi decompression chemiluminescence method. All device models are manufactured by HORIBA Europe GmbH.

In each figure presented below, blocks have been placed around the time intervals seen in table 3 representing the number of vehicles measured during different periods of the day.

The first pollutant which was measured is CO. As shown in figure 6, values are increasing during each of the measured time intervals, with a major increase towards the evening (starting 18:00-21:00).

A reason for this situation might be due to winds. According to a study done in Sao Paulo, Brazil, if wind speed is low, the rate of dispersion of emitted CO from the incomplete combustion process of a vehicles engine is low as well, resulting in a higher amount of CO remaining in the area [12]. For this particular result shown below, it seems that during that time interval of the day, wind speeds were low, resulting in an increase of non dispersed CO levels.

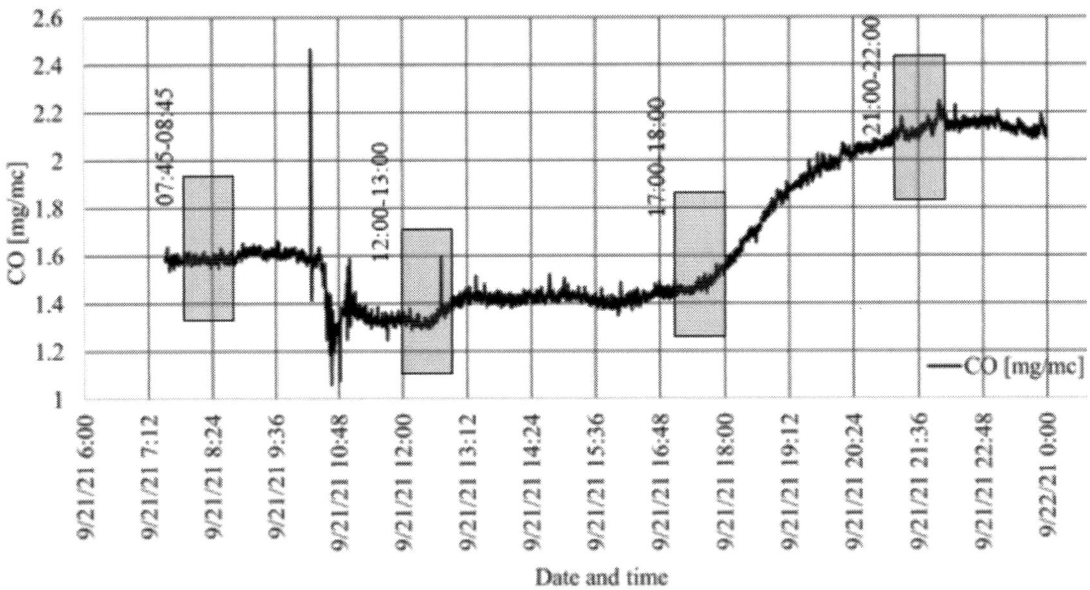

Figure 6. Values of CO shown in comparison to the time interval of high traffic.

For SO_2 levels shown in figure 7, an increase is present at each time interval block, with the highest levels around block 17:00-18:00, then decreasing towards the evening.

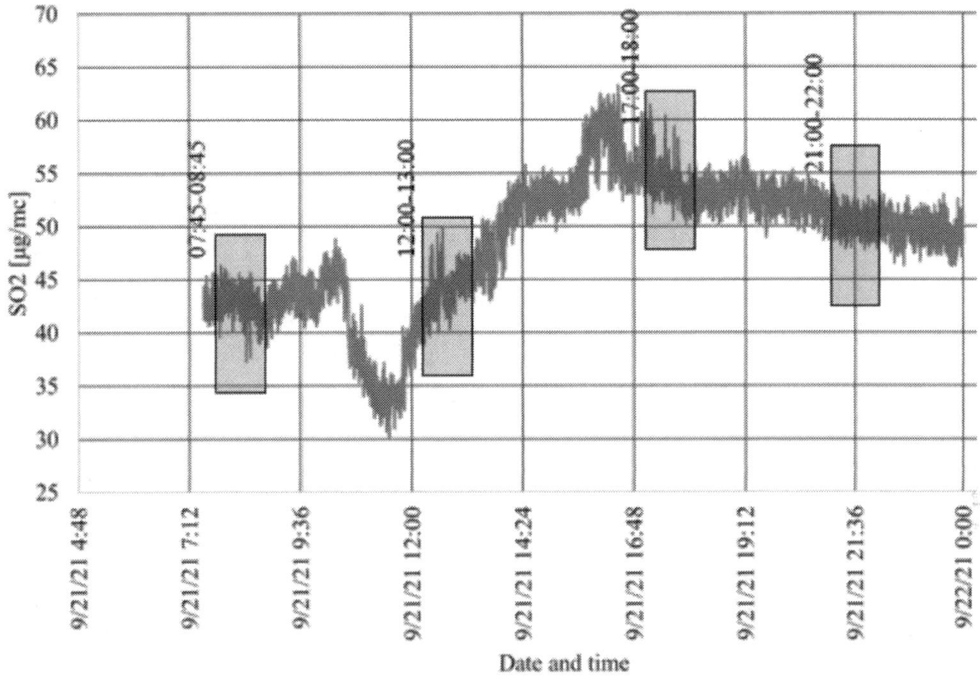

Figure 7. SO_2levels measured.

Finally, for NO_2, levels are shown in figure 8. Mostly the same increasing trend is seen like for SO_2 during each time interval blocks with highest traffic measured. Peaks are shown to be during the morning and evening time interval blocks in comparison to SO_2 when the highest peak was measured during the afternoon time interval block.

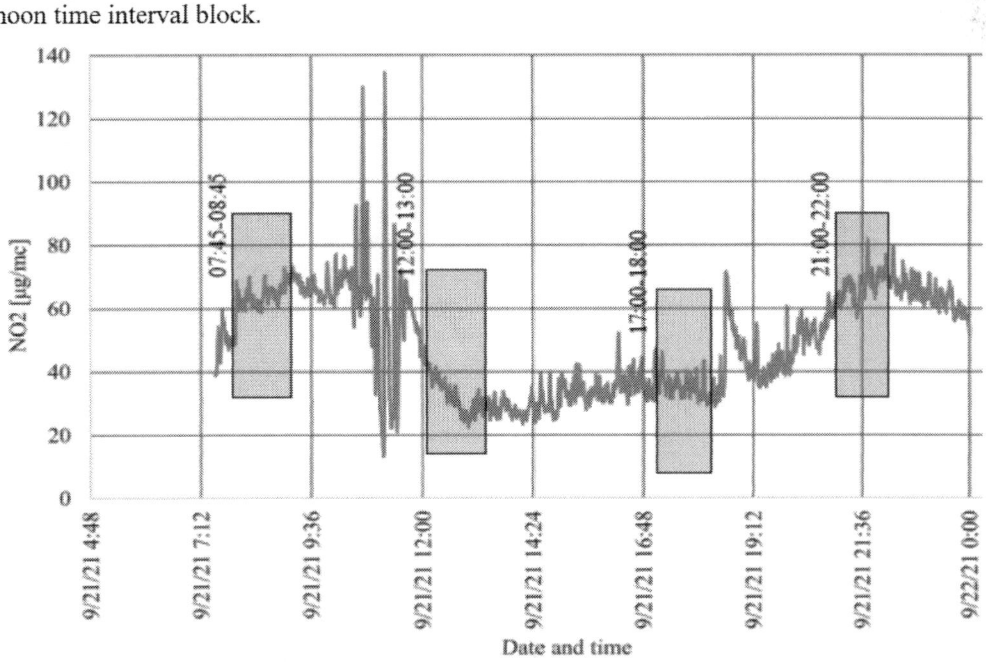

Figure 8. NO_2 measured values [$\mu g/m^3$].

Concentrations for each pollutant are within the accepted values according to Romanian Law No. 104 from 15th of June 2011 and are shown in table 7 [13].

Several improvement ideas have been drawn to reduce the amount of pollution within the area. One of this is referring to an infrastructure project, to double the lane number on the entire road section connecting highway A1 with Timisoara. Thus, an increase of distance between vehicles which reduces congestion will result. Another project is the construction of a belt around Timisoara which restrict high-volume and size vehicles (like trucks) to enter through residential areas.

Table 7. Highest concentration measured versus limit [13].

Pollutant	Averaging time [h]	Highest concentration recorded	Concentration limit
CO	8	1.63 [mg/m^3]	10 [mg/m^3]
SO_2	1	63.22 [μg/m^3]	350 [μg/m^3]
NO_2	1	134.64 [μg/m^3]	200[μg/m^3]

4. Conclusions

Based on the number of vehicles and the time interval blocks (representing rush hour periods with traffic congestions), it can be recognized that pollution levels usually rise together with these two mentioned factors. Recorded values are well within the limit as shown in the results. Even if all concentration levels are in the accepted limit, it must be mentioned that passengers waiting in these public transport stops situated next to highly congested roads are experiencing daily a high and constant number of pollutants which they inhale and may lead to long term exposure diseases.

Acknowledgement
The paper is part of the PhD activity of the first author at the Politehnica University Timisoara.

References

[1] Giavazzi F, Buttini P and Perego C 2007 *Encyclopedia of hydrocarbons* **III** 717
[2] *** https://www.eea.europa.eu/ro/themes/air/intro, accessed on 6th Oct. 2021
[3] Guerreiro C, Gonzalez Ortiz A and de Leeuw F 2017 Air quality in Europe — 2017 report European Environmental Agency 11-12
[4] Metcalfe S and Derwent D2005 *Atmospheric Pollution and Environmental Change*Hodder Headline Group84-85
[5] Dong J, Liu Y and Bao H 2021 *Revalue associations of short-term exposure to air pollution with respiratory hospital admissions in Lanzhou, China after the control and treatment of current pollution* International Journal of Hygiene and Environmental Health **231** https://doi.org/10.1016/j.ijheh.2020.113658, accessed on 10th Oct. 2021
[6] Brumberg H L and Karr C J 2021*Ambient Air Pollution: Health Hazards to Children*Pediatrics **147** 5 https://doi.org/10.1542/peds.2021-051484, accessed on 10th Oct. 2021
[7] Cheng J, Xu Z, Zhang Z, Zhao H and Hu W*Estimating cardiovascular hospitalizations and associated expenses attributable to ambient carbon monoxide in Lanzhou, China: Scientific evidence for policy making* 2019Science of The Total Environment **682** https://doi.org/10.1016/j.scitotenv.2019.05.110, accessed on 10th Oct. 2021
[8] *** https://www.epa.gov/pm-pollution/particulate-matter-pm-basics, accessed on 6th Oct. 2021
[9] Morino Y, Li Y, Sato K, Inomata S, Tanabe K, Jathar S, Kondo Y, Nakayama T, Fushimi A, Takami A and Kobayashi S 2021 *Secondary Organic Aerosol Formation from Gasoline and Diesel Vehicle Exhaust under Light and Dark Conditions*5
[10] *** https://ec.europa.eu/environment/air/quality/standards.htm, accessed on 7th Oct. 2021
[11] *** https://www3.epa.gov/region1/airquality/pm-aq-standards.html, accessed on 7th Oct. 2021
[12] Rozante J R, Rozante V, Alvim D S, Manzi A O, Chiquetto J B, D'Amelio M T S and Moreira D S 2017 *Variations of Carbon Monoxide Concentrations in the Megacity of São Paulo from 2000 to 2015 in Different Time Scales*11
[13] *Law. 104 – 15.06.2011,* http://legislatie.just.ro/Public/DetaliiDocument/129642

Traffic analysis for the reconstruction of the street geometries in the case of the construction of shopping centre in crowded cities

Matei Lucian[1], Dumitru Ilie[2], Racila Laurențiu[3], Tutunea Dragoș[4] and Oprica Alexandru[5]

[1, 2, 4, 5] University of Craiova, Faculty of Mechanics, Department of Vehicle, Transportation and Industrial Engineering, Craiova, Romania
[3] University of Craiova, Faculty of Mechanics, Department of Applied Mechanics and Civil Constructions, Craiova, Romania

racila_laurentiu@yahoo.com

Abstract. The road network in many of the developing cities has remained at a level that allows low-level transport logistics. In this context, and due to the elements of urban development that are currently applied, changes are needed in the road network to meet new road traffic requirements as well as to provide increased road safety. This paper aims to analyse the traffic capacity of the current street network in relation to the introduction of a mall in the study area.

1. Introduction

The goal of this study is to analyze the various types of transport policies and network configurations that could benefit from different outputs, such as speed, pollution, and delay in regards to the complexity of new shopping building constructions (malls).

The travel difficulties encountered by road users in transiting the main road arteries of different medium and large cities in Romania can be categorized into one of the following:

- increase of the car park necessities
- lack of travel alternatives
- impossibility of modifying the existing street tram
- preceding faulty arrangement of the road transport infrastructure without taking into account the traffic flows, as well as the lack of data on these flows

Taking into account the above categories it is necessary to carry out a traffic study in order to assess all the possibilities when changes are necessary [1].

The purpose of the paper is to identify and evaluate all the necessary measures to improve the travel conditions for the car traffic in the area of a new shopping center, taking into account that the main measures [2], those related to the fluidization of traffic, the increase of the passenger transport capacity and the increase of the safety and comfort of the road users, to take into account:

- Average travel speed;
- Number of stops;
- Journey time;
- Fuel consumption;

- Overall efficiency index.

The paper is introducing and evaluate the importance of urban development data in planning and establishing its suitability for future urban development [3]. The steps in achieving this are:

- Define the infrastructure of the studied area;
- Identify and map the real data sets that are used for simulation planning;
- Create a virtual model and simulate the current and new traffic conditions [4].

2. Evaluating the zone infrastructure

The area for the study is located in the eastern part of Slatina city on the street Artileriei. This is a highly circulated zone, taking over a large part of the mobility of travel to work to and from industrial areas [5].

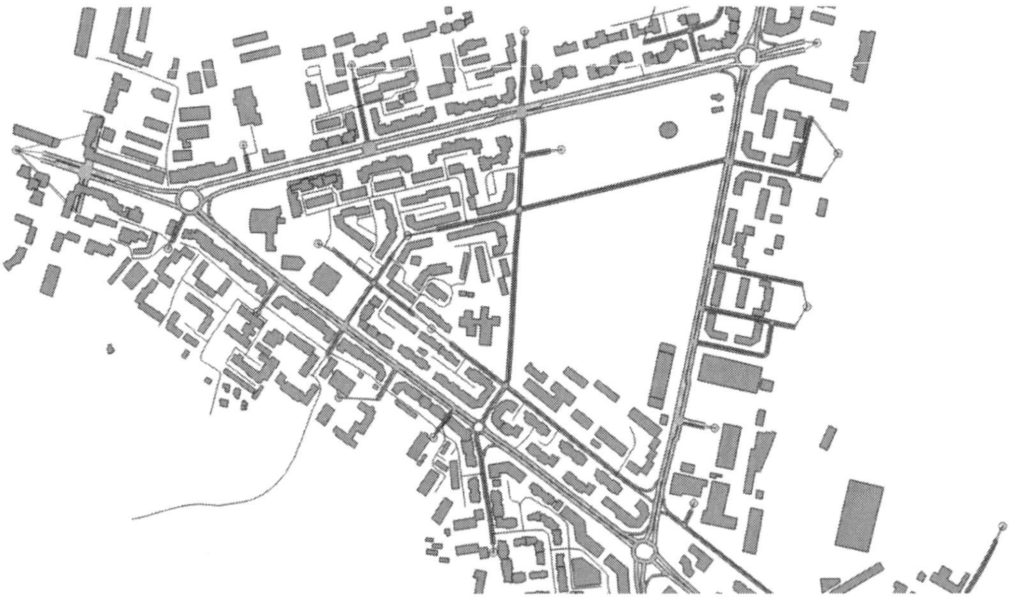

Figure 1. Part of the whole simulated virtual network – overview.

The traffic on the route constituted Artileriei streets is very crowded during peak hours because this is the area used for trips to/from work for citizens working in the industrial park, also is the only street that can have heavy traffic, being considered an exterior ring in what concern the heavy traffic.

On Artileriei Street, near the intersection with Drăgănesti Street, there are located neighborhood shopping centers (Lidl, Kaufland, etc.) that represent points of interest for the nearby residents. At the northern end of the street, to the east, there is the principal railway station of the city, which, by traffic measurements, is not a major traffic generator.

The studied street is delimited in the North and South by roundabouts. The characteristic of the street, on which the new shopping center will be constructed and on which the study and the paper is based, can be seen in the table below:

Table 1. Characteristics of the main street network

Main street	Intersection with main roads	Intersection with secondary roads	Pedestrian crossing	Bus stops
Artileriei	A.I. Cuza, Ecaterina Teodoroiu, Drăgănești, Cireașov, Cazărmii	Nicolae Iorga, Crizantemelor, Piața Gării	4	4

2.1. Public transport evaluation

From the point of view of public passenger transport, the studied area is served by 3 bus lines: line 1 (which also connects the city with the industrial area), line 2, and line 5 which can be seen in the figure below.

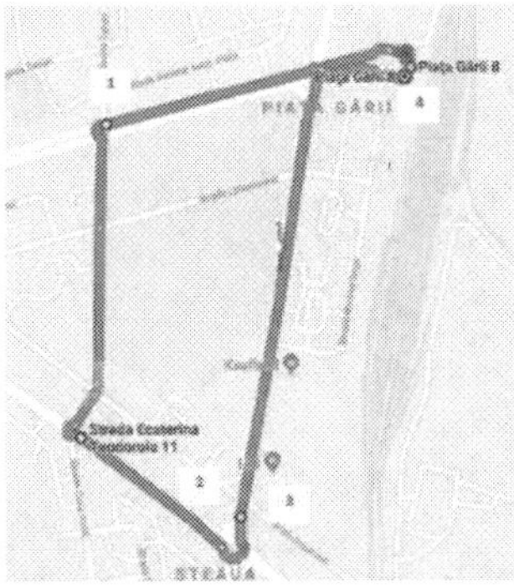

a)

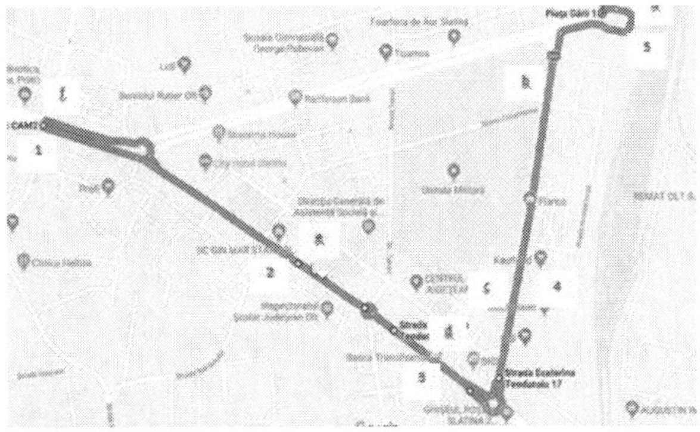

b)

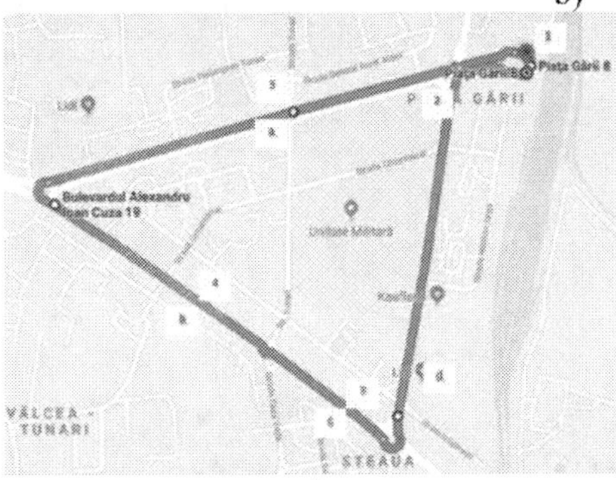

c)

Figure 2. Map of public transport lines in the studied area: a) line 1, b) line 2 and c) line 5.

3. Real data set collection and gathering method

In order to investigate the current situation of road traffic and to validate and calibrate the current transport model as a starting point for forecasting further traffic flows, traffic investigations were carried out [6]. The traffic measurements were made taking into account the recommendations of the normative AND 557/2015 – "Instructions for making road traffic registrations on public roads", approved by the Order of the Minister of Transport no. 481/233.03.2015.

The collection of data on the current characteristics of traffic and mobility of persons and goods was carried out through several harvesting methods:

- Surveys at intersections;
- Origin destination surveys;
- Placement of video equipment for traffic metering at intersections.

Also, the video footage existing in the Traffic Police database in the points of interest for the traffic study was requested and analyzed.

3.1. Retrieving traffic data through dedicated video cameras for traffic measuring

In addition to the classical methods of manual metering of traffic volumes in intersections, in this traffic study, tools dedicated to recording and analyzing road traffic such as video cameras were also used. Their use allows continuous monitoring of traffic in the area and helps to achieve greater accuracy of traffic information, excluding the possibility of human error [7, 8].

The equipment used consists of the Quercus SmartLoop traffic video cameras, which work on the basis of the creation of virtual inductive loops by the user.

Cameras were placed in each intersection on the main streets on the pillars of the lighting network and optimally positioned to retrieve data according to the virtual inductive loops created from the software. The view of the camera was constrained by the location and the angle at which the sensor viewed the intersection. Thus, the images and inductive loops were created based on the viewing angle of the camera, as can be seen in figure 3 below.

Figure 3. Inductive loop setting for traffic measurement.

3.2. Street evaluation

Before making a simulation and analyzing the traffic network it is important to localize on map the street and identify all the existing problems that exist in the studied zone [9].

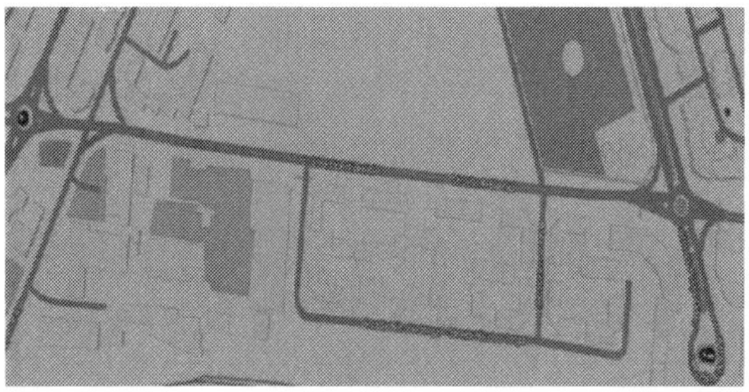

Figure 4. Artileriei street – map localization.

There are a couple of existing traffic problems that can be resolved once are identified. This is important because the problems will be transposed through the simulation of the whole area (mesoscopic simulation) into the small traffic simulation of the roundabout (microscopic simulation).

1. The bus stop next to Lidl store does not have a bus socket for stopping, which limits the flow, when the bus stop at the station, it occupies lane 1, in this context the traffic is carried out only on lane number 2.

Figure 5. Missing socket at Lidl bus stop.

2. Lack of continuous line so that vehicles coming from Drăgănesti Street and enter the roundabout are not allowed the movement; their left movement interrupts the flow of traffic on Artileriei

Street, leading to the increase of blockages in the area of the roundabout with Ecaterina Teodoroiu Street

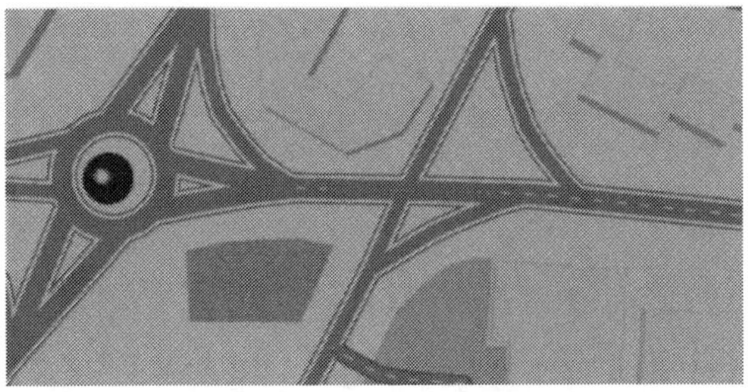

Figure 6. Missing continuous line.

3. The stationary of the vehicles on the Artillery Street practically cancels the use of a traffic lane, which causes the occurrence of the congestion phenomenon

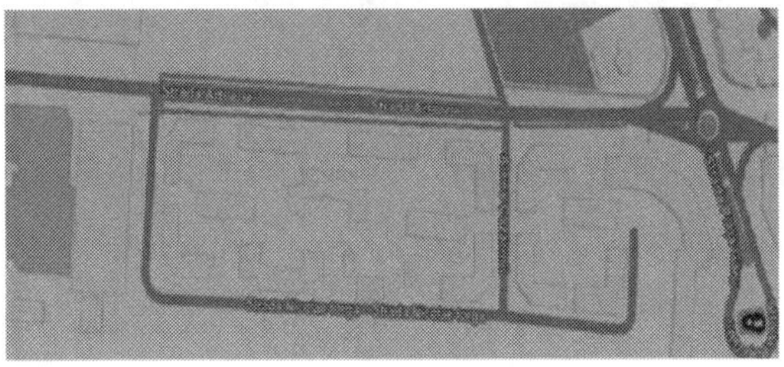

Figure 7. Area where stationary vehicles lead to traffic strangulation.

4. Traffic evaluation in regards to the shopping centre

In order to evaluate the construction of a shopping center (Mall) in the area of the Artillery Street, between the two main roundabouts, it was considered opportune to build a roundabout that would distribute the flows, in a way that is as safe as possible from the point of view of road and pedestrian safety [10].

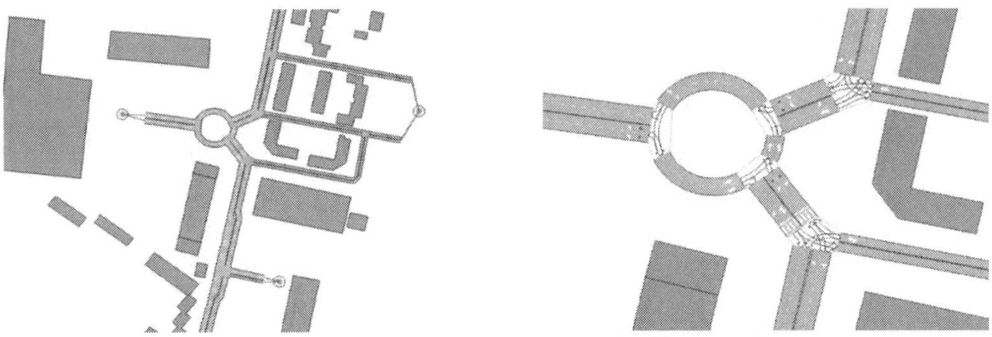

Figure 8. Roundabout shopping center (Mall), overview.

In this regard, only the impact zone of the new roundabout was simulated. For this simulation, the traffic of the entire studied area was taken into account, related to a mesoscopic simulation of the entire area of the study as can be seen in Figure 1 of the paper, and the impact area was simulated microscopically [11]. The traffic data resulted in origin/destination matrix, relative to the current situation.

IOP Conf. Series: Materials Science and Engineering 1220 (2022) 012059 doi:10.1088/1757-899X/1220/1/012059

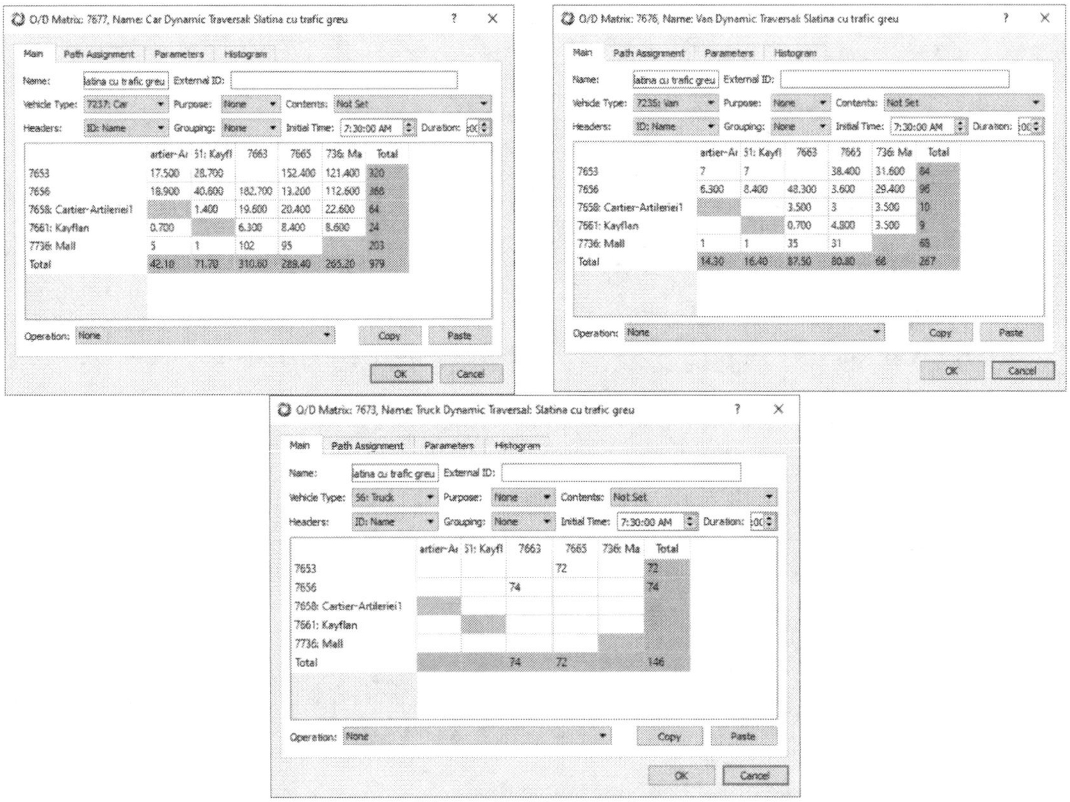

Figure 9. O/D Matrix.

4.1. Traffic evaluation in regards to heavy traffic

Starting from the premise that the traffic is redistributed in order to use the points of interest, the new shopping center, it will be considered that a ratio of 30% to 35% of vehicles will change their destination to the new shopping center. At the same time, a ratio of 35% – 40% of vehicles will exit the commercial area to the rest of the centroids in the simulation.

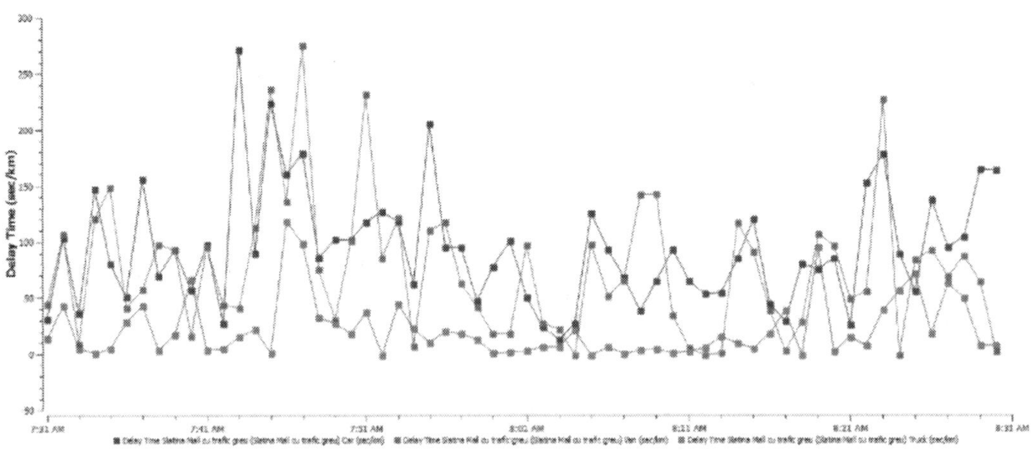

Figure 10. Commercial area simulation delay time.

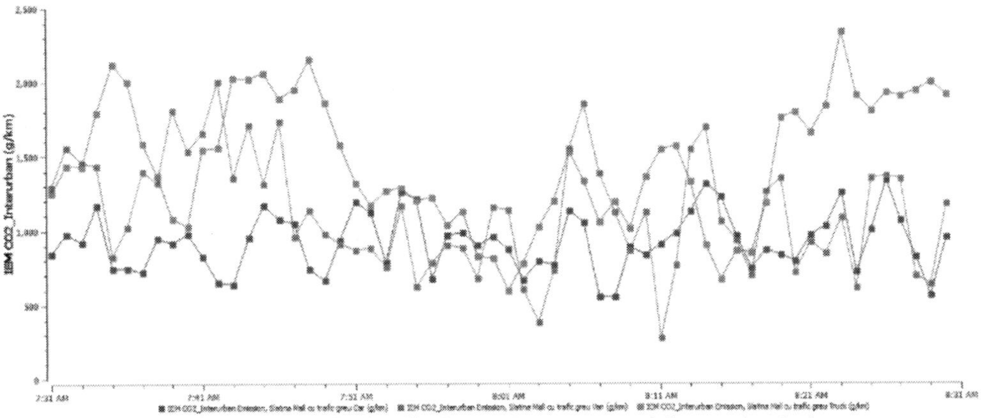

Figure 11. Chemical pollution, CO_2, commercial area simulation.

A slight decrease in about 10% of the delay times in relation to the current situation can be observed. This is due to the introduction of the roundabout so that traffic flows are restructured into new platoons, thus reducing the time lost in the traffic jam area.

4.2. Traffic evaluation without heavy traffic

In the conditions of a ring road, all heavy traffic will be routed on this area so that the traffic will be carried out in much more optimal conditions. This can also be seen from the delay time in the studied area, a time that decreases by more than 50% compared to the initial conditions.

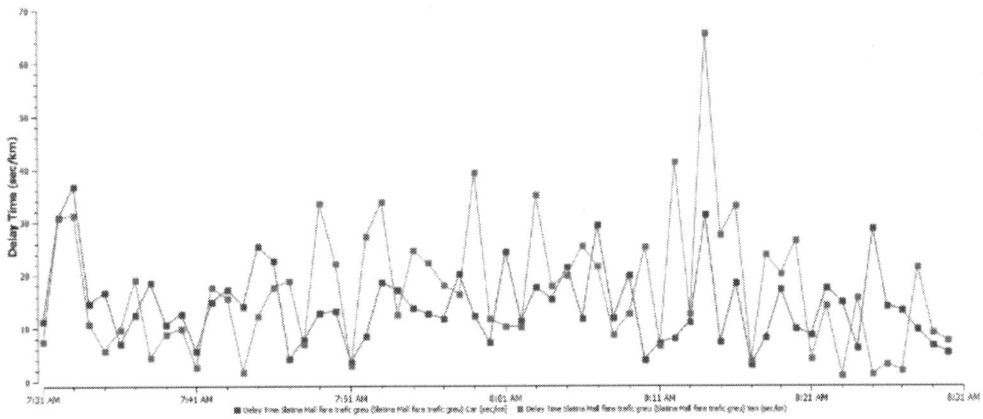

Figure 12. Commercial area simulation delay time without heavy traffic.

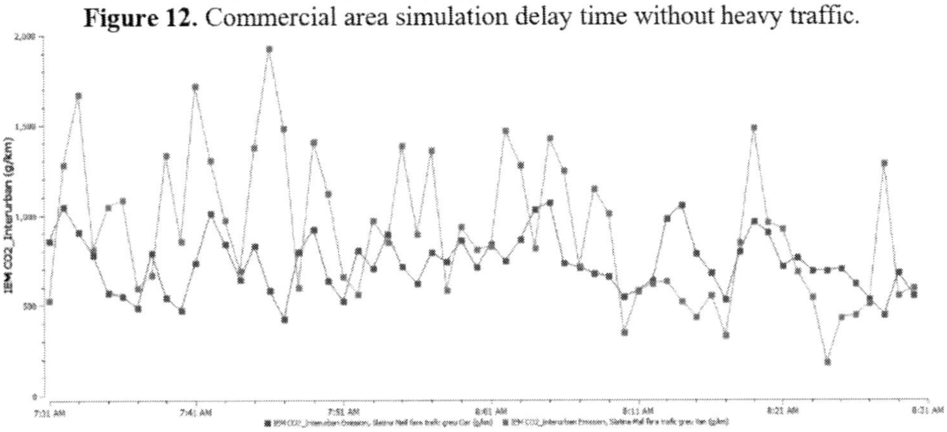

Figure 13. Chemical pollution, CO_2, commercial area simulation without heavy traffic.

5. Conclusion

One of the main important aspects in assessing the traffic flow with regards to a commercial shopping center construction is that all virtual simulations and calculations require as an initial part thorough measurements of traffic conditions by using specialized equipment with a very small error count.

Analyzing the results from the virtual network simulation a couple of conclusions can be seen:

- Evaluating the delay time, a reduction in the fuel consumption area can also be observed, these decreases leading to a much better ratio of pollution parameters;
- Although the assessment of chemical pollution is carried out only in the area of the newly created roundabout due to the reduction of waiting times, there is also a decrease in consumption and polluting elements CO_2, NOx or PM;
- The above reduction also affects the stop times, with a decrease of about 100 sec/km compared to the current situation;
- The downtime is reduced to 4-5 sec/km, thus considerably reducing the possibilities of traffic jams in the case of traffic simulation without heavy traffic compared to the one with heavy traffic.

References

[1] Moridpour S, Majid S, Geoff R - *Lane changing models. A critical review*, Transportation Letters: The International Journal of Transportation Research. 2.,10.3328/TL.2010.02.03.157-173, pp. 157-173, 2010

[2] Virginia Tech Transportation Institute Cambridge Systematics, Inc. - *Microscopic Analysis of Traffic Flow in Inclement Weather*, Publication No. FHWA- JPO-09-066, 2009

[3] Khan Z. H, Aaron T - *A macroscopic traffic model for traffic flow harmonization*, European Transport Research Review. 10. 10.1186/s12544-018-0291-y, 2018

[4] Dumitru I, Dumitru N, Matei L, Racila L - *Saturation flow mathematical model based on multiple combinations of lane groups*, XII Conference on Transport Engineering 7,8,9 June 2016 - Valencia (Spain), 2016

[5] Gipps, P G - *A model for the structure of lane-changing decisions*, Transportation Research B 20(5), pp 403-414, 1986

[6] Dumitru I, Racila L, Tutunea D, Matei L, Dima A and Oprica A - *Experimental investigation of a vehicle behaviour using different complex data acquisition systems*, 2016 CONAT Congress, Brasov, Romania

[7] Matei L. - *Algoritmi de optimizare a circulatiei la nivel de intersectii rutiere*, Ed. Universitaria, Craiova, Romania, ISBN: 978-606-14-1590-2, 2020

[8] Stewart J A, Aerde M V - *An Assessment of Adaptive Co-ordination of Traffic Signal offsets within integration*, Traffic Engineering and Control, Volume 39, 1998

[9] Dumitru I, Matei L, Vinatoru M, Racila L, Oprica T - *Aspects Regarding Priority Settings in Unsignalized Intersections and the Influence on the Level of Service*, Proceedings of the European Automotive Congress EAEC-ESFA, 2015

[10] Matei L, Dumitru I, Racila L., Vînatoru M - *Adaptive traffic signal control on a national road intrsection Applied Mechanics and Materials*, ISSN: 1662-7482, Vol. 822, pp 455-460

[11] Treiber M, Helbing D - *Microsimulation of Freeway Traffic Including Control Measures*, November 2002 - Automatisierungstechnik 49(11).

AUTHOR INDEX